THE VERTEBRATE BODY

By

ALFRED SHERWOOD ROMER
Alexander Agassiz Professor of Zoology,
and Director, Museum of Comparative
Zoology, Harvard University

Illustrated

W. B. SAUNDERS COMPANY
PHILADELPHIA & LONDON
1950

Reprinted January, 1950

Apologia

I HAVE ATTEMPTED to give an introductory account of the vertebrate body, of the varied forms which this versatile and plastic structural type has assumed during its long evolutionary development, and of the functional story associated with this morphological history. A number of admirable texts cover much of the area included here. This book was written for the reason that none, even if excellent in many ways, fully satisfied me in all regards as to the fashion in which the material should be treated. Desirable are:

1. *Fairly adequate illustration.* Even though a student may cover considerable ground in the laboratory, he cannot see all types and structures of interest, and may fail to see the forest for the few trees visible to him.

2. *A truly comparative treatment.* Overemphasis of human structure is not desirable, even for the premedical student. Such a course should be, for him, essentially a "cultural" background, to give him better understanding of the peculiarities and seeming irrational construction of the human body.

3. *Proper paleontological background.* The known facts of vertebrate history should be utilized to give, not only adequate treatment of the skeletal and other systems to whose evolution paleontology contributes, but also a modern phylogenetic point of view.

4. *A developmental viewpoint.* Embryological history is crucial in the establishment of homology, a leitmotiv of comparative anatomy. Further, the time element should be kept in mind in the consideration of any vertebrate body, for an "adult" is merely one stage in a long developmental series.

5. *Inclusion of histological data.* The wielder of the scalpel is liable to forget the basic materials of which gross structures are composed. Such an organ as a stomach is merely a flabby, rather revolting and uninteresting object unless we consider the varied internal epithelia and minute glands which furnish much of its excuse for existence.

6. *Consideration of function.* The almost complete separation of form and function prevalent in instruction today is both unnatural and

unfortunate. It is doubtful if there is such a thing as a nonfunctioning structure, although mention of function is often taboo in morphological works. Nor do functions take place *in vacuo* or without purpose in benefiting the structures which compose the organism, despite the contrary feeling that some physiological treatises imply. Even if attention be held to comparative anatomy in a narrow sense, the study of homology immediately raises the question of the changes of function associated with the changes, often remarkable, undergone by homologous structures.

A critical reader may look askance at the considerable length of the section on the skeleton and suggest that I have been partial to a system of which I am especially fond. I am forced to admit that such a criticism is partly true. But I may plead, in extenuation, that I have been no more partial than many another author; that I have included in that chapter various topics, such as the dermal skeleton and the general history of fins and limbs, that might well have been treated elsewhere; and that more detailed discussion of the skeleton is merited because of the fact that our extensive knowledge of it and our possession of a fine paleontological record gives us, as for no other system, a really adequate evolutionary story.

I have attempted to give a comparative study of the muscular system, rather than the few vague generalities with which the musculature is often (with a shudder) dismissed. I am, however, rather doubtful as to the success of this undertaking.

The attempt to give a treatment of this sort to the story of the vertebrate body has led me into many fields of which I know little. I hope that my friends will deal gently with me and with this book.

Miss Nelda Wright has helped greatly and constantly throughout the preparation of this book. Original drawings are the work of Miss Jeannette Sullivan. I am very grateful to Dr. W. K. Gregory for permission to use a number of valuable figures by the late Bashford Dean, originally published in the latter's "Fishes Living and Fossil." It is my regret that relatively few of the illustrations are original and the reader acquainted with the subject will recognize all too many figures which are all too familiar to him.

I am most especially indebted to Dr. Tilly Edinger for aid in the preparation of the manuscript. She has critically read—and re-read—the entire manuscript to my great profit, and helped in many other ways; many sections are as much her work as mine. Dr. George Wald gave valuable criticism of Chapter IV; Dr. Leigh Hoadley read critically the section on embryology. Dr. Vincent Hall of the University of Illinois has given numerous useful suggestions. Mr. Russell Olsen furnished valuable aid with regard to the derivation of scientific terms.

The original photographs for the illustrations on the cover were supplied through the courtesy of General Biological Supply House, Inc., Chicago, and S. H. Camp & Company, Jackson, Michigan.

While a formal dedication for a book of the limited scope and elementary level of the present work is neither proper nor customary, I may venture to render homage to the two men, now *professores emeriti* at Columbia University, who many years ago instilled in me an interest, which has never flagged, in the story of the vertebrates: Dr. William King Gregory, under whom I studied as a graduate student, and Dr. J. H. McGregor, whom I served as a teaching assistant in vertebrate zoology.

ALFRED S. ROMER

Contents

CHAPTER I

INTRODUCTION

THIS WORK IS designed to give, in brief form, a history of the vertebrate body. Basic will be a comparative study of vertebrate structures, the domain of comparative anatomy. This is in itself an interesting and not unprofitable discipline. Of broader import, however, is the fact that the structural modifications witnessed are concerned with functional changes undergone by the vertebrates—changes correlated with the varied environments and modes of life found in the course of their long and eventful history. The evolutionary story of the vertebrates is better known than that of any other animal group, and vertebrate history affords excellent illustrations of many general biological principles. Knowledge of vertebrate structure is of practical value to workers in many fields of animal biology. To the future medical student such a study gives a broader understanding of the nature of the one specific animal type on which his later studies will be concentrated.

For the most part (Chap. VI-XVI) the present volume is devoted to a consideration, *seriatim,* of the various organs and organ systems. In the present chapter are noted certain introductory matters. Other early chapters discuss general or preliminary topics, including the evolutionary history of the vertebrates and their kin (Chap. II and III); cells as the basic structural elements (Chap. IV); and the embryonic development of the body (Chap. V).

THE VERTEBRATE BODY PLAN

Bilateral Symmetry. A primary feature of the vertebrate structural pattern is the fact that members of this group are bilaterally symmetrical, with one side of the body essentially a mirror image of the other. Vertebrates share this type of organization with a number of invertebrate groups, notably the annelid worms and the great arthropod phylum, which includes crustaceans, arachnids, insects, and so forth. In strong contrast is the radial symmetry of coelenterates and echinoderms, in which the body parts radiate out from a central axis like the spokes of a wheel. The degree of activity of animals appears to be correlated with the type of symmetry which is present. The radiate echinoderms and coelenterates are in general sluggish types, slow-moving or fixed to the bottom or, if free-floating, mainly drifters with the current rather than active swimmers. Vertebrates, arthropods, and marine annelids are, on the other hand, active animals. Activity would seem to have been one of the keys to the success of the vertebrates and is in a sense as diagnostic as any anatomical feature.

Regional Differentiation. In any bilaterally symmetrical animal we find some type of longitudinal division into successive body regions—in the annelid worms, for example, a rather monotonous repetition of essentially similar segments, or in insects a pattern consisting of head, thorax, and abdomen. Vertebrates, too, have well-defined body regions, although these regions are not directly comparable to those of invertebrate groups.

There is in vertebrates a highly specialized *head,* or cephalic region; in this region are assembled the principal sense organs, the major coordinating centers of the nervous system which form the brain, and the mouth and associated structures. Here, as in all bilaterally symmetrical animals (even a worm), there is a strong tendency toward *cephalization* —a concentration of structures at the anterior end of the body.

In all higher, land-dwelling vertebrate groups a *neck* is present behind the head; this is little more than a connecting piece, allowing movement of the head on the trunk. The presence of a neck region is not, however, a primitive vertebrate feature. In lower, water-breathing vertebrates this section of the body is the stout *gill region,* containing the breathing apparatus. The appearance of a distinct neck occurs only with the shift to lung breathing and the reduction of the gills.

The main body of the animal is the next following region, the *trunk;* this terminates in the neighborhood of the anus or cloaca. Within the stout trunk are the body cavities containing major body organs, the viscera. In mammals the trunk is divisible into *thorax* and *abdomen,* the former containing the heart and lungs within their rib basket, the latter enclosing most of the digestive tract; there is, however, no clear subdivision here in lower vertebrates.

In most bilateral invertebrates the digestive tube continues almost the entire length of the body. Among the vertebrates, however, we find, in contrast, that the digestive tract and other viscera stop well short of the end of the body; beyond the trunk there typically extends a well-developed *tail* or *caudal region,* with flesh and skeleton, but without viscera. The tail is, of course, the main propulsive organ in primitive water-dwelling vertebrates. In land animals it tends to diminish in importance, but is often long and stout at the base and well developed in many amphibians and reptiles. In mammals it is generally persistent, but is merely a slender appendage. In birds it is shortened and functionally replaced by the tail feathers, arising from its stump; in some forms—frogs, apes, and man —it is, exceptionally, lost completely as an external structure.

Gills. The presence, in embryo if not in adult, of internal gills, developed in a paired series of clefts or pouches leading outward from an anterior part of the gut (the pharynx) is one of the most distinctive features—perhaps the most distinctive feature—of the vertebrates and their close kin. In higher vertebrates the gills are functionally replaced by lungs, but gill pouches are nevertheless prominent in the embryo. In lower water-dwelling vertebrates they are the primary breathing organs. Among small invertebrates, many with soft membranous surfaces can get enough oxygen through such membranes to supply their wants. But in forms with a hard or shelly surface, and especially in large forms, where the surface area is of course small compared with the bulk of the body, gill structures of some sort are a necessity. Typical invertebrate gills, as seen in crustaceans or molluscs, are feathery projections from the body surface. The vertebrate gill, however, is an internal development, connected with the digestive tube. Water enters the "throat," or pharynx (usually through the mouth), and passes outward to the surface through slits or pouches; on the surface of these passages are gill membranes, at which an exchange of oxygen from the water for carbon dioxide in the blood takes place. In existing lower vertebrates breathing is the primary and practically the sole function of the gills. But in lower chordates and the probable ancestral vertebrates, as we shall see, the gills and gill slits were of importance in food collection—a fact tending to explain the unusual condition of an association of the breathing organs with the digestive tube.

Notochord. In the embryo of every vertebrate there is found, extending from head to tail the length of the back, a long, flexible, rodlike structure —the notochord. In most vertebrates the notochord is much reduced or absent in the adult, where it is replaced by the vertebral column, or backbone. But it is still prominent in some of the lower vertebrates, and is the main support of the trunk in certain lowly relatives of the vertebrates (such as Amphioxus) in which no vertebral column ever forms. So significant is this primitive supporting structure that the vertebrates and their kin are termed the phylum Chordata, a name referring to the presence of a notochord.

Nervous System. Longitudinal nerve cords are developed in various bilaterally symmetrical invertebrate groups. These, however, are frequently paired and may be lateral or ventral in position. Only in the chordates do we find developed a single cord, dorsally situated and running along the back above the notochord or the vertebrae. Invertebrate nerve cords are solid masses of nerve fibers and supporting cells running between equally solid clusters of nerve cells, termed ganglia. The chordate nerve cord is, in contrast, a hollow, nonganglionated structure, with a central, fluid-filled cavity. In various invertebrates the process of cephalization has tended to a concentration of nerve centers in a brainlike structure. Independently, we believe, the vertebrates have evolved a hollow *brain,* with characteristic subdivisions, at the anterior end of the hollow nerve cord. Not exactly matched in any invertebrate group is a series of characteristic *sense organs* developed in the head region of vertebrates—paired eyes and, primitively, a third, median eye; nasal structures, usually paired; paired ears with equilibrium as their primary function.

Digestive System. All metazoans (with degenerate exceptions) have some sort of digestive cavity with a means of entrance to and exit from it. In many of the more primitive metazoans there is but a single opening, serving as both mouth and anus. In vertebrates, as in other, more progressive metazoans, there are separate anterior and posterior openings. The mouth is situated near the front end of the body, typically somewhat to the underside. In arthropods and annelids the digestive tube reaches to the posterior end of the body. In vertebrates, however, this is not the case; the anus is situated at the end of the trunk, leaving, as we have seen, a caudal region in which the digestive tube is absent.

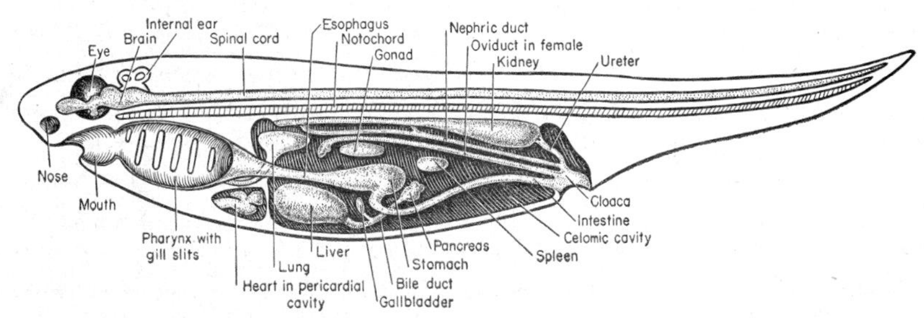

Fig. 1. Diagrammatic longitudinal section through an "idealized" vertebrate, to show the relative position of the major organs.

In most vertebrates the digestive tube is divided into a series of characteristic regions serving varied functions—*mouth, pharynx, esophagus, stomach,* and *intestine* (the last variously subdivided). In lower vertebrates the esophagus may be almost nonexistent, and in some groups the stomach even may be absent. In mammals and certain other vertebrates the digestive tract terminates externally at the *anus*. In most groups, however, there is a terminal gut segment, the *cloaca,* into which urinary and genital ducts also lead.

A *liver* which performs to some extent a secretory function, but is in the main a seat of food storage and conversion, is present in vertebrates as a large, single ventral outgrowth of the digestive tube. Somewhat similar but variable structures are present in many invertebrates.

Kidneys. Among invertebrates some type of kidney-like organs for the disposal of nitrogenous waste and the maintenance of a proper composition of the internal fluids of the body are often present, typically as rows of small tubular structures termed *nephridia*. Nephridia of a special type are still found in Amphioxus among the lower chordates. In true vertebrates, however, the *kidney tubules* serving such a function are of a markedly different type and are characteristically gathered into compact paired kidneys, dorsal in position. *Kidney ducts,* of variable nature, lead to the cloacal region or to the exterior, and a *urinary bladder* may develop along their course.

Reproductive Organs. Male and female sexes are always distinct in the vertebrates, as they are in many invertebrate groups. The organs containing the germ cells—the *gonads*—develop into either *testis* or *ovary.* In all except the lowest vertebrates a duct system leads the eggs or sperm to or toward the surface (frequently by way of the cloaca); and in the female, duct regions may develop for shell deposition or for development of the young.

Circulatory System. In vertebrates, as in many invertebrates, there is a well-developed organ system to circulate a body fluid, the blood, with tubes as vessels and a pump, the *heart,* to circulate the enclosed fluid. The heart in vertebrates is a unit structure, ventrally and rather anteriorly situated. In certain invertebrates the circulation is of an "open" type: the blood is pumped from the heart to the tissues in closed vessels, but is then released and makes its return to the heart by oozing through the tissues without being enclosed in vessels. In the vertebrates, as in some of the more highly organized invertebrates, the system is closed: not only is the blood carried by the *arteries* to the tissues, but the return to the heart, after passing through the tissues in small closed tubes, the *capillaries,* is also made in closed vessels, the *veins.* In most vertebrates *lymph vessels* are present as an additional means of returning fluid from the cells to the heart. Many invertebrates contain in their blood streams pigmented metallic compounds in solution which aid in the transportation of oxygen. Among vertebrates, almost exclusively, the iron compound *hemoglobin* is the oxygen carrier; and further, this chemical is not free in the blood, but is contained in *blood cells.*

In annelids the circulation of the blood is in general forward along the dorsal side of the body, backward ventrally in its return to the tissues. The reverse is true of the vertebrates. The blood from the heart passes forward and upward (primitively via the gills) and back dorsally to reach the organs of trunk and tail, and the major return forward—from the digestive tract at least—is ventral to the gut (although dorsal veins are important).

Celom. In certain invertebrates the internal organs are embedded in the body tissues. In others, however, there develop body cavities—celomic cavities—filled with a watery fluid, in which most of the major organs are found. This latter condition is present among vertebrates. A major body cavity—the *abdominal cavity*—occupies much of the trunk and contains most of the digestive tract; various other organs (reproductive, urinary) project into it. Anteriorly there is a discrete *pericardial cavity* enclosing the heart, and in mammals the lungs are contained in separate *pleural cavities.*

Muscles. Musculature in the vertebrates is of two types, *striated* and *smooth,* or nonstriated, the two differing sharply in minute structure and in distribution in the body. The former, roughly, includes all the voluntary musculature of the head, trunk, limbs, and tail, and the muscles of the gill region; the smooth musculature, more diffuse, is mainly found in the lining of the digestive tract. The musculature of the heart is in various respects intermediate in microscopic structure. The striated musculature of the trunk develops, unlike many other organ systems, as a series of segmental units.

Skeleton. Hard skeletal materials are present in all vertebrates, and in all except certain degenerate or (doubtfully) primitive groups consist in

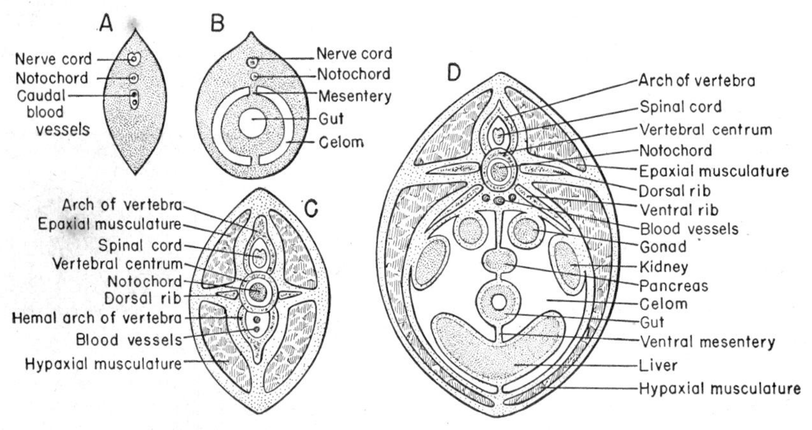

Fig. 2. Cross sections through the body of a vertebrate. *A, B,* Much simplified sections through tail and trunk, to show the essential structure of the trunk as a double tube; in the tail the "inner tube" of the gut is absent. *C, D,* More detailed diagrammatic sections of the tail and trunk to show the typical position of main structures.

part, at least, of bone. Superficial skeletal parts, the *dermal skeleton,* correspond functionally to the "armor" of certain invertebrates, and are typically bony; internal skeletal structures, the *endoskeleton,* are formed in *cartilage,* but are frequently replaced by bone in the adult. Cartilage-like materials are found in invertebrates. *Bone,* however, is a unique vertebrate tissue. It differs in texture and minute structure from the typical chitinous or calcareous skeletal materials of invertebrates, and, in the fact that the salts deposited are calcium phosphate, it differs from most (but not all) invertebrate skeletal structures—in which carbonate is the common calcium compound.

Appendages. Two pairs of limbs, *pectoral* and *pelvic,* are found in most vertebrates in the form of fins or legs and become increasingly prominent in higher members of the group. They are, however, little developed or are absent in the lowest of vertebrates, living and extinct, and hence are not absolutely characteristic. Their structure includes internal skeletal elements, with muscles for their movement arrayed above and below.

Segmentation. The great invertebrate phyla of Arthropoda and Annelida are notable for the presence of *metamerism:* a serial repetition of body parts in a long series of body segments. In annelids this segmentation is readily apparent; in arthropods the metameric structure may be more or less obscured in the adult, but is clearly seen in embryos or larvae.

Vertebrates, too, are segmented, but the segmentation is limited and has obviously developed independently from that of invertebrate groups, in which all structures, from skin inward to the gut, exhibit segmentation. Among the vertebrates neither skin nor gut is segmented; the metameric arrangement is primarily that of the trunk muscles. In relation, however, to the attachments of these muscles and their nerve supply, much of the skeleton and nervous system have taken on a segmental character.

The Body in Section (Fig. 2). We have noted some of the more important body features, with particular regard in many cases to their anteroposterior position in the body. We may now briefly consider the general organization of the body as seen in cross section.

Structurally, the most simple region of the body is that of the tail, strongly developed in most vertebrate groups. The section of a tail (Fig. 2, *A, C*) is typically a tall oval, the surface skin-covered. Somewhat above the center is seen the notochord, or the central region of the vertebrae which typically replace it in the adult; and, above this, a cross section of the nerve tube; the two structures are invariably closely associated topographically. The body cavity and associated viscera are absent in this region; representing them (in a sense) are caudal blood vessels lying below the notochord. Almost all the remainder of the tail is occupied by musculature, usually powerful. This musculature is arrayed in right and left halves, with a median dividing septum above and below.

A typical section through the trunk is more complicated, even when, as in figure 2, *B*, this is represented in its most generalized condition. One may consider the trunk as essentially a double tubular system, roughly comparable in structure to the casing and inner tube of an automobile tire. The outer tube in itself contains all the major elements seen in the section of the tail—notochord and nerve cord, and musculature descending on either side beneath an outer covering of skin. Internally, it is as if we had taken the little area below the notochord in the tail, where only the blood vessels were present, and expanded this to enormous proportions as the celomic cavity of the abdomen. With the development of this cavity the outer "tube" of the trunk now has an inner as well as an outer surface. The surface lining the body cavity is the *peritoneum,* and that part of this lining which forms the inner surface of the outer tube is the *parietal peritoneum.* The part of the outer tube between celomic cavity and the surface of the body is the body wall.

The "inner tube" is primarily the tube of the digestive tract. The outer lining, facing the celomic cavity, is peritoneum—*visceral peritoneum.* The inner lining is the epithelium lining the digestive tract. Between the two, analogous to the musculature in the body wall, are present smooth muscle and connective tissues. In the embryo the gut is connected with the "outer tube" both dorsally and ventrally by *mesenteries*—thin sheets of tissue bounded on either side by peritoneum. The dorsal mesentery—that above the gut—always persists, but the ventral portion frequently disappears for most of its length.

Although we shall treat of the arrangement of the organs in the celom in more detail in a later chapter, we may here go somewhat further in considering the position of the body viscera. (It must be noted that the relative size of the body cavity is never so great as represented in this and other diagrams; the viscera actually expand to fill most of the available space.) In Figure 2, *D* we have indicated the fact that the digestive tube is not a simple tubular structure, but has various outgrowths—most characteristically the liver ventrally and the pancreas dorsally. These are (in theory and in the embryo, at any rate) median structures, and are developed within the ventral and dorsal mesenteries. Further, we may have other organs projecting into the body cavity, but arising from tissues external to it. The kidneys in many groups project into the abdominal cavity at either upper lateral margin, and the reproductive organs—ovaries or testes—typically project into the cavity more medially along its upper border.

DIRECTIONS AND PLANES

Although the vertebrate body is essentially a bilaterally symmetrical structure, there are many exceptions to this general statement. Organs which primitively lay in the midline may be displaced: the heart may be off center; the abdominal part of the gut—stomach and intestine—may be twisted and even convoluted in a complicated asymmetric fashion. Again, in paired structures those of the two sides may differ markedly; for example, in birds but one of the two ovaries (the left) is developed in the adult. A still greater asymmetry is seen in the flounders, where the whole shape of the body is affected by the substitution of the two sides for the normal top and bottom of the animal.

Either in theory or in practice the body of an animal may be sectioned in various ways at various angles. If the body is considered as sliced crossways, as one would cut a sausage, the plane of section is considered *transverse*. If the line of cleavage is vertical and lengthwise, from snout to tail, the plane is a *sagittal* one. Sometimes this latter term is restricted to a cut actually down the midline—the mediosagittal plane—and similar sections to one side or the other are termed parasagittal; but frequently such cuts are considered parts of a series of sagittal sections in a broad sense. The third major plane of cleavage, in the remaining direction, is that of slices cut the length of the body, but horizontally, each going through the width of the body. Such a plane is termed a *frontal* one—that is, one parallel to the forehead.

Direction within the body is of importance in the description of structural relationships and the naming of the various organs. Terms in this category may be considered fixing a position or pointing out a direction.

The head and tail ends of the body are, in most vertebrates, the direction toward which and from which movement of the animal normally takes place. *Anterior* and *posterior* are the common terms of position in this regard; *cranial* and *caudal* are less used, but are essentially synonymous. Upper and lower surfaces—back and belly aspects—are reason-

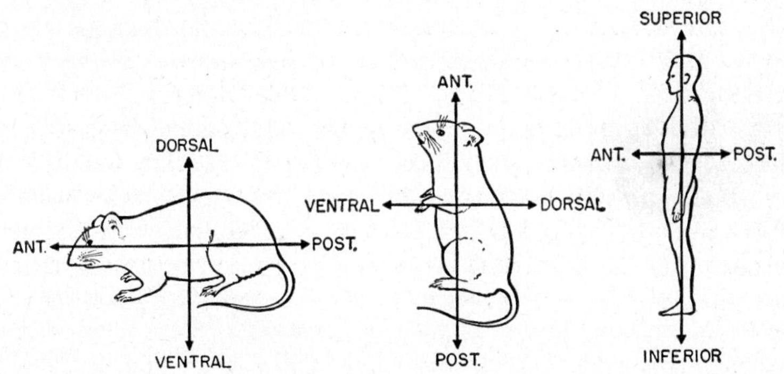

Fig. 3. Diagram to show the contrast in positional terms between normal vertebrates and man.

ably named *dorsal* and *ventral.* Position in the transverse plane is of course given with reference to the midline; *medial* refers to a position toward the midline; *lateral,* a more removed position.

A fourth pair of terms of less positive meaning but of considerable use are *proximal* and *distal.* The former refers in general to the part of a structure closer to the center of the body or some important point of reference, the latter for a part farther removed. These terms are clearly available to limb and tail structures. Within the head and trunk their use is less clear, but we may, for example, speak of proximal and distal parts of a nerve with obvious reference to the spinal cord or brain as a center, proximal and distal regions of arteries with reference to the heart as the assumed center, and so on.

For these adjectives of position there are, of course, corresponding adverbs ending in *ly,* and others (rather awkward) to denote motion in a given direction, ending in *ad,* as *posteriorly, caudad.*

The major directional terms, anterior, posterior, dorsal, and ventral, apply with perfect clarity to almost all known vertebrates. But in man we have an exception, an aberrant form which stands erect—and hence might have different directional terms applied to him.

It is unfortunate that in the terminology currently used in medical anatomy this is the case (Fig. 3). The head and "tail" ends of the body are, in the erect human position, above and below, and are termed *superior* and *inferior,* rather than anterior and posterior. More confusing, however, is the fact that the two latter terms are used—quite needlessly—to replace dorsal and ventral, so that the back side of the human body is termed *posterior* and the belly surface *anterior*. Thus these names have contradictory meanings in special human anatomy and in more normal usage.

General Usage	*Human Terminology*
Anterior	Superior
Posterior	Inferior
Dorsal	Posterior
Ventral	Anterior

This special terminology for the human body causes needless confusion. For example, each spinal nerve has two roots (cf. Fig. 325, p. 535). In a dissecting room these two roots in a human cadaver are termed posterior and anterior. But if a neurologist, working experimentally with (say) rats, tries to use the same nomenclature, he is in an obviously absurd position; one root is not more "anterior" or "posterior" than the other. In both rat and man, however, their designation as dorsal and ventral is reasonable and logical. It is to be hoped that a long-needed revision of anatomical terminology may soon be adopted and that dorsal and ventral will be used for positional terms in the human body.

THE HOMOLOGY CONCEPT

Even in the early days of zoological research it was recognized that within each major animal group there was a common basic pattern in the anatomical plan of the body. The same organs could be identified in many or all members of a group, although frequently much modified in size, form, or even function in correlation with changing habits or modes of existence. With the acceptance by biologists of the principle of evolution in the 60s and 70s of the last century, real significance was given to the concept of *homology:* the thesis that specific organs of living members of an animal group have descended, albeit with modification, slight or marked, from basically identical organs present in their common ancestor. For many decades the tracing out of homologies was a leading motif in zoological research.

Many of the results of such studies were novel and exciting. It was found, for example, that the three little auditory ossicles of our own middle ear (p. 522) were in earlier days part of the jaw apparatus of our piscine ancestors, and appear still earlier to have been part of the supports of the gills of ancestral vertebrates. The muscles with which we smile or frown once helped our fish forbears to pump water through their gills.

Homologous organs are those which are identical—the same—in the series of forms studied. But what do we mean by "the same"? One tends, unthinkingly, to believe that the same actual mass of material, the very same limb or lung or bone, has been handed down, generation by generation, like an heirloom. This is quite absurd, but such a concept has obviously influenced, unconsciously, the minds of many workers. In reality, of course, every organ is re-created anew in every generation, and any identity between homologues is based upon the identity or similarity of the developmental processes which produce them.

The development of the science of genetics has given us a firm base for the interpretation of these processes. They are controlled by hereditary units, the *genes*. These tiny structures are present to the number of some thousands, at least, in every animal cell. The development of the individual is directed by the genes transmitted to the fertilized egg by the parents. Each gene may affect the development of a number of structures or parts of a body; conversely, every organ is influenced in its development by a considerable number of genes. If the genes remain unchanged from generation to generation, the organ produced will remain unchanged (apart from transient environmental effects upon an individual), and the homology is absolute.

Changes, however, do occur in genes, as *mutations;* these mutations produce changes in the structures to which the genes give rise. If the mutations produce effects of small magnitude and occur in only a few of the genes concerned, the organ will be little modified, and its homology with the parent type will remain obvious. If, however, the mutations are numerous and marked in their effects, the organ may be radically modified and its pedigree much less clear. In a sense, a study of organ homology is merely a study of phenomena produced by genes. If the genetic constitution of all animal types were well known, the determination of homology between structures might well rest upon the degree of identity of the genes concerned in their production. But this is not a matter of practical import, for there are few animals whose genetic constitution is adequately known, and it is improbable that our range of knowledge will ever be broadened to the necessary degree.

What are the best criteria for the establishment of homology? Function is no sure guide, for organs which are clearly homologous in two animals may be put to quite different uses. Observation shows that the shape, size, or color of a structure gives little positive evidence of identity. Similarity in general anatomical position and relations to adjacent organs is a more useful clue to identification. Best of all is similarity in developmental history. Embryological processes in vertebrates tend to be conservative, and organs which are quite different in the adult condition may reveal their homology through similarity in early embryonic stages.

Homology is generally applied to structural identity. Some have proposed that the concept be broadened to include functional identity. This suggestion has not, however, met with general acceptance. The term *analogy* is in some regards a parallel, on the functional side, to homology; analogous organs are those which have similar functions. It is, however, somewhat restricted, for as generally used it implies that the organs concerned are not homologous. A lung and a fish gill, for example, are analogous, for both are used for respiration, but the two are quite different structures.

ADAPTATION AND EVOLUTION

The varied modifications which vertebrate structures have undergone and the varied functions which they have assumed have, of course, come about as the result of evolution. One cannot make a comparative study of the vertebrates without the formulation of some concept of the nature of evolutionary processes. Most structural and functional changes in the vertebrate body are quite clearly adaptive modifications to a variety of environments and modes of life. How have these adaptations been brought about? Proper discussion would require a volume in itself; we can here merely indicate the general nature of the problems concerned, and current majority opinion as to their interpretation.

We sometimes speak, thoughtlessly, of adaptation changes, as if the animal "willed" them, or as if its needs or desires in themselves brought new structures or structural changes into being. It would be advantageous, one might say, for a fish to be able to walk on land, and so some fishes made themselves legs; it would be "nice" if the cow's early ancestors developed teeth better able to cope with grain and grass, and so the teeth promptly deepened.

Obviously, such ideas are absurd. They are, however, not far removed from certain "philosophical" theories of evolution which have had, and still have, a certain vogue. These assume that evolution is an unnatural phenomenon; that changes have been brought about by some "inner urge" within the organism, or are the result of the "design" of some supernatural force. Since such theories are nonscientific, they cannot be scientifically disproved; but we are at liberty to look for more reasonable explanations of evolution, based on known facts. If someone tells me that the operation of my automobile engine is controlled by an invisible demon dwelling therein, I cannot prove him wrong. But nothing is gained by adding this hypothetical demon, and I would prefer to attempt to explain the engine's working in terms of known mechanical principles, the nature of electric currents, and the explosive structure of hydrocarbon molecules.

A more plausible attempt at interpretation of structural evolutionary changes was that first advocated over a century ago by Lamarck—a belief in the inheritance of acquired characters through the effect of use and disuse. If the giraffe's ancestors stretched their necks after foliage on high branches, the effects of this stretching would be transmitted to their offspring, generation after generation, and an elongate neck gradually developed in the hereditary pattern. If the snake's lizard ancestors ceased to use their legs in locomotion, the cumulative result of disuse would be the eventual loss of the limbs. This attractive theory seems simple, reasonable, and natural. But its present standing is poor indeed. We may summarize by saying that no one, despite repeated efforts, has been able to furnish any valid proof of any instance of the inheritance of an acquired characteristic. Structures useful to an animal may and often do increase in size or complexity in the course of time, and useless or little-used structures may diminish. But there is not the slightest evidence that the use or disuse of parts by an organism has any effect whatever upon the build of its offspring.

The science of genetics has in recent decades demonstrated that evolutionary changes are due to mutations. These may produce effects of some magnitude, but most cause only minor modifications: a mutation in a fruit fly may, for example, have no greater visible result than the splitting of a single bristle. Of the causes of mutations we still know little, although it is known that x-rays readily induce mutations, and other physical or chemical influences acting on the cell may be important. As for evolutionary theories, however, two things stand out clearly: (1) There is no evidence of "design" or "direction" in mutations. They appear to be quite random, rather than tending in any one direction. Some may well be advantageous; most, however, are obviously harmful, and many are lethal. (2) There is not the slightest evidence that mutations have any relation whatever to use or disuse of body organs; characters acquired by the individual have no specific effect on the nature of mutations of the genes in its sex cells whose effect is transmitted to the offspring.

The process of mutation thus seems to be merely one of blind, random change. But vertebrate evolution certainly appears to be a process which has resulted in changes both useful and adaptive. How can such results have come out of the mutation process?

Much, at least, of the answer was given by Darwin nearly a century ago. He was, of course, quite ignorant of the data now available from genetics, but reasonably assumed that there existed some hereditary mechanisms of the sort with which we are now familiar. Given a supply of random mutations, *natural selection* will act powerfully to eliminate unfit types and preserve the better-fitted forms in which one or a group of useful mutations have occurred in the germ plasm. Even the slowest breeders among animals produce more individuals than can survive. Many are destined to die before becoming adults and reproducing—an act which is nature's standard of success. Which individuals are eliminated is partly a matter of chance. But both observation and experiment indicate that even small mutations in an adaptive direction have a distinct survival value and may become dominant in a species in a short time. This natural selection of such a random series of mutations as have adaptive value would appear to be a major mechanism of evolutionary change.

SURFACE-VOLUME RELATIONS

It is frequently seen that in any group of animals large and small forms differ notably in the relative size of various organs or parts. The reason for many of these proportionate differences lies in a geometrical principle so obvious that it is generally overlooked, namely, the fact that *as the size of an animal* (or any other object) *changes, surfaces increase* (or decrease) *proportionately to the square of linear dimensions, while volumes change proportionately to the cube of linear dimensions.*

This principle is of wide application, for surface-volume relationships are to be found in a variety of structural and functional features of vertebrates. We cite obvious examples: (1) The strength of a leg (like any supporting column) is proportional to its cross section, which varies as the square of linear dimensions, whereas the weight which it supports is proportionate to the cube of linear dimensions. In consequence an elephant cannot have gazelle-like legs. (2) The amount of food which an active animal needs is roughly proportionate to its volume;* the amount of foodstuffs which its intestine can absorb depends upon the area of the intestinal lining. In consequence, large animals have a disproportionate, elongated intestine or one with a complicated structure, resulting in a greater internal surface area for digestion.

NOMENCLATURE

The student of vertebrate morphology is confronted with a bewildering array of unfamiliar names of anatomical structures. This is unfortunate but inescapable. Vertebrate structures are numerous; for many there are no everyday terms. Even where such names are available, they are often vague and not exactly defined in common usage. Further, it is desirable to have some international system of terms understood in the same sense by scientists of every country.

When anatomy was first studied, all "learned" works were, as a matter of course, written entirely in Latin. In consequence, Latin names where already in existence were applied to anatomical structures, and if no term existed, one was manufactured from Greek or Latin roots and cast

* Emphasis on *active;* basal metabolism in a resting condition is quite another thing.

in Latin form. Some notes regarding the formation of anatomical terms are given in Appendix II. Today Latin has ceased to be an international language as far as the general text of scientific books is concerned. Latin anatomical terms, however, are still in vogue. We cannot do without them, although we often use them in a somewhat "anglicized" form—speaking, for example, of the "deltoid muscle" of the shoulder rather than the "musculus deltoideus," or of the "parietal bone" rather than the "os parietalis."

Latin is, of course, an inflected language, and its nouns and adjectives have a variety of endings to express not merely singular and plural numbers, but also a variety of cases and a rather arbitrary system of genders. Until recent decades some knowledge of Latin grammar was part of the equipment of every college student, and the manipulation of Latin terminology presented no difficulty. Today this is not the case; rather unfortunately, for a biologist should at least know enough to avoid such gaucheries as speaking of "humeruses" instead of humeri, "femurs" instead of femora. Fortunately the number of noun and adjective endings ordinarily used in anatomical terms is limited, and these can be readily learned; a brief note regarding them is given in Appendix II.

It is accepted procedure in anatomical nomenclature that where a structure is present in mammals—particularly in man—the name there used be applied to the same structure in other forms. Thus, for example, man and many mammals have a clavicle, or collar bone, and the equivalent element in the shoulder structure should be called by the same name in reptiles, amphibians, or fishes, even though its appearance is radically different. Sometimes, however, too hasty an identification may be made and a name wrongly applied. Teleost fishes have a bone similar in position to the clavicle, and that name was customarily given to it; we now know, however, that the teleosts have lost the true clavicle; the bone present there is a different one (the cleithrum, p. 179). If homologies are in doubt, it is better to use a different name for the structure in question. For example, there is a muscle in the thigh of reptiles which may be homologous with the sartorius muscle of mammals; but since there is some doubt of the homology, it is customary to give the reptilian muscle a different name—the ambiens muscle (p. 273). If a structure encountered in a lower group has no mammalian equivalent, a new name must, of course, be coined.

Although anatomical terminology has been in general a rather stable and uniform system, there arose, quite naturally, a number of differences in terminology between different schools of work and in different countries. Motivated by the laudable desire to achieve uniformity, the German anatomical association, in convention at Basel some decades ago, brought forward a comprehensive scheme of terms which they hoped would receive universal adoption in human anatomy. This terminology, usually referred to as the "BNA," was adopted by many medical schools and is now widely used in medical work.

Unfortunately the results have not been so beneficial as might be desired. We have already noted that the directional terms make many of these names inapplicable to even the ordinary laboratory animals used in medical research. Further, many other BNA names are difficult of application to nonhuman animals, or absurd when they are so applied. Still further, the makers of this terminology actually worked against their professed aims of uniformity and simplicity by unnecessarily coining new terms when appropriate ones already existed, and even applying familiar terms to different structures from those for which they had been generally used (cf., for example, the table of carpal and tarsal names, pp. 202 and 208).

In the 1930s the major imperfections in the system had become apparent, and movements looking toward reform were initiated. The international difficulties of that period, and the war in which they culminated, caused postponement. It is to be hoped that proposals for revision will be revived and that we shall at length arrive at a generally satisfactory nomenclature.

CHAPTER II

THE VERTEBRATE PEDIGREE

Although the present work primarily concerns the vertebrates alone, we must recognize that there exist various animal types lacking a backbone, but closely allied to the vertebrates. Study of these more lowly forms contributes to our understanding of vertebrate structure and history. Further, although (as will be seen) we have little certain knowledge of the early ancestry of

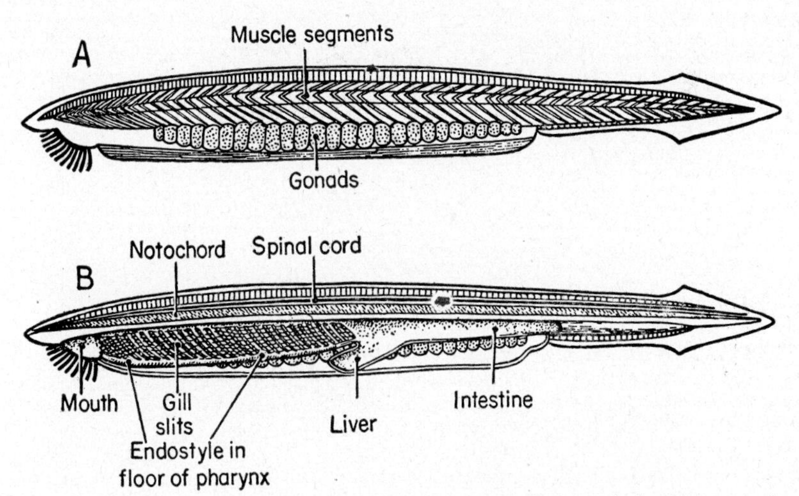

Fig. 4. Amphioxus, a primitive chordate. *A*, As seen through the transparent skin; *B*, a sagittal section. (After Gregory.)

the vertebrates, the subject of their pedigree deserves at least brief consideration.

The vertebrates do not in themselves constitute a major division of the animal kingdom. They are considered merely one subdivision—although by far the largest subdivision—of the *phylum Chordata,* the other members of which are to be briefly considered here. The "lower *chordates*" lack the backbone and many other advanced structures of their vertebrate relatives. They do, however, exhibit basic features characteristic of the vertebrates and not found elsewhere in the animal kingdom. These features indicate that they are truly related to vertebrates and hence properly included in a common group with them. The term Chordata itself implies that a notochord (or chorda), or some structure thought to be equivalent to a notochord, is generally present. Again, a dorsal nerve cord is a common feature. Most characteristic of all is the fact that gill slits are ubiquitous in chordates.

Amphioxus. By most workers the phylum Chordata is subdivided into four subphyla—in roughly ascending order, Hemichordata, Urochordata, Cephalochordata, and Vertebrata. Here we reverse the order and begin our discussion of the lower chordates with the *Cephalochordata,* in which the similarities to the vertebrates are most obvious. The subphylum con-

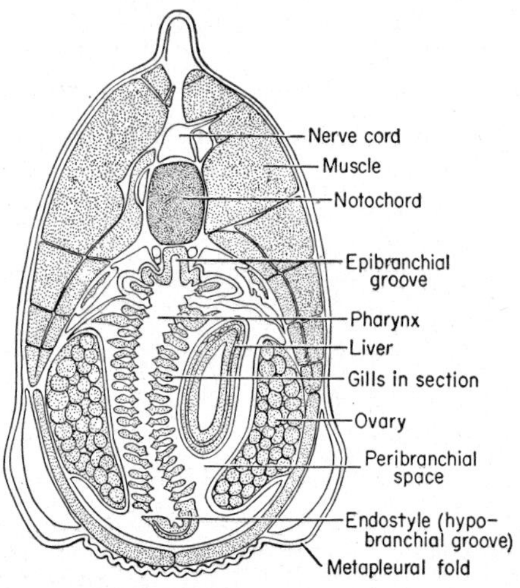

Fig. 5. Cross section of Amphioxus through the pharynx. The peribranchial space surrounding the pharynx, liver, and so on, is, despite its seeming internal position, actually external to the body and is somewhat analogous to the gill chamber of bony fishes, and the like. It is formed by the downgrowth around the pharynx of great metapleural folds meeting one another ventrally, and it connects with the exterior through a posteriorly placed opening. (After Delage and Hérouard.)

sists of a single animal type, usually termed Amphioxus (Figs. 4, 5). This is a small, translucent animal, fishlike in appearance, found in shallow marine waters in various regions of the world, and sometimes locally abundant. As its shape suggests, it can swim readily, but for the most part it spends its time with its body buried in the sands of the bottom, with merely its anterior end projecting.

Despite the piscine appearance, it is obvious that we are dealing with a form far more primitive than any fish. There are no paired fins or limbs of any sort. Slender rods of cartilage-like material stiffen the gills, but no part of the normal vertebrate skeleton of vertebrae, ribs, or skull is to be found. The main supporting structure (apart from connective tissues) is a highly developed notochord, which persists through life and (in contrast to the vertebrate situation) extends clear to the tip of the "nose"—

a feature to which the group name owes its origin. There is a typical hollow dorsal nerve cord; but although the cord is somewhat larger anteriorly, there is no true brain, and there are only dubious traces of sense organs which might correspond to nose or eye.

The digestive tract, except for a liver outgrowth, is simple. Following a mouth and gill region, there is a long slender tube as an intestine; there is no development of subdivisions within the intestine, nor any stomach as such. Internal gills, on the other hand, are highly developed, far more so than in any vertebrate, for there may be as many as fifty or more pairs of gill slits, and each is essentially a double structure. A peculiar hood, the *metapleural folds,* grows downward and backward over the gills, serving as a protection to these delicate structures while Amphioxus is buried in the sand. The high development of gills is presumably related to the mode of life of Amphioxus. Its food is derived from sea water, taken in through the mouth by ciliary action and passed out through the gill slits. Nutrient particles are strained out as the water passes into the gills. These particles are caught in the pharynx in channels in which a sticky mucus is secreted. Ciliated cells in these channels set up a current by which the mucous material and the contained food particles are eventually carried inward to the entrance of the intestine. The pharynx has a major midventral (hypobranchial) groove termed the *endostyle;* there is a corresponding dorsal (epibranchial) groove, and in addition there are smaller channels which run up the walls of the pharynx between the gill slits.

The excretory organs are built on a plan rather different from that of vertebrates; the units of which they are composed are tiny tubes (protonephridia) of a type found in certain invertebrates, and quite in contrast to vertebrate kidney tubules in structure. In the circulatory system the blood, as in vertebrates, courses forward ventrally and back dorsally after passing upward through the gill region; there are numerous, small contractile structures, distributed through the body, which pump the blood, but the single ventral heart characteristic of vertebrates is absent.

In general, then, Amphioxus shows a body plan similar to that of vertebrates, but with a simpler pattern and with certain special features such as the metapleural folds peculiar to this form. It is of interest that in the lampreys, which represent the lowest of existing vertebrate levels, there is generally a larva differing markedly from the adult, and much simpler in structure. This larva shows marked similarities to Amphioxus, a fact which to some demonstrates the ancestral position of the cephalochordates. The majority opinion has been that Amphioxus, although departing to some degree from the directly ancestral pattern, represents a stage in evolution preceding that of vertebrates.

There is, however, a second possible interpretation, namely, that Amphioxus is not really primitive, but a degenerate descendant of primitive vertebrates. We see in many animals the phenomenon known as *neoteny,* in which a larva lingers long before finally changing into an adult, or that of *paedogenesis,* in which the larval animal becomes sexually mature and breeds without ever reaching the adult condition. It is possible that paedogenesis has occurred here and that Amphioxus represents the permanent larva of some early vertebrate type; its simplicity may be due to arrested development.

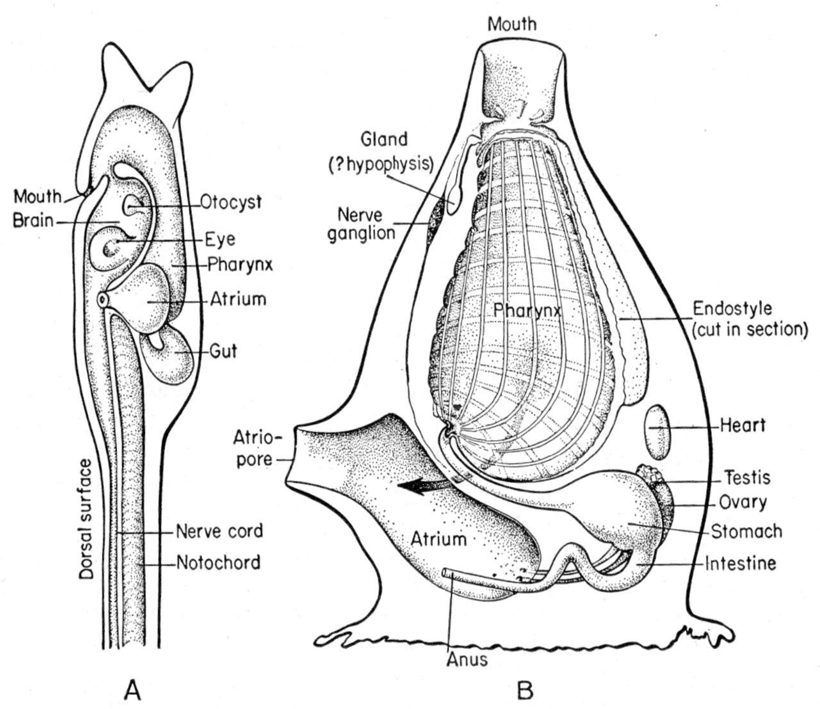

Fig. 6. A solitary tunicate. *A,* Diagram of the structures seen in the free-swimming larva (head end above, and only a short section of tail figured). The otocyst is a simple ear structure. *B,* The sessile adult, formed by elaboration of the structures at the anterior end of the larval body. The original dorsal side lies at the left. The large pharynx is attached to the body wall above and below (left and right in the figure); the atrium (corresponding to the peribranchial space of Amphioxus) bounds it on either side; water passing through the latticework gills of the pharynx enters the atrium and, as indicated by the arrow, streams out through the atriopore. (After Delage and Hérouard.)

Tunicates. A second major group of lower chordates is that of the *Urochordata,* the *tunicates* or *sea squirts* and their relatives. These are rather common small marine organisms. Many are found floating freely in the water, singly or in groups, as tiny barrel-shaped structures; others are attached, either as branching colonies or as individuals. Simplest and perhaps most typical are the solitary tunicates (Fig. 6). As an adult, such an animal is an almost formless sessile object attached to a rock or other underwater object. It is covered with a leathery-looking "tunic."

The only structural features seen externally are an opening at the top, into which water passes, and another lateral opening, through which the water current flows outward. The creature shows no external resemblances to the vertebrates, and internally much of the structure is equally unfamiliar to a student of that group. There is no notochord; nor is there a nerve cord; instead, there is a simple nerve ganglion with a few nerves splaying out from it.

Much of the interior of the animal is occupied by a barrel-shaped structure which serves as the food-gathering device and also as a respiratory organ. The water current, entering it, is strained through slits in the sides of the barrel into a surrounding chamber, the *atrium,* which leads to the lateral, excurrent, opening. On consideration it becomes obvious that the barrel is an exaggerated set of internal gills, constituting the pharyngeal region of the animal. Behind the enormous pharynx, the digestive tube narrows to form esophagus, stomach, and intestine—all of modest size.

We have here, in the pharyngeal gill apparatus, a high degree of development of one of the primary characters of the chordates. For other chordate characters, however, we must turn to the developmental history. In many tunicates propagation takes place by a process of budding. But in some there is a distinct larval form, which has somewhat the appearance of an amphibian tadpole (in one small group of tunicates the larval form persists in the sexually mature state). The "head" of the larva corresponds to the entire body of the adult. The tail is a swimming organ, useful in transporting the young tunicate about in its search for a new home. Once the animal attaches and "settles down" to its sedentary adult existence, the tail dwindles and is absorbed into the body. In this tail, however, are to be found major proofs of vertebrate relationship of the tunicates. There is in the larval tail (as the group name implies) a well-developed notochord, and, above this, a typical hollow dorsal nerve cord. Anteriorly, there is in the larva a rudimentary brain and sense organs. At metamorphosis, notochord and most of the nervous system (unnecessary in the sessile adult) disappear.

The tunicates, thus, are definitely chordates and definitely related to the vertebrates, but represent a lower level of evolutionary development. The majority opinion of workers on the subject is that the tunicates represent a degenerate side branch of the vertebrate ancestral line and that the common ancestor of vertebrates and tunicates was a free-swimming form, somewhat like the larval tunicate. From this, it is suggested, the vertebrates "ascended" by an improved continuation of an active mode of life, whereas the tunicates tended to become degenerate and lost most progressive structural features except for the food-straining gill barrel; became, in fact, fit subjects for evolutionary sermons on the results of slothful living. There is, however, another possible interpretation, namely, that the early chordate was, after all, a sessile food strainer somewhat like an adult tunicate; that the tail first appeared as an adaptation for larval locomotion; and that the development of higher forms came about by the retention of the tail and the free-swimming habit in adult life.

Acorn Worms. The third group of forms definitely related to the vertebrates is that of the *Hemichordata.* Here the key chordate characters are less developed, and some writers would erect the hemichordates into a phylum separate from the Chordata, although related to them.

The typical hemichordates are the *acorn worms,* such as Balanoglossus (Fig. 7). These are not uncommon in tidal flats, where they are present

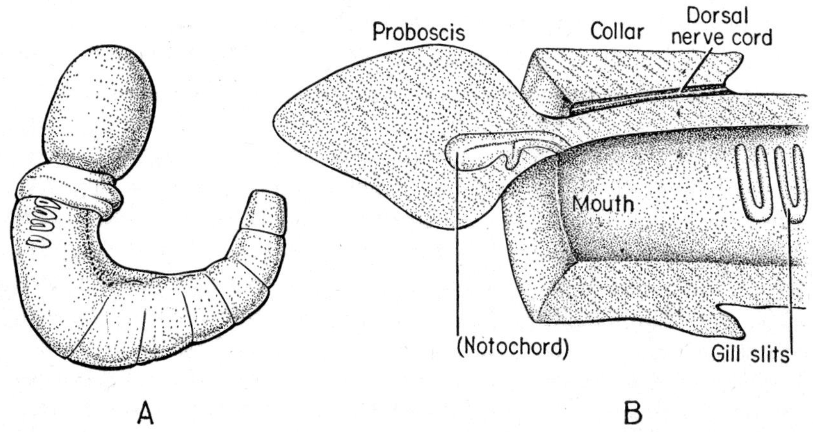

Fig. 7. An acorn worm. *A,* Young individual (the adult body is more elongate), showing successively proboscis, collar region, and an area in which gill slits are developing. *B,* Sagittal section through the anterior part of the body of an adult. (After Delage and Hérouard.)

as mud burrowers. The general body shape is wormlike, but there the resemblance ends, for their structure is not at all comparable to that of ordinary annelid worms. Even externally the acorn worms are distinctive. The body terminates anteriorly in a tough yet flexible "snout" or proboscis, of variable length, which serves as a burrowing organ. Behind the proboscis there is a distinct thickened section of the body, the "collar" region; the name acorn worm is due to the fact that in some forms the proboscis and collar have the appearance of an acorn in its cup.

In most regards the acorn worm shows no especial resemblance to the vertebrates or other chordates. For a short distance—in the collar region—there is a dorsal hollow nerve cord. Over the rest of the body, however, the nerve cells and fibers are rather diffusely distributed in the skin, although there is some development of solid dorsal and ventral strands of nerve tissue. There is no proper notochord, although a stout pouch of tissue at the base of the proboscis has been compared to an imperfectly developed structure of that sort.

But—as in tunicates—we find that a vertebrate type of gill system is present in characteristic and highly developed fashion. The gills are not so pronounced as they are in tunicates, but there is, behind the collar, an extended region of the gut corresponding to the vertebrate pharynx, from which on either side open out gill slits quite comparable—even in details of development—to those of Amphioxus. Vertebrates are not descended from acorn worms as such, but these forms may reasonably be interpreted as a group not distantly removed from our early chordate ancestors.

THE VERTEBRATE ANCESTOR

Attempts to gain a clear picture of the predecessors of the vertebrates from a study of the lower chordate groups are, unfortunately, none too successful. This is to be expected. The fossil story suggests that the vertebrates were passing through the evolutionary stages represented by these lower chordates at a time which is on the order of half a billion years or more ago. There is, however, no direct fossil evidence of such forms, for they were presumably soft-bodied animals and hence unlikely to be preserved in the record of the rocks. That we would have living today animals exactly representing these ancient stages in vertebrate history can hardly be expected.

Putting together the evidence from the living chordates, however, we can get some general ideas as to the type of ancestral chordate from which the vertebrates evolved. The evidence suggests that this ancestor consisted largely of an elaborate food-gathering gill structure, to which were appended a digestive apparatus and reproductive organs, and little else. The gill region, with its function mainly confined to respiration, is of modest size in typical jawed fishes. But in some of the older jawless types (cf. Figs. 15, 16, 190, pp. 38 and 322) the gill chamber was of considerable magnitude, as it is in tunicates and Amphioxus. Although the tunicates themselves are too specialized to be true ancestors, one may imagine the remote vertebrate ancestor to have been a sessile, marine food gatherer. For distribution of the progeny, there may have come about the larval development, as in many tunicates, of a propulsive tail, with notochord, nerve cord, and swimming muscles. The vertebrate line moved from the sea to fresh waters; here, with the necessity of resisting the downward sweep of stream currents, the larval tail may have persisted through life as a useful adjunct. This development, with the addition of skeletal materials, could have led to the early phase of jawless vertebrate life found in the oldest fossil representative of the group. The development of jaws as food gatherers presumably led to the reduction of the gill region and the development of the proportions of gill region to other parts of the body seen in the more typical fishes.

Indications of the high antiquity and ancient importance of the gills are seen in the distinct character of the skeletal structures, muscles, and nerves associated with them (cf. pp. 209–211, 282).

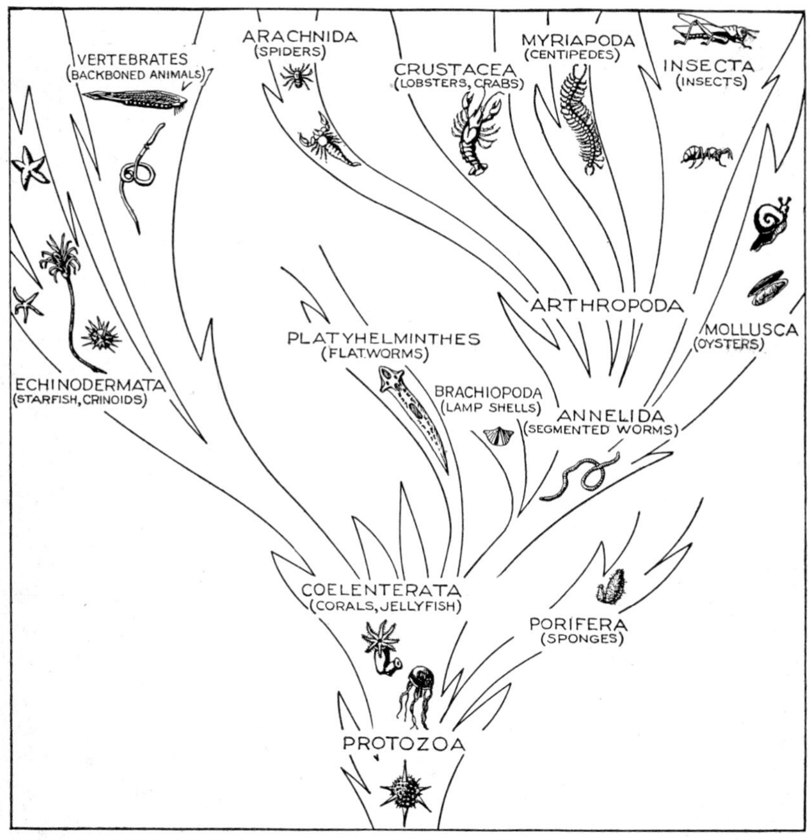

Fig. 8. A simplified family tree of the animal kingdom, to show the probable relationships of the vertebrates. (From Romer, Man and the Vertebrates, University of Chicago Press.)

Invertebrate Phylogeny. We have seen that it is difficult, because of the lack of concrete satisfactory evidence, to plot the course of vertebrate ancestry through the lower chordate levels of evolution; still more difficult and hazardous is the attempt to chart the possible line of development of the chordates as a whole from earlier ancestors among the varied invertebrate phyla.

In recent decades students of invertebrate zoology have tended to agree on a fairly uniform scheme of phylogeny of the major groups of animals without backbones (Fig. 8). The coelenterates, such as jellyfishes, sea anemones, and corals, are regarded as the basal stock of metazoans—all animals, that is, above the level of protozoans and sponges. The coelenterates, however, have a simple, two-layered structure, with little between the "skin"—the ectoderm—and the lining, termed the endoderm, of the inner gut cavity. Above the coelenterate level all animals have a third intermediate body layer, the mesoderm, from which muscular, circulatory, and other systems are formed; the mesoderm becomes—in bulk, at any rate—the most important of the three tissues.

Two contrasting methods of embryonic formation of this third body layer are to be seen. In one type especially characteristic of echinoderms—starfishes, sea urchins, and the like—the mesoderm arises in the form of pouches growing outward from the gut walls; these pouches remain in the adult as closed body cavities. In a second type the mesoderm arises as solid masses of cells budded off from an area near the posterior end of the body, and the body cavities, when formed, arise by cleavage within the masses of mesodermal cells. To this type belong the annelid worms and the molluscs. The great group of jointed-legged animals, the arthropods, appear to belong here as well, although their developmental pattern is much modified; and certain other forms, such as the flat worms, appear to be offshoots from the base of this major stock. We thus have the concept that, above the coelenterate level, the invertebrates form two great branches, in Y-fashion, with the echinoderms at the end of one branch and the great host of familiar advanced invertebrates clustered on the other.

The two stocks contrast not only in the type of formation of the middle body layer, but also in the larval development. Both echinoderms on the one hand and water-dwelling annelids and molluscs on the other grow from the egg into tiny larvae of simple structure, with bands and tufts of cilia arranged in characteristic patterns on the surface of the body. The echinoderm larva has an arrangement of the ciliated bands and other features which differ markedly from those of the larvae of annelid worms and molluscs.

From what point on this family tree of the invertebrates does the chordate (and vertebrate) branch arise? Theories on this subject have been numerous, but have given few positive results.

One solution to the problem might be to suggest a direct origin of chordates from those most primitive metazoans, the coelenterates. Here there are no great difficulties to overcome, for animals on a coelenterate level of evolution have few specialized features which must be lost before starting on the path towards the vertebrates. But in reality, advocacy of coelenterate descent as a solution would seem to be begging the question. A number of basic advances had not been made by the coelenterates, but are found in almost all other invertebrate phyla—development of a middle body layer, presence of both mouth and anus in the digestive tract, and so forth. It seems highly improbable that the vertebrates acquired these progressive features entirely independently of other invertebrate groups. Search seems warranted for possible relatives—if not direct ancestors—on a higher level.

Annelids as Ancestors. The annelid worms offer a possible point of departure; the theory of vertebrate origin from annelids was warmly advocated during later decades of the nineteenth century. The lowly angleworm is none too prepossessing as an ancestor; but there are numerous marine annelids of a more progressive and attractive nature. Annelids have a typical bilateral symmetry, as do vertebrates, and, in correlation with this, are, like typical vertebrates, active animals in contrast with the sessile types common in many invertebrate phyla. Then too, they are segmented forms, as the vertebrates are to at least some extent. As in

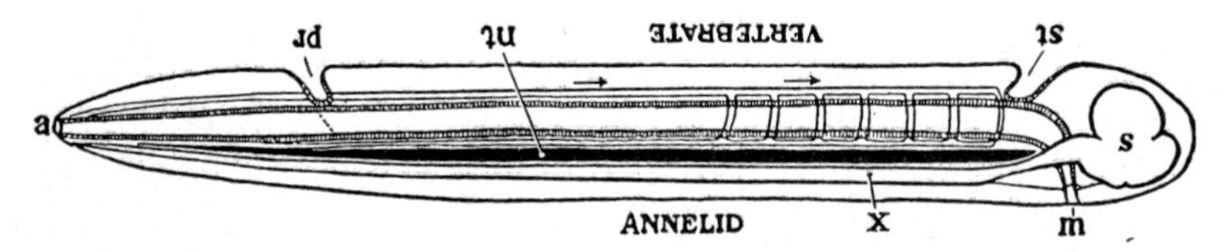

Fig. 9. Diagram to illustrate the supposed transformation of an annelid worm into a vertebrate. In normal position this represents the annelid with a "brain" (*s*) at the front end and a nerve cord (*x*) running along the underside of the body. The mouth (*m*) is on the underside of the animal, the anus (*a*) at the end of the tail; the blood stream (indicated by arrows) flows forward on the upper side of the body, back on the underside. Turn the book upside down and now we have the vertebrate, with nerve cord and blood streams reversed. But it is necessary to build a new mouth (*st*) and anus (*pr*) and close the old one; the worm really had no notochord (*nt*): and the supposed change is not as simple as it seems. (From Wilder, *History of the Human Body*, by permission of Henry Holt & Co., publishers.)

vertebrates, the central nervous system is composed of a brainlike mass at the anterior end of the body and a longitudinal nerve cord.

So far, so good. But beyond this point the comparison breaks down. Even the segmentation is not a perfect argument; for the annelid is segmented in every respect, from skin to gut lining, whereas the segmentation of a vertebrate is primarily confined to part of the middle body layer. The annelid has, it is true, a longitudinal nerve cord. But it is solid, not hollow, and it is ventral rather than dorsal in position. This last point is especially troublesome to advocates of this theory. They have "resolved" the difficulties by assuming that a vertebrate is a worm upside down (Fig. 9). This is hard to swallow (even a worm may have some idea as to which way is up) and involves further perplexities. The worm mouth is on the under side of the head, and so is that of a vertebrate. A reversal of surfaces implies that the old mouth of the worm has closed and been replaced, historically, by a new one. Attempts have been made to find traces of the theoretical old mouth opening in vertebrate embryos—it should pass upward and forward through the brain to the top of the head!—but without convincing results.

Even if this difficulty in orientation be solved, there are more ahead. There is no trace in an annelid of a notochord and none of that seemingly crucial chordate feature—internal gills. And the type of mesoderm formation is in contrast, for in chordates, it would appear, the pouch type of mesoderm formation is the basic pattern. There is, thus, almost no positive reason to believe in a descent of vertebrates from annelids; and there are so many difficulties in the way that there is today little reason to take stock in an annelid theory.

Arachnids as Ancestors. The arthropods, including crustaceans, myriapods, arachnids, and insects, are, it is agreed, descended from annelids or from worm ancestors closely related to them. The arthropods include the most progressive and successful of all invertebrate animals, and it is not unnatural that they have received considerable attention as possible vertebrate ancestors. Among the arthropods it is the arachnids that have been selected as the most likely candidates for vertebrate kinship. Spiders are the commonest of arachnids; but the scorpions are more generalized types. Still more primitive are aquatic arachnids of high geological antiquity; one ancient extinct group, the eurypterids, have been thought by some to be close to the ancestry of the vertebrates.

In arachnids, as in ancestral annelids, there is a ventral nerve cord. This gives us the same problem that was encountered in the worms: the upper and lower surfaces must, it seems, be reversed, a new mouth exchanged for an old one, and so on. Again as in annelids we find conspicuous differences from vertebrates in segmentation; also, there is no trace in arachnids of notochord or internal gills. An added difficulty with arachnids as vertebrate ancestors is the presence of numerous, complex jointed legs of arthropod character. It is impossible that these were transformed into fish fins, and the arthropod legs must be done away with before the development of vertebrate appendages can begin. The old eurypterids were covered with chitinous armor, which in some cases had a superficial resemblance to the bony armor of certain archaic fishes. But even this resemblance is meaningless; for the resemblance is a top-to-top one, whereas, because of the reversal of surfaces, it should be the bottom of the arachnid which resembles the upper side of the vertebrate. To sum up, to make a vertebrate out of an arachnid, the supposed ancestor must have lost almost every characteristic feature he once possessed and reduced himself practically to an amorphous jelly before resurrecting himself as a vertebrate.

An amusing variant of the arachnid theory removes a whole series of difficulties caused by the necessity of turning the arachnid over to make a vertebrate. Under this theory there has been no reversal of surfaces. It is pointed out that the arthropod digestive tube has expansions and subdivisions which rather resemble those of the brain and spinal cord of vertebrates. This particular theory assumes that the original ventral nerve cord of arachnids migrated upward to surround the original digestive tube, which actually became the cavities of the nervous system; and that meanwhile a brand new digestive tract developed by a closing over of the furrows present ventrally between the jointed legs. Technically, this theory solves the difficulties encountered in turning the animal over; but it appears to offer fresh problems as difficult as those solved. There is no adequate explanation for this change from one digestive system to another; and the intermediate stages are, to say the least, difficult to imagine.

Echinoderm Affinities. Unlikely as it seems at first sight, the best clues to chordate origins are to be found in a study of the echinoderms—the starfish, sea urchins, and the like. Several lines of work suggest that, although it is highly improbable that the vertebrates arose directly from

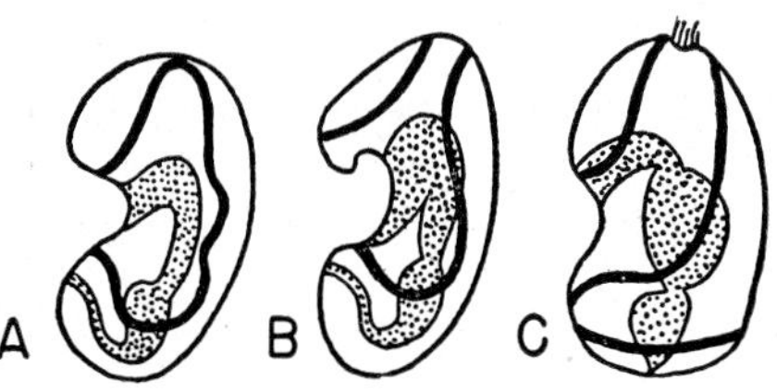

Fig. 10. Diagrammatic side views of the larvae of (*A*) an acorn worm, (*B*) a starfish, and (*C*) a sea cucumber, all much enlarged. The black lines represent ciliated bands. The digestive tract (stippled) appears through the translucent body. Views are from the left side; the larvae are bilaterally symmetrical. (After Delage and Hérouard.)

echinoderms, the two are nevertheless relatives. Adult echinoderms are radially symmetrical, but the larvae exhibit bilateral symmetry—presumably a primitive feature. In most vertebrates mesoderm formation is a complex process, but in Amphioxus we find mesoderm forming from gut pouches just as it does in echinoderms. Another item of interest is the fact that in many of the hemichordates there is a ciliated larva (Fig. 10) of the same type as that of echinoderms—so similar, in fact, that until the life history was known, the hemichordate larvae were thought to be those of starfish!

Even biochemistry helps to establish the case. In all animals with musculature, phosphorus compounds are present which speed up the energy release for muscle activity. In vertebrates the material combined with the phosphate is creatine (cf. Chap. IX); in most nonvertebrate groups another material, argenin, is present instead. Creatine is commonly present in one invertebrate group alone—that of the echinoderms.

It is not believed, under this theory, that vertebrates arose from echinoderms—certainly, at least, not from any "proper" modern echinoderms.

The evidence merely suggests that such kinship as the chordates have lies with that group and definitely not with the annelid-arthropod stock. The common ancestor may have been some small, simply built, bilaterally symmetrical animal of the ancient oceans, from which evolution went onward in two directions—toward the radially symmetrical and essentially sessile echinoderms on the one hand, and on the other, toward the chordates—progressive, persistently active and persistently bilaterally symmetrical.

CHAPTER III

WHO'S WHO AMONG THE VERTEBRATES

THE STUDY OF organs and organ systems and their varied forms and functions—the main concern of the present work—gives us but a one-sided account of the vertebrates. What one should know is not merely the discrete parts, but the total animal; its life and its place in nature. Our present study no more gives us a rounded picture of the vertebrates than the dissection of a cadaver and a course in physiology would give us a complete knowledge of mankind. It is to be hoped that the student will read some work on the "natural history" of vertebrates and thus gain an idea of the living animals whose bodies are verbally dissected in this volume. We shall here give merely a brief survey of the membership of the vertebrate groups in order to place the forms discussed within a phylogenetic framework.

THE GEOLOGICAL RECORD

The fossil record and the extinct animals included in it require attention in this regard. In comparative anatomy we compare the organs of *existing* members of different groups as if one had descended from the other; as if mammals had descended from the existing reptiles, these from existing amphibians and fishes. Obviously, however, this is not the case. A turtle is a reptile, but it is not a mammal ancestor; it has had just as much time to diverge from the common primitive reptilian stock as has the mammal. A frog is an amphibian; but it is definitely not the sort of amphibian from which more progressive land vertebrates were derived. Only through paleontology can we hope to discover the true nature of the common ancestors from which the varied living vertebrates arose.

In discussing fossils, some notion of the geological time scale is necessary (cf. Table I). The earth's total history of upwards of two billion years is divided into a few major time units termed *eras;* these are subdivided into a number of *periods*. For the earlier eras there is little adequate knowledge of life of any sort; the fossil record is almost entirely confined to the last three eras, spanning somewhat over half a billion years of earth history.

The first of these three, the *Paleozoic Era* or Age of Ancient Life, covered about 300 million years, and is divided into half a dozen periods. The fossil record remaining from the seas of the oldest period (Cambrian) contains abundant representatives of almost every major animal group, except the vertebrates. The first faint traces of backboned animals have been found in the rocks of the following Ordovician period, and a modest number of archaic fishes have been found in the Silurian period that follows. The paucity of early vertebrate fossils may be due to the group's having evolved in fresh waters; the sediments of the older Paleozoic periods are mainly marine. In the Devonian period fishes were abundant in fresh-water deposits—so abundant that this period is sometimes termed the Age of Fishes—and some had invaded the seas as well. The continental sediments of the Devonian indicate to the geologists that much of the land was subject to marked seasonal droughts, as are certain tropical regions today; times of abundant rainfall alternated with seasons when streams ran dry and pools were stagnant. These conditions appear to have had a major influence on the history of fishes and in the development of terrestrial life.

TABLE I

Geologic Periods Subsequent to the Time When Fossils First Became Abundant

(The Carboniferous is generally subdivided into two periods, Mississippian [earlier] and Pennsylvanian [later]. The Tertiary period is subdivided into epochs [not given here].)

Era	Period	Estimated Duration (in Millions of Years)	Estimated Time Since Beginning (in Millions of Years)
Cenozoic	Quaternary	1	1
	Tertiary	69	70
Mesozoic	Cretaceous	50	120
	Jurassic	35	155
	Triassic	35	190
Paleozoic	Permian	25	215
	Carboniferous	85	300
	Devonian	50	350
	Silurian	40	390
	Ordovician	90	480
	Cambrian	70	550

At the end of the Devonian appeared the first land vertebrates, the amphibians, and primitive members of that group are common in the swamp deposits that characterize the Carboniferous period, the age during which the earth's major coal seams were formed. Before the end of that period the first reptiles had evolved, and early reptile orders were common land animals in the Permian period, with which the Paleozoic Era closed.

The *Mesozoic Era,* the "Middle Age" of the story of life, is frequently termed the Age of Reptiles, for members of that class dominated the land life of that era, and many types of reptiles now extinct flourished in the seas and in the air as well. The highest of vertebrate groups, further, had their beginnings in the Mesozoic; the oldest mammals appeared about the beginning of the Jurassic, and the oldest known birds are found toward the end of that period, but both groups remained inconspicuous till the end of the era.

The *Cenozoic Era* is the Age of Modern Life or Age of Mammals. Toward the end of the Mesozoic the reptile hordes died out, to leave that class of vertebrates in its modern impoverished phase. Modern bird types appeared early in the Cenozoic, and, most conspicuously, the mammals rapidly evolved into the varied progressive groups which dominate the land today.

VERTEBRATE CLASSIFICATION

The backboned animals constitute the subphylum Vertebrata of the phylum Chordata. The next step in classification is a division of the varied vertebrates into a series of classes. The distinguishing features of certain of these classes are obvious to anyone who has the slightest familiarity with nature. The *class Mammalia* includes the mammals, the familiar warm-blooded, hair-clothed animals among which man himself is to be included; the birds, *class Aves,* are readily distinguished by the presence of feathers and wings and by their possession, equally with mammals, of a high, controlled body temperature. The *class Reptilia,* lacking the progressive features of the birds and mammals, represents a lower level of land life, with lizards, snakes, turtles, and crocodiles as representatives. A fourth group is that of the *class Amphibia,* including frogs, toads, and salamanders—four legged animals, but reminiscent of fishes in many respects.

One commonly lumps the remaining lower vertebrates as "fish," and these forms (or most of them) are sometimes included in a single vertebrate class—the attitude being that, after all, they seem to be built on a common plan, as water dwellers with gills and locomotion performed by fins rather than limbs. But this is a rather personal, human viewpoint. An intellectual and indignant codfish could point out that this is no more sensible than putting all land animals in a single class, because, from his point of view, frogs and men, as four-limbed lung-breathers, are much alike. Actually, when we look at the situation objectively, a codfish and a lamprey, at two extremes of the fishy world, are as different structurally as amphibian and mammal. The fishes are perhaps best arranged in four classes of lower vertebrates: *class Agnatha* for jawless vertebrates, such as the living lampreys and fossil relatives; *class Placodermi* for primitive jawed fishes of the Paleozoic, now extinct; *class Chondrichthyes,* cartilaginous fishes, sharks, and their relatives; *class Osteichthyes,* the higher bony fishes which today constitute most of the piscine world.

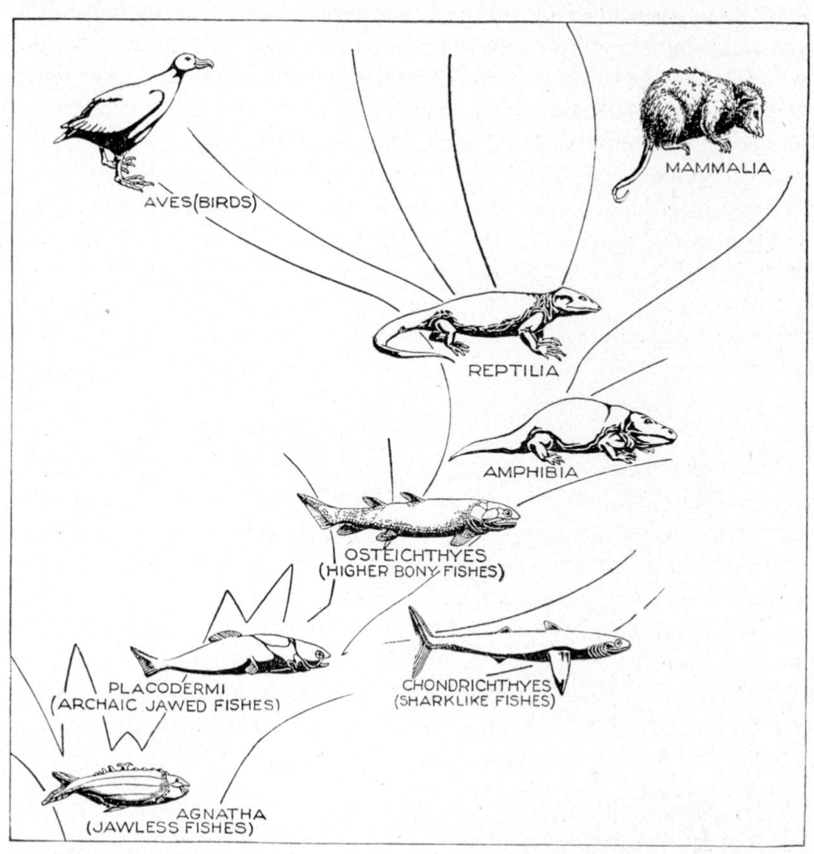

Fig. 11. A simplified family tree of the classes of vertebrates. (From Romer, Man and the Vertebrates, University of Chicago Press.)

If we wish to group these eight classes, we may for convenience consider the four higher land groups as constituting a *superclass Tetrapoda,* or four-footed animals, the fishes as making up a *superclass Pisces:*

Superclass Pisces	Class Agnatha
	Class Placodermi
	Class Chondrichthyes
	Class Osteichthyes
Superclass Tetrapoda	Class Amphibia
	Class Reptilia
	Class Aves
	Class Mammalia

Another, and in one respect a more scientific grouping, is to consider the three highest classes as forming a group termed *Amniota,* the remaining five constituting the *Anamniota*. This is based upon the fact that the lower types generally have a rather simple mode of reproduction, with eggs laid in the water and young developing there, whereas reptiles, birds, and ancestral mammals evolved a shelled egg, laid on land, within which a complex sort of development (described in a later chapter) takes place. The name Amniota is derived from the amnion, one of the membranes surrounding the embryo in this pattern of development.

A synoptic classification of vertebrates is given at the end of the book. In its simplest form the phylogeny of the vertebrate classes (Fig. 11) may be diagrammed thus:

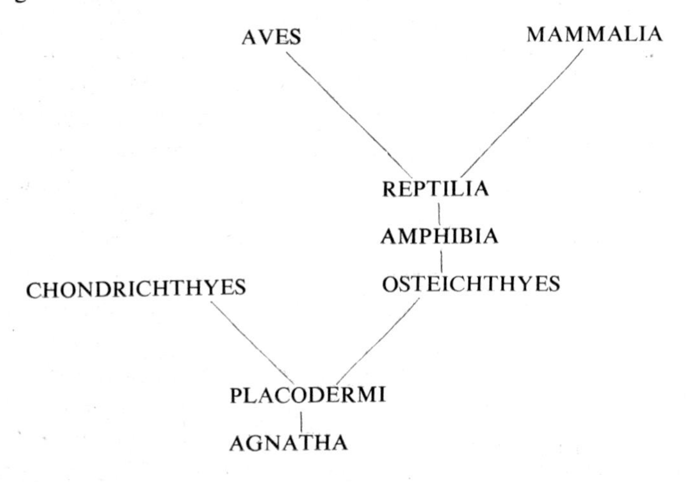

JAWLESS VERTEBRATES

Living *lampreys* and *hagfishes,* together termed *cyclostomes* (Figs. 12, 13), are representatives of a lowly group, the *class Agnatha*—jawless ver-

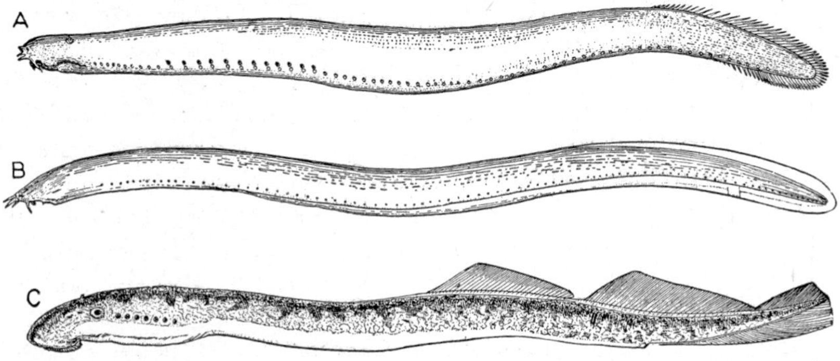

Fig. 12. The three main types of cyclostomes. *A*, The slime hag, Bdellostoma; *B*, the hagfish, Myxine; *C*, the lamprey, Petromyzon. (From Dean.)

tebrates. These fishes are eel-like in appearance, but much more primitive in their structure than true eels (which are highly developed bony fishes).

They are soft-bodied and scaleless and, though having a feeble skeleton of cartilage, lack bones entirely. There are no traces of paired fins, and, most especially, jaws are totally lacking. Adult cyclostomes are predaceous; a rasping tonguelike structure substitutes for the absent jaws and enables them to attack the flesh of higher fishes upon which they prey. A curious feature is that there is but a single nostril, which in hagfishes is at the tip of the snout, in lampreys high up on the top of the head. In lampreys there is a distinct larval stage in which body structure is much simpler than that of the adult and in fact resembles that of Amphioxus in many regards. Like Amphioxus, and unlike the adult cyclostome, the larva is a relatively sessile food strainer.

It is generally agreed that the absence of jaws and, probably, that of fins are primitive features of cyclostomes. Other characters, however, are more dubiously primitive. There is considerable reason to regard the absence of a bony skeleton as a degenerate feature; the predaceous habit can hardly have been present in ancestral vertebrates (mutual cannibalism is not, to say the least, advantageous), and the tongue rasp is a cyclostome specialty. Cyclostomes represent a primitive level of vertebrate development; they are not, however, in themselves ancestral vertebrates.

When we look into the fossil record, we find that the oldest and most primitive of fossil vertebrates, found in Ordovician and Silurian freshwater deposits and surviving into the Devonian, were small fishlike creatures known as *ostracoderms* (Figs. 15, 16). Superficially there is little resemblance to the cyclostomes, but study shows that ostracoderms were jawless ancient representatives of the class Agnatha. In a majority of ostracoderms there was, as in cyclostomes, but a single nostril, and that high atop the head. Many ostracoderms (like cyclostomes) lacked paired appendages, although in some there were paired spines or peculiar flaps projecting from the body behind the head.

The ostracoderms were, in the absence of jaws or other biting or rasping structures, nonpredaceous. We find that in most of them there was a greatly expanded "head" region, most of which was occupied by large gill chambers (Fig. 190, p. 322). It seems obvious that these oldest vertebrates, like their chordate ancestors and like the larval lampreys of today, made their living by straining food materials through their gill system. Many of them, although capable of locomotion with a fishy tail, were much flattened and must have been relatively sluggish animals.

A major contrast with the modern cyclostomes lies in the skeleton. All ostracoderms were covered by a good bony armor, and in certain of them there was developed an internal bony skeleton as well. It was formerly assumed that the primitive vertebrates were (like the living cyclostomes and sharks) boneless forms, with a skeleton of cartilage only. This may have been true of the still older ancestral chordates and immature ostracoderms. But the prevalence of bone in the oldest known fossil vertebrates, and evidence of reduction instead of increase in ossification in the later history of many fish groups, suggest that the early vertebrates were armored as adults, and that absence of bone in the lower living vertebrates is a degenerate rather than a primitive characteristic. As to the reasons for this early development of bone, we are none too certain. One suggestion lies in the fact that we find in association with the oldest vertebrates, in the stream deposits in which they are found, remains of eurypterids—ancient water scorpions. These forms were voracious and, on the average, considerably larger than the little ostracoderms amidst which they lived. It may be that in their earliest phases the vertebrates were the underdogs in the world of fresh-water life; bony armor may have been a defense against these invertebrate enemies. Later, as vertebrates became larger, speedier, and themselves predaceous, the eurypterids vanished from the fossil record.

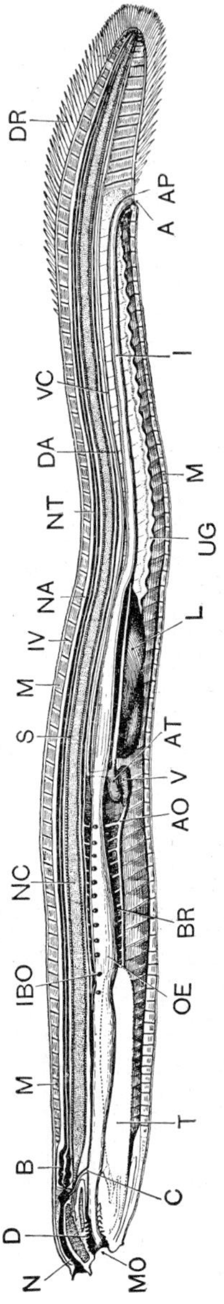

Fig. 13. Longitudinal section of a slime hag, Bdellostoma. *A*, Anus; *AO*, ventral aorta; *AP*, abdominal pore; *AT*, atrium of heart; *B*, brain; *BR*, gill pouch; *C*, duct from nasal pit to throat; *D*, horny, toothlike structures; *DA*, dorsal aorta; *DR*, dorsal fin rays; *I*, intestine; *IBO*, internal gill openings; *IV*, septum between muscle plates; *L*, liver; *M*, muscle segments; *MO*, mouth; *N*, nostril sac; *NA*, sheath of spinal cord; *NC*, notochord; *NT*, neural tube (spinal cord); *OE*, pharynx; *S*, sheath of notochord; *T*, extrusible "tongue;" *UG*, urogenital organs; *V*, ventricle of heart; *VC*, posterior cardinal vein. (From Dean.)

PLACODERMS

The ostracoderms were at the peak of their development during the Silurian. At the end of that period there appeared somewhat more advanced fish types which were exceedingly prominent in the following Devonian period, but became extinct before the close of the Paleozoic. These were mostly grotesque forms, quite unlike any fishes living today. They are currently grouped as a special class of vertebrates, the *Placodermi,* the name referring to the fact that most were, like ostracoderms, covered to a variable degree by armor plating.

All had jaws. This represents a major advance over the ostracoderms, one which opened up new avenues of life to fishes and enabled them to become more active and wider-ranging animals. The term *gnathostome*—"jaw-mouthed"—is often applied to placoderms and all higher vertebrates in contrast with the Agnatha. The jaws of placoderms, however, are frequently of peculiar types and are seemingly primitive or aberrant in build—nature was still "experiment-

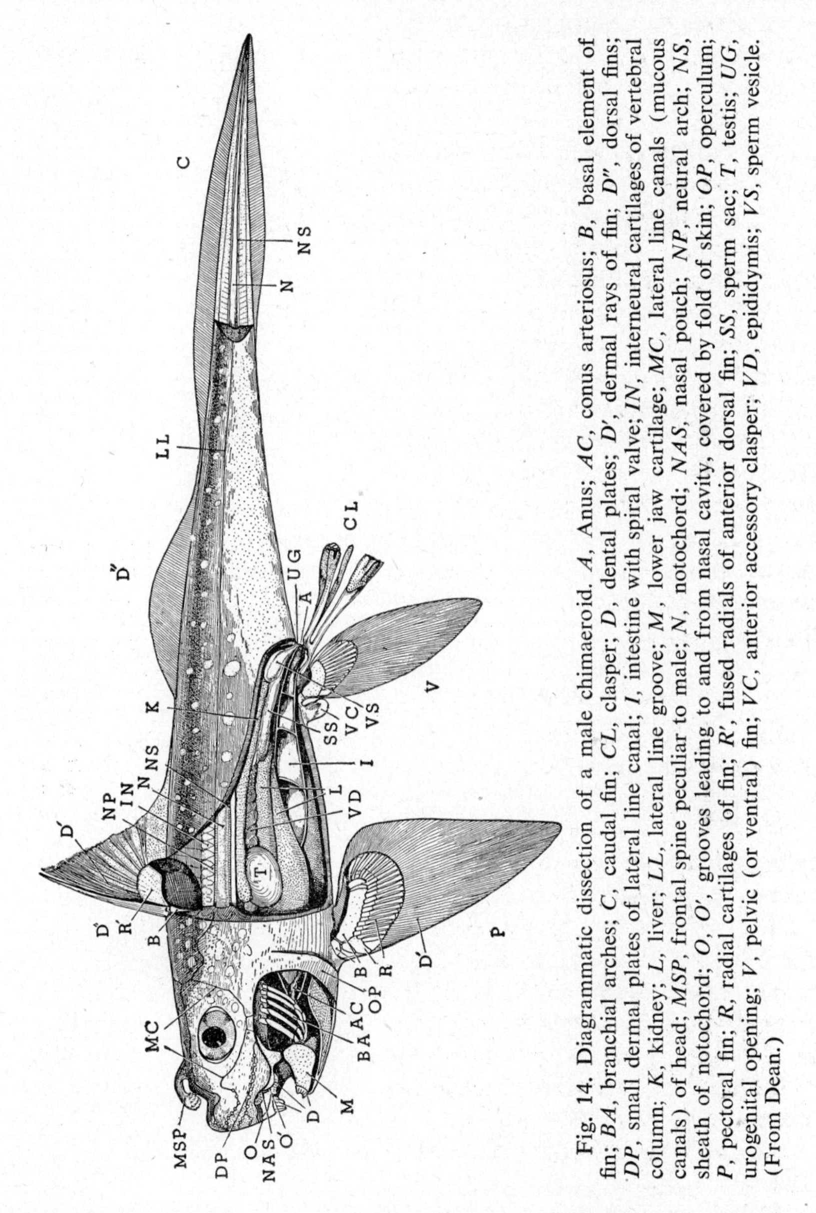

Fig. 14. Diagrammatic dissection of a male chimaeroid. *A*, Anus; *AC*, conus arteriosus; *B*, basal element of fin; *BA*, branchial arches; *C*, caudal fin; *CL*, clasper; *D*, dental plates; *D′*, dermal rays of fin; *D″*, dorsal fins; *DP*, small dermal plates of lateral line canal; *I*, intestine with spiral valve; *IN*, interneural cartilages of vertebral column; *K*, kidney; *L*, liver; *LL*, lateral line groove; *M*, lower jaw cartilage; *MC*, lateral line canals (mucous canals) of head; *MSP*, frontal spine peculiar to male; *N*, notochord; *NAS*, nasal pouch; *NP*, neural arch; *NS*, sheath of notochord; *O*, *O′*, grooves leading to and from nasal cavity, covered by fold of skin; *OP*, operculum; *P*, pectoral fin; *R*, radial cartilages of fin; *R′*, fused radials of anterior dorsal fin; *SS*, sperm sac; *T*, testis; *UG*, urogenital opening; *V*, pelvic (or ventral) fin; *VC*, anterior accessory clasper; *VD*, epididymis; *VS*, sperm vesicle. (From Dean.)

ing" with these new structures. Fins, too, were developing, in connection with the new freedom which fishes were acquiring, but these structures also were variable and often oddly designed (from a modern point

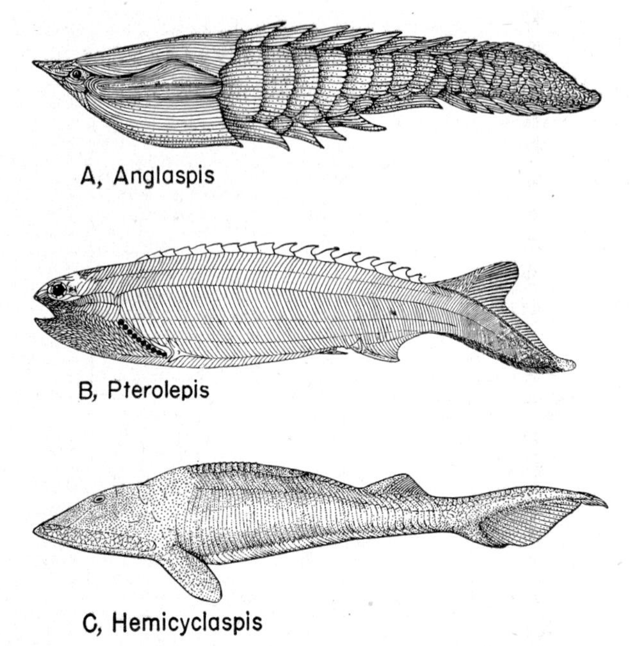

Fig. 15. Fossil ostracoderms. *A*, Anglaspis, representative of a group (Heterostraci) in which the nostrils were presumably paired and ventral in position. The gills opened by a common slit behind the main armor plates on either side. *B*, Pterolepis, member of a group (Anaspida) of relatively active swimmers. The tail is tipped downward, the reverse of the shark condition. The gill openings are rounded circles on the flank; small spines are present at the point where paired pectoral and pelvic fins develop in higher fishes. The nostril opening lay dorsally between the eyes. *C*, Hemicyclaspis, a member of the Cephalaspis group (order Osteostraci). The head and gill region are enclosed above in a solid, bony shield. The nostril opening is, again, placed dorsally between the orbits. The gills opened ventrally beneath the broad "head" (cf. Fig. 190, p. 322). Flipper-like structures are seen in the position of pectoral paired fins. (*A* and *C* after Heintz; *B* after Kiaer.)

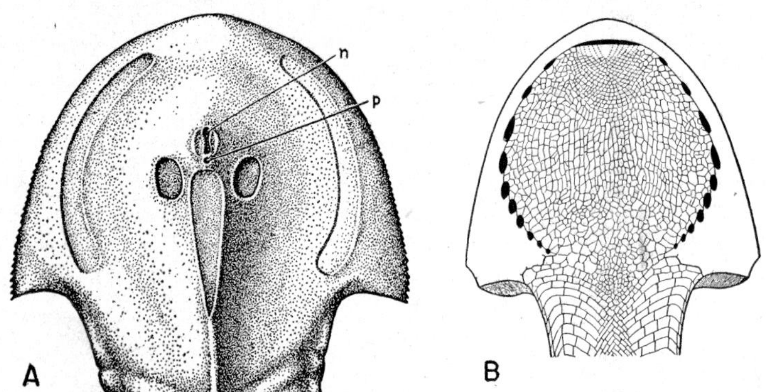

Fig. 16. *A*, Dorsal and *B*, ventral views of the head region of a fossil ostracoderm of the Cephalaspis type (cf. Fig. 15, *C*). Dorsally are seen openings for the paired eyes, median eye (*p*), and a median slit (*n*) for nostril and hypophyseal sac. Ventrally, the throat was covered by a mosaic of small plates, beneath which lay an expanded set of gill pouches (Fig. 190). Round openings on either side are for the gills; the mouth is a small anterior slit. (After Stensiö.)

of view). The early placoderms were essentially fresh-water dwellers, like the ostracoderms before them; but during the Devonian, certain of them had invaded the seas.

Of placoderms, the most "normal" in appearance were the *acanthodians* (Fig. 17, *A*), usually termed "spiny sharks." The general body proportions were sharklike, but the acanthodians were most unsharklike in other respects—particularly the fact that they were fully clad in well-

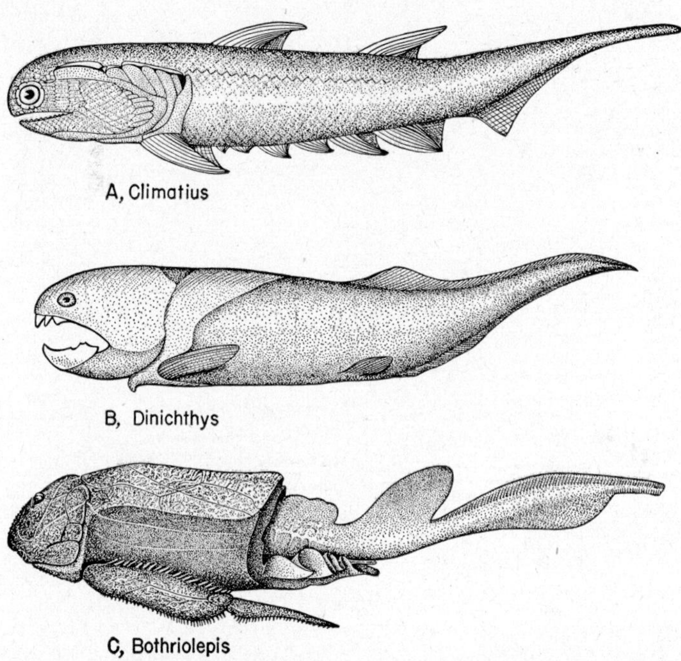

Fig. 17. Fossil Devonian placoderms. *A,* An acanthodian (spiny "shark"), with large fin spines and accessory fins between pectoral and pelvic pairs. *B,* Giant arthrodire with "head" and thoracic armor plating and a naked body. *C,* An antiarch with peculiar bony "flippers" in the place of pectoral fins. (*A,* data from Watson; *B* after Heintz; *C* after Patten.)

developed bony scales comparable to those found in some of the higher bony fishes. The fins mainly consisted of spines—sometimes of large size—with, apparently, but a small web of skin behind them. More common in much of the Devonian were the *arthrodires*—jointed-necked fishes (Fig. 17, *B*). In these the head and gill region was covered by a great bony shield, and a ring of armor sheathed much of the body, the two sets of armor connected by a movable joint. Peculiar bony plates served the function of jaws and teeth. The posterior part of the body was quite naked. In some forms there have been found true paired fins, but in the most primitive arthrodires all we find is a pair of enormous, hollow, fixed spines projecting outward from the shoulder region—some sort of hold-fast or balancing structures. A third group of placoderms was that of the *anti-*

archs of the Devonian (Fig. 17, *C*)—grotesque little animals which had two sets of armor like the arthrodires, but small heads, tiny nibbling jaw plates and, for limbs, a pair of jointed "flippers" projecting from the body like bony wings. Still other (but poorly known) placoderms had reduced armor and more normal fin development and rather suggest a transition from armored ancestors to shark types.

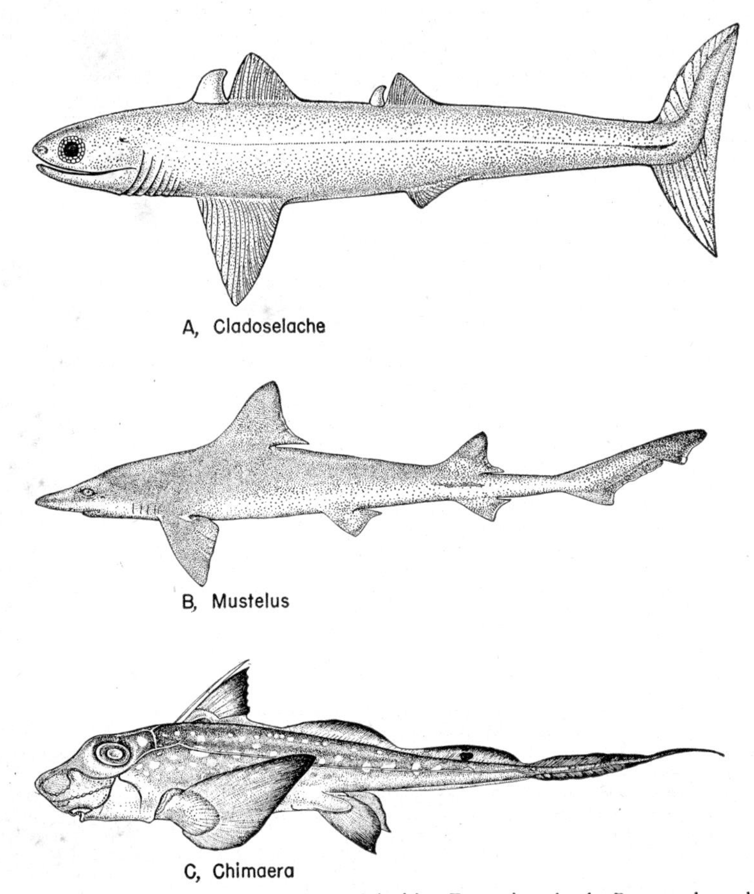

Fig. 18. Cartilaginous jawed fishes. *A*, Primitive Devonian shark; *B*, a modern shark type; *C*, a chimaera. (*A* after Dean and Harris; *B* after Garman; *C* after Dean.)

Most of the placoderms were obviously far off the main lines of vertebrate evolution, and few, if any, of the known types may possibly be regarded as actual ancestors of later vertebrates. As a group, however, they appear to represent nature's first essay in the development of jawed vertebrates. Most of these "experimental models" were not, in the long run, successful; others—poorly known or still unknown—gave rise to the two classes of more advanced fishes.

SHARKLIKE FISHES

The modern sharks are the typical representatives of a major surviving group of jaw-bearing marine fishes—the *Chondrichthyes.* This name, "cartilaginous fishes," refers to the fact that bone is unknown in any member of the group. It seems probable that the absence of bone in sharks is due to a process of reduction; the toothlike denticles present in the shark skin and the spines sometimes present on the fins are the last remnants of the armor which once sheathed their placoderm forbears. Sharks first appear in the latter part of the Devonian, and Cladoselache (Fig. 18, *A*) of that period is a form which may be roughly ancestral to many,

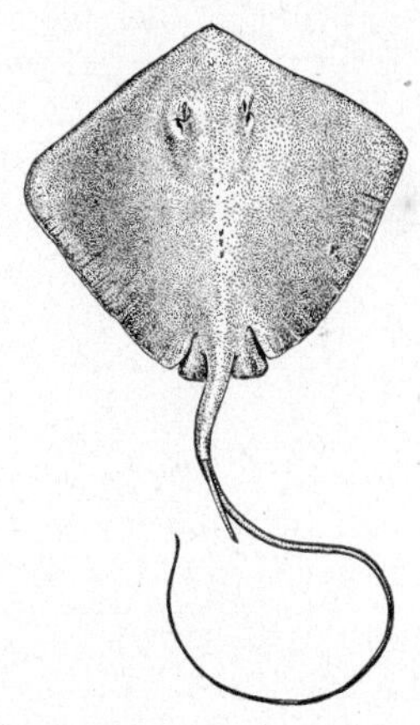

Fig. 19. A skate, Dasybatus. The pectoral fins are enormously expanded, the tail is reduced to a whiplash. The spiracle is placed just behind the eye. (After Garman.)

if not all, of the later cartilaginous fishes. There was a variety of shark forms in the seas of the late Paleozoic, and toward the end of the Mesozoic we find shark types similar to those in modern oceans (Figs. 18, *B,* 20). In the Mesozoic, too, there appear the first of the *skates* and *rays* (Fig. 19), forms derived from sharks; they have taken to a mollusc-eating diet and a bottom-dwelling mode of life with which is correlated the flattened body shape of these unattractive fishes.

A distinct group of cartilaginous fishes is that of the *chimaeras* or *Holocephali* (Figs. 14, p. 37; 18, *C,* 20), which are relatively rare oceanic forms. These are, like the skates, mollusc eaters; but the body is not depressed, and their peculiarities include, among other features, the development of peculiar tooth plates and upper jaws which (in contrast to those of sharks) are solidly fused to the braincase. It is presumed that the chimaeras were derived from the early sharks; the fossil record, however, is imperfect.

BONY FISHES

The class *Osteichthyes* includes the vast majority of fishes. As the name implies, they are forms in which (in contrast to the sharks) a bony

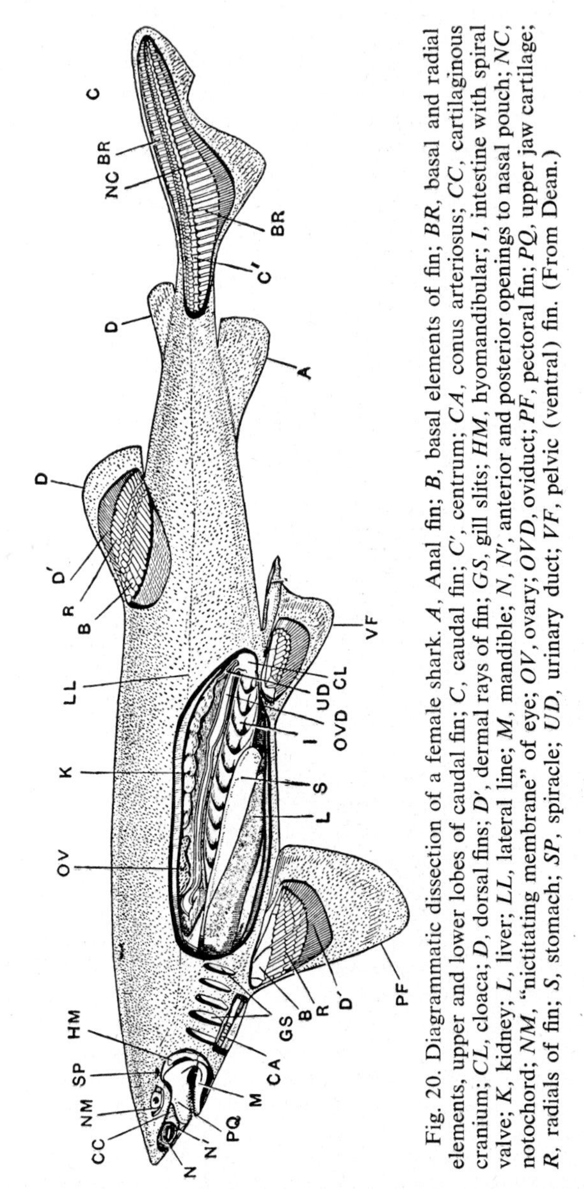

Fig. 20. Diagrammatic dissection of a female shark. *A*, Anal fin; *B*, basal elements of fin; *BR*, basal and radial elements, upper and lower lobes of caudal fin; *C*, caudal fin; *C′*, centrum; *CA*, conus arteriosus; *CC*, cartilaginous cranium; *CL*, cloaca; *D*, dorsal fins; *D′*, dermal rays of fin; *GS*, gill slits; *HM*, hyomandibular; *I*, intestine with spiral valve; *K*, kidney; *L*, liver; *LL*, lateral line; *M*, mandible; *N*, *N′*, anterior and posterior openings to nasal pouch; *NC*, notochord; *NM*, "nictitating membrane" of eye; *OV*, ovary; *OVD*, oviduct; *PF*, pectoral fin; *PQ*, upper jaw cartilage; *R*, radials of fin; *S*, stomach; *SP*, spiracle; *UD*, urinary duct; *VF*, pelvic (ventral) fin. (From Dean.)

skeleton has been retained, and improved upon. A characteristic bone pattern is to be found, with variations, in most members of the group, in skull, jaws, gill coverings, and in a set of bony scales covering the body.

It was once believed that these fishes were descendants of sharklike forms and that bone was in them a new acquisition. It now, however, appears

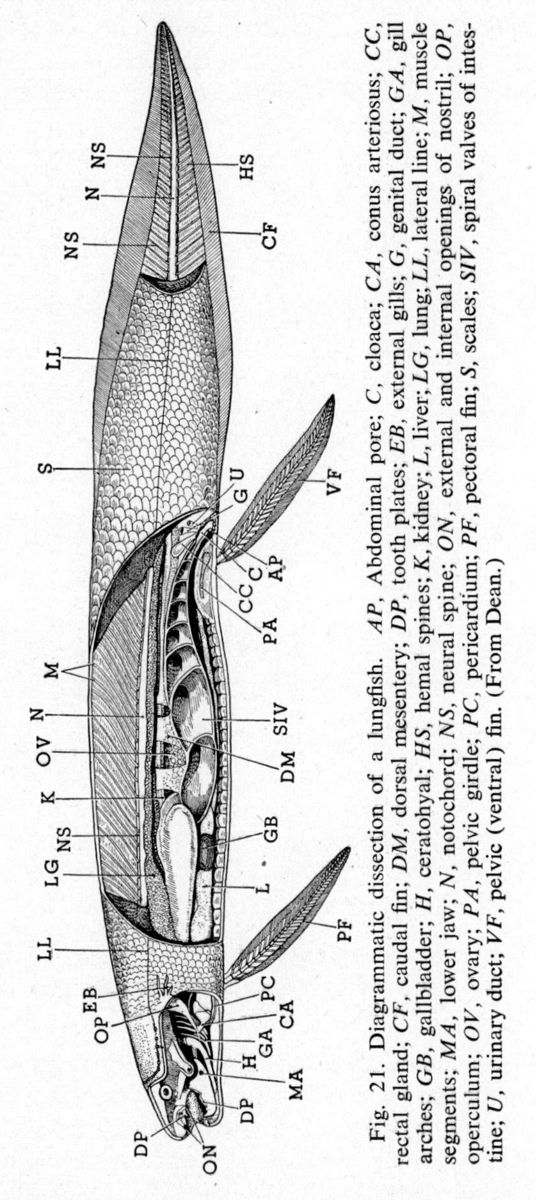

Fig. 21. Diagrammatic dissection of a lungfish. *AP*, Abdominal pore; *C*, cloaca; *CA*, conus arteriosus; *CC*, rectal gland; *CF*, caudal fin; *DM*, dorsal mesentery; *DP*, tooth plates; *EB*, external gills; *G*, genital duct; *GA*, gill arches; *GB*, gallbladder; *H*, ceratohyal; *HS*, hemal spines; *K*, kidney; *L*, liver; *LG*, lung; *LL*, lateral line; *M*, muscle segments; *MA*, lower jaw; *N*, notochord; *NS*, neural spine; *ON*, external and internal openings of nostril; *OP*, operculum; *OV*, ovary; *PA*, pelvic girdle; *PC*, pericardium; *PF*, pectoral fin; *S*, scales; *SIV*, spiral valves of intestine; *U*, urinary duct; *VF*, pelvic (ventral) fin. (From Dean.)

more probable that the bony skeleton here is simply a retention and perfection of that which was present in ostracoderms and placoderms.

The first Osteichthyes are found in rock of early Devonian age; the class is, thus, much older than the sharks. By the middle of the Devonian, bony fishes were already the dominant forms in fresh waters; and they were present in great variety and abundance in later Paleozoic periods. In the Mesozoic they invaded the seas as well, and marine waters became the headquarters of the class. Today 95 per cent or more of all fishes are members of the Osteichthyes. Lungs appear to have been present in all primitive bony fishes, although today such structures usually have been lost or converted into a swim bladder. Lungs were presumably an aid to

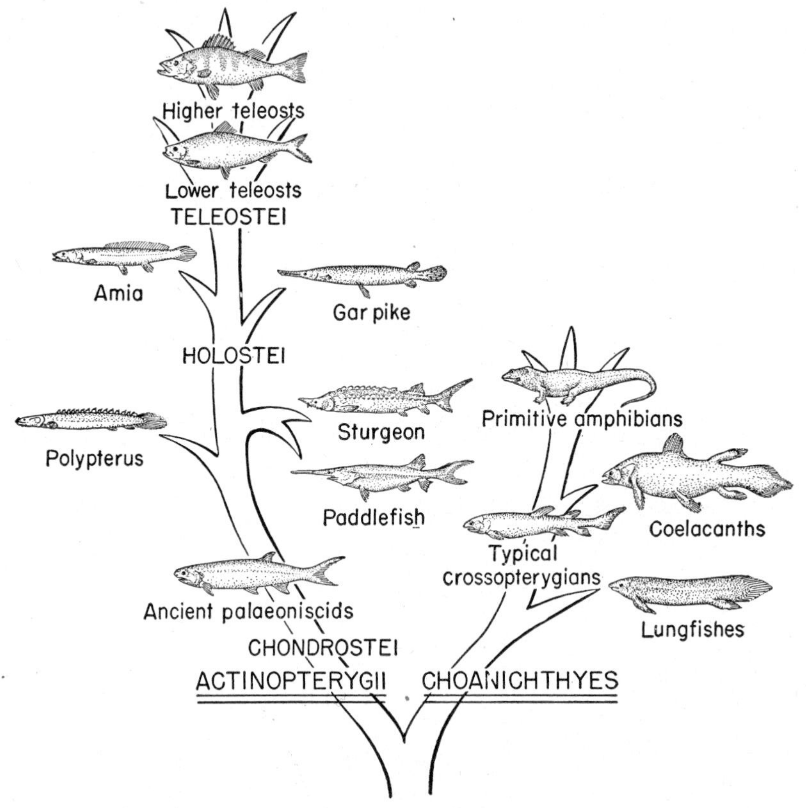

Fig. 22. A simplified family tree of the bony fishes, to show their relations to one another and to the amphibians.

survival under conditions of seasonal drought; such conditions may have been present in the fresh waters in which the ancestral Osteichthyes lived.

The phylogeny of the bony fishes is complicated, but must be reviewed in order to keep in mind the position on the family tree of many interesting and anatomically important types (Fig. 22). At the very beginning of their known history the Osteichthyes had already subdivided into two major groups, termed the subclasses Choanichthyes and Actinopterygii.

Choanichthyes. From the viewpoint of the descent of land animals the Choanichthyes are the more important of the two, for they contain the order Crossopterygii, from which land vertebrates appear to have descended, and the order Dipnoi, the lungfishes, which are surviving cousins of our piscine ancestors. The Choanichthyes have, in contrast to the other subclass, internal nostrils associated with lung breathing (the name refers to this feature), fleshy-lobed fins (suitable for development into land limbs), and, as a technical character, scales which in early forms were of a structure quite distinct from those of the actinopterygians (cf. p. 146).

Crossopterygii. In the middle and later Devonian the commonest of bony fishes were crossopterygians (Fig. 23, *A*), aggressive, predaceous fishes which in most regards show features of a sort to be expected in the

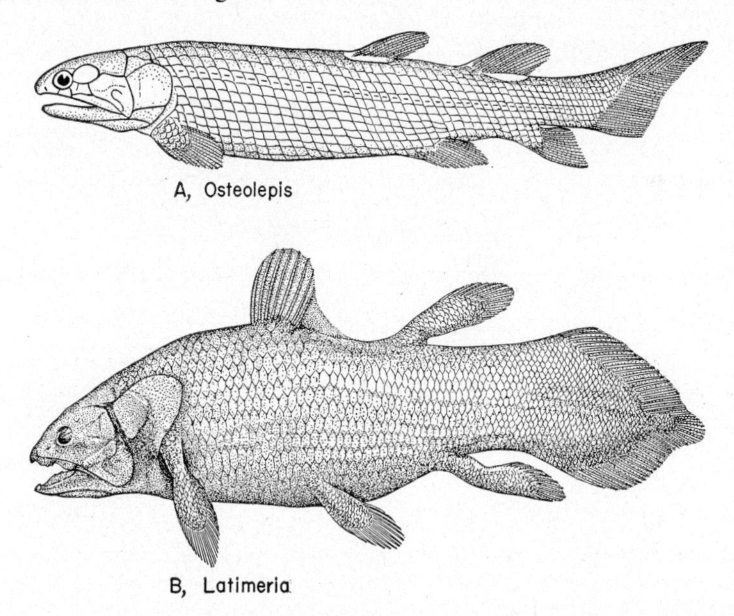

Fig. 23. Crossopterygians. *A*, Typical Devonian form; *B*, the living coelacanth. (*A* after Traquair; *B* after Smith.)

ancestors of the amphibians. In the Carboniferous, however, they became relatively rare, and typical crossopterygians were extinct by the close of the Paleozoic.

Meantime, however, a peculiar side branch of the crossopterygians had developed, termed the coelacanths (Fig. 23, *B*). These were forms with stub snouts, feeble jaws and teeth, which migrated into the Mesozoic seas. The last fossil coelacanths are found in Cretaceous rocks, and it was long taught that our crossopterygian relatives had been extinct since the days of the dinosaurs. In 1939, however, to the surprise of science, a strange fish caught off the coast of South Africa proved to be a coelacanth! This single specimen was, unfortunately, incompletely preserved; it is to be hoped that further specimens will be obtained in future years. Knowledge of the anatomy of this fish is greatly to be desired. But it must be kept in mind that coelacanths are highly specialized fishes, not too representative of the ancestral crossopterygians from which the tetrapods were derived; coelacanth structure may not be too good a guide to a knowledge of the nature of the Paleozoic ancestors of land vertebrates.

Lungfishes. The *Dipnoi,* or lungfishes (Figs. 21, 24), are represented today by three genera, living, one each, in tropical regions of Australia, Africa, and South America. In many anatomical features and in their mode of development the lungfishes closely resemble the amphibians, and they were once thought by many to be actual amphibian ancestors. But it is now more reasonable to believe that these features were present as well in their relatives, the ancestral crossopterygians, and that the lungfishes are to be regarded as "uncles" rather than the actual progeni-

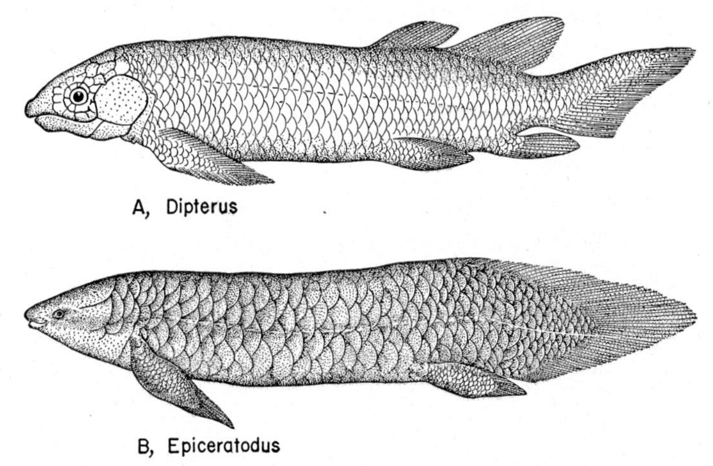

Fig. 24. Lungfishes. *A,* The oldest Devonian fossil type; *B,* Epiceratodus of Australia. The median fins have changed greatly during the history of the group. (*A* after Traquair; *B* after Dean.)

tors of land vertebrates. The skull structure of lungfishes, living and fossil, is of a peculiar type obviously unlike that of a proper amphibian ancestor; and, in connection with a diet of hard invertebrates, there are present in all lungfishes specialized fan-shaped toothplates. It is of interest that the lungfishes have survived only in regions where we find today conditions of seasonal drought similar to those which we believe to have been present in the Devonian.

Actinopterygians. As types ancestral to higher vertebrates, the Choanichthyes are of major interest; but as successful fishes, the *Actinopterygii,* or *ray-finned fishes,* are vastly more important. From Carboniferous times on, these have been the dominant fishes. In contrast with the choanate orders, internal nostrils are never present; and except in a few primitive forms there is never a fleshy lobe to the fins. Instead, as the name implies, the paired fins are webs of skin supported by horny rays.

The actinopterygians have long been divided into three groups—here considered as superorders—which are, in ascending order: the Chondrostei, Holostei, and Teleostei. The names are not particularly appropriate, but may be retained for convenience.

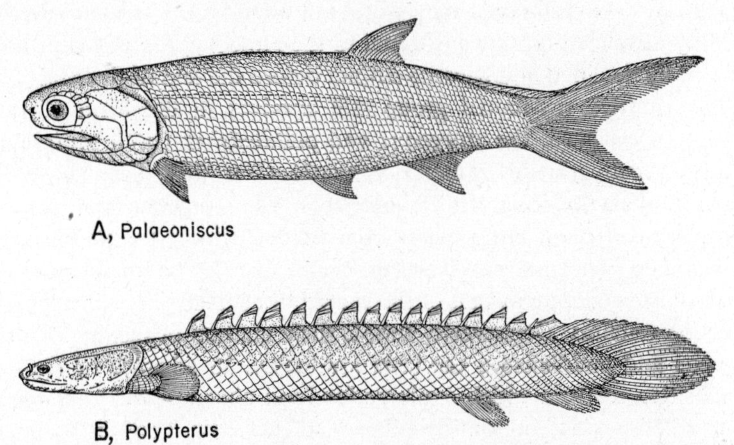

Fig. 25. Primitive ray-finned fishes. *A*, An early Paleozoic type; *B*, a living representative of the ancient palaeoniscoids, with modified fin structure. (*A* after Traquair; *B* after Dean.)

Chondrostei. In the Paleozoic the ray-finned fishes were represented by abundant genera of Chondrostei known as *palaeoniscoids* (Fig. 25, *B*). These were generally fishes of small size, with rather uptilted sharklike

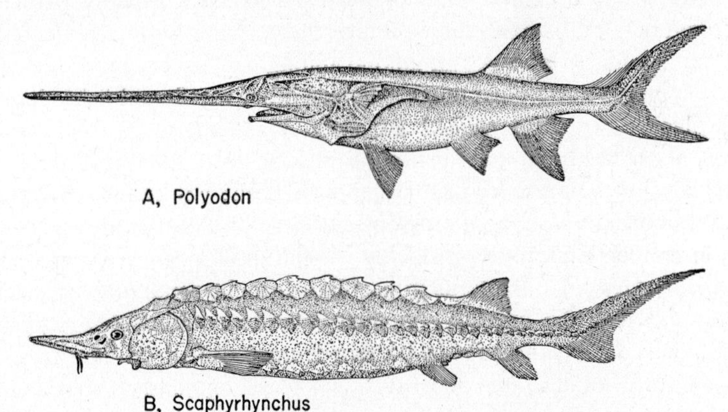

Fig. 26. Chondrosteans. *A*, The paddlefish or "spoon-billed cat" of the Mississippi; *B*, a sturgeon. (After Goode.)

tails (the heterocercal type, cf. Fig. 97, *A*, p. 171) and with scales covered by a shiny material known as ganoine (cf. p. 146). The term "ganoid" is sometimes applied to fishes of this group, but is to be avoided, for it is usually applied indiscriminately and variably to any old fish with shiny scales. In the earliest days of bony fish history, primitive ray-finned forms were outnumbered by crossopterygians and lungfishes, but by the Carboniferous they were far more numerous than their early rivals and swarmed in the ancient lakes and streams in immense numbers and variety. In the Triassic they were still abundant, but mainly represented by advanced types transitional to the Holostei; the palaeoniscoids then rapidly declined and became extinct before the end of the Mesozoic.

This primitive ray-finned group still survives in the form of three aberrant types. Two, the *sturgeons* and *paddlefishes* (both represented in North America), are rather degenerate (Fig. 26). They have lost the ganoid scale covering of their ancestors. The sturgeons have retained several rows of plain bony plates as a partial armor; the paddlefish is nearly naked. The internal skeleton, highly ossified in their ancestors, is nearly as degenerate as that of the sharks, for it is mainly cartilaginous; little bone remains. Degenerate, too, is their method of feeding. In both sturgeons and paddlefishes the jaws are feeble. In advance of the jaws is a sensitive rostrum which explores for food ahead of them; sturgeons and paddlefishes are bottom-dwelling mud grubbers. Only in their persistently sharklike tail fin is there much evidence of the trade-marks of the older palaeoniscoids.

The third type of chondrostean survivor is *Polypterus* (Fig. 25, *B*) of Central Africa, which lives in much the same environment as the lungfish of that continent.* In its fins Polypterus is much modified from the ancestral type. Its tail fin has become essentially symmetrical; its dorsal fin is split up into a series of small sail-like structures (to which its name refers), and its paired fins, unlike those of any proper actinopterygian, have a fleshy lobe. Again unique among actinopterygians is the fact that Polypterus has typical lungs, whereas other ray-finned fishes have in their place a structure termed the swim bladder (cf. p. 325), which almost never has breathing functions and is, instead, a hydrostatic organ.

Because of the presence of lungs and the fleshy paired fins, Polypterus was long considered to be a crossopterygian. But closer study shows this to be incorrect. Lungs were probably present in all primitive bony fishes, and the survival of Polypterus in peculiar drought conditions appears to be due (as with the lungfish) to their retention. Although the fins are fleshy, they differ markedly in pattern from those of crossopterygians. The anatomy of the animal is as a whole in agreement with that of actinopterygians rather than that of the Choanichthyes, and the scales are of the true ganoid type, contrasting strongly with those of the Choanichthyes. Polypterus is surely a somewhat modified descendant of the ancient palaeoniscoids.

* Calamoichthys is a closely related form, but with a more elongate eel-like shape, from the same region. I have made no reference to this genus in later sections, but comments on Polypterus usually apply to Calamoichthys as well.

Holostei. Succeeding the chondrosteans as dominant fishes in the middle Mesozoic were the *holosteans.* In them the old, long, upturned sharklike tail had become shortened, the jaws tended to have a shorter gape, and the scales, in many cases, tended to lose their shiny ganoid covering. Another trend, too, was apparent at this time: the ray-finned fishes were invading the seas. The major center of actinopterygian evolution from the Jurassic period onward was the ocean.

The only two surviving holosteans, however (the group became rare in the Cretaceous), are fresh-water forms, both present in North America. The gar pikes (Lepidosteus, and the like, Fig. 27, *A*) are fast swimming fishes fairly representative of the ancestral holosteans in many ways, but

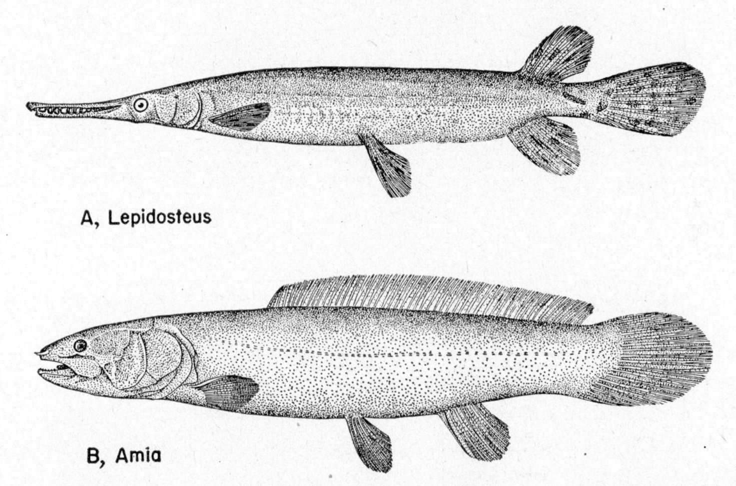

Fig. 27. Holosteans. *A,* The gar pike; *B,* the bowfin. Both are inhabitants of North American fresh waters and are the only survivors of a stage antecedent to the teleosts. (After Goode.)

specialized in their elongated jaws associated with predaceous habits. A more advanced type is *Amia* (Fig. 27, *B*), a lake and river fish of the Midwest and South, popularly termed the "dogfish" or bowfin. In these holosteans the internal skeleton is not at all degenerate, but in Amia the scales have lost their ganoine covering, and the tail is much like that of the teleosts.

Teleostei. The teleosts, as the name suggests, form the end group of the ray-finned fishes, and are the fishes dominant in the world today. They appear to have originated from the holosteans in the Mesozoic oceans, and before the close of the Cretaceous had replaced the older group as the most flourishing of fish types. In teleosts the originally sharklike tail has been reduced, and the tail fin has a superficially symmetrical appearance. The scales have lost all trace of the original shiny ganoid covering, and are generally thin, flexible, bony structures.

In the oceans where they originated the teleosts constitute, despite the presence of sharks and skates, the vast majority of all piscine inhabitants. They have invaded every possible marine habitat from the strand line to the abyssal depths. They reinvaded fresh waters as well, where they constitute the entire fish population except for a few species of lampreys and the half-dozen bony fish types mentioned earlier.

The herrings and similar forms (Fig. 28, *A*) appear to represent a primitive group of teleosts; salmon and trout are related. The carps and

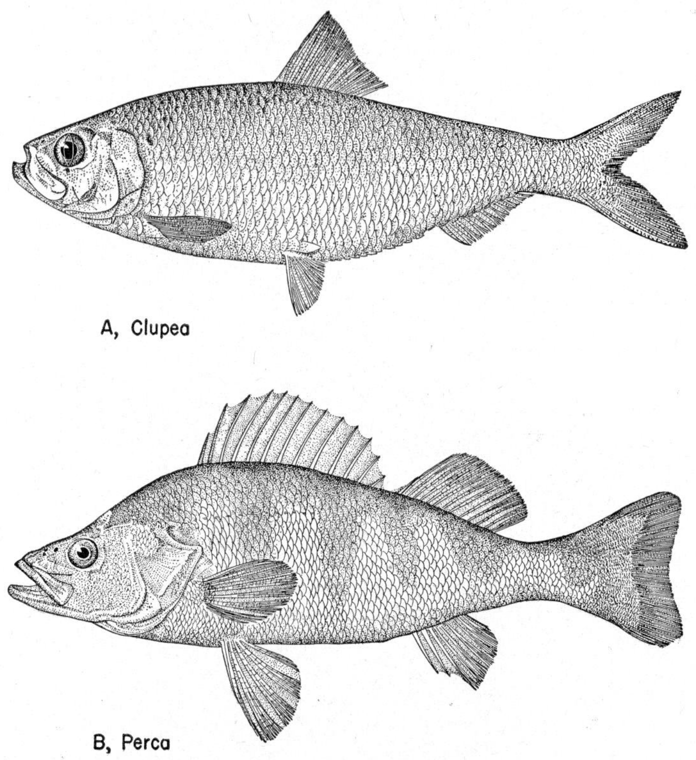

Fig. 28. Teleosts. *A*, Primitive type, the herring; *B*, an advanced, spiny teleost, the yellow perch. (After Goode.)

catfishes are characteristic of a major fresh-water division of the teleosts. More progressive teleosts, almost all marine, are the spiny-finned forms—the perch (Figs. 28, *B*, 29) is typical—in which parts, at least, of the fins are supported by stout spines rather than softer rays.

In any comparative study of vertebrate anatomy or physiology in which the teleosts are involved, their ecologic history must be kept in mind. Since land vertebrates come from fresh-water fishes, one tends to assume that structures or functions seen in fresh-water teleosts may be representative of those once present in tetrapod ancestors. But in addition to the fact that our common fishes belong to a different branch of the fish family tree from that from which land animals took origin, it must be kept in mind that modern fresh-water teleosts have not been, in all

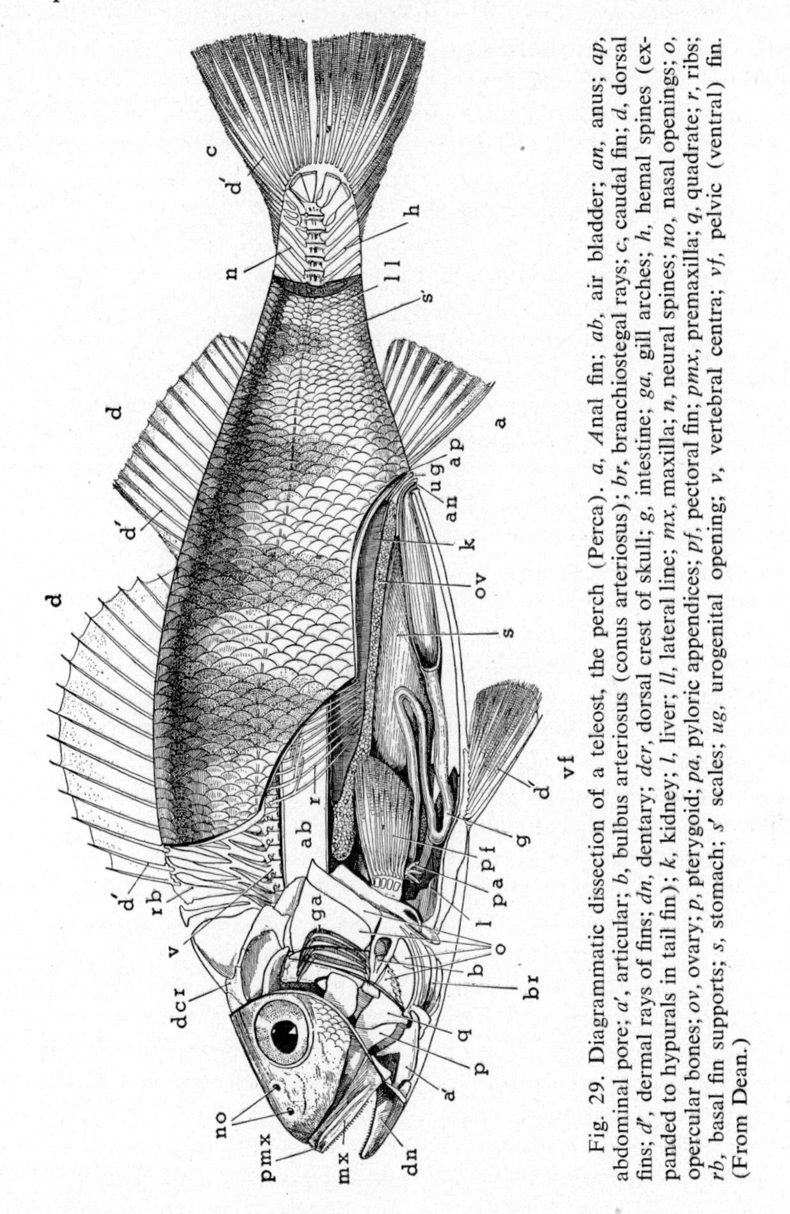

Fig. 29. Diagrammatic dissection of a teleost, the perch (Perca). *a*, Anal fin; *ab*, air bladder; *an*, anus; *ap*, abdominal pore; *a′*, articular; *b*, bulbus arteriosus (conus arteriosus); *br*, branchiostegal rays; *c*, caudal fin; *d*, dorsal fins; *d′*, dermal rays of fins; *dn*, dentary; *dcr*, dorsal crest of skull; *g*, intestine; *ga*, gill arches; *h*, hemal spines (expanded to hypurals in tail fin); *k*, kidney; *l*, liver; *ll*, lateral line; *mx*, maxilla; *n*, neural spines; *no*, nasal openings; *o*, opercular bones; *ov*, ovary; *p*, pterygoid; *pa*, pyloric appendices; *pf*, pectoral fin; *pmx*, premaxilla; *q*, quadrate; *r*, ribs; *rb*, basal fin supports; *s*, stomach; *s′* scales; *ug*, urogenital opening; *v*, vertebral centra; *vf*, pelvic (ventral) fin. (From Dean.)

probability, continuous residents in that environment since the early days of fish history. Between that time and the present there intervened a long marine phase.

AMPHIBIANS

Greatest, perhaps, of all ventures made by the vertebrates during their eventful history was the development of tetrapods in the invasion of the land—a step which involved major changes in function and resulted in profound structural modifications. The shifts, from swimming to four-footed walking, and from gill breathing to the dominance of lungs, are the most obvious of the modifications necessary in this step. But analysis shows that functional and structural changes were necessitated in almost every organ or organ system of the body.

The basic group of land vertebrates is the class Amphibia. There are three living orders: the frogs and toads (*Anura*); the newts and salamanders (*Urodela*); and some wormlike burrowers (*Apoda*). Commonest are the anurans, familiar to us in temperate regions and represented in great variety in the tropics. Their specialized nature is obvious in their jumping habits, which are responsible for many modifications in their structure, particularly in the skeletal system. Much more generalized in appearance, at least, are the salamanders of the order Urodela, retiring but not uncommon dwellers in moist habitats in the temperate zone. The Apoda will not be familiar to many readers of this work, for they include only a few genera of tropical burrowers which look not unlike earthworms.

In their external form the salamanders resemble the ancestral amphibians which first sprang from ancestral fishes. There is a fairly elongate but stoutly built body, with powerful trunk musculature, and a well-devoloped tail which is an aid in swimming. The median fins of fishes are gone, but the paired fins have developed into the typical land limbs which are the trade-mark of the tetrapods. In salamanders, and in the other modern amphibian orders as well, we find various internal features which bridge structural gaps between the crossopterygian fishes and the higher classes of land vertebrates.

Amphibian structure and function are important in comparative studies. But due caution must be exercised. Frogs and salamanders and apodans are amphibians, and the amphibians are the most primitive of tetrapod classes; but we must not complete a false syllogism by concluding that these modern amphibians are really primitive tetrapods. Even in its external anatomy, a frog or apodous form shows obvious signs of specialization and degeneration, and there is much evidence that many internal structures are likewise aberrant from the primitive tetrapod pattern. A frog is, in many ways, as far removed structurally from the oldest land vertebrates as a man, and even a salamander must be regarded with suspicion.

For the actual ancestors of the land dwellers as a whole, we must turn to the fossil record of the late Paleozoic when, in the Carboniferous and early Permian, there lived numerous and varied amphibians of a more primitive nature. Two major groups were then distinguishable.

One included a series of small animals, termed lepospondyls, of which a diagnostic feature lies in the fact that the central portions of the backbone segments were spool-shaped. The vertebrae of modern urodeles and apodans are built on essentially the same plan (cf. Fig. 85, *H*), and it is highly probable that these living orders are remnants of the archaic lepospondyl stock.

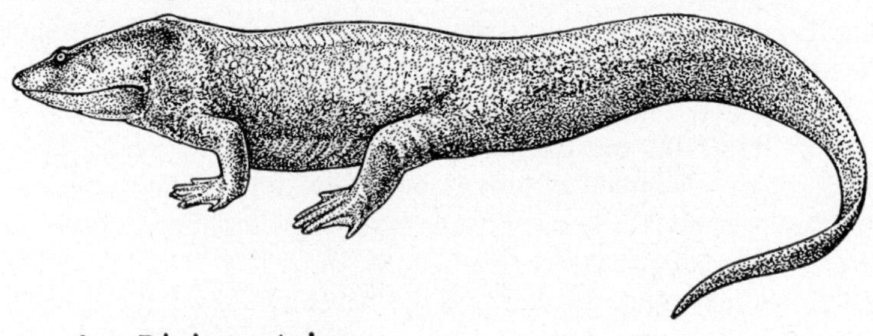

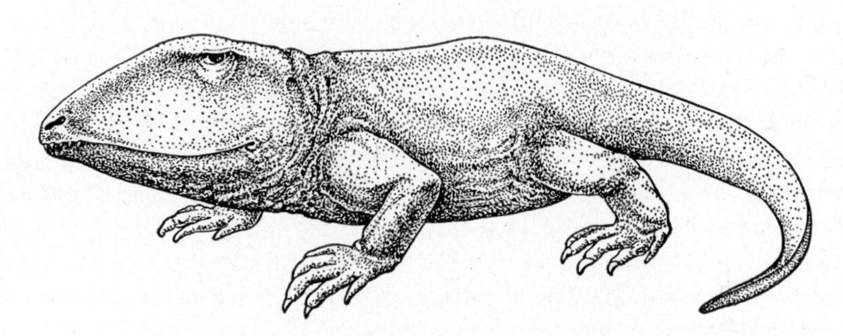

Fig. 30. *A,* Diplovertebron, a primitive Paleozoic amphibian (labyrinthodont); *B,* Ophiacodon, an early Permian reptile, representative of primitive members of that class, although showing indications of relationship to mammalian ancestry. (*A* after Gregory.)

It seems likely, however, that the older lepospondyls were, despite their antiquity, a side branch, if an early one, of the basal stock of land animals. The true base is to be sought among a second early group of amphibians, the *labyrinthodonts* (Fig. 30, *A*). These animals, of variable size, were in general considerably larger than contemporary lepospondyls; some soon attained crocodile proportions. Their vertebral construction was a diagnostic feature and was one from which that of reptiles and higher vertebrates could have been readily derived. Except for the absence of median fins and the presence of short but sturdy legs developed from paired fins, many features of the early labyrinthodonts are highly comparable with those of the crossopterygians from which they came. They were the first vertebrates to walk on land.

In the late Paleozoic and earliest Mesozoic, labyrinthodonts were abundant and varied. From them at an early stage came, presumably, the lepospondyl stock from which the salamanders and Apoda have descended. The frogs are believed, on current evidence, to have arisen from the labyrinthodont "main line" at a somewhat later date, and hence, despite their peculiar specializations, may be rather closer, phylogenetically, to the higher land groups than are the other two orders.

Before the close of the Paleozoic the labyrinthodonts had given origin to the reptiles. With the rise of that more progressive class, the amphibians rapidly dwindled in importance. The labyrinthodonts vanished at the close of the Triassic, and the surviving amphibians play but a modest role in modern vertebrate history.

One sometimes tends to think of the development of the early land vertebrates as the result of some "urge" toward terrestrial life among their fish ancestors. This is, of course, absurd; the evolution of the earliest amphibians capable of walking on land seems to have been essentially a happy accident. The amphibians appear to have evolved from crossopterygian ancestors toward the close of the Devonian, an age during which seasonal droughts were, it seems, common over much of the earth. Lungs are an excellent adaptation for use under stagnant water conditions. But when a stream or pool dries up completely, a typical fish is rendered immobile. Some further development of the fleshy fins already present in crossopterygians would give their fortunate possessor the chance of crawling up or down the stream bed (albeit with considerable pain and effort at first) and enable him to reach some surviving water body where he could resume a normal piscine existence.

Legs, the diagnostic feature of the tetrapod, may thus have been, to begin with, only another improvement for an aquatic life. The earliest amphibian was little more than a four-legged fish. Life on land would have been the farthest thing from his thoughts (had he had any). It was probably only after a long period of time that his descendants began to explore the possibilities of land existence opened out to them through their new locomotor abilities. And even today those of his descendants which have remained amphibians have never capitalized fully on these potentialities.

The term amphibian implies the double mode of life exhibited by many members of this class. Some toads spend much of their lives on good dry land, but most amphibians do not venture far from the stream banks, and some modern forms are still essentially water dwellers like their ancestors. The typical amphibian mode of development, as exemplified by the familiar frogs and toads of northern temperate regions, is still essentially that of the ancestral fishes. The eggs are laid in the water, and develop there into water-dwelling, gill-breathing tadpoles. Only when adult size is neared do lungs replace gills, limbs develop, and terrestrial life becomes possible. An amphibian is chained to the water by his mode of development and the necessity of returning to that element periodically for reproductive purposes. Although numerous modern amphibians have adopted a variety of adaptations to avoid this complication, none has been a complete success as a fully terrestrial form.

REPTILES

The reptiles are the descendants of the ancient amphibians which solved this reproductive problem and became the first fully terrestrial vertebrates. The "invention" of the amniote egg (with the associated developmental processes; see Chap. V) is the major diagnostic feature which distinguishes reptiles from amphibians.

The reptilian egg can be laid on land; thus there is avoided the necessity of any adaptation for water existence in either young or adult. This egg type is the familiar one preserved in the reptiles' avian descendants. The shell offers protection. A large yolk furnishes an abundant food supply so that the reptilian young (unlike the tadpole) can hatch out, at a fairly good size, as a miniature replica of the adult and thus avoid the necessity of prematurely foraging for its food. Of membranes developed within the egg shell, one sheathes externally both young and yolk. A second forms a lunglike breathing mechanism for the oxygen which penetrates the porous shell. A third (the amnion, from which the egg type gets its name) encloses the developing embryo in a liquid-filled space—a miniature replica of the ancestral pond.

It is probable that the oldest reptiles were still amphibious in their habits, and that the amniote egg was merely an adaptation parallel to, but better than, other adaptations seen in modern amphibians—a device which removed the eggs from the dangers of drought, and of enemies present in the ancestral waters. Here again, a major advance in the evolution of vertebrates may well have been a happy accident.

From the ancestral reptiles there rapidly developed, beginning in the late Paleozoic, a great array of types (Fig. 31). The reptiles became a widespread and prolific group during Mesozoic times, and, further, gave rise during that period to the birds and the mammals. Both these more progressive groups retained the developmental plan found in the reptiles (although it is modified with the introduction of viviparous habits in mammals), and the three classes are often styled collectively the amniotes.

Most of the reptiles which flourished in Mesozoic times have since become extinct; only four orders survive. We shall briefly recount the history of the group as a whole.

Although most reptiles utilized their potentialities and became fully terrestrial types, several groups failed to do so. The *turtles*—order *Chelonia*—were apparently an early offshoot from the stem reptiles; once

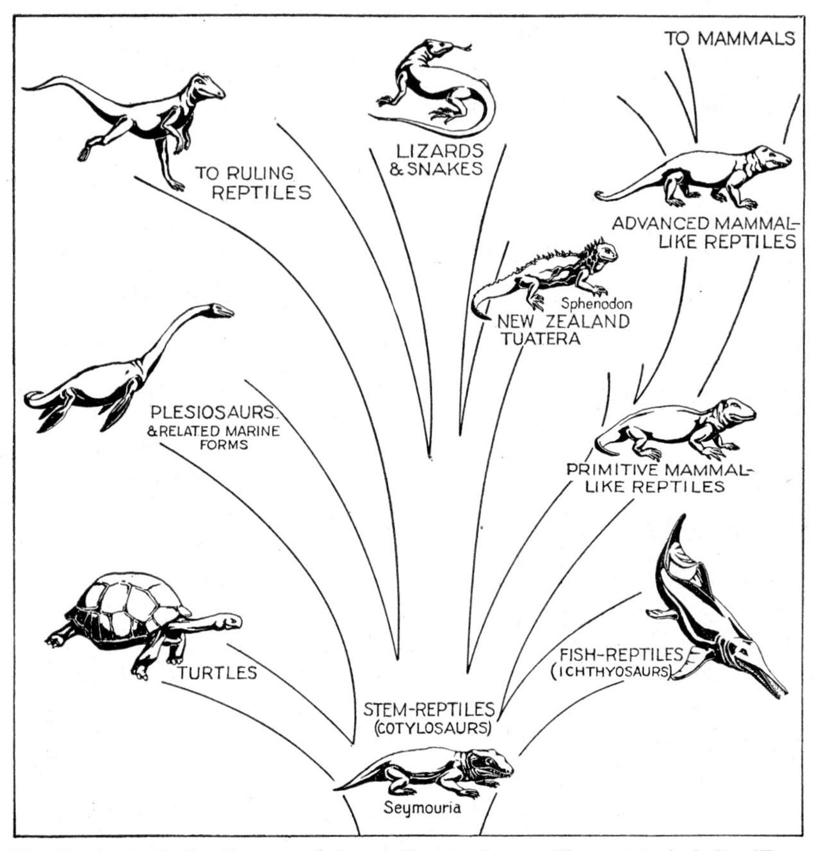

Fig. 31. A simple family tree of the reptiles (ruling reptiles not included). (From Romer, Man and the Vertebrates, University of Chicago Press.)

encased in a protective armor, they have remained a conservative group. The chelonians are still mainly amphibious and have in some cases reverted completely to an aquatic existence. Two extinct orders of reptiles which flourished in Mesozoic days became highly specialized marine forms. Most extreme in their specializations were the *ichthyosaurs,* which completely lost the power of terrestrial locomotion, turned their legs into short, finlike steering appendages and resumed a "stream-lined" fishlike shape, with the redevelopment, in piscine fashion, of back and tail fins. Not so extreme in their specializations, but equally successful, were the *plesiosaurs;* the stocky body was apparently incapable of redeveloping the fish type of locomotion, and was propelled by two pairs of great oar-shaped limbs.

The temple region of the skull affords diagnostic features useful in reptilian classification. A majority of reptile orders are characterized by the perforation of this region by two openings (fenestrae, cf. p. 230), each

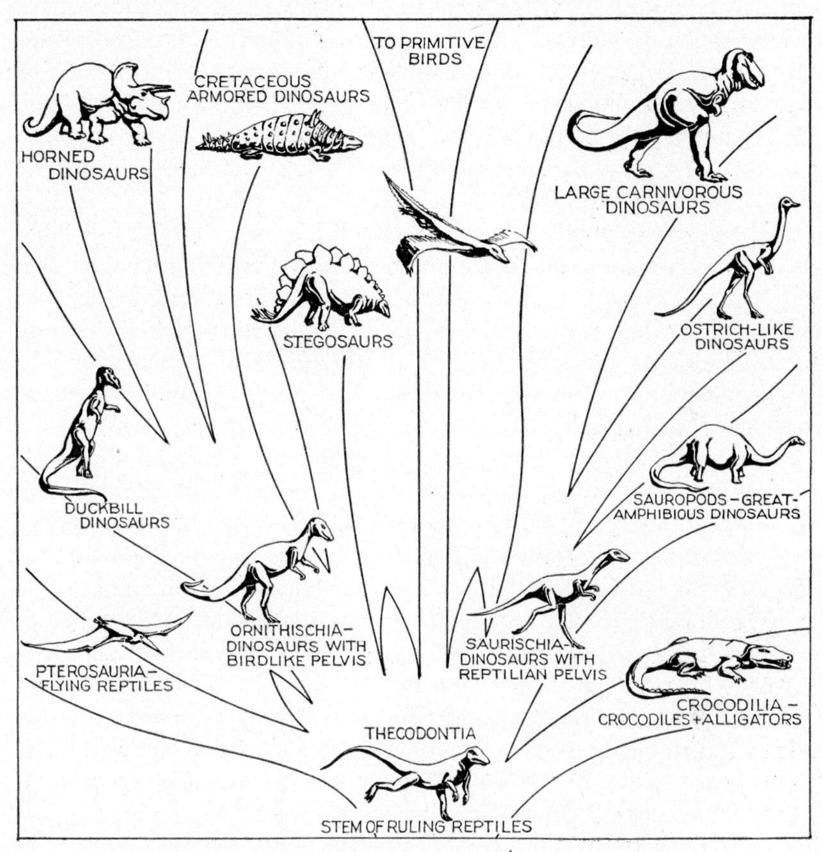

Fig. 32. A simple family tree of the ruling reptiles. (From Romer, Man and the Vertebrates, University of Chicago Press.)

bounded by an arch below it. A two-arched (i.e., diapsid) group which appeared early in the Mesozoic and has persisted to the present day, although never prominent, is the order *Rhynchocephalia,* represented in modern times by Sphenodon, surviving in isolation on a few New Zealand islets.

In appearance and in all structural features except the temple region this Mesozoic relic resembles the lizards; and the fossil record indicates that the Sphenodon group and the order *Squamata*—the *lizards* and *snakes*—are closely related. These latter types can be technically separated from the rhynchocephalians by the fact that the lower arch of the temple region has disappeared (cf. Figs. 135 and 136, p. 230). The order Squamata was the last major group of reptiles to appear in the fossil record. Lizards did not appear in any numbers until close to the end of the Age of Reptiles; they are today abundant and varied in all the warmer regions of the world. In several groups of lizards there is a tendency for reduction of the limbs and a return to the ancestral pattern of locomotion by body undulation; a successful limbless group of lizard descendants is that of the snakes, which differ further from their lizard ancestors in adaptations of the skull and jaws for swallowing large prey. Snakes are entirely unknown before the Cretaceous, last of the Mesozoic periods, and seem to have continued their evolution to their present state during the Age of Mammals.

An exceedingly important group with the same type of skull build as Sphenodon is the great subclass *Archosauria,* the ruling reptiles (Fig. 32). Today they survive only in the form of the rather aberrant crocodiles and alligators; but most of the dominant land reptiles of the Mesozoic were archosaurs, and the birds are descendants of this group.

The basal stock of the archosaurs is found in the Triassic in the shape of rather small, slenderly built predaceous reptiles included in the order *Thecodontia.* Elongate hind legs, a modified hip structure and other features suggest that they were becoming adapted to a bipedal mode of life. From these modest beginnings came the dinosaurs. These are popularly considered as constituting a single group of gigantic reptiles. This concept is far from correct, for although many dinosaurs were large, some were small (one was no bigger than a rooster). There were two major dinosaur stocks, not particularly closely related to one another, although descended in common from thecodont ancestors.

In one group, termed the *Saurischia,* or *reptile-like dinosaurs,* the ancestral forms were bipedal carnivores. Some of the smaller and more primitive of these bipeds can scarcely be distinguished from their thecodont ancestors. Others grew to immense size; Tyrannosaurus was the most ponderous flesh eater the earth has ever seen. Derived from early bipeds of this sort were the amphibious (sauropod) dinosaurs, which changed to a herbivorous mode of life, reverted from a bipedal pose to four-footed walking and grew to such giants as Brontosaurus and Diplodocus. It would appear that these great reptiles spent much of their lives in lagoons where soft vegetation abounded; such was their weight (one is estimated to have reached 50 tons) that it seems improbable that their stocky limbs could have supported them effectively on land.

A second major group was that of the *Ornithischia,* or *birdlike dinosaurs,* in which the hip girdles (but not other anatomical features) were comparable to those of birds. Like their saurischian cousins, primitive members of this group were bipeds; but in contrast with that other dinosaur stock, all the birdlike forms were herbivores. Best known of the bipeds of this group are the duckbills (hadrosaurs), which were abundant in the closing days of the Age of Reptiles. As in the saurischians, so in the ornithischians there was reversion to a quadrupedal pose. There were, in fact, three distinct types of quadrupeds developed in this order, all with some sort of defense against the great carnivores of the day. These types are exemplified by such popular museum exhibits as those of Stegosaurus, whose backbone is capped with defensive plates and spines; Ankylosaurus, low and flat and heavily armored on back and tail; and the horned dinosaurs, such as Triceratops, with horns—usually a trio of them—and a great frill of bone protecting the neck.

Still another group of extinct archosaurs was that of the order *Pterosauria,* the *winged reptiles*. In them the front limbs had one finger (the fourth) enormously elongated. From it there was extended, in somewhat batlike fashion, a great wing membrane.

Sole survivors today of the archosaurs are the alligators and crocodiles comprising the order *Crocodilia.* Although many of their structural features are indicative of a bipedal ancestry and descent from the thecodonts, the crocodilians, like many of their dinosaurian relatives, have reverted again to a quadrupedal gait and have become amphibious. The crocodiles are phylogenetically remote from the base of the reptilian family tree; their anatomical features are hardly to be considered characteristic of reptiles as a whole, and, as might be expected, they show numerous features found in the birds, whose ancestors were archosaur relatives of the crocodiles.

BIRDS

Birds have been aptly termed "glorified reptiles." We customarily treat of them as a separate class, *Aves;* but in many regards they are little farther removed from the general reptilian stock than are some of the ruling reptiles from which they sprang. Within that group, as we have noted, was included one series of flying forms, the pterosaurs; the birds are not descended from pterosaurs, but are a second archosaur flying type, in which, instead of a membrane, feathers—diagnostic of the class—form the wing surfaces of the modified pectoral limbs. In certain respects, notably bipedal adaptations, the birds are similar to their dinosaurian relatives; and almost every notable bird character is an adaptation to flight. The maintenance of a high and constant body temperature and improvements in the circulatory system are associated with the need of a high metabolic rate in flight; lightening of the body in various ways (particularly by the development of air sacs and hollow bones) is also associated with flight, as are modifications in the brain and sense organs.

A happy accident of preservation has given us knowledge of two skeletons of ancestral birds, usually termed Archaeopteryx, from deposits of late Jurassic age. These birds, still retaining teeth, in which the wing still has clawed fingers, and in which there still persisted a long reptilian type of tail, so nearly split the differences in skeletal structure between ruling reptiles and modern birds that their systematic position would be a matter of doubt were not their feathers preserved with them.

Birds, mainly owing to the delicacy of their skeletons, are relatively rare in the fossil record. However, there is evidence that before the close of the Cretaceous there had evolved birds quite modern in every structure except the retention of teeth. And by the dawn of the Cenozoic, even these ancestral structures had disappeared, to give place to a horny bill.

Although taxonomists divide the birds into a considerable series of orders, the structural differences between them are for the most part small; they are no greater than those which distinguish families in some of the mammalian orders.

There is, however, one partial exception to this; for there does seem to be a distinction between two groups, representing primitive and advanced stages in the evolution of modern birds. Technically, the two can be defined by details of palatal structure (with which we need not concern ourselves) which give rise to the terms "palaeognathous" and "neognathous" to distinguish them. Most birds with which the reader is ordinarily familiar—indeed, the majority of all living birds—are members of the latter, higher group. To the lower assemblage, often termed *ratites,* belong the ostrich and similar forms, such as the cassowary and emu of Australia, the rhea of the South American pampas, the extinct moas and the little kiwi of New Zealand.

Most of the palaeognathous birds are flightless, a fact that has given rise to the claim that they represent a primitive stage in bird evolution in which flight had not been attained. But anatomical study strongly indicates that this is not the case, and the ratites are probably degenerate descendants of once flying types. Most of them are found on islands where there are few terrestrial enemies, or on continents (Australia, South America) where, the fossil record tells us, the same was true at the time that the native ratites evolved. If ground-dwelling enemies are absent, much of the "point" of flying has been lost.

MAMMALS

Mammal-like Reptiles. The mammals are descended from reptiles; but the fossil record shows that the reptilian line leading to them diverged almost at the base of the family tree of that class. Their relationship to the existing reptilian orders is exceedingly remote.

Oldest of the reptile line leading to mammals were the *pelycosaurs* (Fig. 30, *B,* p. 53), a group which flourished in the early Permian. They were in most regards exceedingly primitive reptiles, but certain characters in skull structure indicate that they represent a first stage toward the evolution of mammals. Succeeding them in late Permian and early Triassic days were the *therapsids,* progressive mammal-like forms which were the commonest animals of their day (Fig. 33). The characteristic therapsids were flesh eaters, active four-footed runners in which, as in their mammalian descendants, elbow and knee had been swung in toward the body, making for better support and greater speed. In the advanced Triassic members of the group many features of skull, jaw, dentition, and limbs approach closely the mammalian pattern.

The evolution of mammal-like reptiles was the major feature of early reptilian evolution. But in the Triassic period other reptile groups became prominent—notably the dinosaurs. It appears that the therapsids could not compete successfully with them, and rapidly dwindled and disappeared from the scene. There survived, however, small forms from which there evolved the oldest mammals, sparse remains of which are found

Fig. 33. A mammal-like reptile (Lycaenops) from the late Permian of South Africa. (After Colbert.)

in later Mesozoic deposits. Living as they did for tens of millions of years as contemporaries of the dinosaur dynasties, our small Mesozoic ancestors were seemingly insignificant in the life of their times.

Intelligent activity may be reasonably regarded as the keynote of mammalian progress. With activity may be correlated not only the efficient locomotor apparatus characteristic of mammals, but also (as in birds) circulatory improvements and high body temperature (with which the development of hair is related). There is no evidence of marked increase of relative brain size in therapsids; but even the stupidest of mammals is an intellectual giant compared with any reptile. The habit of bearing the young alive—characteristic of all except the most primitive forms—and the development of the nursing habit are mammalian innovations which have resulted in giving a long period for the development and elaboration of delicate nervous and other mechanisms before the young is sent out into the world.

Monotremes. A few early mammals are known from sparse Jurassic remains. It must have been at least this early that there separated from the ancestors of the higher mammals the ancestors of the monotremes now living in Australia—the duckbill and the spiny anteater. These curious animals are definitely mammalian in many features, but primitive in such regards as the fact that, alone of mammals, they still lay a shelled egg like their reptilian ancestors. They are so specialized in many ways that they cannot be regarded as in themselves ancestral types. Their survival in Australia may be due to the relative isolation of that area. Unfortunately we know almost nothing of their history.

Marsupials. Rare finds from the later Cretaceous tell us that there had by then evolved two more progressive mammal groups—the marsupials and the oldest placental mammals. The marsupials, or pouched mammals, owe their name to the fact that although the young are born alive, they are born at a tiny and immature stage; typically, the female marsupial carries on her belly a pouch in which the newborn young are kept and nourished for a further period after birth. The common opossum is characteristic of the group and is a primitive mammal in many regards. In most regions of the world the marsupials have not been able to compete successfully with more progressive mammals. But South America proved a haven for many marsupials during Tertiary times, when that continent was long isolated; and Australia is a region where these forms have flourished greatly. The geological evidence suggests that the latter continent became separated from the rest of the world at the close of the Cretaceous and has since remained isolated. No placental mammals, it appears, had reached Australia at the time of separation, and few have been able to reach it since. There the marsupials had no opposition, and expanded and diversified to fill almost every type of adaptive niche which placental mammals have occupied in other regions. The kangaroos are the most grotesque of these forms, but there are in addition marsupial "squirrels," "cats," a "wolf," and so on—even a pouched imitation of the moles of other regions.

Placental Mammals. The major, progressive group of mammals includes the host of living forms properly termed the *Eutheria,* the "true mammals," but usually called placentals. The latter name is due to the fact that, in contrast with most marsupials, there is an efficient nutritive connection, the placenta, between mother and embryo; as a result the young can develop to a much more advanced stage before birth. On the extinction of the dinosaurs, these highly developed mammals were already in existence; they rapidly expanded into a host of types, many of which have continued down to modern times. In some other groups of vertebrates the family "tree" is actually treelike, with a main trunk or at least major branching limbs. That of the placental mammals, however, is comparable to a great bush: the various orders are difficult to assemble into groups and appear, for the most part, to have branched out independently of one another in early times. We may briefly note some of the main components of the placental assemblage.

The ancestral placentals—indeed, the early mammals as a whole—seem to have been small shy animals which were potential flesh eaters, but were forced to live, owing to their size, on small prey, such as insects, grubs, and worms, with presumably a small percentage of softer vegetable materials. The shrews today live similar lives. These, with a variety of other related forms, such as the moles and the hedgehogs of the Old World, form the order *Insectivora,* a group which in some ways may be regarded as the least modified descendants of the early placental stock. The *bats,* order *Chiroptera,* which alone of the mammals have acquired true flight, are apparently fairly direct descendants of the insectivores. Also connected rather closely with primitive insectivores is the order *Primates,* including lemurs, monkeys, apes, and man. Primitive mammals are thought to have been to some degree arboreal; this mode of life was emphasized in the early primates and appears to have been correlated with the development of many primate features in the limbs and—most notably—the brain. A few primates, such as men and baboons, have become ground dwellers, but still carry a strong imprint of their arboreal ancestry in their bodily build.

Some orders appear to have diverged rapidly from the insectivore stock at an early stage and have acquired structural features far removed from the more common patterns. Such are, for example, the so-called toothless mammals (most of which have teeth, but poorly and oddly developed). The order *Edentata,* in a technical sense, is essentially a South American group, including armadillos, tree sloths, and the anteaters of that continent and—among fossil forms—the armored glyptodonts and the giant ground sloths. A second important group of aberrant forms is that of porpoises and whales, the order *Cetacea,* with major adaptations enabling them to invade the seas in much the same fashion as the ichthyosaurs of the Mesozoic. A third series of forms which departed far and rapidly from the basal stock is that of the gnawing animals or *rodents.* Most are included in the order *Rodentia,* by far the most successful of all mammals if variety and numbers are accepted as criteria. A pair of chisel-like front teeth, followed after a gap by good grinders in a cheek series, are diagnostic features. Squirrel, rat, and guinea pig are typical members of three major groups of living rodents. The hares and rabbits were once included in this same order; but though they are rather similar to the true rodents in dental equipment, they are now thought to be quite distinct (they retain, for example, two pairs of upper gnawing teeth) and are placed in a separate order, the *Lagomorpha.*

The insectivores were potential flesh eaters. With the development of numerous mammals of the more harmless varieties there soon arose varied true *carnivores* capable of preying upon them. Some of the first "experiments" of the placentals along such lines were relatively ineffective, slow, clumsy, and (it would seem) stupid, but more progressive members of the order *Carnivora* soon spread over the world. The civets, mongooses and related forms of the Old World tropics seem to be rather close to ancestral types. The dog and cat families are two major and contrasting carnivorous groups. Bears and raccoons are apparently related to the dog group; the weasel family includes a variety of forms which are mainly found in north temperate forests. A notable side branch of the carnivores is that of the forms termed pinnipeds, which, as the seals and walruses, have paralleled the cetaceans in reverting to an aquatic life.

Ungulates. Notable in all stages of placental evolution has been the development of large herbivorous forms. Most of these tended to develop hoofs and hence are termed ungulates. This term, however, is to be avoided if any precision is desired, for the ungulates are not a single order, but a series of orders representing different "experiments" of nature in exploring the possibilities of the herbivorous mode of life.

The two major living ungulate orders are the Perissodactyla and Artiodactyla; horse and cow, respectively, are familiar examples of the two. In the *Perissodactyla,* to which the rhinoceroses and tapirs belong as well as the horse tribe, one key tendency is that for reduction in the number of toes about the third as an axis, so that the trend has been for the development of a three-toed condition and, in the late members of the horse group, a single-toed state. The *Artiodactyla,* on the other hand, have the axis of the foot between the third and fourth toes. In this foot type the loss of the first digit results in a four-toed foot; the lateral elements are usually reduced or lost and the two-toed "cloven hoof" results. Pigs, peccaries, and hippopotamuses are among the less progressive artiodactyls. The others—much more numerous—are cud-chewing types such as the camels, deer, giraffes, prongbuck, cattle, sheep, goats, and the numerous types of antelopes.

Also to be included in the ungulate category, in a broad sense, are three minor orders thought to be related to one another and often termed *subungulates;* Africa may have been their early common home. In some ways the most primitive of the three is the order *Hyracoidea,* in which there are included only the little African "rock rabbits" or conies. The elephants are the only living members of the order *Proboscidea;* in older times the group was much more numerous, including a variety of mammoth elephants and more primitive relatives, the mastodons. Although it is a contradiction in terms, there is included in this assemblage the order *Sirenia,* the sea cows, a herbivorous group of marine mammals whose most primitive fossil representatives show relationship to proboscideans and hyracoids.

CHAPTER IV

CELLS

ALTHOUGH anatomical and physiological studies of the vertebrate body generally deal with gross structures—organs and organ systems—it must never be forgotten that these structures are composed of tissues, and these in turn of cells, and that the cells are the basal living units from which the entire complex body is built. Cytology, histology, and cellular physiology are special fields of study and research which deal with cells and tissues. Here we shall but briefly review our knowledge of these topics as a background for a better understanding of structures of greater complexity with which the remainder of this work is mainly concerned.

The Cell. A century ago, with the development of microscopical methods, evolved the important concept that the body of any animal (or plant) was basically composed of cells, each a highly organized but tiny living unit. All organic materials are composed of cells or of materials formed by cells. Each cell lives its own life; but each is dependent upon other cells and tissues for its existence, and each in turn makes its contribution to the welfare of the total organism.

Within the body the shapes, structures, and functions of the myriads of cells which it contains are highly varied. There is no such thing as a "typical" cell, for each is specialized for a particular role in the vertebrate organism; a muscle cell, a nerve cell, a gland cell, all differ widely from one another in many regards. Nevertheless, numerous basic features common to all may be discussed at this point.

The gross morphology of a cell is readily discernible. Centrally situated is a nucleus, normally spherical in shape, and usually constituting but a relatively small proportion of the volume of the cell. The nucleus is bounded externally by a well-defined membrane. Without this membrane lies the body of the cell, composed of living material, the cytoplasm, which may exhibit various inclusions. External again to the cytoplasm is the cell membrane, typically thin. Some of the structural features will be discussed later. For the moment we shall concern ourselves with the chemical materials of the cell.

Cell Chemistry. The basic material of cells is *protoplasm,* the stuff of life itself. This forms the main content of the cell body, the cytoplasm, and, in modified form, the substance of the nucleus. Its exact nature we do not know (when, if ever, we do, we shall have solved many of the fundamental problems of biology), but we are even now familiar with many of its attributes. As seen in the living state, protoplasm is (except for inert, or relatively inert, inclusions) a seemingly clear, homogeneous material. Chemical and physical studies, however, show that it is a complex colloidal solution of variable consistency which contains a great variety of chemical materials.

Much of vertebrate structure and function is related to the collection, transformation, and transportation to the cells of the basic chemicals needed to form and maintain the protoplasm and enable it to play its proper part in the work of the body and, again, to the disposal of wastes formed by its activity. In consequence, it is important that we understand the fundamentals of cellular chemical structures and processes. This topic will be reviewed in simple—rather, oversimplified—fashion.

Water, of course, is the major constituent of protoplasm. A vertebrate, for all its seeming solidity, is an exceedingly watery structure; perhaps three-quarters of the cell contents are water. In solution in the watery portion of the protoplasm may be found a variety of ions of *inorganic salts*. Most of these ions are identical with those found in the extracellular body fluids. It is notable that sodium and chloride ions, abundant in those fluids, are generally in relatively low concentration in the cells themselves. Calcium, phosphorus, potassium, and sulfur (as sulfate) are the other common elements which may be present in inorganic form. There are further half a dozen "trace" elements generally present in minute quantities, but even so are known to be absolutely vital requisites in various vertebrates. These include iodine and the metals iron, copper, zinc, manganese, and cobalt. In addition there may be present various simple compounds in process of inclusion in the chemical structure of the cell, or on their way out as end products of metabolism. Most important, however, are the numerous and varied organic compounds which appear to be responsible for the vital cellular structure.

In basic chemical analysis these all contain a considerable proportion of carbon, that versatile element which, with its four bonds for ready chemical union with other elements, is the prime feature of *organic compounds*. The carbon "skeleton" of organic compounds most frequently takes the form of chains of carbon atoms: —C—C—C—C—C—C— (each C with a free bond above and below) to the free bonds of which hydrogen, oxygen, or nitrogen atoms or hydroxyl groups (OH) are most commonly appended. Also characteristic are rings of carbon atoms, typically with six members, and often including oxygen or nitrogen as well as carbon.

Oxygen and hydrogen are almost as abundant in organic compounds as carbon, and nitrogen, present in smaller amounts, completes the quartet of basic organic elements. Sulfur and phosphorus are also present, to a lesser degree, in many of these compounds, and a variety of other elements are generally present in still tinier quantities, particularly sodium, potassium, calcium, magnesium, and chlorine.

Cellular organic compounds may be for the most part grouped into three categories—carbohydrates, lipoids, and proteins. Simplest in structure of the three are the *carbohydrates,* which, as their name implies, consist of a combination of carbon with water, mostly in the form of H and OH groups attached to one or more ring-shaped "skeletons" formed by carbon and oxygen atoms. Sugars and glycogen are the characteristic animal carbohydrates. The common simple *sugar* found in the vertebrate body is *glucose,* with the following structure:

```
               OH
               |
             H—C—H
               |
               C—————————O
              /|          \
      H      / H           H
      |     /              |
      C                    C
      |   \               /|
      OH   \  OH      H  / OH
              |       | /
              C———————C
              |       |
              H       OH
```

This is the sugar present in the blood stream and found to enter the cell, but as such it is hardly an integral part of the "vital" materials of the protoplasm. Rather, it is to be considered a source of immediate energy supply for cell life processes; in the presence of oxygen this sugar readily breaks down into water and carbon dioxide with a considerable release of energy. A molecule of glucose easily combines end-to-end with other similar molecules (the bond forming with the elimination of a molecule of water at the point of union) to form more complex sugars (polysaccharides); with still greater numbers and complexity in the union of glucose molecules we reach the stage of starches. The characteristic starch of animal cells is *glycogen,* composed of many thousands of glucose units. Glycogen derives its name from the fact that it re-forms glucose on its dissolution. Carbohydrate energy is characteristically stored in cells in this form. But even so, carbohydrates are hardly to be regarded as vital parts of the protoplasm, since they tend to be stored, as fuel reserves, in solid granules.

The *lipoids,* or fatlike substances, include a considerable variety of materials. Most characteristic and familiar are the true *fats.* The major bulk of a fat consists of molecules of *fatty acid.* These have a monotonous structure—a long chain of carbon atoms tied (except at one end) to an unvarying series of hydrogen atoms; for example:

```
    H  H  H  H  H  H  H  H  H  H  H  H  H  H  H  H    OH
    |  |  |  |  |  |  |  |  |  |  |  |  |  |  |  |   /
 H—C—C—C—C—C—C—C—C—C—C—C—C—C—C—C—C
    |  |  |  |  |  |  |  |  |  |  |  |  |  |  |  |   \\
    H  H  H  H  H  H  H  H  H  H  H  H  H  H  H  H    O
```

These chains are stable under most conditions, but in the presence of appropriate catalysts are oxidized with a great release of energy (petroleum fuels, the hydrocarbons, seem to be essentially somewhat modified "fossil" fatty acids with similar properties). The various fats are formed by tying three of these elongate molecules into a single package by their union with a molecule of *glycerin* (glycerol to the chemist), a simple compound rather similar to a simple sugar:

```
    H  H  H
    |  |  |
 H—C—C—C—H
    |  |  |
    O  O  O
    H  H  H
```

The fatty acids take the place of the —OH groups, with elimination of water in the process.

Fats are a general constituent of cells and are sometimes present in great volume; but they too appear to be in the main fuel reserves in storage, used when carbohydrates are exhausted, and play a relatively small part in the living protoplasm. Related to this is the fact that fats are insoluble in water; they are hence not an integral part of the watery protoplasm, and tend to occur as separate droplets. Despite the seeming inertness of fats, however, it is notable that the presence of four specific fatty acids appears to be as necessary for the maintenance of life as the presence of certain specific inorganic materials noted above, or of specific vitamins.

Somewhat less inert are the *phospholipids,* or phosphated fats, which differ from true fats in that but two fatty acids are attached to the glycerin molecule; the third position is occupied by phosphoric acid bound also to another organic molecule. The phospholipids are partially soluble in water, and are, in relation to this fact, a more integral part of the active protoplasmic materials than their fatty cousins. They may represent a modification of fat activated for transport by the watery fluids of the body and more available for cellular use.

Classed with the lipoids as a catch-all (primarily because we do not have any other category in which to place them) are various substances broadly termed the *sterols,* small in quantity in protoplasm, but important in protoplasmic activity and general vital processes. Like the fats they are composed almost exclusively of hydrogen atoms hung on a carbon "skeleton;" however, this framework is not composed merely of simple chains, but includes carbon rings in complex formations. A ubiquitous cell component of this group is cholesterol, the function of which is still unknown; many highly active substances, such as the antirachitic vitamin and male and female sex hormones also belong in this general category.

Most distinctive of vital materials, however, are the *proteins*. A basic difference from the carbohydrates and fats lies in the presence of nitrogen. Protein molecules are giants in the chemical world. They are believed to include in many cases hundreds of thousands of atoms and may approach, if not reach, the limit of visibility of the ordinary microscope. The basic structure of proteins is well understood. They are formed by a compounding of simple chemical structures, *amino acids*. The formula of such an acid is invariably in the following pattern:

$$
\begin{array}{ccccccccc}
 & & \mathrm{H} & & \mathrm{H} & & & & \\
 & & | & & | & & & & \\
\mathrm{H} & - & \mathrm{N} & - & \mathrm{C} & - & \mathrm{C} & - & \mathrm{OH} \\
 & & & & | & & \| & & \\
 & & & & \mathrm{R} & & \mathrm{O} & &
\end{array}
$$

R designates the one variable in the system; there may be found here either a hydrogen atom or a variety of atom complexes. About two dozen variants are known, and hence that number of different amino acids.

Protein materials enter the cells in the form of these units, simple in nature and limited in variety. Within the cells of the body, however, these building stones are combined into molecular structures of almost infinite complexity and variability. End-to-end combinations of acids occur, forming elongate structures; unions in various directions may yield complex networks; and although the number of basic types of material is limited, variations in the numbers and arrangement of these units render possible uncountable numbers of differing proteins. The nature of the proteins present varies greatly from one animal species to another, and may even vary from tissue to tissue; there is a "protein specificity."

A special class of proteins is that of the *nucleoproteins*. These consist of varied proteins combined with nucleic acid; this acid in turn consists of a combination of phosphoric acid and further organic bases. Nucleoproteins are present to some degree in the cytoplasm of the cell body. Of especial interest, however, are the types of nucleoproteins confined to the nucleus. There they are found in the chromosomes, the bearers of the hereditary materials, and it is possible that these large molecules may be the actual genes of the geneticist.

Still another special category of protein compounds are the *enzymes*. These ubiquitous and absolutely essential materials are catalysts, which vastly speed up reactions, typically reversible, involving organic materials; they function much as do simpler catalysts in simpler chemical processes. These substances are proteins, to some of which is attached a simpler compound which may be the active agent in their catalytic activity. Some enzymes, as will be noted later, function outside the cells, in the digestive tract. But all are produced intracellularly, and most of them (dozens are known, and undoubtedly many more are as yet undiscovered) play specific parts in cell metabolism.

Finally, of chemical materials essential to the well-being of cells we may mention the *vitamins,* prominent in medical and popular thought in recent decades. These substances have been discovered in the course of the study of dietary deficiencies in man and higher vertebrates, and are organic compounds essential, although in extremely small amounts, to the operation of the body. Their discovery and study have been in relation to the needs of the body as a whole, but it is obvious that in general they effect their influence through the activities of the cells and therefore are part of the vital cell materials. Certain of the vitamins are definitely known to join with proteins to form enzymes, and it is not unreasonable to assume that most, if not all, function in this fashion. They differ, however, in one important respect from the typical enzymes, which are manufactured by the cells from simpler chemical materials. With the vitamins "home manufacture" is not possible. The fact that the vertebrate cell cannot, in general, manufacture vitamin molecules led to their discovery. The vitamins, or similar materials readily transformable into vitamins, must be supplied in the food, sometimes from other animal sources, but ultimately from plants which manufacture them. The cells cannot live unless supplied with these materials, any more than they can exist without a supply of specific inorganic salts, specific fats, and specific amino acids.

The emphasis on dietary and medical aspects of the topic obscures the general biological picture of the vitamin story. Almost all studies have been made on man and other mammals (to a lesser extent on birds), and we know almost nothing of vitamin needs of lower vertebrates. Further, we know little of their basic action or influence; our knowledge of their function is confined almost entirely to diseases or body defects of man and laboratory animals caused by vitamin deficiencies.

Protoplasmic Structure and Activity. Protoplasm is a colloid, typically a fluid watery "sol" with particles of "living" organic materials in solution in it, but varying greatly in viscosity and sometimes reaching the "gel" state. Fixed and stained cell material frequently appears to show structural features in the form of granules or a network, even in cells in which no major structural elements are found in the cytoplasm. Many of these appearances are probably artificial; nevertheless, there is every reason to believe that the arrangement of cytoplasmic materials is not haphazard, and that basic structural features exist in life, owing presumably to the patterns present within the protein molecules and to their mutual arrangement. Amino acids readily join end-to-end, and long protein strands appear to be common formations. In addition, however, there are frequent side-to-side bonds between these chains, so that protein molecules may take the form of a latticework; still further, the lattice may become three-dimensional.

These structures are not, however, static. Constant chemical and physical activity is characteristic of living protoplasm; a cessation of metabolic activity is almost synonymous with death.

Every cell must produce energy for its particular function in the body (whether secretion, nervous conduction, muscular movement, or what not); but in addition it must constantly undergo chemical change in the maintenance of its internal economy. A cell, like a body, has, so to speak, its own "basal metabolism," necessary for the support of its vital activities.

In part these metabolic processes are constructive. Glucose is taken in and built up for storage as glycogen. Lipoid materials are received and stored or transformed. Simple amino acids are accepted and built up into complex proteins. Further, although these three classes of materials are, in general, discrete, there is evidence of a certain amount of transformation of one into the others; of carbohydrates into fats and vice versa, and even, to some degree, an interchange of materials between the proteins and the "simpler" organic elements. Despite the seeming stability of proteins, once formed, and of fat and glycogen in reserve storage, it appears that these large molecules are constantly dropping off fractions consisting of amino acids, fatty acids, and glucose, respectively, and substituting others from "pools" of these substances present in the cell.

Still further, though animal cells depend on food materials (and fundamentally on plants) for the elementary organic substances, they themselves have, to a high degree, the ability to manufacture from these simple materials a great variety of complex organic compounds.

The energy needed for the cell's own maintenance and for the performance of its role in bodily activities is gained by the breaking down of organic materials into simpler compounds. Some of this breakdown needs no oxygen immediately. It is a fermentation comparable to that produced by yeast on sugar. Much as yeast fermentation breaks down carbohydrate into alcohol and carbon dioxide, so animal cell enzymes may cause glycogen to break down into carbon dioxide and a relatively simple compound, lactic acid.

In the long run, however, most of the cellular energy is produced by a "burning" process of oxidation, requiring large amounts of oxygen in constant supply. The carbohydrates are readily oxidized with a high energy release; the fatty materials are good energy producers, but more in the nature of a reserve. Both types of material, being composed entirely (or almost entirely) of carbon, oxygen, and hydrogen, give as end products for elimination from the cell simply water and carbon dioxide (and further excess "metabolic water" is produced in the constructive metabolic processes).

Even the proteins tend to be steadily consumed. The breakdown here, however, is complicated by the presence of nitrogen, and hence a nitrogen compound is an element of the waste as well as water and carbon dioxide. The nitrogenous waste material leaves the typical vertebrate cell in the simple form of ammonia, NH_3. Large quantities of ammonia are, however, toxic when present in the body liquids, and much of the ammonia is combined, in the liver cells or elsewhere, with waste carbon dioxide to yield more harmless compounds—urea, $CO(NH_2)_2$, or the more complex and insoluble uric acid, $C_5H_4N_4O_3$.

"Gross" Cell Structures. In addition to its submicroscopic structure, cytoplasm usually contains more highly organized components or inclusions visible under the microscope. Many of these are specific for the cells of specialized tissues; others are commonly found and may be noted here. Visible droplets of fat and granules of glycogen may be present. Widespread are mitochondria, small structures which may be mere dots or long filaments, but characteristically are short rods. Special stains reveal in many cells an internal reticular apparatus (frequently named after its discoverer, Golgi) which forms an irregular network of variable extent. Mitochondria and reticular apparatus have been thought to be related to secretory activities, but the question of their function is far from settled. Fibrils are often seen; they are most highly developed in muscle and nerve cells in relation to special functions, but are also present in other cases where they are thought to aid in attaining relative stability and rigidity of the cell. Frequently a pair of tiny dots is seen in a specialized cytoplasmic area close to the nucleus. These are the centrioles, which play a part in cell division.

The *nucleus* is usually relatively small, spherical, centrally situated, and surrounded (except at the time of cell division) by a relatively stout membrane. Its protoplasm, as seen under the microscope, is characterized by the presence of chromatin, material given this name because of its characteristic staining capacity and probably consisting essentially of the nucleoproteins which chemical analysis shows to be characteristic of the nucleus. Most of the time during which the cell is performing its normal activities, the chromatin appears to be irregularly diffused in a network throughout the nucleus. At cell division, however, the chromatin (and associated materials) arranges itself in the shape of elongate threads or rods, the *chromosomes,* for equitable distribution to the daughter cells.

A vast amount of work in the field of genetics has shown, without the possibility of doubt, that the nucleus plays a dominant role in inheritance and in development. The greater part of the "work" of the cell is done by the cytoplasm external to the nucleus. This cytoplasm is extremely sensitive to stimuli acting on it from without, not merely in the form of general environmental conditions in its surroundings, but also by way of nervous and hormonal controls. It is obvious, nevertheless, that the nucleus has an important and continuous influence upon cytoplasmic activity, for no cytoplasmic mass can long exist if the nucleus be removed from the cell. The nature of this influence is in great measure a matter of speculation rather than of accurate knowledge. We may, however, imagine the nucleus playing somewhat the part of a general staff, issuing various "directives" to the cytoplasm—directives which, however, are subject to sharp modification as the result of variations in external influences. In the absence of this directive force, organization disappears, and chaos and cell death result.

Cell division in the vertebrate body is most active in the embryo. In most tissues, however—the nervous system is a conspicuous exception—cell division continues at a slower rate in later stages for replacement of used-up cells or for further growth. As far as we can be sure, this is always the normal process of *mitosis*. This process includes a doubling of the chromosome number; the formation of a spindle and the division of the chromosomes into two equal groups which withdraw to form two nuclei; and the subsequent division of the cytoplasm of the cell body to form two new cells. The maturation of germ cells involves a more complicated process, as a result of which the chromosome number eventually present in egg and sperm is half that present in normal body elements.

Although it may be exceedingly thin, the *cell membrane* is a definitely organized structure, as shown by the fact that its presence prevents the free entrance into the cell of certain materials found in the surrounding medium. It is partly permeable, however, and permits, selectively, the entrance of substances needed for cell activity and the exit of wastes. It presumably consists of a thin layer of molecules; if injured, it is in general rapidly replaced by materials drawn from the deeper protoplasm. Most foods needed by the cells and most wastes leaving them are water-soluble materials of small dimensions which appear to pass through the interstices between the molecules or groups of molecules of which the membrane is composed. Part of the membrane appears to consist of lipoid droplets; and fatty molecules soluble in such lipoids, even if of large size, may readily get into the cell through these droplets by entering into solution in them.

Cell Environment; Interstitial Fluid. A major duty of the organism is to provide the environmnt, physical and chemical, required by its cellular units. These requirements are rather rigidly fixed for vertebrate cells.

Although the organism as a whole may be able to exist over a fairly wide range of external temperatures, the temperature to which the cells in the interior of the body can be subjected and still survive is limited. In general, lower, primitively water-dwelling forms have a temperature range lower in the scale than do terrestrial vertebrates. With perhaps some exceptions, even fishes are unable to stand internal temperatures much below the freezing point—the formation of ice crystals in the cells and the consequent disruption of the protoplasmic structure being, it would seem, the major factor in cell death. Thirty degrees centigrade is about the upper limit for fish cells, 45° C. or so for the higher, terrestrial vertebrate classes. Beyond such a point "heat death" occurs; proteins undergo irreparable coagulation.

A cell can live and avoid desiccation only if bathed in a watery liquid medium. Such a material, the *interstitial* (or intercellular) *fluid,* pervades the body. But, as amply demonstrated by work on tissue cultures, water itself is not enough. This liquid must contain a considerable amount of material in solution; otherwise, through osmotic pressure (see below), swelling and disruption of the cells may take place. Still further, it is found that, in or out of the body, cells flourish only if the materials in solution, mainly inorganic salts, adhere rather closely to the formula actually present in interstitial fluid normally found in the body of vertebrates. This liquid contains considerable amounts of sodium and chloride ions, lesser quantities of potassium, calcium, and magnesium ions, and small amounts of other elements.

Reasons for the requirement of this salt formula (quite different from that in the watery constituents of the cells themselves) are complex and poorly understood. It is of interest, however, that with one major exception (the absence of sulfates), the materials in the interstitial fluid are similar to those found in sea water, although rather more dilute. The suggestion has been made that this is no mere accident; that the ancestors of the vertebrates were simple animals bathed in and permeated by the waters of the early Paleozoic ocean; that their cellular physiology was evolved with this type of environment as a basic feature; and that when, as complex organisms, they developed an independent internal environment, this salty interstitial liquid persisted as, so to speak, a remnant of the ancient seas. The interstitial fluid is in communication with the similar liquid constituting the bulk of the blood plasma, and it is therefore through the blood circulation that there is accomplished the even distribution through the body of the fluid materials necessary for cell survival.

It is, of course, through the interstitial fluid that the cell is supplied with the materials which it needs for its existence and through this same liquid that it is relieved of its waste materials. These materials have been referred to in earlier discussion, and are summed up here.

For its existence the cell must be supplied with (1) water, (2) oxygen, (3) simple salts and "trace" elements, (4) glucose, (5) fatty materials, (6) amino acids, (7) vitamins.

From the cell the surrounding fluids receive as waste (1) carbon dioxide, (2) excess water produced by metabolic processes, (3) nitrogenous waste, as ammonia, (4) small amounts of still other discarded materials.

The interstitial fluid must be in intimate contact with the circulatory system so that food materials may be constantly supplied and waste constantly taken away. Further, the digestive and respiratory systems must supply, directly or indirectly, the necessary "food" materials; the excretory and respiratory organs must eliminate the wastes. And, indeed, in a broader view, almost every body organ is involved in some fashion in the acquisition of the food supply for the multitude of body cells and the maintenance of the proper environment for them. The cells are dependent upon the proper functioning of the body as a whole; the body is dependent for its existence upon the proper functioning of its individual cells, and labors to that end.

Osmosis. The nature of the cell membrane, and the necessity of there being a sufficiency of materials in solution in the interstitial fluid to balance the concentration within the cell, bring to our attention for brief review the basic physical-biological phenomenon of osmosis—a phenomenon of importance in a great variety of biological activities and hence influential in the molding of structure and function in many parts of the body.

If two bodies of liquid—say, of water with materials in solution—are separated by a membrane, man-made or organic, a variety of conditions may exist. The membrane may be so impermeable that little or no exchange of materials may take place, or so tenuous that it may be readily penetrated by all the materials of the liquids present, which may thus exchange freely. Intermediate conditions, however, may exist. It may be that the submicroscopic interstices in the membrane are so small that large molecules or even large ions cannot pass through. Such membranes, common in the vertebrate body, are termed *semipermeable.*

We conceive of the molecules and ions of a liquid, no matter how quiet it may appear to the eye, as being in rapid motion, dashing violently to and fro. If two liquids are separated by a semipermeable membrane, that membrane is constantly bombarded by particles on both sides. If both liquids consist of pure water, a large proportion of the particles will pass through the gaps in the membrane, but since the passage is equally free from both sides, no change in quantity will occur. Again, if both liquids contain in solution equal quantities of materials which cannot pass through the mesh, conditions will likewise remain stable; a given percentage of particles of the large ions or molecules strike the membrane and fail to penetrate it; but since the percentage of large particles which cannot pass, and water particles which can, is the same on both sides, quantities and qualities of the two liquids are undisturbed.

But when the two solutions are not in balance, we see a different and striking result. Let us take the extreme case where a thick solution of materials is present on one side, and pure water on the other. From the solution side, only water particles are able to penetrate the membrane, and these constitute only a fraction (if a large one) of the liquid. But from the pure water side, every particle is water and has a chance to pass. In consequence the net result is a steady, one-way flow of liquid from the pure water side into the solution. In a laboratory experiment with simple liquids it can be seen that the solution becomes increasingly diluted, owing to the incoming water, and expands, even against considerable pressure; body cells, unsuited for normal life under such conditions, are seen under the microscope to expand when placed in pure water, obviously have their protoplasm diluted to a nonviable stage, and finally burst.

Cellular membranes of variable degrees of permeability, noted in later chapters, are present in numerous situations in the organs of vertebrates. It is obvious that osmosis is a phenomenon which enters notably into the determination of the architecture of the body.

Epithelia. With a few exceptions (as of cells in the blood stream), the body cells are not isolated entities; they are part of formed *tissues*—organized associations of cells of similar origins and functions. In some cases—notably the connective tissues and skeletal structures, to be described later—the aggregation of cells may be relatively diffuse, and sometimes ill defined or amorphous. In many instances, however, the cell association is that termed an *epithelium*—a regular and compact arrangement of cells in a sheet, which borders on one of its aspects the surface of the body or one of its cavities. In the embryo, most epithelia are simple and diagrammatic in appearance. Later, however, by thickening and modification they may lose much of the early epithelial appearance. Thus, for example, the thick "gray matter" of the human brain, or a mass of liver tissue, has little of the appearance of an epithelium; but if their embryological history be traced, they will be seen to have been derived from simple epithelial linings of the brain or gut tube.

For convenience, various classificatory terms are used for different shapes of epithelial cells. *Squamous cells* are those much reduced in height, appearing in transverse section as thin lines. *Cuboidal cells* are those in which height and width are about equal, so that in section they have a square outline. In *columnar cells* the height is in excess of the width.

Embryologically, epithelia often consist at first of a single layer of cells, and this arrangement is retained in various adult cases as a *simple epithelium*. In such an epithelium any one of the three cell types noted may be present. Embryonic cells tend to be rounded; when closely packed they tend to assume a cuboidal shape, in a simple epithelium; this type is not uncommon in the adult. A thin squamous type is characteristic, for example, of the lining of blood vessels. Surface cells with secretory or other important functions tend to be of a deeper, columnar type.

When an epithelium shows two or more layers of cells, it is termed a *stratified epithelium*. The stratified types are named from the nature of the cells on their surface; these may be squamous, columnar, or (more rarely) cuboidal. The deeper cells of a stratified squamous epithelium may be cuboidal or columnar, and the underlying cells of a columnar type may be much flattened.

On surface view, or in section parallel to the surface, the cell boundaries of an epithelium may be sometimes demonstrated to be irregular, but they are typically polygonal in arrangement, frequently with the hexagonal outline, which (as the bee knows) gives the best compromise between the ideal circular individual shape and the necessities of a compact arrangement. The adjacent cells in an epithelium may be tightly packed together, but generally are separated by small spaces filled by the interstitial fluid. These spaces may be bridged by numerous tiny bars of protoplasm connecting cell with cell. As a rule blood vessels are absent in epithelia, and nutrient material reaches the cells by passing upward in the interstitial fluid from vessels beneath the base of the epithelium. In most cases epithelia are bounded below by connective tissues; between the two there is typically a *basement membrane* formed by condensed connective tissue material.

In a stratified epithelium there is frequently a constant loss or destruction of surface cells; these are replaced from below, and the lower cells of the epithelium form a *germinative layer* capable of making replacements by cell division (cf. Fig. 59, p. 117). The major activities of an epithelium are in general associated with its outer, free surface. As suggested by these facts, it is natural to find that in many stratified epithelia the cells are progressively more specialized as the surface is approached. We further find that in an epithelium of any type, the individual cells are frequently more or less polarized, the proximal (lower) and distal (upper) ends being different in nature.

The free surface of an epithelium frequently shows marked specializations. The cells may secrete a *cuticle,* a more or less solid layer of surface material covering the epithelium. More frequently, however, the surface of the protoplasm itself may be modified. The free surface may exhibit striations indicating the presence of hairlike surface extensions; still larger processes of the same sort may give a structure like that of a miniature hairbrush. Most specialized of surface structures are *cilia,* found on the surface of a variety of embryonic and adult epithelia of vertebrates as well as invertebrates. These are slender, mobile organs, homogeneous in apparent structure, but with a readily staining basal corpuscle beneath. Cilia beat in but one direction; the beat in a ciliated surface progresses in wavelike fashion so as to carry mucous or other materials along in a constant unidirectional stream.

In certain cases a simple epithelium may present a seemingly stratified appearance—a *pseudostratified epithelium.* If, for example, a simple but tall epithelium contains cells of two sorts, with nuclei at different levels, or with cell bodies expanded at different levels, one gets the impression that two layers are present. The term is better justified in certain epithelia where two types of cells are present at the base of a simple epithelium, but only one type reaches the surface. (Incidentally, it will be realized that if a simple epithelium is cut at an angle, several cell layers may appear, deceptively, to be present in microscopic section.)

In a broad sense the term epithelium may be applied to any tissues of the sort described here, no matter where they are found in the body. By many authors, however, the term is restricted and sometimes confined to tissue layers on the outer surface of the body on the one hand, and the gut and its derivatives on the other, or to cavities obviously originating from these outer or inner surfaces. The theoretical consideration behind this is the fact that (as will be seen in the next chapter) such epithelia are derived during individual development from the surfaces of the early embryo. If such a restriction is made, one or more alternative terms are used for tissues lining cavities formed secondarily in the deeper layers of the body. The name *endothelium* is generally used for the epithelial material lining the vessels of the circulatory system. Sometimes this term is applied as well to the lining of the body cavities; *mesothelium* is here an alternative name.

Glands. Secretory activity—the production of materials to be discharged from the tissue—is characteristic of a variety of cells; cells in which this is a dominant characteristic are termed *gland cells*. Frequently their nature is obvious in microscopic preparations, owing to the presence of granules of the secreted material or vacuoles filled by it in the cell body. Usually gland cells are long-lived, but in some cases part or all of the cell body is destroyed in the secretory process.

Isolated gland cells may be found scattered in an epithelium which serves other functions. Frequently, however, concentrated masses of epithelium are present which have exclusive secretory functions and form organized glands.

In some cases these structures secrete into the interstitial fluid or blood vessels; these are glands of internal secretion—*endocrine glands*. In such deeply placed structures there is usually a rich network of circulatory vessels, making for efficient conduction of the secretion. Much more commonly, glands open to the surface of an epithelium and hence discharge their products on the body surface or into some internal cavity or tube, as *exocrine glands*.

Two common types of gland cells are mucous and serous cells. The *mucous cells*, usually occurring in simple epithelia, produce mucin or *mucus*, a thick, slimy "lubricating" secretion such as that characteristic of the mammalian nose or throat. The secretion frequently accumulates in the cells in the form of large, distended drops, causing them to be termed *goblet cells*. *Serous cells* produce a watery secretion, in which there are frequently specific, useful enzymes.

A glandular area may be simply part of the surface of an epithelium, all its cells discharging separately into the cavity served, as is true in certain regions of the digestive tract. More frequently, however, we find that in the course of embryonic development glandular tissues withdraw from the general epithelial surface to form special tubes or pockets opening to the surface only by narrow ducts; the lining of the duct usually is nonsecretory. Such glands are of variable complexity; they may conveniently be classified into simple and compound types, either of which may be a tubular or an alveolar gland (Fig. 34).

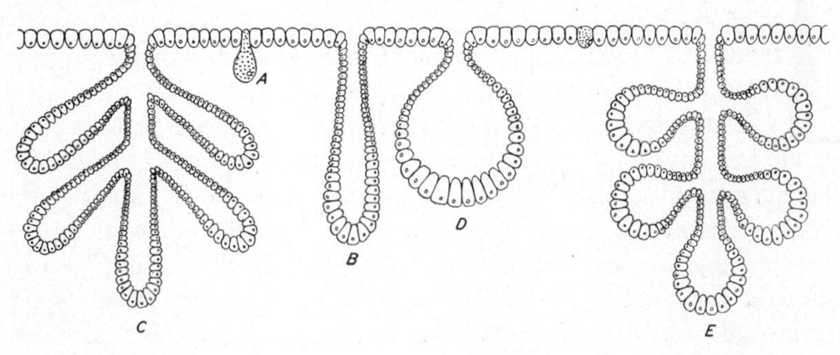

Fig. 34. Diagrams to show different types of glands. *A*, Unicellular gland; *B*, simple tubular gland; *C*, compound tubular gland; *D*, simple alveolar gland; *E*, compound alveolar gland. (From Turner.)

A *simple tubular gland* may be the simple tube which the name implies; however, the tube may be coiled (as in a sweat gland) or even somewhat branched in its inner secretory portion, as long as the branches seem to form a single compact unit. A *compound tubular gland* is characterized by a branching duct, with a cluster of tubules about each. A *simple alveolar gland* consists of a single secretory pocket or *alveolus,* or a few closely grouped pockets opening by a duct to the surface; a *compound alveolar gland* consists of a number of complex pocket areas.

CHAPTER V

THE EARLY DEVELOPMENT OF VERTEBRATES

THE ADULT PHASE of an organism is merely the last major stage in a long series of stages through which it passes during the lifetime of the individual; these earlier stages are more transient, but none the less important. In later chapters the development of the various specific organs and tissues will be noted. Here we shall discuss briefly the early developmental history of the vertebrate from the egg to the point where the major organ systems have differentiated and the basic ground plan of the body has been established. In doing so we shall simplify the story in somewhat diagrammatic fashion, and omit many features of interest to the embryologist.

EGG TYPES

The vertebrate egg, exceedingly variable in size, is typically spherical and exhibits little of structural features to foreshadow its remarkable potentialities. It contains a nucleus (with which that of the sperm fuses after fertilization), an amount of clear cytoplasm, and a variable quantity of relatively inert yolk (lecithin) which furnishes nutritive material for the developing embryo.

The quantity of yolk present varies greatly; it determines the size of the egg and is crucial as regards the pattern of early development. In some eggs—those of Amphioxus and mammals, for example—little yolk is present, and that rather evenly distributed through the cytoplasm. Such an egg is called *isolecithal* (or homolecithal). Actually it is doubtful if a truly isolecithal egg exists, for yolk material is heavier than ordinary protoplasm and tends to be more plentiful in the (topographically) lower part of the egg.

A second vertebrate egg type is that which may be termed *mesolecithal* (or mixolecithal); the egg is somewhat larger, containing a moderate amount of yolk which tends to settle into one hemisphere. The frog has a characteristic mesolecithal egg, and the same type is found in urodeles, lungfishes, lower actinopterygians, and lampreys; it is so widespread in lower aquatic forms that it is reasonable to conclude that the mesolecithal egg was characteristic of the ancestral vertebrates.

In the sharks and skates on the one hand, and reptiles and birds on the other, we find eggs of large size—the *telolecithal* type—with yolk constituting most of the volume of the cell, and with the relatively small amount of cytoplasm concentrated at one pole. So overwhelming is the yolk mass that in the kitchen the cell body of the hen's egg is simply termed the "yolk," to the neglect of the tiny amount of clear cytoplasm which it contains. In the common modern bony fishes—the teleosts—the egg is usually small, but heavily loaded with yolk. It behaves in development somewhat like that of a shark or bird, but with peculiarities that need not be considered here.

The tendency for concentration of the yolk in the lower hemisphere brings to light the first evidence of organization in the egg—a polarity, with an "animal" pole in the relatively clear cytoplasmic area above, a "vegetal" pole in the yolk region below. The egg thus has a radially symmetrical structure. In many invertebrates the axis connecting these two poles becomes the anteroposterior axis of the body, the vegetal pole becoming the posterior end. In the vertebrates and lower chordates this is also true, but with qualifications. Related to the greater complexity of development, the adult axis in Amphioxus lies about 45 degrees off the egg axis, so that (to put it crudely) the animal pole slants down beneath the prospective chin of the adult, and the vegetal pole slants upward and posteriorly toward the back of the animal. To make a bilaterally symmetrical rather than a radial animal, we need, in addition to an anteroposterior axis, a medial plane separating left and right halves of the future body. In some cases the point of entrance of the sperm and the path of its nucleus to a union with that of the egg appears to determine this plane; in other cases, however, the symmetry appears to be already established in the unfertilized egg.

The seemingly inert and featureless egg contains within itself all the potentialities needed for the complete development of the adult. The sperm supplies nuclear material which is of importance in future heredity and in the later phases of development, but which has little effect on the earlier phases of differentiation. When ripe, the egg is "set" and ready, waiting only a proper stimulus to begin the developmental story. Physical or chemical stimulation is capable in many cases of beginning this. Under normal conditions, however, the trigger effect is produced by the entrance of the sperm.

To illustrate the varied patterns seen in the early development of vertebrates we shall select eggs of the three different types discussed above. For an almost diagrammatically isolecithal egg we shall, in fact, leave the true vertebrates and resort to Amphioxus, a lower chordate relative of the vertebrates. The frog or urodele egg is a characteristic mesolecithal type. That of the shark or skate is illustrative of an extremely heavy yolked telolecithal type, and the bird's egg is similar in nature. The mammalian egg is tiny and almost devoid of yolk; but mammals have descended from reptiles with large-yolked eggs, and their developmental pattern is of a peculiar type with many "reminiscences" of that of telolecithal forms.

CLEAVAGE AND BLASTULA FORMATION

Amphioxus. As a first major sequence of events in development we shall describe the process of *cleavage,* leading to the stage known as the *blastula.* This process is most diagrammatically seen in Amphioxus.

In this form (Fig. 35) the first cleavage is longitudinal, extending from pole to pole (much as one cuts an apple into two portions), and results

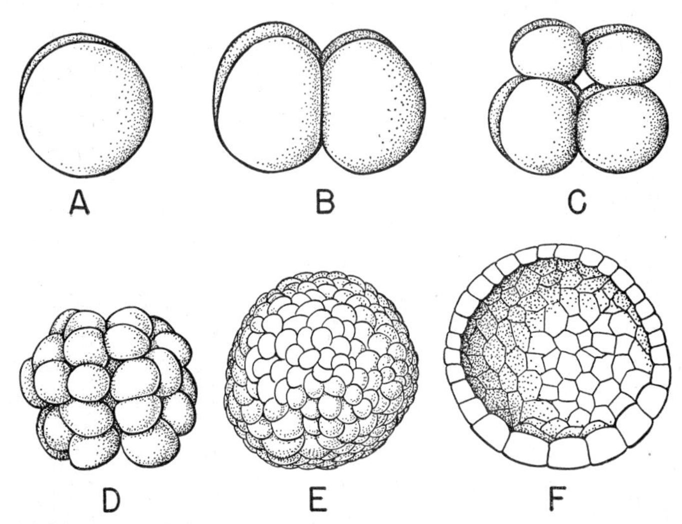

Fig. 35. Cleavage and blastula formation in an isolecithal egg—that of Amphioxus (cf. Figs. 36 to 38). *A,* First cleavage; animal pole of uncleaved egg is at top of figure. *B,* Second cleavage, to four-celled stage. *C,* Third cleavage; cells of animal hemisphere are somewhat smaller. *D,* After about two further cleavages. *E,* Blastula. *F,* Section of blastula, to show segmentation cavity in interior, and single-layered surface, with cells at vegetal pole rather larger. (Mainly after Cerfontaine.)

in the formation of two cells of equal size, which are destined, in normal development, to become the right and left halves of the body. The two adhere to one another, but nevertheless each tends to round up and approach a spherical shape—a tendency repeated in later divisions. A second cleavage is, like the first, longitudinal in direction, from pole to pole; it is at right angles to the first, and the process is similar to that of cutting an apple into quarters.

The third division is at right angles to both those preceding, and is essentially a cut around the equator of the egg, dividing each of the four cells then present into upper and lower components and resulting in an eight-celled stage. By this time, however, it is manifestly difficult for each of these cells to maintain a spherical shape and yet remain in contact with all former neighbors. In consequence there tends to develop a central cavity inside the sphere of cells—a cavity which becomes increasingly large as division proceeds, and is known as *segmentation cavity* or *blastocele*.

We have spoken of the Amphioxus egg as isolecithal, but pointed out that the yolk distribution is not absolutely uniform; there is a bit greater concentration toward the vegetal pole. A cell tends to divide, not through the center of its gross mass, but through the center of its living protoplasm, without regard for such relatively inert materials as yolk. In consequence

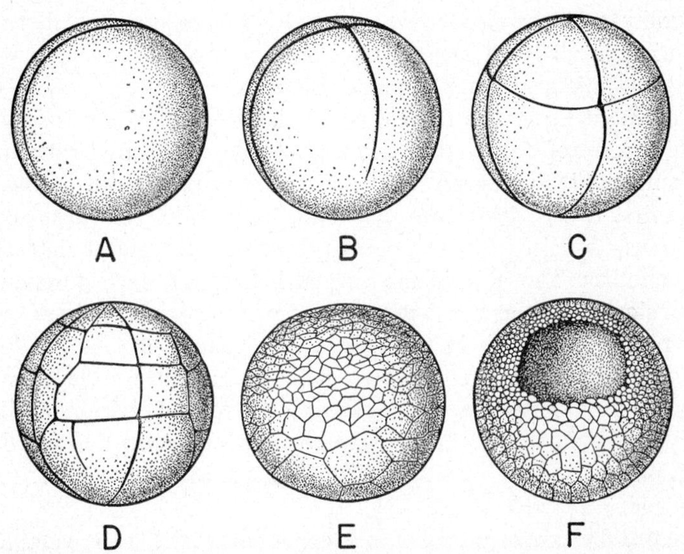

Fig. 36. Cleavage and blastula formation in a mesolecithal type of egg found in amphibians (cf. Figs. 35, 37 and 38.) *A,* First cleavage. *B,* Second cleavage. *C,* Third —meridional—cleavage, with smaller cells in animal hemisphere. *D,* About 36-celled stage; cleavages irregular, but slower and with larger cells in vegetal hemisphere. *E,* Blastula, with strong contrast between cells at two original poles. *F,* Section of blastula, showing segmentation cavity of restricted size; blastula a number of cell layers in thickness. Yolky mass at vegetal pole has cleaved, but slowly and into a mass of large cells.

the equatorial division which results in the eight-celled stage is not exactly through the equator, but slightly above this plane. As a result we find that the upper quartet of cells is slightly smaller than that below, and that the concentration of yolk is slightly greater in the ventral quartet.

The next set of cleavages is, as would be expected, longitudinal; each cell of both quartets is subdivided into two, to give a sixteen-celled stage. Then follows a paired set of cleavages in both animal and vegetal hemispheres to give a thirty-two-celled stage with, from pole to pole, four (rather irregular) rings of cells around the circumference, and with distinct (if small) size differences between these rings as we travel from pole to pole. Beyond the thirty-two-celled stage, since the process is a geometrical progression, the number of cells increases rapidly, to about sixty-four, 128, and so on; now, however, there is much less regularity in the divisions and a less synchronous cleavage. Within a short time the original egg cell has given rise to several hundred cells, arranged as a single-layered sphere about a central cavity. This sphere, the result of segmentation, is termed a *blastula*. Within is a large cavity, the blastocele, the initiation of which was noted above. The cells of the blastula are not too dissimilar from one another, but observation reveals the presence of smaller cells toward the original animal pole, and larger and somewhat more yolky cells toward the vegetal pole. In consequence there is a differentiation, although an imperfect one, of two types of cells which, as will be seen later, have a different fate in the embryo.

Mesolecithal Eggs. The normal amphibian egg (Fig. 36) as seen in a typical frog or urodele is of the mesolecithal type, with a moderate amount of yolk more highly concentrated toward the vegetal pole. In such an egg the first two divisions are essentially comparable to those seen in Amphioxus—longitudinal, and cutting from pole to pole at right angles to one another. The yolk in the vegetal hemisphere has, however, the effect of slowing up the cleavage of this part of the egg. The first cleavage begins promptly enough in the animal hemisphere. But as the furrow progresses downward, it encounters difficulty in cleaving the inert yolk materials, and is slowed up. In consequence the second division may begin dorsally before the first has been completed. This tendency for retardation of division in the lower portion of the egg persists through the period of cleavage.

The third division of an egg of this type is equatorial, as in Amphioxus. This cleavage tends, as there, to divide the protoplasm in each of the four cells into equal halves; but since there is considerably more yolk and less protoplasm in the vegetal hemisphere, the line of cleavage is well above the equator (almost as far north, one might say, as the Tropic of Cancer). In consequence we find in the frog or urodele egg, from this time on, a marked difference between the small cells of the animal hemisphere and the large, yolky cells of the vegetal hemisphere.

As the end result of the process of cleavage in an egg of this sort, we arrive at a blastula which in certain regards resembles that of Amphioxus, but in which there is a greater disparity between the cells of the two hemispheres in size and yolk content. Further, the blastocele, lying within the sphere of the blastula, is not so extensive as in Amphioxus; the yolk-filled protoplasm toward the vegetal pole has failed to divide promptly in the frog or urodele and a mass of large yolky cells partially fills the internal region of the sphere.

Telolecithal Eggs. Although the amphibian egg contains a considerable amount of yolk, it is, nevertheless, as is that of Amphioxus, *holoblastic;* that is, there is a complete cleavage of the entire egg in blastula formation.

The *teloblastic* type of cleavage (Fig. 37) is found in such forms as the elasmobranchs, reptiles, and birds, in which most of the egg substance consists of a great, inert yolk mass. The protoplasm is confined to a small area capping the yolk at the animal pole; cleavage and blastula formation are confined to this clear protoplasmic area; the yolk does not cleave at all.

As in eggs of other types, first and second cleavages pass through the center of the animal pole. They are not, however, complete cleavages of the egg; they affect only the materials of the protoplasmic cap and do not extend to the yolk. There follow "equatorial" divisions, separating a cluster of cells at the pole from more peripheral members. Further

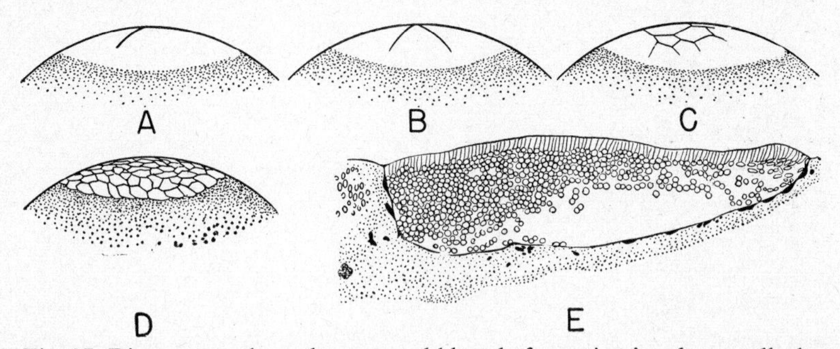

Fig. 37. Diagrams to show cleavage and blastula formation in a large-yolked egg, as that of a shark, reptile or bird (cf. Figs. 35 and 36). In *A* to *D* is figured only the animal pole of the egg, containing an area of clear protoplasm on top of the large, inert yolk mass. *A* to *D* show cleavage stages comparable to those in *A* to *D* of the two preceding figures; the result of cleavage is not a sphere, but a flattened plate of cells. In *E* is shown, at higher magnification, a section through the formed blastula of a shark. The blastula is a flat plate, a number of cells in thickness, with an irregular segmentation cavity lying below it, but above the unsegmented yolk mass. (*E* after von Kupffer.)

subdivisions, generally rather irregular, occur, until the entire protoplasmic area at the animal pole region is divided into a considerable number of cells, which we must regard as a blastula, and may term a *blastodisc*. In Amphioxus the blastula was only one cell thick. In a mesolecithal egg, however, the surface of the blastula is generally a thicker structure, several or many cells deep; and in telolecithal eggs the entire blastula region of the egg may be a thick mass of cells. This cell mass tends to be slightly lifted off the underlying yolk; the resulting space is considered to be equivalent to the blastocele.

But the blastula is essentially a flat sheet, not a sphere. Its margin, bounded all about by the yolk, consists of cells which in less yolky eggs would lie in the vegetal pole region, but are here unable to attain such a position. In cartographic terms, the blastula is a sphere flattened down into a two dimensional "map" on a north-polar projection.

Mammals. The early stages in mammalian development (Fig. 38) are quite specialized and unlike those of any other vertebrates. Except for the monotremes, mammals bear their young alive, and the nutritive materials for growth of the embryo are derived from the blood stream of the mother through the instrumentality of connecting tissues termed the *placenta.* The placenta is formed through a modification of embryonic membranes found in all amniotes, and much of the developmental process is comparable to that seen in a reptile or bird. But the necessity for the rapid development of the placenta is responsible for an early embryonic history that is unique.

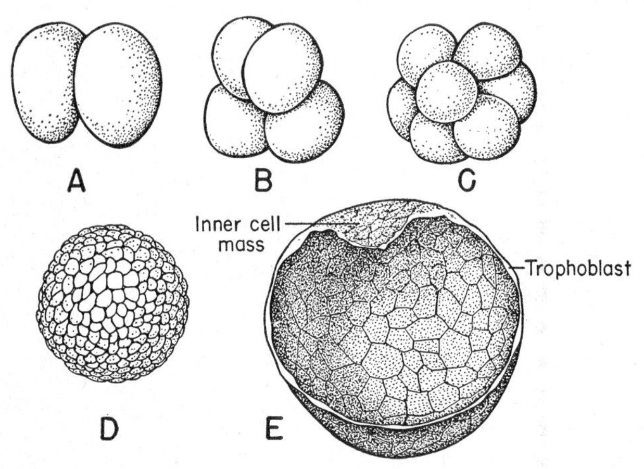

Fig. 38. Cleavage and blastula formation in a mammal. The small, almost yolkless egg cleaves (*A* to *D*) in a fashion similar to that of Amphioxus. The formed blastula (seen in section at *E*) has a deceptive resemblance to that of Amphioxus. Actually, however, the thin external sphere is the trophoblast, which forms a connection with the uterine wall, and the true blastula is merely the inner cell mass. This is a sheet of cells placed above the internal cavity much as the blastula of a telolecithal egg (Fig. 37, *E*) is situated atop the yolk mass. (After Streeter.)

Related to the fact that food materials are later accessible from the maternal blood stream, the mammalian egg is a tiny object practically devoid of yolk. In consequence, total cleavage, on the mode of Amphioxus, is possible. This cleavage results presently in the production of a little cluster of sixteen cells or so.

There now occurs a process unparalleled in other vertebrates. With further division, there differentiates (1) a spherical group of cells termed the *inner cell mass,* and (2) an expanded sphere of cells external to it, the *trophoblast.* As will be seen later, the trophoblast becomes an important part of the placenta, and its early development is a functionally vital embryonic adaptation. The egg is fertilized as it enters the oviduct, and reaches the uterus, in the walls of which it becomes embedded, within a few hours or days; the trophoblast must develop rapidly to make contact with the maternal tissues as soon as the uterus is reached.

GASTRULATION AND GERM LAYER FORMATION

Amphioxus. We have seen as the result of cleavage and blastula formation the development of the egg into an early embryo which consists in most types of a single body layer in the form of a sphere or sheet of cells. The next major event is gastrulation—the transformation of the blastula into a two-layered embryo, with an outer surface roughly corresponding

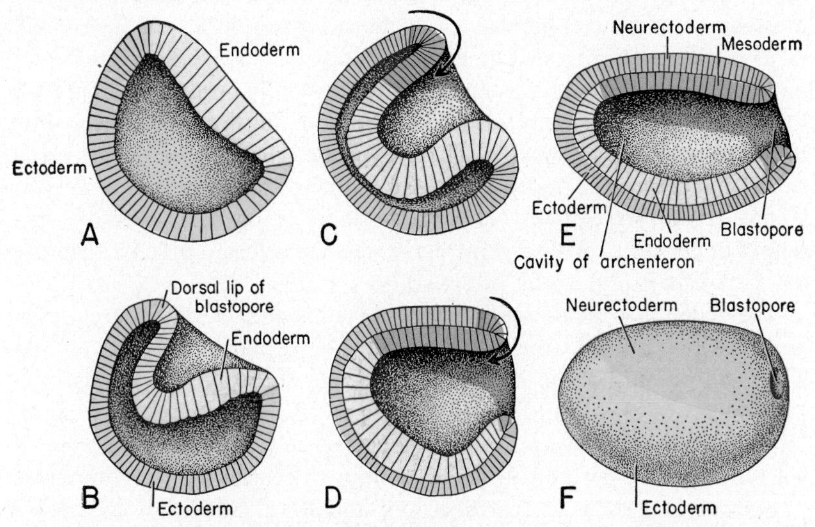

Fig. 39. Gastrulation in Amphioxus. *A* to *E,* Median sagittal sections showing successive stages. (The embryo has been rotated from the original egg position to that of the future adult, with head end at left, the former vegetal pole region at posterior end.) In *A* and *B,* invagination has occurred; *C* to *E,* growth by epiboly and involution, mainly at dorsal lip of blastopore (indicated by arrows). *F,* Surface view of late gastrula, seen from left. (After Hatschek, Cerfontaine.)

In this and subsequent figures the following colors are used to distinguish germ layers: skin ectoderm, blue; neurectoderm, green; mesoderm, red; endoderm, yellow.

to the skin surface of the adult, and an internal surface roughly corresponding to the adult gut lining. This stage of the embryo is termed the *gastrula.* It is comparable in many ways to the adult of the lowest of metazoan animals, the coelenterates, which never develop beyond a two-layered condition.

Amphioxus presents the simplest and most diagrammatic type of gastrula formation (Fig. 39). The first major event in the gastrulation of this animal is the pushing in of the vegetal hemisphere of cells into the blastocele (which it almost completely occludes), much as one might, with a thumb, push one side of a soft rubber ball into the other. The result is a double-layered cup, with the smaller cells of the original animal hemisphere lying outside, the larger and yolkier cells of the vegetal hemisphere lining the inner surface. The inner lining of the cup is basically comparable to the gut of the adult animal, and is termed the *archenteron*, or primitive gut. The opening from the archenteron to the outside is termed (somewhat confusingly) the *blastopore*.

In many invertebrates this process of inpushing of the vegetal cells, known as *invagination*, constitutes the whole act of gastrula formation. The cells of the original animal hemisphere in such cases constitute the *ectoderm*, or outer germ layer of the late embryo and adult; the inner cells are the *endoderm*, or inner germ layer. In the chordates, however, gastrulation is no such simple process. The chordate body is a complex structure. The body materials of the simple coelenterate cup would produce in a chordate little more than the superficial part of the skin and the inner lining of the digestive tract and its glands. There is needed an additional part of the superficial ectodermal layer for the formation of the complicated nervous system. And—exceedingly important—there must be formed the materials of the third major germ layer, the mesoderm, which constitutes the greater part of the bulk of the body.

The materials which form the nervous system and mesoderm are present as cells which lie on the outer surface of the early blastula (Fig. 40, *A*, *B*). They can be marked or stained at this stage and followed through into later stages of the development of the gastrula. But as yet they exhibit little or no differentiation from neighboring cell areas; and most are as yet far from the position they must attain to realize their potentialities.

Further stages in gastrulation and the development of the neural ectoderm and of mesoderm are accomplished by activity and growth in the region of the blastopore, particularly at its dorsal "lip," which acts as a major center of growth and organization within the embryo at this stage in its career. Beginning at the dorsal lip, and later spreading laterally and ventrally, there is a rapid multiplication of cells around the blastopore margins. With multiplication, the external tissues tend to extend backward, partially covering the originally large opening of the blastopore. This process is termed *epiboly* (a "throwing over"). But since the continuity of inner and outer layers is always preserved at the curved edge of the lip, there must be continually an inward rolling—*involution*—of cells at the lip from outer to inner layers.

With continuation of this growth, the gastrula is converted from a hemisphere into a somewhat elongate hollow cylinder, closed anteriorly, but still open posteriorly through a much reduced blastopore. The earlier cup-shaped gastrula was asymmetrically disposed, dorsoventrally, with regard to the prospective anteroposterior axis, the center of the primary ectoderm lying rather ventrally and anteriorly. The greater posterior growth from the dorsal blastopore lip results in a more symmetrical definitive gastrula. The disposition of materials, however, is still asymmetrical at this stage. The ectoderm, in a narrow sense, is still confined to the anterior and ventral parts of the surface of the gastrula. Dorsally and posteriorly we find a newly expanded area of cells produced by growth around the original blastopore lips. Although not visibly distinguishable from the ordinary or *skin ectoderm,* later development shows this oval cell area to be the source of the future nervous system; it may be regarded as essentially a separate germ layer, a *neurectoderm.*

Internally, the gastrula shows a division of the gut lining into two parts, a division roughly comparable to that seen externally. Ventrally and anteriorly lie the invaginated materials of the original vegetal hemisphere of the blastula. These form the definitive gut lining, the endoderm, and occupy a position internally which corresponds roughly to that of the normal ectoderm outside. The dorsal and posterior parts of the primitive gut, however, are lined with smaller cells which were originally on the outer surface dorsally and laterally, but have been rolled inward in the process of involution. These are seen in later development to be not part of the definitive gut, but the source of the future middle body layer, the *mesoderm* (including the notochord). We may thus conceive of the Amphioxus gastrula as a two-layered structure, with each layer divided topographically into two components: externally, definitive ectoderm and neurectoderm; internally, endoderm and mesoderm.

Although not properly a part of gastrulation, the next stage in the development of the mesoderm will be described at this point; this is the formation of mesodermal pouches (Figs. 47, 48, p. 98). Of the mesoderm, that part which forms a longitudinal band along the dorsal midline of the archenteron roof is not involved in pouch formation, and constricts to form the notochord. At the sides, however, the mesoderm folds outward into a pair of longitudinal grooves. At the anterior end of each groove a spherical pouch pinches off as a *mesodermal somite.* This contains a cavity, formerly continuous with that of the archenteron, which is destined to become part of the celom; its walls later differentiate into mesodermal tissues. Posteriorly, successive pairs of somites are budded off the length of the body. As these pouches are budded off, the endoderm, originally lateral and ventral only in position, comes to extend upward internal to them and across beneath the notochord to form a continuous lining to the definitive gut. This pouch type of mesoderm formation is highly comparable to the process seen in acorn worms and echinoderms and is a prime basis for the belief that chordates are related to the echinoderm phylum.

Mesolecithal Types (Fig. 41). In such types as the urodeles and frogs, with mesolecithal eggs, gastrulation takes place in a fashion similar to that of Amphioxus. But the presence of a large mass of yolk cells filling much of the interior of the blastula reduces the amount of invagination that can take place; in consequence, much of the formation of the gastrula occurs by growth of the blastopore lips. At the start of gastrulation there is a certain amount of invagination in the region of the future dorsal lip of the blastopore (Fig. 41, *B*); but it is a physical impossibility for all of the massive material of the vegetal hemisphere—the potential

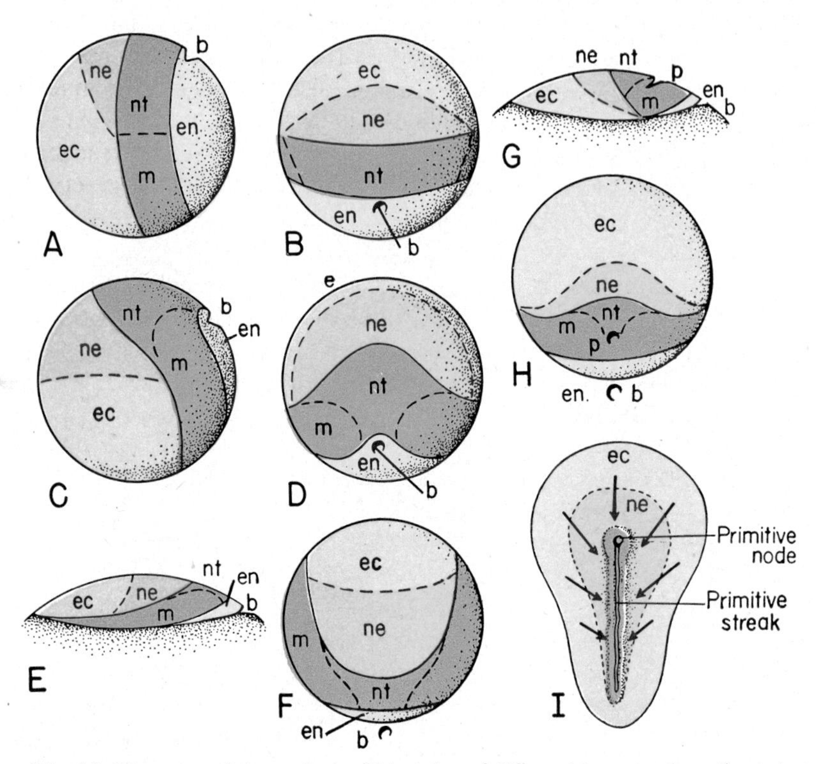

Fig. 40. Diagrams of the surface of blastulae of different types to show the potential fate of various regions in normal development. *A, B,* Left side and dorsal views of the blastula of Amphioxus (cf. Fig. 35, *E*). *C, D,* Similar views of an amphibian egg (cf. Fig. 36, *E*). In these figures the embryo has been rotated from the position in which the egg originally floated to that assumed by the gastrula; the shift is such that the vegetal pole region, originally ventral, has rotated posteriorly and upward to essentially the position from which the endoderm develops. The position at which the blastopore develops is arbitrarily indicated by an indentation. In *E, F,* and *G, H,* similar lateral and dorsal views are shown for the flattened blastula of telolecithal eggs of shark and bird. Note that in all forms the pattern of potential germ layer areas is similar. *I,* Stage in bird development beyond *H;* the embryo is elongating the endoderm is already invaginated, and mesoderm and neural ectoderm are moving inward (as indicated by arrows) to the primitive streak. *b,* Position at which blastopore or equivalent develops; *ec,* future skin ectoderm; *en,* endoderm; *m,* mesoderm; *ne,* neural ectoderm; *nt,* notochordal region of mesoderm. *p,* position in which primitive node and streak make their appearance. (Data from Conklin, Vogt, Vandebroek, Pasteels.)

endoderm—to fold itself neatly within the mantle of ectodermal cells formed from the animal portion of the egg. The archenteron as formed by invagination is a modest pocket; the remainder of the process of gastrula formation basically results from epiboly and involution, owing to the growth of the blastopore lips.

As in Amphioxus, the future mesodermal areas can be outlined on the outer surface of the blastula (Fig. 40, *C, D*), and the mesodermal materials roll over the blastopore lips on to the inner surface (Fig. 41, *C*). There is, however, from here on a marked difference from the Amphioxus condition. In that form the mesoderm remains for a time in continuity with the endoderm and only later folds outward, in the form of pouches, into its definitive position. In amphibians the dorsal band of mesodermal tissue (the future notochord) remains for some time in contact anteriorly

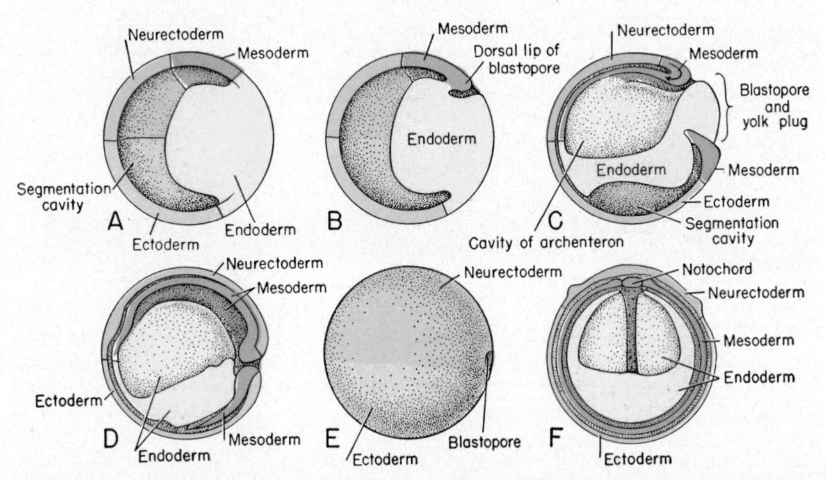

Fig. 41. Gastrulation in an amphibian type of egg (cf. Fig. 40). *A* to *D* are views comparable to *A* to *E* of Figure 39. The presence, however, of a large mass of yolk restricts invagination to the extent shown at *B;* the remainder of gastrulation is performed by epiboly and involution, as indicated for dorsal lip by arrow in *C*. *E*, Gastrula from left side. *F*, Transverse section, looking anteriorly. As can be seen from *D* and *F*, the mesoderm folds inward between ectoderm and endoderm (cf. Fig. 49, *A, B*, p. 99). (After Hamburger.)

with the endoderm (Fig. 41, *C, D*). But at either side, the infolded mesoderm pushes out between endoderm and ectoderm and grows downward as a thin sheet which eventually reaches the ventral midline (Figs. 41, *F*, 49, *A, B*, p. 99). As this sheet of mesoderm is laid down, the endoderm (as in Amphioxus) grows upward to form a roof for the definitive gut. A furrow sometimes seen at the point at which the mesoderm folds outward from the archenteron suggests an "attempt" at pouch formation. One may believe that the process of mesoderm formation seen in mesolecithal eggs is basically similar to that of Amphioxus; but here the mesoderm is at first a continuous solid sheet, and it is only later that segmental conditions and celomic cavities appear.

Elasmobranchs (Fig. 44, p. 94). Obviously, from the nature of the blastula, "typical" gastrulation in a telolecithal vertebrate is an impossibility. There is merely a flat plate of cells, a blastodisc; no sphere to be invaginated. Epiboly and involution, however, are much in evidence.

The primary event in normal gastrulation is the formation of an endoderm beneath the ectoderm; and part, at least, of endoderm formation should be by the involution of materials at the blastopore lips, and most markedly at the dorsal lip. But where are the blastopore lips in this flat plate? The most reasonable answer is that they lie at the margins of

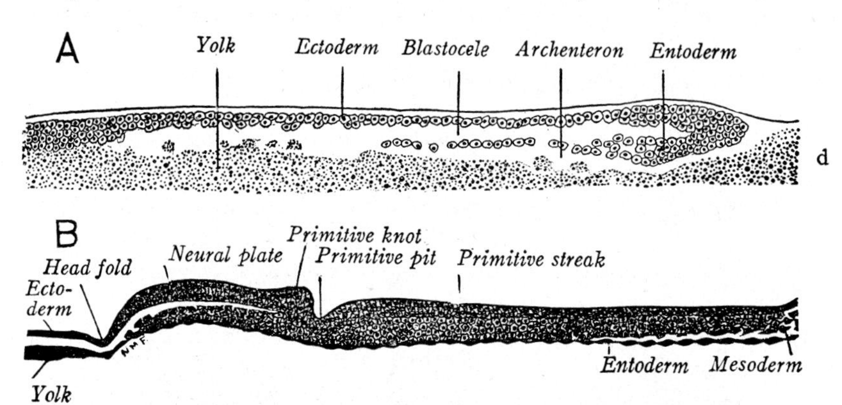

Fig. 42. Two successive longitudinal sections of a bird embryo, to show gastrulation. *A,* Figure comparable to Figure 44, *B*; the endoderm is turning in posteriorly to form a roof to the archenteron. This area of inrolling is comparable to a blastopore (cf. Fig. 39, *C*). *B,* Later stage, comparable to Figure 40, *I;* the section runs through the streak from the primitive pit backward; cells are turning down inward from the surface (in the plane of the paper) and rolling forward and laterally to form notochord and typical mesoderm. (After Arey.)

the plate; and the dorsal lip, the most active area, should, by comparison with nontelolecithal forms, lie at the posterior end of the forming embryo.

In such a form as a shark or skate the comparison with a typical gastrula can be made without too great difficulty. A center of activity is early seen at one part of the disc margin (Fig. 54, *A*, p. 105), and the components of the formed embryo have been found distributed over the disc surface (Fig. 40, *E, F*) in such a fashion that this area can be safely regarded as the dorsal lip region. At this point, and to either side of it, there is rapid growth, backward extension in epibolic fashion, and active involution. There is first an inturning of endoderm (Fig. 44, *B, C*), which spreads forward beneath the germ disc to transform it into a two-layered, flattened equivalent of a gastrula. Following the endoderm, mesodermal elements are folded under to spread out between ectoderm and endoderm (Fig. 44, *D*). The resulting embryo has thus the essential structure of an amphibian gastrula, with notable differences, however; for the endoderm does not surround the yolk of the egg, but is merely spread out over its upper surface. Similarly, although the mesoderm spreads on either side between ectoderm and endoderm, it extends outward, not downward. The embryo is, so to speak, unbuttoned ventrally.

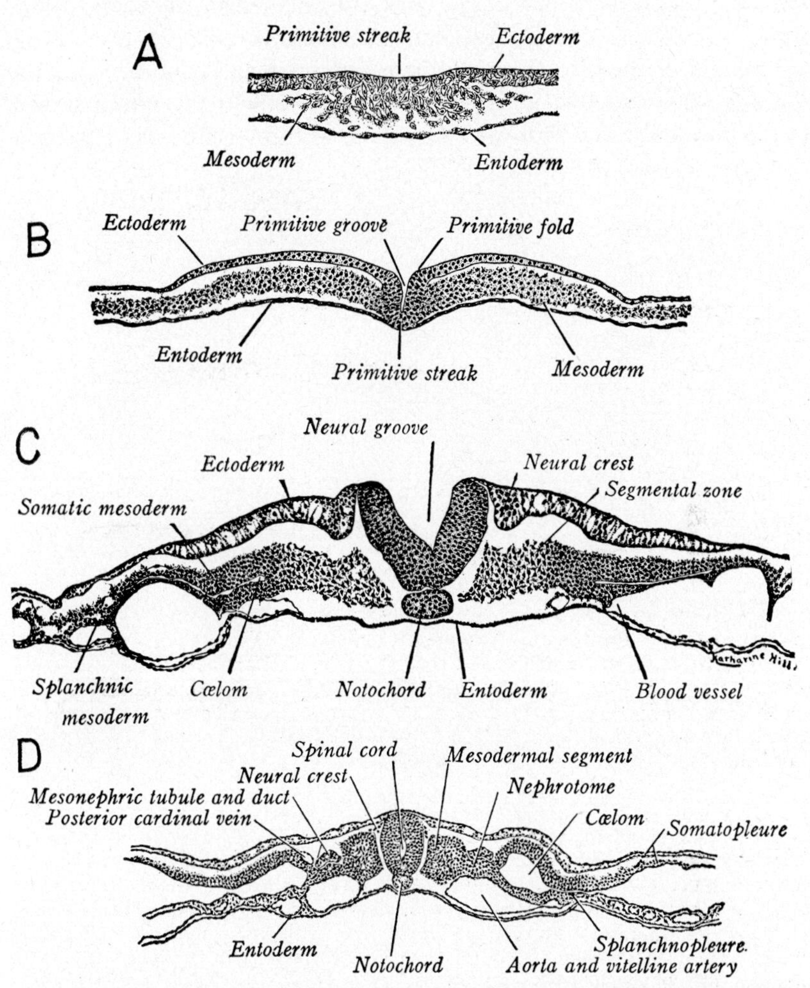

Fig. 43. Cross sections of chick embryos to show successive stages in development of mesoderm and neural tube. *A,* Stage in which the inturning of mesoderm in the primitive streak has barely begun. *B,* The mesoderm has spread widely on either side between ectoderm and endoderm, but has not differentiated further. *C,*The celom has begun to appear, splitting the lateral part of the mesoderm into outer—somatic—and inner—splanchnic—parts. Centrally, the notochord has separated from the remainder of the mesoderm, and neural folds and crests are appearing on either side of a neural groove. *D,* The neural folds have closed to form a tube, the spinal cord. The mesoderm has divided into somites, nephrotomes, and a lateral plate in which a celom separates inner and outer layers of the body—splanchnopleure and somatopleure. (From Arey.)

Reptiles and Birds (Figs. 42, 43). The egg-laying amniotes—reptiles and birds—have a still more specialized type of gastrulation. The potential embryonic regions have a distribution on the flat blastodisc similar to that in the shark (Fig. 40, *G, H*). Over most of its periphery the disc is rather tightly adherent to the yolk. In one region, however, which proves to be the future posterior end of the embryo (Fig. 55, *A*, p. 106), the margin of the disc is for a time a free-growing lip, at which, as in sharks, cells from the dorsal surface roll inward to the under surface of the disc and extend forward beneath it to form a thin sheet of endo-

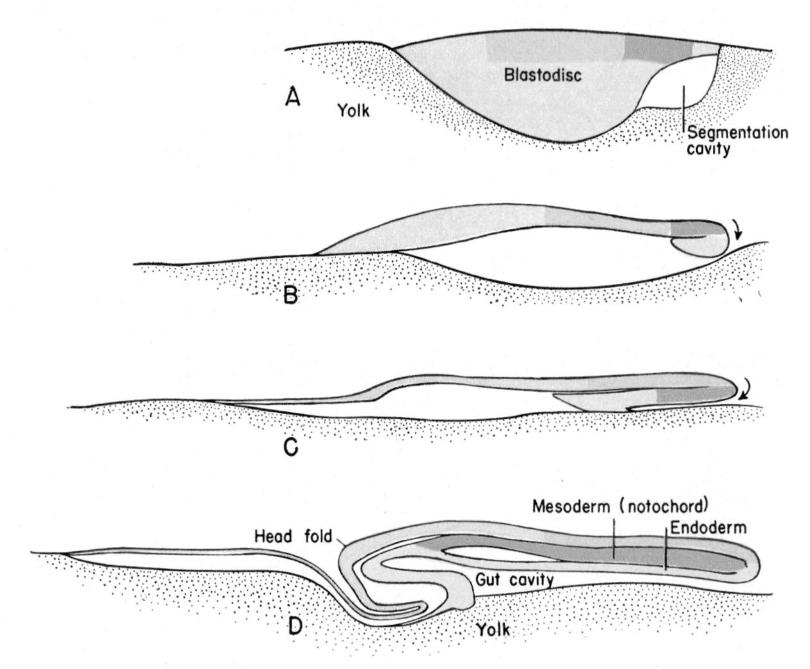

Fig. 44. Longitudinal sections of successive stages in gastrulation of an egg of telolecithal type, as seen in an elasmobranch. Only the disc of the blastula and the neighboring part of the yolk are shown. Anterior end at left. *A*, Blastula (cf. Fig. 37, *E*); *B*, involution of endoderm at posterior end of disc, corresponding to blastopore; *C*, continued process of epiboly and involution, with inturning of mesoderm; *D*, mesoderm separated from endoderm; gut cavity formed, open below, roofed by endoderm. (After Vandebroek.)

derm (Fig. 42, *A*). There is thus performed the primary act of gastrulation, with the production of a two-layered disc. The free posterior margin of the disc, where overgrowth and involution take place, is here, as in sharks and skates, comparable to the dorsal lip of the blastopore.

There is, however, no inturning of mesoderm in this process. For mesoderm formation a different procedure is adopted. Extending forward from the region of the blastopore lip, there shortly appears on the dorsal surface of the disc the *primitive streak* (Figs. 40, *I*, 55, *B*, p. 106). This consists of parallel ridges with a groove between them; at the front end of the groove there develops a pit which in many amniotes extends forward and downward as a canal to pierce the roof of the underlying gut cavity. Observation and experiment show that the primitive streak is a region of crucial activity in the formation of the embryo. On the dorsal surface of the disc there is a steady movement of cells medially and

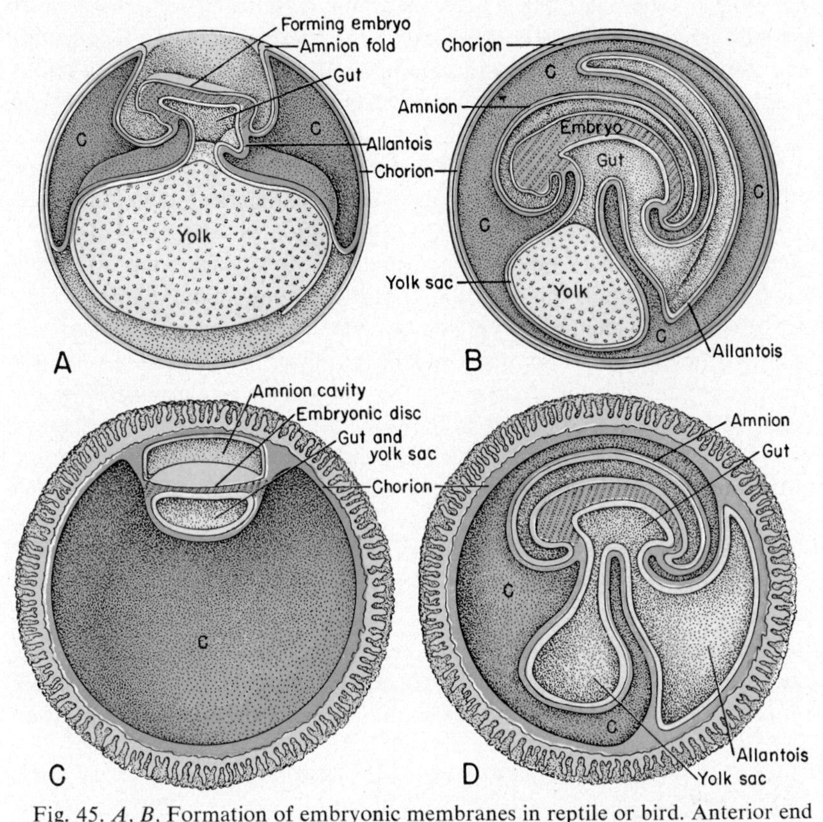

Fig. 45. *A, B,* Formation of embryonic membranes in reptile or bird. Anterior end at left. *A,* An early stage. The embryo has been lifted somewhat off the yolk, but gut cavity proper and yolk sac are broadly connected. Yolk sac incompletely formed; amnion folds and chorion incomplete; allantois barely indicated. *B,* Later stage; embryonic membranes formed and yolk already partially reduced. *C, D,* Comparable views of mammalian type of development as seen in primates. *C,* Stage beyond the blastula seen in Figure 38, *E.* The inner cell mass has split ventrally to produce a gut cavity—constituting the major act of gastrulation—and split dorsally to produce an amnion cavity. Between the two cavities the embryo forms a disc in which primitive streak formation occurs much as in a reptile or bird. Mesoderm has already appeared, and chorionic villi are establishing connections with the surrounding uterine wall. *D,* Later stage in mammalian development, corresponding to *B. c,* Celomic cavity.

posteriorly into the margins of the primitive streak; in the walls of the streak the sheets of cells move downward and then fan outward to interpose themselves between ectoderm and endoderm (Figs. 42, *B,* 43). Anteriorly, cells taking part in this movement push forward from the pit region to form the notochord; laterally, the material moves outward to form the somites and other mesodermal structures.

It is obvious that the primitive streak, although not an open blastopore, performs the functions of that structure in forming mesoderm; the cell movements which take place here are comparable to those performed by the mesoderm at the blastopore margins in a mesolecithal egg. The primitive streak persists for a considerable period during embryonic development as the embryo elongates and mesoderm is formed for more and more posterior regions (Fig. 55, *C*).

Mammals (Fig. 45, *C*). Gastrula formation in mammals is a unique process. In later stages the mammal comes to be identical in most respects with its amniote relatives, but until gastrulation is completed it is still quite atypical; it has not yet recovered, so to speak, from its initial vagaries. The details of gastrulation vary among mammalian groups; described here is that characteristic of primates.

The so-called blastula consisted of an external trophoblast, which makes contact with the uterine tissues, and an inner cell mass. The first stage in further development is the appearance in the inner cell mass, above and below, of cavities which expand to leave between them a flat, two-layered plate of cells. The upper cavity, lined with ectoderm, is that of the amnion; the lower, a yolk sac with an endodermal lining. The cavities and the materials lining them are parts of the amniote membrane system, to be described later, the flat two-layered plate between them is a *blastodisc,* in which the embryo is to arise. The essential act of gastrulation, the formation of endoderm, is directly accomplished by the cleavage of the yolk sac cavity out of the under margin of the inner cell mass; older processes of invagination or involution of endoderm have here been completely abandoned.

Beyond this point, however, embryo formation proceeds as in other amniotes. A primitive streak arises on the blastodisc (Fig. 56, *A,* p. 107); as in many reptiles and birds, a canal bores inward and forward from a primitive pit at the anterior end of the streak, bearing mesoderm cells with it; mesoderm rolls inward and downward along the course of the primitive streak to push outward below, between ectoderm and endoderm. Relatively little is known about the organization of the mammalian blastodisc, but presumably ectodermal and mesodermal components are arranged on it in a fashion similar to that of a bird or reptile.

NEURAL TUBE FORMATION

After the completion of the process of germ layer formation and the development of a gastrula of one type or another, there occur formative processes in the neural ectoderm which result in the development of a neural tube and the progression of the embryo to a stage termed the *neurula.*

In Amphioxus the neural ectoderm occupies a large oval area on the dorsal and posterior surface of the gastrula (Fig. 39, *E, F*, p. 87). We presently find on either side a folding upward of tissue at the junction of the future body ectoderm and neurectoderm areas (Fig. 47). In Amphioxus (not in true vertebrates) the two tissues separate as this fold forms. The ectodermal margins from right and left sides grow over the neural region and finally meet to form a complete layer of "skin" over the top of the body. Beneath this layer the lateral margins of the neurectoderm sheet gradually roll upward, meet, and form a circular tube—the neural tube (Figs. 47, *F*, 48). The tube first closes midway of its length; the process gradually progresses both forward and backward. For some time the anterior end still opens to the surface as a *neuropore*. More curious is the situation at the posterior end of the canal. Here the folds cover over

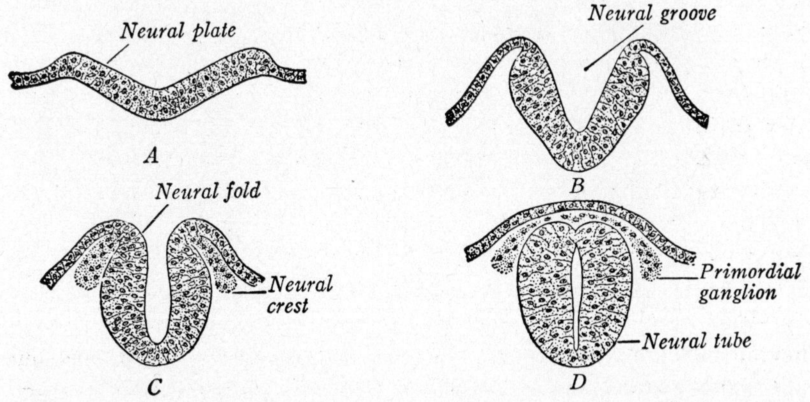

Fig. 46. Formation of the neural tube and crest as seen in a typical vertebrate (mammal); a series of transverse sections at successive embryonic stages. (From Arey.)

the blastopore (Fig. 48). This remains open, but now opens, not onto the surface, but into the posterior (and lower) end of the neural tube. The primitive gut thus remains in connection with the surface; but the connection is indirect, running up into the neural canal and then forward to the neuropore. The connecting piece between gut and neural tube is termed the *neurenteric canal*. In later development, as the tail sprouts out, it closes, and neural and gut tubes become discrete structures.

Most vertebrates show a type of neurectoderm differentiation contrasting with that seen in Amphioxus (Figs. 43, *C, D*, 46). The ectoderm tissues fold upward on either side as a *neural fold*. Eventually the two folds meet one another dorsally; ectodermal and neural tissues in each fold separate and meet their "other numbers" from the opposite side, and skin surface and neural tube are completed. The end result is much the same as in Amphioxus, although attained in different fashion. During the folding process, as the folds form high *neural crests* on either side, masses of cells are pinched off into the interior. Most of these crest cells are destined to become part of the nervous tissues; others, noted in later chapters, have a more varied history. In the head region neural crest materials may free themselves from the neighboring ectoderm before the

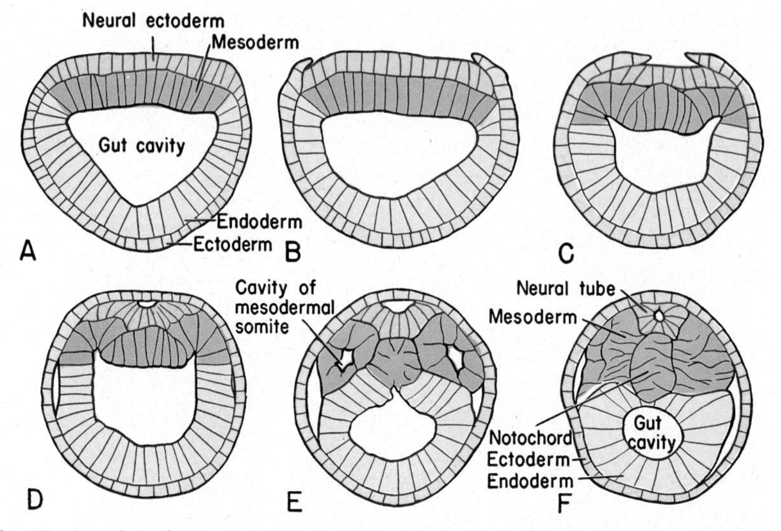

Fig. 47. A series of cross sections to show formation of mesodermal pouches and neural tube in Amphioxus. (After Cerfontaine.)

neural folds are well formed; still other future nervous system elements and sensory structures may arise as *placodes*—thickenings of the embryonic ectoderm lateral to the neural tube region, which detach themselves from the under surface of the future skin.

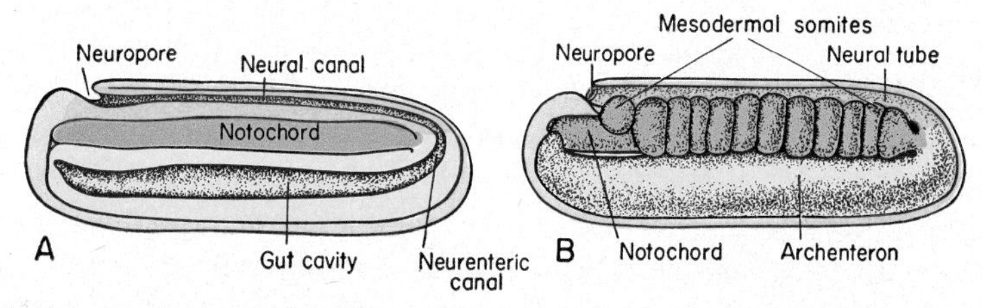

Fig. 48. Amphioxus embryos at a stage in which the neural tube has formed and mesoderm is differentiating. *A,* Sagittal section. *B,* Longitudinal view with skin ectoderm sectioned medially, but internal structures preserved intact. (After Cerfontaine and Conklin.)

MESODERM DEVELOPMENT

The mesoderm forms the greater part of the bulk of the body. Except for the brain and spinal cord, the ectoderm forms little but the superficial portion of the skin. Except for a mass of liver and pancreas tissue, the endoderm forms little but a thin film of epithelium lining the gut. Practically all the rest of the body is derived from the mesoderm—muscles, connective tissues, skeleton, circulatory, urinary, and genital tissues. If comparison be made with a house, the ectoderm corresponds to the paint on the outside and the wiring system; the endoderm to the floor varnish, wall paper and perhaps the kitchen stove. All the rest—frame, plumbing, sheathing, flooring, even the floor boards, lath, and plaster—is comparable to the mesoderm derivatives.

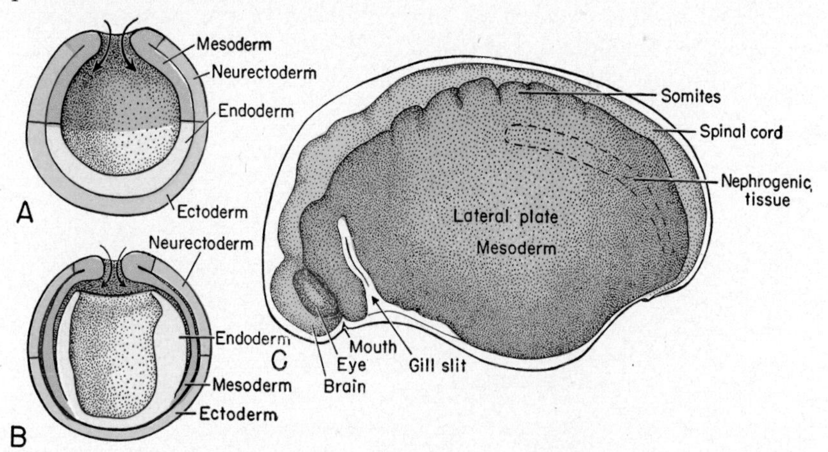

Fig. 49. Mesoderm formation in an amphibian. *A,* Section of a urodele gastrula, cut transversely through the blastopore, showing involution of mesoderm into the lateral walls of the archenteron. This is essentially similar to the situation in Amphioxus at the stage of Figure 39, *C* or *D*. *B,* Later stage; the mesoderm, instead of forming hollow pockets, as in Amphioxus (Fig. 47), attains its intermediate position by pushing downward and forward between ectoderm and endoderm. *C,* A later embryo of an amphibian, after the neural tube is formed, seen in side view. The mesoderm forms a long, continuous sheet on either side of the body. The dorsal part is beginning to subdivide into somites. The part of the mesoderm which will later form kidney tissue is indicated by broken lines. The lateral plate is being broken up anteriorly by the formation of gill clefts. (*A* and *B* after Hamburger; *C* after Adelman.)

Mesodermal Divisions. In Amphioxus the mesoderm (apart from the notochord, discussed later) forms a paired series of segmentally arranged somites, which from the beginning contain a celomic cavity (Figs. 47, 48). In true vertebrates, as we have seen, there is marked modification of this pattern. There is at first no segmentation in the mesoderm, which pushes out on each side as a longitudinally continuous sheet between ectoderm and endoderm; nor does the mesoderm at first contain any cavities of celomic type. In mesolecithal eggs the mesoderm of either side grows as a hemicylinder, following the curve of the formed body wall downward and then inward to the ventral midline of the belly (Figs. 41, p. 91; 49). In telolecithal eggs, where the body is at first spread out flat above the yolk mass (and in mammals where the same pattern is followed despite the absence of yolk), the mesoderm spreads out laterally as a flat plate (Fig. 43, p. 93) and continues beyond the region of the body proper to contribute to the development of the extra-embryonic membranes. In these latter types, we noted, the body is at first "unbuttoned" ventrally, so that it is only at a late stage that the mesoderm growths of the two sides gain contact ventrally, and the body (and mesodermal) contours come to resemble those of mesolecithal forms.

There presently appears, in all vertebrates, a differentiation, from the dorsal midline outward, of three divisions of the mesoderm, each extending the length of the trunk (in head and tail the mesoderm is more restricted in its development). Next to the neural tube and notochord

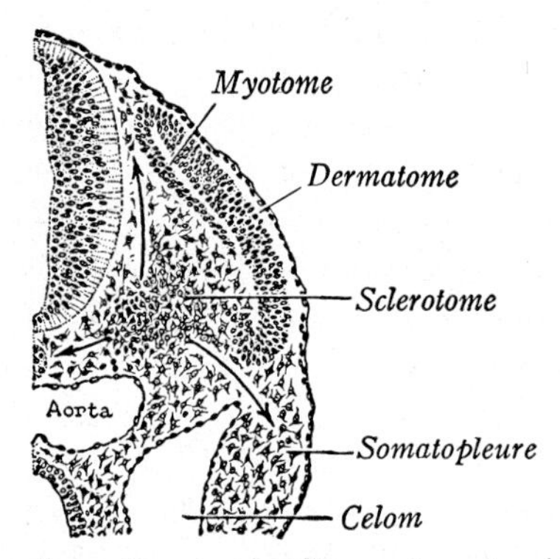

Fig. 50. Hemisection through a mammalian embryo to show the subdivision of the somite into myotome, dermatome, and sclerotome. Arrows show directions in which mesenchyme grows from sclerotome to form vertebra and rib. The small notochord is present above the aorta, and part of the gut wall is shown below that large vessel. (From Arey.)

the mesoderm thickens and subdivides on either side into a longitudinal row of blocklike structures—the *mesodermal somites* (Figs. 43, *C, D,* 49, 55, *C, D,* 56, *C,* pp. 106, 107), comparable to the somites of Amphioxus. These are the first indications of true segmentation in the vertebrate body, and it appears that (apart from the independently derived serial arrangement of the gill structures) the segmentation seen in other vertebrate organs is due to the influence of the segmental arrangement of the mesodermal somites.

Within each somite there is generally for a short time a development of a small celomic cavity comparable to that which exists from the first in Amphioxus. Soon differentiation appears within the somite (Fig. 50). Medially, there is a loss of its epithelial nature, and there is a great proliferation of cells from its ventral medial corner. These cells form an area of loose embryonic tissue, of the type termed mesenchyme, which expands around the nerve cord and notochord and seems to form much of the axial skeletal structures; in relation to this, the medial part of the somite is termed the *sclerotome*. The epithelial layer on the outer side of the somite disintegrates, and proliferates mesenchyme cells which appear to form connective tissues of the skin; hence the name *dermatome* given to this part of the somite. After the loss of these two areas, medial and lateral, the remaining portion of the somite, termed the *myotome,* gradually differentiates and expands to form the axial musculature, the history of which shall follow in a later chapter.

Lateral or ventral to the somites a relatively small intermediate region of the mesoderm develops in the trunk into materials, the *nephrogenic tissue,* from which are derived the kidney tubules and their ducts, and the deeper tissues of the gonads as well. This region may exhibit a segmentation comparable to that of the somites adjacent to them and form *nephrotomes,* but frequently it develops for most or all of the length of the trunk as a continuous band (Fig. 49, *C*).

Beyond the nephrogenic region—curving ventrally in mesolecithal forms, but at first extending straight laterally in other types—is a great sheet of mesoderm termed the *lateral plate* (Fig. 49, *C*). Never is there any segmentation in this plate. At first the plate is a solid band; later, however, it cleaves, and a *celomic cavity,* the cavity which in adult life surrounds most of the viscera, develops within the plate (Fig. 43, *C, D,* p. 93). In addition, however, the lateral plate proliferates further supplies of mesenchyme.

Mesenchyme. During much of embryonic development there exist, between the epithelia and tissue masses of the major organs, relatively empty spaces filled with fluid. Scattered through these spaces, however, we find diffused star-shaped cells which are believed to have in many cases the properties of ameboid movement. These cells compose the mesenchyme, the embryonic connective tissue.

We have already cited areas of origin of this type of tissue. Much of it is formed by proliferation from the somites, and, indeed, at one time it was thought that all mesenchyme had this type of origin. In addition, however, as just mentioned, the lateral plate is a source for mesenchyme. These areas of origin are both from the mesoderm, and certainly most mesenchyme is of mesodermal origin. But, as noted elsewhere, the ectoderm in certain instances gives rise to materials of a mesenchyme-like nature, and the endoderm, too, it seems, may produce tissues of a sort normally derived from mesenchyme; its production is not confined to one region or one germ layer alone.

Though the mesenchyme acts as an embryonic connective tissue, its activities are not confined to such an essentially passive and transient function. It is a most versatile material. It gives rise to the connective tissues of the adult. Vertebrate skeletal materials, both cartilage and bone, are of mesenchymal derivation; they are essentially modified and strengthened types of connective tissue. The mesenchyme gives rise to the circulatory system in toto—to the blood vessels and to the blood corpuscles as well. Much of the musculature of the body is of mesenchyme derivation, including all smooth musculature, the specialized cardiac musculature, and even a part of the striated musculature.

Connective Tissues. The most direct adult products of the mesenchyme are the connective tissues, which form the "stuffiing" of the body and reinforce the epithelia of many of the body organs. The simplest form of connective tissue is a loose, *reticular* type. Much of its bulk consists of a gelatinous ground substance, presumably secreted by the sparsely distributed connective tissue cells. It contains a loose network of long and

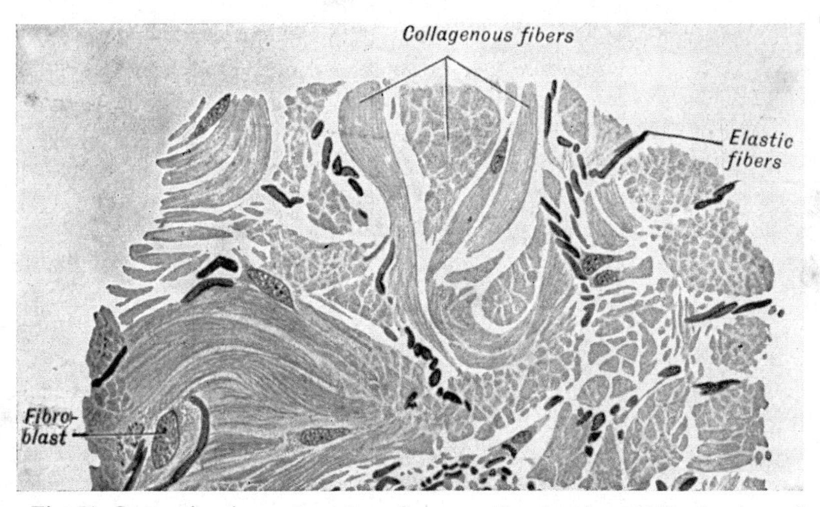

Fig. 51. Connective tissue. A section of mammalian dermis, × 500, showing collagenous fibers of ordinary connective tissue, sectioned elastic fibers, and fibroblast cells. (From Maximow and Bloom.)

slender *connective tissue fibers,* flexible but inelastic structures which are formed by these cells. A more common type of connective tissue is a *compact* one (Fig. 51) in which there is a densely packed mass of interlacing fibers, producing a feltlike structure. Most connective tissues contain a small percentage of coarse, yellow *elastic fibers;* in some instances the elastic type of fiber is dominant, producing *elastic tissues*. *Tendons,* forming the attachment of many muscles, consist of parallel bundles of connective tissue fibers; *ligaments* are comparable structures uniting skeletal elements; *fasciae* are sheets of connective tissue investing muscles or other objects. Fatty or *adipose* tissue (Fig. 52) is developed within loose connective tissues of the embryo or adult. Fat droplets form and flow together within modified mesenchyme cells to fill most of their distended volume.

Notochord. The notochord is an ancient structure, present even in lower chordates, in which many of the characteristic vertebrate organs fail to make their appearance. As has been seen, the notochord is derived from cells which fold in at the dorsal lips of the blastopore or at the corresponding position at the anterior end of the primitive streak, and, pushing forward, come to lie beneath the neural ectoderm. It soon differentiates, along the median axis, from adjacent regions of the mesoderm.

The notochord is a prominent structure in the embryo of every vertebrate, and in many lower vertebrates it continues little changed throughout life. It is prominent in adult cyclostomes (where vertebrae are little developed) and there forms, as in Amphioxus, the major supporting structure of the trunk and tail (Figs. 13, 84, pp. 36, 154). Anteriorly in all vertebrates it extends to the region just behind the hypophysis and

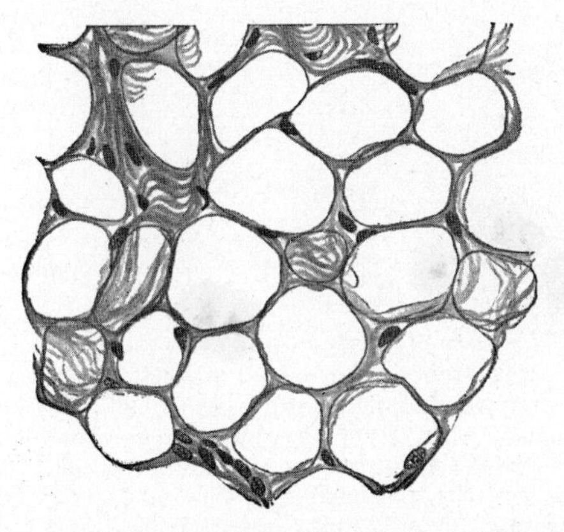

Fig. 52. Fat tissue. (From Maximow and Bloom.)

infundibulum of the brain; posteriorly it extends to the terminus of the fleshy part of the tail. The notochordal cells are soft and gelatinous; they are, however, surrounded by a cylindrical sheath and an external membrane, apparently of their own making. In lower vertebrates with a large and persistent notochord this sheath is of considerable thickness; it renders the whole a relatively strong yet flexible supporting structure, resembling a slender, elongate sausage.

In more advanced vertebrates the notochord is progressively replaced functionally by the vertebrae, which develop around it and give greater strength, if lessened flexibility, to the back. As this replacement occurs, we find a correlated reduction in the adult notochord. In many fishes and in ancient fossil amphibians and reptiles the notochord is present in the adult as a continuous, unbroken structure running the length of the body. It is, however, generally greatly restricted by the vertebrae; it may expand between successive vertebral centra, but constrict within each segment, so that its contours resemble those of a series of hour glasses set end to end. In most tetrapods even this tenuous connection between successive segments fails, and the notochord is represented in the adult only by gelatinous materials which may persist between successive centra of the vertebral column. Even in amniotes the notochord is a large and prominent structure in early stages of the embryo; later, however, it ceases to grow and hence becomes relatively smaller and smaller.

BODY FORM AND EMBRYONIC MEMBRANES

Amphioxus and Lower Vertebrates. Except for certain aspects of mesodermal differentiation, we shall not in this chapter carry onward in any detail the further development of the embryo. We may note here, however, in a general way, the gradual assumption of definitive body shape during

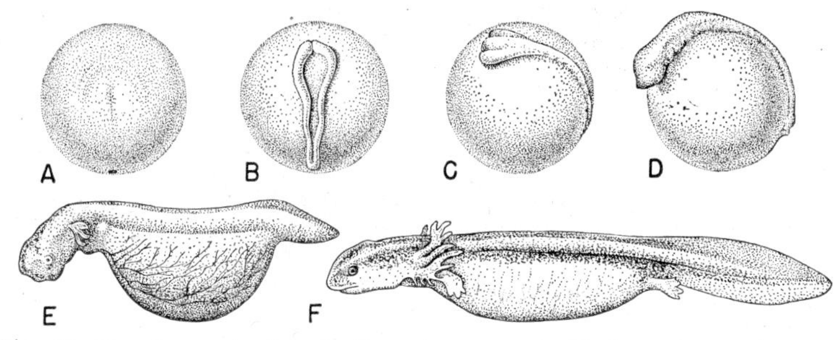

Fig. 53. Development of body form in a mesolecithal egg type—the urodele Necturus (the mud puppy). *A*, Late gastrula, seen from above, head end at top. *B*, Neural folds forming. *C*, View from left side, neural tube formed, brain bulging upward above sac partly filled with yolk. *D*, Head and trunk taking shape dorsally. *E, F*, Steps in reduction of yolk-filled belly sac and assumption of normal form. External gills and eye appear in *E*, limbs in *F*. (After Keibel.)

embryonic growth and the nature of the embryonic membranes which are important in the development of large-yolked eggs.

In Amphioxus we left the embryo at the neurula stage as a rather short cylinder. The remainder of the developmental story is, superficially, mainly one of bodily elongation. Anteriorly, there is a development of a mouth and of a complex gill region; posteriorly, the neurenteric canal closes, and there buds out a tail in which develops a continuation of spinal cord and notochord and numerous mesodermal somites.

In vertebrates with a moderate amount of yolk, including various fishes and characteristic amphibians (Fig. 53), the neurula is likewise spheroid. Dorsally, there is a rapidly growing nervous system; below this, internally, a gut cavity, its endodermal cells distended with yolk. With brain growth the head region expands in an anterior direction; posteriorly, there is a major development of a tail beyond the anal region, much as in Amphioxus. Soon there is attained a body shape not far removed from that of the later fish or amphibian larva, except for a distended gut region which is gradually reduced as the yolk material is absorbed.

Bodily form is slower of attainment in large-yolked eggs. In the elasmobranchs (Fig. 54) the neurula is little more than a flat sheet overlying a thick yolk mass, with the neural tube, above the notochord, marking out the "main line" of body organization. Anteriorly, the growing brain extends forward over the yolk surface, with the ectoderm folded beneath it; the anterior end of the digestive tract assumes a tubular form, separate from the yolk cavity; the mesoderm sheets of either side now grow downward to meet ventrally. At the posterior end, too, there is a tendency for

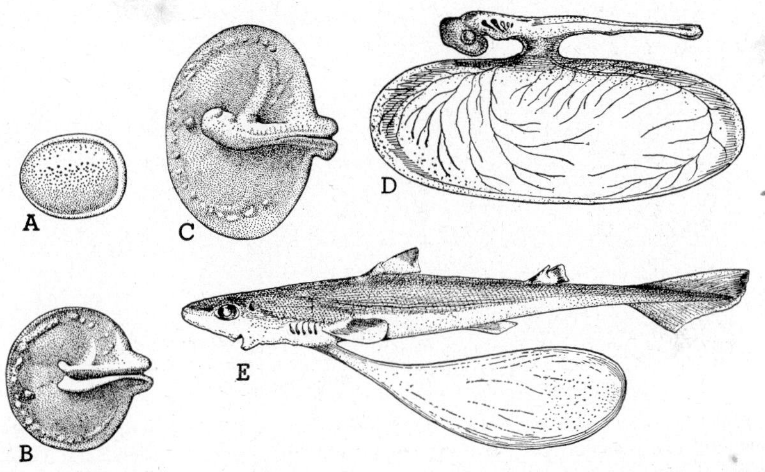

Fig. 54. Development of body form in a shark. *A* to *C* are dorsal views of the cellular disc from which the embryo forms; the underlying yolk is omitted in these figures. *A*, The embryonic disc at gastrulation; the endoderm is rolling under at the thickened posterior and lateral margins (cf. Fig. 44, *B*, p. 94). *B*, The disc is enlarging, and the neural folds are developing on the upper surface. *C*, The neural folds are closed except at the growing posterior end; the body of the embryo is lifting off the yolk, and head region and somites are visible. *D*, The yolk sac is completely formed and the embryo connected with it by a stalk; eyes and gill slits are visible. *E*, Nearly normal shape has developed except for retention of a relatively small yolk sac. (After Ziegler, Dean.)

the body to lift itself off the yolk as the tail buds out; only near the midlength of the body does the gut still remain connected with the yolk mass. Meanwhile the endoderm still lying on the yolk surface, covered with mesoderm and with an outer surface of ectoderm, continues to expand over the yolk and eventually completely encloses it in a *yolk sac* which is essentially an extra-embryonic part of the gut. The yolk is gradually digested and carried to the embryo by blood vessels developed in the walls of the gradually dwindling sac.

Reptiles and Birds. The shelled amniote egg has evolved as a structure in which, in contrast to the egg of lower vertebrates, development can take place on land rather than in the water. Apart from the protection of the shell, we find that the embryo becomes surrounded by a series of membranes which afford it protection and aid its metabolic activities. These membranes include a yolk sac, amnion, chorion, and allantois (Fig. 45, *A*, *B*, p. 95).

There is in amniotes a development of a yolk sac somewhat as in sharks. The endoderm, to begin with, is a flat sheet of cells capping the yolk; much as in sharks, endoderm, with accompanying mesoderm, gradually grows downward to enclose the yolk in a sac, and the connection of the sac with the growing embryo is eventually constricted to a stalk.

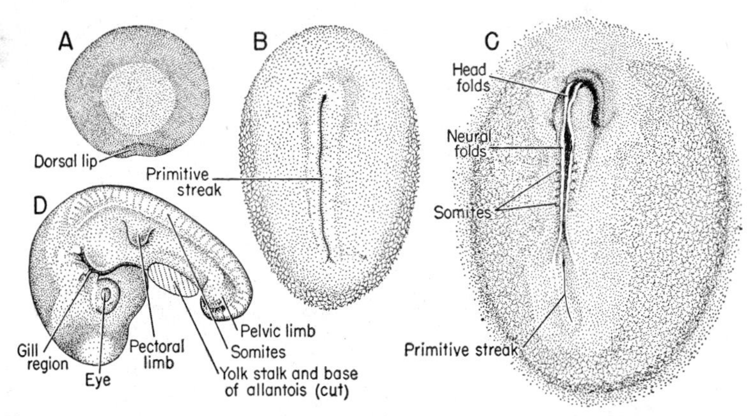

Fig. 55. Some stages in amniote development as seen in reptile or bird. *A*, Small germinal disc situated on the upper surface of the yolk; the endoderm is rolling under at the thickened posterior (lower) edge, which corresponds to the dorsal lip of the amphibian blastopore (cf. Fig. 42, *A*). *B*, Formation of primitive streak and elongation of germ disc (cf. Figs. 40, *I*, 42, *B*, and 43). *C*, The embryo is enlarging to cover more of yolk; the head region is lifting off the yolk surface; neural folds and somites are appearing; the primitive streak, now relatively small, is still active in formation of posterior part of body. *D*, Side view of a considerably later stage, comparable to Figure 45, *B*. The embryo is separated from the yolk except by a stalk (cut). Many head and body structures are formed, and limb buds are appearing. (*B* and *C* after Huettner.)

In sharks the ectoderm lying outside the region of the embryo grows downward around the yolk sac. In amniotes the situation is more complicated; this extra-embryonic ectoderm (with accompanying mesoderm) does not directly envelop the yolk sac, but gives rise to other important membranes. Folds grow upward around the embryo to form a closed sac, the *amnion;* its liquid-filled cavity furnishes a miniature replica of the former aquatic environment for development of the embryo. Externally, the ectoderm sheet, with mesodermal reinforcement, expands to enclose the entire set of embryonic structures in a protective membrane, the *chorion.*

With the restriction of the yolk sac connection to a stalk, the remaining part of the gut cavity has assumed a tubular form. At a relatively late stage there grows out ventrally from the posterior end of the gut a stalk of endoderm plus mesoderm. This rapidly expands into a large sac which underlies and is attached to the chorion over much of its area. This is the *allantois*. When the kidneys of the embryo begin to function, the allantoic cavity becomes an embryonic bladder. Much more important, however, is the function of the allantois as a breathing organ. The combined chorionic and allantoic membranes operate as a lung surface for absorbing the oxygen which enters through the porous shell; blood vessels in the allantoic walls carry oxygen in to the embryo and carry outward carbon

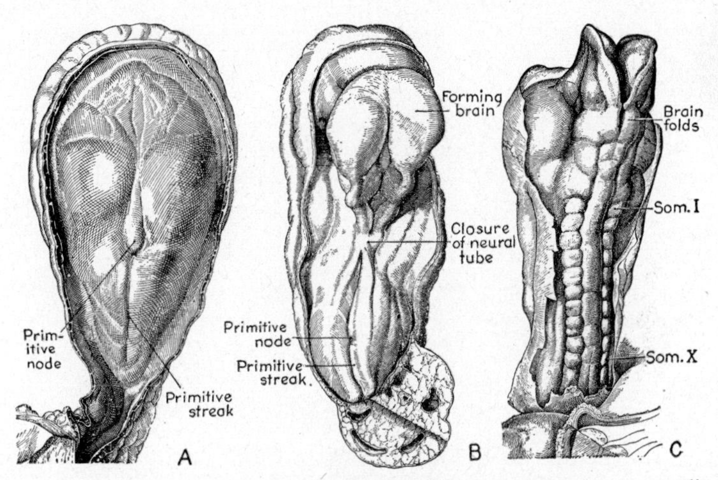

Fig. 56. A series of early human embryos, to illustrate stages in mammalian development. All three are dorsal views of the embryo, with the embryonic membranes cut away. *A,* Primitive streak stage, comparable to Figure 55, *B,* for a bird or reptile. *B,* Later stage in which the primitive streak is still active posteriorly, but the neural tube is forming more anteriorly. This stage is comparable to Figure 54, *C,* for the shark, not quite so advanced as Figure 55, *C,* for the bird. *C,* More advanced stage, with neural tube nearly completely closed and somite formation well advanced. (After Heuser, West, Corner.)

dioxide in exchange. With these membranes formed, the embryo takes shape and grows within the expanding cavity of the amnion (Fig. 45).

Mammals (Figs. 45, *D,* p. 95; 56, 57). We have noted that in placental mammals as a preliminary to gastrulation there appear, by processes of separation within the inner cell mass of the developing egg, cavities lying above and below the embryonic disc. These cavities are exactly comparable to those of the amnion and yolk sac of reptiles and birds, but because of a precocious development of the trophoblast, the order of events is quite different in mammals. The blastodisc, present between amniotic and yolk sac cavities, is connected posteriorly with the trophoblast by a stalk of mesodermal cells. Along this stalk there pushes out a gut outgrowth which expands beneath the chorion into an allantoic sac. Mesodermal tissue comes to reinforce the endoderm in its extension as a yolk sac and its allantoic development, converts the amniotic ectoderm into a formed amnion, and sheathes the trophoblast internally to form a chorion comparable to that of reptiles. As an end result we find that the mammal has, by a different route, arrived at the formation of a series of embryonic membranes highly comparable to that of its reptilian forbears (cf. Fig. 45, *B, D,* p. 95).

There are, of course, notable differences; the yolk sac contains no yolk, and the external membranes, as noted below, are transformed into a placenta. As in reptiles and birds, the growing embryo takes form by growing upward and gradually separating its body from the yolk sac. Eventually its only ventral connection is by a narrow area, the *umbilical*

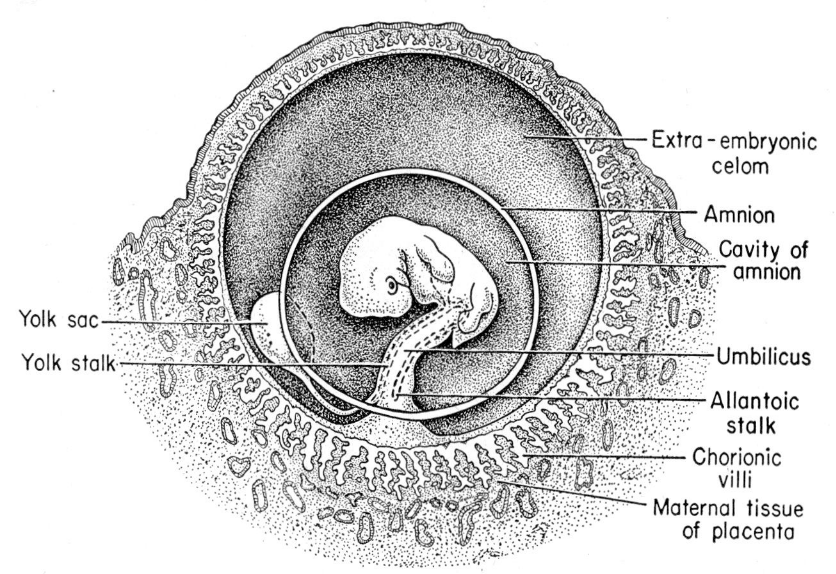

Fig. 57. Diagram to show the development of a mammalian embryo inside its membranes. The stage represented is one at which the embryo, though well formed, is still of small size.

cord. This cord is surrounded by the ectoderm of the amnion and contains stalks connecting embryo with yolk sac and allantois, together with mesodermal tissues and blood vessels associated with those structures.

A major difference between a typical mammal and its amniote relatives is the development of a *placenta,* through which the embryo receives food and oxygen from the mother and sends back waste materials in return. This characteristic mammalian structure is formed by an intimate union of the embryonic membranes with the surrounding maternal uterine tissues. The basic embryonic component of the placenta is the chorion, formed from the trophoblast plus the mesodermal tissues which come to reinforce it. On its outer surface there develop finger-like processes, which project into the maternal tissues. These are highly variable in size, disposition, and the degree of complexity with which they interlock with the uterine walls. A number of different placental types, which need not concern us here, are found in the various mammalian groups.

The placenta is richly supplied with blood vessels from both mother and embryo, and in it there is a constant exchange of materials between the two organisms. It is, however, important to note that the two blood streams are never in contact; they are everywhere separated by a membrane through which small molecules may pass readily, but never larger bodies such as blood proteins or blood corpuscles.

The chorion, however, is only part of the placental structure supplied by the embryo. This membrane is not in close contact with the embryo within. Some means must be supplied through which the blood may freely flow from embryo to placenta and return. In typical marsupials the yolk sac adheres to the under surface of the chorion and develops blood vessels which serve this purpose; to some extent a yolk sac type of placenta persists in some higher mammal groups. In nearly all higher forms, however, the dominant structure is an allantoic placenta. In reptiles and birds the allantois extends far around the outer surface beneath the chorion and becomes closely attached to that structure. Allantoic blood vessels are important in those groups in carrying oxygen and carbon dioxide from surface to embryo, and vice versa. In most mammals the allantois is similarly developed beneath the chorion as an integral part of the placenta; its vessels are vital as carriers of food as well as oxygen to the embryo.

Larvae. In amniotes and the large-yolked eggs of the sharklike fishes, in which an abundant food supply is available, development proceeds rather directly toward adult structure; the young, at birth, is essentially a sturdy little replica of the adult, soon capable, in a modest way, of setting about its business of making a livelihood in the fashion of its elders. Not so in many water-dwelling lower vertebrates—lampreys, many bony fishes, amphibians. In these forms the supply of yolk available for embryonic growth is limited; the young, when hatched, is of tiny size. At such a stage it is in many cases unable to take up adult habits and is liable, because of its size, to dangers to which the adult is not subjected. Under such conditions it is to be expected that frequently the young of such forms assumes a mode of life quite different from that of the adult, often lives in a different environment and develops, temporarily, structures suited to its livelihood which may differ markedly from those of the adult. A *larval stage* is thus interjected into the life history. With growth, larval features are eventually lost and adult structures and habits assumed—the process of *metamorphosis*. The tadpole stage of the frog or toad is a familiar larval form. Aquatic water-breathing larvae are found in the Urodela; most metamorphose into a somewhat more terrestrial gill-less adult stage, but some fail, partially or entirely, to metamorphose—the common mud puppy, Necturus, of Midwest streams and the axolotl of Mexico are examples. Many teleosts have larval stages; an extreme type is the translucent, leaf-shaped larva of the eel, found in midocean, whose relationship to the adult was long unsuspected. The adult lampreys are large predaceous forms, mainly oceanic; the larva is a tiny food strainer, sessile in the mud of stream bottoms.

DEVELOPMENTAL MECHANICS

In the earlier sections of this chapter we have given a description of the orderly series of events which take place during the developmnt of a vertebrate, but have said little as to the "why?" of these occurrences.

The answer to this question is a major interest of embryologists today. The development of the individual from seemingly simple egg to complex adult is a miracle so common that we regard it as commonplace. When some accident occurs in the normally well-regulated process, we tend to be puzzled or disturbed over the abnormality that results. Rather, we should marvel that the process of development normally proceeds so effectively, that development takes place at all. Most of the major events in vertebrate embryology are well known, but the mechanisms of development are in most regards still a mystery, and the solution of the chemical and physical problems involved is the chief concern of modern embryology.

The external environment is, of course, influential in development. It is, for example, obvious that a typical anamniote egg must have for proper development water of appropriate salinity; that mammalian development depends upon the reception of proper nutritive materials through the placenta. Again, the hereditary factors carried by the chromosomes can be considered basically responsible for the entire developmental story. But their influence is not, in general, direct; it can, for example, be proved in certain cases that the sperm nuclear material introducing half the hereditary factors for the new individual has little effect on the embryo until a relatively late stage in its development. The hereditary materials from the female are presumably responsible for the organization of the egg. Once organized, however, the egg is essentially beyond their control; the nature and distribution of its cytoplasmic materials are the factors mainly responsible for development. Once development is "triggered" by the entrance of the sperm, growth and differentiation proceed inexorably toward their determined goal.

A considerable degree of organization is already present in the ripe egg. We have noted the polarity associated with yolk distribution; and in certain cases it can be proved that despite its seemingly radial symmetry the chordate egg already has an established median plane, with the consequent attainment of bilateral symmetry. Still further, it can be shown in some cases that a distribution of materials within the egg has determined the future anteroposterior axis.

The successive cleavages of the egg appear to depend upon mechanical factors. For example: a cell tends to divide, as noted earlier, through the middle of its protoplasmic mass; the orientation of the plane of cleavage is correlated with the position of the mitotic spindle on which the chromosomes divide. As differentiation progresses, chemical and physical conditions in the various areas may be responsible for differences in the rate of cell division and consequent differences in the number or sizes of cells in these areas.

In some invertebrate groups the fate of the various cells into which the egg divides is determined at an early stage in cleavage; each cell is destined to form one particular part of the adult and can form no other. In the vertebrates the pattern is much less rigid. For example, each of the two cells formed by the first cleavage normally forms half the body, but in various types, from Amphioxus to mammals, it can be shown that either of the two, if separated from its fellow, can alter its normal destiny and form a complete embryo. However, every embryonic cell or region sooner or later loses its original broad capacities and becomes increasingly limited to a more and more narrow range of possibilities.

In the amphibians, in which studies of this sort have been most extensively carried out, this determination of the future fate of cell areas becomes marked during the process of gastrulation. We have noted, for example, that future skin ectoderm and neurectoderm areas can be marked out at the blastula stage (Fig. 40, p. 90). The cells of these areas, however, are at this time exactly the same in nature, and their fates are as yet undetermined; if bits of tissue are exchanged between the two regions, they follow the fate of their new locations. The same holds true for an exchange at an early gastrula stage. But by about the end of gastrulation these two areas are irrevocably determined as to their general fate; potential neural tissue will remain such if transplanted into the future skin area; potential skin will develop as skin even when included in the neural tube. Still later, narrower areas of potentiality may be determined. It is as if each cell passed along a series of ever-branching pathways; at each fork it must take one path or the other, and its future possible goals are constantly more and more narrowly restricted.

This progressive determination appears to be brought about through progressive chemical changes within the cell cytoplasm. In part, it would seem, such changes take place as a process of self-differentiation within a group of cells, a *"field"* in some well-marked-off region of the embryo. For the most part, however, determination appears to be due to the chemical influence of some adjacent region already determined. Such a region is generally termed an *inductor;* it induces or evokes a response toward differentiation from a cell mass previously showing little evidence of determination.

In various mesolecithal eggs the major *"organizer"* in early embryonic stages is the region of the dorsal lip of the blastopore. This appears to play a major role in the growth and infolding of the tissues of the area; for if transplanted to another region, the blastopore lip can initiate processes of gastrulation similar to those which it would have produced in its home territory. We have noted that the primitive streak of amniotes is a modified blastopore; as might be expected, this region is a major center of organization in amniote embryos. Later in development there are many known examples of induction of a more limited nature, such as the process of lens formation mentioned in Chapter XV.

ONTOGENY AND PHYLOGENY

Even in the early days of the scientific study of embryology, a century ago, it was noticed that animals vastly different as adults are similar in structure and appearance as embryos and that the embryos of "higher" vertebrates often exhibit conditions quite different from those of the adult, but similar to those seen in the adults of "lower" groups. From such observations came the idea of a *biogenetic "law,"* which proclaimed that individual development—*ontogeny*—repeats the history of the race—*phylogeny;* that, in other words, an animal in its development climbs its own family tree, successive embryonic stages representing the adult stages of ancestral types.

This "law" was for decades an important factor in the stimulation of embryological work and in the study of homology. But further consideration has shown that it is only a half truth. A mammalian embryo at an early stage is fishlike in many regards; it has, for example, prominent "gills" which are, of course, later reduced or modified. But there is actually no resemblance to an adult fish; the "gills" are the gill pouches seen in the embryonic fish, and they fail to open to the surface as do the gill slits seen in the mature fish. Actually, it is the fish embryo, not the adult fish, which the mammalian embryo resembles. As evolution has progressed, there has been no progression ontogenetically of piling one adult stage on top of another. New forms, it would seem, have arisen by diverging at some point from old developmental sequences and thus arriving at new goals in the adult.

Development tends to be a conservative process, for departure from the old, tried and true methods will usually result in failure and death. In consequence, much of the developmental pattern in forms only remotely related to each other may be similar, and the mode of development may follow a devious course, with indications of structures which were once present in the adult of ancestors, but which now never reach maturity. Ontogeny repeats many important steps in the developmental pattern of ancestral forms; it is especially likely to repeat them when they are structurally or functionally necessary or useful in the derived type's own development.

But embryos as well as adults may be modified in relation to their environment, and such modifications may result in the creation of embryonic structures never seen in adults at any lower level. The earlier amniotes developed a shelled egg with a large yolk mass as food and with embryonic membranes as devices for improved development; no amniote ancestor ever had, of course, an adult stage with a pendant yolk sac, nor was there ever an adult amniote ancestor enclosed in a "caul"—the amnion.

Although, in general, ancestral developmental patterns are conservatively followed or are at the most subject to gradual modifications, there may be drastic modifications of the pattern, and later embryonic stages may be reached by a different route from that followed by ancestral types —presumably in relation to powerful adaptive requirements. We have seen that large yolked amniotes have abandoned the older methods of gastrulation and mesoderm formation, although reaching the same goal by another path. Another striking example of this is seen in the early development of the mammal. Other amniotes have a neat, "logical" method of forming the embryonic membranes. The mammal departs radically from its reptilian ancestors in the process of membrane formation, but arrives at exactly the same end result.

THE GERM LAYERS

In our account of development we have laid emphasis upon the early differentiation and segregation of the major structural elements of which the adult body has been built. The theory of the germ layers was an early concept in embryology and a most useful and fruitful one. In the early embryo two layers, ectoderm and endoderm, may be distinguished as comparable in general to the outer and inner layers which alone constitute the entire body of a coelenterate; soon there is developed a third, intermediate, germ layer, the mesoderm, from which in vertebrates, as in all invertebrates above the coelenterate level, the greater part of the body substance is derived. We have in this chapter adhered to the germ layer concept, although emphasizing the early separation between body and neural portions of the ectoderm.

In the adult vertebrate, tissue components of organs and organ systems can for the most part be sorted out readily as regards their derivation from the germ layers. Details and exceptions will be found in later chapters; major derivatives are listed here.

From the body ectoderm: the superficial portion (epidermis) of the skin of the body surface together with its extensions into both ends of the digestive canal (mouth, cloacal region); epidermal skin structures, such as hairs, feathers, skin glands, and so forth; sensory epithelia of nose and internal ear; eye lens.

From the neural ectoderm: the entire nervous system; the eye retina;

and from the neural crest certain non-neural structures (gill cartilages, pigment cells).

From the mesoderm: connective and skeletal tissues; the musculature; the vascular system; most of the urinary and genital systems; the lining of the coelomic cavities; the notochord.

From the endoderm: the lining of the major part of the digestive tract and the substance of organs (liver, pancreas) connected with it; much of the gill system of lower vertebrates and all the lining of the lung-breathing apparatus of higher forms; parts of the urinary and reproductive tracts.

So generally do the patterns of origin of various adult tissues adhere to tabulations of this sort that beliefs as to the absolute specificity of origin of such tissues completely dominated the minds of many embryologists for many decades. In more recent times, however, further observation and experimental work have shaken such a faith. It is found, for example, that though most skeletal material arises in orthodox fashion from mesodermal mesenchyme, certain cartilages in the head and throat come from cells of the neural crest, otherwise mainly devoted to the development of nervous system structures; that the gill lining may come from either ectoderm or endoderm; that cell areas which normally produce nerve cord will, if transplanted, produce skin, and vice versa. There has been, in consequence, a tendency on the part of some to abandon the germ layer concept as meaningless. This, however, is a counsel of despair. What a cell becomes depends, as we have noted, on the potentialities present in it at any given stage of development and on the external influences to which it is subjected. But in general we find that in normal development the embryonic cells and tissues do follow a consistent regional pattern; and, if nothing more, the germ layer terminology is useful as a description of the topography of development. It is, however, more than this. Experimental work has shown that, although in early stages there may be little differentiation between various regions of the embryonic germ layers, there is increasingly, in later stages, a limitation of capacities in different regions. The prospective fate in normal development of the germ layers and subsidiary areas of these layers is in general accord with the experimentally deduced story of their prospective potencies.

CHAPTER VI

THE SKIN

SKIN FUNCTIONS. Forming a covering for the entire body, the *integument*—the skin with its accessory structures—is an organ system performing varied and important functions, many of which are protective. The tough "hide" of many vertebrates—sometimes reinforced with dermal scales or bones—is a protection against mechanical injury and against the attack of predaceous enemies. The skin is a continuous unbroken line of defense against the invasion of bacteria and other micro-organisms. As a sheath of tissue isolating the internal structures from the exterior, the skin may ward off physical or chemical influences disturbing the inner economy. It may aid in regulation of the water content of the body, preventing too great a loss of water in marine or terrestrial vertebrates, too great an influx of water in fresh-water dwellers. Skin pigment prevents the intake of injurious amounts of light. The skin itself, in addition to auxiliary structures such as hair or feathers, insulates the body against too great a loss (or gain) of heat and plays a major part in temperature regulation in birds and mammals.

The skin may further play an active physiological role in the absorption or elimination of materials through moist skin membranes or glandular structures. Breathing—the absorption of oxygen and release of carbon dioxide—is a function of the skin in many forms, and in some vertebrates the skin has become the major breathing organ. The skin may, particularly through its glands, function as an accessory to the kidney in the elimination of wastes.

As the region of the body in immediate contact with the outer world, the skin would appear to have been the area of origin of sensory and nervous structures in ancestral metazoans. In the vertebrates the nervous tissues have become a separate organ system withdrawn from the surface; but embryologically, we have seen, the nervous system still arises in continuity with the skin ectoderm, and the skin is still the seat of abundant sensory structures.

The skin is not a single structural entity, but consists of two parts, *epidermis* and *dermis,* which differ greatly in their nature. The epidermis is superficial in position and is essentially a cellular material, an epithelium derived from the ectoderm of the embryo; the deep-lying dermis has primarily a fibrous structure, with relatively few cells, and is derived from the embryonic mesenchyme, of mesodermal origin. The epidermis is generally thin, the dermis much thicker. The epidermis gives rise to a host of differentiated structures, such as hair, feathers, various glands; the dermis, on the other hand, has, except for bone formation, a relatively simple and uniform composition.

Epidermis. This, the superficial body covering, is usually much the thinner of the two skin layers. Derived directly from the embryonic ectoderm, it forms a continuous epithelium over the entire body. It tends to give rise in higher groups to varied special skin structures (described in later sections).

In Amphioxus the epidermis is in the simplest possible condition; it consists merely of a single layer of columnar cells, covered by a thin film of *cuticle* which they have secreted. Except for the cyclostomes, the cuticle

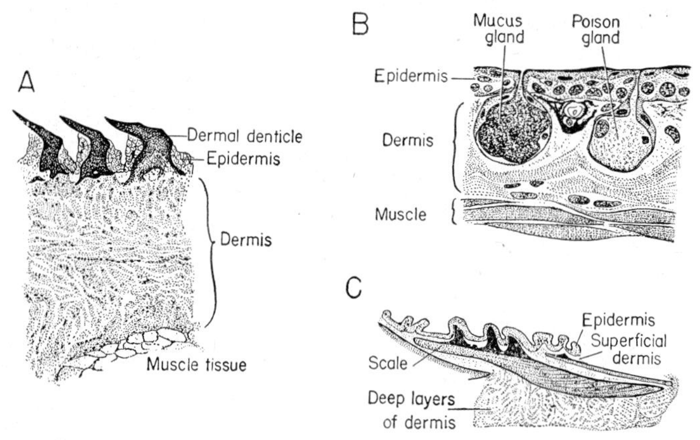

Fig. 58. Sections of the skin of *A*, a shark; *B*, a salamander; *C*, a teleost. (After Rabl.)

is absent in the adult vertebrate, and in every case the epidermis of true vertebrates is a stratified epithelium.

In fishes and in water-dwelling amphibians the epidermis (Fig. 58) remains in general a simple structure, apart from the presence of glandular elements. Nerves and blood vessels are seldom if ever present in the epidermis in any vertebrate group. Pigment may be present in the vertebrate epidermis in the form of *melanin,* an organic compound which in various concentrations gives shades of brown and black. However, in lower vertebrates coloration is mainly due to color-bearing cells situated in the dermis (and described later).

In fishes and amphibians generally the entire thickness of the epidermis consists of "live" cells, containing a normal protoplasm. At the surface of the epidermis, cells are, however, lost by wear or injury, and there is a constant replacement from below. The basal layer of cells, usually more or less columnar in shape, is the vitally important part of the epidermis in any vertebrate. This is the "matrix" of the epidermis, the layer from which successive generations of cells are budded off to form the outer portions of the epithelium. Superficial damage to the epithelium is readily repaired. But if, through major injury or serious burns, a large area of this basal matrix is destroyed, a re-covering of the flesh by skin becomes difficult, if not impossible.

With the assumption of a terrestrial life by certain of the amphibians and the amniotes, the nature of the epidermis is changed. In land dwellers, where water loss through the skin is a serious matter, the outer part of the epidermis becomes differentiated from that beneath. As the surface is approached, the layers of cells are successively flatter, and the protoplasm

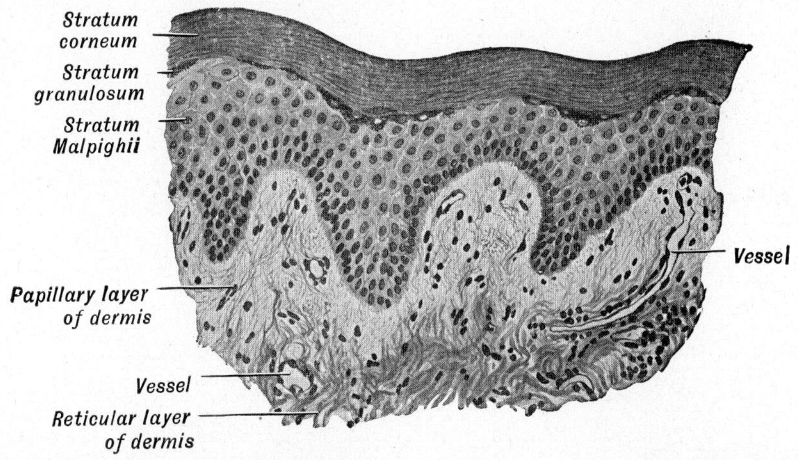

Fig. 59. Section of the skin of the human shoulder, × 125. In addition to the germinative layer (stratum Malpighii) and stratum corneum, there is an intermediate granular layer in the epidermis; in some mammalian situations there is further a transparent layer (stratum lucidum) between horny and granular levels. (Maximow and Bloom.)

increasingly modified and lifeless. The surface of the skin is dry and covered by thin, "dead" cells which may be rubbed off or lost piecemeal (dandruff is a familiar example) or shed seasonally as a whole (as in snakes).

Notable is the development in the superficial cells of large amounts of *keratin,* a protein material abundant in the horn sheaths of cattle, finger nails, and so forth.* Absent in most water dwellers, keratin is found in the superficial skin cells of a limited number of bony fishes, and in certain amphibians—notably the more terrestrial types of toads—and is highly developed in the outer skin layers of all amniotes. In many cases, described below, keratinized epidermis forms a variety of structures—scales, scutes, claws, hoofs, horns, feathers, hair, and so on.

* *Keratos* and *cornu* are, respectively, the Greek and Latin words for horn, whence are derived English words used for this substance.

Even in the absence of such specializations, however, the amniote skin always shows a contrast between deep and superficial portions of the epidermis (Fig. 59). Below is the zone of "live" cells, a *stratum germinativum;* superficially, with an abrupt or gradual transition, is a *stratum corneum,* of flattened, deadened, and cornified cells. (In certain thickened regions of the mammal skin, one or two intermediate layers may be distinguished: a *stratum granulosum* overlying the germinative zone, followed by a *stratum lucidum* of translucent cornified cells.)

Keratin Skin Structures. Throughout the higher vertebrates keratin-filled epithelium develops into a variety of special structures. Simplest perhaps are thickenings or swellings of the stratum corneum, such as in

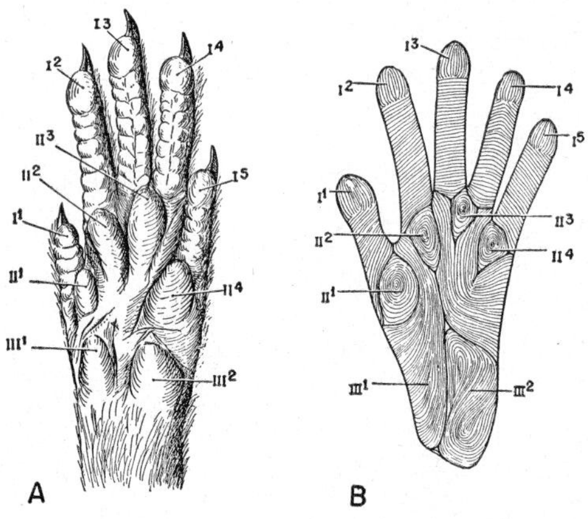

Fig. 60. Palm surface of the manus of *A*, an insectivore; and *B*, a monkey (macaque). The insectivore shows a presumably primitive mammalian structure, with thick pads on either side at the proximal end of the palm (III^1, III^2), pads between the bases of successive digits (II^1 to II^4) and pads at the tip of each toe (I^1 to I^5). In higher primates these pads are replaced by patterns of friction ridges. (After Whipple.)

the "warts" of toads, and the like. Such thickenings are frequently present on surfaces subject to wear, as the under surface of the feet. In mammals generally we find a characteristic arrangement of *foot pads* (Fig. 60, *A*), including, in five-toed types, a pair of pads on the proximal part of palm and sole, one on the base of "thumb" or big toe, a series between the bases of the other toes, and one on each toe tip. In higher primates, where the hand and foot grasp the tree limb in locomotion, we find palm and sole covered instead by a pattern of *friction ridges* (Fig. 60, *B*) which aid in obtaining a firm grip. These epidermal ridges, in the position of the pads of other groups, form complex series of loops and whorls. As is well known, the pattern of the fingertips is almost infinitely varied and affords a ready means of identification of a human person.

In reptiles, thickening and hardening of the cornified epidermis results in the formation of *horny scales* or *scutes*. In crocodiles and turtles they form a pattern of flat plates; in the latter (as the tortoise "shell" of commerce) they overlie the bony dermal skeleton of the back and belly. In lizards and snakes (Fig. 61) overlapping scales are commonly present, and these are in snakes highly developed as aids to locomotion, giving a "hold-fast" in the absence of limbs. Though such structures are termed scales, it cannot be too strongly emphasized that these horny epidermal scales of reptiles (and other amniotes) are not at all homologous with the bony dermal scales most characteristically developed in fishes.* Reptilian horny scales develop embryologically as an outpushing of the epidermis containing a papilla of mesodermal tissue; the broad upper surface of the papilla becomes the intensely cornified scale.

In mammals and birds the horny scales once present in their ancestors have for the most part disappeared. They persist, however, on the legs of birds, and are present on the legs and tails of a variety of mammals, notably

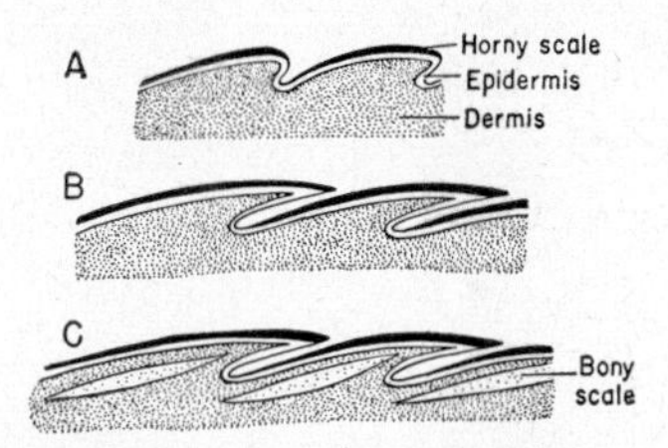

Fig. 61. Diagrammatic sections of reptile skin to show scale types. *A,* Lizard skin with simple, horny epidermal scales, gently overlapping; *B,* deeply overlapping horny scales of snake type; *C,* type of scale present in many lizards, with bony scale underlying horny element. (After Boas.)

rodents. The pangolin of the Old World tropics is notable as a mammal which has redeveloped a complete body covering of large horny scales.

In many amniotes in which teeth are reduced or absent, the skin on the jaw margins may cornify as a substitute, producing a *bill* or *beak*. Such structures are, of course, characteristic of the birds as a group, are well developed in the turtles, and are found in a few mammals, such as the monotremes.

Claws, nails, and hoofs are keratinized epidermal structures tipping the digits of amniotes (Fig. 62). The claw is the basal type; nails and hoofs are mammalian modifications. A typical *claw* forms a protection for the top, sides, and tip of a terminal toe joint; an inverted V in section, the claw becomes increasingly narrow distally and curves downward beyond the tip of the toe. The base of the claw includes a germinative layer protected by a fold of skin; from this matrix the keratinized claw epithelium continually grows outward over the dermis to be as continually worn away at the claw tip. On the under surface is a pad of softer, less

* In some lizards bony scales may be present as well as horny scales, the latter, of course, being more superficial in position (Fig. 61, *C*).

cornified tissue, a *subunguis,* which effects a transition between claw and normal epidermis.

A *nail,* as developed in the arboreal, grasping type of locomotion of the primates, is essentially a broadened and flattened claw restricted to the upper surface of the finger or toe. *Hoofs* are characteristic of the various ungulate mammals, which have undergone toe reduction and walk on the tips of the remaining digits. The original claw has shortened and broadened to become essentially a hemicylinder sheathing the toe tip; it is the curved or V-shaped distal end of the hoof which rests on the ground, the subunguis forming a pad within the curve of the hooftip.

Horns and hornlike structures are widespread in distribution, particularly among mammals. Though actual horn—i.e., keratin—is not always present in these structures, the general picture may be discussed here.

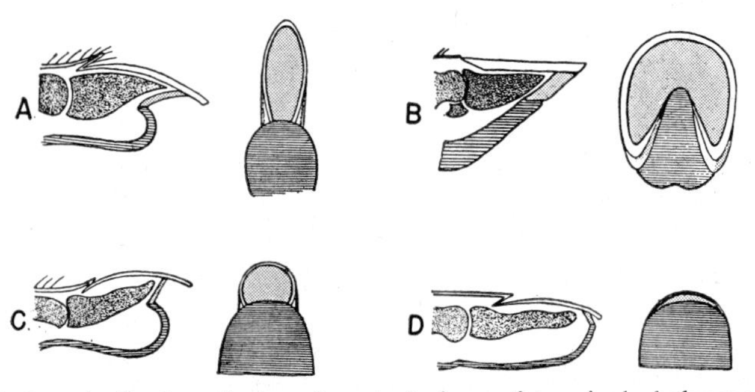

Fig. 62. Longitudinal sections and ventral views of terminal phalanges of mammals to show the build of claw, nail, and hoof. Toe phalanges, stippled; subunguis, fine stipple; epidermis of ventral surface of foot, hatched; epidermis of upper surface and horny material of claw, clear. *A,* Claw of carnivore type; *B,* a horse's hoof; *C,* a nail of a typical primate; *D,* a human nail. (After Boas.)

A true *horn* is seen in cattle and present in other members of the cattle family—sheep, goats, and antelopes. The core of the horn is a spike of bone arising from the skull. Sheathing and extending this, however, is a hollow cone of true horn substance, formed by keratinization of skin epidermis. Neither core nor sheath is ever shed, and it will be noted that these typical horns, variously curved, are never branched.

Although often called a horn, the *antler* of the deer is quite a different structure. When matured it consists solely of bone; during growth only it is covered by skin in the form of "velvet"; no actual horn substance is present. As further points of contrast we may note that an antler is branched, and is shed annually.

Still other types of "horns" are found among mammals, and comparable structures are seen, although less commonly, in reptiles and even in birds. Among mammals, for example, may be mentioned the horn of the prongbuck, which, like that of cattle, has a horn-sheathed bony core; but in contrast, the horny sheath is branched and is shed annually. Again, the horns of rhinoceroses are formed entirely of keratinized epidermis, but this is a fused mass of modified hairlike material, rather than a typical horn.

Feathers. The possession of feathers is the distinguishing mark of the bird. Primarily horny epidermal structures derived, it is believed, from reptilian scales, feathers perform two major functions in the avian economy. As a body covering they form an effective insulation, aiding in the maintenance of the high body temperature which is as characteristic of birds as of mammals. Bird flight is rendered possible through the development of large feathers which form the wing surface and the tail "rudder."

Three types of feathers may be distinguished (Fig. 63): the down feather, the contour feather, and the filoplume. As the most familiar (if

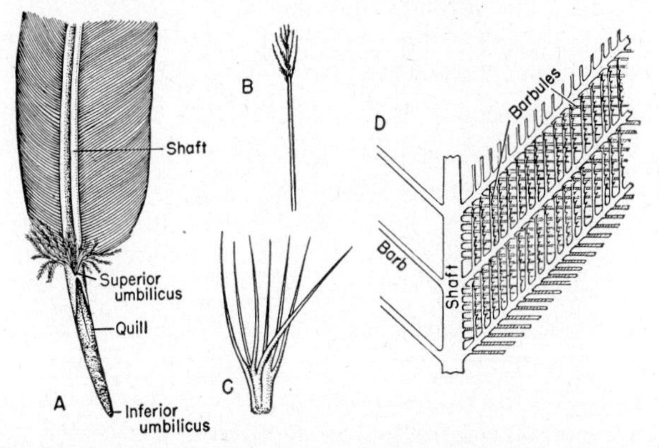

Fig. 63. Feathers. *A,* Proximal part of a contour feather; *B,* filoplume; *C,* down feather; *D,* diagram of part of a contour feather, to show interlocking arrangement of barbules. (After Gadow, Bütschli.)

the most complicated) type, *contour feathers* may be first described. These are the large feathers which sheath the body surface and (as the term will be used here) include as well the feathers of the wing and tail. A mature feather is formed entirely from greatly modified and cornified epithelial cells. The basal portion of the feather is the *quill* or calamus, a hollow cylinder with its cavity more or less filled by a pithy material—the remains of mesodermal tissue which was present here during the development of the feather. At the base of the feather is a small opening into this cavity, an *inferior umbilicus,* and a similar *superior umbilicus* is present at the distal end of the quill. The quill lies in a *follicle,* a deep cylindrical pit surrounded by a sheath of epidermal tissue, sunken into the dermis of the skin.

Beyond the quill lies the exposed and expanded portion of the feather, the *vane*. The axis is continued by the *shaft* or *rachis* (there is in some feathers a smaller, secondary axis at the plume base, an *aftershaft*). Unlike the quill, the shaft is a solid rather than a hollow structure, its interior filled with a sponge of horny, air-filled cells. Obliquely out from either side of the shaft extend the major shaft branches, *barbs*. These are so closely connected with one another as to appear to the naked eye to form a continuous sheet. Closer inspection, however, shows that the continuity of the feather surface is effected by tertiary elements, *barbules,* which arise in rows on either side of each barb. By a complicated system of hooks and notches each barbule interlocks with neighbors on the next barb. Any disruption of this arrangement can be repaired by the bird's preening action. In nonflying ostrich-like birds, where smooth feather contours are unnecessary for wing surfaces or for streamlining of the body, there is little development of barbule hooks, and the contour feathers may be fluffy plumes.

Basically similar, but simplier in build, are *down feathers*. These form the entire body covering of the chick and underlie the contour feathers over much of the body of the adult, forming the main insulation. A proximal quill region is present in these small feathers as in the contour type. Distally, however, beyond the quill termination, there is no shaft, but instead a spray of slender simple branches.

Filoplumes, "pinfeathers," are still simpler in appearance. These are small, hairlike feathers with a slender shaft continuing out from the body; they may terminate in a tiny tuft of barbs.

Much of the beauty of birds lies in their varied and attractive coloration. In part this is due to pigments present in the feathers. Much of the color, however, and the iridescent sheen of feathers are not attributable to pigments, but to the refraction of light from the horny feather substances.

Feathers are not, of course, uniformly distributed over the bird body. Contour feathers occur in definite feather tracts, *pterylae*. Most prominent is a long row of large feathers along the back of the "forearm" and "hand," the *remiges,* which form the wing surface; another group of large feathers, the *retrices,* forms the tail. Over a number of body areas no contour feathers arise, but the arrangement of the pterylae is such that the entire body is completely and smoothly ensheathed.

In its initial stages the development of a feather is comparable to that of a reptilian scale (Fig. 64). Mesodermal tissues, including small blood vessels, gather beneath the site of the prospective feather to form a feather papilla. Above this the epidermis extends outward into a cone-shaped structure with an epithelial surface and a pulp-filled center.

From this point onward, development diverges from that of the horny reptilian scale. The papilla does not remain on the surface; instead, the epidermis surrounding it sinks inward to form the feather follicle, out of which the papilla grows in further development into a mature feather.

Taking up first the development of a down feather, we find that the papilla becomes a greatly elongated tubular structure. The living mesodermal tissues of the papilla, a nutritive pulp, extend the length of the tube during development. The epidermal surface gradually takes on a

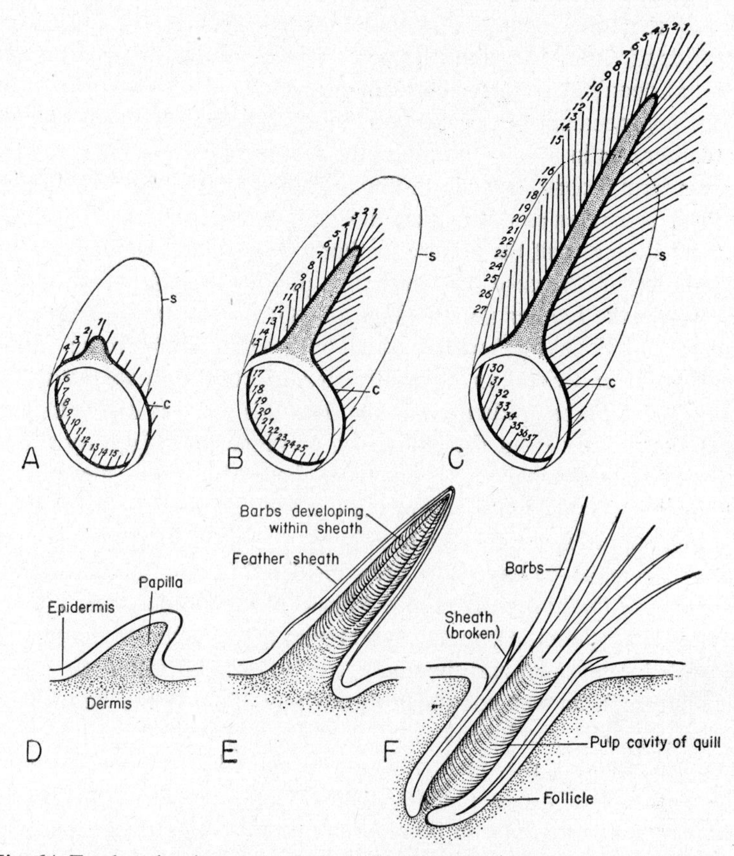

Fig. 64. Feather development. *D, E, F,* Diagrammatic sections through successive stages in the development of a down feather. Development begins in the form of a mesodermal papilla; later the structure sinks into a follicle. An outer layer of the ectodermal covering separates as a thin feather sheath. The remainder of the ectoderm forms basally a hollow tube which becomes the quill. More distally it divides into a number of parallel columns. With rupture of the sheath these are freed as the barbs. *A, B, C,* Diagrams to show the development of a replacement contour feather; the basic pattern is comparable to that of a down feather. Feather growth begins at a basal ectodermal collar (*c*), from which develop, as in the down feather, parallel columns of tissue within the feather sheath (*s*). One exceptionally strong upgrowth (stippled) becomes the shaft; the parallel columns of tissue migrate successively (as shown by the numbering) on to this to become the barbs. (*A* to *C* from Lillie and Juhn.)

horny character. Proximally, in the future quill region, the epidermal tube is a simple cylinder. Distally, however, a thin outer layer separates as a conical sheath from the inner regions of the epithelium. The inner mass develops as a series of thickened longitudinal ridges. When feather growth is completed, the mesodermal tissue of the distal part of the feather is resorbed, the thin sheath breaks down, and the ridges are freed to become the spreading distal filaments of the down feather. Proximally, however, the tube remains intact as the quill sector. The mesodermal tissue filling it dries to a pith; the two umbilici are the original proximal and distal openings of this tube segment.

More complicated, but basically similar, is the development of a contour feather. As in a down plumule, the feather rudiment takes the form of an epidermal cone, mesoderm-filled, in the distal part of which a surface layer separates as a thin sheath. One might expect the complicated branching structure of the mature feather to grow in treelike fashion. This is not, however, the case, for the entire process of differentiation takes place within the feather sheath (Fig. 64, *A-C*). The future barbs arise, like those of a down feather, as parallel outgrowths from a basal "collar." At one point, however, an especially strong process grows out to form the shaft. The barbs gradually migrate onto this shaft as it develops, and further barbs form from the collar and continue the migratory process. At a later stage the barbules form from the barbs. The entire developmental process takes place within the confines of the cylindrical sheath; when this ruptures, the feather has simply to unroll to attain its mature state. Feathers, as we have said, are considered to be a modification and elaboration of the horny scales of the birds' reptilian ancestors. The early stages of development are in agreement with this conclusion, but the later phases of development and the adult structure are far different.

Feather replacement continues throughout life. At the base of each follicle the papilla and its epidermal covering persist as a feather matrix; from these materials new feathers form. In many birds, particularly those of temperate and arctic regions, feather replacement is the seasonal phenomenon of *moulting;* in others there may be a gradual replacement throughout the year.

Hair. As an insulating device formed of keratinized epidermis, hair is a mammalian analogue to the avian feather. In other regards, however, the two structures are in contrast. As will be seen, they develop in different fashion, and, unlike a feather, a hair is a purely epidermal structure, without the mesodermal component seen within the feather during its development. Hairs, unlike feathers, are not modifications of horny scales, but are new structural elements of the skin. It is probable that hairs had developed before scales were lost by our reptilian forbears. In certain instances where horny scales are retained in mammals, hairs are found growing in definite patterns between the scales; even when (as usual) scales are absent, the same arrangement of hairs may persist (Fig. 65).

A typical hair (Fig. 66) includes two portions, the projecting *shaft,* and the *root,* which lies in a pit sunk within the dermis, termed the *hair follicle*. Both shaft and root consist (except at the very base) of essentially dead and heavily keratinized epidermal cells. At its base, however, the root expands into a hollow bulb. Beneath is a dermal *papilla* containing connective tissue and blood vessels. Through the latter come nutrient materials for the sustenance of the bulb cells and thus for the growth of the hair.

In the bulb alone the hair includes typical "living" cells, with a normal protoplasmic content—the hair *matrix*. Growth and division of these cells results in the development of the hair. As cells are budded off distally in the bulb region and pushed outward by the development of further cells beneath, they become part of the root, and eventually of the shaft section of the hair, and gradually assume a "dead" keratinized structure as typical hair cells.

Within the hair two or three layers may be distinguished in section. Externally is a thin *cuticle,* consisting of a single layer of transparent

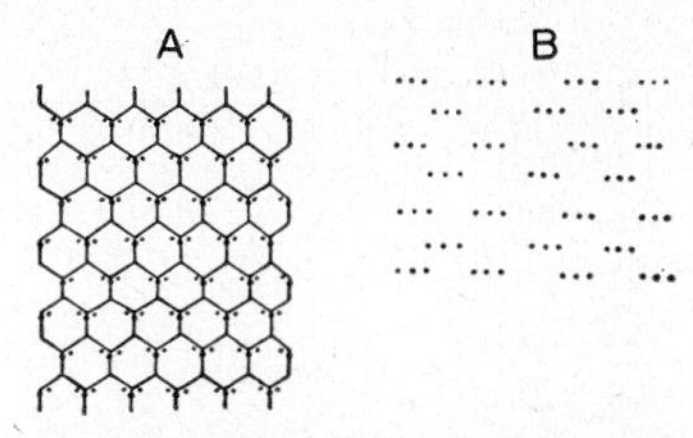

Fig. 65. Hair patterns of mammals to show presumed derivation from structures developed in the interstices between scales. *A*, Part of the scaly tail of a tree shrew, with the hair (represented by dots) in this position; *B*, skin of a marmoset, with the hairs arranged in a similar pattern despite the absence of scales. (After De Meijere.)

cells which often overlap in shingle fashion. Beneath the cuticle is the main substance of the hair, a dense horny material of modified cells containing a variable amount of pigment and air vacuoles. In stout hairs, a thin central area of shrunken cells and large air spaces may be distinguished; when present, this is considered a *medulla,* and the main mass of the hair substance termed a *cortex*.

Lining the follicle is the hair sheath, which includes several layers. Internally is a thin, cornified epithelial layer, an *inner root sheath,* past which the hair pushes outward in its growth. The thicker *outer root sheath* is a living epithelium continuous with the deeper layers of the epidermis on the skin surface. Still more externally placed is a *dermal sheath* of connective tissue.

Adjacent to the follicle and emptying its oily lubricating material into it, may be found a sebaceous gland (cf. p. 129). Each hair is further provided with a small *arrector pili muscle,* composed generally of smooth muscle fibers. Attached at one end to the superficial part of the dermis, the muscle slants inward to insert into the hair follicle deep down at the side toward which the hair inclines. Contraction of such a hair muscle (under the control of the "involuntary," autonomic nervous system) brings the hair erect (and by its pull on the skin causes "gooseflesh"). Of

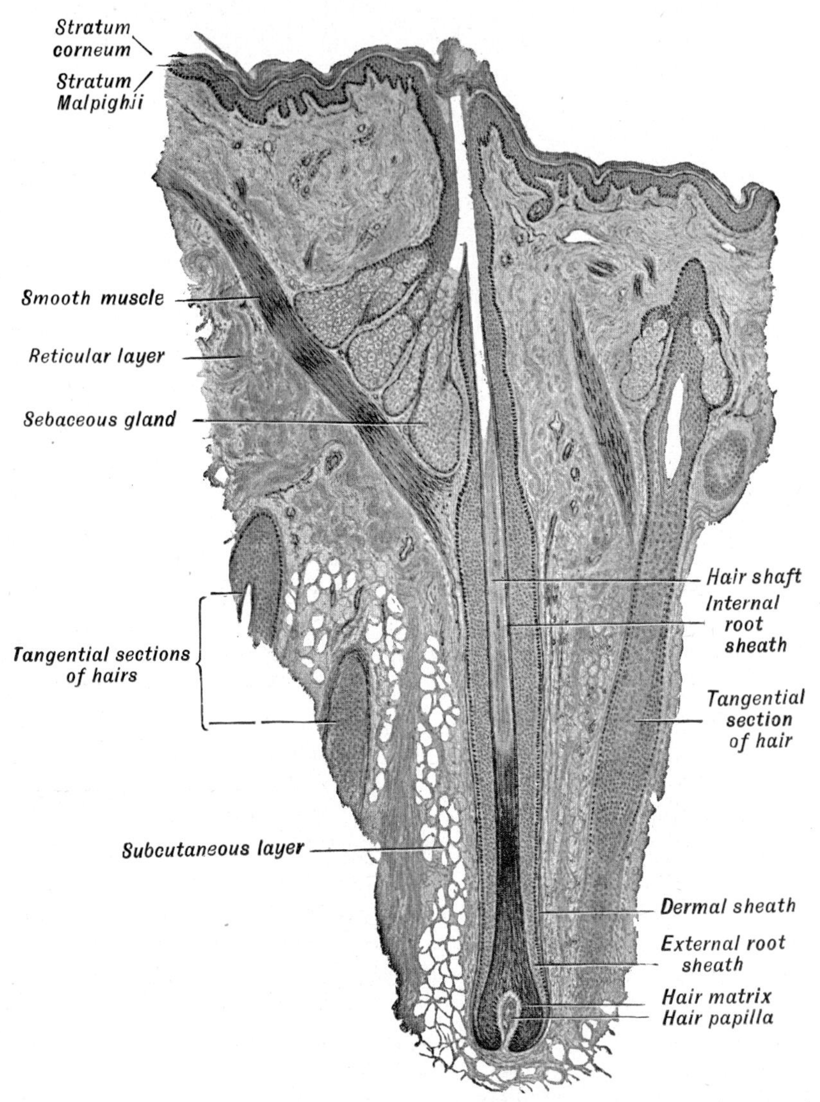

Fig. 66. Section of mammalian skin, to show hair structure. (From Maximow and Bloom, after Schaffer.)

little avail in such a thinly haired form as man, this potentiality of raising and depressing the hair can bring about, in a thickly furred animal, marked changes in the insulating powers of the coat.

In the development of a hair in the mammalian embryo there is no formation of a mesoderm-filled papilla such as we find initially in feather development. Instead, we see the growth of a solid column of epidermal cells downward into the dermis. At the bottom of this column a thick mass of these epithelial cells develops as a hair germ; a hollowing of its base allows the development of a vascular papilla. Above, the column of cells rising to the surface forms the epithelial walls of the follicle, and growth of the hair upward from its germinal base takes place in a tube hollowed out within this column.

Once developed, a hair is not (any more than a feather) a permanent structure; throughout life most hairs are cast off and replaced, either as a gradual process or as a seasonal shedding of much or all of the entire coat. Resorption takes place at the bulb, and a new hair develops in the same follicle from the matrix cells surrounding the papilla.

Although hair may serve a variety of special functions, its primary use is as an insulating and thermal regulating material. It is perhaps difficult for us, as members of a mammal group in which hair is much reduced, to appreciate this fact fully; but it must be remembered that in mammals generally the hair forms a thick covering over almost the entire body. It is a furry coat, containing numerous "dead" air spaces which give it many of the attributes of commercial rock-wool insulating material. In connection with its insulating qualities we may note the seasonal shedding of hair in mammals of temperate and arctic regions, whereby a thicker coat is grown for winter, a thinner one for milder summer weather.

All hairs contain pigment to some degree. Melanin is the common substance, producing in various concentrations various shades of brown and black; a related pigment is responsible for reddish tinges. Air bubbles present in the hair will lighten the intensity of coloration; when abundant, air spaces will turn darker colors to gray and in the absence of pigment will produce a silvery white.

Loss of hair color with age in an animal is due both to reduced pigmentation and to increase in the air content of the hair. Coat color may change notably with the seasons when the hair is shed and replaced, as is witnessed by the white winter coats of various arctic animals, admirably adapted to concealment in the snowy landscape and contrasting sharply with the dark summer pelage of the same species.

Mammalian hair is, of course, highly variable in a number of regards. It varies from form to form, between one area of the body and another, from one season to another, from youth to old age. The hairy covering in many mammals is a thick fur. In others, such as the higher primates, which are primarily tropical forms, it is a much thinner and uneven coat. In whales, which depend on fat rather than hair for insulation against the cold of sea waters, it may be entirely absent. An animal may have a double pelage, consisting of a sparse outer set of long coarse hairs and an insulating undercoat of much more numerous finer hairs.

Specialized types of hair are found in many instances. Various head regions may sprout *vibrissae,* long stout hairs with sensory structures at their bases, which function as tactile organs—the cat's "whiskers" are typical. Eyelashes are on the same order. Hairs which are rounded in cross section tend to be straight and erect, and if stoutly built may develop as bristles or spines. Hairs which have an oval or flattened section are more readily bent and may result in the production of a curly or woolly covering.

Skin Glands. Glandular structures are developed in the epidermis in every vertebrate class. In fishes they generally occur only in the form of discrete, mucus-secreting cells which are widely distributed over the surface of the body. These mucus cells are responsible for the slimy feel of

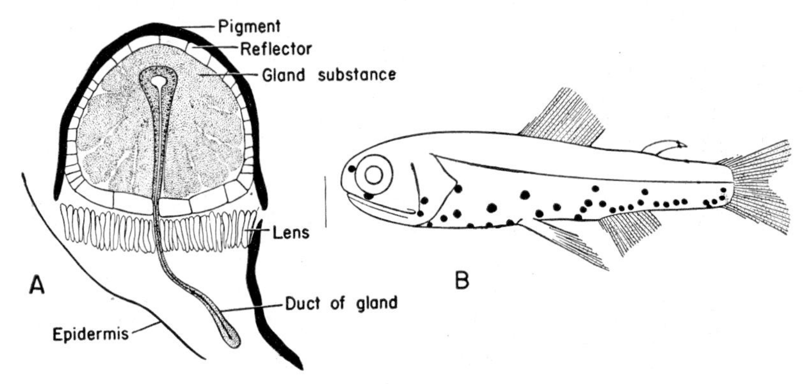

Fig. 67. *A,* Section through a light organ of a teleost (Cyclothone). Part of the duct leading in from the body surface is seen. *B,* Light organs of a small teleost (Myctophum). The organs are in black. (After Brauer.)

fish skin; they are excessively abundant in the hagfishes, and it is said that a single one of these animals can turn a whole bucket of water into a jelly of slime.

In a few cases, however, formed glands which serve special functions are present in the fish skin. *Poison glands* are found associated with spines on the fins, tail, or gill cover in several sharks and chimaeras and a number of teleosts. A most unusual development in various deep sea fishes is that of luminous organs, *photophores,* formed by modified glands (Fig. 67). In some instances the light is produced by the symbiotic presence of phosphorescent bacteria in the organ; in others complicated chemical processes of oxidation in the glandular material appear to result in the production of a "cold light." In many teleosts these organs develop accessories, including a pigmented reflector behind the light organ and a lens in front, giving the photophore much the build of an automobile headlight.

In water-dwelling amphibians, mucus production persists. Here, however, we are dealing not with individual mucus cells, but with formed *mucus glands,* although glands of a simple alveolar nature (Fig. 58, *B*, p. 116). A second type of gland is also frequently present in amphibians: the *granular glands* owe their name to the granulation of the protoplasm of the secreting cells. The varied product is poisonous, ranging from mildly irritating substances to alkaloids which may be highly toxic.

When the reptilian stage is reached, the primitive mucus cells have disappeared, and glands of any kind are rare in the hard dry skin. In birds, too, skin glands are rare; the only conspicuous gland is the uropygial or *preen gland,* a large, compound alveolar structure situated on the back above the base of the tail. Its function appears to be the supplying of an oily "water-proofing" material used by the bird in preening its plumage.

Fig. 68. A sweat gland from the mammalian skin, much enlarged. At top, epidermis with coiled duct of gland. The coiled gland is deeply sunk in the dermis.

In mammals, glands of new types make their appearance. Alveolar *sebaceous glands* are associated with hair follicles (Fig. 66), but may persist in regions where hair is absent; their oily secretion is a lubricant for hair and skin. *Sweat glands* (Fig. 68) have a simple tubular structure, but with a much elongated tube which grows down to coil in complicated fashion in the dermis (Oliver Wendell Holmes, Sr., who was an anatomy professor as well as author, compared them to the intestine of small fairies). The watery secretion of the sweat glands contains salt, urea, and other waste products. These structures thus act as accessory kidneys, and we are familiar with the fact that excessive sweating results in a considerable loss of salt from the body. The evaporation of sweat on the skin surface is a cooling device important in mammalian temperature regulation. In carnivores, generally, sweat glands are much reduced in number; the panting of a dog utilizes salivary evaporation on the tongue to a similar end.

Still another mammalian gland type—one to which, in fact, the class owes its name—is that of the *mammary glands,* inactive in males (except in monotremes) but in females secreting a nourishing milk for the young. No antecedents for these are found in lower classes; it is thought, from their construction, that they are specialized derivatives of the sweat glands.

There is considerable variation in the structure of mammary glands. In monotremes there are simply two bundles of discrete glands in the abdominal wall which discharge a sticky secretion onto the belly surface, to be licked off by the young. In all higher mammalian groups there are well-formed mammae, the gland openings concentrated in projecting nipples or teats whence the fluid is sucked by the young. Their embryonic development begins with the appearance of a pair of longitudinal swellings, *mam-*

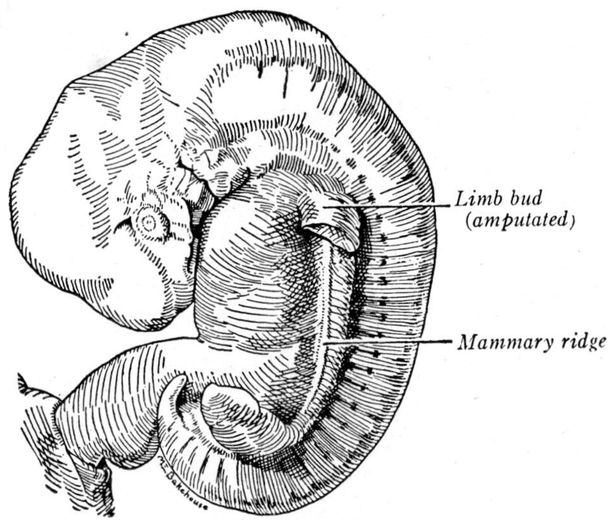

Fig. 69. Mammalian embryo to show mammary ridge or "milk line." (From Arey.)

mary ridges (Fig. 69), running along the length of the trunk on either side. Concentrations of tissue at specific points on either line result in the eventual development of the mammary glands. They develop further in the female when sexual maturity is reached and become functional at the time of parturition, the development being under hormonal control. The number is generally correlated with the number of young produced at a birth; they range from a single pair up to half a dozen pairs or more. Their position is likewise variable. In marsupials they are general enclosed within an abdominal pouch in which the young are carried after birth, each one tightly clamped to a nipple. In many ungulates the mammae are abdominal in position; in higher primates, on the other hand, they are in the pectoral region; forms with large litters (as pigs, many carnivores) usually develop two long rows of nipples.

Dermis. Thicker but less varied in structure than the epidermis is the deeper skin component, the dermis. Basically this consists, in most vertebrate groups, of a tightly felted fibrous mass of connective tissue, derived from embryonic mesenchyme. It is this part of the skin of various animals which, after appropriate treatment, becomes commercial leather (a fact which the alternative name of *corium* for this layer implies). The dermis is an effective insulating material, and its flexible yet unyielding nature makes it a major defense against injury.

In most bony fishes, however, we find a different kind of dermis. A certain amount of connective tissue is present, but in considerable measure it is replaced by mesenchyme derivatives of a more specialized sort in the form of bony dermal scales or plates (Fig. 58, *C*, p. 116). These structures, forming a defensive armor of a harder and stiffer type than that offered by the normal connective tissues, are part of the external skeleton, and as such are described in the chapter following. Except for the skull region, this dermal armor is much reduced or absent in most land vertebrates (turtles form a conspicuous exception), and it is likewise absent in the cyclostomes and sharklike fishes (except for the small, toothlike denticles embedded in the skin of the latter).

Offhand, one would assume that the more common condition of the dermis in modern vertebrates—that in which it is fibrous—is primitive and that the presence of bone in the dermis is secondary. The history of vertebrates shows, however, that the reverse is the case. The evidence is definite in the land vertebrates; tetrapods are descended from fishes in which the dermis was occupied by thick bony scales, and stages in the reduction of the bony armor to the modern fibrous condition of the skin are observed in the paleontological record of early amphibians and reptiles. For lower fish groups, the evidence, although not absolutely conclusive, indicates that here too the ancestors had bony dermal scales, for the oldest vertebrates had an armored skin. A bone-filled dermis was, it would seem, the primitive vertebrate condition; the leathery type prevalent today has been arrived at as the result of dermal degeneration.

Though connective tissue fibers form the main bulk of the typical dermis, other materials are present here as well. A certain (though small) percentage of the fibers are of the elastic type. Cellular components are present among the fibers, notably the fibroblasts, to which the tissue formation is due. The deeper layers of the dermis are a major locus for the development of fat tissues. Fat is an excellent insulating material; in whales, as the thick "blubber," it substitutes for the absent hair in this regard. Down into the dermis projects, in higher vertebrates, the basal portion of glands and of feather or hair follicles. Smooth muscle fibers may develop in the dermis, and striated muscle tissues, derived from underlying body muscles, may attach to the under surface of the skin. The sensitivity of the skin is due to the presence in the dermis (relatively seldom the epidermis) of nerve fibers; though some end freely, most terminate in sensory corpuscles, often numerous, situated in the outer part of the dermal layer.

Circulatory vessels are, in general, abundant—lymphatics, small arteries and veins, and capillary networks of complicated pattern. The blood vessels supply nutriment not merely to the dermis, but also, by diffusion of materials through the tissues, to the overlying epidermal cells. In forms with moist skins the presence of this rich vascular supply gives the possibility of exchange of materials with the surrounding medium. Notable is the utilization of the skin as an auxiliary breathing organ in many amphibians and some bony fishes; certain salamanders and eels, indeed, are able to supply their entire oxygen need this way.

Temperature Regulation. Proper functioning of the body in vertebrates can take place only over a restricted temperature range. In mammals and

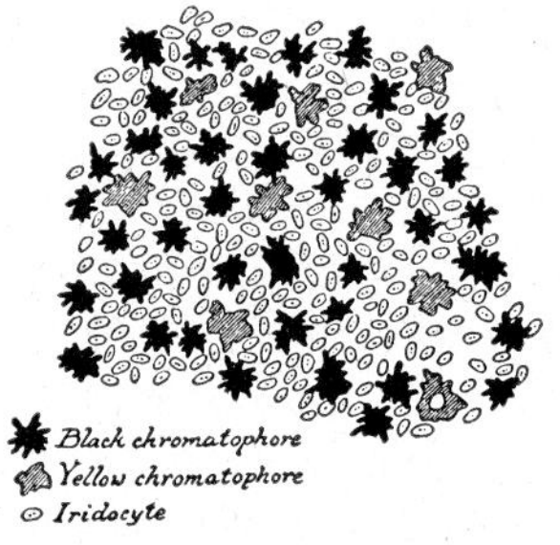

Fig. 70. Enlarged surface view of a piece of the skin of a flounder, seen by transmitted light, to show the three types of chromatophores present—pigmented melanophores, lipophores, and crystalline guanophores (or iridocytes). (After Norman.)

birds body temperatures are so regulated that the internal temperature varies little from a norm which is generally within a few degrees of 100° F. This is a *homothermous* condition, contrasted with the *poikilothermous* state of lower vertebrate groups, in which body temperature tends to vary in relation to external temperature. Regulation is under the control of a center in the hypothalamic region of the brain stem which acts as a neural "thermostat." Internal temperature depends upon the amount of heat created by metabolic processes and upon the amount lost (or gained) by the body. Most of the heat lost (some 70 per cent in man) is lost through the skin, which is thus of the greatest importance in regulation. It may be noted that in a small animal the skin surface is relatively larger in proportion to body bulk than in a larger form; size, therefore, is a factor in heat conservation, and in general the larger members of any mammalian genus are those which live in colder climates.

The skin is to some extent a static insulating material in every vertebrate, but in birds and mammals the heat loss can be regulated by various devices. We have already noted that feathers and hair are adjustable insulators produced by the epidermis, and evaporation of liquid from the sweat glands produces a cooling effect. The dermis supplies an important factor in regulation in its abundant blood vessels. Under control of the autonomic nervous system, the skin capillaries can be dilated, the skin flushed, and heat lost rapidly; with constriction of the capillaries (and a blanched skin) a major saving in heat is effected.

Chromatophores. The color of animals is generally influenced to some degree by pigmentation in epidermal cells, and in birds and mammals feathers or fur is mainly responsible for surface coloration. In lower vertebrate groups, however, skin color is due almost entirely to special color-bearing cells, the *chromatophores,* located in the outer part of the dermis (Fig. 70). The chromatophores are typically stellate, with elongate cell processes, and contain numerous granular elements. Common types include (1) *melanophores,* with the dark brownish pigment melanin, (2) *lipophores,* with red or yellow carotenoid pigments, (3) *guanophores* (or *iridocytes*). The last contain not pigment, but crystals of an organic substance, guanine, which by refraction may alter the effect of the pigmented materials present. Nearly all the varied colorations seen in fishes, amphibians, and reptiles are due to chromatophores of these three types, present in varied numbers and in varied arrangements. Lying as they generally do in the dermis, the chromatophores have been regarded as derivatives of the embryonic mesenchyme. Recent work, however, has shown that (curiously) amphibian chromatophores are derived from cells of the neural crest—a structure which for the most part contributes to the nervous system—and it is possible that this is the source of chromatophores in other groups as well.

The chromatophores are capable of producing striking color changes in a variety of fishes, amphibians, and reptiles (particularly lizards). The chameleon is proverbial in this regard, and the flounders are equally remarkable in the variety of color and color patterns which they are able to display. In part the color changes are attributable to shifts in the relative position of the chromatophore types in the skin, one or another becoming more broadly exposed at the dermal surface or more fully masked by neighboring cells. In the main, however, the color changes are due to changes in the distribution of the color granules within the individual cell. If the granules are dispersed throughout the cell, the effect is maximum; if they are concentrated in a tight cluster, little color appears.

Chromatophore changes may result from the direct stimulus of light or heat, but are in general under the control of the nervous system, and for the most part appear to be changes toward the color scheme of the environment, based on information reaching the brain through the eyes. The paths through which the nervous influence reaches the chromatophores are varied. In some cases nerve fibers reach the chromatophores to give a direct stimulation; in others the effect is due to hormones from glands of internal secretion, especially the pituitary. In still other cases chromatophores are rendered active by nerves situated at some little distance from these cells. It appears that here the stimulation is due to the diffusion through the skin of some chemical material given off by the tips of the nerve fibers—a type of substance which has been termed a *neurohumor*.

CHAPTER VII

THE SKELETON

FROM A physiological or biochemical point of view the skeleton is a relatively inert organ system. From a broader functional viewpoint, however, it is of the greatest importance. Evolved from the connective tissues, the hard skeletal structures are vital in welding together the softer organs and helping in support and maintenance of proper body form. Almost all the striated musculature attaches to the skeleton, which is hence the agent through which bodily movement is accomplished. Still further, the more superficial skeletal elements have a protective function, acting as a shield for softer or more vital structures beneath them.

SKELETAL TISSUES

Cartilage. Two skeletal tissues are characteristic of vertebrates—cartilage and bone. Both are specialized derivatives of the connective tissues

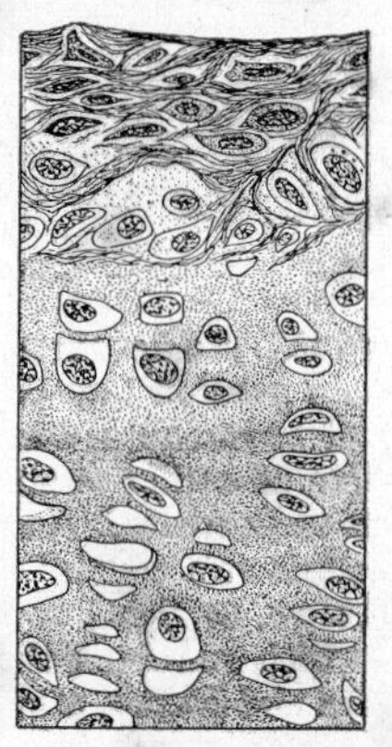

Fig. 71. Section through part of a cartilage (from the sternum of a rat). The surface layers (at the top) show a fibrous condition transitional from the perichondrium. (After Maximow.)

and arise from mesenchyme, but they differ markedly in nature, in mode of origin, and (frequently) in position.

Typical *hyaline cartilage* (Fig. 71) is a flexible, rather elastic material, with a semitransparent, glasslike appearance. Its ground substance, or matrix, is a complex protein (a chondromucoid); through this is spread a network of connective tissue fibers. Throughout are spaces containing cartilage cells. These are usually rounded and without the branching processes characteristic of bone cells; they are isolated within the matrix which they have secreted. There are normally no blood vessels within a cartilage; the nutriment supplied to the cells must thus reach them by

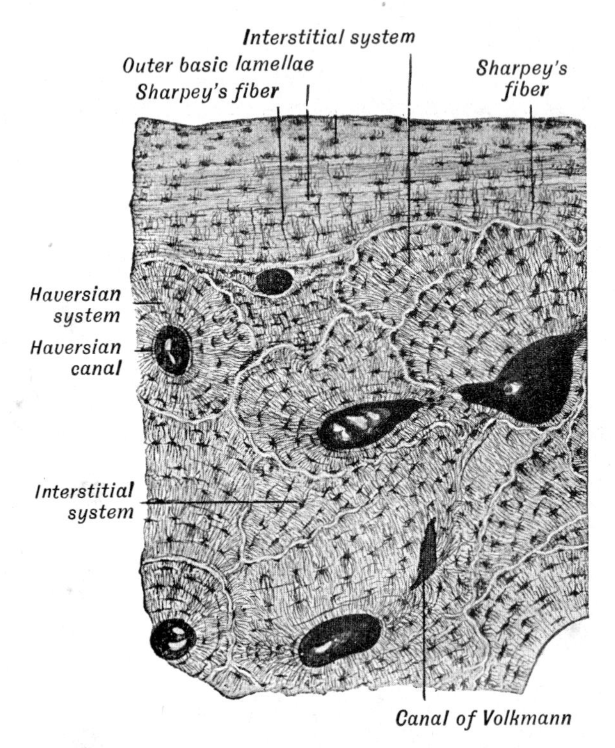

Fig. 72. Bone structure. A ground thin section through a mammalian metacarpal. Toward the top (outer) margin are parallel lamellae of bone formed from the periosteum; within are a number of haversian systems cut at various angles. The "interstitial system" includes remains of earlier formed bone layers not destroyed when the present haversian systems were created. A cementing substance binds the different bone areas together. "Volkmann's canals" carry blood vessels from the surface or bone marrow cavity (below, at right) to haversian systems. Sharpey's fibers are connective tissue fibers which run from the periosteum inward through the bone substance. (From Maximow and Bloom, after Schaffer.)

passing through the ground substance. The outer surface of a cartilage is covered by a layer of dense connective tissue, the *perichondrium*.

There are numerous variants from ordinary, hyaline cartilage. In animals in which the adult skeleton is largely cartilaginous, there is a strong tendency for the development of *calcified cartilage;* a deposition of calcium salts in the matrix produces a relatively hard, brittle, and opaque substance superficially somewhat similar to bone. *Elastic cartilage* (seen characteristically in the external ear of mammals) gains flexibility through the presence of many elastic fibers in the ground substance. *Fibrocartilage* is a form of material transitional between a dense connective tissue and cartilage, and is often present in the neighborhood of joints and the attachments of ligaments and tendons. *Mucous cartilage,* found in various instances in lower vertebrates, is a relatively soft and diffuse type, with branching rather than rounded cells.

Embryologically, cartilage is derived from mesenchyme. The irregularly shaped mesenchyme cells round up, and develop between themselves the interstitial ground substance and fibers. At first closely packed, the cells tend to separate as more material is laid down. Frequent cell divisions are seen in growing cartilage; we may find cells grouped in pairs or quartets, indicating their origin by division from a single parent cell. A cartilage may grow by the addition of new cells to its outer surface from the perichondrium. But it is also capable of growth by internal expansion —a unique and important character of cartilage as contrasted with bone.

Cartilage is essentially an *internal* skeletal structure. With few exceptions (as the mammalian ear pinna) it is never present in the skin or near the surface of the body but is, rather, characteristic of deep-lying parts of the skeleton. It is always abundant in the embryo and young of vertebrates. In the higher groups of living vertebrates and in many of the older fossil types of lower vertebrates, however, the adult skeleton is in great measure bony, and cartilage is relatively reduced. Only in living lower groups—cyclostomes and Chondrichthyes—is cartilage the major skeletal material in the adult.

Cartilage is a deep tissue, an embryonic tissue, a relatively soft and pliable and readily expandable tissue.

Bone (Fig. 72). Bone is the dominant skeletal material of the adult in most vertebrate groups. Like cartilage, it is a derivative of the mesenchyme and consists of transformed mesenchyme cells enclosed in a ground substance containing connective tissue fibers. The two materials differ, however, in numerous regards. The bone matrix is, from its first deposition, a hard, opaque, calcified material quite different from the ground substance of cartilage; laid down in it are salts in which phosphate and carbonate are combined with calcium in complex fashion. Enclosed within this matrix are the bone cells which have secreted it. They differ markedly from those of cartilage, for they (and the lacunae in the bone in which they are situated) are irregular, star-shaped, branching structures. Their branching processes continue outward from the lacunae in all directions, in tiny canaliculi, to reach neighboring bone cell spaces. In contrast with cartilage, bones are penetrated by blood vessels. The solid matrix of the bone is apparently impervious to nutritive materials, and the cells receive sustenance by the transfer of food materials from the blood vessels via the canaliculi. The external surface of bone is covered (like that of a cartilage) by a layer of dense connective tissue, the *periosteum.*

Whereas a cartilage is generally rather uniform in appearance when seen in section, bones have a complicated internal structure. Many bone areas (particularly bone surfaces) consist of *compact bone,* which forms a hard white mass in which openings of microscopic size are present. *Spongy bone,* generally present in the interior of large skeletal elements, contains a framework of bars of bony material, enough to preserve the shape and rigidity of the structure; but much of its area is occupied by vascular, fatty, or other tissues which form a *bone marrow*. The development of this type of bone aids in lightening the skeleton, and the marrow itself has positive functions (fat storage, blood cell formation).

Much of the substance of any bone was laid down originally in the form of layers of bone—*lamellae*—and this arrangement persists in many

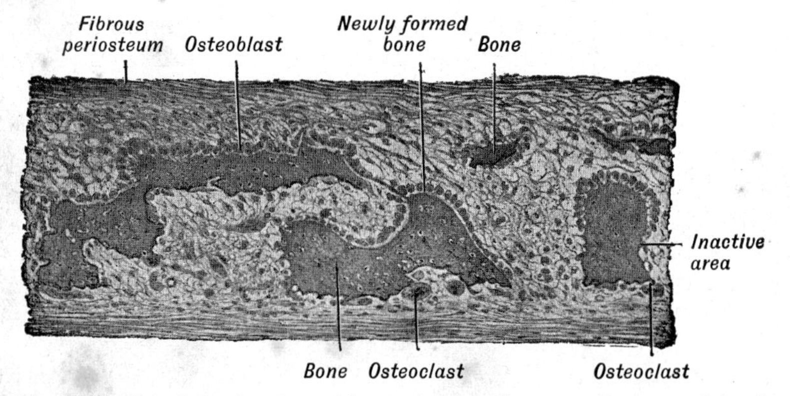

Fig. 73. Section through a dermal bone of the skull at an early stage of development. Thin bits of bones are already formed, with bone-forming cells—osteoblasts—around them. Bone-destroying cells—osteoclasts—are also present. (From Maximow and Bloom, after Schaffer.)

cases in the adult. However, bones are reworked as they grow by resorption of material and redeposition of new bone, so that the histological pattern is often highly complicated. Characteristic is the usual development of haversian systems, consisting of small *haversian canals* containing blood vessels and nerves, which branch through the bone. They are surrounded by concentric cylinders—bone layers, the cells of which obtain their nutriment from the canal. Haversian systems are formed in general by the destruction of the bone materials formerly present in a given region; tubular channels are formed, and bone is redeposited in concentric layers inside their walls.

Bone development. Bone arises through the activity of mesenchyme cells termed *osteoblasts,* which form about themselves a thick substance containing numerous fiber bundles. The deposition of calcium salts in this intercellular material completes the formation of bone, and the osteoblasts, their main activity over, become the bone cells—*osteocytes*.

Two radically different modes of bone formation—i.e., *ossification*—are to be seen in the vertebrate embryo. The simpler of the two is the formation of *membrane bone* (Fig. 73). In this case the bone forms simply and directly from mesenchyme. Membrane bone formation is seen in the dermal bones formed in the skin. The bone develops first as a thin, flattened, irregular bony plate, or membrane, in the dermis. It gradually expands at its margins and becomes thickened by the deposition of successive layers of additional bone on inner and outer surfaces. In bony fishes dermal bones are usually formed over nearly the whole of the body (including the mouth cavity), in the form of large plates anteriorly, and

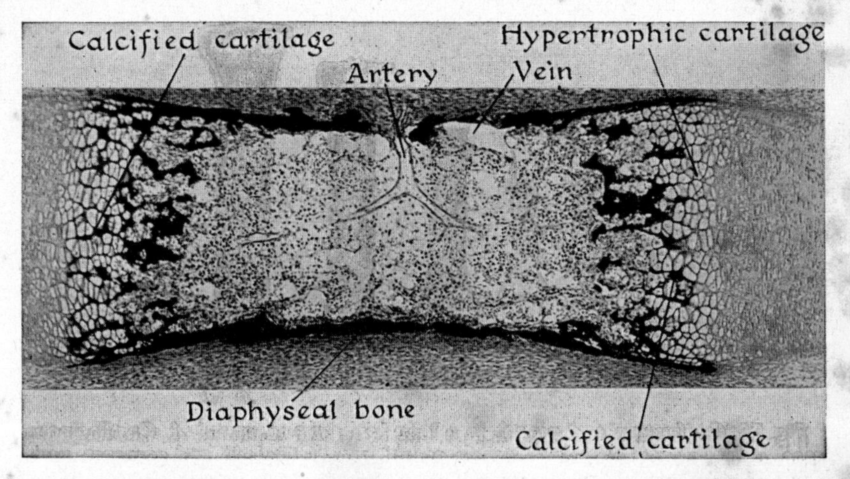

Fig. 74. Section of a metapodial of a rat embryo in which ossification is taking place in the shaft of a bone (bone material stained black). Perichondral ossification is present superficially. At either end is normal cartilage; toward the center of the shaft the cartilage is swollen, and becomes calcified and replaced by endochondral bone. (From Maximow and Bloom.)

of bony scales over most of the trunk. In higher vertebrates the extent of dermal bone formation is much restricted, for bony scales tend to disappear, and in birds and mammals dermal bones are normally found only in the skull, jaws, and shoulder girdle.

Quite different and much more complicated is the formation of *endochondral bone* (Figs. 74, 75). This is primarily the replacement of an embryonic cartilage by an adult bony structure. But, as will be seen below, actual replacement of cartilage by bone is only part of the process; much of the bone is really laid down directly in intramembranous fashion external to the embryonic cartilage.

In such typical internal structures as the long bones of the limbs of tetrapods the cartilage tends to assume the definitive form of the adult bone at an early stage and tiny size. Presently the cartilage begins to undergo modification and degeneration, particularly near the middle of its length; the cartilage cells begin to multiply and arrange themselves in longitudinal columns: and the material between these columns calcifies. Blood vessels break into the cartilage from the surface, and destruction of the cartilage of this central area takes place. Osteoblasts which enter with the blood vessels lay down bone in place of the destroyed cartilage. Ossification proceeds from the center toward either end of the element.

If the cartilage failed to grow, the replacement of the cartilage by bone would be accomplished in a short time, and a completely ossified bone of tiny size would result. But this does not occur, for the cartilage grows at either end about as fast as it is destroyed centrally. The cartilage, so to

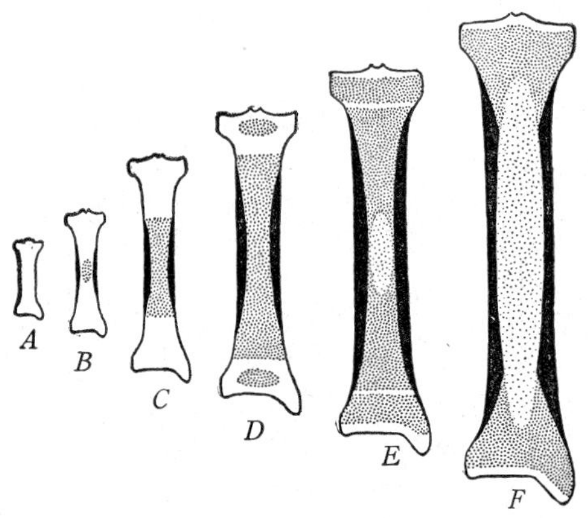

Fig. 75. Ossification and growth in a long bone of a mammal. *A,* Cartilaginous stage. *B, C,* Deposit of spongy, endochondral bone (stippled) and compact, perichondral bone (black). *D,* Appearance of an epiphysis at either end. *E,* Appearance of the marrow cavity (sparse stipple) owing to resorption of endochondral bone. The growth of the bone is confined to the thin strips of actively growing cartilage between shaft and epiphyses. *F,* Union of epiphysis with shaft, leaving articular cartilages on end surfaces; enlargement of marrow cavity by resorption of periosteal bone, centrally, as deposition continues peripherally. (From Arey.)

speak, leads the osteoblasts a long "stern chase;" the process of ossification does not catch up with the cartilages at either end of the element until adult size has been attained. With complete ossification, growth stops; for at their ends, internal skeletal elements are usually articulated with their neighbors, and bone cannot be readily added to the articular surfaces without damage to the joint areas.

In lower vertebrates, generally, internal bones ossify from a single center—the shaft region of a long bone, for example—and even in the adult the articular ends are often tipped by cartilage. But in mammals (and to a much more limited extent in certain reptiles) accessory ossifications—*epiphyses*—are found (Fig. 75). These are characteristically developed at either end of a long bone, and on prominent processes for muscular attachment. In epiphyses there develop ossification centers similar to those of the shaft region, but more limited in extent. These new centers may ossify the articular region of the bone long before the growth of the shaft is completed and thus allow the element to function properly despite its incomplete ossification. Between epiphysis and shaft there is a long-persistent band of cartilage. This would seem at first sight to be a relatively weak, inert, and functionless region. Actually, of course, it is highly important. This is a zone of growth; the cartilage here is continually growing, and as continually being replaced by bone from shaft and epiphysis. Once ossification does away with this band of cartilage, shaft and epiphysis unite; bone growth is over, and definitive adult size has been attained.

Though much of the growth of an internal "cartilage bone" takes place by the replacement of cartilage, this is not, as one will readily see upon reflection, the whole story. When the center of the shaft ossified, the bone had a small diameter. As the cartilage grows at either end, these ends gradually expand in width toward adult size; but the bone, if formed solely by internal replacement, would be shaped rather like an hour glass, with a constricted middle portion. With the result of rectifying this imperfection, we find that, in addition to internal replacement, cartilage bones have a major increment produced by the direct formation of bone on the surface of the cartilage as *perichondral bone,* formed in successive concentric layers rather after the fashion of membrane bone. These layers are thickest in the originally thin middle part of the shaft. This process of superficial addition of bony material may continue after the underlying shaft region has been ossified; at this stage the term *periosteal bone* rather than perichondral is more appropriate.

The process of endochondral bone formation in higher vertebrates was long believed to be an example of the ontogenetic recapitulation of the history of the race. The replacement of cartilage by bone in embryonic development was thought to repeat the phylogenetic story of the vertebrate skeleton, under the assumption that the cartilaginous condition found in the adult cyclostome or shark was a truly primitive one. However, the fossil evidence strongly suggests that this is not the case; that, on the contrary, the living lower vertebrates are skeletally degenerate and that cartilage in the adult is a retention of an embryonic condition.

The question may, however, be fairly asked: Why this roundabout way of forming bones if there is no phylogenetic story concerned?

The answer may be deduced from the fact that not all bones are preformed in cartilage. Those which lack that stage are the dermal elements. It is significant that these are usually simple and platelike and are capable of free growth at every surface until adult size is reached, whereas most of the deeper elements, which are preformed in cartilage, have complicated articulations with their neighbors and important muscular attachments. The vertebrate skeleton at an early embryonic stage is an almost perfect if tiny model of that of the adult. If the internal elements, with their complicated muscular attachments and articular relations, were formed directly in bone, growth to adult size would be impossible; for bone can grow only at its surface, and surface growth would disrupt the relations of the bone with surrounding structures. Some type of moldable material, capable of growth without disturbance of surface relations, is needed for the growth of deep-lying elements. Cartilage, with its capability of growth by internal expansion, is an ideal embryonic adaptation for this purpose. Except where degeneration has occurred, bone is the normal adult skeletal material in a vertebrate, cartilage its indispensable embryonic auxiliary.

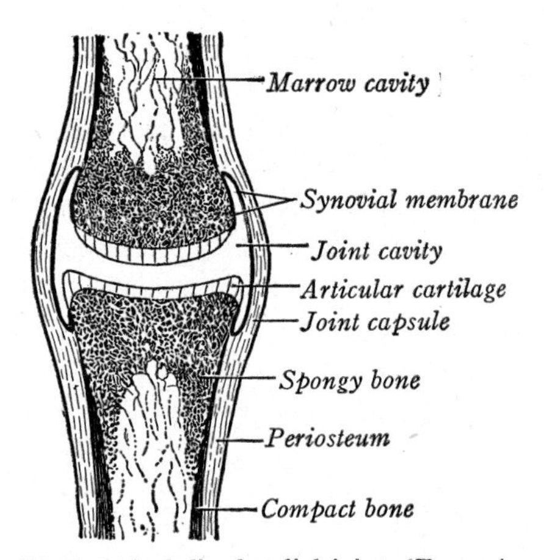

Fig. 76. A typical diarthrodial joint. (From Arey.)

Joints. Bones and cartilages are joined to one another by structures of varied types. In such cases as the bones of the skull, where one abuts directly against another, there is usually little if any possibility, or desirability, of movement. The connection between such elements—termed a *synarthrosis*—may be formed by thin intervening sheets of cartilage or connective tissue, or by direct contact of bone to bone, so that the lines of apposition, or *suture,* may be obliterated in the adult.

In freely movable joints between adjacent elements, one speaks of the joint as a *diarthrosis* (Fig. 76); the adjacent surfaces are typically covered by a film of cartilage even if the articulating elements are bony in major structure, and are bound together by stout connective tissues, frequently enclosing a liquid-filled joint cavity.

Classification of Skeletal Elements. The skeleton includes a wide variety of elements of varied form, structure, function, position, and embryonic origin, associated in variable combinations and showing major modifications from group to group. Classification of skeletal elements is difficult, and no classification is absolutely satisfactory. We here adopt the following scheme of major subdivision:

- Dermal skeleton
- Endoskeleton
  - Somatic
    - Axial
    - Appendicular
  - Visceral

A primary distinction may be made between the *dermal skeleton,* consisting of bony structures—plates and scales—developed in the skin, and the more deeply placed skeletal parts, the *endoskeleton.* Dermal elements are never preformed in cartilage. On the other hand, the remaining deep-lying skeletal structures here grouped as endoskeleton are preformed in cartilage and may remain cartilaginous throughout life.

Within the class of endoskeletal elements, a distinction may be made between two categories of unequal size, the somatic and visceral groups (as noted in Chapter IX, the associated muscular system may be similarly subdivided). As the *visceral skeleton* we may classify the elements which lie between and operate the gill openings, and structures (such as the jaw cartilages and ear ossicles) derived from them. The *somatic skeleton* includes all the remaining internal structures. As further discussed in other sections, the gill region is distinctive as regards its musculature and nerve supply; its skeleton is equally distinctive; and, as will be seen, its embryological origin is radically different from that of most other skeletal elements.

The somatic skeleton includes in its more primitive, axial portions the vertebrae, the ribs, and other related elements of the trunk and tail segments, and, anteriorly, the braincase, which forms a major element of the skull. These structures constitute the *axial skeleton.* The limb girdles and the elements of the free appendages, forming the *appendicular skeleton,* may be considered derivatives of the axial skeleton.

In certain instances we find that structural units of the adult skeleton contain elements derived from two or more categories. Thus, for example, the shoulder girdle of many vertebrates contains both dermal and endoskeletal components; the lower jaw of several vertebrate classes includes both visceral and dermal elements. Most complex of all is the skull, which in bony fishes and land vertebrates includes dermal, axial, and visceral structures in its formation.

DERMAL SKELETON

Fishes. The skin over most of the body of many living vertebrates contains no hard skeletal parts, but dermal bony structures are usually present in the head region, at least, and it seems certain that ancestral vertebrates were completely ensheathed in a bony armor. Although there may be a superficial covering of other hard skeletal materials, the major component of skin plates and scales is bone, formed directly in membrane. The primitive armor is seen highly developed in the ancient ostracoderms of the Paleozoic era (Fig. 15, p. 38), and such armor was preserved in numerous placoderms and higher bony fishes. In the ostracoderms the trunk and tail were flexibly sheathed in a series of scales or small plates which allowed for the body movements necessary in swimming. The anterior part of the body, in contrast, was typically encased in a series of solid, immovable plates. One tends to think of these structures as forming a skull. Their extent in ostracoderms was much greater, however, for this "head" armor included the large gill region and, frequently, part of

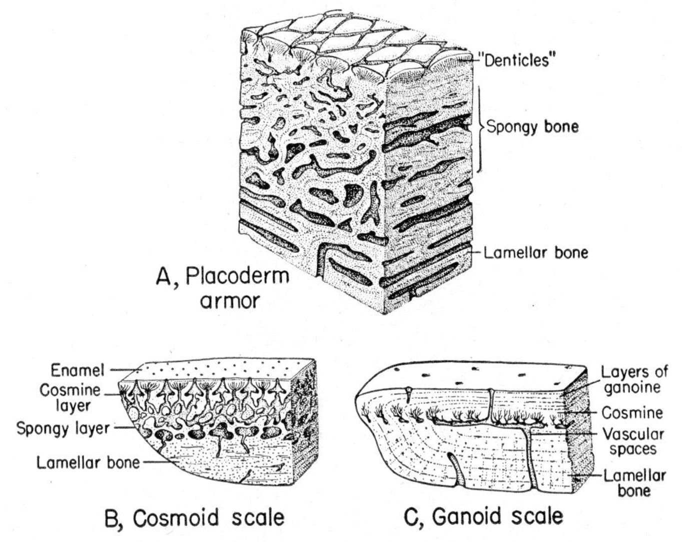

Fig. 77. The structure of dermal plates and scales in primitive vertebrates. *A*, Devonian placederm. *B*, Primitive crossopterygian with a cosmoid scale type. *C*, A Paleozoic ray-finned fish with a ganoid scale. (After Kiaer, Goodrich.)

the trunk as well. In some instances most or all of the adult armor was a solidly fused shield; in other cases there were a number of discrete elements. The pattern of the bony elements of these ancient forms cannot be readily homologized with the dermal bones of more advanced vertebrates. Probably there was great variation in the number and arrangement of plates developed ontogenetically in the head region in early vertebrates; their exact arrangement mattered little as long as they fused together in the adult to form an adequately solid structure. It is only in bony fishes and land vertebrates that the plates of the head region have tended to "settle down" to a consistent pattern of cranial elements that can be identified from group to group; this pattern will be studied in detail in a later section.

In ostracoderms and other ancient forms (as in higher bony fishes) scales and head plates are similar in microscopic structure (Fig. 77, *A*).

The inner layer consisted of compact bone. Above this was a layer of spongy bone, with numerous spaces presumably containing blood vessels. Toward the outside was a further compact layer. The surface of the plates and scales was frequently ornamented with tubercles or ridges. Beneath each tubercle lay a cavity presumably filled with a "pulp" of blood vessels and connective tissue. The body of the tubercle was formed of a substance resembling the dentine of a tooth, and the surface of the tubercle and, indeed, of the plate as a whole, was frequently covered by a thin layer of hard material comparable to tooth enamel. The entire tubercle, in fact, resembled a tooth in structure, and it is extremely probable that teeth are actually derived from such structures.

The extinct placoderms were, as the name implies, also armored; but the armor was variable. The grotesque arthrodires, for example (Fig. 17, *B*, p. 39), had a heavy armor over the top and sides of the head and gill region and a second set of armor plates encircling much of the trunk,

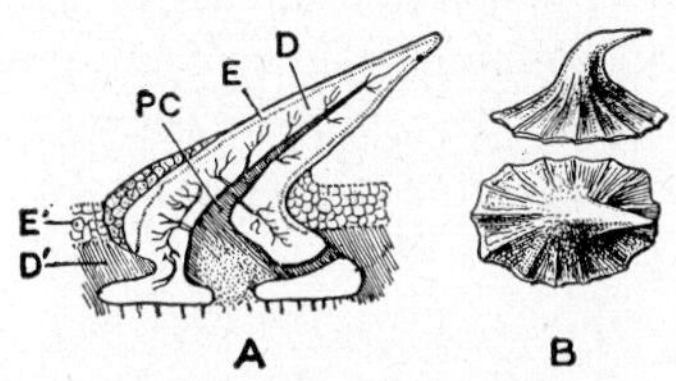

Fig. 78. Shark dermal denticles. *A*, Section through a denticle: *D*, dentine; *D'*, dermis; *E*, hard, enamel-like surface of denticle—vitrodentine; *E'*, epidermis; *PC*, pulp cavity. *B*, Side and surface views of denticle. (From Dean.)

but the tail appears to have had a naked skin. The "spiny sharks" (Fig. 17, *A*) were sheathed in scales similar in structure to the ganoid scales of certain bony fishes mentioned below. In some placoderms the armor was much reduced, presumably a degenerate condition and one suggesting a transition to that of the sharks.

The living cyclostomes are entirely devoid of hard dermal structures. It is highly probable that this condition is secondary and degenerate. Among the Chondrichthyes there are present, in chimaeras and some sharks, fin spines of a dentine-like material, but the skin is otherwise naked except for the *dermal denticles,* or placoid "scales" (Figs. 58, *A*, p. 116; 78). These are small, isolated structures, often conical in form, which resemble simple teeth (cf. Fig. 172, p. 298). The base contains a pulp cavity; a mass of dentine forms the bulk of the tubercle; and there is a hard shiny surface film which resembles tooth enamel, but appears, in sharks, to be merely a specialized superficial layer of the dentine itself. It was formerly assumed that the presence here of dermal denticles marked the first phylogenetic appearance of dermal armor and that bony scales and plates appeared later as the result of fusion of such denticles. It seems, however, probable that the reverse is actually the case. Dermal denticles are the last remnants of the ancestral armor; the bone proper has disappeared, leaving only its superficial tubercles.

Early members of the bony fish group (Osteichthyes) were, like the older and more primitive ostracoderms and placoderms, completely armored in bony scales. In general the scaly covering has been retained, but it has undergone many modifications, and there are some teleosts in which scales are absent. In bony fishes, in addition to body scales, there is a well-defined cranial shield which forms part of a typical skull; dermal bony plates are further associated with the lower jaw and the inner surfaces of the mouth, and there is a dermal shoulder girdle. These structures will be discussed later. We may here, however, note briefly the presence, between head and shoulder, of a series of *opercular elements* lying in the flap of skin covering the gill region (Figs. 131, 132, pp. 224, 225). There is typically a major plate, the *operculum,* on either side behind the cheek region of the skull. Other, smaller elements may be present adjacent to the main operculum (preoperculum, suboperculum, and so on). The opercular structures follow the contours of the gill chamber downward and forward beneath the jaws. Here there was primitively a series of plates termed *gulars;* in teleosts these have been reduced to a series of parallel rods termed *branchiostegal rays* (Figs. 29, 99, pp. 51, 172).

In early bony fishes the scales, overlapping shingle-fashion, were thick, shiny structures covered by an enamel-like material. On sectioning them it is seen that radically different types were present in the two major fish groups. In the early Choanichthyes the *cosmoid* scale (Fig. 77, *B*) was characteristic. In this there was the three-layered structure noted above for older fish groups, with a surface layer of tubercles covered by thin enamel. The main substance of the tubercles consisted of a material, *cosmine,* similar to the dentine of a tooth (cf. p. 298), but differing in that the tiny canals penetrating it from the pulp cavity are branching structures. Such scales were characteristic of fossil crossopterygians and the oldest lungfishes. In recent lungfishes, however, the scales have degenerated to a simple, rather fibrous and leathery type of bony structure; cosmoid scales are nonexistent in living animals. In primitive actinopterygians there was present quite a different type of scale, the true *ganoid* scale (Fig. 77, *C*), still found little modified in Polypterus and the gar pikes. This differs from the cosmoid type in that, instead of a single layer of enamel above the cosmine, there is layer after layer of a shiny enamel-like material termed *ganoine.* Hosts of Paleozoic and early Mesozoic actinopterygians, the palaeoniscoids, had scales of this type. In most later actinopterygians, however, the ganoine was reduced and lost, and the cosmine layer as well, leaving in teleosts only relatively thin, pliable scales of a bonelike material in which, however, bone cells are rare or absent (Fig. 58, *C,* p. 116). In early fishes the scales, arranged in diagonal rows, generally had a V-shaped, free posterior border or, less commonly, a rounded one. In teleosts the basic type of scale is that with a rounded—*cycloid*—border. In advanced teleosts the posterior margin is frequently frayed out to give a comblike—*ctenoid*—border, but the difference is trifling.

We may here (parenthetically) discuss the nature of the rays which stiffen the peripheral portions of fish fins, medial or paired (Fig. 79). In some fishes with a scaly covering, scales like those of the body sheathe the fins, but become smaller and more attenuate distally. In higher bony fishes these fin scales tend to be modified into elongated rays, termed *lepidotrichia,* each of which represents a row of slender scales. The very tip of the fin in bony fishes may be additionally stiffened by tiny horny rods developed in the dermis of the skin—*actinotrichia*—and larger rays of this sort—termed *ceratotrichia*—are the sole support of the shark fin web.

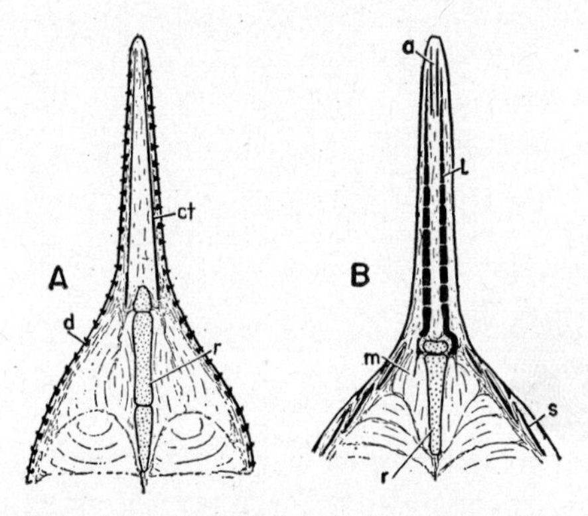

Fig. 79. Sections through the dorsal fin of *A,* a shark; *B,* a ray-finned fish; to show the nature of the fin supports. *a,* Actinotrichia; *ct,* ceratotrichia; *d,* dermal denticle; *l,* lepidotrichia; *m,* muscle tissue at fin base; *r,* skeletal supports of fin; *s,* bony scales. (From Goodrich.)

Tetrapods. Land vertebrates are descended from fishes with a complete bony covering of plates and scales. The dermal bones of the head are retained as part of the skull and jaw structures, and a dermal shoulder girdle generally persists. The remainder of the dermal armor, however, has tended strongly to reduction and loss.

With the reduction of the gills in land vertebrates, the series of opercular bones vanishes. The bony body scales may persist, but only in a much reduced or modified fashion. In the older fossil amphibians the original fish scales were usually retained in V-shaped rows on the belly and flanks, perhaps as a protection to the low-slung body in land locomotion; more rarely are bony scales found over the back. In modern amphibians scales are completely lacking except for functionless vestiges buried in the skin of the Apoda.

In some of the oldest reptiles the ventral bony scales persisted in the shape of a series of jointed V-shaped rods on the belly, termed abdominal ribs or *gastralia*. These structures are dermal and should not be confused with ventral extensions of the true ribs, which may lie beneath them, but are, of course, part of the endoskeleton. Abdominal ribs persist today in lizards, crocodilians, and Sphenodon; but in birds and mammals these last representatives of the old fish bony scale system have vanished.

The skin, however, persistently retains its potentialities of forming dermal bone, and in many reptilian types we find a redevelopment of bony

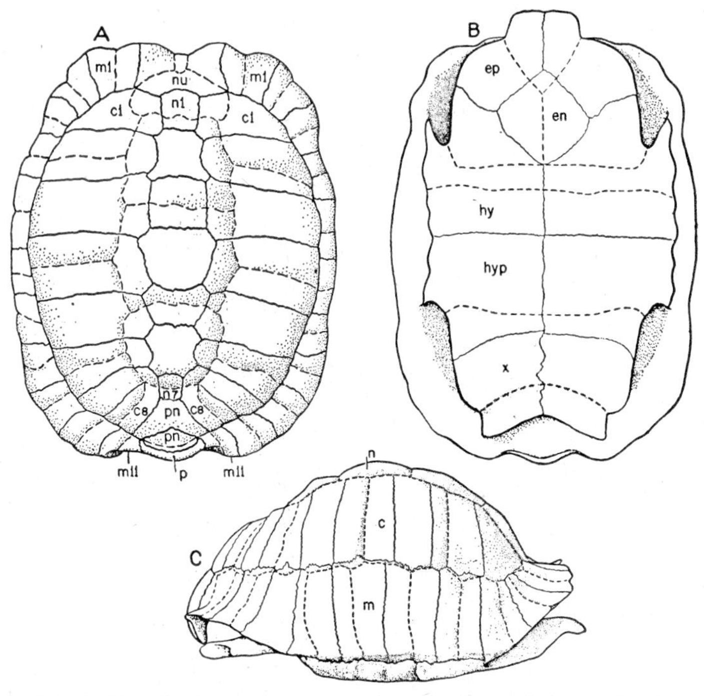

Fig. 80. *A,* Dorsal, *B,* ventral and *C,* lateral views of the shell of a tortoise (Testudo). Sutures between bony plates in solid line, outlines of horny scutes in broken line. On carapace, *c, c1* to *c8,* costal plates; *m, m1* to *m11,* marginals; *n, n1* to *n7,* neurals; *nu,* nuchal; *p,* pygal; *pn,* postneurals. On plastron: *en,* entoplastron; *ep,* epiplastron; *hy,* hyoplastron; *hyp,* hypoplastron; *x,* xiphiplastron.

nodules, scales, or plates as a protective armor. Examples of this sort are common in extinct reptiles; in the Cretaceous Period, for example, one group of clumsy four-footed dinosaurs developed a massive armor covering the back. In lizards there is frequently a secondary development of bony scales underlying the horny scales in the epidermis (Fig. 61, *C,* p. 119), and in the crocodilians there develops an armor of subquadrate bony plates covered by horny skin.

Redevelopment of dermal bone is not found in birds, but does occur in a few mammalian types. Most of these belong to the edentate stocks native to South America. Extinct ground sloths reinforced their tough hide with lumps of dermal bone buried in the dermis. In the armadillos and their extinct relatives, the glyptodonts, bony plates form a complete carapace covering the back and the top of the head and in glyptodonts sheathe the tail as well. In the glyptodonts the dome-shaped dermal shield was a solid structure composed of small polygonal plates. In armadillos the plates are arranged in transverse bands; the more anterior and posterior rows tend to fuse together to a variable degree into solid structures, but those of the middle of the back remain separate, giving some flexibility and allowing, in some cases, a curling up of the body into a tight ball in which the soft belly is concealed.

Turtle Armor (Fig. 80). Although in extinct reptiles there are many interesting examples of armor, that of the living chelonians is unequalled for completeness and efficiency. We have noted earlier the presence of horny scutes over the surface of the turtle body; beneath these scutes there is in most chelonians a stout and effective bony shell. This consists of a convex oval dorsal shield, the *carapace,* and a flat belly slab, the *plastron.* At the sides the two parts are connected by a bony bridge; front and back, the shell is open for the head, limbs, and tail.

The carapace consists essentially of three sets of plates. Circling the margin of the shield is a set of *marginal* plates; the most anterior and posterior marginals have the special names of *nuchal* and *pygal* plates. From front to back in the midline is a series of subquadrate plates, most of which are fused to the vertebrae beneath them and termed *neurals;* the last two or so are not so attached and are termed *postneurals.* Transversely placed between marginals and neurals are eight paired *costal* plates, solidly bound to underlying ribs. The elements of the carapace are firmly fused in many cases, with obliteration of sutures; the lines of separation of the horny surface scutes, however, are often impressed on the surface of the bony carapace. For the most part the boundaries between the horny scutes of the carapace (and of the plastron as well) alternate with the sutures between the bony plates, a condition which lends strength to the shell structure as a whole.

The plastron of typical chelonians consists of four paired plates and an anterior median element; in primitive types additional paired elements may be present. The diamond-shaped unpaired plate is the *entoplastron;* the paired elements are, in order, *epiplastra, hyoplastra, hypoplastra,* and *xiphiplastra.* The three plates at the anterior margin are actually elements (clavicles and interclavicle) of the dermal shoulder girdle described later in this chapter; the remaining plastral elements, and all the carapace plates, appear to be new developments in turtles, although there is evidence that dermal abdominal ribs have been incorporated in the plastron structure.

AXIAL SKELETON

Vertebrae—General. The host of skeletal structures remaining for description are endoskeletal, lying (in contrast to those of the dermal skeleton) in the interior of the body; they are first formed in the embryo as cartilages, the cartilage being usually replaced and supplemented by bone. The great majority of endoskeletal elements belong to the system here termed somatic; they are formed (in contrast to those of the visceral system) from mesenchyme of mesodermal origin. Apart from the special category of limb supports, the somatic skeletal elements can be classed as axial.

The major axial structure is the vertebral column, which replaces the notochord as the main longitudinal support of the vertebrate body. For a small water dweller the notochord may afford sufficient support to the back, support combined with ideal flexibility. But with greater size,

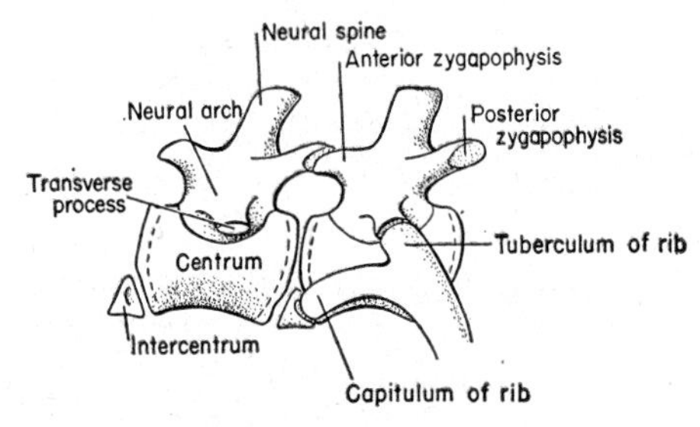

Fig. 81. Two vertebrae of an early generalized reptile.

greater strength appears to have been a necessity, and the need for strength is emphasized in land vertebrates in which the back must suspend the weight of the body. Cartilage and, especially, bone give the needed strength, and the jointed nature of the backbone affords a considerable degree of motility. As a second major function, the vertebral elements extend upward from the notochordal region to surround and protect the spinal cord adjacent to it.

As an introduction to vertebral structure we may first consider the relatively simple situation seen in typical vertebrae of amniotes (Fig. 81). Each vertebra, formed of bone in these forms, includes two principal portions, a neural arch and a centrum. Ventrally placed is the *centrum,* which functionally replaces the notochord. In many vertebrate groups this is a spool-shaped cylinder articulated at either end with the adjacent centra. In more primitive amniotes the centrum may be hollowed at both ends, a condition termed *amphicelous* (Fig. 82, *A*). The cavity thus present between successive centra is filled with soft material which may be, in part at least, derived from the embryonic notochord, and there may be a tiny hole, piercing the centrum lengthwise, through which the notochord may extend as a continuous (though much constricted) structure.

In most living amniotes, however, the notochord has disappeared except for, at the most, thin pads of tissue lying between successive centra, and the centra have become solid structures, whose end faces are closely apposed to those of their neighbors. Those vertebrae which have flat terminal faces, as is common in mammals, are termed *acelous* (Fig. 82, *D*). In many cases one end of the centrum expands in spherical fashion, to be received in a hollowed socket in the neighboring centrum. If the concave socket is at the anterior end of the centrum, a condition found in most living reptiles, the vertebra is termed *procelous* (Fig. 82, *C*); if posteriorly placed, *opisthocelous* (Fig. 82, *B*). Since the nature of the articular faces may change regionally along the column, various irregularities and combinations of structure may occur, and there are numerous variants from these orthodox types. Notable is the development in the neck region of birds of *heterocelous* vertebrae, with saddle-shaped articular faces between the centra, and the development of complicated central connections in the retractable neck of turtles.

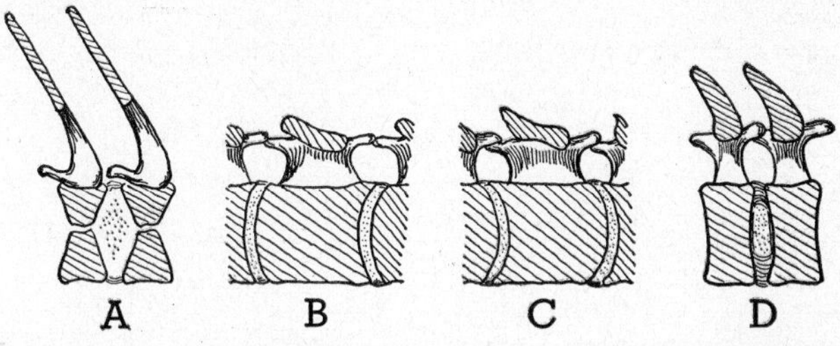

Fig. 82. Longitudinal sections through amniote vertebrate of various types. *A*, Primitive amphicelous type; the form represented is pierced through the centrum by an opening for a continuous notochord. *B*, The opisthocelous type. *C*, The procelous type. *D*, Essentially an acelous type, but with centra slightly biconcave and with room for an intervertebral disc. (After Gregory.)

The rib head may attach in amniotes to the centrum at a rounded facet, typically near the anterior margin, and in mammals may articulate, intercentrally, with the posterior edge of the adjacent centrum as well; the attachment may occur at the tip of a projecting process termed a *parapophysis* (*apophysis* is a term used for any projecting vertebral structure).

In reptiles, and in the tail of mammals, there may be present in the trunk small elements, crescentic in end view, which are wedged in ventrally between successive centra. These are *intercentra* (or hypocentra), remnants of "central" elements important, as will be seen, in lower vertebrates. In the tail the intercentra have as ventral extensions the *hemal arches* or chevrons, which extend downward, joining to form a spine, in the septum between the muscles of either side of the tail. The hemal arches, like the neural arches above, are forked at their bases as seen in end view, with a *V* or *Y* shape; between the two branches lie the major blood vessels of the tail.

The upper surface of the centrum forms the floor of the canal carrying the spinal cord. At either side rises the broad base of the *neural arch;* centrum and arch are firmly fused in most adults. Above the spinal cord the two arches fuse to form a *neural spine,* which runs upward between the dorsal muscles of the two sides; successive spines are usually bound together by an elastic ligament. At the base of the arch on either side there is frequently a stout *transverse process* (*diapophysis*) to which attaches the upper head (tuberculum) of a two-headed rib.

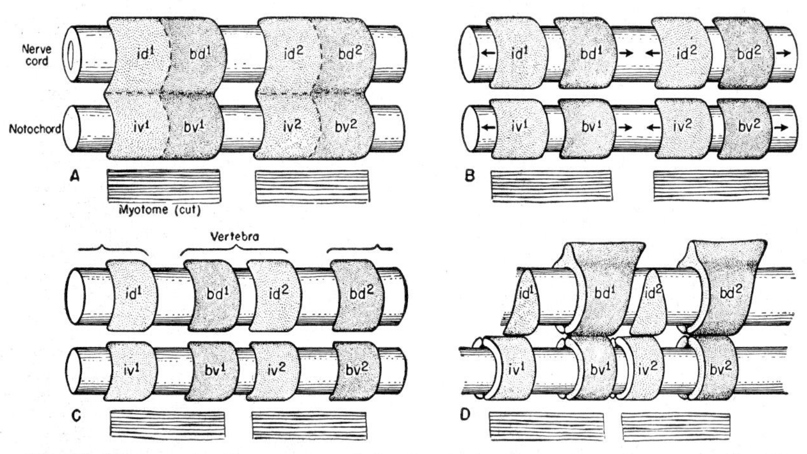

Fig. 83. Diagram to show theory of development of vertebrae from arcualia. Two mesodermal segments of an embryo seen from the left side. *A,* Skin and most of the muscle segments removed to show mesenchyme of sclerotome lying lateral to nerve cord and notochord. This is already showing indications of separating into anterior and posterior parts. *B,* Four pairs of precartilaginous tissue blocks are formed in each segment; anterior and posterior parts are separating, as indicated by arrows. *C,* The tissue blocks (arcualia) are re-forming into vertebrae which (as shown by comparison with the muscular tissue) alternate with the original segmentation. *D,* In the center, a vertebra formed from parts of two segments; in front and behind, parts of two other vertebrae formed from other parts of the two segments figured in *A*. *bd,* Basidorsal; *bv,* basiventral; *id,* interdorsal; *iv,* interventral.

In land vertebrates (and in a few bony fishes) strength is added to the column by the presence of articulations between successive neural arches. At either end there develop paired "yolking" processes, or *zygapophyses.* The anterior zygapophyses terminate in rounded surfaces which generally face upward and inward; to them are opposed surfaces facing downward and outward on the posterior zygapophyses of the next anterior arch (for mnemonic purposes one may recall that in any procession, those at the front of the parade are up and in—those at the rear, down and out). There are sometimes, in tetrapods, additional articulating processes for firmer vertebral union.

Vertebral Development (Fig. 83). The structures described are those normally found in the higher vertebrate classes; in amphibians and the various fish groups there is a great variety of vertebral types. Before discussing them, however, we should consider the embryological development of vertebrae. This is by no means uniform, and there are many variable and debatable features, but the general pattern seems to be approximately as follows.

Three stages may be encountered in the development of vertebrae: (1) mesenchyme condensations, (2) cartilaginous blocks, and (3) ossifications. In many lower vertebrates all or part of the structure may remain as cartilage; in forms where ossification occurs, part of the vertebra may ossify directly without passing through a cartilaginous stage.

In the discussion of mesoderm differentiation it was noted that a considerable amount of mesenchyme proliferates from the medial side of each somite—the area known as the sclerotome (Fig. 50, p. 100). Its major function is the formation of vertebrae; the sclerotome cells spread out in a sheet applied to the sides of both notochord and spinal cord opposite each somite.

The number of vertebrae in the adult corresponds to the number of somites from which their materials are derived, and one might assume that in consequence each vertebra would lie opposite the muscle segment likewise produced by each somite. Actually this is not the case; the vertebrae alternate with the muscle segments, and the mesenchyme from which a vertebra is composed is derived, half and half, from each of the two adjacent somites. In many vertebrates there early appears a distinction between the anterior and posterior portions of the mesenchyme sheets produced by each segment. As development proceeds it is found that the front half of a vertebra is formed from the posterior part of one segmental mesenchyme sheet, the back half from the anterior part of the mesenchyme of the next segment.

On reflection it is seen that this mode of mesenchyme division is not a senseless developmental complication, but a functionally useful one. In lower vertebrates, particularly, the segmentally arranged axial muscles are powerful and are important in locomotion. To exert their force they depend in great measure on attachment to the vertebrae and ribs. Their attachments, to be effective, must be to two successive vertebrae; in consequence, vertebral structures must alternate with muscular ones. It will be noted that the primary segmentation, that of the somites, is retained by the axial muscles; the vertebral segmentation, developing later in both phylogeny and ontogeny, is subsequent to that of the muscles and alternating with it.

The mesenchymal materials which form a vertebra thus consist of anterior and posterior parts on either side. Further, each of these four masses tends to divide into dorsal and ventral halves, the dorsal components lying alongside the spinal cord, the ventral ones beside the notochord. We may thus conceive of the potential vertebra as consisting in an early stage of eight pieces, four on either side, each quartet including anterior and posterior components both dorsally and ventrally.

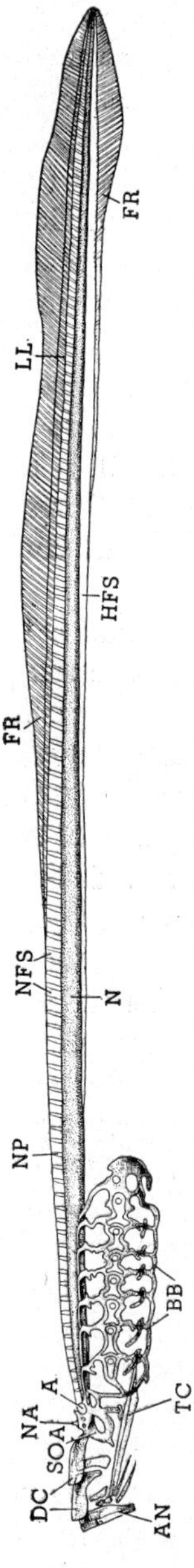

Usually there is a transformation of the mesenchymal masses into cartilaginous structures; these may be retained in lower vertebrates as adult vertebral elements or may be replaced by bone. These cartilaginous blocks tend to arch around notochord, below, or spinal cord, above, and are termed *arcualia* (cf. Fig. 83). The members of the anterior group may be termed the *basal arcualia,* those of the posterior group *interarcualia;* the four elements of one side of an embryonic vertebra are *basidorsal, basiventral, interdorsal, interventral.* Ventral arcualia may fuse with their mates of the opposite side to form central elements; dorsal pairs may fuse to form neural arches. In some cases it is difficult to interpret adult vertebral structures in terms of known (or assumed) embryological arcualia. In general, however, the basidorsals appear to form the neural arch; the basiventrals are responsible for the formation of the intercentra of land vertebrates, one or both of the interarcualia for the definitive centrum of higher land classes. It must be noted, however, that even where a cartilaginous arch forms the basis for an adult bony structure, much of this may develop directly from mesenchyme and that, as will be seen, there may be, in amphibians, direct development of centrum from mesenchyme without any intervention of a cartilaginous stage.

Fig. 84. Skeleton of the lamprey, Petromyzon. *A,* Otic capsule; *AN,* ring cartilage surrounding mouth; *BB,* cartilages of branchial basket; *DC,* dorsal cartilages of mouth region; *FR,* dermal fin rays; *HFS,* fibrous sheath around dorsal aorta; *LL,* longitudinal ligament connecting tips of neural processes; *N,* notochord; *NA,* openings of nasal capsule; *NFS,* fibrous sheath of spinal cord; *NP,* neural processes; *O,* openings of gill chambers; *SOA,* cartilaginous arch around orbit; *TC,* tongue cartilage. (From Dean.)

Fish Vertebrae. The cyclostomes are hardly to be classed as vertebrates in a strict technical sense, for the main support of the back is the persistently large notochord with its thick fibrous sheath (Fig. 84). Hagfishes have no vertebral elements at all, although, posteriorly, cartilages form part of the fin supports.

Lampreys have no ventral elements, but there are two pairs of small dorsal cartilages alongside the spinal cord in each segment.

In elasmobranchs (Fig. 85, *A*) there are typically developed all four pairs of arch cartilages. The dorsal pairs form neural arches and *inter-*

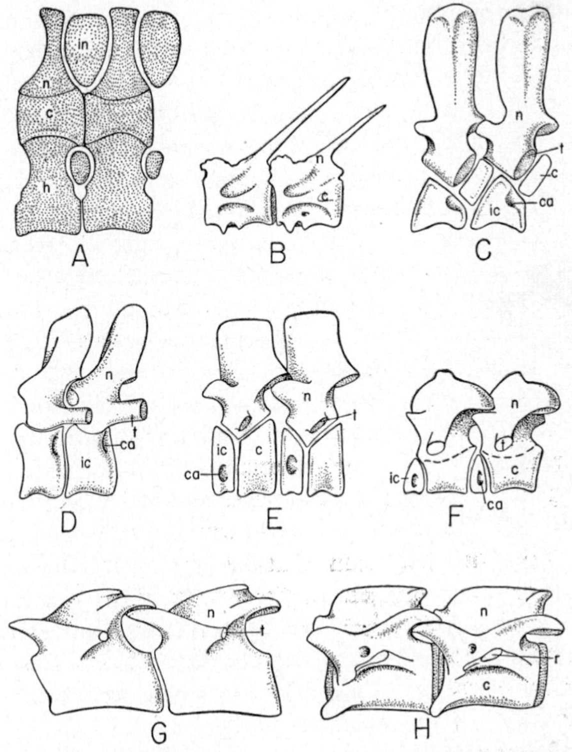

Fig. 85. Vertebrae of lower vertebrate groups. *A*, Shark, showing accessory cartilages in addition to the typical arcualia. *B*, Teleost. *C* to *F*, Vertebrae of primitive tetrapods, the extinct labyrinthodonts. *C*, The rhachitomous type, presumably primitive. Neural arch and intercentrum represent basidorsal and basiventral; the pleurocentrum is formed from interarcual elements. *D*, The stereospondylous type. *E*, Embolomere type. *F*, Type approaching that of reptiles (seen in Seymouria and others) with large true centrum (= pleurocentrum) and reduced intercentrum. *G*, Frog vertebrae; the ventral elements are reduced, and the neural arch ossification forms the central region. *H*, The urodele type; the centrum is formed mainly by a perichordal ossification, but ventral cartilages may persist and form the tips of the centrum. *c*, Centrum (pleurocentrum in labyrinthodonts); *ca*, attachment of capitulum of rib; *h*, hemal arch; *ic*, intercentrum; *in*, interneural element; *n*, neural arch; *r*, rib attachment in urodeles; *t*, transverse process.

neurals as well; below are formed centra, and both dorsally and ventrally there are often accessory cartilages. The most prominent element of the vertebra is a large disc-shaped centrum. The embryonic notochord has an extremely thick fibrous sheath, which is invaded in every segment by cells from the four adjacent basalia and eventually transformed, with the superficial addition of further cartilage, into a single cartilaginous structure in which the contributions of the separate arcualia are indistinguishable. Usually the cartilage of the centrum becomes more or less strongly calcified; the calcium deposition takes place in a variable but usually complicated pattern, so that concentric rings or star-shaped structures are seen when a vertebra is sectioned. In the tail region there is an interesting condition—that of *diplospondyly*—in which two complete vertebrae instead of one may develop for each muscle segment.

The chimaeras differ from the elasmobranchs in that instead of a single disc-shaped centrum there may be as many as five thin and feebly developed rings about the notochordal region in each segment. In adult lungfishes the notochord is persistently large. Two pairs of arcualia develop as cartilages and may invade the notochordal sheath, but never form complete centra. Though the main mass of the arch materials remains cartilaginous, its dorsal portions ossify as neural arches and spines, and hemal spines also ossify in the tail.

In none of the remaining vertebrates is there any invasion of the notochordal sheath by arch elements; the formation of centra is a variable process, but always takes place external to the notochord. Among the actinopterygians, sturgeons, paddlefishes, and many fossil forms have unossified centra, but there are well-formed, spool-shaped centra in Polypterus, the living holosteans, and the whole horde of teleosts (Fig. 85, *B*). Arcualia are prominent in the embryo of ray-finned fishes, particularly the basidorsals, which form neural arches. But centrum formation is essentially a process independent of the arches. A bony ring is laid down around the notochordal sheath without being preformed in cartilage; such a structure is a *perichordal centrum*.

Primitive Tetrapod Vertebrae. The basal stock of the tetrapods appears to be included in the Labyrinthodontia, a great ancient amphibian group, now extinct, whose vertebral structure is hence of importance. Though there is some doubt as to the most primitive vertebral structure in this group, the *rhachitomous vertebra* (Fig. 85, *C*) may well be the basic type. In this there are three sets of elements, ossifications in cartilaginous arcualia. Dorsally there is a well-developed neural arch derived from basidorsal cartilages. Below, the notochord is continuous, but much constricted by the development of two types of central structures. Anteriorly is the *intercentrum,* wedge-shaped in lateral aspect, but seen in end view to be a crescent curving ventrally around the notochord. The intercentrum, it is agreed, is formed by the fusion of the pair of basiventrals (the switch here between "basi"- and "inter"- prefixes is, regrettably, a stumbling block to the student). Posteriorly there is a pair of wedge-shaped elements, termed *pleurocentra;* these curve downward on each side around the notochord from the upper rim of the central region; the pleurocentra are the structures from which, by expansion, the true centra of amniotes have evolved. They are derived from the embryonic interarcualia, but whether from dorsal or ventral components, or both, is not clear. In any event, we have but three sets of elements, rather than the primitive four, to deal with in land vertebrates.

The modern amphibian groups (Fig. 85, *G, H*) have departed widely from this primitive type. In no living amphibian order do the ventral arcualia take any major part in the formation of the centrum. These cartilages may be present as vestigial nubbins at the ventral edge of the centrum in some anurans; they are responsible for hemal arches in the urodele tail and for a ring of cartilage which forms between successive centra in urodeles and to some extent in the Apoda. But in all three existing orders the centrum forms directly by perichordal ossification, without being preformed in cartilage. In the frogs the ossification spreads downward from the neural arch. In the urodeles and Apoda the major part of the centrum

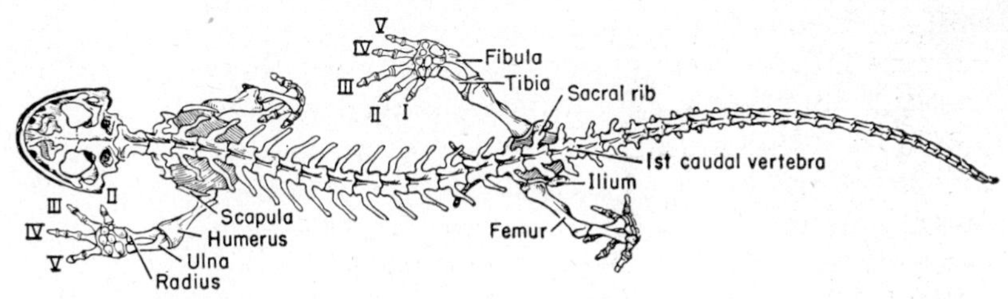

Fig. 86. The skeleton of a salamander, as seen from above. (From Schaeffer.)

ossifies perichordally in absolute independence of either neural arch or ventral cartilages.

Among the Labyrinthodontia the presence of (1) the rhachitomous type of vertebra, in which there are two incomplete rings in the region of the centrum, has already been described. There are a number of derivatives of this structure. (2) In the *stereospondylous* type (Fig. 85, *D*) the pleurocentra dwindle and disappear, while the intercentra form complete discs. (3) In *embolomerous* vertebrae (Fig. 85, *E*) both intercentra and pleurocentra form complete checker-like discs, each pierced for the persistent notochord (some actinopterygian fishes show a parallel development). (4) In a fourth type (Fig. 85, *F*) the intercentra remain as ventral crescents; the pleurocentra, on the other hand, fuse and expand to form stout centra.

From this last type the vertebral structure of all amniotes appears to have been derived. The vertebrae of many of the oldest reptiles were of just this sort, and can scarcely be distinguished from those of related amphibians. In later amniote history the centrum has generally elongated to become a spool rather than a short disc, and the intercentra are reduced and frequently lost. The embryological development of amniote vertebrae is often complicated, but the general phylogenetic picture seems clear; basidermal and interarcualia appear to be primarily responsible for neural arch and centrum, respectively.

Regional Variations in Vertebrae. It is obvious that there may be considerable variation along the length of the vertebral column. In higher vertebrates vertebral regions may be defined according to the variations in the nature and presence or absence of ribs. In fishes regional distinctions are difficult to make. One can, however, distinguish between trunk and tail vertebrae through the presence of hemal arches in the caudal region. In land vertebrates one or more vertebrae carry sacral ribs for

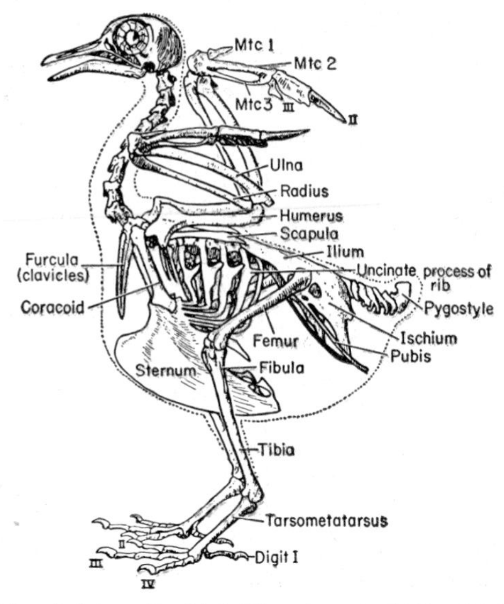

Fig. 87. The skeleton of a bird (the pigeon). (After Heilmann.)

attachment to the pelvic girdle. There is thus distinguished (unless the girdle is secondarily lost) a *sacral* region of the column, behind which is the *caudal* region. In primitive land forms ribs may be present on every segment from the neck to the base of the tail. However, the longest ribs are generally found in the "chest" region, the thorax; the vertebrae carrying such ribs are distinguished as *thoracic,* in contrast to those of the neck, the *cervical* region, and to a *lumbar* region between thoracic and sacral groups. In birds and mammals free ribs disappear from cervical and lumbar vertebrae, making for clearer distinction between regions.

In primitive land vertebrates there is a variable vertebral count, but thirty or so presacrals, of which about seven were cervicals, appears to have been a fairly common number in early tetrapods. In amphibians there is normally but a single sacral. Primitively the tail was elongated, with perhaps half a hundred or more caudals. There is great variation in modern amphibian vertebral numbers; the elongated Apoda may have 200 or so vertebrae; typical frogs, on the other hand, have reduced their backbone to nine vertebrae plus a rodlike *urostyle* representing a number of fused caudals.

The primitive reptiles appear to have had about twenty-seven presacrals and a long tail; two or more sacrals are generally present. There is great variation within the reptiles in vertebral counts.

A curious condition is seen in the tail of Sphenodon and some lizards, in which there is a "breaking point" in the middle of each caudal at which the tail may be shed. The appearance is like that of some fishes with double caudal centra or the embolomerous amphibians, but the structure here is probably a specialized, not a primitive, one.

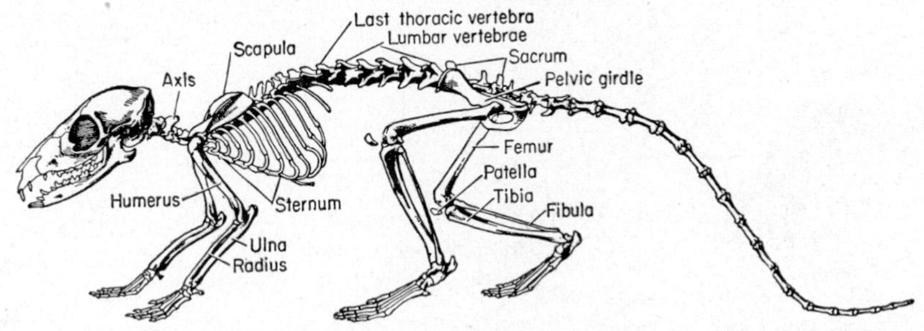

Fig. 88. The skeleton of a generalized mammal, the tree-shrew, Tupaia. (Cf. Figs. 84, 98, 99.) (After Gregory.)

In birds the distinct cervical region has a highly variable number of vertebrae. Most of the thoracic vertebrae tend to be fused together, as an aid to the effective bracing of the wings in flight. The posterior thoracic, lumbar, and proximal caudal vertebrae are generally fused with the original sacrals into an elongated *synsacrum* supporting the pelvic girdle; there is almost no motility in the avian column beyond the neck region. The bony tail is much shortened, and is capped by a *pygostyle* formed of fused vertebrae.

Ancestral mammals appear to have had, much like early reptiles, about twenty-seven presacral vertebrae, two or three sacrals, and a tail which, however, tended to become relatively short and weak. With remarkable consistency the number of cervicals, which lack free ribs, remains at seven. Only in a few tree sloths, with six or nine cervicals, is this rule violated, and a neckless whale and a giraffe both have seven cervicals, appropriately abbreviated or elongated. The remaining presacrals can be considered a *dorsal* series, divided into *thoracic* and *lumbar* subgroups, the former rib-bearing, the latter ribless. The number of dorsals is usually in the twenties and tends to remain rather constant within the limits of many mammalian families or orders. However, the number of ribs is variable; hence the proportion of thoracics to lumbars may be variable. Thus, in the cow-antelope family of ungulates, the number of dorsal vertebrae is almost always nineteen, but thoracics may vary from twelve up to fourteen, and the lumbars, correspondingly, from seven down to five.

Atlas-Axis Complex (Fig. 89). In fishes, head and trunk move as a unit. In land vertebrates independent movement of the head is important, and modifications of the most anterior vertebrae and their skull articulations facilitate this. The articular surface at the back of the skull, the condyle, is in fishes a single rounded structure, comparable to the end of a centrum, and the oldest amphibians still retained this type.

In most amphibians, living or fossil, the single condyle has divided into a pair, one at either side, and the articular surface of the first vertebra has become correspondingly divided. This allows the head to swing up and

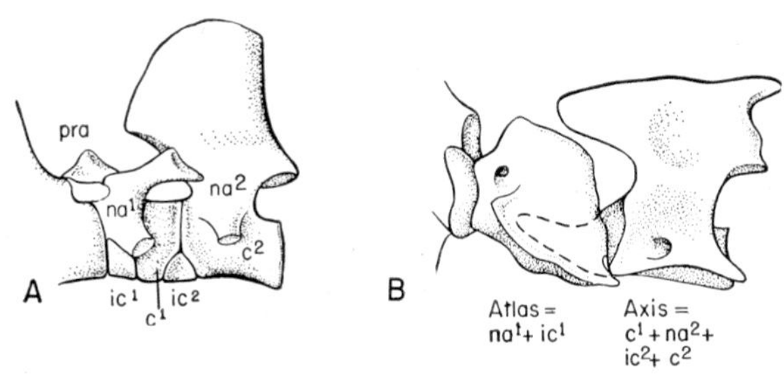

Fig. 89. The atlas-axis complex. *A*, The occipital condyle and first two vertebrae in a primitive reptile (Ophiacodon). *B*, The same region in a typical mammal, showing fusion of elements. The proatlas in *A* is the neural arch of a "lost" vertebra of which the centrum has fused with the occiput. The broken line in *B* shows the position of the odontoid process (c^1), which runs forward inside the ring of the atlas. c^1, c^2, centrum of first and second vertebrae; ic^1, ic^2, intercentra; na^1, na^2, neural arches; *pra*, proatlas.

down in hinge fashion, but gives little scope for lateral movement. In most reptiles and in birds the condyle remains single, but the first two vertebrae, termed the *atlas* and *axis,* have been modified to permit considerable freedom of movement. In the atlas the intercentrum and neural arch tend to form a ringlike structure (in many reptiles there may be, incidentally, a small, extra neural arch in advance of that of the atlas—a *proatlas*). The centrum of the first vertebra tends to remain independent of the atlas proper and to associate itself with the second vertebra, the axis, which typically has a large neural spine for the attachment of head ligaments. In mammals, as in amphibians, the condylar articulations with the skull are double. The axis bears anteriorly a process projecting from its centrum into the ventral part of the atlas ring; the embryological story shows that this *odontoid process* includes the centrum of the atlas segment as well as the intercentrum of the axis itself. The head may move upward and downward on the atlas; the major component in sideways movement of the head is movement of atlas on axis.

Ribs in Fishes. Although the vertebral column is the main skeletal structure upon which the axial muscles play, little musculature is directly attached to the vertebrae. For the most part muscular force is exerted, in fishes, upon the connective tissue septa—*myocommata*—between successive muscle segments; ribs, formed at strategic points in these septa, connect with the vertebrae and render the muscular effort more effective.

In most fishes each muscle segment in both embryo and adult is divided into dorsal and ventral parts by a longitudinal septum running the length of the flank (Figs. 2, p. 6, and 90). A logical place for rib formation is at the intersection of this septum with the successive myocommata; in

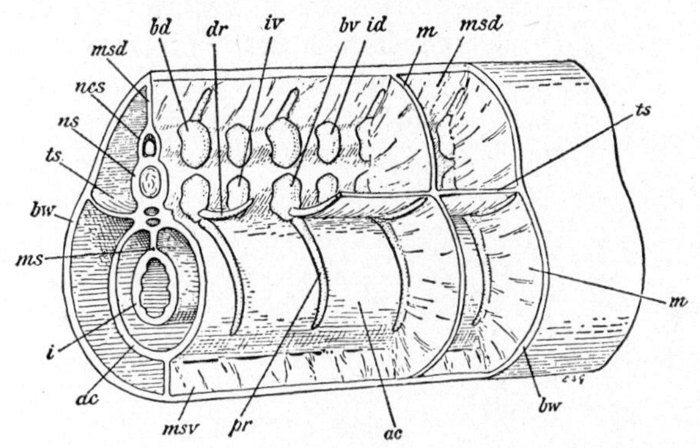

Fig. 90. Diagram of a section of the trunk of a vertebrate to show the connective tissue system and the axial skeletal elements. A view from the left side, as if partially dissected out and the muscles removed from between the septa. Vertebral elements develop in the connective tissue sheath surrounding the spinal cord and notochord; ribs, dorsal or ventral, develop where the transverse septum intersects the horizontal septum or celomic wall. *ac,* Outer wall of celomic cavity; *bd,* basidorsal; *bv,* basiventral; *bw,* body wall, cut; *dr,* dorsal rib; *i,* intestine; *id,* interdorsal; *iv,* interventral; *m,* transverse septum (myocomma); *ms,* dorsal mesentery; *msd,* median dorsal septum; *msv,* median ventral septum; *nes,* neural tube; *ns,* notochordal sheath; *pr,* ventral or pleural rib; *ts,* horizontal septum. (From Goodrich.)

many fishes *dorsal ribs* are present at these points. A second position in which ribs may develop is that where the myocommata reach the walls of the celomic cavity ventrally and internally. Ribs in this position are termed pleural or *ventral ribs.* In addition to these two types of characteristic ribs, riblike intermuscular bones may be found in teleosts (the shad is all too good an example) at other points in the myocommata, making for greater muscular efficiency. The ribs, like the connective tissues in which they are found, are derived from mesenchyme and usually are preformed in cartilage even if later ossified. Proximally, close to the vertebrae, rib mesenchyme may be derived from the sclerotomes, more distally from the lateral plate of the mesoderm. In fishes both dorsal and ventral ribs attach to some point on the centrum of the vertebra.

The hemal arches of the tail are essentially homologous with the ventral ribs of the trunk region, although the hemals are fused with the basiventrals (or intercentra) from which they both originate, while the ventral ribs, although arising from the basiventrals, are discrete elements. In some fishes a transition is seen between the two structures; passing backward along the trunk, the ventral ribs of the two sides come to extend more directly ventral and approach one another more closely as the posterior end of the celom is reached; where the celom ends, the two processes meet ventrally and become, in the caudal vertebrae, the hemal arches.

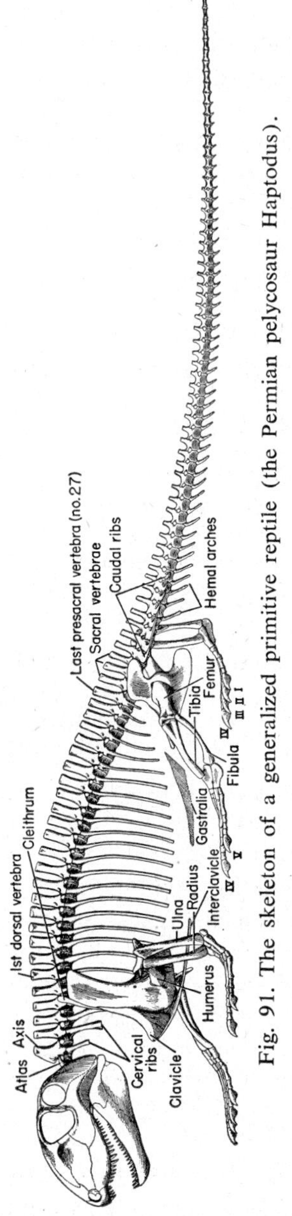

Fig. 91. The skeleton of a generalized primitive reptile (the Permian pelycosaur Haptodus).

There are no ribs in cyclostomes. In sharks, short ribs may be present; they are directed rather ventrally, but lie in the longitudinal muscular septum and hence appear to be, by definition, dorsal ribs. In many bony fishes only ventral ribs are present, but in Polypterus, on the one hand, and in numerous teleosts on the other, both dorsal and ventral ribs are to be found.

Tetrapod Ribs. In tetrapods but one type of rib is present in the much reduced muscular wall of the body. Although doubts have been expressed, it is fairly generally agreed that these ribs correspond to the dorsal ribs of fishes. As in fishes, ribs were primitively borne by every vertebra from the atlas to the base of the tail (cf. Fig. 91). The more anterior—cervical—ribs are short, in relation to the development of a flexible neck region. The thoracic region is characterized by stout and elongated rib structures which are, through muscular connections, concerned in supporting the trunk on the pectoral limb girdles. These thoracic ribs are in most groups of tetrapods bound to a median ventral structure, the sternum. In primitive tetrapods the vertebrae of the lumbar region may bear ribs, but these are relatively short. One or more highly developed sacral ribs, not developed in fish, connect the backbone with the pelvic girdle. Beyond the pelvis, ribs, diminishing in size posteriorly, may be present in the longitudinal muscle septum of the tail.

In ancestral tetrapods the ribs were characteristically double-headed structures on the anterior vertebrae at least, although the two heads may be connected by a thin web of bone (Fig. 81, p. 150). The *capitulum*—the head proper—was attached, in early forms, to the intercentrum. With reduction of the intercentrum in advanced tetrapods this attachment tends to shift to the anterior edge of the adjacent centrum or to the intercentral space. The second head, the *tuberculum,* is essentially a short process from the curved proximal part of the rib, and primitively attached to the transverse process of the neural arch. In the posterior part of the column the two heads tend to come close to one another, and the capitular attachment may shift upward and backward onto the centrum; the heads, however, usually remain separated by a groove through which passes a small blood vessel. In some amphibians and reptiles the ribs near the scapula, from which strong muscles (serratus ventralis) pass to that element, are broadly expanded. The sacral rib or ribs are powerfully built, with a broad head, a short shaft, and a distal expansion applied to the inner surface of the dorsal element of the pelvic girdle. The sacral ribs are immovably attached to the vertebrae, and the lumbar and caudal ribs, where present, tend likewise to be incapable of movement.

In modern amphibians the ribs, like the vertebrae, have departed more radically than those of early reptiles from the primitive tetrapod type. They are always much reduced and never reach the sternum. In most frogs and toads they are entirely absent except for a sacral. In urodeles they are present but short, sometimes forked distally, and proximally attach by two heads to a transverse process of a peculiar type, not comparable to that in any other group.

Within the reptiles there is much variation in rib structure; particularly notable is the variation in vertebral attachment. The thoracic ribs are typically formed in two segments, a proximal, ossified, rib proper, and a distal *sternal rib* (Fig. 92) which almost always remains cartilaginous. A joint between the two segments allows for flexibility necessary in the expansion and contraction of the chest and consequent filling and emptying of the lungs. In turtles the ribs are reduced in number; eight of them are firmly fused to the overlying costal plates of the carapace. Reptilian ribs may remain two-headed, but with both heads sometimes attaching to the centrum or to the neural arch; or there may be but a single head, with a similar choice of origins. In the Squamata the single-headed ribs articulate with the centrum. In the crocodiles the ribs are two-headed; in the cervical region the capitulum arises from the centrum, but in the trunk both heads attach to the transverse process.

In birds the cervical ribs are fused to the vertebrae; free ribs are confined to the compact thoracic region. These carry, as do those of some living reptiles and early fossil tetrapods, expansions in the form of separate *uncinate processes* for the attachment of muscles supporting the shoulder blade. In monotremes short, immovable ribs are visible on the cervical vertebrae. In higher mammals they appear to be absent, but the embryological story shows that the so-called transverse processes of the cervical vertebrae actually include short, fused, two-headed ribs. Ribs are present on all the thoracic vertebrae in mammals and are, indeed, diagnostic of that region; most curve forward ventrally to reach the sternum, but the shorter posterior members of the series are bound to that struc-

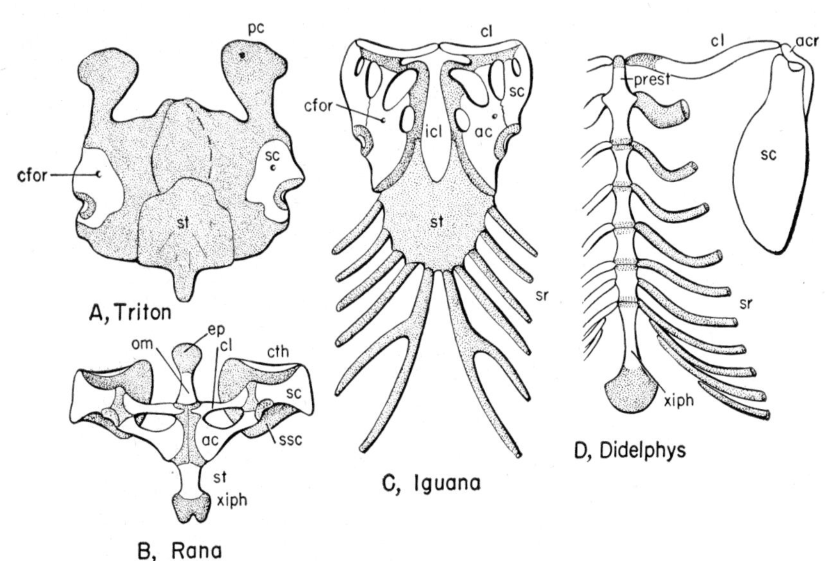

Fig. 92. Ventral views of the shoulder girdle and sternal apparatus in various tetrapods. For lateral views, cf. Figures 101 and 102 (pp. 177, 180). *A*, Salamander; *B*, a frog; *C*, a lizard; *D*, a mammal (opossum). Anterior end at the top of the figures. In *A* and *C* the dorsally turned scapulae are invisible. In *A* the two coracoid cartilages overlap, as indicated by the broken line. *ac*, Anterior coracoid element; *acr*, acromion; *cfor*, coracoid foramen; *cl*, clavicle; *cth*, cleithrum; *ep*, episternum; *icl*, interclavicle; *om*, omosternum; *pc*, precoracoid region of coracoid plate; *prest*, presternum; *sc*, scapula; *sr*, sternal ribs; *ssc*, suprascapula; *st*, sternum; xiph, xiphisternum. Cartilage stippled. (*A*, *C*, and *D* after Parker.)

ture only via ligamentous connections with their longer anterior neighbors. The lumbar vertebrae have long transverse processes, but no ribs, and caudal ribs are never developed in mammals.

Sternum (Figs. 92, 102, p. 180). A generalized condition of the sternum is that seen in many reptiles. It is a ventrally placed, shield-shaped cartilage which articulates anteriorly with the shoulder girdle, and posterolaterally connects on either side with the ventral ends of the thoracic ribs to form a complete enclosure of the chest region, the thorax. Although there is (owing to the usually cartilaginous nature of the structure) little fossil evidence, the sternum was probably developed in early tetrapods; no such structure is found in fishes. Among reptiles the sternum has been lost in snakes and certain snakelike lizards and in the turtles.

In modern amphibians, with reduced ribs, the sternum is atypical. In urodeles there may be developed a cartilaginous plate posterior to the shoulder girdle in the ventral midline; in frogs there is found a plate or a rod posterior to the girdle ventrally, and a second element anterior to the girdle.

In birds (Fig. 102, *F,* p. 180) the sternum is an enormous ossified structure, its size being due to the attachment to it of most of the mass of the chest muscles, important in flight; except in the ostrich-like birds, in which flight has been lost, the bone bears a huge keel for additional muscle attachment.

In typical mammals the sternum, usually ossified, has a different type of structure from that of either reptiles or birds. It has the form of an elongated, jointed rod with which the ribs are articulated at the "nodes."

Although the sternum is associated topographically with the shoulder girdle, it is generally considered part of the axial skeleton because of its close association with the rib system. Its embryological development, however, shows that in both reptiles and mammals it arises independently of the ribs.

Braincase. The braincase, of cartilage or bone, forms the anterior end of the axial skeletal system, here greatly modified in relation to the brain and specialized sense organs of the head region. In most vertebrates the braincase is fused with dermal and visceral arch skeletal materials to form a definitive skull structure of compound nature. In cyclostomes and the sharks, however, the braincase, owing to the absence of dermal bones—presumably a secondary condition—is a discrete skeletal element. The shark braincase shows a structure which, with modification, is repeated in all the lower jawed vertebrates, and in further modified form is represented in the skull of even the highest of vertebrate groups. In sharks it retains its embryonic cartilaginous nature; in most other forms it is more or less replaced by bony elements.

As seen in sharks, the braincase (Fig. 93) is a troughlike structure which forms a floor and side walls for the brain and a partial roof. The roof, however, is never complete; one or more dorsal gaps, *fontanelles,* may be present. This lack of a complete roof is repeated in most other vertebrates; in forms with an ossified skeleton, as presumably in the shark's ancestors, a dermal roof was present overhead, and a complete braincase roof was unnecessary as a protective measure.

Four major braincase regions may be distinguished, from back to front: occipital, otic, orbital, and ethmoidal. The most posterior part of the braincase, the *occipital region,* is of relatively narrow width. Posteriorly there is an opening, the *foramen magnum,* through which passes the spinal cord. Below this is a rounded *occipital condyle* which abuts against the first vertebra; the condyle is typically concave. The notochord in the embryo, and sometimes in the adult, may extend forward in the floor of the braincase from the condyle to a point near the pituitary body pendant from the brain. Anterior to the occipital region the braincase expands laterally in the otic region; there is here incorporated on either side an *otic capsule* enclosing the sacs and canals of the internal ear. Still

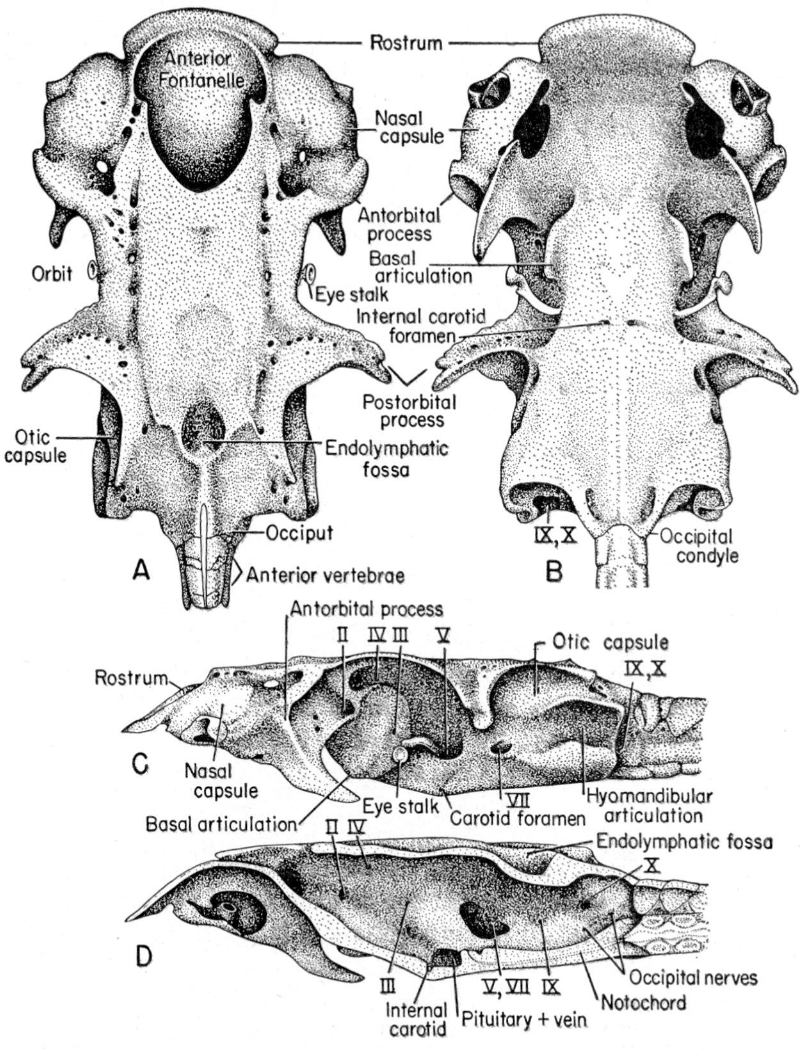

Fig. 93. The braincase of the shark Chlamydoselache; *A*, dorsal, *B*, ventral and *C*, lateral views, and *D*, sagittal section. Nerve exits in Roman numerals. (After Allis.)

farther forward the braincase narrows in the *orbital region* to aid in forming a socket, the orbit, for the eyeball and its musculature. Here a median depression in the braincase floor may mark the position of the pituitary. Still farther forward the braincase expands in an *ethmoid* region to terminate in a rostrum; on either side, in front of a preorbital ridge (antorbital process), are the *nasal capsules* which enclose the olfactory organs.

Numerous openings—foramina—are present in the braincase for the cranial nerves and blood vessels. Into the orbit emerges the optic nerve (II) of the eye, and in variable fashion, three smaller nerves (III, IV, and VI) for the eye muscles. Near the posterior end of the orbit there is typically a large foramen or gap for the trigeminal nerve (V), and the facial nerve (VII) usually leaves the braincase close by. Behind the otic capsule there are openings for nerves associated with the gill region, the

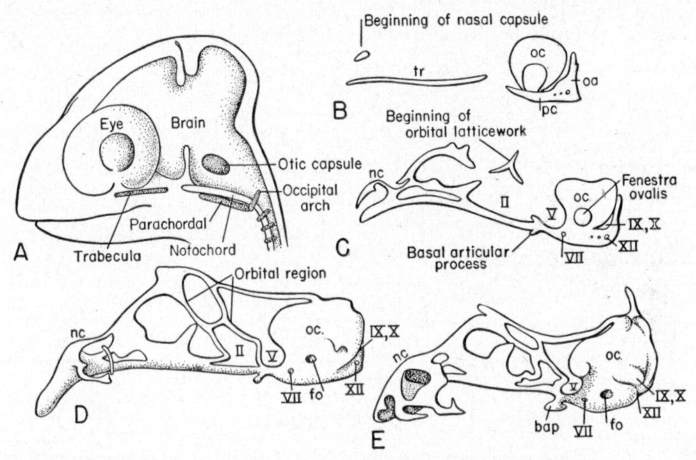

Fig. 94. Stages in the embryonic development of the braincase of a lizard. In *A* the outlines of the head, brain, and notochord are given for orientation. The main elements of the braincase structure are appearing—trabeculae, parachordals, otic capsule, and occipital arches (the nasal capsule appears later); further development consists in great measure of the growth and fusion of these elements. In the lizard the orbital region grows as a complicated latticework, rather than a plate. Nerve positions are indicated by Roman numerals. *bap,* Basal articular process with upper jaw cartilage; *fo,* fenestra ovalis; *nc,* nasal capsule; *oa,* occipital arch; *oc,* otic capsule; *pc,* parachordal; *tr,* trabecula. (After DeBeer.)

glossopharyngeal (IX) and vagus (X), and, posterior to the last, for small occipital nerves. In sharks the endolymphatic ducts from the internal ear (cf. Chap. XV) open to the surface, typically into a median depression, or fossa, in the braincase roof. Ventrally, in the pituitary region, an opening is present for the carotid arteries, which supply blood to the brain, and other arterial and venous openings are present laterally.

A specialized gill bar element, the hyomandibular, articulates rather loosely with the outer surface of the otic capsule, and the cartilages which form the shark upper jaws also articulate with the braincase (Fig. 128, p. 213). The common connection in sharks is one in which the upper jaws are loosely joined anteriorly with the under surface of the braincase; in primitive sharks there is an additional articulation of the upper jaws with the skull posterior to the orbit. In higher fishes and early land vertebrates this last connection tends to be reduced or absent, and there is instead a strong *basal articulation* of the upper jaw structures with the base of the braincase in the orbital region (Figs. 94, *E*, 130, p. 218; 132, p. 225; 144, p. 241).

In the embryo vertebrate (Fig. 94) the brain, and the notochord extending forward beneath it to the pituitary region, are already far advanced in development before skeletal structures appear. Braincase development begins ventrally, then continues laterally; because of the large size of the brain, it is difficult, so to speak, for the braincase to surround it, and the roof, as we have noted, often remains incomplete. The two basic embryological braincase structures are the parachordals and trabeculae. The *parachordals* are a pair of cartilages which develop beneath the posterior part of the brain on either side of the notochord; farther forward there is in lower vertebrates a similar pair, the *trabeculae* (in mammals, however, a single median structure represents the trabeculae). Presently these paired elements fuse with their mates and with each other to form a nearly solid braincase floor. There may, however, be a persistent median opening for the hypophyseal sac, which grows upward from the roof of the mouth to form part of the pituitary body.

Alongside the posterior part of the brain, the sacs and canals of the internal ear develop at an early stage. A shell of cartilage, the *otic capsule*, forms around the outer surface of these structures and tends to enclose them nearly completely. Farther forward, the eye, of course, is not enclosed, but there soon appears, in one fashion or another, a plate or latticework of cartilage internal to the eye socket, which often forms an *orbital plate*. The nasal region is relatively slow to develop, but there eventually appears a *nasal capsule* of complicated structure. The posterior boundary of the vertebrate skull is somewhat variable from group to group. The occipital region appears to be formed from one or several modified vertebrae. In agreement with this, we find that in the embryo there appears behind the otic capsule a variable number of *occipital arches* somewhat resembling the vertebrae behind them.

As development proceeds, the ventral elements of the braincase fuse with the lateral structures and with the occipital arches to form side walls as well as a floor to the brain (cf. Fig. 125, *D-F*, p. 210). Most of the major openings for nerves and vessels lie in spaces between the major components. Thus nerves V and VII have their exits in openings in front of the otic capsule; nerves IX and X emerge between otic capsule and occipital region. Although a roof is slow to form, there is always a union of the two otic capsules above the foramen magnum.

The description above of the adult braincase was mainly based on that of sharks. In the other lower vertebrates in which the braincase is a separate entity—cyclostomes and chimaeras—its structure is degenerate or highly modified and need not be considered in detail. In cyclostomes

(Figs. 84, p. 154; 125, *A*, p. 210) it consists of little more than a rather simple brain trough, a pair of otic capsules, and a capsule for the single nostril. The chimaera braincase (Fig. 95) is basically like that of the sharks, but short and high, and specialized in various features, of which the most noticeable is the fusion of the short upper jaw cartilages to the braincase. This peculiarity (repeated in certain other fishes and in land vertebrates) is here associated, apparently, with the stronger support necessary for the shell-crushing activity of these mollusc-eating fishes.

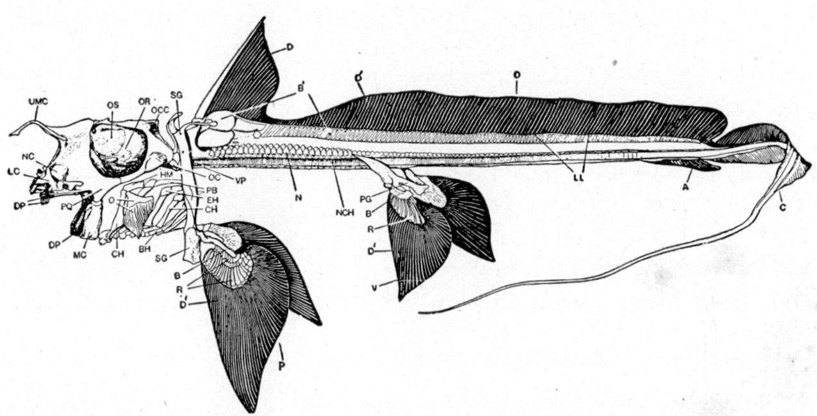

Fig. 95. The skeleton of a female chimaera. *A,* Anal fin; *B,* fin basals; *B′,* dorsal fin supports; *BH,* basibranchial; *C,* caudal fin; *CH,* ceratohyal, ceratobranchial; *D,* dorsal fin, with spine; *D′,* dermal fin rays; *DP,* dental plates; *EH,* epibranchial; *HM,* hyomandibular; *LC,* cartilages in nasal region of "lips;" *LL,* ligament connecting fin supports; *MC,* lower jaw (Meckel's cartilage); *N,* cartilages of neural arch regions; *NC,* nasal capsule; *NCH,* notochord, with ring calcifications of sheath; *O,* cartilages in operculum; *OC,* occipital condyle; *OCC,* crest on occiput; *OR,* orbit; *OS,* interorbital septum; *P,* pectoral fin; *PB,* pharyngobranchial; *PG,* pelvic girdle; *PQ,* palatoquadrate, fused with skull; *R,* radial fin supports; *SG,* shoulder girdle; *UMC,* upper median cartilage of snout (this is not the clasper of the male); *V,* pelvic fin; *VP,* plate formed of fused anterior vertebrae. (From Dean.)

The braincase of more highly developed fishes and tetrapods, to be described as part of the skull structure, is usually well ossified in the adult, but always develops in cartilage in a fashion similar to that of sharks. In sharks the braincase floor is flat, and this *platybasic* condition is repeated in the degenerate modern amphibians; in most other vertebrates, however, the braincase tends to be relatively high and more constricted ventrally in the orbital and ethmoidal regions—a *tropibasic* type.

Median Fins. The primitive vertebrate was a water dweller. As the builder of ships is aware, the most efficient shape, that offering the least resistance to the water, is a fusiform one, in which the surfaces are as smoothly contoured as possible—streamlined—and in which, following a pointed anterior tip, the greatest width is not far from the anterior end of the body. Most fishes approximate this shape, although there are conspicuous variations. In general, the posterior part of the body tends to be flattened from side to side. This flattening is associated with the nature of the propulsive force; forward movement is due to lateral muscular movements of the body, which effect a backward pressure of trunk and tail upon the surrounding water (Fig. 96, *A*). Tension of the muscles, first on one side of the body and then on the other, produces lateral curva-

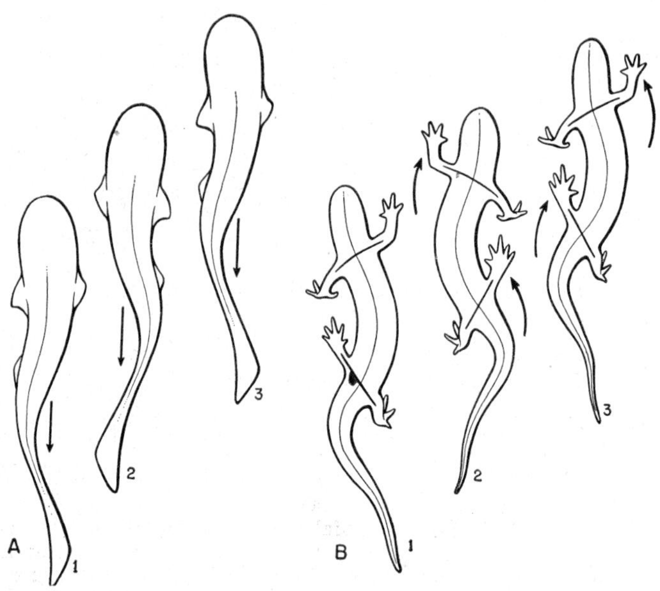

Fig. 96. *A,* Dorsal views of a fish swimming, to show the essential method of progression by the backward thrust of the body on the water, resulting from successive waves of curvature traveling backward along trunk and tail. The curve giving the thrust (indicated by arrow) in *1* has passed down the tail in *2* and is replaced by a succeeding wave of curvature; the thrust of this wave carries the fish forward to the position seen in *3,* and so on. *B,* Dorsal views of locomotion in a salamander. Although limbs are present, much of the forward progress is still accomplished by throwing the body into successive waves of curvature. In position *1,* the right front and left hind feet are kept on the ground, the opposite feet raised; a swing of the body in *2* carries the free feet forward, as indicated by arrows. If these feet are now planted and the other two raised, a following reversed swing of the body will carry them forward another step, as seen in *3.*

ture; and such curves, successively produced, travel back along the trunk and tail, pushing the body forward as a result of their backward thrust. The "snap" of this whiplike motion is accentuated at the posterior tip of the body, and an expanded caudal fin is usually present and greatly developed to gain the maximum effect from the push at this point.

These propulsive movements would tend to be unregulated (as are those of a tadpole) in the absence of fins as aids in stabilization and steering. Rotary, rolling motions are checked in part through the broad plane of the caudal fin, and further prevented by the development of other median fins: dorsal fins—usually one or two of them—above, and an anal fin below, posterior to the anus.

Some believe that originally all the median fins formed a continuous fold down the back, around the tail, and forward to the anal region. Some fish larvae exhibit a continuous median fin, and dorsals and anals are often more broadly based in the embryo than in the adult. These facts tend to support such a theory; but the oldest known fossil vertebrates, with a few exceptions, show just as discrete median fin structures as any modern fish.

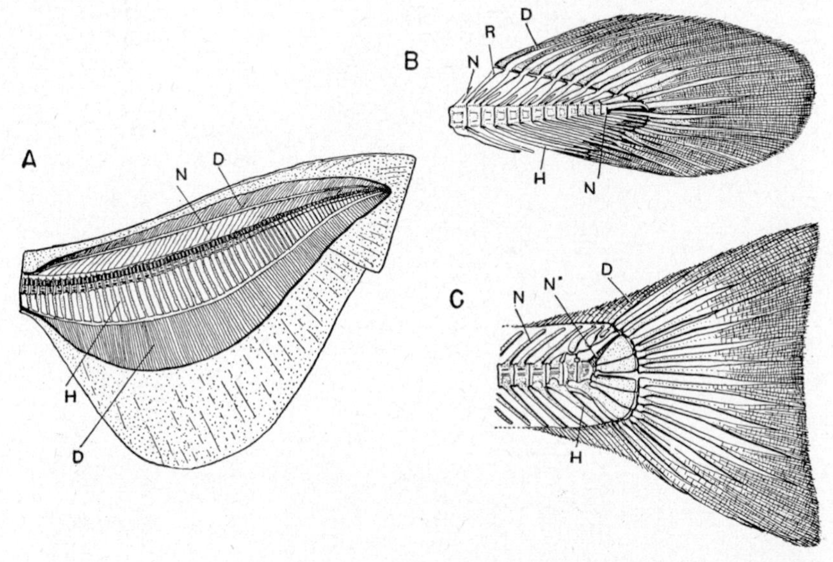

Fig. 97. Caudal fins. *A*, Heterocercal type seen in sharks, sturgeons, paddlefish; *B*, diphycercal type, as seen in Polypterus; *C*, homocercal type of teleosts; *D*, Dermal fin rays; *H*, hemal spines; *N*, neural arches; *N'*, tip of notochord; *R*, fin radials. In *C*, enlarged elements beyond *H* are hypural bones. (From Dean.)

The skeleton of median fins is formed in the median dorsal septum in which the neural spines are formed, and in the tail also in the ventral septum in which the hemal arches occur. In the tail region neural and hemal spines may themselves contribute directly to the support of the caudal fin (Fig. 97). This, however, is never the case with the other median fins—the dorsals and anal (Figs. 95, p. 169; 98, 99). Here the fin supports are most commonly rods of cartilage or bone, the *radials,* which may articulate with neural or hemal elements of the column, but are often separated by a gap from these structures. The radials are sometimes a series of simple parallel bars, but are often broken up into proximal and distal rows. It is probable that the radials originally corresponded in number to the body segments, but frequently in teleosts double the number is present. Primitively, it appears, the radials extended well into the fin, as is the case today in sharks; in actinopterygians, however, they hardly extend at all into the free fin, which is supported mainly by fin

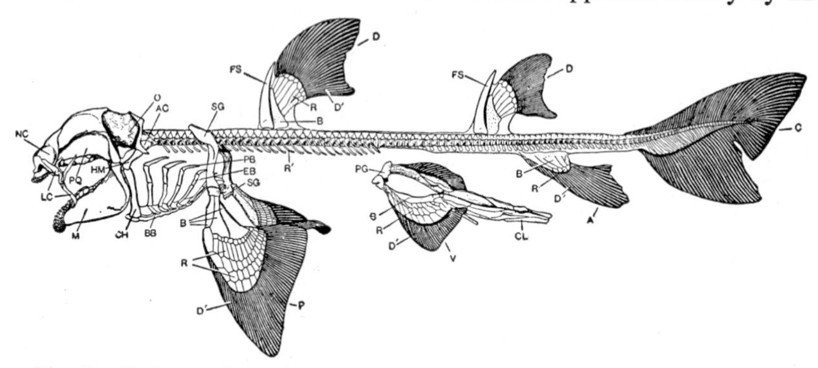

Fig. 98. Skeleton of a shark (Cestracion). *A*, Anal fin; *AC*, auditory capsule; *B*, basal elements of fin; *BB*, basibranchial; *C*, caudal fin; *CH*, ceratohyal; *CL*, claspers of male; *D*, dorsal fins; *D′*, dermal rays of fin; *EB*, epibranchial; *FS*, fins spines; *HM*, hyomandibular; *LC*, labial cartilages; *M*, mandible; *NC*, nasal capsule; *O*, orbit; *P*, pectoral fin; *PB*, pharyngobranchial; *PG*, pelvic girdle; *PQ*, palatoquadrate; *R*, radial elements of fin; *R′*, ribs; *SG*, pectoral girdle; *V*, pelvic (ventral) fin. (From Dean.)

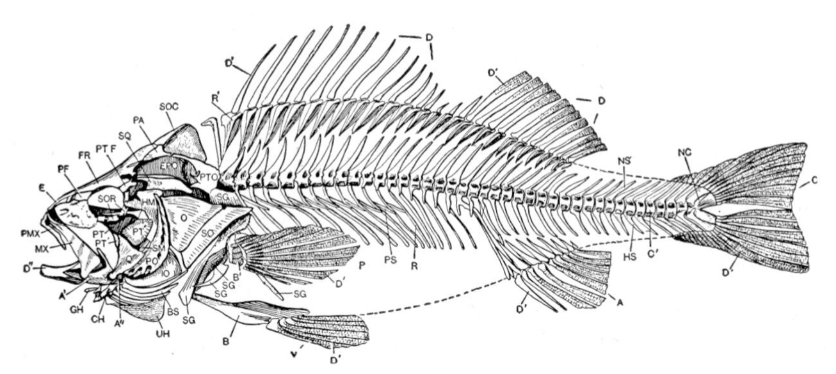

Fig. 99. Skeleton of a teleost (Perca). The interpretation of certain skull elements differs from that used in Figure 132, but is one commonly used in works on teleosts. *A*, Anal fin; *A′*, articular; *A″*, angular; *B*, pelvic girdle; *B′*, skeleton of pectoral fin: *BS*, branchiostegal rays; *C*, caudal fin; *C′*, centrum of vertebra; *CH*, ceratohyal; *D*, dorsal fins (anterior one supported by stout dermal spines); *D′*, dermal rays of fins; *D″*, dentary; *E*, ethmoid; *EPO*, epiotic; *FR*, frontal; *GH*, hypobranchial (glossohyal); *HM*, hyomandibular; *HS*, hemal spine; *IO*, interopercular (part of opercular series); *MX*, maxilla; *NC*, tip of notochord in tail; *NS*, neural spine; *O*, opercular; *P*, pectoral fin; *PA*, parietal; *PF*, prefrontal; *PMX*, premaxilla; *PO*, preopercular (part of opercular series); *PS*, supporting processes of ribs; *PT*, *PT′*, *PT″*, bones of palatal region; *PTF*, postfrontal; *PTO*, pterotic; *Q*, quadrate; *R*, ribs; *R′*, supports of dorsal fins; *SG*, various portions of shoulder girdle; *SM*, symplectic (binding hyomandibular to skull); *SO*, subopercular; *SOC*, supraoccipital; *SOR*, suborbitals; *SQ*, squamosal; *UH*, urohyal (ventral element of branchial skeleton); *V*, pelvic fin. (From Dean.)

rays (lepidotrichia) articulating with the ends of the bony radials. The radials generally extend outward as parallel structures, but there may be, as in many sharks, a concentration of elements at the fin base. With such concentration there may appear larger elements termed *basals* (Fig. 98), presumably representing a fusion of radials.

In a number of sharks and the chimaeras the median fins carry anteriorly spines which act as cutwaters, and spines were present on nearly all the fins in the fossil acanthodians (Fig. 17, *A*, p. 39). In some of the lowly placoderms and in ostracoderms we find rows of spines as stabilizers in the absence of fins. It is not impossible that spines were the basic structures from which dorsal and anal fins evolved; a web of skin may have developed behind projecting spines, with a later strengthening of the structure by skeletal elements and fin rays.

Dorsal fins are normally one or two in number. Ostracoderms, placoderms, and Chondrichthyes all show both conditions in various cases; among bony fishes the actinopterygians primitively have a single dorsal (Fig. 25, *A*, p. 47), early Choanichthyes two (Fig. 23, p. 45). Dorsal fins may be reduced, or fused with the caudal as in modern lungfishes; on the other hand, there may be a secondary increase in numbers, as in Polypterus (Fig. 25, *B*).

A single *anal fin* is almost always present ventrally behind the anus. It may be lost in such forms as the bottom-dwelling skates and rays.

Caudal Fin. In the caudal fin (Fig. 97) the major skeletal supports are the arches, neural and hemal, of the distal elements of the vertebral column. Additional radial elements are sometimes present, but these are seldom well developed; the fin proper consists of a stout web of skin supported by powerful finrays.

Three main types of caudal fins are to be found in fishes—heterocercal, diphycercal, and homocercal. The *heterocercal fin* type is that familiar in sharks (Figs. 18, *B*, p. 40; 98); the tip of the body turns upward distally, and the greater part of the fin membrane is developed below this axis. In the *diphycercal fin* the vertebral column extends straight back to the tip of the body, with the fin developed symmetrically above and below it; the living lungfishes and Polypterus are examples of fishes with fins of this type (Figs. 24, *B*, 25, *B*). The *homocercal* fin is characteristic of teleosts (Figs. 28, p. 50; 99). It is superficially symmetrical, but dissection shows that the backbone tilts strongly upward at the tip; the fin expanse is purely a ventral structure, despite its appearance, and is supported by enlarged hemal spines known as *hypurals*.

With seeming logic one tends to arrive at the conclusion that the diphycercal caudal fin is the primitive type, from which the others have been derived. This is not the case. In most instances diphycercal tails are shown by fossil history to have arisen from heterocercal ones, and the homocercal type is likewise of heterocercal origin.

The oldest of vertebrates, the ostracoderms, have tails of asymmetrical build, generally of a "reversed heterocercal" type (Fig. 15, *B*, p. 38), in which the backbone tilts downward rather than upward posteriorly. The modern cyclostomes have as adults a diphycercal type of tail, but the larval lamprey shows some development of the reversed heterocercal condition. In the oldest jawed vertebrates of the Devonian in which the tail structure is known, the heterocercal type is dominant (Figs. 17, p. 39; 18, *A*, p. 40; 23, *A*, 24, *A*, p. 45; 25, *A*, p. 47). The tails of all well-known placoderms had this structure; the heterocercal type is found in the oldest Chondrichthyes and in the oldest bony fishes of all types as well; it is clearly the ancestral form in all these fish groups.

In later Chondrichthyes the heterocercal tail remains dominant and is seen in typical form in modern sharks. However, skates and rays, with depressed bodies, and chimaeras (Figs. 18, *C*, 19, p. 41; 95) tend to reduce the tail to a rather whiplike structure, although the basal fin pattern is that of an attenuated heterocercal form.

Among the actinopterygians the typical Paleozoic forms, the palaeoniscoids (Fig. 25, *A*), had a good heterocercal tail fin. The sturgeons and paddlefishes have retained it (Fig. 26, p. 47), but Polypterus, although primitive in many other ways, has modified the tail to a diphycercal type (Fig. 25, *B*). In Mesozoic days the holosteans, which form the middle group of ray-finned fishes, exhibit a tail technically heterocercal, but an abbreviate modification; the distal extension of the column into the fin is much shortened. This condition is retained in the gar pikes and Amia, the modern holostean representatives (Fig. 27, p. 49). Their tails are superficially symmetrical and appear much like those of teleosts; but it is not until we reach this last group that further shortening of the tip of the column and development of large hypural bones gives us the true homocercal type with its deceptively simple and seemingly primitive appearance (Figs. 28, p. 50; 99).

In the Choanichthyes, the earliest lungfishes had typical heterocercal caudals (Fig. 24, *A*, p. 46). Soon, however, in geological history the tip of the tail tended to straighten and the dorsals and anal fins to lengthen out in the direction of the caudal. In modern genera (Fig. 24, *B*) the tail fin is diphycercal; it reaches far anteriorly both dorsally and ventrally, owing apparently to the incorporation in it of dorsal and anal fins. The most primitive of Devonian crossopterygians likewise had a heterocercal tail fin (Fig. 23, *A*, p. 45). In this group, however, there was a rapid trend toward straightening of the tip of the vertebral column, and most crossopterygians show a characteristic type of three-lobed, diphycercal caudal (Fig. 23, *B*).

In land vertebrates, even in amphibians, the fish median structures have been completely abandoned; a tadpole or salamander exhibits a tail expanded dorsoventrally as a good swimming organ, but we fail to find any evidence of internal structures characteristic of fish fins. Many amniotes have returned to an aquatic mode of life. Some, such as the turtles, the extinct plesiosaurs, and the seals, have never redeveloped the tail as a propulsive organ, and rely instead upon limbs for locomotion. The extinct ichthyosaurs among reptiles and the whales and sea cows among mammals have redeveloped median fins, but these are never the same as those of their fish ancestors, long since lost. Ichthyosaurs and cetaceans redeveloped dorsal fins, but skeletal supports are absent. The caudal fins of whales and sirenians are expanded transversely, not vertically, and lack true skeletal supports. The ichthyosaurs have come closest to the redevelopment of a fish type of caudal fin supported by the axial skeleton; the backbone extends into the sharklike fin, but extends into its lower, not into its upper, lobe.

Heterotopic Bones. Besides the normal cartilages and bones which are welded into the comprehensive pattern of the vertebrate skeleton, there may be mentioned here a variety of bones which are not part of the proper skeleton, but arise as accessories to bodily organs in one vertebrate group or another. Generally these elements take the place of fibrous connective tissues otherwise present in the same situation; they may form directly from such tissues or pass through an intermediate cartilaginous stage. Except for sclerotic plates to be described in connection with the eye, such heterotopic bones appear later than proper skeletal elements; they are skeletal "afterthoughts."

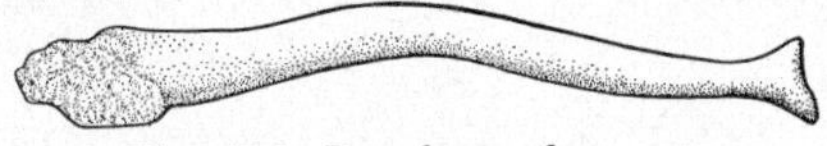

Fig. 100. Baculum of an otter.

A *baculum* (os priapi, os penis, Fig. 100) is a heterotopic bone, which is the skeleton of the penis, found in all insectivores, bats, rodents, and carnivores, and in all primates except man; its widespread distribution indicates that it must have developed early in the evolution of mammals. It forms in the fibrous septum between the corpora cavernosa and above the urethra. In shape it is the most varied of all bones; it may be a plate or a rod of varied form—straight, bent, or doubly curved; round, triangular, square, or flat in section; simple, pronged, spoon-shaped, or perforated; long or short.

Other heterotopic bones have a more limited distribution in special groups. We may mention, as examples, the *os palpebrae* of crocodilians, a plate embedded in the eyelid; the *os cordis* formed in the heart of deer and bovids; *rostral bones* found in the muzzles of some mammals; the *os falciforme,* a sickle-shaped element in the digging "hand" of moles, functioning as the skeleton of a supernumerary digit.

APPENDICULAR SKELETON

The girdle and limb elements constituting the appendicular skeleton definitely belong to the general somatic system of endoskeletal structures;

their history, however, has been distinctive. Normally two pairs of appendages are present, as fins in fishes and as the limbs of tetrapods. The anterior, *pectoral appendages* are situated just behind the gills in fishes; in land vertebrates in an equivalent position at the border of neck and chest regions. The posterior, *pelvic appendages* are placed at the back end of the trunk region, just anterior to anus or cloaca.

Origin of Paired Fins. Paired appendages were not present in the ancestral vertebrates; they were developed during the course of early fish evolution. They are absent in the living jawless vertebrates, and were rarely present in the ancient ostracoderms; the only known case is that of pectoral "paddles" in cephalaspids (Fig. 15, *C,* p. 38)—structures difficult to compare with normal vertebrate limbs.

The origin of paired fins has been much debated. Many decades ago a famous anatomist of the day suggested, ingeniously, that these fins are modified gill structures. The limb girdles were compared with gill bars; the fin was regarded as developed from the flap of skin which, in a shark, forms the outer margin of the gill structure; its skeleton was supposed to have been derived from a concentration and extension of the small cartilages which may stiffen the shark gill flap. A host of embryological and morphological facts show that this idea is purely fanciful, but a reminissence of this theory remains in the name archipterygium given to a leaf-shaped fin type (Figs. 107, *H,* p. 189; 108, *D,* p. 191) which was supposed, under this theory, to be primitive (but apparently is not).

In reaction against this theory arose a rival *fin fold theory* of the origin of paired fins. Its advocates pointed out that paired fins are similar to median fins in essential structure. In both there is a fold of skin extending outward from the body, a centrally placed set of skeletal supports, and, on each surface of the skeletal structures, a sheet of musculature which moves the fin. If the structure of median and paired fins is similar, may they not have had a similar origin? If, as we believe, median fins have evolved in simple fashion as stabilizing keels growing outward dorsally or ventrally from the trunk, may not the paired fins have developed from the flanks in a comparable way?

The functional aspect of paired appendages may be considered in relation to this problem. From the viewpoint of tetrapods we tend to think of paired limbs as active propulsive organs. This is, however, not generally a function of the paired fish fins. In many modern fishes they have relatively narrow bases and are capable of a free movement which is of aid in steering, but has little to do with propulsion. But in fins which seem to be of relatively ancient patterns there is generally a broad base which makes the fin relatively immobile. Primitively, it would seem, the paired fins were little more than horizontal stabilizing keels, which aided, as do the median fins, in the prevention of rolling movements, and had the further function of preventing fore and aft pitching. Both structure and function suggest that, like the median fins, the paired fins were developed as stabilizing aids in the aquatic locomotion of early vertebrates.

We have noted earlier the possibility that median fins may have developed in connection with spine structures; recent paleontological work

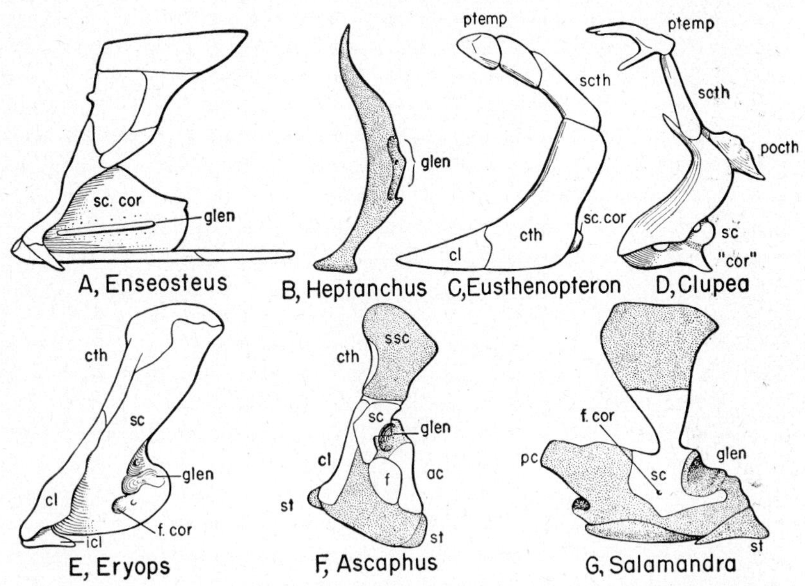

Fig. 101. The shoulder girdle in fishes and amphibians. *A,* Devonian placoderm; *B,* a shark; *C,* a Devonian crossopterygian; *D,* a teleost (herring); *E,* a Paleozoic primitive amphibian; *F,* a frog; *G,* a salamander. Cartilage stippled. In all except *B* and *G* a dermal girdle is present; in fishes *A, C, D,* this is the most prominent part of the girdle, including all parts except that labeled scapula and coracoid, and in the placoderm (*A*) the dermal girdle is the lateral part of an extensive thoracic armor. In amphibians the dermal girdle is reduced or absent. Except in the shark the endoskeletal girdle is relatively small in fishes and partially hidden beneath the dermal elements. In amphibians the endoskeletal girdle is expanded, but generally ossifies from a single center, comparable to the scapula of amniotes; the frog, however, has a coracoid element. Much of the endoskeletal girdle is cartilaginous in living amphibians. *ac,* Anterior coracoid element; *cl,* clavicle; *"cor",* coracoid of teleost (homology with that of land forms doubtful); *cth,* cleithrum; *f,* foramen in coracoid plate of frog; *f.cor,* coracoid foramen for nerve and blood-vessels; *glen,* glenoid cavity, point of fin attachment in fishes; *icl,* interclavicle; *pc,* precoracoid process of coracoid plate; *pocth,* postcleithrum; *ptemp,* posttemporal; *sc,* scapula; *sc. cor,* single scapulocoracoid ossification of fishes; *scth,* supracleithrum; *ssc,* suprascapula; *st,* sternum. Cartilage stippled. (*A* after Stensiö; *C* after Jarvik; *D* and *F* after Parker.)

suggests that the same may be true of the paired fins. Since the extinct placoderms are the oldest of gnathostomes, one would, a priori, expect in them paired fins of diagrammatically simple structure. This is the exact opposite of the actual findings. The fins of placoderms were "experimental models," most of which differed as much from the more orthodox paired fins of later date as did many early (and unsuccessful) flying machines from the modern plane. Notable is the frequent presence of spinous structures as the main element in placoderm fins. In some placoderms (Fig. 17, *C*, p. 39) the entire paired "fin" is a bony structure; in the acanthodians (Figs. 17, *A*, 107, *A*, p. 189) the fins are more normal in appearance, but are supported by a stout (and fixed) spine.

Typical living vertebrates have two pairs of appendages; but other phyla of animals may have other numbers of limbs, and there is no reason to think that there is anything sacrosanct about four as a limb count. In this regard, as in others, the placoderms experimented. Some had well-developed pectorals, but no pelvic fins. Some acanthodians went to the other extreme and show additional fin pairs (Fig. 17, *A*); in one "spiny shark" as many as seven pairs are present. Above the placoderm level, however, supernumerary appendages are never seen; it is vain (if amusing) to speculate as to the possible results had they survived in higher vertebrates.

Pectoral Girdle—Dermal Elements (Figs. 101, 102). Each vertebrate limb includes not only skeletal elements lying within the free portion of the appendage, but a basal supporting structure, the limb girdle. This lies within the substance of the trunk, acts as a stable base for the motions of the fin, and, in land vertebrates, forms an intermediary through which the weight of the body is transferred to the limb.

The pectoral girdle is, in all major groups except the Chondrichthyes, a duplex structure including both dermal and endoskeletal elements. The latter form the fin supports; the dermal bones, however, give added strength and help to tie the endoskeletal girdle to the body.

As we have noted, the anterior part of the fish body was early armored by solid plates of bone, in contrast to the flexible scale system found more posteriorly. In varied placoderms, particularly the arthrodires, this armor extended backward to enclose part of the trunk (Figs. 17, *B*, p. 39; 101, *A*). The side walls of the thoracic "chest" armor, behind the gill opening, can be considered a *dermal shoulder girdle,* although differing greatly from that in higher fishes; an endoskeletal girdle was present beneath and behind the dermal girdle. The absence of dermal elements in the Chondrichthyes is presumably correlated with the general reduction of the bony skeleton in that group.

In the more primitive bony fishes (Fig. 101, *C*) we find a dermal pectoral girdle with a pattern basic to that of all remaining vertebrate types. On either side is a vertical band of superficially placed bone extending dorsoventrally along the back edge of the gill opening and turning inward at its front edge to form part of the back wall of the gill chamber; the endoskeletal girdle, usually of modest dimensions, is bound to its inner and posterior surfaces. The main element in the fish dermal girdle is a vertically placed paired *cleithrum;* below it is the smaller *clavicle,* which curves downward and forward beneath the gill chamber and is expanded ventrally, where the two clavicles have a union—a symphysis—with one another. Above each cleithrum there are usually additional elements—typically a *supracleithrum* and *post-temporal,* and sometimes other bones as well—which curve upward and forward above the gill chamber and anchor the dermal girdle to the skull.

This general type of structure is found in a great variety of fishes—all known crossopterygians and lungfishes and all chondrosteans, living and extinct. In the holosteans and teleosts (Fig. 101, *D*), however, the clavicle is lost, leaving but a single large bone on either side—the cleithrum.

In early fossil amphibians (Fig. 101, *E*) the dermal girdle is retained, but with notable changes. Dorsally, the connection with the skull is lost, a condition which allows the head to move more freely on the trunk. For part of their length cleithrum and clavicle are rather slender dermal rods lying along the front margin of the much enlarged endoskeletal girdle. The upper end of the cleithrum may expand to cap the shoulder blade; ventrally, the clavicles typically expand in triangular shape; between them there is a diamond-shaped median *interclavicle*—a new element, although foreshadowed in some crossopterygians.

Among modern amphibians (Fig. 101, *F, G*) the dermal girdle has disappeared completely in the skeletally degenerate urodeles. In the frogs it is much modified. There is no interclavicle, but the cleithrum is present dorsally as a slender sliver of bone. The clavicle forms a bar connecting scapula and sternum. In some frogs the clavicle is developed in connection with a bar of cartilage. This is a highly unorthodox procedure, of course, for a dermal bone; it is probable that this cartilage is an intrusion from the underlying endoskeletal girdle.

In primitive fossil reptiles (Fig. 102, *A*) the dermal girdle was like that of early amphibians. The cleithrum, however, disappeared at an early reptilian stage and is not present in any living amniote. Clavicles and interclavicle are found little changed today in Sphenodon and lizards (Fig. 102, *B, C*), but have been subject to change and loss in other groups. In turtles (Fig. 80, p. 148) they are incorporated in the plastron. In crocodilians the clavicles have disappeared, but the interclavicle persists ventrally (Fig. 102, *E*). In birds the clavicles and interclavicle are present (Fig. 102, *F*) in fused form as the *furcula,* the wishbone.

As witnessed by the living monotremes, the primitive mammals retained the general reptilian pattern of interclavicle and paired clavicles (Fig. 102, *I*), but above the monotreme level the former has vanished (Fig. 102, *J*). The clavicles remain in many of the more generalized marsupials and placentals, articulating ventrally with the sternum. They are, however, frequently lost or reduced (as in the cat), particularly in running or bounding

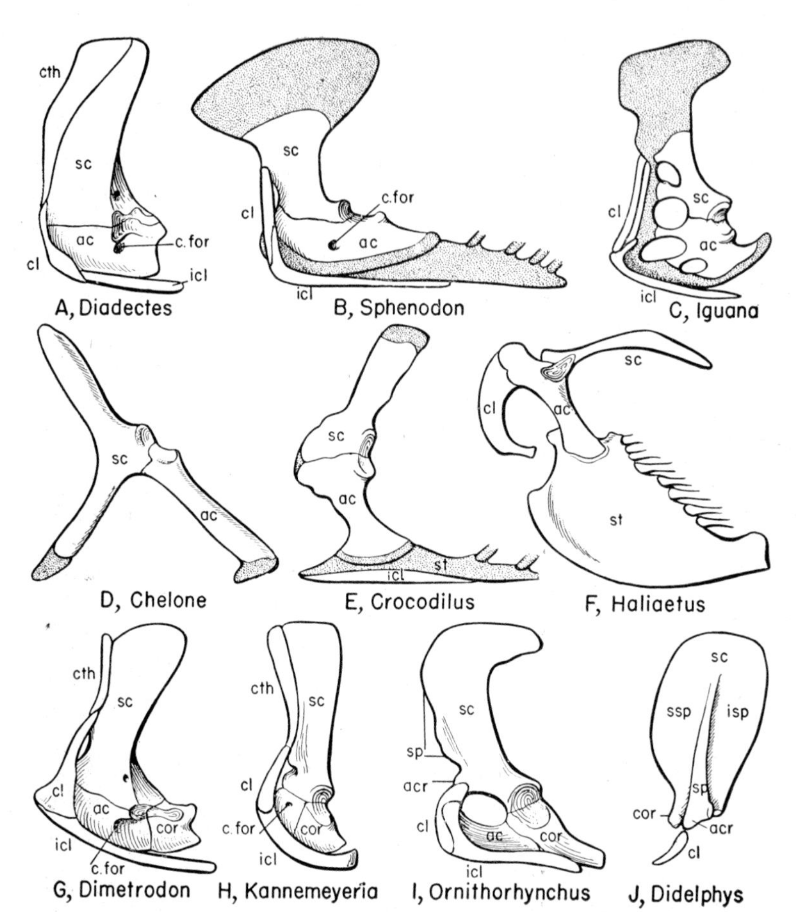

Fig. 102. The shoulder girdle in reptiles and mammals. *A*, "Stem reptile" (cotylosaur); *B*, Sphenodon; *C*, a lizard; *D*, a turtle; *E*, a crocodile; *F*, a bird; *G*, a pelycosaur (primitive mammal-like reptile); *H*, a therapsid; *I*, a monotreme (duckbill); *J*, opossum. In *B*, *E*, and *F* the sternum is shown. In most reptiles and birds only one (anterior) coracoid element is present; in mammal-like forms the true coracoid appears and persists, despite the loss of the coracoid plate area. The borders of the scapula and coracoid are often cartilaginous in reptiles. In lizards, scapula and coracoid are often fenestrated at points of muscular origins. The cleithrum has vanished in all living amniotes, but persisted long in mammal-like forms (*G*, *H*); its position is represented by the scapular spine, which lies at the front edge of the scapula in monotremes, but back of the new supraspinous fossa in higher mammals. *ac*, Anterior coracoid bone of reptiles, birds, monotremes; *acr*, acromion; *c for*, coracoid foramen; *cl*, clavicle; *cor*, true coracoid; *cth*, cleithrum; *icl*, interclavicle; *isp*, infraspinous fossa; *sc*, scapula; *sp*, spine of scapula; *ssp*, supraspinous fossa; *st*, sternum. (*C* and *I* partly after Parker; *H* after Pearson.)

types where complete freedom of the shoulder blade from the body skeleton is desirable to relieve the jars otherwise transmitted from front legs to body.

Endoskeletal Shoulder Girdle (Figs. 101, 102). Functionally, the endoskeletal girdle is much more important than the dermal girdle, for it always carries the limb articulation and is used as the major base of attachment of limb muscles. It is bound into the trunk by the attachment of axial muscles to its inner surface and, in tetrapods, by its connection ventrally with the sternum. This more deeply placed girdle component is cartilaginous in the Chondrichthyes and partly or entirely cartilaginous in some of the more degenerate bony fishes; otherwise it is an ossified structure, although cartilage may persist in peripheral areas in many fishes, amphibians, and reptiles.

Centrally situated on the endoskeletal girdle on either side is the limb articulation. In land animals with but a single proximal appendicular element this is an articular socket termed the *glenoid fossa;* in most fishes, however, a number of skeletal fin elements articulate with the girdle. The area above the limb articulation may be termed in general the *scapular blade;* that below it, the *coracoid plate.* (These regional names are derived from those of bony elements found in these areas in land vertebrates, but we shall use the terms here merely as descriptive of girdle regions, without reference to ossifications.) In most fishes the endoskeletal girdle is relatively small. The scapular region forms a buttress above the fin articulation and is firmly attached, anteriorly, to the inner surface of the dermal girdle. The coracoid region forms a plate below the glenoid area and may extend forward beneath the clavicle. The dorsal muscles which elevate the fin attach generally to the scapular blade, the opposed ventral muscles to the coracoid plate; both girdle areas are sometimes excavated in fishes for better accommodation of these muscles.

In the Chondrichthyes, as we have noted, the dermal girdle is not developed, and in compensation, apparently, the endoskeletal cartilages (Fig. 101, *B*) are enlarged, those of the two sides fusing ventrally to produce a U-shaped structure. In bony fishes the girdle is ossified in variable fashion. In teleosts as many as three bony elements may be present; it is useless, because of our lack of knowledge of connecting types, to attempt to identify these bones with specific elements present in land vertebrates.

In land vertebrates, in which the dermal girdle tends to undergo reduction, the endoskeletal girdle is, in contrast, much expanded. This expansion is obviously related to the larger size of the land limb, as compared to the fish fin, and the more massive nature of the limb musculature attaching to the girdle. Above the glenoid fossa there is in primitive tetrapods (Fig. 101, *E*) an elongate yet broad scapular blade; below, an expanded coracoid plate with a characteristic *coracoid foramen* for a nerve and blood vessels. In primitive amphibians the entire endochondral girdle may be ossified, but each half ossifies as a single element. The modern urodeles (Fig. 101, *G*) likewise have but a single ossification; much of the girdle remains unossified, and there is an elongate cartilaginous anterior process from the coracoid plate region. In some amphibians, living and fossil, and in all early reptiles there are two ossifications on either side. One, the *scapula,* centers in the scapular blade. The other, a ventral ossification, is sometimes called the coracoid, but, since it is not homologous with the mammalian ossification of that name, is better termed the *anterior coracoid.* The frogs (Fig. 101, *F*) have both scapula and anterior coracoid. In urodeles and primitive frogs the lower margins of the two coracoid plates may overlap ventrally. In some frogs a *firmisternal* condition develops (Fig. 92, *B,* p. 164) in which the margins of the two coracoid plates are butted together ventrally, bracing the animal against the jar of landing in its leaping mode of progression.

In many reptiles (Fig. 102, *A-C*) there persists a girdle structure highly comparable to that of primitive tetrapods, with a well-developed scapular blade and coracoidal plate, each with a single ossification; Sphenodon retains essentially this pattern, as do primitive lizards, except that the coracoid plate may be fenestrated at the areas of attachments of fleshy muscles. The chelonians (Fig. 102, *D*) show a triradiate girdle structure, with two ventral prongs. It is sometimes said that there are here two coracoid bones; but actually the anterior ventral process is a downward extension of the scapula, which thus retains its connection with the clavicle (here a plate situated ventrally in the plastron). In crocodilians (Fig. 102, *E*) and dinosaurs both scapula and coracoid are rather slender elements which meet at an angle at the glenoid fossa, the scapular blade slanting upward and backward, the coracoid back as well as down. This construction is in general retained in birds (Fig. 102, *F*), although the coracoids are powerful elements in the "ratites."

The earliest of mammal-like reptiles, the pelycosaurs (Fig. 102, *G*), had a girdle similar in many ways to that of the most ancient reptiles and amphibians. There was, however, the addition of a third ossification not present in other reptiles. Most of the coracoid plate was still ossified as an anterior coracoid, but behind it there appeared a new element, equivalent to the true *coracoid* of mammals. This bone was small in pelycosaurs; in therapsids (Fig. 102, *H*) it grew forward at the expense of the anterior coracoid to occupy the greater part of this ventral plate. Meanwhile, excavation of the front edge of the coracoid plate had the result that the point of attachment of the clavicle to the scapula became a distinct process, the *acromion.*

In the most primitive of living mammals, the monotremes, the endoskeletal girdle (Fig. 102, *I*) greatly resembles that of the therapsids. With the step upward from monotremes to marsupials and placentals, however, there is a major change (Fig. 102, *J*). The entire coracoid plate has vanished; the anterior coracoid has disappeared, and the coracoid has dwindled to a process like a crow's beak (as its name implies), attached to the lower margin of the scapula. The girdle is practically reduced to a dorsally situated scapular blade, with the glenoid cavity below it. Even the scapula has radically changed, for, instead of being a simple plate, it has a ridge, the *scapular spine,* running down its length which separates *supraspinous* and *infraspinous fossae;* the acromion is situated not at the front end of the bone, but at the base of the spine.

This marked change in girdle structure is related to the shift in limb posture in mammals as compared to reptiles, and consequent changes in limb musculature (cf. p. 273). Much of the musculature which once arose from the coracoid plate has shifted upward to the scapula, rendering this plate useless. The spine and acromion on the scapula actually represent the original anterior margin of that bone; the supraspinous fossa is a new shelf built out in front of the erstwhile front margin, to accommodate part of the musculature which has migrated upward.

Pelvic Girdle (Figs. 104, 105, p. 185; 106, p. 188). The pelvic girdle, a purely endochondral structure, differs markedly from the pectoral throughout the vertebrate series, even more markedly, perhaps, in lower than in higher types. In fishes (Fig. 108, p. 191) each half of the girdle is a small and simple ventral plate, often triangular, embedded in the muscles and connective tissues of the abdomen, typically just anterior to the cloaca. Never is there in fishes any connection with the vertebral column; the ventral tissues seem to be sufficient to anchor girdle and limb in place. The half-girdles of the two sides are usually in contact anteromedially. This contact, a primitive *pelvic symphysis,* affords further mutual support, and in the sharks and lungfishes the two half-girdles are fused to form a single element. Along its posterolateral margin the girdle affords a point of attachment for the fin skeleton. Although the half-girdle of either side is usually a single, compact structure, the sturgeons and the early shark Cladodus show evidence of subdivision, suggesting that the pelvis arose from a fusion of originally independent basal fin elements. In the Chondrichthyes, lungfishes, and a few ray-finned forms the girdle remains in a cartilaginous condition in the adult; in other fish groups each half of the girdle ossifies as a single element.

With the presence, in tetrapods, of large limbs which support much of the weight of the body, the pelvic girdle changes in radical fashion. Greater areas are needed on the girdle for the attachment of the limb musculature; the support of the body necessitates that the girdle be tied more closely into the body (Fig. 103).

In the process of tetrapod development the original, ventral part of the girdle on either side becomes a large plate of bone lying in a tilted position in the flank. This ossifies from two centers, the *pubis* anteriorly and the *ischium* posteriorly, and may be termed the *pubo-ischiadic plate.* Externally, it offers an area of origin for limb muscles; it is pierced by the *obturator foramen,* carrying a nerve supply to part of these muscles. Above the pubo-ischiadic plate there develops the *acetabulum,* a large rounded socket which receives the head of the proximal limb bone, the femur. Pubis and ischium enter into the formation of the acetabulum; its upper margin is occupied by the third of the pelvic elements, the *ilium,*

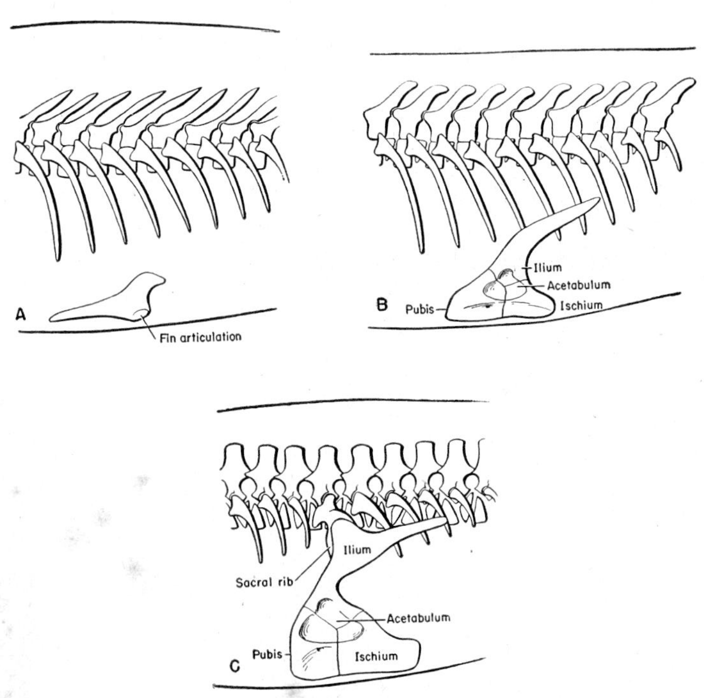

Fig. 103. Diagrams to show the development of the pelvic girdle and sacrum in the evolution of amphibians from fishes. *A,* Left lateral view of the pelvic region of a fish, with the vertebral column and ribs above, the small pelvic girdle placed ventrally. *B,* Primitive tetrapod stage, found in some early fossil amphibians. The girdle has expanded, with the three typical bony elements. The ilium extends upward, but was presumably connected with the column only through ligaments binding it to the neighboring ribs. *C,* The girdle has grown further, and the ilium is firmly attached to an enlarged sacral rib.

which forms the dorsal part of the girdle. In fishes there is sometimes a small dorsal girdle process; in even the most primitive of known amphibians this had grown far dorsally and posteriorly as a rodlike structure which presumably connected by ligaments with the ribs in this region of the body. In most amphibians there early developed an iliac blade in addition to the primitive rod; beneath this blade lies an expanded sacral rib, which ties girdle and body together and is the major factor in the support of the body on the limbs.

The inner surface of the girdle offers, ventrally, some area of origin for limb muscles, but for the most part lies adjacent to the walls of the celomic cavity. The two halves of the pelvic girdle are broadly apposed ventrally in a pelvic symphysis; usually the union between the two pubic areas is the stouter. The cloaca lies posterior to the pelvic girdle; hence the girdle and the ribs and vertebrae with which it joins form a ring of bone bounding a *pelvic outlet,* through which materials in genital, urinary, and digestive systems must pass before leaving the body; the size of the pelvic

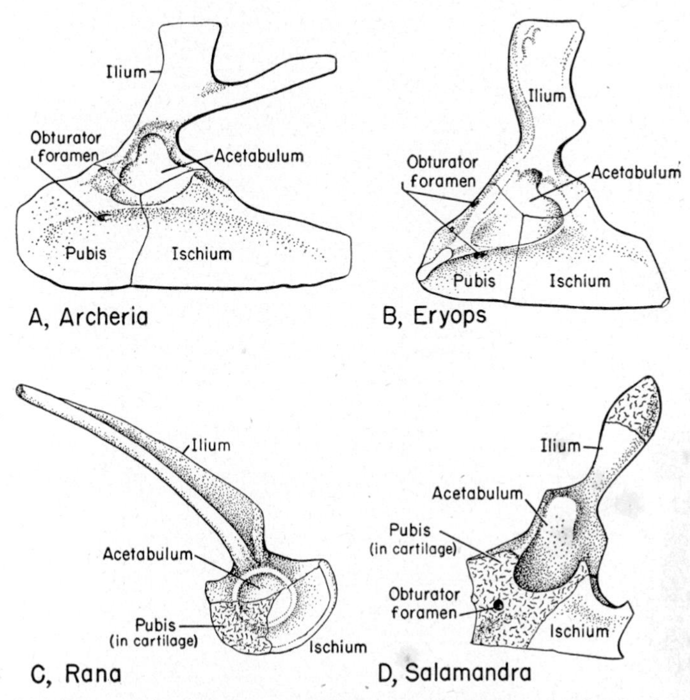

Fig. 104. The pelvic girdle in amphibians. *A,* A primitive fossil labyrinthodont; *B,* a typical later labyrinthodont; *C,* a frog; *D,* a urodele. A posterior process of the ilium, present primitively, is retained in many reptiles (cf. Figs. 105 and 106), but is lost in most Amphibia (represented by a prong in *B*). In the anurans the ilium is a specialized and elongate rod. The pubis was primitively well ossified, but remains cartilaginous in many fossil forms and all modern amphibians.

outlet is an important factor in forms laying large eggs or bearing the young alive.

There are marked variations among amphibians in the structure of the pelvic girdle (Fig. 104). The posterior projection of the ilium was usually reduced even in ancient times, and this bone became only a narrow blade apposed to the sacral rib. In many of the older amphibians the pubis failed to ossify, and it is never ossified in modern forms. The urodele pelvis (Fig. 104, *D*) is basically similar to that of older types except for the development anteriorly of a median cartilaginous prepubic process (not figured), which serves as an added support for the belly wall. In the anurans (Fig. 104 *C*) the girdle is highly specialized, in correlation with the hopping gait and shortening of the vertebral column; the ilium is a long slender bone extending far forward as well as upward from the acetabulum.

The pelvic girdle of early reptiles (Fig. 105, *A*) was little changed from that of the early amphibians, except that the blade of the ilium tended to be more expanded for the attachment of more extensive limb muscles on the outer surface, and for the attachment medially of two or more sacral ribs rather than the single rib usual in amphibians. Within this class, however, there occurred great variation in pelvic structure.

Much of the outer surface of the pubo-ischiadic plate was primitively occupied by the broad, fleshy origin of a muscle (obturator externus; cf. Fig. 160, p. 272) running to the under side of the femur. An opening, or fenestra, covered by a membrane, tended to develop beneath the origin of the muscle, separating pubis and ischium and turning the pelvic girdle into a tripartite structure. The *pubo-ischiadic fenestra* (or *thyroid fenestra*) is present in turtles, Sphenodon, and lizards (Fig. 105, *B*).

In the archosaurians (Fig. 105, *C-E*) the girdle becomes highly modified in relation to the development of bipedal habits in the group. In these reptiles and their avian descendants the acetabulum usually becomes open at its base for the better reception of the head of the femur. The pubo-ischiadic plate tends to be reduced, and the ventral limb muscle attachments moved fore or aft towards the ends of pubis or ischium, gaining a more advantageous leverage. Both these bones curve downward at their ends, and in all except the most primitive fossil members of the Archosauria the symphysis becomes restricted to their tips. In bipedal archosaurs and in birds the ilium is elongate. This is mainly correlated with a considerable increase in the number of sacral vertebrae, which must transfer the entire body weight to the limbs via the ilium.

In the so-called reptile-like dinosaurs, the saurischians, the pubis and ischium tend to become rodlike, giving a triradiate type of pelvis (Fig. 105, *D*). In the crocodiles (Fig. 105, *C*) the pelvic structure is basically similar to that of saurischians, except that the ischium has expanded and excluded the pubis from the acetabulum. The ornithischians (Fig. 105, *E*), the so-called birdlike dinosaurs, get their name from the fact that (as in birds) the slender main shaft of the pubis has swung far back to parallel the ischium; the reasons for this peculiar condition are far from clear. Anteriorly, the pubis sends out, as a belly support, a stout process in much the position of its original shaft.

In birds (Fig. 105, *F*), as in ornithischians, the pubis has swung back to a posterior position. There is, however, little development of an anterior pubis process; the greatly developed sternum forms a bony sheath over most of the ventral surface of the body, and belly support by such a proc-

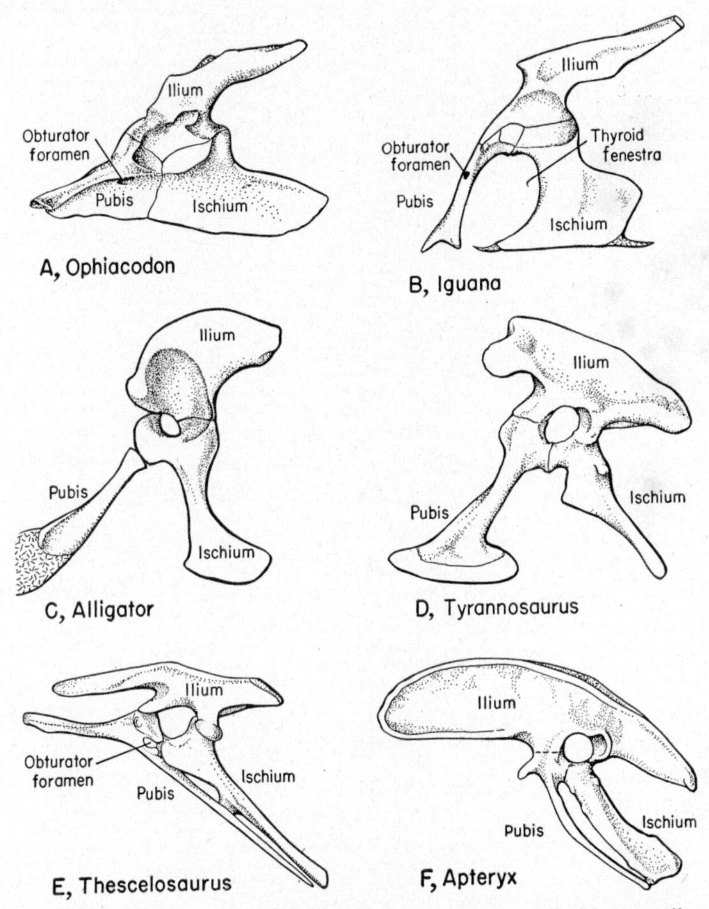

Fig. 105. The pelvic girdle in reptiles and birds. *A,* Primitive reptile; *B,* a lizard; *C,* a crocodilian; *D,* a reptile-like dinosaur (Saurischia); *E,* a birdlike dinosaur (Ornithischia); *F,* a bird (the kiwi). In *A* the ilium is a low blade, and the same is true in lizards; in other forms this structure is more expanded. In the dinosaurs and birds, with a bipedal type of locomotion, the ilium has grown forward somewhat as in mammals (cf. Fig. 106). In the archosaurs (*C, D, E*) and the birds descended from them there is typically an open bottom to the acetabulum for the better reception of the head of the femur. In the primitive reptile the puboischium is a solid plate. In lizards there is a large thyroid fenestra developed between pubis and ischium, from which a large muscle to the femur takes origin (Sphenodon and turtles are similar). This fenestra is comparable to one seen in mammals, but there the obturator foramen is concerned in the development of the fenestra. In such archosaurs as the alligators and saurischians there appears at first sight to be a similar structure. Actually, however, this is not the case; the pubis and ischium are twisted downward, and the true ventral margin of the girdle is the curving lower margin of pubis and ischium. In *C* and *D* the pelvis is triradiate, with a simply built pubis; in *E* the pubis is two-pronged, and the pelvis tetraradiate. In the bird the anterior process of the pubis is reduced or absent. In the alligator the pubis is excluded from the acetabulum by the ischium and a nubbin of calcified cartilage; the pubic region of the girdle extends forward along the belly as a fibrous cartilage.

ess is little needed. Except in the ostrich, the pelvic symphysis has disappeared, and the two halves of the girdle are widely separated ventrally, a situation correlated with the relatively large size of the bird egg and the consequent necessity for a large pelvic outlet. Some strengthening of the ventral parts of the girdle is, however, afforded by a fusion between the

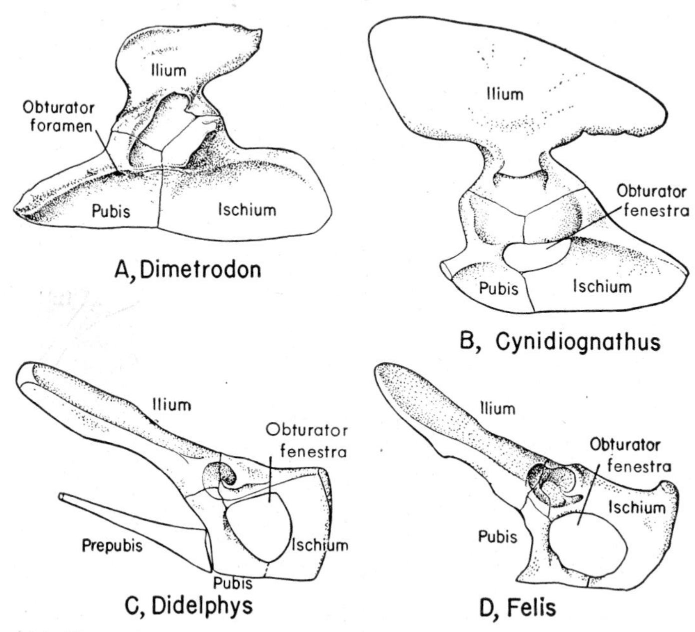

Fig. 106. The pelvic girdle in mammal-like reptiles and mammals. *A,* Primitive pelycosaur; *B,* a therapsid; *C,* the opossum; *D,* a cat. The Dimetrodon pelvis is of a primitive reptilian type. In the therapsid the ilium has grown forward dorsally, pubis and ischium have swung back ventrally, and the obturator foramen has expanded (comparable to the situation in many modern reptiles) into an obturator (or thyroid) fenestra. The opossum and cat show a typical mammalian structure, with a large obturator foramen, a shortened ischium and a slender ilium (secondarily broadened in many heavy mammals, however). The opossum, like other marsupials and the monotremes, has a pair of "marsupial bones" not found in other groups. (In the cat, as in certain other mammals, an accessory element is seen in the acetabulum.)

posterior ends of all three elements of either side in many birds; the tips of pubis and ischium usually become connected, and the ischium may become continuous posteriorly with the iliac blade.

In the mammal-like reptiles and mammals (Fig. 106) marked changes associated with changed limb posture and changed musculature have occurred. The muscles and the bones from which they originate undergo a marked rotation, counter-clockwise as seen from the left side. The ilium, which primitively grew mainly backward dorsally, comes to extend anteriorly as it reaches upward to the sacrum. The pubis and ischium, on the contrary, have moved posteriorly so that the ventral plate of the girdle extends hardly at all anteriorly to the acetabulum.

The mammalian ilium is primitively a rather slender rod, triangular in section; however, in heavy-bodied ungulates (as horses, cattle, elephants)

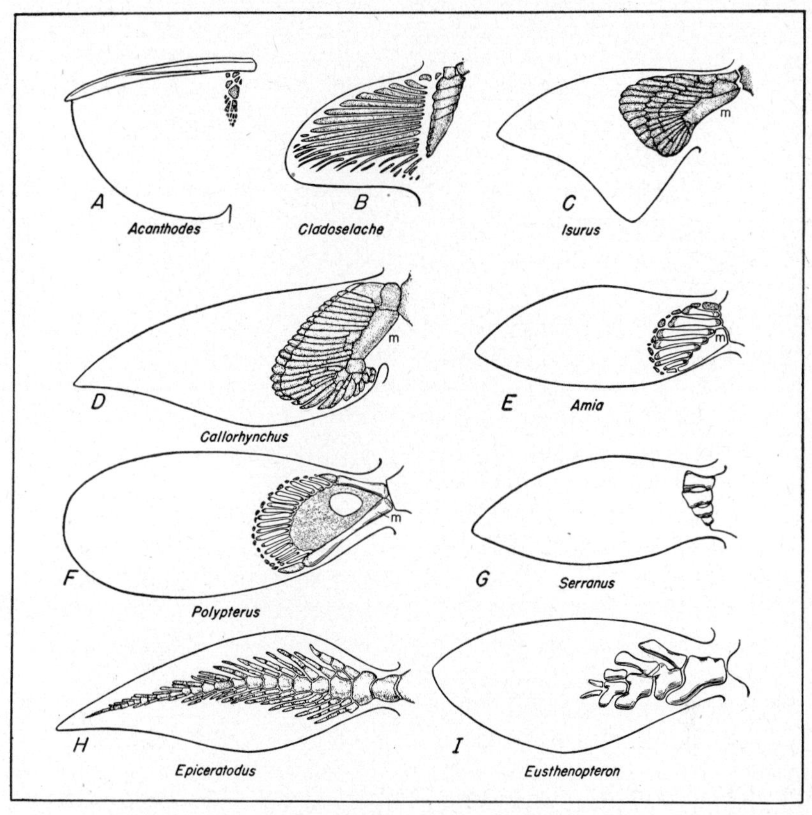

Fig. 107. Pectoral fins of fishes. All are fins of the left side, viewed from the upper surface, so oriented that the long axis of the body is vertical on the page, the anterior end above. The outline of the complete fin is indicated, and (except for *A*) the articular region of the girdle is included at the right. *A,* Fossil acanthodian, with the fin skeleton little developed, and a spine forming a cutwater and main fin support. *B,* Primitive fossil shark, with a parallel-bar type of fin. *C,* Modern shark with a narrow-based flexible fin and a basal concentration of bars with formation of a metapterygial axis (the metapterygium is the elongate posterior basal element, *m*). *D,* Comparable type found in chimaeras. *E,* Primitive actinopterygian type, with parallel-bar construction, but metapterygial axis present. *F,* An aberrant modification of the last, found in the archaic actinopterygian, Polypterus. *G,* Teleost (sea bass) with a much reduced skeleton. *H,* The typical archipterygium of the Australian lungfish. *I,* The abbreviate archipterygium of a fossil crossopterygian. (*A* after Watson; *B* after Dean; *C* and *D* after Mivart.)

or bipeds (as man), in which there are powerful gluteal muscles running from ilium to femur, this bone may be much expanded. As in many reptiles, the pubo-ischiadic plate is fenestrated. Here, however, in contrast to reptiles, the foramen for the obturator nerve is included in this opening, termed the *obturator fenestra*. The symphysis between the two halves of the girdle is usually stoutly developed between the pubes and may be absent posteriorly. In connection with viviparity, the pelvic opening is often broader in the female, and in some forms the symphysis is loosened (under hormonal control) at the time of parturition. In monotremes and marsupials there is a pair of "marsupial bones" (prepubes), extending forward from the pubis and supporting the body wall. They were not present in the ancestral reptiles and are absent in placentals; the reason for their appearance in these two groups of primitive mammals is not clear.

Paired Fins in Fishes (Figs. 106, 107). We noted earlier some of the theories and problems connected with the origin and nature of fish fins. Apart from various peculiar early fossil types, we see in the general run of living and fossil fishes two basic types of paired fin skeletons, together with various intermediates. One type is the *archipterygium,* typically developed in the lungfish Epiceratodus (Figs. 107, *H,* 108, *D*). The skeleton of the leaf-shaped, narrow-based fin consists of a jointed central axis and side branches which are usually better developed on the anterior (preaxial) margin of the fin. Typical archipterygia are found in fossil lungfishes and in some early crossopterygians; hence the archipterygium may be a basic type in the Choanichthyes. But except for an aberrant family of extinct sharks it is unknown in other fish groups, ancient or modern, and hence is unlikely to have been antecedent to other fin types. In most crossopterygians we find an abbreviated type of archipterygium (Figs. 107, *I,* 108, *E*) in which the axis is short and the branches are generally confined to the anterior margin—a fin pattern of great interest as one antecedent to the skeletal plan of the tetrapod limb.

In strong contrast to the archipterygium is the finfold type of fin seen in ancient Paleozoic sharks such as Cladoselache (Figs. 107, *B,* 108, *A,* Cladodus). Here the fin had a broad base, and hence was little more than a fixed horizontal stabilizer; there is no fin axis, and the skeleton consisted of parallel bars of cartilage. This type of fin persists with modification in modern cartilaginous fishes (Figs. 107, *C, D,* 108, *B, C, F*). The major change today is that the base of the fin has generally become much narrowed, making it more movable and a more effective steering device. With this constriction of the base, the fin cartilages have, of necessity, become crowded together; there is a tendency for one of the most posterior cartilages to become particularly prominent and act as an axis on to which many of the other bars articulate. This axial element is frequently termed the *metapterygium.** It is obvious that the modern type of shark fin shows a condition intermediate between the finfold type and the archipterygium;

* It is sometimes stated that the shark fin has typically three elements attaching basally, termed the pro-, meso-, and metapterygium. This situation is seen at times, but actually there is great variation in elements other than the metapterygium.

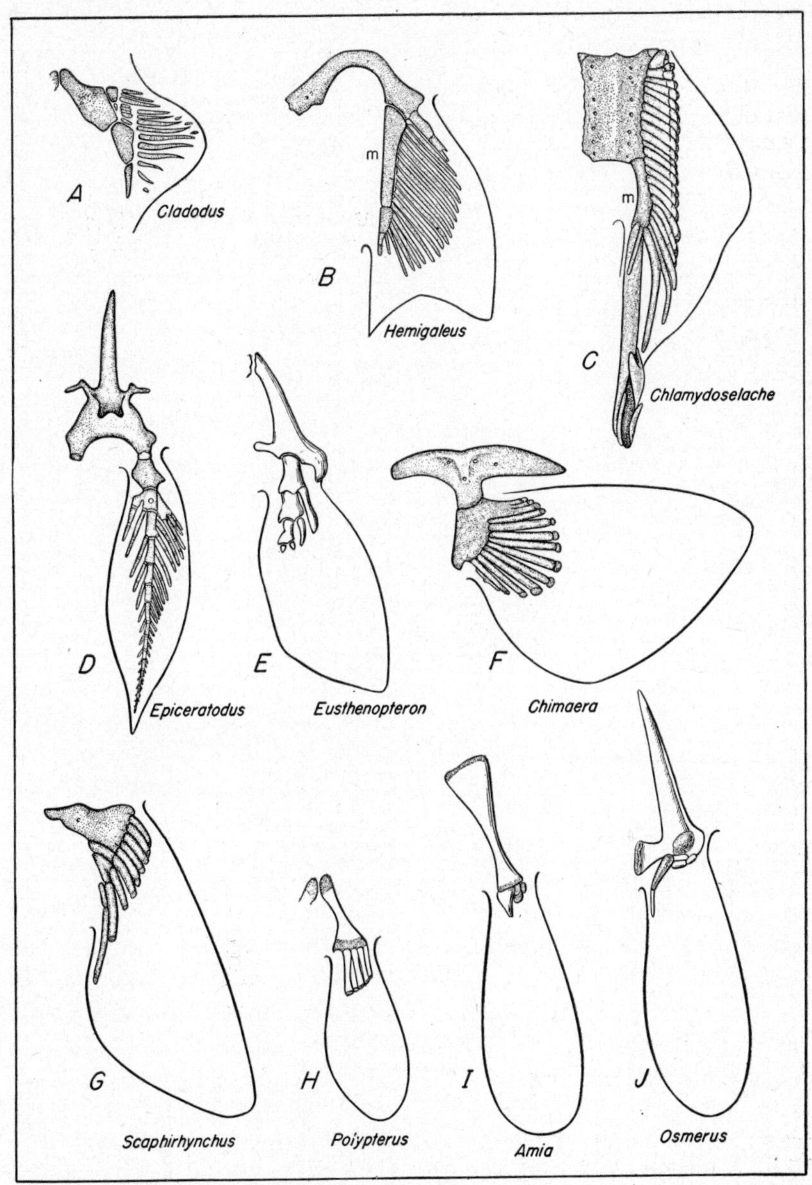

Fig. 108. Pelvic fins of fishes. All are of the left side, viewed from below; the midline of the body is at the left, the anterior end above. The fin outline is indicated in each case. The left half of the pelvic girdle is included, or the whole girdle if the halves are fused. *A,* Primitive Carboniferous shark, with a broad-based fin and parallel bars as fin support. *B,* Modern female shark, with similar construction. *C,* Male shark, with additional cartilages supporting the clasper. *D,* The archipterygium of the Australian lungfish. *E,* The abbreviate archipterygium of an ancient crossopterygian. *F,* Female chimaera fin, essentially similar to that of sharks. *G,* The sturgeon, also with a sharklike construction. *H,* Polypterus; *I,* the bow-fin; *J,* a teleost; showing nearly complete reduction of the bony fin skeleton. (*A* after Jaekel; *B* after Garman; *C* after Goode; *D, H, I* after Davidoff; *E* after Gregory.)

there is here a main fin axis with side branches, although the branches are almost all on the exterior or anterior margin of the axis. In the shark group, at least, the parallel-barred finfold type is primitive, suggesting that the archipterygial type is in general a secondary, not a primitive, type of fin.

The actinopterygians are defined as a group of fishes in which, in general, skeleton and flesh extends hardly at all into the fin, which is mainly a web of skin supported by horny rays. Consideration of a series of actinopterygian fins indicates rather surely that they have been derived from the finfold type, and have evolved in a fashion comparable to that of sharks. The basic structure is a series of parallel bars; in such a primitive

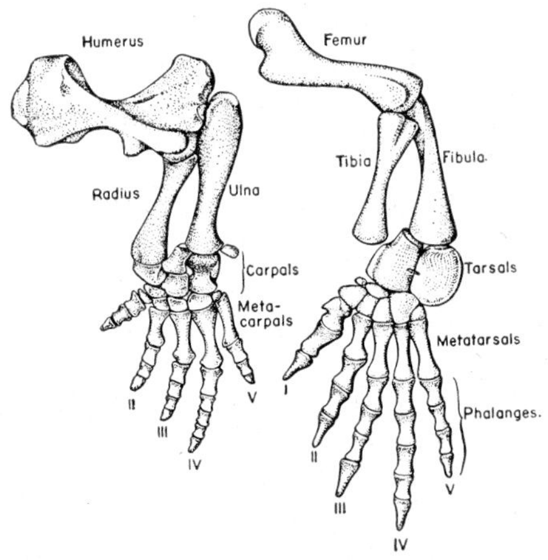

Fig. 109. Left front and hind limbs of a primitive reptile (Ophiacodon) to show the general pattern of limb construction in early tetrapods.

ray-finned form as the sturgeon the pelvic fin (Fig. 108, *G*) is broad-based, with numerous bars extending into the fin rather as in a primitive shark. As in later sharks, there is in ray-finned forms a trend toward a narrow fin base and greater flexibility, with a consequent development of a metapterygial axis in some cases (Fig. 107, *E*).* In teleosts (Figs. 107, *G*, 108, *J*) the bony elements are much reduced; there typically remain in the pectoral fin only a few short parallel bars, and in the pelvic fin these may be further reduced to nubbins of bone or cartilage.

In fishes generally the pectoral fins are the larger of the two pairs and have a more highly developed skeleton. Pectorals are practically universal in jawed fishes of every sort, but the pelvics may be reduced or may, in many teleosts, move forward to a position beneath the shoulder region (Fig. 28, *B*, p. 50) or even forward beneath the "chin." In the Chon-

* Polypterus (Fig. 107, *F*) shows a peculiar and overelaborated development of the pectoral fin skeleton, but the primary structure is nevertheless one of parallel bars.

drichthyes, associated with internal fertilization (cf. Chap. XII), the finger-like *claspers* are supported by extra posterior cartilages on the pelvic fins (Fig. 108, *C*).

The Primitive Tetrapod Limb. The limbs of land vertebrates appear at first sight strikingly different from fish fins, but although the greater size and greater complexity of the land limb tends to obscure the relationship, the two are readily comparable in basic structure. The numerous muscles that sheathe the tetrapod appendage can be resolved on comparative and embryological grounds into two series comparable to the simpler muscle masses on the upper and lower surfaces of the fish fin, and the limb elements of land forms are comparable in essence to the fin supports of crossopterygian fishes with abbreviate archipterygia.

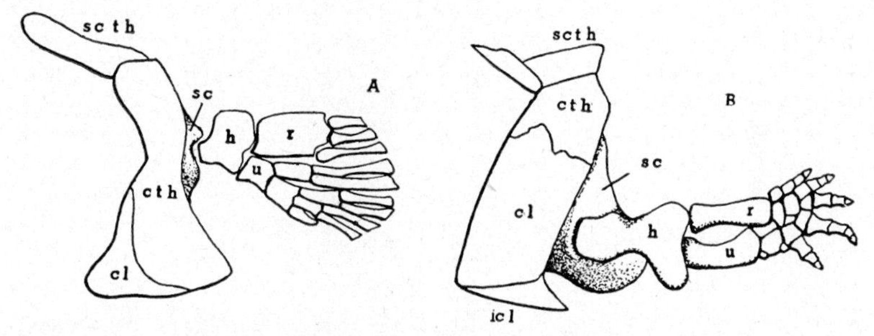

Fig. 110. The shoulder girdle and pectoral fin of a crossopterygian, *A*, and the same structures in an ancient fossil amphibian, *B*, placed in a comparable pose to show the basic similarity in limb pattern. *h, r,* and *u,* Humerus, radius, and ulna of the tetrapod and obvious homologues in the fish fin. *cl,* Clavicle; *cth,* cleithrum; *icl,* interclavicle; *sc,* scapula; *scth,* supracleithrum.

The limb of a primitive land vertebrate is composed of three major segments (Fig. 109). In both front and hind limbs—corresponding, respectively, to pectoral and pelvic fins—the proximal segment projects laterally from the body and includes but a single bony element, *humerus* or *femur,* which primitively moves fore and aft in an essentially horizontal plane. Beyond the elbow or knee region a second segment is present which lifts the body off the ground. (Trackways of ancient amphibians, by the way, prove erroneous the popular belief that these first tetrapods crawled on their bellies.) Here movement is rotary in a vertical plane parallel to the long axis of the body, capable of increasing the progression caused by the fore and aft movement of the proximal elements. Two bones are present in this segment—*radius* and *ulna* in the forelimb, *tibia* and *fibula* in the hind, the first named in each case being the anterior or medial element of the pair.

The third segment in either pectoral or pelvic limb is that of the foot, termed the *manus* in the pectoral, *pes* in the pelvic, limb. This includes proximally the wrist or ankle region—*carpus* or *tarsus*—which forms a more or less flexible adjustment between the foot and the more proximal part of the limb, and distally the toes, or *digits,* placed upon the ground. The proximal element of a toe (contained in palm or sole) is a *metapodial* —a *metacarpal* or *metatarsal;* the remaining joints form the *phalanges.*

This pattern of limb skeletal construction is, as mentioned, comparable basically to that found in crossopterygian relatives of the tetrapods (Fig. 110). In such fishes there is a single proximal element articulating with the shoulder girdle and equivalent to humerus or femur; in a second segment are two elements comparable to the radius and ulna of pectoral limb or tibia and fibula of the hind leg. Beyond this point the fin skeleton of known crossopterygians shows an irregular and variable branching ar-

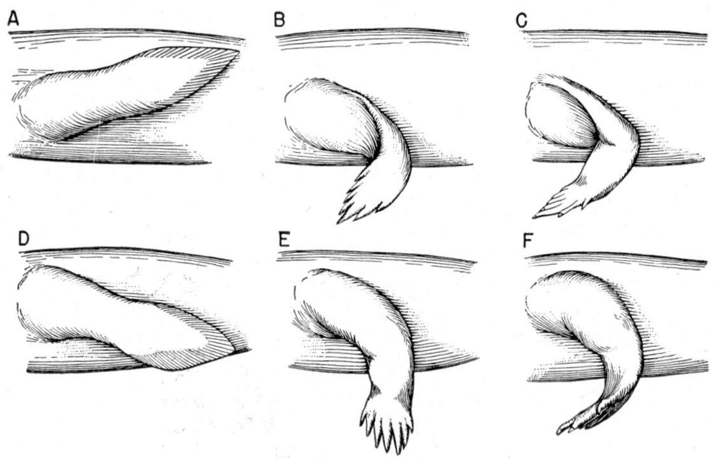

Fig. 111. Diagrams to show the postural shift in the paired limbs in the transition from fish to amphibian. *A* to *C,* Pectoral limb; *D* to *F,* pelvic limb; *A, D,* fish position; *B, E,* transitional stage; *C, F,* amphibian position (cf. text).

rangement; attempts at detailed homologies of these distal elements with particular bones of the manus or pes seem for the most part to be valueless.

From the first there were notable differences between the front and hind legs of land vertebrates, despite the basic similarity of their bony patterns, and many of these differences are retained in later tetrapods. We note particularly the differences in the major joints between limb segments. Below the elbow, radius and ulna rotate rather freely on the humerus, but the comparable knee joint is invariably a simple untwistable hinge. Wrist articulations are essentially hingelike; but although the ankle in mammals is a hinge, the foot could be readily rotated on the shin in reptiles and early amphibians. This contrast in joint structure can, it seems, be traced back to a contrast in the position of the pectoral and pelvic fins in the ancestral fishes (Fig. 111). Presumably, as in living lungfishes, the pelvic fin of the fish forebears of tetrapods extended outward and downward from the body in simple fashion, with the future sole surface down; the pectoral fin, on the contrary, appears to have been turned back along the flank with the future palm turned outward. For the pectoral fin a major twist at the future elbow is necessary to bring the limb into proper position, but once that twist is made, the "hand" points forward properly, and only a simple hinge is needed in the future wrist region. In the pelvic fin the future limb, projecting laterally, needed only a hinge at the knee to bring its two long segments into proper position; the "foot," however, would point laterally, and sharp rotation at the future ankle is needed to bring it into a fore and aft position.

Limb Function and Posture. The development of terrestrial habits brought about a major revolution in vertebrate locomotion. In fishes, propulsion was accomplished by undulatory movements of the trunk and tail. Undulatory movement appears to have long remained important in early land animals and is still important today in salamanders (Fig. 96, p. 170). Here the feeble limbs play only a minor role as positive propulsive agents; they are in great measure stationary organs through which the push of body undulations may be exerted on the ground. In tetrapods generally, however, the limbs have taken over a positive and dominant role in progression.

Primitive land vertebrates, both amphibians and reptiles, had a sprawling posture, with the limbs extending far out from the side of the body. This primitive pose is still preserved in living urodeles, and with little modification in turtles, lizards, and Sphenodon. But this pose is wasteful of energy, for much muscular effort is used up in merely keeping the body off the ground; most land vertebrates have modified this primitive structure in various ways. The frogs have become specialized for a hopping gait, with highly modified hind legs. In early Ruling Reptiles a strong trend toward a bipedal mode of life occurred, and there existed in Mesozoic days a variety of bipedal dinosaurs. The hind limbs were strengthened and the knees brought around forward to a position essentially beneath the body, thus transferring much of the burden of weight support directly to the limb bones. Many of the giant reptiles, however, reverted to a quadrupedal gait, and the living archosaurs, the crocodilians, are likewise quadrupedal.

With the front limbs freed from ground locomotor function, the development of wings and flight became possible among Ruling Reptiles. One group, the pterosaurs, evolved a wing membrane supported by one elongate finger. Birds, with feathered wings, developed from another group of bipedal archosaurs.

A second type of major improvement in land locomotion, one in contrast with bipedalism, was the development of the efficient four-footed gait characteristic of mammals. The knee is turned forward as in the bipeds, the elbow backwards. In this pose the entire limb movement is a fore and aft swing, making for a longer stride and greater speed; further, muscular efficiency is gained through the fact that most of the body weight is carried directly by the bones of the four limbs. This pattern of locomotion was developed by the therapsids and remains relatively unchanged in most modern mammals. There are, however, many variations, such as the flying mechanisms of the bats, the digging specializations of moles, the bipedalism of man, and so forth.

Many amniotes, despite their fitness for terrestrial life, have returned to the water, with consequent modification of limb structure. In general, the limbs of aquatic tetrapods tend to reassume a finlike shape, with

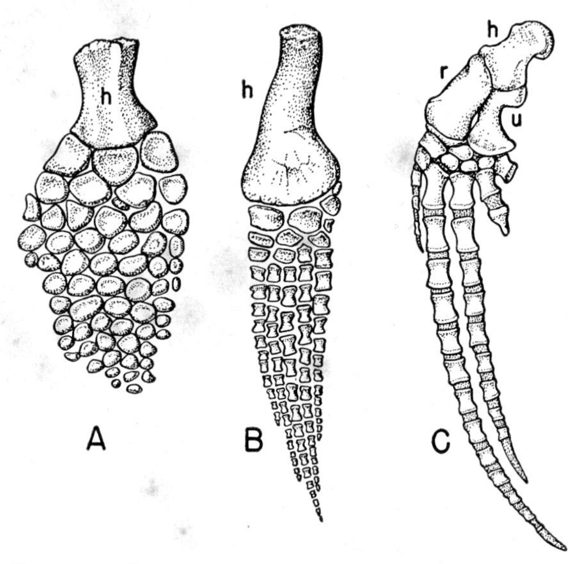

Fig. 112. Examples of tetrapod appendages retransformed into fishlike paddles in marine forms. Pectoral limbs of *A*, an ichthyosaur; *B*, a plesiosaur; *C*, a whale. In all there has been a shortening and broadening of the proximal bones and a multiplication of the phalanges. *h*, Humerus; *r*, radius; *u*, ulna. (*A* and *B* after Williston; *C* after Flower.)

shortened proximal segments and an expanded "foot" area. In less specialized aquatic types, such as seals, the proximal elements may remain little modified, but in advanced types, such as the whales and the extinct ichthyosaurs and plesiosaurs, the whole limb may be modified into a flipper-like structure (Fig. 112). Aquatic birds usually use webbed feet for swimming, but penguins and auks use the powerful wings as swimming organs.

The Apoda among amphibians, the snakes, and certain lizards, have abandoned limbs and reverted to an essentially piscine type of locomotion based on body undulation. In whales and sirenians the pelvic limbs are reduced or absent, and among bipedal forms the "arm" is reduced in certain dinosaurs and nonflying birds.

Humerus (Fig. 113). In many tetrapods the two ends of the humerus are much expanded and seemingly "twisted" on one another. This apparent twist is, however, a primitive feature; the proximal end of the bone is essentially horizontal, the distal end tilted so that its lower surface faces the forward-slanting forearm. Proximally, stout processes are developed for attachment of the pectoralis and deltoid muscles in the mammalian humerus for the muscles from the scapula—the *greater tuberosity*. In primitive forms the proximal articular surface is an elongate oval cap-

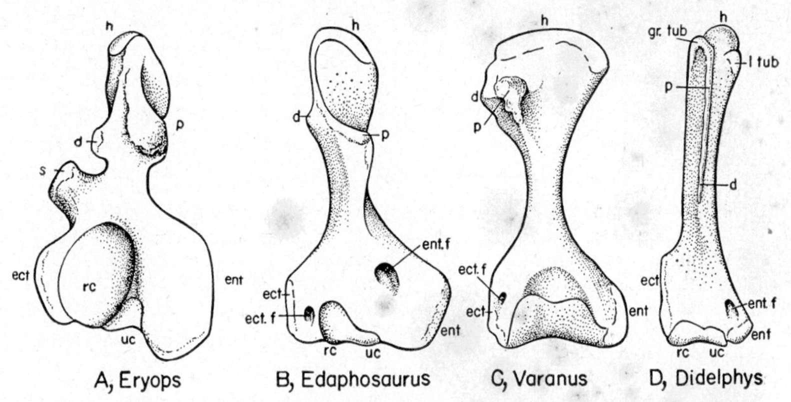

Fig. 113. Humeri of *A*, a Paleozoic amphibian; *B*, a primitive fossil reptile; *C*, a lizard; *D*, the opossum, seen from the under surface. Primitive humeri were short, practically without a shaft, and much expanded at both ends (in *A* and *B* the proximal end is twisted about 90 degrees with the distal and hence appears thin). In all a prominent crest is present to which is attached the pectoralis and deltoid muscles. In later types the bone became relatively long and slender, particularly in small animals. In primitive reptiles, foramina developed distally on the inner or posterior side (entepicondylar foramen, *ent f*) and on the outer or anterior margin (ectepicondylar foramen, *ect f*). The former foramen persists in various mammals and in Sphenodon, the latter in many reptiles. *d*, Deltoid crest; *ect*, ectepicondyle, for attachment of extensor muscles of forearm; *ent*, entepicondyle, for attachment of flexor muscles of forearm; *gr tub*, greater tuberosity, for attachment of supraspinatus and infraspinatus muscles; *h*, head; *l tub*, lesser tuberosity, for attachment of subscapular muscle; *p*, pectoral crest; *rc*, radial condyle; *s*, process (supinator) which in reptiles aids in formation of ectepicondylar foramen; *uc*, ulnar condyle, or trochlea.

ping the end of the bone; in later types it tends to become spherical and distinct from the shaft. Primitively little if any shaft was present between proximal and distal expansions. In many small reptiles and in most birds and mammals the bone is more slenderly built and elongate.

Distally, the humerus bears on its under surface a rounded *radial condyle;* posterior to this the end of the bone is notched for a pulley-like ulnar articulation, the *trochlea*. Expansions on either side distally, for forearm muscle attachment, are the *ectepicondyle* on the front (or outer) side, and the larger *entepicondyle* on the back (or inner) margin. Universally in early reptiles (seldom in amphibians) the entepicondyle is pierced by a large *entepicondylar foramen* for a nerve and blood vessels.

Proximal to the ectepicondyle there is primitively a projecting *supinator process;* in some reptiles this fuses with the ectepicondyle, bridging an

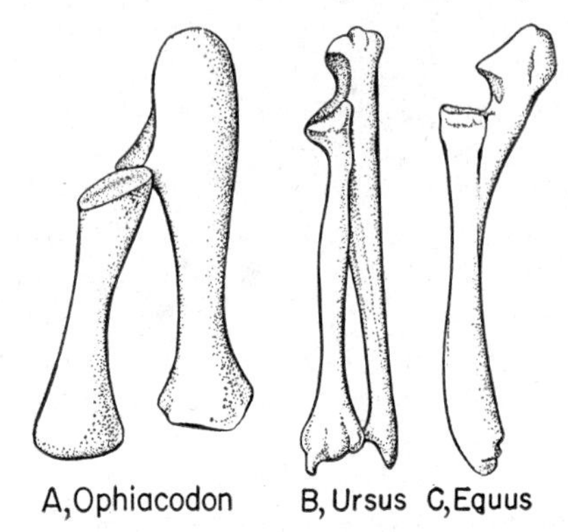

Fig. 114. Radius and ulna, seen from the anterior or extensor surface in *A,* a primitive reptile; *B,* a bear, representing a typical mammalian condition; and *C,* a horse. The humerus articulates with the curved surface of the notch in the ulna and the adjacent head of the radius; above, the projecting olecranon of the ulna serves for the attachment of the powerful triceps muscle, which extends the forearm. In many mammals, as in the horse, the lower part of the ulna is reduced and fused with the radius.

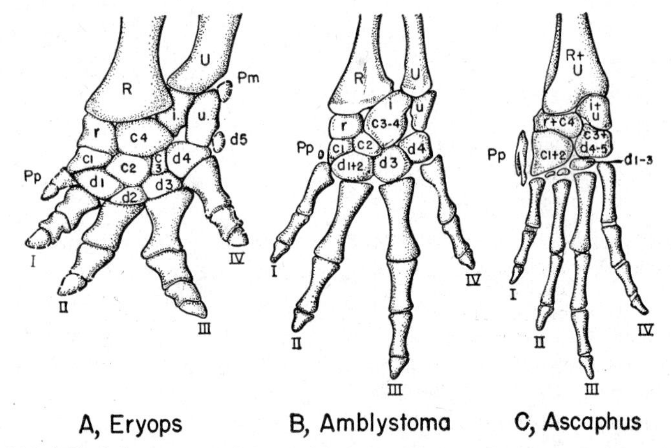

Fig. 115. The left manus of amphibians, including an early labyrinthodont, a urodele, and an anuran. Restored elements in *A* in broken line. All twelve elements thought to have been present in the primitive carpus are found in Eryops; in modern amphibians fusions of various sorts have occurred. As in most amphibians, four toes are present in the forms figured, although all have some development of a digit medial to the pollex, and an element in Eryops may have been the stub of an extra toe beyond the reduced fifth digit. The phalangeal count of two or three is usual in amphibians. *c1* to *c4,* centralia; *cu.* cuboid; *d1* to *d5,* distal carpals; *i,* intermedium; *l,* lunare; *m,* magnum; *m1, 3, 5,* metacarpals; *p,* pisiform; *Pm,* postminimus digit; *Pp,* prepollex; *R,* radius; *r,* radiale; *s,* scaphoid; *td,* trapezoid; *tm,* trapezium; *U,* ulna; *u,* ulnare; *un,* unciform; *I* to *V,* digits. (After Gregory, Miner, and Noble.)

ectepicondylar foramen. The ectepicondylar foramen persists in many turtles and lizards and in Sphenodon, and the entepicondylar opening is present in the last form and in many of the more primitive mammals; but in many reptiles and mammals, and in birds as a whole, both foramina have disappeared.

Radius (Fig. 114). The radius is a columnar-shaped bone which supports the body on the front foot, and articulates at its upper and lower

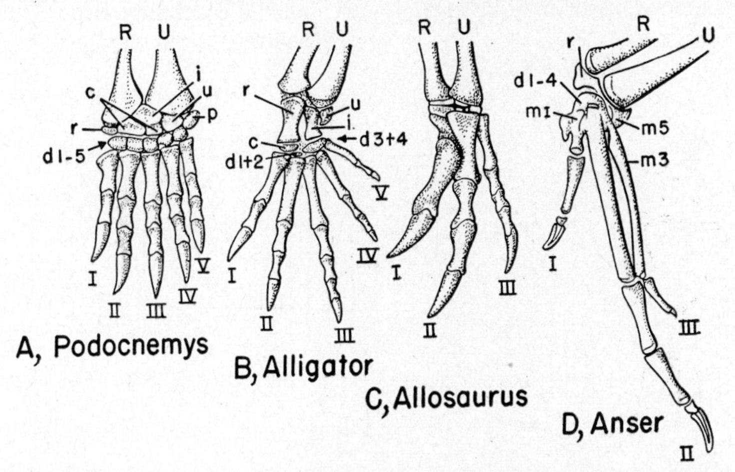

Fig. 116. The manus in *A*, a turtle; *B*, the alligator; *C*, a carnivorous dinosaur; *D*, the goose. Abbreviations as in Figure 115. The bird manus includes the fused elements of the first three digits; a somewhat comparable structure is seen in certain dinosaurs. (*A* after Williston; *C* after Gilmore; *D* after Steiner.)

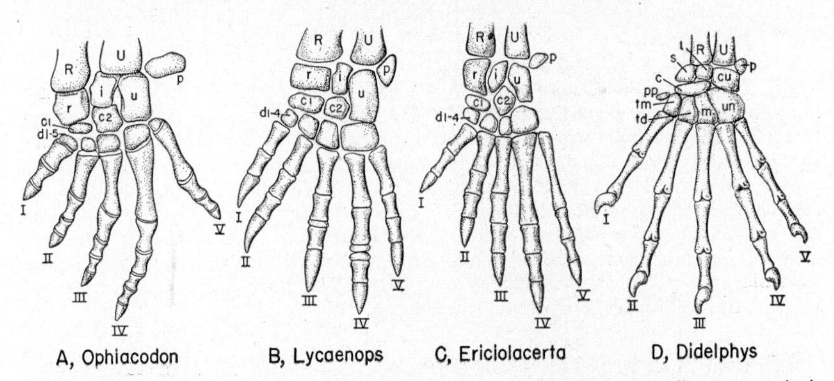

Fig. 117. Evolution of the mammalian manus. *A*, Primitive reptile; *B*, a primitive therapsid; *C*, an advanced therapsid; *D*, a primitive mammal, the opossum. Abbreviations as in Figure 115. In the carpus there is a loss of the fifth distal element, a reduction from two centralia to one; distally, "supernumerary" phalanges are lost from digits III and IV. (*B* after Broom; *C* after Watson.)

ends with humerus and carpus, respectively. Except in extreme aquatic specialization, there is little variation in its structure.

Ulna (Fig. 114). The ulna lies lateral to the radius in the forearm. It seldom bears any notable part of the weight, but is important for muscular attachment. Below, it articulates with the carpus; above, at the elbow, it articulates by a notch with the distal edge of the humerus. Above this notch the ulna projects as the *olecranon,* the "funny bone;" to this attaches the main extensor muscle of the upper arm, and the pull of this muscle on the olecranon is the major force extending (i.e., opening out) the forearm. Particularly among mammals the shaft of the ulna may fuse with the radius or may be entirely lost, but the olecranon is persistently present.

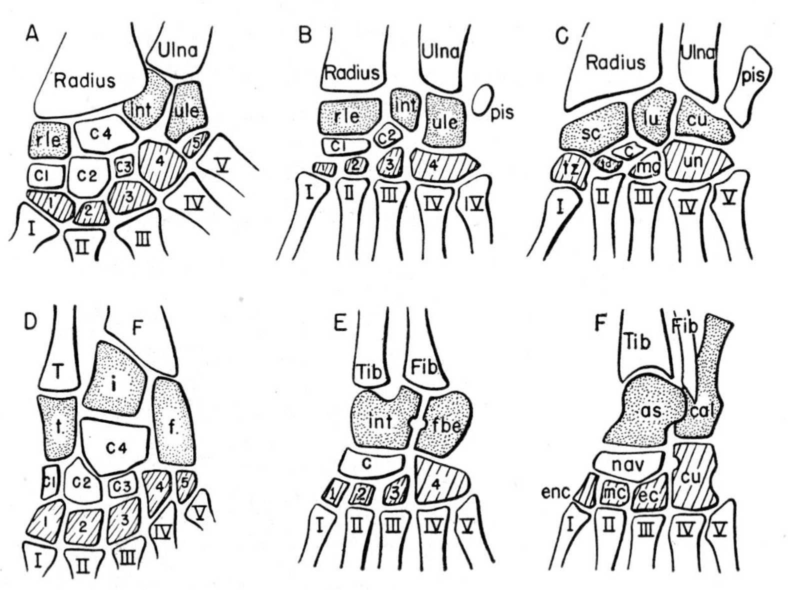

Fig. 118. Diagram of carpus (*A* to *C*) and tarsus (*D* to *F*) to show essential homologies between primitive tetrapod (*A, D*), primitive reptile (*B, E*), and mammal (*C, F*). Proximal row of elements stippled; central row (and pisiform) unshaded; distal row hatched. Digits indicated by Roman numerals; distal carpals and tarsals by Arabic numerals. *as,* Astragalus; *c, c1* to *c4,* centralia; *cal,* calcaneus; *cu,* cuneiform in carpus, cuboid in tarsus; *ec,* external cuneiform (ectocuneiform); *enc,* internal cuneiform (entocuneiform); *f, fbe,* fibulare; *F, Fib,* fibula; *i, int,* intermedium; *lu,* lunar; *mc,* middle cuneiform (mesocuneiform); *mg,* magnum; *nav,* navicular; *pis,* pisiform; *rle,* radiale; *sc,* scaphoid; *t,* tibiale; *T, Tib,* Tibia; *td,* trapezoid; *tz,* trapezium; *ule,* ulnare; *un,* unciform.

Manus (Figs. 115–119). The proximal part of the hand, or manus, consists of a series of small elements which form the *carpus,* a region of flexible adjustment between the arm region and the digits. Primitively, as seen in early tetrapods, there appear to have been nine elements in the carpus, arranged in three series:

1. Three proximal elements, one beneath each of the two forearm bones, a third intermediate in position. These three are reasonably termed (from the inner edge outward) the radial carpal or *radiale, intermedium,* ulnar carpal or *ulnare.*

2. A series of central carpals or *centralia,* wedged in between this proximal set and a distal one. A row of three centrals was primitively present, running outward from the inner, radial margin, and a large fourth element lay in a central and rather proximal position.

3. A row of *distal carpals.* Typically, these are subquadrate or rounded elements, one lying opposite the head of each toe, or digit. Since five digits appear to have been a primitive number, five distal carpals are typically present in the primitive carpus.

It is impossible in any limited space to recount all the variations undergone by the carpal elements in the history of land vertebrates. There are

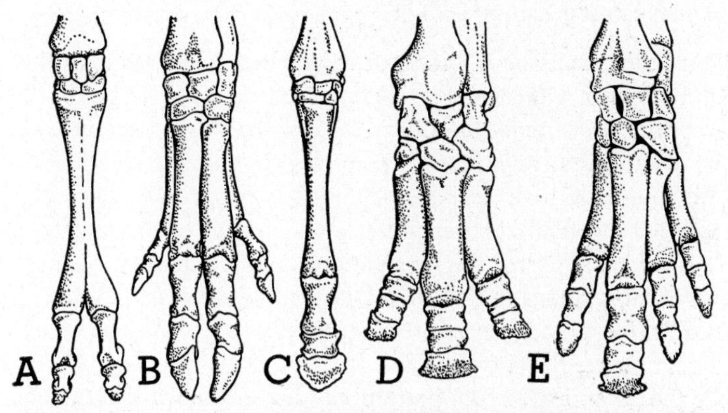

Fig. 119. Left front feet of ungulates—in order, camel, pig, horse, rhinoceros, and tapir. The first two are artiodactyls, in which the axis of symmetry lies between the third and fourth toes. In the pig (*B*) lateral toes, 2 and 5, are complete but small. In most artiodactyls, as in the camel (*A*), the two main metapodials are fused into a cannon bone. The three remaining forms are perissodactyls, in which the axis runs through the third toe. In the tapir (*E*) the pollex is lost, but the other four toes remain; in modern rhinoceroses (*D*) the fifth toe has disappeared; in modern horses (*C*) the second and fourth are reduced to splints. (After Flower.)

numerous reductions and fusions. The three proximal elements persist in most cases. The centralia, on the other hand, are almost always reduced. Even in the most primitive reptiles there are never more than two; a single centralia is a common reptilian and mammalian condition. Distally, there is a general trend for the loss of a fifth distal carpal, even when the fifth digit is retained. With toe reduction there is generally a loss of corresponding distal carpals. In reptiles and mammals is found an additional small bone, the *pisiform,* attached to the outer margin of the carpus and forming a point of attachment for the tendons of muscles along this aspect of the limb.

The terminology of the carpals in lower tetrapods is relatively simple. In mammalian anatomy, unfortunately, this is not the case; each was early given a distinct name, still used in comparative studies (Fig. 118);

and (to make matters worse) still another series of names, not useful elsewhere, have been applied to these elements in the medical school dissecting room. We list this plethora of terms without comment:

General Terminology	*Mammalian Anatomy*	*Human Anatomy*
Radiale	Scaphoid	Os naviculare
Intermedium	Lunar (or semilunar)	Os lunatum
Ulnare	Cuneiform	Os triquetrum
Pisiform	Pisiforme	Os pisiforme
Centrale	Centrale	(Absent)
Distal carpal 1	Trapezium	Os multangulum majus
Distal carpal 2	Trapezoid	Os multangulum minus
Distal carpal 3	Magnum	Os capitatum
Distal carpal 4	Unciform	Os hamatum

Beyond the carpus lie the toes, or *digits*. For each free digit there is a proximal segment which in life is contained in the flesh of the "palm." These proximal elements in either front or hind limb are termed *metapodials;* more specifically the term *metacarpal* applies to those of the forefoot. In mammals with speedy locomotion, particularly the ungulates, there may be a considerable increase in the length of the metapodials of such toes as are present; this elongation has the effect of adding essentially an extra segment to the limb. Distal to the metapodials are the elements of the free toe, the *phalanges;* distal phalanges may be much modified to bear claws, nails, or hoofs.

Since the complicated foot pattern had no antecedents in the ancestral fish fin, it is probable that considerable variation occurred in foot structure in the earliest land vertebrates. Unfortunately, however, complete feet are seldom preserved in fossil form. The presence of five toes in both front and hind feet is a condition common and apparently primitive in amniotes and found as well in some of the oldest amphibians. However, there are never more than four toes in the front foot of living amphibians, and the same number is found in many fossil forms. Though we have never found more than five toes in any early tetrapod, there is embryological evidence suggesting that as many as seven might have been present in ancestral types.

Five digits tend to persist in the manus in most reptiles. In ichthyosaurs (Fig. 112, *A*, p. 196) there is great variation in the number of digits present in the "fin;" there may be as few as three or as many as eight longitudinal rows of disc-shaped phalanges. In the bipedal archosaurs the "arms" are partly or completely freed from duty as locomotor organs, and there may be much modification of the hand (Fig. 116, *B, C*). Frequently there is a reduction of the outer elements; in some dinosaurs only the three inner fingers remain, and in an extreme case only two. The flying reptiles, the pterosaurs, lost the fifth digit; the three inner ones are feeble, clawed structures, but the fourth is enormously elongated as a support for the wing membrane. The birds appear to have descended from Ruling Reptiles in which only the three inner fingers remained; these three digits are preserved, in modified fashion, in the bird wing (Fig. 116, *D*).

Five toes were present in the manus in the ancestral mammals and are found in many forms today (Fig. 117). Early mammals, it is believed, were more or less arboreal, and the first digit, as a thumb or *pollex,* became more or less opposable to the others to give a grasping power useful in tree locomotion. Most mammals abandoned arboreal life, however, at an early day; this divergent digit was of little use in terrestrial locomotion and is frequently reduced or lost.

Digital reduction and modification occur in a great variety of mammals. We may note, for example, the development in bats of a wing membrane which, in contrast to that of pterosaurs, is supported by all the digits except the "thumb," and on the other hand the modification of the whale manus into a structure somewhat comparable to that of ichthyosaurs (Fig. 112, *C*).

The development of the ungulate foot is accompanied by a reduction and elimination of digits (Fig. 119). In most ungulates the "thumb" was early lost, and a four-toed manus was characteristic of forms as dissimilar as ancestral horses and cows. Further reduction, however, followed two different paths. In one group the main axis of the limb extends down into the third digit, which enlarges at the expense of the others. This is the pattern followed by the Perissodactyla, the horses and their relatives, or odd-toed ungulates. Most of the extinct horse genera were three-toed; digits 1 and 5 were lost, 2 and 4 were small, and 3 became increasingly prominent. In modern horses splints of the metapodials of 2 and 4 are present beneath the skin, but there is a functionally one-toed condition, with toe 3 the survivor in the elimination contest. The Artiodactyla, or even-toed ungulates, followed another pattern. In them the primitive manus was four-toed, with the axis of the limb passing down between toes 3 and 4. These two toes tend to dominate jointly, and become the so-called "cloven hoof" of pigs, camels, and numerous familiar ruminating animals. The side toes (2 and 5) may persist, but are usually reduced to vestiges; the two major metapodials may fuse to form a "cannon bone."

A *phalangeal formula* indicates the number of phalanges in each digit in succession from inner to outer toes. In amphibians the number of phalanges is in general low, usually two or three. In primitive reptiles, however, there was established a formula of 2.3.4.5.3 for the manus. This persists with little change through a great variety of reptilian groups. As stated in this formula, the number of phalanges (and the length of the toes) increases steadily from pollex to fourth toe; the fifth toe is reduced in size and phalangeal count, and tends to diverge laterally from the other four. This situation is perhaps due to the fact that in the sprawling limb posture of primitive tetrapods the foot strikes the ground at an angle. The higher, outer side must be longer to get a proper "footing;" the fifth toe abandons the attempt, so to speak, of gaining sufficient length and extends out sideways as a lateral prop.

In chelonians there is frequently a reduction in phalangeal formula (Fig. 116, *A*). On the other hand, the extinct marine plesiosaurs and ichthyosaurs greatly increased the number of phalanges in each toe; each individual phalanx was, however, much abbreviated (Fig. 112, *A, B*).

In the mammal-like reptiles the earlier types (Fig. 117, *A, B*) retained the reptilian formula. Advanced forms, however, reduced to 2.3.3.3.3, which is the generalized formula for primitive mammals (Fig. 117, *C, D*). One joint has been lost from the third toe, two joints from the fourth;

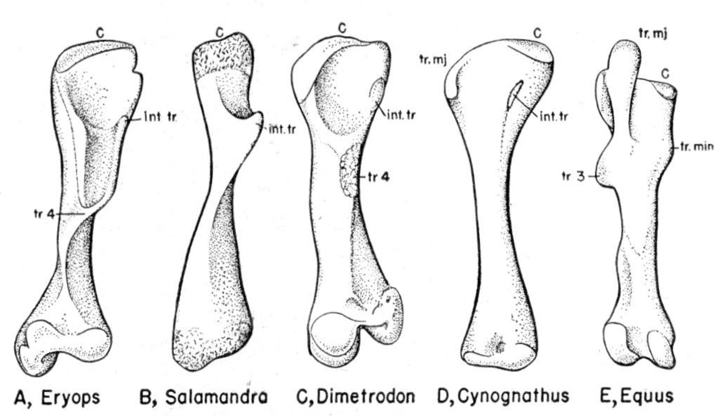

Fig. 120. Femora seen from the ventral surface. *A*, Primitive fossil amphibian; *B*, a urodele; *C*, a primitive reptile; *D*, a mammal-like reptile; *E*, the horse. The proximal end above; distally, the articular surfaces for the tibia. *c*, Head (capitulum); *int. tr.*, internal trochanter of primitive forms, for attachment of obturator externus muscle or equivalent; *tr. 3*, third trochanter of perissodactyls for part of gluteal muscles; *tr. 4*, fourth trochanter, to which are attached tail muscles pulling the femur backward in many amphibians and reptiles; *tr. mj.*, trochanter major of mammals, for gluteal muscles; *tr. min.*, trochanter minor, for iliopsoas muscles of mammal.

some therapsid genera show a transitional stage in which the joints to be lost are present but vestigial. This change in formula may be related to the changed limb posture; with the limb in a straight fore and aft plane, all the toes strike the ground equally well and hence tend to be subequal in length. Among mammals, as we have noted, there is frequently a loss of digits, but seldom a reduction in phalangeal count in any persisting toe. In the Cetacea, alone among mammals, we find an increase in the phalangeal count to as high as thirteen or fourteen in a digit; the whales here parallel the extinct marine reptiles (Fig. 112, *C*).

Femur (Fig. 120). The thighbone in primitive land vertebrates is essentially a cylindrical structure with expanded ends. Proximally there is a terminal head fitting into the acetabulum. Beneath the head is a cavity, or fossa, into which some of the short ventral muscles insert, and at the anterior (or inner) margin of this fossa is seen ventrally a strong *internal* *trochanter* for further ventral muscle attachment. Along the length of the shaft ventrally is an *adductor crest* for the attachment of strong adductor muscles. Near the proximal end of this ridge is frequently seen the so-called *fourth trochanter,* to which attach powerful muscles running onto the thigh from the tail. Distally, the femur expands, with rugosities on either side for shank muscle origins; there is a deep groove along the dorsal midline distally for the tendon of the extensor muscle of the thigh, which extends to the tibia. Below, at the distal end, there is a broad, double articular surface for the head of the tibia, and along the posterior (or lateral) margin of the femur an adjacent area for the fibular articulation.

The general nature of the femur remains little modified in a majority of tetrapod groups except for the head of the bone. This usually develops

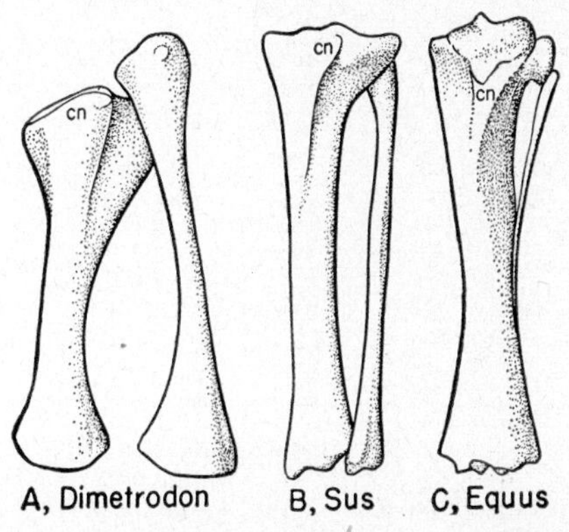

Fig. 121. The left tibia and fibula, seen from the extensor (dorsal) surface, of *A,* a primitive reptile (the Permian Dimetrodon) with a fibula of good size; *B,* the pig, showing a primitive mammalian condition, with the fibula complete, although slender; *C,* the horse, exemplifying a type with reduced fibula. *cn,* Cnemial crest.

into a more or less spherical structure set off from the shaft. In forms in which the hind limbs are rotated forward (archosaurs, birds, mammals) the femur head faces inward for better reception in the acetabulum, and a *greater trochanter* may develop on the upper end of the shaft. In mammals the internal trochanter disappears, and a new *lesser trochanter* develops in much the same region, but for different muscular attachments. The fourth trochanter is absent in mammals, where the muscles once attaching to it have dwindled or disappeared. The adductor crest may persist or be represented by rugose lines, the *linea aspera.*

Tibia (Fig. 121). The tibia, corresponding to the radius in the front leg, is the main supporting element of the lower leg segment; it is always stoutly developed, articulating broadly below with the inner side of the tarsus, and above by a much expanded head, triangular in section, with almost the entire under surface of the distal end of the femur. The front margin of the head bears a strong ridge, the *cnemial crest,* to which attaches the terminal tendon of the main extensor muscle of the thigh. Primitively the attachment of the tendon was a direct one. However, the sharp curve which this tendon must make down over the end of the femur is disadvantageous. In mammals there generally develops the *patella,* a nubbin of bone which rides over the knee joint; the extensor tendon inserts into this ossification, which is, in turn, tied to the cnemial crest.

Fibula (Fig. 121). This bone is in some respects comparable to the ulna in the forearm; it bears little weight and, in consequence, tends to

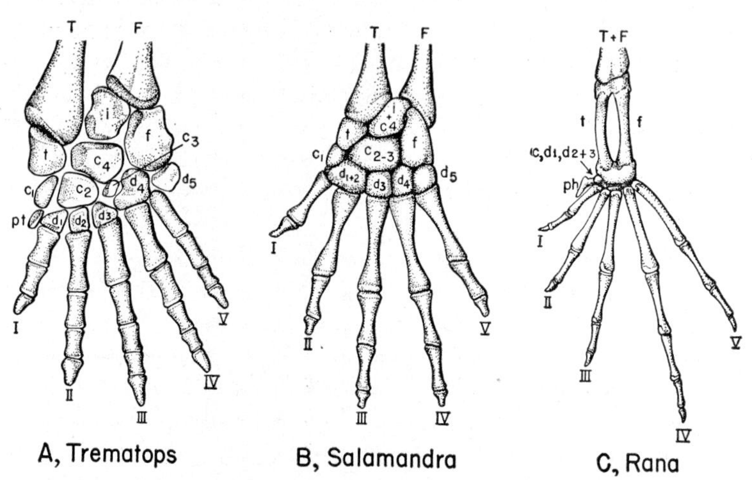

Fig. 122. The pes of amphibians, including *A*, an early labyrinthodont; *B*, a salamander; and *C*, a frog. The tarsus of Trematops includes all elements presumed to be present in ancestral tetrapods, as well as an additional pretarsal bone. In the urodele some fusion of tarsal elements has occurred. In the frog, tibiale and fibulare are so elongated as to constitute an additional limb segment; except for a few small distal elements, the fate of the other tarsal bones is not clear. Five toes are typically present, as contrasted with four in the manus, and the count of phalanges is generally higher in the pes than in the amphibian manus (cf. Fig. 115). The frog has a developed prehallux. *c1* to *c4*, centralia; *d1* to *d5*, distal tarsals; *F*, fibula; *f*, fibulare; *i*, intermedium; *ph*, prehallux; *pt*, pretarsal element; *T*, tibia; *t*, tibiale. (*A* after Schaeffer; *B* after Schmalhausen; *C* after Gaupp.)

undergo reduction. Since in the hind leg the extensor tendon attaches to the inner element, the tibia, the fibula never has any proximal development comparable to the olecranon. The fibula is articulated proximally with the posterior (or outer) aspect of the end of the femur, distally with the outer side of the tarsus. In many living groups the fibula has lost its femoral articulation, and that with the tarsus may be much reduced or lost. It usually persists, although as a slender structure, but its ends may fuse with the tibia, or the shaft may be lost.

Pes (Figs. 118, p. 200; 122–124). The structure of the ankle region, the *tarsus,* was primitively quite similar to that of the carpus; the same number of bones was present in much the same arrangement, and a terminology similar to that of the carpus may be used. A proximal series includes elements beneath tibia and fibula and one intermediate in position; these are logically termed *tibiale, intermedium,* and *fibulare.* There

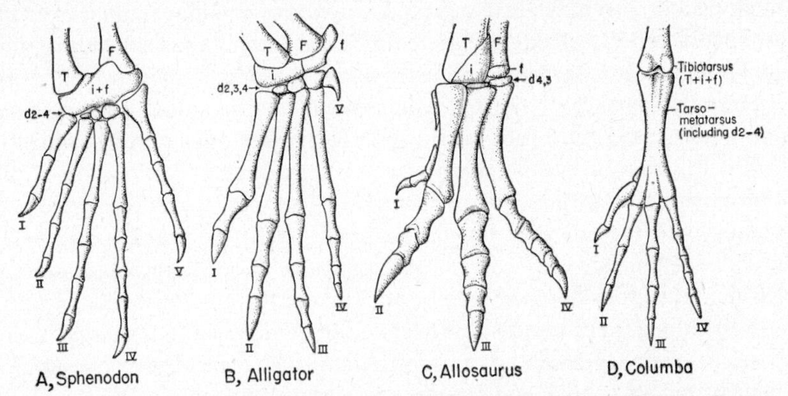

Fig. 123. The pes in reptiles and birds. *A,* Sphenodon; *B,* the alligator; *C,* a carnivorous dinosaur; *D,* the pigeon. In all forms there have been various types of reduction and fusion of the tarsal elements and a trend toward the development of a main joint within the tarsus between proximal and distal elements. In the birds all the tarsals are fused with the tibia proximally or with the fused metatarsals distally. In archosaurs, such as the alligator and dinosaurs, there is a strong trend toward the loss of the fifth toe, and it has disappeared in the dinosaur figured and in the bird. Except for the lack of fusion of the metatarsals, the Allosaurus foot is quite similar to that of birds. Abbreviations as in Figure 122. (*B* after Williston; *C* after Gilmore.)

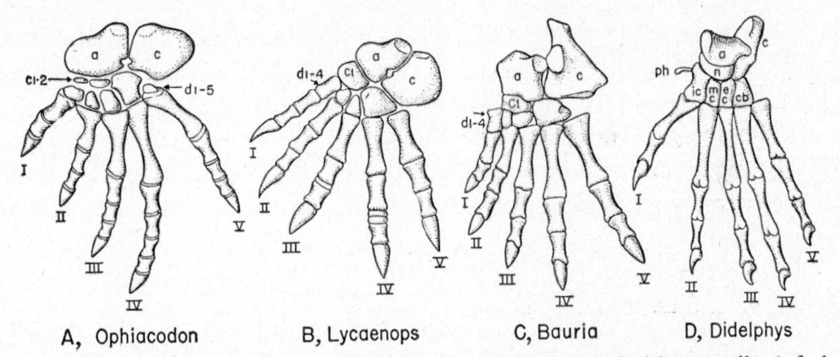

Fig. 124. Evolution of the mammalian type of pes. *A,* Primitive reptile (of the early Permian); *B,* a primitive therapsid (of the late Permian); *C,* an advanced Triassic mammal-like reptile; *D,* the opossum. Principal changes include development of a pulley surface on the mammalian astragalus, development of a heel on the calcaneum (in *C* and *D*), loss of two small tarsal elements, and reduction in phalangeal count (a transitional stage in *B*). *a,* Astragalus; *c,* calcaneum; *c1* to *c4.* centralia; *cb,* cuboid; *d1* to *d5,* distal tarsals; *ec,* external cuneiform; *ic,* internal cuneiform; *mc,* middle cuneiform; *ph,* prehallux (exceptional in mammals). (*B* and *C* after Schaeffer.)

were, as in the manus, four *centralia* in primitive types and five *distal tarsals.* As in the carpus, the central elements are reduced to two in primitive reptiles and to one in mammals, and the fifth distal element tends early to be reduced and to disappear.

There are, however, notable differences in the history of the tarsus as compared with the carpus. There is no development of any element corresponding to the pisiform. More important, however, is the change which occurs in reptiles in the proximal row; for the inner, tibial element disappears. The two remaining elements are often called by their mammalian names of *astragalus* and *calcaneum;* the tibia comes to be applied to a rounded surface on the astragalus. We have noted that in the original limb position the hind foot would tend to project out laterally from the body if it were not twisted sharply forward on the leg; elimination of the tibial tarsal aids in the forward turning of the foot. In reptiles generally the tibia can rotate freely on the astragalus. In most reptiles and early amphibians the calcaneum is a broad, flat plate of bone, the outer margin of which is important as an area of attachment for muscle tendons.

In the bipedal archosaurs there tended to develop an "intratarsal" joint between lower leg and toes; astragalus and calcaneum were closely applied to tibia and fibula, and reduced distal tarsals were functionally joined to the metatarsals. In birds, actual fusion has taken place, so that we speak of a *tibiotarsus* as a single limb element, succeeded distally by a *tarsometatarsus.*

In mammals, astragalus and calcaneum are much modified. The astragalus develops a head as a keeled rolling surface which fits into a corresponding set of grooves in the distal end of the tibia. The joint, contrasting with that of reptiles, is a hinge, without possibility of any degree of twisting, but allowing a great amount of extension and flexion of the foot. Characteristic of artiodactyls is an astragalus in which the lower end also has a keeled rolling surface, so that an extreme amount of backward and forward movement is possible. The calcaneum has little or no articulation with the more proximal limb elements in mammals, but develops a powerful heel projection to which attach the calf muscles through the "tendon of Achilles."

In the tarsus, as in the carpus, we find a complicated series of names applied to the mammalian elements:

General Terminology	*Mammalian Anatomy*	*Human Anatomy*
Intermedium	Astragalus	Talus
Fibulare	Calcaneum	Os calcis
Centrale	Navicular (or scaphoid)	Os naviculare pedis
Distal tarsal 1	Internal cuneiform	Os cuneiforme primum
Distal tarsal 2	Middle cuneiform	Os cuneiforme secundum
Distal tarsal 3	External cuneiform	Os cuneiforme tertium
Distal tarsal 4	Cuboid	Os cuboideum

The description given for the digits of the manus applies in general to the pes, but the proximal elements are, of course, termed *metatarsals* rather than metacarpals. The phalangeal formula of the pes in amphibians and reptiles frequently corresponds closely to that for the manus, except that in reptiles there is usually a fourth phalanx in the last toe. In bipedal archosaurs there was often, in relation to speedier locomotion, an elongation of the metapodials comparable to that noted in ungulate mammals. There was, further, a reduction in the number of digits in these reptiles; the divergent fifth toe often disappeared, and the first toe, or *hallux,* was turned back as a prop for the foot. There remained, in the more "advanced" bipedal types, three forwardly turned toes, symmetrically disposed, the center one the longest. These toes were, of course, the second, third, and fourth. Despite the modification seen here in their comparative length, the primitive phalangeal count had been retained; the second and fourth toes, although subequal in length, had three and five phalanges, respectively; the middle (third) toe, four. The birds have descended from bipedal archosaurs and have exactly the same toe structure; so similar are they that dinosaur footprints when first discovered were reasonably thought to be those of gigantic birds.

In mammals the history of the hind foot is comparable in general to that of the manus. We may note, however, that in the horses digital reduction proceeded more rapidly in the hind foot than in the forefoot; the earliest known fossil horses had already attained a three-toed stage.

VISCERAL SKELETON

Between the gill openings of lower vertebrates is a series of cartilaginous or bony bars—arches which aid in the support and movements of the gill apparatus. Such structures are ancient in nature, for a cartilage-like substance is found here even in Amphioxus, in which no other skeletal parts of any sort are present. These gill arches are the basic components of a set of structures termed the visceral skeleton. Except in agnathous forms, the elements of the visceral system never form more than a small proportion of the skeletal materials of the body. They are, however, notable for their versatility, the degree of modification of form and function which they have undergone in the course of vertebrate history. Primitively, to be sure, they were merely gill supports. But early in fish history they played a leading role in the development of the jaws; and although gills as such have disappeared in amniotes, persistently surviving visceral skeletal elements are to be found even in mammals in such varied areas as the braincase, the auditory ossicles, and the larynx.

At first sight it seems difficult to justify classifying these usually modest structures in a special category contrasting with all other endoskeletal elements. Consideration, however, shows that such distinction is well warranted. The gill region, or pharynx, although relatively insignificant in land vertebrates, is, if traced downward in the chordate series, increasingly prominent, important, and in every way distinctive. In the nature of its muscles and nerve supply the gill region is in strong contrast, primitively, with any other body region, and its skeleton is likewise unique.

Other endoskeletal structures form in the "outer tube" of the body;

those of the gill region are formed in the walls of the gut, and are truly visceral in position. In addition, they are unusual in their embryological origin. The visceral cartilages are derived from mesenchyme; but this mesenchyme is not of mesodermal origin—a situation embarrassing to

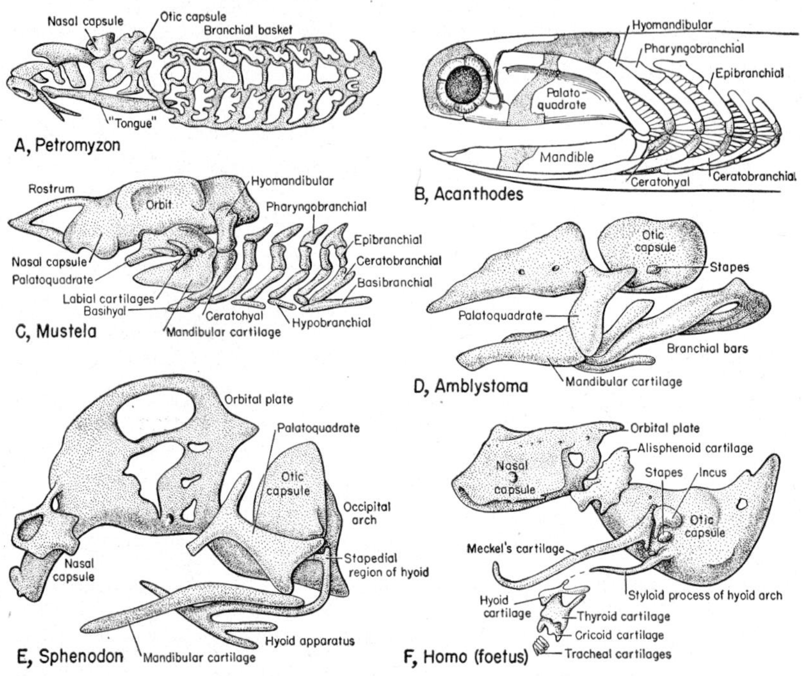

Fig. 125. Visceral skeleton and braincase in representatives of six vertebrate classes; *A* to *C*, adult forms, *D* to *F*, embryos; cartilage stippled, bone unstippled. *A*, The lamprey, with a peculiar braincase and associated cartilages not readily comparable to those of other groups; the branchial basket is fused anteriorly to the braincase. *B*, Paleozoic spiny "shark," a placoderm in which jaws are developed, but the hyomandibular is unspecialized (the sclerotic plates of the orbit are figured, as is a slender dermal bone below the mandible). *C*, Shark, with hyostylic jaw suspension. *D*, Salamander; the palatoquadrate is reduced, and the branchial arches are reduced even in the embryo or larva. *E*, The reptile Sphenodon. The braincase is incompletely formed (cf. lizard in Fig. 94*D*), and hyoid and branchial bars are reduced to a hyoid apparatus and stapedial cartilage. *F*, Human foetus. The braincase develops only ventrally and anteriorly around the brain. The palatoquadrate is reduced to alisphenoid and incus (= quadrate). The lower jaw (Meckel's cartilage) is reduced, the proximal part becoming the malleus in the adult. Other visceral elements include the hyoid and its styloid process, stapes, laryngeal and tracheal cartilages. (*A* and *C* after Goodrich; *B* after Watson; *D* after Hörstedius and Sellman; *E* after Howes and Swinnerton; *F* after Gaupp, Macklin.)

the embryologist who lays down rules for embryos to follow. It is, instead, derived from the ectoderm.

We have noted in our embryological review the development of the neural crest—cell masses from which elements of the nervous system are major derivatives. But in certain fishes and amphibians part of the neural crest cells of the head have a different fate, and the same situation may well hold true in other vertebrate groups. They migrate downward and inward, form a mesenchyme material similar to that elsewhere derived from the mesoderm, and ultimately form skeletal structures. This mesenchyme, sometimes termed *"mesectoderm,"* appears to form the most anterior part of the braincase; its major function, however, is the formation of the gill bar system.

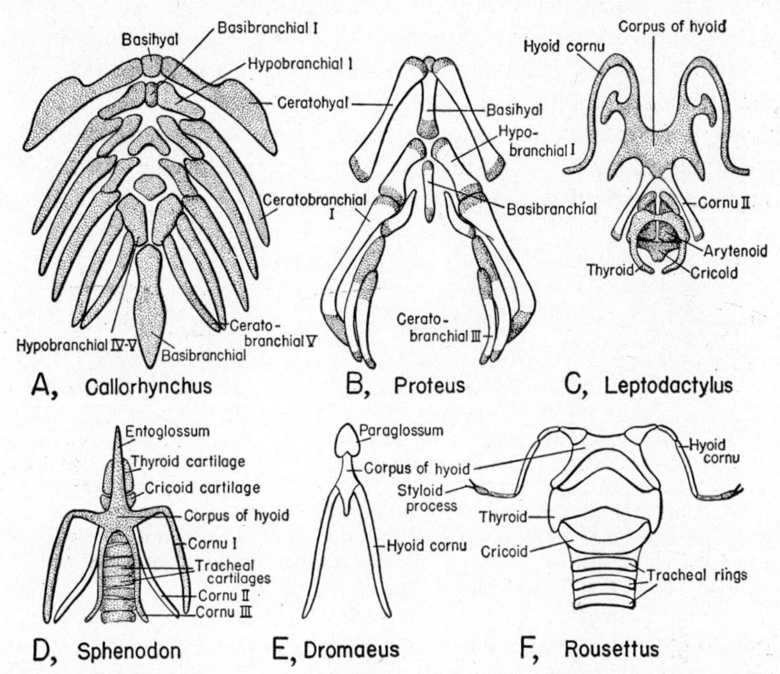

Fig. 126. Gill bars and tetrapod derivatives, in ventral view, in *A*, a chimaera; *B*, a water-dwelling salamander; *C*, a frog; *D*, a reptile, Sphenodon; *E*, the cassowary; and *F*, a bat. In *A* the dorsal arch elements are not included. In *C*, *D*, and *F* laryngeal cartilages are included. The entoglossum of *D* and paraglossum of *E* are tongue-supporting anterior developments from the body of the hyoid.

The Gill Skeleton. In the cyclostomes (Figs. 84, p. 154; 125, *A*) certain of the gill structures are highly specialized, serving to stiffen the rasping "tongue" which has evolved in these predaceous forms as a jaw substitute. The remainder of the visceral skeleton forms a continuous latticework surrounding the gill region and connecting in lampreys with the braincase. In the cephalaspid ostracoderms, some of the oldest of fossil vertebrates, the skeleton of the gill pouches was fused above with the rest of the "cranial" skeleton (Fig. 190, p. 322). These facts suggest the probability that in the ancestral vertebrates the gill supports were not discrete elements, but were fused with one another and with the cranial region. Our knowledge is, however, too imperfect as yet to be certain of the true ancestral situation, and in all gnathostomes the gill skeleton consists (except for cases of secondary fusion) of numerous separate movable structures.

The gill skeleton in jaw-bearing fishes consists of jointed bars which follow one another in sequence along the walls of the pharynx, forming arches between successive gill slits. These *visceral arches* are serially arranged structures, as are the gill openings themselves and the muscles and nerves of the gill region. It must, however, be emphasized that no proof has been found, despite many attempts, that this segmentation is in any way related to the segmentation present in other parts of the body and is there founded on the segmentation of the mesodermal somites. The pharyngeal segmentation is based, it seems sure, on the seriation of the endodermal gill pouches, and this has no proved relationship to the series of mesodermal somites seen in the trunk or head.

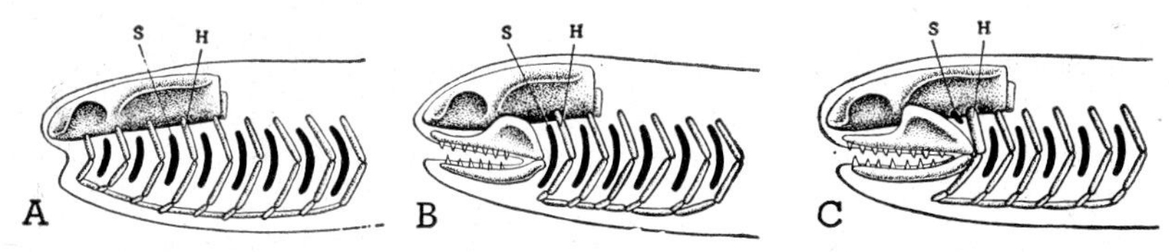

Fig. 127. Diagrams to show evolution of jaw and hyoid region. Gill openings in black. *H,* Hyomandibular; *s,* spiracular gill slit. In *A,* a primitive jawless condition. *B,* Jaws formed from a pair of gill arches (two anterior arches and slits may have been lost in the process); spiracular gill slit unreduced, and hyomandibular not specialized. This condition may have prevailed in the extinct placoderms. *C,* Condition seen in most jawed fishes; the hyomandibular has become a jaw support, and the intervening gill slit reduced to a spiracle.

Each typical gill bar of a jawed fish (Figs. 125, *C,* 126, *A,* 162, p. 284) includes on either side a major dorsal and a major ventral element, the two bent somewhat forward on each other. The dorsal elements are the *epibranchials;* the ventral ones, *ceratobranchials.* There are generally *pharyngobranchials,* turned inward over the pharynx, at the upper end of each normal arch. Below, the successive arches are tied to one another and with their mates of the opposite side by elements which extend downward, forward, and inward to the midline. There is a great variation here, but frequently there are short paired *hypobranchials* below the ceratobranchials, and median ventral structures—*basibranchials* or *copulae.* The gill bars generally bear on their inner margins *gill rakers,* and on their outer surfaces a row of *gill rays* which stiffen the gill septum or gills (Fig. 186, p. 314). In most jawed fishes five sets of typical gill bars are present, in addition to the specialized jaw and hyoid arches now to be described.

Jaw Development. Perhaps the greatest of all advances in vertebrate history was the development of jaws and the consequent revolution in the mode of life of early fishes. In this process the visceral skeleton played a leading role, for transformed gill bars are basic structures in jaw formation and, indeed, form the entire jaw structure in the Chondrichthyes.

In the sharks, without the complication of investing dermal bones, the jaws are formed solely of paired upper and lower cartilages which seem surely to represent transformed gill arch elements (Figs. 125, *C*, p. 210; 128). The shark upper jaw element, running fore and aft below and beside the braincase, is the *palatoquadrate* (or pterygoquadrate) *cartilage;* it is comparable to the epibranchial of a normal gill arch. Posteriorly, it articulates with a lower jaw or *mandibular cartilage,* presumably derived from the ceratobranchial of the same arch as the upper jaw. It appears that in the development of jaws, a pair of gill bars lying adjacent to the expanding mouth cavity became armed with teeth and enlarged to function in a new capacity as biting jaws (Fig. 127). Whether the gill arch

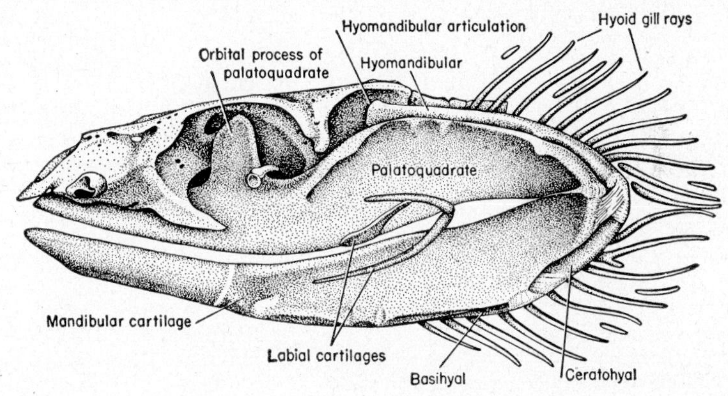

Fig. 128. The cranial skeleton of a shark, Chlamydoselache: braincase, jaws, hyoid arch. (After Allis.)

which became modified into jaws in the ancestral gnathostome was the most anterior one present in the jawless ancestor is doubtful; it is probable that one or two anterior arches were destroyed in the expansion of the mouth cavity. It has been suggested that the small *labial cartilages* present in the angle of the mouth in sharks (Fig. 128) represent such arches, but this is doubtful.

In forms other than sharklike fishes, dermal elements are also present in the complex structure of upper and lower jaws (cf. Fig. 129), and the part played in jaw formation by gill arch elements is much reduced. The jaw articulation, however, remains persistently that between upper and lower visceral bars in every group except the mammals, in which (as discussed later) the entire structure of the jaws and palate is taken over by dermal elements. Even in mammals, however, relics of the gill bar jaw structures persist, although, as will be seen, in curiously transformed fashion as ear ossicles and a component of the brain capsule.

Jaw Suspension. It is believed that in placoderms the upper jaw structures articulated directly with the brain case in simple fashion. In most more advanced fishes, however (Figs. 127, *C*, 128), the second or *hyoid arch,* lying behind the first gill opening (the spiracle) is specialized as an aid to jaw support. The ventral part of this arch is normal; there is here on either side a major element termed the *ceratohyal* and usually a smaller *basihyal.* In the upper part of this arch, however, the epibranchial segment is an element, often of considerable size, termed the *hyomandibular.* This is generally braced dorsally against the otic region of the braincase. Below, it runs to the region of the jaw articulation and is bound by ligaments to the jaw structures. This mechanism is an effective support for the jaws, propping them on the braincase. In tetrapods the upper jaw and the palatal structures are satisfactorily bound to the remainder of the

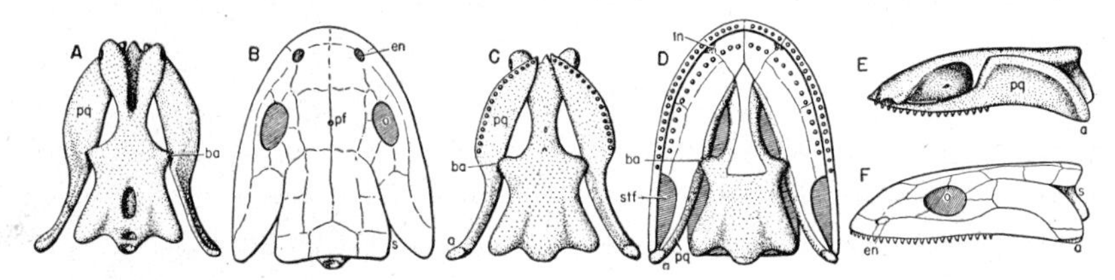

Fig. 129. Diagrams to show the components of the skull. *A, C, E,* Dorsal, ventral, and lateral views of endoskeletal structures of braincase and palate (or upper jaw) as present in a shark or an embryo of higher jawed vertebrates; *B, D, F,* similar views with dermal elements added. In *B* and *F* a dermal roof shield binds together and covers braincase and endoskeletal jaw structures; in *D,* dermal palatal structures are seen reinforcing the endoskeletal palate, and a further dermal element underlies the braincase. *a,* Articulation with lower jaw; *ba,* basal articulation of braincase with palatoquadrate; *en,* external naris; *in,* internal naris; *o,* orbit; *pf,* parietal foramen; *pq,* palatoquadrate; *s,* notch in skull for spiracle or ear drum; *stf,* subtemporal fossa for jaw muscle.

skull without the aid of the hyomandibular support. In relation to this we find that the hyomandibular has been transformed into an auditory bone, the stapes; its history is discussed in connection with the ear.

The nature of the hyomandibular serves to introduce the general problem of the suspension of the jaws on the skull in fishes. If, as is the case in most modern fishes, the upper jaw structure has no direct connection with the braincase (except far anteriorly) and the jaw joint is braced entirely by the hyomandibular, the condition is said to be *hyostylic.* In a few sharks, however, there is support both from the hyomandibular and from a direct articulation of upper jaw and braincase; this is an *amphistylic* condition. In still other fishes the hyomandibular is nonfunctional as a supporting structure, and the upper jaw region is entirely responsible for its own suspension; this is termed *autostylic.* In chimaeras (Fig. 95, p. 169) and lungfishes the autostylic condition is associated with the fact that the upper jaw itself is fused tightly to the braincase. Land vertebrates are all autostylic, for the hyomandibular never functions as a jaw support.

Tetrapod Gill Arch Derivatives (Figs. 125, *D–F*, p. 210; 126, *B–F*, p. 211). In larval amphibians gills are still functional, and in relation to this fact the gill bars remain prominently developed. There is already, however, some tendency toward reduction; there never are more than four arches present behind the hyoid arch, instead of the five customary in fishes, and these are often continuous bars, without clear-cut divisions into the various elements typical of their fish ancestors.

In metamorphosed amphibians and in amniotes functional gills have been done away with; visceral arches nevertheless persist (apart from the jaw and ear ossicles) as two sets of structures formed in connection with new tetrapod developments in the former gill region. (*a*) A tongue, never characteristically present in fishes, arises in the floor of the mouth where there once lay the more anterior and ventral parts of the visceral bars; a reduced and transformed remnant of the anterior gill bars serves, as the *hyoid apparatus,* to support the tongue. (*b*) With increased importance of the lungs, skeletal structures derived from or related to the more posterior part of the visceral skeleton are developed about the larynx and trachea which form the lung entrance. These structures consist of a laryngeal skeleton, the "Adam's apple" of man, and ring-shaped tracheal cartilages.

The tetrapod hyoid apparatus typically includes a main body, the *corpus,* formed from one or more of the median ventral arch elements (copulae or basals), and remnants of the hyoid arch and of one or two branchial arches extending outward and upward as "horns" or *cornua* (singular, cornu). The body of the hyoid is situated in the floor of the throat in the base of the tongue, ventral to the anterior end of the windpipe. In some reptiles and in birds the corpus of the hyoid develops a long anterior process for better tongue support. The horns extend upward on either side of the windpipe, and although the most anterior—"principal"—horns represent only the lower half of the original hyoid arch (the upper part forms the stapes), they may nevertheless gain a slender secondary connection with the otic region of the skull, either directly or by ligaments. In mature amphibians and in reptiles there are frequently three pairs of cornua; in birds generally a single pair; in mammals typically two pairs, representing the hyoid and first branchial arch. In mammals additional ossifications may occur in the slender process of the hyoid horn, reaching up to the ear region, and in some mammals, including man, the distal end of this horn may fuse with the skull as a *styloid process.*

But the hyoid apparatus is only one part of the tetrapod derivatives of the fish gill supports; skeletal structures supporting the windpipe itself have developed from the visceral arch system. Just beyond the entrance to the windpipe, the glottis, there typically develops in tetrapods an enlargement in the tube as the larynx (cf. Chap. X). The basal part of the hyoid apparatus generally lies close to the front end of the floor of this expansion; additional cartilages, at least part of which appear to be modified visceral arch structures, tend to form a complex *laryngeal skeleton.* When fully developed, these include unpaired *thyroid** and *cricoid cartilages* in line ventrally behind the body of the hyoid and, dorsally, paired *arytenoids;* these cartilages are not infrequently ossified. In addition, we find in mammals a cartilage developing in the flap covering the glottis—the epiglottis. In most tetrapods the windpipe (trachea), running down from the larynx, tends to be strengthened by ring-shaped *tracheal cartilages*. These last structures are not readily comparable to any specific fish elements, but are reasonably to be regarded as new developments of the visceral skeletal system.

* The term "thyroid" means simply shield-shaped; the thyroid gland, also shield-shaped, happens to lie in this general region of the throat, but the topographic juxtaposition is pure coincidence.

CHAPTER VIII

THE SKULL

THE TERM "SKULL" is used in somewhat variable fashion. In a broad way it refers to any type of skeletal structure found in the head region. In this sense we may consider the lamprey as having a skull composed of a braincase and the peculiar tongue cartilages, the shark as having a skull consisting of braincase and isolated upper and lower jaw bars. But in common parlance, and as we shall use it here, the term has a somewhat different meaning. The familiar skull of every form from a bony fish to a mammal is a fused, unit structure, in which braincase and endoskeletal upper jaws are welded together by series of dermal bones; the lower jaw is not included.

In the older Agnatha and in the extinct placoderms there appears to have been present a fused head skeleton which included dermal as well as endoskeletal materials and is hence to be considered a skull in the present sense. But in most of those archaic fishes the structures seem to have been aberrant, and they are as yet too incompletely known to warrant their being considered here. In the higher bony fishes and all tetrapods there is a well-constructed skull, with many common features present throughout these groups. Even so, it is difficult to give a generalized description which will hold in all cases. And the task is further complicated by the fact that in most living groups considerable degeneration and specialization have taken place, obscuring the true phylogenetic story.

In the study of soft anatomy we suffer under the handicap of knowing nothing of the actual ancestral types from which the organs of modern animals have been derived; we can, at the best, merely make guesses (which we hope are intelligent) at the structural conditions in such ancestors, extinct these many millions—or hundreds of millions—of years. In the case of skeletal history, however, we are better off, for there are frequently preserved as fossils the bones of the actual forbears. As regards the skull, we have a fairly complete knowledge of the structure of the head in the ancient labyrinthodonts of the later Paleozoic, whose skulls are of the very type from which that of all later tetrapods has been derived. Further, though bony fish history is still obscure in some respects, the skull pattern of the higher fishes is not far removed from that of the ancient amphibians. We shall, therefore, give at this point a rather full account of this central skull pattern and then discuss the major modifications

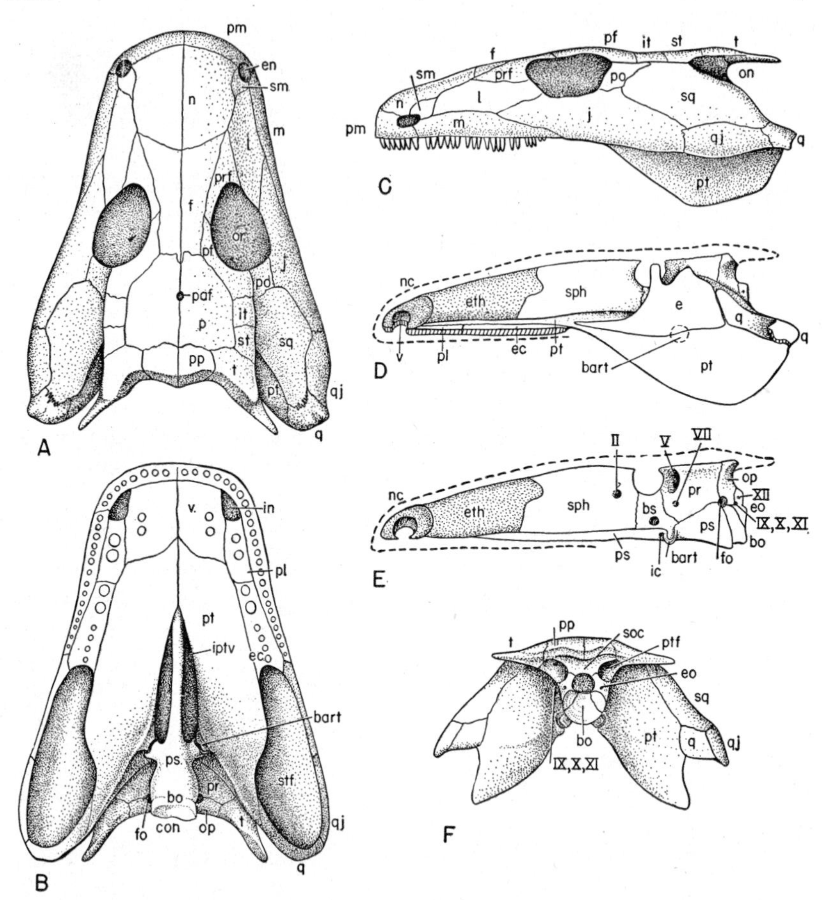

Fig. 130. The skull of an ancestral land vertebrate, based primarily on the Carboniferous labyrinthodont, Palaeogyrinus. *A,* Dorsal view of dermal skull roof; *B,* palate; *C,* lateral view; *D,* lateral view with dermal skull roof removed (outline in broken line); palatal bones—dermal and endoskeletal—of left side are shown; deep to them, the braincase. The hatched area is the sutural surface of palatal bones against maxilla. *E,* Lateral view of braincase; *F,* posterior view. Roman numerals, foramina for cranial nerves. *a,* Angular; *ab,* auditory bulla; *a proc,* angular process; *ar,* articular; *as,* alisphenoid; *bart,* basal articulation of palate and braincase; *bo,* basioccipital; *bs,* basisphenoid; *c,* coronoid; *con,* condyle; *c proc,* condylar process; *d,* dentary; *do,* postparietal (dermal supraoccipital); *e,* epipterygoid; *ec,* ectopterygoid; *en,* external naris; *eo,* exoccipital; *esl esm,* extrascapular, lateral and medial; *et,* entotympanic; *eth,* ethmoid region; *ett,* ethmoturbinal; *f,* frontal; *fm,* foramen magnum; *fo,* fenestra ovalis; *g,* gulars; *gl,* jaw articulation (glenoid fossa); *hy,* hyomandibular articulation; *ic,* foramen for internal carotid; *in,* internal naris; *ina,* internasal; *inf de for,* inferior dental foramen; *iof,* infraorbital foramen; *iptv,* interpterygoid vacuity; *it,* intertemporal; *j,* jugal; *l,* lacrimal; *ls,* laterosphenoid; *m,* maxilla; *me,* mesethmoid; *m for,* mental foramen; *mt,* maxillo-turbinal; *n,* nasal; *nc,* nasal capsule; *nt,* nasoturbinal; *ntc,* notochord (restored); *o,* opercular; *oc,* otic capsule; *occ,* occipital bone; *on,* otic notch; *op,* opisthotic; *or,* orbit; *os,* orbitosphenoid; *p,* parietal; *pa,* prearticular; *paf,* parietal foramen; *pap,* palpebral; *pd,* predentary; *pe,* periotic; *pf,* postfrontal; *pin,* pineal foramen; *pl,* palatine; *pm,* premaxilla; *pn,* postnasal; *po,* postorbital; *pop,* preopercular; *popr,* paroccipital process;

seen in the various fish and tetrapod groups. The roof pattern and general proportions of the skull figured (Fig. 130) are those of the embolomerous amphibian Palaeogyrinus; the internal structures, not fully known in that genus, are restored after the pattern of certain other primitive labyrinthodonts.

Skull Components. For analysis the skull may first be resolved into a number of major components, each including a variety of ossifications (Fig. 129, p. 214). We may distinguish:

A. Dermal structures, including
   1. A dermal roof—a shield of membrane bone covering thoroughly the top and sides of the head to the jaw margins, unbroken by openings save for the nostrils (*external nares*), eyes (*orbits*), and a small *parietal foramen* for a third, median, eye. At the sides the shield is primitively notched behind the orbit for the spiracular gill opening, or for the middle ear and eardrum region which replace it in tetrapods. The lower margins of the dermal shield bear the marginal teeth.
   2. Dermal palatal bones, formed in the skin of the mouth roof on either side. In the Choanichthyes and early land vertebrates there is a lateral gap anteriorly on either side for the *internal nares,* or *choanae*. Posteriorly, the palatal elements are broadly separated from the edge of the shield by the *subtemporal fossae,* through which descend the temporal muscles which close the jaws.
   3. A median dermal ossification forms in the skin of the mouth roof on the under surface of the braincase.

B. In all jawed vertebrates the visceral arch series contributes a cartilaginous bar (palatoquadrate cartilage), or bones replacing it, to the upper jaw structure; in vertebrates with a formed skull this is welded to the dermal palatal bones to form a palatal complex. This visceral structure primitively bears the articulation with the lower jaw, and may articulate with the braincase at one or more points.

C. The braincase, formed in cartilage, but usually ossified to a considerable degree.

The Primitive Amphibian Skull. *Dermal Roof* (Fig. 130, *A, C*). The dermal skull roof of an early amphibian includes a considerable number of bony elements, suturally connected to form a practically solid armor

---

por, postrostral; *pos,* postsplenial; *pp,* postparietal; *pr,* prootic; *prf,* prefrontal; *prp,* preparietal; *prs,* presphenoid; *ps,* parasphenoid; *pt,* pterygoid; *ptf,* post-temporal fossa; *q,* quadrate; *qj,* quadratojugal; *r,* rostral; *s,* stapes; *sa,* surangular; *sc,* sagittal crest; *se,* sphenethmoid; *sm,* septomaxilla; *so,* supraorbital; *soc,* supraoccipital; *sop,* subopercular; *sp,* splenial; *sph,* sphenethmoid; *sq,* squamosal; *st,* supratemporal; *stf,* subtemporal fossa; *t,* tabular; *tl,* temporal lines indicating edge of original skull roof and upper boundary of temporal muscles; *tm,* tympanic; *v,* vomer. (Mainly after Watson.)

shield. In general these bones are paired; occasionally we find as variants unpaired median ossifications, usually variable and of little importance. It is a strain to commit to memory the names of the many paired elements present. Many of them, however, occur in bony fishes and in all tetrapods and are important to the student of osteology, of taxonomy, and of paleontology. As an aid to memory we may group them into several series:

(*a*) Tooth-bearing marginal bones. At the tip of the skull there is on either side a small *premaxilla* bearing the most anterior teeth. The remaining marginal teeth are carried on the elongate *maxilla*. The suture between premaxilla and maxilla lies below the external naris.

(*b*) A longitudinal series along the dorsal midline. From the premaxilla back to the occipital end of the skull we find in order the following bones: nasal, frontal, parietal, postparietal. The *nasals* lie medial and posteromedial to the external nares. The *frontals* lie between the orbits, but primitively do not enter into their margins. The *parietals* are generally large and occupy the central part of the skull table—the flattened roof area behind and between the orbits; the parietal foramen is situated here. The *postparietals* (dermal supraoccipitals) are at the back rim of the roof and may descend as flanges on to the posterior surface of the braincase.

(*c*) A circumorbital series of five bones—prefrontal, postfrontal, postorbital, jugal, and lacrimal—which primitively form a complete ring around each orbit. *Prefrontal* and *postfrontal* form the upper rim of the orbit; the *postorbital* occupies a place at the upper back orbital margin. The *jugal* is usually a large bone centered beneath the orbit. The *lacrimal* extends forward from the orbit to the nasal region; the lacrimal (tear) duct follows this course in tetrapods, and the bone derives its name from the fact that in amniotes it carries a canal for this duct on its inner surface. (We may mention here in passing the *septomaxilla,* a small and obscure dermal element which may be present at the back margin of the external naris adjacent to the lacrimal, or appear as a small shell of bone within the nasal cavity; it is described occasionally in tetrapods of one group or another, but its history is poorly known.)

(*d*) Temporal series. The notch in the skull behind the orbit is known as the *otic notch* in early tetrapods, since the eardrum is believed to have been placed here. Extending back above the notch at the margin of the skull table is a row of three small bones—*intertemporal, supratemporal,* and *tabular*—which tend to be reduced and lost in most later tetrapods.

(*e*) Cheek bones. Most of the cheek region, behind the orbit and below the otic notch, is occupied by a broad, platelike bone, the *squamosal;* below it there is primitively a smaller marginal element, the *quadratojugal.*

In primitive tetrapods the skull shield usually had a "sculptured" surface pattern of ridges or pits to which the dermis was strongly attached. This pattern is repeated in primitive bony fish skulls, but is generally replaced by a smooth surface in living types. Fishes and water-dwelling amphibians bear sensory lateral line organs (cf. Chap. XV) in a pattern of canals or grooves on the head. In fishes the canals are usually sunken into the substance of the roofing bones as closed tubes with occasional pores opening to the surface; in amphibians they are more superficial and indicated at the most by grooves in the roofing bones. Similar canals or grooves are found in the lower jaws and in the operculars of fishes.

Palatal Complex (Fig. 130, *B, D*). In sharks, lacking dermal bones, the only "skull" structure lateral to the braincase is the cartilage which functions as an upper jaw. This, however, is not comparable to the jaw margins of bony fishes or tetrapods, for in them this element is combined with dermal bones to form a complex palatal structure (Fig. 129, *D*, p. 214), and the jaw margins are formed by bones of the skull shield.

The major part of the palatal complex consists of four pairs of dermal bones developed in the skin lining the roof of the mouth. The largest and most medially situated is the *pterygoid,* which primitively runs the greater part of the length of the skull; it is applied to the superficial—that is, lower—surface of the palatal cartilage or the ossifications formed in it, and in part replaces that structure as an element of strength in palatal construction. Lateral to the pterygoid is a row of three further dermal elements: the *vomer,* which meets its partner anteriorly and extends back medial to the internal naris; the *palatine,* extending back from that opening along the palatal margin; and the *ectopterygoid,* terminating at the subtemporal fossa. Teeth may be present on any or all of the four dermal elements of the palatal complex in bony fishes and early tetrapods, and the three lateral bones sometimes carry a series of teeth even larger than those of the jaw margins (shark teeth are perhaps comparable to this set, rather than to the marginal teeth of bony types).

These dermal palatal bones serve many of the functions performed by the shark upper jaw cartilage; but that structure, or bones replacing it, persists, although in somewhat reduced form, in the skull of all bony fishes and tetrapods as the *palatoquadrate* (or pterygoquadrate) cartilage. In bony fishes a number of ossifications, variously named, may be found along the length of this bar; they may be termed as a group, *suprapterygoids,* since they are found in general above or internal to the pterygoid bone. In tetrapods but two ossifications are found in the palatoquadrate cartilage. Posteriorly, the *quadrate* bone forms the articular surface for the lower jaw. More anteriorly a second ossification, the *epipterygoid,* lies, as the name implies, above the pterygoid. This bone articulates at the basal articulation with the sphenoid region of the braincase; a movable articulation at this point between palate and braincase was characteristic, it would seem, of early bony fishes and early land vertebrates. A vertical bar of the epipterygoid extends upward toward the skull roof; more posteriorly, the epipterygoid may extend backward to meet the quadrate. Anteriorly, the palatoquadrate cartilage originally extended, it seems, on toward the nasal region, but this area is reduced and cartilaginous in early tetrapods.

The anterior part of the palatal complex is an essentially horizontal plate of bone, notched laterally by the internal nares, but otherwise tightly apposed to the marginal tooth-bearing dermal bones. The palatal plates of the two sides meet anteriorly in a long suture; more posteriorly, they separate, leaving interpterygoid vacuities and exposing the ventral surface of the braincase. The posterior half of the palatal complex is, by contrast with the anterior part, a vertical sheet of bone, bounding medially the subtemporal fossa for the jaw muscles. At the junction of anterior and posterior parts of the palatal complex is the basal articulation with the braincase. Behind this point there is an open cleft between the vertical palatal plate and the side walls of the braincase. Through this cleft pass important blood vessels and nerves; through it also passes in primitive bony fishes the spiracular gill slit, and in amphibians the ear passage which replaces the spiracle.

The Braincase (Fig. 130, *B, E, F*). In bony fishes and all lower tetrapods a median dermal element, the *parasphenoid,* forms in the skin of the roof of the mouth beneath the braincase and is applied closely, sometimes indistinguishably, to its lower surface. The anterior part of the parasphenoid is primitively a slender and sometimes rodlike structure; more posteriorly, from the region of the basal articulation, it spreads broadly over the expanded ventral surface of the braincase.

The braincase is well ossified in most bony fishes and tetrapods. However, the region of the nasal capsules never ossifies in land vertebrates, and in the skeletally degenerate living amphibians braincase ossification is greatly reduced. In the better ossified forms the bony elements are frequently fused in the adult, so that it is difficult to delimit the areas of the individual bones.

In the occipital region a ring of four bones—median *basioccipital* and *supraoccipital* elements, and paired *exoccipitals*—surrounds the foramen magnum. In fishes and primitive tetrapods the occipital condyle is a single concave circular structure, mainly formed by the basioccipital, but with the exoccipitals entering into its upper and lateral borders. Basioccipitals and exoccipitals are formed, embryologically, from fused occipital arch elements and hence (not unexpectedly) tend to resemble the centrum and neural arches of a vertebra. The supraoccipital, on the other hand, arises in a cartilage connecting the otic capsules of the two sides. The basioccipital may extend well forward beneath the floor of the brain cavity. The exoccipitals, however, are relatively small. Except in the degenerate modern amphibians, they are pierced in all tetrapods—and in crossopterygian fishes as well—by openings for the twelfth (hypoglossal) nerve. The vagus and accessory nerves (X and XI) and usually a vein draining part of the brain pass through a large foramen (jugular) lying anterior to the exoccipital and representing the embryological gap between occipital arch and ear capsule; nerve IX may use this exit or pierce, more anteriorly, the ear capsule.

The roof of the braincase is generally open between the two otic capsules, but the region is, of course, protected by overlying dermal elements. Two ossifications are present in the otic region, the *prootic* and *opisthotic,* but the two are generally firmly fused and difficult to distinguish in the adult. The sacs and canals of the internal ear are buried within these ossifications of the capsule and may extend into the supraoccipital as well. In fishes there is no external opening from the otic region, but the head of the hyomandibular bone abuts against its outer surface. In land animals the *fenestra ovalis* is present at this point to receive the footplate of the sound-transmitting stapes (evolved from the hyomandibular). Nerve VIII enters the auditory capsule from within, for the reception of its sensory stimuli; nerve VII usually pierces the prootic near the anterior end; and the large trigeminal nerve (V) emerges through one or more openings at the anterior margin of the prootic. There is frequently an opening, the *post-temporal fenestra,* running in forward from the back surface of the skull above the otic capsule and below the dermal bones of the skull roof.

Forward from the otic region the braincase rapidly contracts in width to the interorbital or sphenoid region. A major element here is the *basisphenoid,* an unpaired bone which has its center of ossification in the braincase floor. At either side the basiphenoid sends out a *basipterygoid process,* which forms the basal articulation with the palatal complex. A pocket, the *pituitary fossa* (or sella turcica, not seen in the figures) extends down into this element from the floor of the brain cavity, and contains the pituitary body. Nearby there is, ventrally, an opening from below on either side through which the internal carotid artery enters the braincase, and just behind the pituitary there is primitively a canal containing a vein which crosses from one orbit to the other. The lateral walls of the interorbital region are pierced by an opening for the optic nerve (II) and tiny foramina for the eye muscle nerves (III, IV, VI).

A median ossification, the *sphenethmoid,* sheathed below by the anterior process of the parasphenoid, is present in the narrow ethmoid region. Within it the two olfactory nerves diverge toward the nasal capsules. The more anterior part of the ethmoid area and the nasal capsules remain unossified in early tetrapods; they are ossified in some early fishes, but the nature of the ossification centers here is unknown.

The Skull Roof in Bony Fishes. Before following through the story of skull evolution in later tetrapods, we shall first compare the skull pattern seen in early tetrapods with that of their bony fish relatives.

Tetrapod ancestors are to be sought among the Choanichthyes, either the crossopterygians or forms closely related to them. We have considerable knowledge of the cranial structure of Paleozoic crossopterygians; their skulls show a resemblance to the early tetrapods in many features.

The dermal shield of these forms (Fig. 131, *A, C*) is as a whole comparable to that of tetrapods, and, despite considerable confusion in past interpretations, many of the individual dermal elements can be directly homologized. It is, however, interesting to note the marked evolutionary

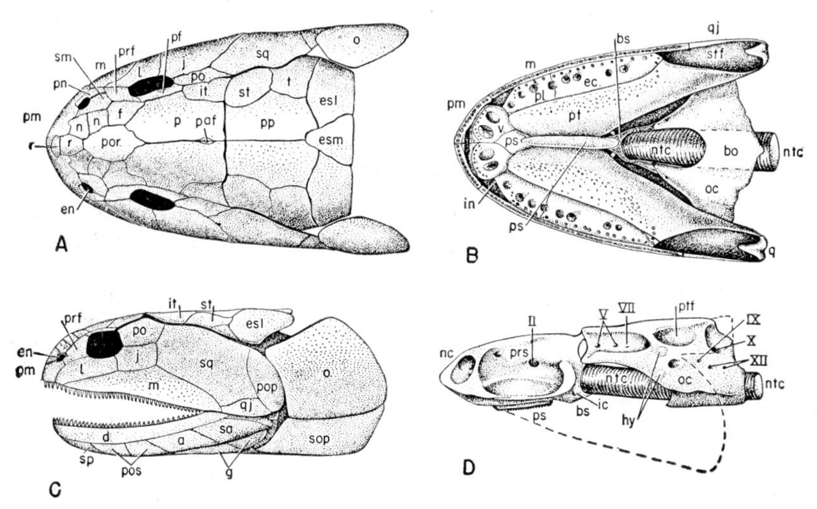

Fig. 131. The skull of a Paleozoic crossopterygian (composite), for comparison with that of a primitive land vertebrate (cf. Fig. 130). *A*, Dorsal view; *B*, palatal view; *C*, lateral view; *D*, lateral view of braincase. The labyrinthodont dermal covering differs primarily in loss of opercular elements and posterior row of bones on roof; relative reduction of length of posterior part of skull and elongation of "face" region; reduction of small elements in rostral and nasal region. The palate is similar in the two. The labyrinthodont braincase is less completely ossified, formed in one piece instead of the two present in crossopterygians, and lacks the greatly expanded notochord of the latter. Abbreviations as in Figure 130. (*A* and *B* based on Eusthenopteron; *C* based on Osteolepis; *D* based on Ectosteorhachis; data from Jarvik, Romer, Säve-Söderberg, Stensiö.)

change in proportions between front and back regions of the skull. In crossopterygians the paired eyes and parietal foramen are far forward on the head; there is a long skull table and cheek region and a short "face." In early land vertebrates the facial region is much more developed, and the posterior part of the head relatively short. This suggests that the early ancestors of the bony fishes were very short-faced and that the facial region is in process of expansion in the crossopterygians. This idea appears to be borne out if we examine the dermal roof pattern. In the back part of the head most of the bony elements are directly comparable to those of tetrapods. Anteriorly, however, the small elements present are variable, and difficult in many cases to homologize with those of land vertebrates;

the face, it would seem, was a new, growing development, in which stability had not yet been attained.

The dipnoans are regarded as an aberrant offshoot of the primitive stock of the Choanichthyes. The skull roof is certainly aberrant. In modern lungfishes the skull roof is incomplete and includes but a few large elements not readily comparable to those of crossopterygians or amphibians. In the oldest fossil lungfishes the roof shield consisted of numerous small polygonal bones which varied not merely from form to form, but even from one individual to another. In later genera there was a trend toward reduction in the number of these elments and for the preservation of a relatively few large plates such as those present today.

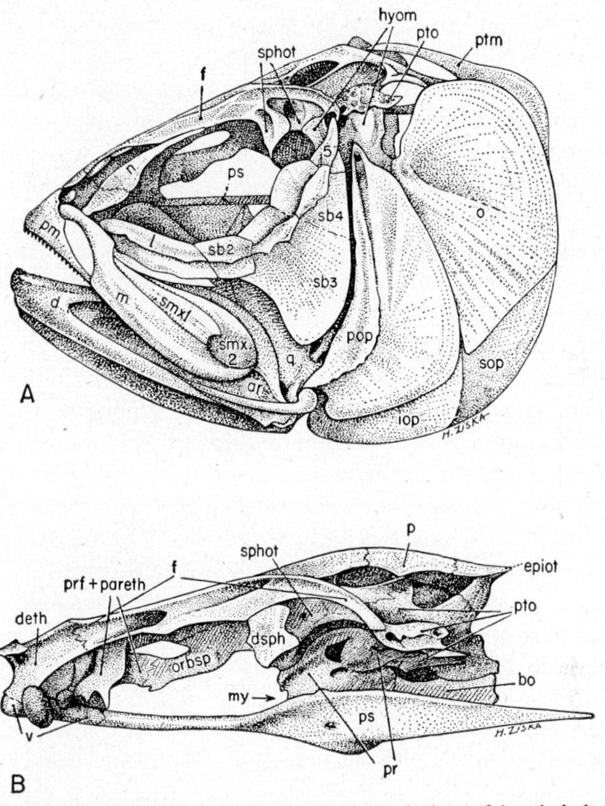

Fig. 132. The teleost Clupea (herring). *A*, Lateral view of head skeleton; *B*, lateral view of braincase. The structure is different from that of tetrapods and crossopterygians. *deth*, Dermethmoid; *dsph*, dermosphenotic; *epiot*, epiotic; *hyom*, hyomandibular; *iop*, interopercular; *my*, opening of myodome indicated by arrow; *orbsp*, orbitosphenoid; *pareth*, parethmoid; *ptm*, post-temporal; *pto*, pterotic; *sb2* to *sb5*, suborbitals; *smx1*, *smx2*, supramaxilla; *sphot*, sphenotic. Other abbreviations as in Figure 130. (From Gregory.)

The presence of a variable pattern of numerous small bony plates over the entire roofing shield in early lungfishes and on the snout region of crossopterygians suggests that in the early bony fish ancestors in general bony centers were numerous, small, and variable in the head shield, and that only later was there a tendency at various times for a reduction in the number of elements present and for the stabilization of the pattern. Since it is improbable that this process of elimination and stabilization would have been the same in all cases, it is perhaps too much to expect that homologies can be firmly established between individual roofing bones in the various groups of bony fishes.

The actinopterygians (except the paddlefishes) have a well-developed dermal roof in which many of the elements are quite stable and placed in positions comparable to those of certain of the bones in crossopterygians and tetrapods (Figs. 99, p. 172; 132, *A*). In consequence, these elements are customarily given familiar names, such as parietals, frontals, and so on; we have, however, little guarantee of true homology. A notable difference from crossopterygians and tetrapods is that in ray-finned fishes there is seldom much development of a cheek region behind the orbit and no development, normally, of a bone there corresponding to the squamosal. Except in a few early forms there is no parietal foramen. In actinopterygians the orbits are generally large, and the facial region anterior to them persistently short and undeveloped. The jaws were primitively long, and the maxillae well developed. In higher actinopterygians, however, the gape is reduced and the jaw articulation moved forward; the marginal dentition tends to disappear in teleosts, and the maxilla may be eliminated from the borders of the mouth.

The Palate in Bony Fishes. In crossopterygians the palatal structure (Fig. 131, *B*) is quite comparable to that of their tetrapod descendants; the dermal elements are almost identical in every regard. In the endochondral component, however, the crossopterygians show some differences which are probably primitive, for instead of the two replacement bones seen here in amphibians there appears to be a whole row of suprapterygoid bones, of which epipterygoid and quadrate are the tetrapod survivors.

In lungfishes the marginal teeth of the mouth are lost, and the remaining upper dentition includes only a pair of large, fan-shaped tooth plates fused to the pterygoids and a much smaller anterior pair on the vomers. In correlation we find that the dermal pterygoid bone is well developed, and that there is a small vomer; the other palatal bones, however, have been lost. There is no ossification of any sort in the palatoquadrate cartilage, which is firmly fused to the braincase in autostylic fashion.

In actinopterygians the palatal construction appears primitively to have been similar to that of crossopterygians, with several replacement bones forming in the palatoquadrate cartilage. In most living actinopterygians the basal articulation of palate and braincase is absent, and in compensation the hyomandibular is large and well ossified (Figs. 99, p. 172; 132, *A*); an extra bone, the *symplectic,* appears in this area to bind further the hyomandibular to the upper jaw complex.

The Braincase in Bony Fishes. In lungfishes the braincase, like other parts of the skeleton, is degenerate, persistently cartilaginous except for a pair of exoccipitals; in the sturgeons and paddlefishes much of the braincase has likewise degenerated to a cartilaginous state. In the ancient crossopterygians and older fossil members of the actinopterygians the whole structure was thoroughly ossified, but most sutures were obliterated during growth, and hence little can be said as to the individual bones present. In modern higher ray-finned forms, including the teleosts, the well-ossified braincase generally shows discrete bones; many appear to be the same as in tetrapods, but the otic region often includes as many as five bones which are not directly comparable to those of land animals (Fig. 132, *B*).

In many regards, even in small details, the crossopterygian braincase structure (Fig. 131, *B, D*) is quite similar to that of early land animals. There is, however, one striking difference. Whereas the braincase of all other known vertebrates is built, in the adult, as a single compact unit, that of the crossopterygians so far studied is built in two quite separate pieces. The anterior segment is rather firmly bound to the upper jaw and palatal structures; it includes the nasal, ethmoid, and sphenoid regions, back to the area of the pituitary fossa and the basal articulation. The posterior segment includes the otic and occipital parts of the braincase. Embryologically, the front segment includes the trabeculae and associated structures (cf. p. 168); the back segment is formed from the parachordals, otic capsules, and occipital arches. The two parts are movably articulated with one another, and a further specialization tends to give them a flexible union; the notochord runs forward, without constriction, through a canal in the floor of the posterior moiety to terminate in a socket on the posterior surface of the anterior segment.

This unusual type of braincase construction appears to have had a functional value in reducing the shock of the jaw snap in a predaceous fish. The divided braincase may be a specialization within the group. In that case we must look for the ancestors of the tetrapods not among known crossopterygians, but among ancient relatives of theirs, as yet undiscovered, which were similar in structure, but lacked this specialized feature. On the other hand, it is quite possible that this crossopterygian condition is ancestral, and that a two-part braincase was present in the forbears of the tetrapods.

In actinopterygians the eyes and orbits are usually large, and in relation to this fact we find that between the orbits the braincase may be reduced to a membranous septum. In most actinopterygians there develops behind the pituitary region a large excavation in the floor of the braincase, a *myodome,* in which the rectus muscles of the eyeball take their origin.

History of the Tetrapod Skull Roof. Rather than discuss seriatim the changes rung on the primitive tetrapod skull pattern as a whole in group after group of higher vertebrates, we shall follow through separately the history of the major components.

As regards the elements of the dermal skull roof, their later history is almost exclusively one of loss and degeneration. Almost never is there any development of a new element; always there is, in time, a greater or lesser

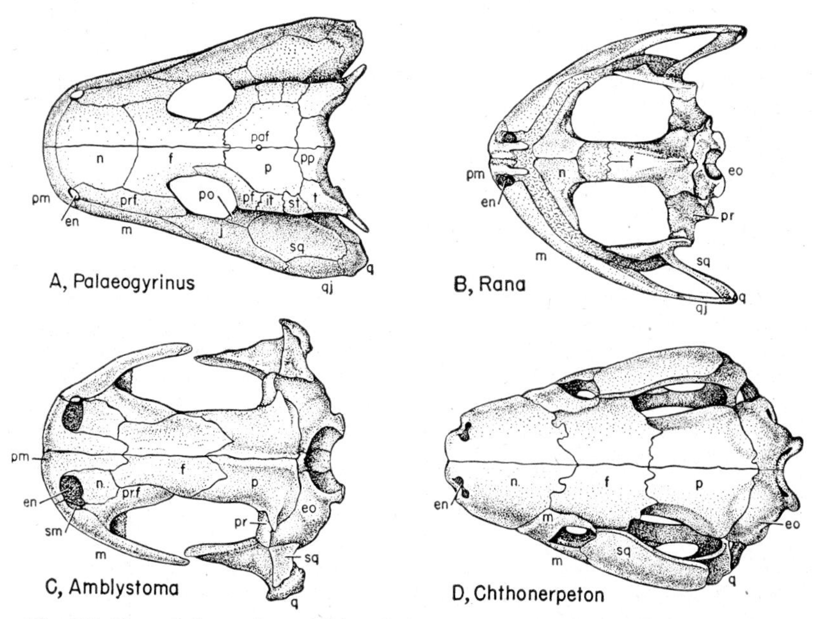

Fig. 133. Dorsal views of amphibian skulls. *A,* The labyrinthodont, Palaeogyrinus; *B,* a frog; *C,* a salamander; *D,* an apodan. Abbreviations as in Figure 130. (*A* after Watson; *D* after Marcus.)

degree of reduction. No living tetrapod has retained in full the pattern of its early ancestors, and few have preserved a solid roof covering.

Although a great variety of Paleozoic and early Mesozoic amphibians retained the primitive head shield pattern with little change, the modern amphibians show a greater degeneration of this structure than do many of the amniotes. In the broad, flat skull of a modern frog or salamander (Fig. 133, *B, C*) only a small proportion of the ancient dermal roof elements is preserved. Premaxilla, maxilla, frontal, parietal, and squamosal form the usual complement of bones; in urodeles a prefrontal is also present, but in anurans the parietal is absent (replaced by the expanded frontals), and in some urodeles maxilla and nasal are absent. All other elements have vanished, and much of the upper surface of the head is bare of dermal bone. In the Apoda (Fig. 133, *D*) the skull roof is a solid structure, but there is a similar reduction in the number of bones present; this solidity is probably a secondary condition associated with the burrowing nature of those wormlike forms.

In the stem reptiles (cotylosaurs) (Figs. 134, *A*, 135, *A*) the primitive pattern of the skull roof was retained with practically no change from primitive amphibians; indeed, the only diagnostic difference was the disappearance of the little intertemporal bone. In later reptiles (Figs. 134,

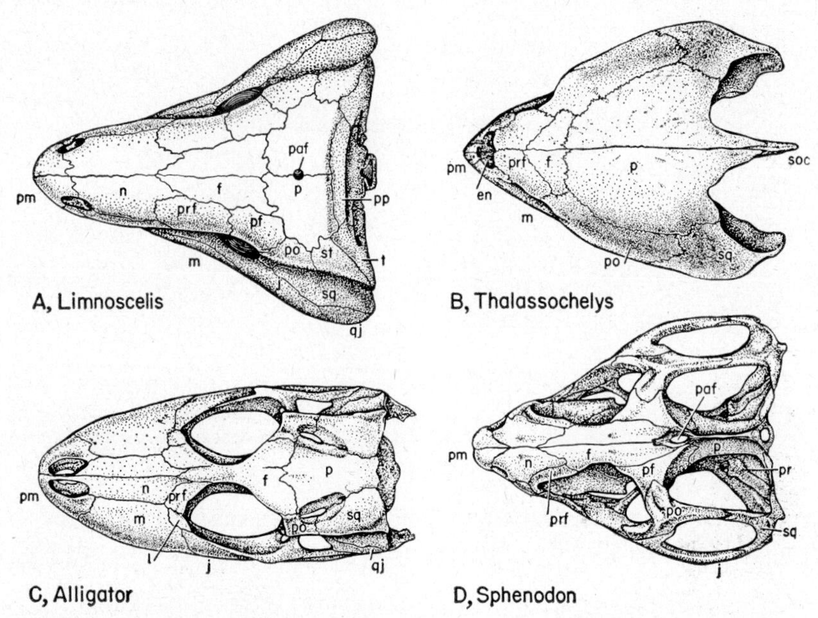

Fig. 134. Dorsal views of reptilian skulls. *A,* Stem reptile of the Paleozoic; *B,* a sea turtle; *C,* a young alligator; *D,* Sphenodon. Abbreviations as in Figure 130.

B–D, 135, *B–F*), however, there are notable losses and structural modifications. The elements which most generally disappear are those in the temporal region and the back margin of the skull—supratemporal, tabular, and postparietal; no one of the three is positively identifiable in any modern reptile. The pineal opening became unfashionable, it would seem, in many reptiles; Sphenodon and many lizards retain it, but it has vanished in turtles, snakes, and crocodiles (and extinct Ruling Reptile groups as well), and it is absent, too, in birds and mammals.

Many major changes in the skull roof of reptiles are associated with the development of *temporal fenestrae* (Fig. 136). In earlier bony vertebrates the temporal muscles closing the lower jaw took origin beneath the solid plate of bone covering the temple and cheek origin. This was the situation in the stem reptiles, and the chelonians have apparently retained the primitive, solidly roofed—*anapsid*—condition. In most reptiles, how-

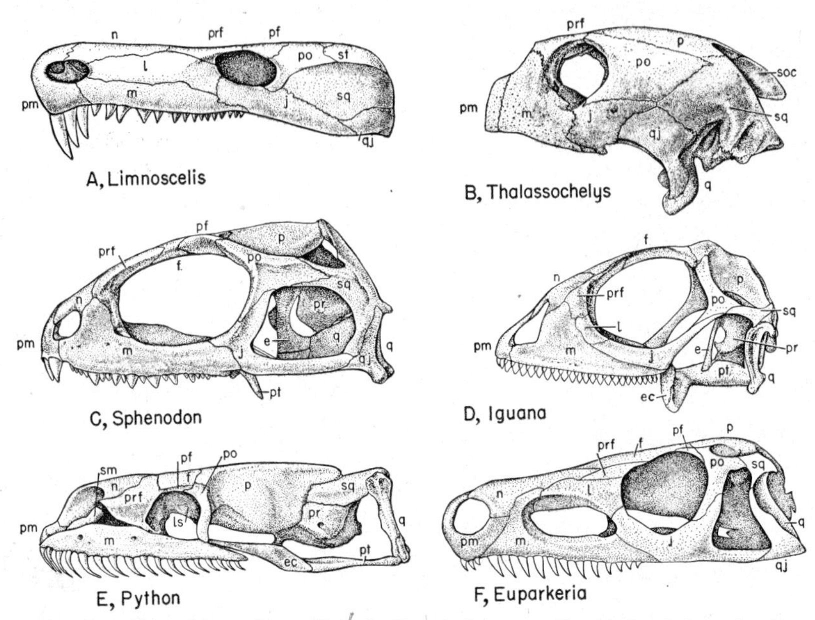

Fig. 135. Side views of reptilian skulls. *A*, Stem reptile of the Paleozoic; *B*, a sea turtle; *C*, Sphenodon; *D*, a lizard; *E*, a python; *F*, a primitive ruling reptile of a type from which birds, dinosaurs, and crocodilians have descended. Abbreviations as in Figure 130. (*F* after Broom.)

ever, one or two openings on each side have been made in the cheek or temple region to allow freer play for the jaw musculature. Two openings

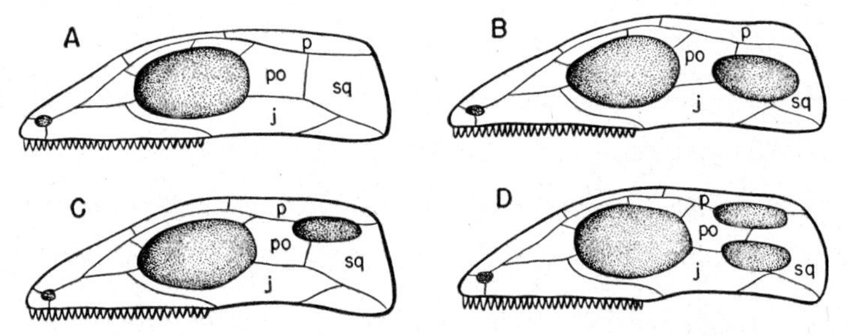

Fig. 136. Diagrams to show types of temporal openings in reptiles. *A*, Anapsid type (stem reptiles, turtles); *B*, synapsid type (mammal-like reptiles); *C*, parapsid type (extinct plesiosaurs, and so forth; *D*, diapsid type; rhynchocephalians, ruling reptiles, lizards, and snakes derived by loss of one or both temporal arches. *j*, jugal; *p*, parietal; *po*, postorbital; *sq*, squamosal.

have developed in many extinct and recent reptiles, such as rhynchocephalians and all the Ruling Reptile stock. This is termed the *diapsid* condition. One opening is present high up toward the skull roof, a second well down the side of the cheek; a horizontal arch of bone between the two openings connects the postorbital and squamosal elements. The Squamata are derived from early diapsids; the bar closing the lower fossa has, however, been lost in lizards, leaving but a single arch and closed opening, and in snakes the upper bar has been lost as well, leaving the whole cheek clear of dermal bone. By this process of reduction the quadrate has been left as a relatively free element, giving a high degree of motility (useful in swallowing large prey) to the jaw structures. In other (extinct) types, sometimes called *parapsid* (such as the plesiosaurs), the single opening was dorsally situated, the postorbital and squamosal bones meeting below the fenestra.

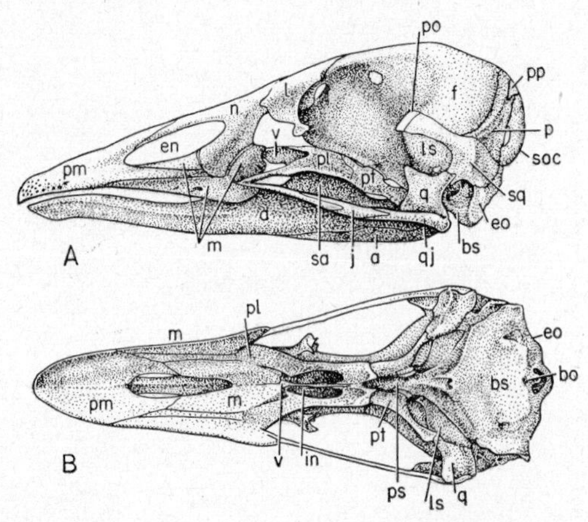

Fig. 137. *A*, Lateral and, *B*, ventral views of the skull of a duck (Anas). Abbreviations as in Figure 130. (After Heilmann.)

The chelonians (Figs. 134, *B*, 135, *B*) are technically anapsids, for there is no true temporal opening. In many turtles, however, the skull roof has been, so to speak, eaten away from behind, and is often strongly scalloped out on either side of the parietals; in some cases the entire temporal region has been excavated, and there is no dermal roof over the cheek region from the orbits to the back edge of the skull. Posterolaterally, the turtle skull is uniquely specialized for the eardrum and middle ear structures.

In Aves (Fig. 137, *A*) the lightly built skull has such a close fusion of elements that sutures are in general obliterated. Birds have evolved from archosaurian reptiles, and the dermal roofing pattern is essentially that found in generalized members of that group. However, the expanding brain and large orbit have, so to speak, crowded the cheek region, originally diapsid in structure; much of the bars behind the orbit and between the two fenestrae have disappeared.

In the evolution of mammals (Figs. 138, 139) a first change, seen in

Permian pelycosaurs (Figs. 138, *A*, 139, *A*), was the development of a small temporal opening on the side of the cheek; this is termed the *synapsid* condition (cf. Fig. 136, *B*). In later mammal-like reptiles, the therap-

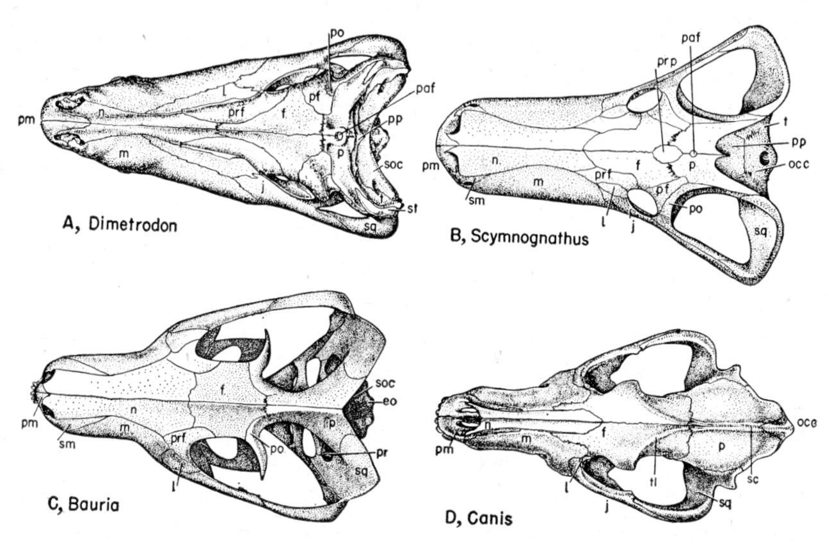

Fig. 138. Dorsal views of skulls to show the evolution of the mammalian skull roof. *A*, Primitive early Permian mammal ancestor (pelycosaur); *B*, a later Permian therapsid; *C*, a progressive Triassic therapsid; *D*, the dog. Abbreviations as in Figure 130 (*B* after Watson; *C* after Boonstra.)

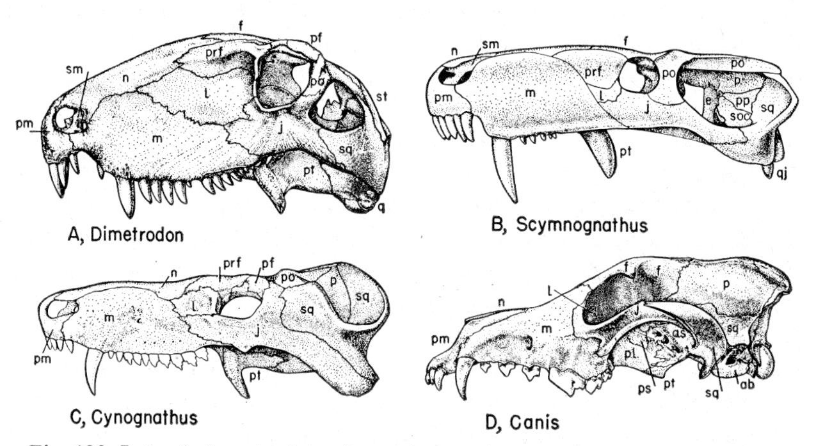

Fig. 139. Lateral views to show the evolution of the mammalian skull. *A*, Primitive early Permian mammal ancestor (pelycosaur); *B*, a late Permian therapsid; *C*, a progressive Triassic therapsid; *D*, the dog. Abbreviations as in Figure 130. (*B* after Watson; *C* after Broili and Schroeder.)

sids (Figs. 138, *B*, *C*, 139, *B*, *C*), this fenestra enlarged greatly toward the top of the skull; the openings of the two sides almost meet dorsally, leaving only a narrow strip of the original parietal surface between them.

Ventrally, only a bar, often narrow, formed by the jugal and squamosal, remains below the temporal opening as a *zygomatic arch* (cf. Fig. 144, p. 241). The bar between orbit and temporal opening disappeared in some advanced therapsids and is absent in primitive mammals; its upper end may remain as a *postorbital process* of the frontal. Farther forward, the maxilla (bearing the large canine teeth) expands upward and reduces the size of the lacrimal. The two external nares crowd toward the midline, where they form (as also in turtles) a single bony orifice. Various elements tended toward reduction and were lost by the time the definitive mammalian stage was reached; these include the septomaxilla, prefrontal, postfrontal, postorbital, supratemporal, and quadratojugal. The originally paired postparietals form a small median *interparietal* in therapsids; this persists in mammals, but is commonly fused with the occipital bones.

The primitive pattern of the mammalian skull roof is preserved, in the main, in such a form as the dog (Figs. 138, *D*, 139, *D*). There are, however, numerous variations in proportions and disposition of skull parts in the different groups of mammals, most of which are apparent in the dermal bone arrangement. The superficial bar behind the orbits was lost in the ancestral mammals, but it has been rebuilt in a number of different groups, and in higher primates a complete, deep, bony wall between orbit and temporal fossa has developed. In some forms (notably among insectivores and edentates) the zygomatic arch may be incomplete. The face is much lengthened in insectivores and is especially elongate in anteaters of several orders; on the other hand, it is shortened in higher primates and withdrawn, so to speak, beneath the expanded brain vault. The external narial opening may move backward over the roof of the face region in forms which develop a flexible snout (as the tapirs) or trunk (elephants), and in the aquatic sirenians and whales. In the cetaceans the nasal opening, the "blowhole," is at the top of the skull; there has been a backward movement of the nasal region and a forward push of the occipital bones as well, so that the original skull roof is practically eliminated. In rodents the masseter muscles of the jaw (cf. p. 288) are highly developed; the side of the face may be deeply excavated for their origin, and a canal leading forward from the orbit on to the snout may be opened out into a large fenestra for their accommodation. In small animals within any group the brain is relatively large, and in consequence the dorsal surface of the skull appears swollen; larger species or races, with relatively small brains but heavy temporal muscles, are more likely to develop a median longitudinal *sagittal crest* for muscular attachment. A transverse *nuchal crest* may form across the back margin of the roof for muscles and ligaments supporting the head. The tall, domed skull roof of the elephant is not due to brain expansion; largely filled with air sinuses, it provides a broad surface for attachment of the neck muscles and ligaments which support the heavy head and tusks.

The Palatal Complex in Tetrapods. The elements originally present in the earliest tetrapods persisted in later labyrinthodonts, but the structural arrangement became considerably modified in the palate of the broadened and flattened skull of these fossil forms and in the modern urodeles and anurans (Fig. 140). The basal articulation has lost its motility, and palate and braincase are broadly fused here. Anterior to this point of union the gap between the dermal bones of the palate and the base of the braincase is widened out into a pair of broad *interpterygoid vacuities*. The ectopterygoid is absent in both urodeles and anurans, and in the

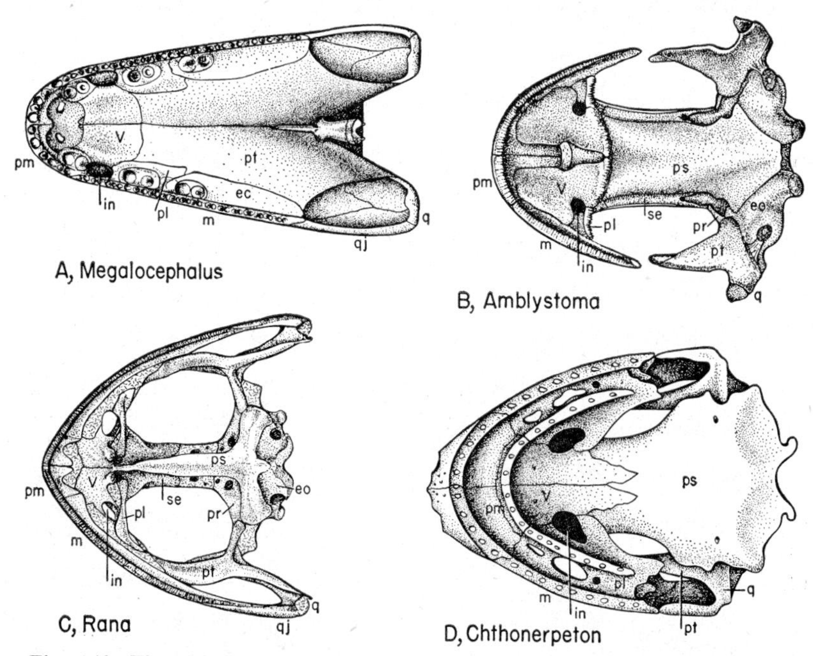

Fig. 140. The palate of amphibians. *A*, Paleozoic labyrinthodont; *B*, a salamander; *C*, a frog; *D*, an apodous amphibian. Abbreviations as in Figure 130. (*A* after Watson; *D* after Marcus.)

former the palatine is also small or absent, so that the palate is poorly developed laterally. Further, ossification is reduced in the palatal cartilage of modern amphibians. The epipterygoid is never ossified, and in urodeles the quadrate region remains cartilaginous as well; here, as elsewhere, the modern amphibian skeleton is definitely degenerate. The apodan palate has a more compact structure, but is essentially comparable to that of other modern amphibians, and the ossifications present are the same as in the frog. In the oldest tetrapods the jaw gape was long, and the quadrate articulation consequently well back of a plane drawn transversely through the occiput. In later amphibians the jaws are much shortened and the quadrates level with or in advance of the occiput.

The reptiles (Fig. 141) tended to hold to the primitive type of palate more persistently than did the amphibians; the primitive construction is still seen little changed in Sphenodon and lizards. A prominent lateral flange, often tooth-bearing, developed on the pterygoid in early reptiles, and is still preserved in Sphenodon and lizards. As noted previously, the quadrate has become a relatively free and movable element in lizards. In snakes the palate is still more flexible; indeed, the entire skull structure, except for the compact braincase, is pliable.

In the turtles the palate has become fused to the braincase through the reduction of the originally free basal articulation; the epipterygoid has

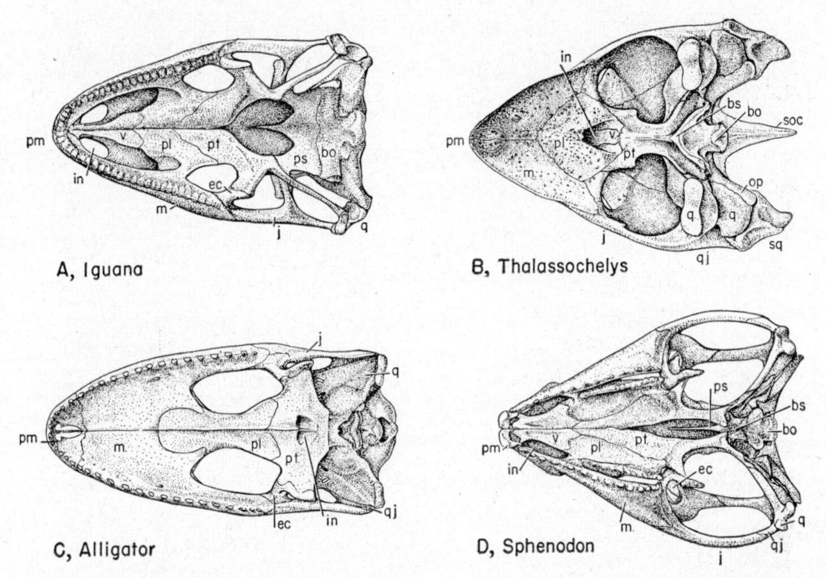

Fig. 141. The palate of reptiles. *A*, Lizard; *B*, a sea turtle; *C*, a young alligator; *D*, Sphenodon. Abbreviations as in Figure 130.

disappeared in the process. In the chelonians the pterygoids have fused with the lateral margins of the braincase floor; the quadrate is massive and is broadly in contact with the otic region of the braincase, thus increasing the solidity of the skull structure. The ectopterygoid is lost. The internal nostrils have moved some distance back in the roof of the mouth; they are generally situated in a deep dorsal pocket, which may be to some extent covered ventrally by bone. We see here the beginning of a *secondary palate,* a structure developed to a much greater degree in crocodiles and in mammals.

In crocodilians palatal mobility is likewise lost. The two pterygoids have gained a firm median union beneath the floor of the sphenoid region and have welded themselves onto the braincase. A further notable change from that of primitive forms is seen in the crocodilian skull in the presence of a highly developed secondary palate. Beneath the original palatal roof the maxillary, palatine, and pterygoid bones have built a secondary shelf of bone; between these two layers canals run back from the external nares to open into the mouth far back, above the opening to the windpipe. A flap of skin just in front of this point can close off the pharynx from the mouth so that breathing may be accomplished under water, even with the mouth open, so long as the tip of the nose is above the surface—a useful adaptation in a predaceous water dweller.

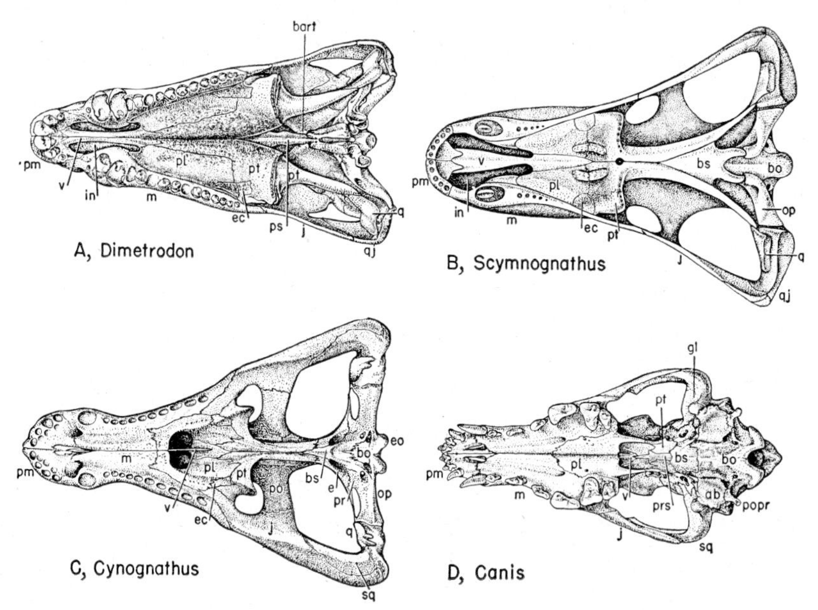

Fig. 142. A series of skulls in ventral view to show the evolution of the mammalian palate. *A*, Primitive early Permian pelycosaur; *B*, a late Permian therapsid; *C*, an advanced Triassic therapsid; *D*, the dog. Principal changes include development of secondary palate in *C* and *D*; loss of movable basal articulation and fusion of braincase and palate in *B* and later figures; reduction of pterygoid; loss of quadrate from skull structure and development of new jaw joint in *D;* addition of auditory bulla in *D*. Abbreviations as in Figure 130.

In birds (Fig. 137, *B*, p. 231) the palatal structures are lightly built and flexible. The vomers are small bones, centrally placed between the large choanae; the ectopterygoids and epipterygoids are lost. The palatines run lengthwise lateral to the choanae to terminate at the movable basal articulation. The pterygoids are short bars which extend back from this point to the quadrates, which are movably socketed on to the side of the braincase. The structure as a whole has much of the flexibility seen in the palate of lizards.

In the most primitive of mammal-like reptiles (Fig. 142, *A*) the palate was still essentially that of their early reptilian and amphibian ancestors. In the therapsids, however, there appear a series of modifications leading to the mammalian type of palate (Fig. 142, *B–D*). Anteriorly, a major change is the gradual development, in mammal-like forms, of a secondary palate rather comparable to that of the crocodiles. The maxillae and, farther back, the palatines fold ventrally to produce an elongate shelf lying below the original roof of the mouth; channels are thus created through which the air passes far back from the original position of the internal nares before entering the mouth. The vomers, originally paired, fuse into a single element (they maintain their position in the original roof of the mouth and hence are almost concealed in ventral view). They aid in forming a floor beneath the presphenoid and ethmoid and, further, form a partition between the two nasal cavities. The bony secondary palate does not extend so far back as that of crocodilians, but is continued by a fold of skin as a "soft palate." The origin of the secondary palate in mammal-like reptiles and mammals may be reasonably associated with the development of the constant body temperature characteristic of mammals. Continuous breathing is practically a necessity, and the secondary palate is an aid in the maintenance of breathing while the mouth is functioning in eating.

In the mammalian line, as in many lower tetrapods, movable articulation of palate and braincase has been abandoned (Fig. 144, p. 241); the pterygoid bones fuse on either side to the sphenoid region of the skull. In this process the pterygoids, once large elements, become much reduced in size. The lateral flange on the palate disappears (and with it the little ectopterygoid bone), and the posterior arm of the pterygoid, once running backward and outward to the quadrate, gradually dwindles. In a mammal the pterygoids are little more than small wings of bone projecting ventrally on either side below the sphenoid region of the braincase and forming the walls of the posterior end of the air passage.

The original endochondral elements of the palate have changed markedly in mammals and are no longer present as part of the palatal structure. The mammals have acquired a new jaw joint; the skull portion of it, the *glenoid fossa,* is formed by the squamosal, and the quadrate has thus lost its major function. The quadrate was already small, in therapsids, and rather loosely connected with other skull bones. In mammals it has disappeared as a proper skull element; it is, however, present, but as a tiny auditory ossicle. After the fusion of palate with braincase, the epipterygoid lost much of its original function. The epipterygoid, besides forming a socket for the basal articular process of the braincase, had primitively a vertical process extending up to the skull roof lateral to the braincase. In mammals this has fused into the braincase walls as the *alisphenoid* bone (p. 242).

The Braincase in Lower Tetrapods. In the later labyrinthodonts and modern amphibians the braincase tends to become relatively broad and flat—platybasic—and to become greatly reduced in its degree of ossification. On its under surface it is covered by a broad parasphenoid (Fig. 140, p. 234); but in the braincase itself little ossification remains. There are paired exoccipitals which may extend into the otic region; often a prootic bone, and, anteriorly, a sphenethmoid. Other ossifications have disappeared. In most fossil amphibians and all modern types the condyle, primitively single, is paired. The skull appears to have been shortened in the occipital region, and no foramen for a twelfth nerve is present.

In reptiles the braincase tends to remain in a better ossified condition than in amphibians. There is none of the flattening tendency seen in amphibians; the occipital region remains well developed; there is always a foramen for the twelfth nerve, and the condyle, a convex structure, remains single (except in the line leading to the mammals). The ossifications present are generally those seen in early tetrapods. Fusion of the palatal structures with the braincase in turtles, crocodiles, and various fossil reptiles results in a modification of braincase form, including a loss of the basal process of the basisphenoid and a union of the otic region with the quadrate, and so forth; but usually the basic structure is readily discernible.

There is, however, one notable difference between the structure of the reptilian braincase and that of typical amphibians. In those lower forms the braincase area between the orbits, in front of the otic region, is relatively thick and has well-ossified walls. In primitive reptiles, however, the braincase is here reduced. For much of the height only a membranous septum is present between the orbits, from the front margin of the otic region forward to that of the sphenethmoid; the brain primitively lay entirely back of this point. In certain reptiles, however—the snakes and the crocodilians—there appears to have arisen a functional need for a better enclosure of the brain cavity here. This enclosure is accomplished by the development of a new paired element above and in front of the otic bones. It is sometimes called an alisphenoid, but does not appear to correspond to the mammalian bone of that name and is better termed a *laterosphenoid* (Fig. 135, *E*, p. 230)—a new braincase ossification.

In birds the brain is much expanded, and the braincase is a swollen structure completely surrounded by bone (Fig. 137, *B*, p. 231). There has occurred here a process somewhat similar to that described below for mammals. Much of the brain expansion is sheathed by new extensions of the old roofing elements—frontals, parietals, and squamosals. In addition, however, a laterosphenoid element is present.

The Mammalian Braincase. In the mammal-like reptiles the braincase elements were those seen in early amphibians and reptiles. When the mammalian condition is attained, notable changes have occurred (Figs. 143, 145). The occiput region is basically the same as that of early reptiles. There is, however, a division of the originally single condyle into a paired structure analogous to that seen in amphibians. In the adult mammal the elements of the occipital region are usually fused into a single *occipital bone,* although ossification centers of all four primitive elements are seen in the embryo. There are usually added to this element dorsally several dermal bone centers which appear to be remnants of the post-parietal and (questionably) tabular bones of the ancestral skull roof.

Primitively the otic capsule was, we noted, formed from two centers (prootic and opisthotic). In mammals, however, there is a variable num-

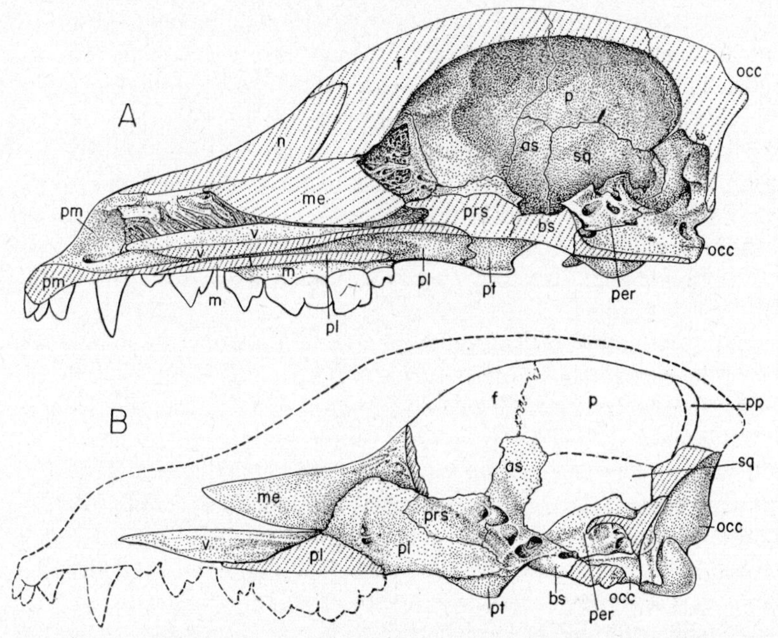

Fig. 143. *A,* Medium sagittal section of the dog skull. Diagonal lines indicate sectioned bones. *B,* Diagrammatic dissection of the dog skull in lateral view. All dermal bones of the skull roof are removed (including the dermal component of the occipital), leaving a series of bones comparable to those in the primitive amphibian of Figure 130, *D.* Bones present include palatal elements (*v, pl, pt* and *as*—the last a cartilage replacement bone) and braincase elements (*me, prs, bs, per, occ*). The upper margin of the brain cavity is shown in outline, and the contributions of the dermal roof bones (*f, p, pp, sq*) are indicated; the braincase itself forms little more than a floor to the brain cavity. (For other views of the dog skull, see Figures 138 *D,* 139, *D,* 142, *D,* pp. 232, 236.) *per,* Periotic, other abbreviations as in Figure 130.

ber of centers which in the adult fuse into a compact mass, the *periotic.* In reptiles the otic capsule occupies a relatively large area on the side wall of the braincase; in mammals, however, it is relatively small, and is situated low down at the lateral braincase margins. For the most part the capsule is buried deeply beneath the surface of the skull. Its posterior end, however, may be exposed as the *mastoid process*; this is frequently adjoined, posteriorly and medially, by a *paroccipital process* of the occipital bone.

In the otic region we find in most mammals a new addition to the skull in the shape of the *auditory bulla* (Figs. 142, *D*, 145). Just outside the periotic region lies the middle ear cavity, containing in mammals the three small and delicate auditory ossicles and, at its outer margin, the eardrum (cf. p. 522). In the earliest mammals, as in most marsupials today, this region had no satisfactory protection against injury from without; in most placentals this protection is supplied by the development of the bulla as an ensheathing capsule. Frequently present here is a *tympanic* bone. This, in its most primitive form, is merely a bony ring surrounding the eardrum; but in other cases this bone expands inward to form part or all of the bulla. The tympanic is a dermal element, and embryological evidence suggests (unlikely as it seems at first) that it is a modification of the angular bone of the reptilian lower jaw, which lies close to the eardrum in reptiles (cf. Fig. 314). In addition to the tympanic, there forms in many mammals a second bone, an *entotympanic,* which may surround the middle ear cavity and may fuse in the adult with the tympanic to form a compound bulla. The entotympanic is formed in cartilage. This cartilage, however, is not part of the original tetrapod skull structure; we have here the unusual feature of the addition of an entirely new skull component. The otic elements—periotic and new bulla—remain discrete in some forms; in others they fuse with one another and with the adjacent squamosal bone of the skull roof to form a compound *temporal* bone.

The basisphenoid forms the braincase floor in the region of the pituitary. In many cases it fuses with neighboring elements—alisphenoid, presphenoid, orbitosphenoid—to form a complex (if small) element termed the *sphenoid.* The parasphenoid, which sheathed the braincase floor in lower forms, has disappeared as a separate ossification in mammals.

The sphenethmoid of lower tetrapods persists in mammals, anterior to the basisphenoid, as the *presphenoid;* its lateral walls in the orbit are sometimes distinct and termed the *orbitosphenoid,* but the bone is essentially a unit structure. In some mammals, including various primitive types and both major ungulate orders, the presphenoid is the most anterior braincase element; but in other orders, including primates, rodents, and carnivores, a new ossification, the *mesethmoid,* is present still farther anteriorly. When both are present, the presphenoid is restricted to the anterior part of the braincase floor, appearing externally in the roof of the choanae ventrally and (as the orbitosphenoid portion of the bone) in the orbital walls anterior to the alisphenoid. In mammals the nasal pouches have extended backward so that they are separated from the expanded brain cavity only by a transverse plate of bone, the *cribriform plate.* The mesethmoid, when present, ossifies in this area, and extends forward as a median septum between the two nasal cavities. In these cavities variable scrolls of cartilage or bone, *turbinals,* which represent the original nasal capsule, develop in the embryo; covered with nasal mucosa, they increase the surface area of the nasal passages.

The turbinal system (cf. p. 492), when typically developed in mammals, includes a *maxilloturbinal,* a *nasoturbinal,* and usually several *ethmoturbinals,* formed, as the name suggests, in connection with adjacent skull elements. The maxilloturbinal is of great extent, occupying most of the ventral and anterior parts of the nasal chamber. Above it lies the smaller nasoturbinal. In the true olfactory region, placed dorsally and posteriorly, are the ethmoturbinals, often numerous, running forward from the mesethmoid region along the lateral wall of the nasal cavity.

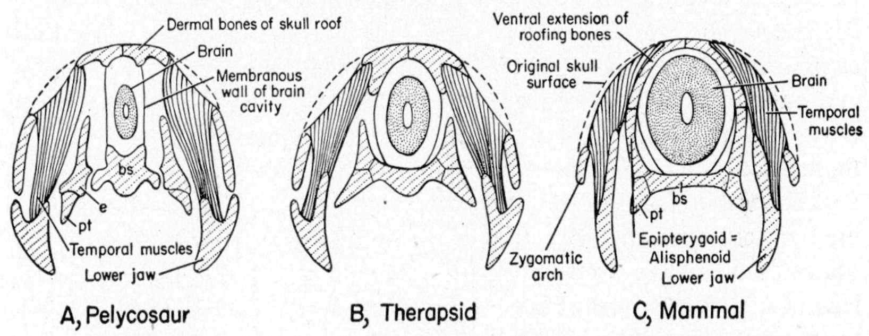

Fig. 144. Diagrammatic cross sections of the skull and jaws of *A*, a primitive mammal-like reptile; *B*, an advanced mammal-like form; and *C*, a mammal; to show features in the development of the skull roof and braincase. (1) Part of the lateral walls enclosing the expanding brain were originally membranous. This lateral area comes to be enclosed by ventral extensions of the roofing bones and by the incorporation of the epipterygoid of the palatal structure as the alisphenoid. (2) This ventral extension of the roofing bones downward around the brain gives the appearance in mammals of being the original skull surface; the original surface, however, lay external to the temporal muscles, as indicated by the broken lines. (3) The upper jaw elements (*e*, epipterygoid; *pt*, pterygoid) originally articulated movably with the braincase (*bs*, basisphenoid); the two structures are fused in *B* and *C*.

The bones so far named include the entire roster of elements formed in the braincase proper. But if we examine the actual "braincase" of a typical mammal (Fig. 143, *A, B*), it will be seen that they form little but the floor and the back wall of the brain cavity. The mammalian brain has expanded to such proportions that the braincase proper has been unable, so to speak, to keep up with its growth; most of the walls of bone surrounding the brain in the adult placental mammal have been derived from other skull materials.

A parallel to this situation was noted in the birds. As there, the elements of the dermal roof have been drawn upon for most of the roof and walls of the brain capsule. The parietals and frontals in early reptiles formed a broad roof over the brain region. With the development of the temporal fenestrae their surface area was much reduced (Fig. 144, *A*). They tended, however, to send flanges downward, deep to the jaw muscles, along the walls of the brain cavity beneath. With the great expansion of the brain in mammals (Fig. 144, *B, C*), these flanges became much enlarged and spread out laterally over the swollen brain surface to form much of the roof and side walls of the brain box; the squamosal bone from the side of the dermal roof has also been drawn into the work of forming a covering for the brain. These newly expanded areas of the dermal roofing elements give the appearance of forming part of the original surface of the skull. They are not, however. They lie beneath the temporal jaw muscles (the original skull roof was external to these muscles). The upper limits of temporal muscle attachment on either side are marked by the *temporal lines,* which may become confluent posteriorly as the sagittal crest (Fig. 138, *D*, p. 232). Only the area of the frontals and parietals which lies between these lines, or forms the sagittal crest, represents the original skull surface; the remainder of the expanse of these bones is a new, deeper, internal growth.

Still another addition has been made to fill out the walls of the expanding braincase: the incorporation of the epipterygoid bone of lower vertebrates as the alisphenoid (Figs. 143, *A, B,* 144, *C*). We have noted the lack of ossification in the reptilian braincase wall between the orbits. The epipterygoid in reptiles is mainly a vertical rod which, as part of the upper jaw structure, lies external to this unossified interorbital region. In therapsids this bar was expanded to a plate, but still lay external to the braincase. In mammals, however, it has been incorporated into the braincase wall, filling the gap above the basisphenoid and so firmly fused with this last bone that (as its new, mammalian name implies) it appears to form a wing of the combined sphenoid element.

In sum, the brain capsule of mammals is a composite affair. The original braincase is able to do little more than form its floor; dermal elements plus the alisphenoid are called into play to form the walls and roof of this expanded structure.

The interior of the mammalian braincase is partially subdivided by one or two ridges of bone extending downward from the roof. The transverse *tentorium* above the chamber containing the cerebellum and brainstem separates it from the more anterior cavity mainly occupied by the large cerebral hemispheres; this latter is in the majority of mammals partitioned dorsally by the median longitudinal *falx cerebri,* a brain sheath in which ossification may occur. Anteriorly, at the junction of the cranial and facial segments of the skull, there is frequently considerable space between the brain cavity and the surface contours; instead of being filled by solid bone, these areas, in such bones as the frontals, maxillae, presphenoid, are often occupied by air-filled *sinuses,* connecting with the nasal cavities. They are advantageous in reducing skull weight.

Mammalian Braincase Foramina (Fig. 145). We have mentioned the various openings for nerves and vessels found in the braincase of lower tetrapods. The situation may be reviewed for the mammals, for although many of the openings are the same, the nomenclature differs; further, the incorporation of the alisphenoid and auditory bulla into the braincase has modified the nature of the openings.

The olfactory nerves in mammals take the form of numerous small filaments which pierce the ethmoid bone in the sievelike cribriform plate. The optic nerve usually enters the brain cavity through an *optic foramen* in the orbitosphenoid.

In early reptiles the area posterior to the ethmoid region, on the side of the braincase as far back as the ear capsule, was open for the exit of nerves III to VI. The filling in of the braincase wall by the alisphenoid has changed this situation. We find a large opening in front of the alisphe-

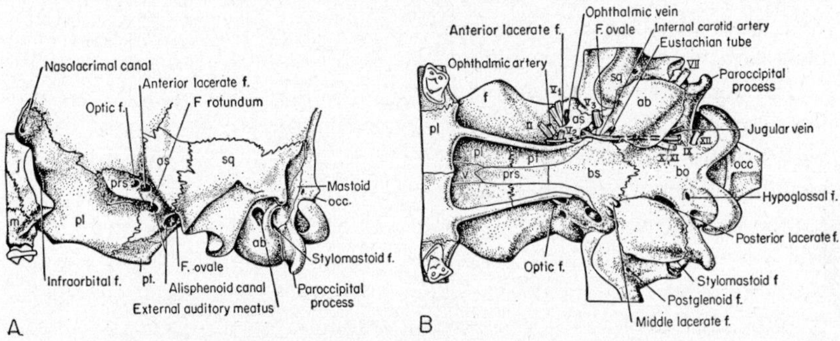

Fig. 145. Braincase region of a dog in *A*, lateral view, *B*, ventral view, to show foramina. In *B* are indicated the main nerves, the course of the internal carotid artery and its palatine branch and the jugular vein.

noid, the *anterior lacerate foramen* (sphenoidal or orbital fissure); two openings—*foramen rotundum* and *foramen ovale*—usually piercing the alisphenoid; and a *middle lacerate foramen,* lying at the back edge of the alisphenoid. Typically, the eye muscle nerves and the most anterior division of the trigeminal nerve pass through the anterior lacerate foramen; maxillary and mandibular branches of the trigeminal use the round and oval foramina; the middle lacerate foramen is of little functional importance except that the internal carotid artery enters the braincase here or close by.

Nerves VII and VIII enter the ear capsule in an *internal auditory meatus*. The former nerve primitively emerged at the surface toward the front end of the ear capsule; in mammals, however, in relation to the complicated middle ear structure, it remains buried in the otic bones until it emerges back of the bulla region at the *stylomastoid foramen*. At the back end of the bulla, in a gap between the otic and occipital regions, is a *posterior lacerate foramen* (jugular foramen) through which nerves IX, X, and XI and the internal jugular vein draining the skull make their exit.

Nerve XII usually occupies a separate *hypoglossal foramen,* sometimes multiple, in the floor of the occiput.

The auditory bulla has, of course, an external opening, the *external auditory meatus,* and at its front end an opening for the *eustachian tube* from the throat. The internal carotid artery originally ran forward on the under surface of the skull past the middle ear region. This area is covered in mammals by the bulla, beneath which the artery runs forward in a *carotid canal* to a point near that at which it enters the braincase. Near this latter point there is sometimes present an *alisphenoid canal* through which a branch of the artery continues forward toward the palate.

Among other skull openings and ducts not connected with the braincase we may mention the *incisive foramen* near the front of the palate, connecting the mouth with Jacobson's organ (cf. p. 493); the *infraorbital foramen,* sometimes enlarged to a canal, carrying blood vessels and nerves forward from the orbit to the surface of the snout; the *nasolacrimal canal* from orbit to nasal cavity, for the tear duct; variable canals in the palate for the forward passage of nerves and small blood vessels.

Lower Jaw (Figs. 146, 147). The lower jaw, *or mandible,* is in most vertebrates a complex of both dermal and endoskeletal structures. Presumably, dermal elements were present in the jaw of the ancestral gnathostomes; in the living Chondrichthyes, however, they are absent, and the lower jaw of sharklike fishes is, like the upper, simply a tooth-bearing bar of cartilage derived from a gill arch element (Figs. 125, *C,* p. 210; 127, 128, p. 213). This mandibular cartilage, also termed *Meckel's cartilage* after its discoverer, is present in the jaw of the embryo in bony fishes and all tetrapods (Fig. 125, *D-F*). In all these types except the mammals the cartilage, or a bone derived from it, persists as part of the adult jaw structure. Its anterior part may disappear in the adult or remain as a cartilage buried deep within the jaw structure; rarely (as in common frogs), an anterior part of the cartilage may ossify. The posterior portion, which articulates by a socket with the upper jaw, is a cartilage in degenerate forms, but usually ossifies as the appropriately named *articular* bone. Anterior to it is a *prearticular fossa* opening downward into the jaw from above, which affords an area of insertion for muscles closing the jaw and offers a place of entrance into the interior for nerves and blood vessels.

The lower jaw is sheathed both laterally and medially by dermal bones. Early fossil amphibians and bony fishes had a complex series of such bones; their descendants have simplified the pattern in various ways. The primitive pattern will be described first.

On the outer surface the major element is the *dentary,* which bears the marginal teeth and corresponds to the maxilla of the upper jaw. Below the dentary there is primitively a row of dermal bones which includes anteriorly two small *splenials,* followed by a larger *angular,* and a *sur-*

angular; the last runs to the back end of the jaw ramus, sheathes the articular externally, and rises to the upper margin of the jaw behind the dentary. The symphysis between the jaw rami is mainly formed by interlocking rugosities on the inner surface of the dentaries, but the anterior splenials usually enter into the symphysis as well. All these external elements may curve inward ventrally to appear also on the lower part of the inner surface of the jaw.

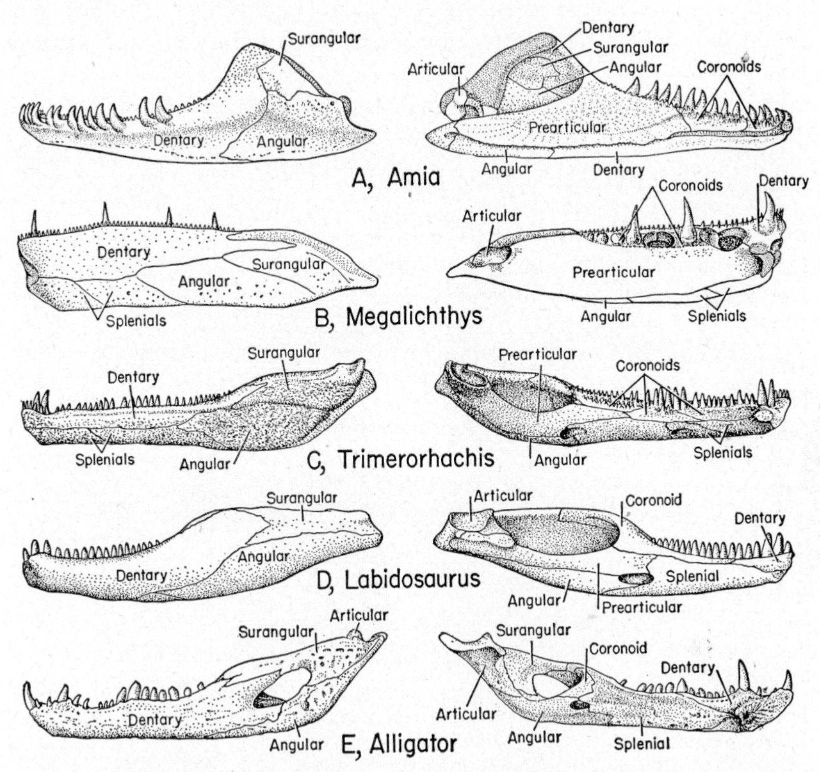

Fig. 146. Left lower jaws, outer views at left, inner views at right, of *A,* a ray-finned fish, the bow fin; *B,* a primitive crossopterygian; *C,* a primitive labyrinthodont; *D,* a primitive reptile; *E,* an alligator. The jaws of modern teleosts, amphibians, and reptiles, in which the number of elements is reduced, have been derived from the types shown in *A, C,* and *D,* respectively.

The internal sheathing elements of the primitive jaw were four in number. A major dermal bone is the elongate prearticular, which runs forward from the articular. Toward the upper margin there is a series of three slender bones, the *coronoids.* They bear, in early types, teeth of variable size, and teeth may be present on the prearticular as well. It is of interest that the inner aspect of the lower jaw is primitively very much a mirror image of the upper jaw and palate (cf. Figs. 140, *A,* 146, *C*). The prearticular, running forward from the articular, internal to the prearticular fossa, is comparable to the pterygoid, running forward from the quadrate internal to the subtemporal fossa; the three coronoids are comparable to the three lateral palatal elements.

The structure described above is that found in both early crossopterygians (Fig. 146, *B*) and early amphibians (Fig. 146, *C*). In other groups of vertebrates the changes from this primitive condition are almost entirely a matter of fusion or, more frequently, loss of elements.

In ray-finned fishes loss and fusion occur (Fig. 146, *A*), leading to a condition in which, in teleosts, but three bones are recognized, termed

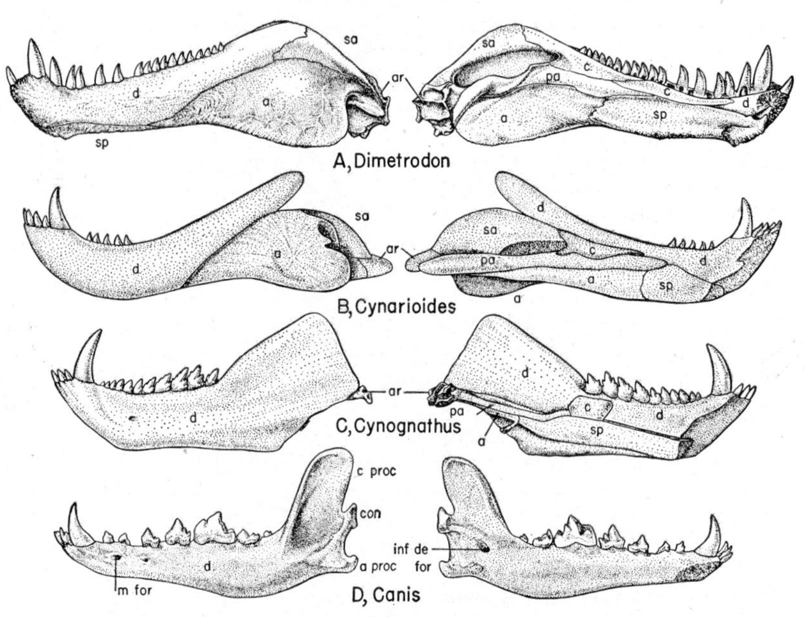

Fig. 147. Left lower jaws of mammal-like reptiles and mammals, illustrating the reduction of jaw elements. Outer views (*left*) and inner views (*right*). *A,* Primitive mammal-like reptile (pelycosaur); *B,* a primitive therapsid; *C,* an advanced therapsid; *D,* a typical mammal (dog). Abbreviations as in Figure 130.

dentary, articular, and angular. In sturgeons, paddlefishes, and in lungfishes, skeletal degeneration leads to the retention in the adult of a large unossified Meckel's cartilage and a strong reduction in dermal ossification.

In modern amphibians, too, ossification is much reduced. Typical urodeles have three dermal bones, the dentary and two bones on the inner surface which are probably the prearticular and a coronoid. In frogs there is, apart from the dentary, but a single other element, presumably the prearticular. In reptiles (Fig. 146, *D*) there is never more than one splenial, and almost never more than a single coronoid, but a general retention of other elements—articular, dentary, angular, surangular, and prearticular. In crocodilians (Fig. 146, *E*) and other archosaurs a large lateral jaw fenestra is usually developed in the angular region. In birds the characteristic reptilian elements (except, apparently, the coronoids) are retained in a fused condition, and the archosaurian lateral fenestra is present.

In mammal-like reptiles (Fig. 147, *A-C*) can be seen stages in a notable transformation of the jaw structure which is completed in mammals (Fig. 147, *D*). The dentary increases steadily in importance; posteriorly, it extends to a position beneath the squamosal close to the old jaw articulation, and more anteriorly an ascending process grows upward beneath the temporal region for the attachment of temporal muscles. Concomitantly the other jaw ossifications are reduced in size. In mammals we find that these other structures have been lost from the jaw and, in so far as they are still present, are to be looked for in the auditory apparatus. The jaw consists of the dentary bone alone. This includes a *ramus,* the original tooth-bearing part of the bone; a *coronoid process* to which temporal muscles attach; a new backward projection which may bear ventrally an *angular process* (often prominent); posteriorly, a transversely elongate rounded *condyle* which fits into the glenoid fossa of the squamosal bone.

CHAPTER IX

MUSCULAR SYSTEM

THE MUSCULAR SYSTEM, judged quantitatively, at least, should loom large in any animal study of the present sort, for muscle tissue constitutes from a third to a half of the bulk of the average vertebrate. Functionally, too, the musculature is of highest importance. The activity of the nervous system—even the highest functioning of a human brain—has little mode of expression other than the contraction of muscle fibers. From locomotion to the circulation of the blood, the major functions of the body are caused by or associated with muscular activity.

Smooth Muscle. Histologically, two major types of muscular tissue are distinguished—smooth and striated fibers. Smooth muscle fibers are the

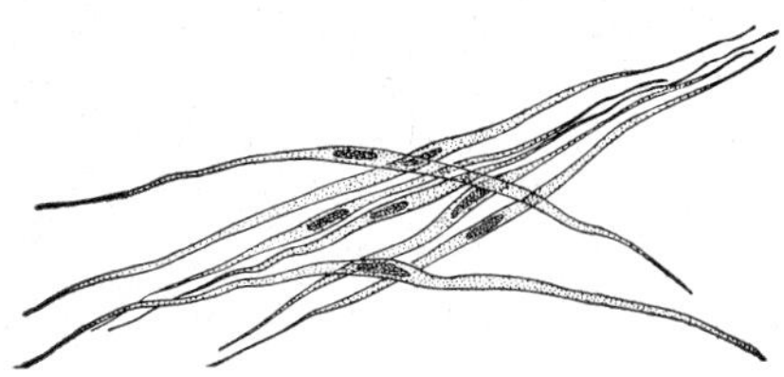

Fig. 148. Smooth muscle fibers from the wall of the stomach of a cat. × 220. (After Maximow and Bloom.)

simpler of the two. These are derived from the embryonic mesenchyme and hence developed in association with the connective tissue cells. The main seat of smooth muscle is in the lining of the digestive tract. Certain other sites, such as the ducts of the glands associated with the gut, the bladder, and the trachea and bronchi of the lungs, are outgrowths of this tract. Still other loci for smooth muscle, however, are independent of the gut system, these including circulatory vessels, genital organs, and the connective tissue of the skin and other areas.

A typical smooth muscle fiber (Fig. 148) is a slender, spindle-shaped body averaging a few tenths of a millimeter in length. There is a single, centrally situated nucleus; the cytoplasm appears homogeneous, but with special chemical treatment tiny fibrils can be seen running the length of the cell. As the name implies, the fibrils appear to be simple, and there is no cross-banding such as is seen in striated muscle cells. Smooth muscle fibers are sometimes scattered. More generally, however, they are arranged in bands or bundles, with interspersed connective tissue fibers uniting them into an effective common mass (cf., for example, Figs. 205, 217). In part, at least, smooth muscle cells are stimulated by nerve fibers; but they may apparently be stimulated by the contraction of their neighbors as well, and, particularly in the digestive tract, one may find a wave of contraction passing along a band of smooth muscle tissue.

Cardiac Muscle. In the heart is found a special type of musculature not present elsewhere (Fig. 149). This tissue shows a cross-banding similar to that seen in striated fibers, but in other regards cardiac muscle differs markedly from striated muscle. There are no separate fibers; instead, the whole heart musculature is a continuous network of dividing and recombining strands, with nuclei at intervals along them. Prominent cross bands, termed *intercalated discs,* more or less separate this

Fig. 149. Mammalian cardiac muscle, showing intercalated discs. About 450 ×. (From Maximow and Bloom.)

network into short units. Embryologically, heart muscle is a special type of smooth musculature in which the cells have united as a syncytium; its histological peculiarities are presumably associated with the important functions of this constantly active tissue.

Striated Muscle. Fibers of this type form the "flesh" of the body, the voluntary muscles, derived in great measure from the myotomes of the embryo; they are in general arrayed as formed muscles which attach to and move skeletal structures. Striated muscle fibers (Fig. 150) are large cells, elongate and cylindrical, with lengths which vary from about a millimeter to a number of centimeters. These cells are multinucleate, with numerous nuclei scattered along the entire length. Primitively, the nuclei may be situated in the interior of the fiber; in mammals they generally assume a superficial position. The fiber surface is covered by a thin sheath, the *sarcolemma.* As in smooth muscle, the fiber contains a large number of longitudinal fibrils closely packed within the cell. Its striated appearance under the microscope is due to the fact that each fibril consists of alternating light and dark portions; these occur at the same point on each fibril and hence appear as cross striations on the fiber as a whole. The major impression is that of alternating light and dark bands; closer study, however, frequently shows in the middle of the light band a thin dark disc, and still other banding effects may be seen in the contracted phase. Each cell is innervated by a nerve fiber ending in close contact with it. At some point along the outer surface there is an area without fibrils and containing a number of nuclei; this is a *muscle plate,* into which penetrate the coiled ends of nerve fibrils.

Striated muscle cells are arranged in parallel fashion to form muscles. Connective tissue fibers run between the muscle fibers and bind them

Fig. 150. Two striated muscle fibers. One fiber is crushed, and shows the sarcolemma, × 250. (From Maximow and Bloom.)

together, form sheaths for fiber bundles—*fascicles*—and, further, form an external sheath, the *perimysium,* for the muscle as a whole.

Muscle Fiber Function. The force of a muscle is exerted through a contraction of its individual fibers. There is, of course, no reduction in total bulk; in consequence, the shortened fibers, and the muscle as a whole, become thicker with contraction. In smooth musculature the contraction is relatively slow, and the amount of contraction less than in a striated muscle. Studies indicate that the fibrils contain parallel chains of stretched-out protein molecules specific for muscle cells. In contraction these chains are crumpled and shortened; in striated fibers changes in the striations are an associated visible phenomenon. A considerable and speedy release of energy is required for this contraction. In the long run the muscular energy is supplied by consumption of carbohydrates present as glycogen in the muscle fibers. However, it appears impossible to oxidize glycogen with sufficient rapidity, and special chemical mechanisms are developed.

Muscle fibers are notable for the presence in them of considerable amounts of phosphate in combination with certain simple organic compounds—adenylic acid and creatine. A rapid breakdown of these substances occurs in anaerobic fashion (i.e., without need of oxygen), with a release of most if not all of the energy necessary for contraction. Before a second contraction can take place, these phosphate compounds must be rebuilt, and energy must be supplied for the rebuilding process. Here the normal method of carbohydrate oxidation as an energy source comes into the picture. The glycogen stored in the muscle cell is burnt, and the energy used not so much directly for fiber contraction as for rebuilding the phosphate compounds. If, however, activity of the fiber is too great and too continuous, efficient utilization of the glycogen does not take place; "fatigued" fibers contain considerable amounts of lactic acid, a product resulting from incomplete oxidation of the glycogen.

In sum, carbohydrates are basically responsible for the energy needed for muscle contraction; but for "explosive" use, much of this energy is stored in rapidly destructible organic phosphate compounds.

A striated muscle as a whole may contract slightly or strongly, briefly or for a considerable period of time; the result varies according to the number of fibers stimulated and the number and rapidity of the nerve stimuli. Individual muscle fibers, however, work on an "all or none" basis; each fiber either contracts as fully as possible or fails to contract at all. A single sharp contraction of a striated fiber and the more gradual relaxation which follows occupy altogether but a tenth of a second or so. Normally, however, muscle contraction is due not to a single stimulation, but to a continuous rapid tattoo of nerve impulses. A second impulse given before the effect of the first has worn off will increase the contraction, a third will further increase it, and a long series may bring the muscle fibers to a state of maximum contraction or tension (tetanus).

Classification of Muscle Tissues. Attempts to range the muscular tissues of the body in major categories have been based on a variety of criteria; some contradictions are encountered, no matter what criterion is considered fundamental. One obvious suggestion is a classification according to histological structure, with two main divisions consisting, on the one hand, of smooth muscles (including cardiac muscles) and on the other of the striated muscles. This seems at first sight reasonable. The striated muscles are for the most part under voluntary control, whereas the smooth muscles are under the influence of the involuntary or autonomic nervous system; the striated muscles are in general large, well-formed structures, whereas the smooth muscles are often diffuse and may be incorporated in the substance of other organs; the striated muscles are mostly formed in the "outer tube" of the body, while the smooth musculature is in great measure associated with the gut tube; much of the striated muscle is derived from the mesodermal somites, while the smooth musculature arises from mesenchyme.

One prominent group of muscles, however, is in many ways anomalous and ruins the seeming simplicity of such a classification. This is the *branchial system* of striated muscles which in primitive vertebrates is associated with the gill bars (the visceral skeletal system) and which remains prominent in the head and "neck" region in all vertebrates. The branchial muscles and their derivatives are striated, but exhibit strong contrasts with all other striated muscles and, curiously, show affinities in diagnostic characters with the smooth musculature. Muscles of this group are not derived from somites as are other striated muscles, but arise from mesenchyme like smooth muscle fibers (Fig. 153, p. 255). Other striated muscles are innervated from brain or spinal cord, by fibers of a specific type—the somatic motor system (p. 540). In contrast, the gill arch muscles are innervated, through special cranial nerves, by a different category of neurons, which belong to the visceral motor system; the smooth muscles are similarly innervated. Finally, while "ordinary" striated muscles lie in the "outer tube" of the body, the gill arch muscles, although often superficial in higher vertebrates, are, like the smooth muscles, primarily associated with the digestive tube—that is, its anterior, pharyngeal region.

All this suggests the picture of the primitive vertebrate as having had two discrete sets of musculature. The first, which we may term the *somatic musculature,* was that forming the muscles of the "outer tube" of the body; this musculature is derived from myotomes, is universally striated, and is innervated by somatic motor neurons. The second was the *visceral musculature,* connected mainly with the gut tube; muscles of this group are all derived from mesenchyme (never from myotomes) and innervated by visceral motor fibers. In this second group of muscles those of the more posterior segments of the gut retained a simple structure as smooth muscle; but, in the pharyngeal region, important in early vertebrates for food gathering and breathing, the striated condition was assumed.

It appears, then, that despite the seeming unnaturalness of associating in one major group all smooth and some striated elements, the most natural classification of muscles is as follows:

Somatic	Axial	Trunk and tail Eyeball
	Appendicular	
Visceral	Branchial	
	Smooth (gut, and the like)	

(As will be seen later, the muscles of the eyeball are a special category of the axial musculature. The limb muscles are, historically, derivatives of the axial system, but have assumed such special importance as to merit separate treatment.)

The smooth muscles are, in general, component parts of various organs and need no separate consideration here. We shall in this chapter discuss only the formed muscles of striated type—the various somatic muscle groups and the branchial muscles of the visceral system.

Muscle Terminology. As with other organ systems, muscles are given, as far as possible, names used in human anatomy. Unfortunately the comparative study of musculature is still in its infancy, and we are in doubt in many cases as to the homologues of human muscles in lower vertebrates. Elementary students in zoology, for example, frequently dissect the thigh of a frog and apply to its muscles the names of those found in the human thigh; but it is doubtful if many of the muscles given like names are really homologous. Another procedure, often followed, is to give muscles of lower vertebrates of which the homologies are, or have been, in doubt, names which are simply descriptive of their general position or attachments. Thus we have in reptiles an iliofemoralis muscle running from ilium to femur; it may be more or less homologous with some of the gluteal muscles lying in this region in man and other mammals (cf. p. 273), but since exact homologies are doubtful, it is safer to give the reptile muscle a name that avoids a definite commitment.

There are several terms frequently used to describe muscles, particularly limb muscles, according to the type of action they perform. An *extensor* muscle is one which acts to open out a joint; a *flexor* closes it. An *adductor* draws a segment inward; an *abductor* does the reverse. A *levator* raises a structure, in contrast to a *depressor*. A *rotator* twists a limb segment; a *pronator* or *supinator* rotates the distal part of a limb toward a prone or supine position of the foot (i.e., with palm or sole down, or vice versa). *Constrictor* or *sphincter* muscles are those which surround orifices (as gills, anus) and tend to close them when contracted.

Muscles most often attach to skeletal elements at either end. One attachment is usually the more stable, and is considered the area of *origin,* the other the *insertion;* in limb muscles the proximal end is always considered to be the point of origin. The fleshy mass of a muscle is its *belly.* Muscle fibers never attach of themselves to a bone or cartilage; the attachment is always mediated by connective tissue fibers. In many cases the muscle fibers approach the bone closely, and hence we speak of a "fleshy" attachment; in others, the muscle terminates in a *tendon,* or a flat sheet of connective tissue—an *aponeurosis* or *fascia.* A muscle may include two or more bellies; and more than one muscle may attach by a common tendon.

Muscle Homologies. The comparative study of musculature is a difficult procedure because of the variability of muscles and the apparent ease with which their functions may alter. A mass of muscle tissue which is a unit in one animal may be split into two or more distinct muscles in another animal, and there appear to be other cases in which originally separate muscles have fused. The general pattern of their arrangement may be of aid in an attempt to sort out individual muscles, but the pattern may be obscured by the fact that a muscle's origin or insertion may shift in relation to differing functional needs.

Embryological origin is here, as ever, an important criterion for identification of homologies. We find that, in many cases at least, groups of

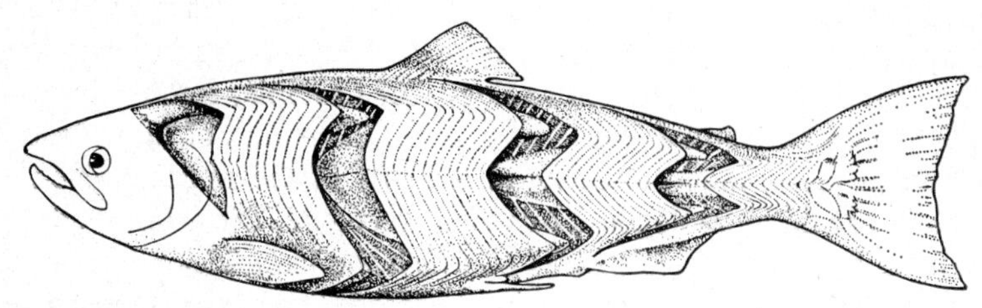

Fig. 151. Dissection of a salmon to show axial musculature. In four places a series of myomeres has been removed to show the complicated internal folding of these segmental structures. Within the body each V projects farther anteriorly or posteriorly than it does at the surface. The lateral septum is visible, cutting the main, anterior-pointing V. (After Greene.)

individual muscles in the adult can be traced back to larger aggregations of muscular or premuscular tissue in the embryo (cf. Fig. 157, *B*, p. 266) and that the mode of breaking up these mother masses gives valuable evidence of homologies. Unfortunately, only a limited amount of work has been done on muscle embryology.

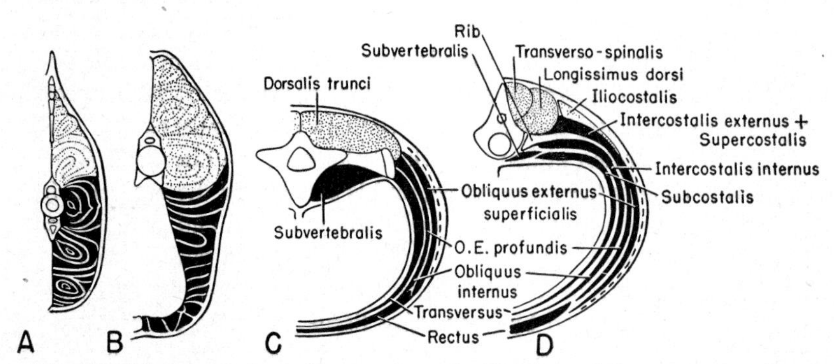

Fig. 152. Diagrammatic sections to show the divisions of the trunk musculature in *A*, a shark tail; *B*, the shark trunk; *C*, a urodele; *D*, a lizard. The epaxial muscles are stippled, hypaxial muscles in black. In *D* a rib is assumed to be present dorsally, and the adjacent parts of the hypaxial muscles are labeled as in the rib-bearing region; more ventrally the names are those of the corresponding abdominal muscles. (Mainly after Nishi.)

The motor innervation of muscles provides valuable clues. Limb muscles, for example, are generally innervated by a series of trunks and branches arising out of a nerve plexus associated with the limb. (Fig. 328). The pattern of the plexus and the nerves extending into and from it tend to remain rather uniform, and muscles known to be homologous usually receive their nerve supply from like branches of the plexus. In consequence there arose in the minds of many workers a belief that there is an unalterable phylogenetic relation between a given nerve and muscle. Embryology, however, gives no indication that there is any mysterious affinity between specific nerve fibers and the specific muscle fibers which form a given muscle, and in some cases it seems quite certain that the innervation of a muscle is different in different animals. Nevertheless, actual practice indicates that, as the embryological picture suggests, the nerve supply to a particular mass of muscle does tend to remain constant, and that innervation affords an important clue to the identity of the muscles.

AXIAL MUSCLES

Trunk Musculature in Fishes. The major part of the somatic muscle division in fishes is the axial musculature, which forms much of the bulk

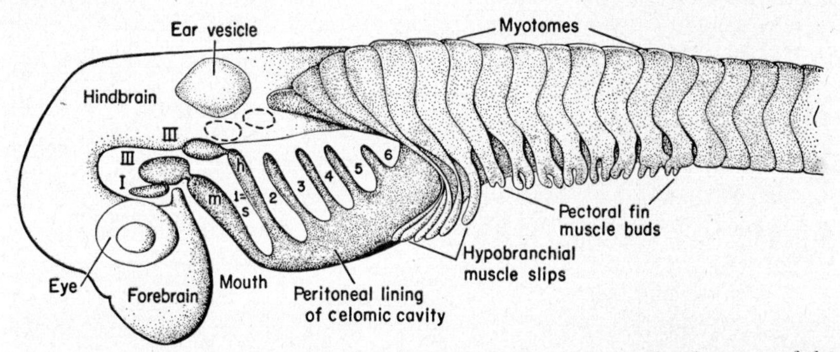

Fig. 153. A diagrammatic view of a shark embryo to show the development of the muscles. Skin and gut tube removed; the brain and eye and ear vesicles are included as landmarks. Posteriorly, the myotomes have extended downward to form myomeres; in the region of the pectoral fin, paired buds are forming from neighboring myotomes as potential fin muscles. Anterior to this, buds from anterior myotomes extend ventrally to form hypobranchial muscles. In the ear region, myotomes (broken lines) are rudimentary or absent, but farther forward three myotomes (I to III) persist to form eye muscles. The position of the spiracular slit (*s*) and normal gill slits (2 to 6) is indicated. These interrupt the continuity of the celom and its peritoneal epithelium. Buds of this project upward between the gill slits; from these arise the visceral muscles of the mandibular arch (*m*), the hyoid arch (*h*), and the more posterior gill arches. (In part after Braus.)

of the body (Figs. 151, 152, *A*, *B*). Arranged for the most part in segmental masses along either flank, it forms a major locomotor organ of a fish. By rhythmic, alternate contractions of the muscles of the two sides the fish's body is thrown into waves which, traveling posteriorly and gaining a cumulative effect at the tail, thrust backward against the resistance of the water and propel the fish forward (cf. Fig. 96, p. 170).

Dorsally, the axial muscles form a thick mass extending outward and upward on either side the full distance from the region of the vertebrae to the skin. In the tail there is a corresponding ventral development of the musculature; in the trunk, of course, the body cavity occupies much of the ventral region of the body, and the ventral trunk musculature takes the form of an enveloping sheath.

This axial musculature is of direct myotomic origin. In the embryo one can observe parallel strips of musculature growing down the body on either side as extensions of the dorsally placed myotomes (Fig. 153). The segmental arrangement is for the most part preserved in the adult fish, so that the major part of the trunk musculature consists of successive segments—the *myomeres*—running along each flank, corresponding in number to the vertebrae. The muscle fibers are oriented in an anteroposterior position in each myomere. Few reach skeletal parts directly; between successive myomeres are stout sheets of connective tissue, the *myocommata,* into which the muscle tissue is bound. The myocommata reach inward to tie into the vertebral column, and it is in these segmentally arranged septa that ribs, and the extra intermuscular bones of many teleosts, develop to give further support (Fig. 90, p. 161).

The myomeres develop initially as simple vertical bands; they do not retain this pattern, however, but fold in a complicated zigzag fashion which appears to promote muscular efficiency. This folding is especially prominent in sharks, and in fishes in general the folding is greater dorsally than toward the belly. Seen on the surface, each myomere is generally in the form of a *W* with the upper edge turned forward. There is thus a forwardly projecting *V* half-way down the flank, and backwardly projecting *V's* above and below. (There may also be one or two small extra jogs above.) Deep to the surface, the angles of the zigzag become increasingly sharp, so that each *V* of a myomere has the shape of a plow or engine cowcatcher. Each myomere folds forward or backward beneath its neighbors, so that any vertical cross section through trunk or tail cuts a considerable number of myomeres; their tissues are often seen in section to be grouped in a number of complex folded masses, each of which corresponds to one of the *V's* seen on the surface (Fig. 152, *A, B*).

In fishes above the cyclostome level there develops on either flank, usually just below the tip of the anterior-pointing *V,* a *horizontal septum* of connective tissue dividing every myomere into dorsal and ventral portions. It is in this septum, at the point where it intersects successive myocommata, that dorsal ribs are developed (Figs. 2, *D*, p.6; 90, p. 161).

As a result of the formation of the septum, the axial muscles of gnathostomes may be divided into two major groups: the dorsal or *epaxial musculature,* lying above the septum and external to the dorsal ribs (when developed), and the *hypaxial musculature* of the flanks and belly, lying below the septum and in the main internal to the dorsal ribs.

Although most of the fish trunk musculature is involved in the major structural organization of myomeres, there may nevertheless be minor units separated from the embryonic myotomes as buds or cell masses which form discrete muscles. Most important of these are the fin muscles.

In the median fins—dorsals, anal, and caudal—there are small, symmetrical muscle slips placed on either side of the radials supporting them. Paired fin muscles have a similar phylogenetic origin, we believe, but because of the importance of the limb musculature in tetrapods, the muscles of the paired appendages will receive consideration in a special section.

The general description of fish axial musculature just given applies primarily to that found in the main part of the trunk region. Anteriorly and posteriorly, specialized regions may be found. The shoulder girdle

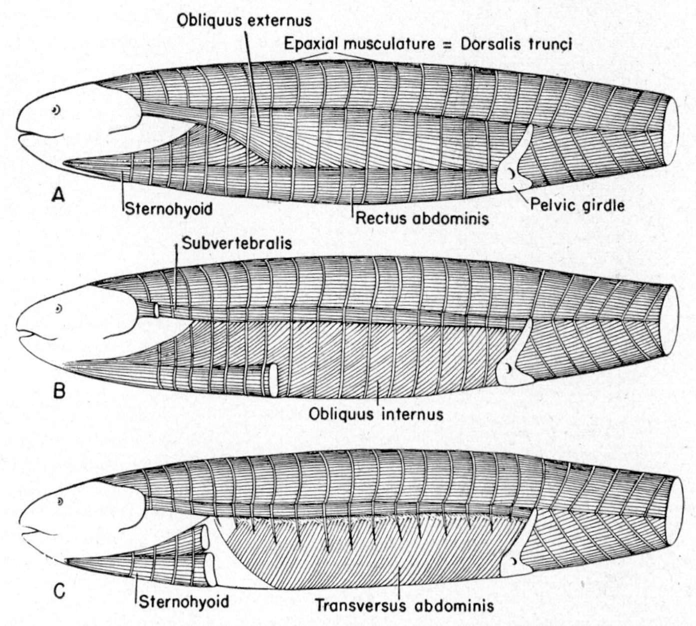

Fig. 154. Lateral views of the axial musculature of a urodele. *A*, Surface view (a thin superficial sheet of the external oblique, however, has been removed). *B*, The external oblique and rectus have been cut to show the internal oblique and subvertebral muscles. *C*, The internal oblique has been removed to show the transversus. (Modified after Maurer.)

may partially interrupt the course of these muscles; farther forward, the gill slits block the lateral part of the trunk muscles, but special dorsal and ventral axial masses pass forward to the head region. Posteriorly in fishes there is little interruption of the axial musculature as it passes backward from the trunk past the pelvic fins and cloaca to the tail, but in land vertebrates the tail muscles may show a different condition from those of the trunk.

In the further treatment of axial muscles here, we shall first follow the history of the two major trunk muscle groups—epaxial and hypaxial—from fishes upward through the tetrapods, and then return to pick up the story of more specialized anterior and posterior regions.

Epaxial Trunk Muscles. The dorsal musculature has led a relatively uneventful phylogenetic career. In fishes (Figs. 151, 152, *A, B*) generally it is a massive column of segmented muscle, lying lateral to the vertebrae and the medial septum, which runs along the back of the body, from the

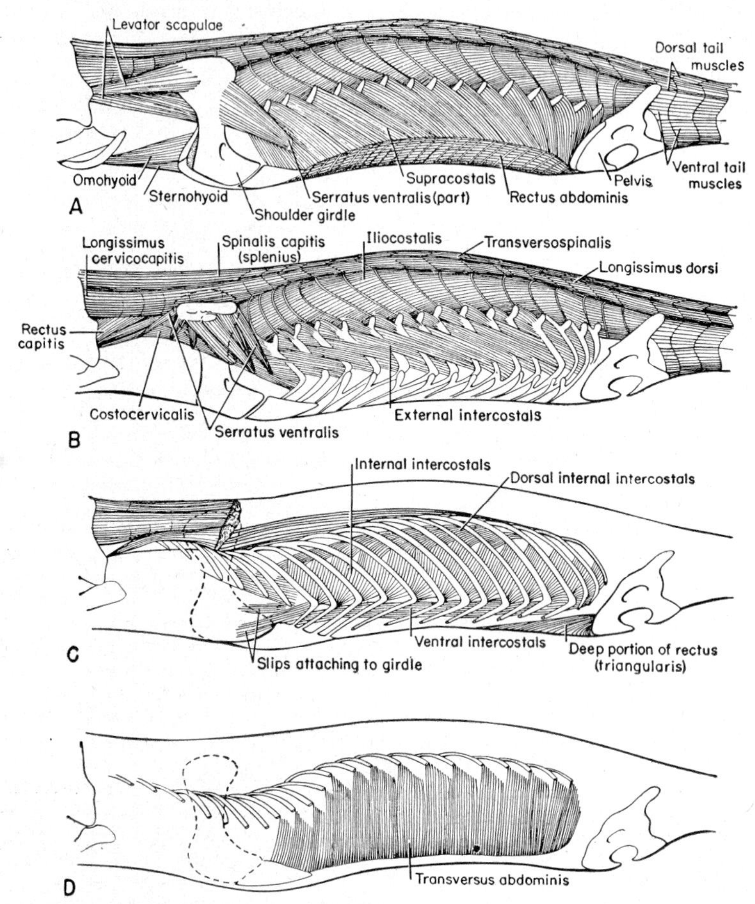

Fig. 155. A series of diagrammatic dissections of Sphenodon to show the anatomy of the axial muscles. In *A* a thin superficial sheet of the external oblique has been removed. In *B* the supracostals, rectus, throat muscles, and more superficial muscles to the scapula have been removed. In *C* the epaxial muscles are cut posteriorly, and the internal intercostals and triangularis (not shown in the last figure) are indicated. In *D* the ribs are cut, and all other muscles removed to show the transversus. (After Maurer and Fürbringer.)

skull or braincase to the end of the fleshy tail, without, in most instances, any great degree of regional differentiation. We may simply consider the whole mass as a single dorsolateral trunk muscle, the *dorsalis trunci.*

In land vertebrates (Figs. 152, *C, D,* 154, 155) this dorsal musculature —indeed, the trunk musculature as a whole—is much reduced in relative volume. This reduction is, of course, associated with the fact that in land vertebrates the limbs have taken over the propulsive duties of the axial muscles. The axial muscles still function, however, in such lateral movements as continue. In addition, they are responsible for dorsoventral bending of the column, a type of movement practically unknown in fishes.

In most land vertebrates there is little height to the back above the level of the neural arches, and the dorsal muscles are restricted to a channel, of limited dimensions, running the length of the body between the neural spines medially and the transverse processes laterally. In amphibians the dorsal musculature may still be considered a single muscle, the dorsalis trunci, which is still segmental and shows little in the way of distinct subdivision.

In reptiles the dorsal trunk muscles are still metameric in composition in most cases, but tend to have a complex build. Laterally, a thin sheet of dorsal musculature, the *iliocostalis,* extends downward on to the flank, external to the ventral muscles, and attaches laterally to the ribs. A *longissimus dorsi,* the main member of the dorsal series, lies above the transverse processes of the vertebrae. Most medially, between the longissimus and the vertebral spines, we find a complicated crisscross of little muscles which tie together successive vertebral spines and transverse processes. Although they pass under a variety of names, they may be termed collectively the *transversospinalis system.* In snakes, where the axial muscles have resumed the major propulsive function, the dorsal muscles are highly developed and extremely complicated. In turtles, with the development of a solid shell, the dorsal trunk muscles (and the ventral ones, as well) are, of course, much reduced, and in birds, in which the trunk vertebrae are in great measure fused, they are also weakly developed.

In mammals the three reptilian divisions of the dorsal musculature are still present, although little of the original metameric arrangement is preserved. Much of it consists of series of small muscle slips to which a plethora of names has been given. Along the lumbar region the two more medial divisions are united into a strong *sacrospinalis,* which helps preserve the arch of the backbone.

Hypaxial Trunk Muscles. In fishes the hypaxial musculature lies below the transverse septum and extends downward around the outer body wall to the ventral midline. It is essentially a unit in most cases, but there are frequently signs of dorsoventral subdivision.

In land vertebrates (Figs. 152, *C, D,* 154, 155) the hypaxial muscles, like their epaxial colleagues, are considerably reduced in volume and for the most part form but thin muscle sheets around the belly and flanks. They have, however, taken on a new function in land life in the support of the trunk viscera—a function in which they are aided by the development in all tetrapods (except the modern amphibians) of a powerful rib system. The presence of ribs adds to the complexity of the hypaxial musculature of tetrapods through a tendency for the development of individual slips for each rib from various muscle sheets. Somewhat arbitrarily we may divide the hypaxial trunk muscles of land vertebrates into three groups arranged dorsoventrally as follows:

1. Subvertebral muscles dorsally and medially;
2. A lateral series of superimposed muscle sheets along the flanks;
3. A rectus group ventrally.

Subvertebral musculature is little developed in fishes. In land vertebrates a deep longitudinal band of hypaxial muscle tissue forms at the sides of the vertebrae below the transverse processes, acting as an opponent of the dorsal musculature in dorsoventral movements of the spinal column. The subvertebral musculature is often continuous with the deepest layer of the flank muscles.

Complicated and varied are the *flank muscles,* which follow the curve of the trunk over the general area extending from the transverse processes of the vertebrae to the ventral territory held by the rectus system. Three superimposed major sheets of segmentally arranged muscle may be generally found; each of these, however, may be subdivided in various regions and areas.

In urodeles the absence of ribs gives us a relatively simple picture of these muscles (the fact that the absence of ribs and the resulting simplicity is a degenerate condition need not cause us undue concern). The most external sheet is the *external oblique* muscle, the fibers of which run in general anteroposteriorly, but tend to slant upward anteriorly in each segment (a thin, external subcutaneous sheet is generally present in addition to the main muscle). A middle layer is that of the *internal oblique;* the fibers are again essentially longitudinal in orientation, but slant upward posteriorly. The third, deepest layer is that of the *transverse* muscle, whose fibers are oriented, in contrast to the obliques, in a transverse, dorsoventral direction. None of the three layers is of any great thickness, but the diverse orientation of the three muscle sheets makes for a body sheath of considerable strength.

In amniotes the presence of ribs on part or all of the trunk vertebrae creates a more complicated situation, except in the lumbar region, where short abdominal sections of the muscle sheets may be present in relatively primitive fashion. The external oblique is split into two layers, superficial and deep (in addition to a subcutaneous sheet in reptiles), and the transversus may be split into two layers as well. In the major rib-bearing region of the trunk the outer layer of the external oblique lies in general superficial to the ribs and produces *supracostal* muscles, and the transversus is for the most part transformed into a *subcostal* series. The intermediate muscle layers, however, form *intercostal* muscles—external intercostals arising from the deep part of the external oblique, internal intercostals from the internal oblique. In most amniotes the transversus tends to persist internally as a continuous if thin sheet; the outer layers, however, tend to become discontinuous, covering only restricted parts of the trunk. From form to form and from one region of the body to another, portions of the various layers of the lateral trunk muscles are given a bewildering series of different names; to follow them through would be as tiring to the author as to the student.

Ventrally, the *rectus abdominis* runs anteroposteriorly from the shoulder region to the pelvis; the portions from each side of the trunk are frequently separated by an intervening tendinous area, the *linea alba.* A superficial portion may be more or less continuous with the external oblique; for the most part, however, the rectus appears to be more closely associated with the internal oblique group. With the growth of the sternum, the rectus is restricted in length, and in mammals is a purely abdominal muscle.

The components of the axial musculature of the tetrapod trunk may be classified, in somewhat simplified fashion, as follows:

- Epaxial ....
  - Transversospinalis system
  - Longissimus system
  - Iliocostalis system
- Hypaxial......
  - Dorsomedial...Subvertebralis
  - Lateral
    - Obliquus abdominis externus
      - (Subcutaneous)
      - Superficialis = supracostals
      - Profundus = external intercostals
    - Obliquus abdominis internus = Internal intercostals
    - Transversus abdominis = Subcostals
  - Ventral...Rectus abdominis

Trunk Muscles of the Shoulder Region (Figs. 154, 155). In many urodeles the ventral trunk musculature extends past the pectoral region almost without break to continue forward beneath the throat. This situation, however, is probably not primitive, but is the appearance in the adult of an embryonic condition, due to the loss of the dermal shoulder girdle. Even in sharks there is a partial interruption of the ventral muscles here, and in bony fishes and tetrapods other than urodeles there is a complete break. The rectus system and the lower margins of the flank muscles typically terminate anteriorly in muscle slips attaching to sternum and coracoid. The more anterior ventral trunk muscles reappear in front of the girdle as the hypobranchial muscles (p. 262).

Whereas the pelvic limb bears body weight by a solid sacral connection with the backbone, there is no such connection in the shoulder region. In tetrapods, in contrast with fishes, the pectoral girdle and limb have no skeletal connection with the backbone other than that gained indirectly via the sternum. The body is suspended between the two scapular blades in slings formed by muscles which on either side run from the top of the scapular blade downward to insert mainly on the ribs. The utilization of muscles for this supporting system furnishes an elasticity which makes up in the shoulder region for the lack of the bony strength present in the pelvis; body and limb movements can be pliably adjusted to one another, and (analogous to the use of springs in automobile construction) the body is eased of much of the jolts and jars of locomotion.

The muscles which form those slings are derivatives of the external oblique muscle which have been pressed into service for this special purpose. Seen in side view (Fig. 155, *A, B*), the major elements of this system spread out fanwise fore and aft from the under surface of the scapular blade to attach posteriorly to the chest ribs, anteriorly to cervical ribs or transverse processes. Because of their jagged appearance, elements of this group are known as the *serratus* muscles (serratus ventralis or anterior); the most anterior part is termed the *levator scapulae*. In addition, we find in mammals a more dorsally placed sheet, the *rhomboideus,* which tends to keep the scapular blade in place by a pull toward the midline.

Neck Muscles (Figs. 154, 155, *B*). In fishes the gill region interrupts anteriorly the course of the axial muscles; they continue forward, however, above and below the gills, as *epibranchial* and *hypobranchial* groups. In tetrapods the same situation persists (the effect is somewhat masked by the presence, superficially, of visceral striated muscles of the trapezius group; cf. p. 285; Fig. 163, *B*, p. 284). A dorsal axial group runs forward as neck muscles, which serve for the support and movement of the head. The epaxial series extends forward here in unbroken fashion, although with a special arrangement of small muscle slips connecting head and cervical region. A minor contribution to the neck musculature is made by the hypaxial muscles; small muscles of the subvertebral series continue forward beneath the column to the skull base.

Throat Muscles (Figs. 154, 155, *A*). The anterior ventral part of the fish axial muscles, the hypobranchial musculature, runs forward along the "throat" beneath the gills. The main mass of this musculature is known collectively as the *coracoarcuales,* since it originates posteriorly from the coracoid region of the shoulder girdle and in its forward course to the jaw area sends off slips which insert on the lower segments of the adjacent gill arches. (Note that these muscles are *not* part of the special group of branchial muscles situated in the gill region.) In land animals this musculature persists in its original location, in reduced form, as a series of ventral slips running from sternum and shoulder girdle to the thyroid cartilage of the larynx and to the hyoid, and from these elements to the region of the jaw symphysis (*sternohyoid, omohyoid* muscles, and so forth). We find one remarkable development of the hypobranchial musculature in tetrapods. The tongue develops in the floor of the mouth from the region of the base of the hyoid arch. As it develops, it carries with it a mass of the hypobranchial muscle fibers present in this region; these constitute the flesh of the tongue.

The embryology and nerve supply of the hypobranchial muscles are not without interest. As axial muscles, they are of myotomic derivation;

however, the development of the gill slits separates the throat region from direct connection dorsally with the myotomes of the occipital and neck regions from which we would expect them to come. In some embryos, slips from these myotomes are seen in process of migrating circuitously backward above the gills, down behind the gill chamber, and then forward in the throat to form the hypobranchial (and tongue) muscles (Fig. 153, p. 255). We have noted earlier that there tends to be a constant nerve supply to a given mass of muscle, even if it has migrated far from its original position. In correlation with this fact we find that the hypobranchial muscles in fishes are innervated by nerves, from the occipital region of the skull and the anterior part of the cervical region, which follow the same path of migration as did the muscle tissue, around the back of the gill chamber and forward along the throat. In amniotes comparable nerves form the hypoglossal nerve and cervical plexus; in the embryo these nerves follow the ancestral route back and down behind the embryonic gill pouches, and even in the adult they pursue a roundabout course to the throat and tongue (cf. p. 556).

Diaphragm Muscles. Mammals are notable for the development of the *diaphragm,* a partition separating thoracic and abdominal cavities and of functional importance in the expansion (and contraction) of the lung cavities (cf. pp. 336, 482). The movement of the diaphragm is performed by a series of striated muscles which converge from all sides toward its center—from the sternum ventrally, the ribs laterally, and the lumbar vertebrae dorsally. These muscles appear to be derivatives of the rectus abdominis system of axial muscles. They are innervated by a special (phrenic) nerve which arises in the cervical region. This strongly suggests that the muscles of the diaphragm originated from an anterior part of the rectus musculature, perhaps in the shoulder region, and migrated posteriorly in an early stage of tetrapod history in conformity with a backward movement of the heart and a backward expansion of the lungs.

Caudal Muscles. In fishes the axial musculature continues into the tail without interruption, except for the cloacal or anal opening. The epaxial musculature of the tail is simply a continuation of that of the trunk. Ventrally, however, in the absence in the tail of the body cavity and its contained viscera, the hypaxial muscles change from a series of sheetlike structures to a compact pair of ventral bundles similar to the epaxial muscles above (Fig. 152, *A*, p. 254).

In tetrapods the great development of the pelvic girdles and the limb muscles arising from them has tended to break the continuity of the axial muscles between trunk and tail. The epaxial muscles are relatively little affected; in many amphibians and reptiles they extend backward on to the tail with little or no interruption. The hypaxial muscles, however, are completely, or nearly completely, interrupted at the pelvis. In the more distal regions of the tail they form a relatively uniform cylindrical muscle bundle occupying a position on either side below the transverse processes of the tail vertebrae similar to that of the dorsal muscle above. Anteriorly, approaching the pelvic region, the hypaxial tail muscle mass becomes constricted and sends terminal fibers downward and forward to insert on to the ischium, or upward and forward to the ilium. In urodeles and in reptiles the basal region of the tail is expanded to contain, ventrally, beside the tail muscles proper, large muscles which run forward and outward to the femur; these, however, are parts of the limb musculature, not axial muscles. Needless to say, there is a reduction or loss of caudal musculature in forms in which the tail is reduced. Remnants of tail musculature may still be found attached to the bony tail stump of birds, frogs, and the like. In primitive mammals the tail is still moderately elongate, but slender, and the caudal muscles are correspondingly reduced in volume. From the ventral muscles just behind the pelvic girdle there usually develops in

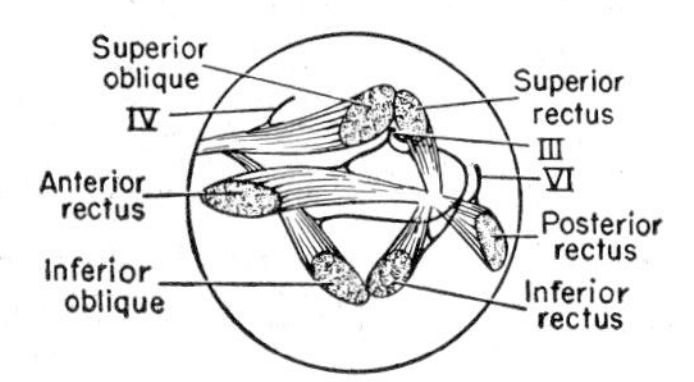

Fig. 156. Eye muscles. A lateral view, with the eyeball (in outline) removed; the ovals are the muscle attachments. The three eye muscle nerves are shown (III, IV, VI). (After Goodrich.)

land vertebrates a sphincter muscle which can close the cloaca; in mammals there is a corresponding anal sphincter.

Eye Muscles. The muscles which move the eyeball form a far-flung anterior outpost of the axial musculature. In embryonic cyclostomes there is present a series of mesodermal somites continuous from the front part of the head back to the trunk. In most vertebrates the series is interrupted, for with the expansion of the braincase in the ear region the somites there tend to be vestigial or absent. More anteriorly, however, in the region of the developing eye socket, three small somites persist in every vertebrate class (Fig. 153, p. 255). They are in great measure separated at an early stage from the lateral mesoderm by the development of the gill slits, and play little if any part in the formation of skeletal or connective tissues. They are, however, competent as myotomes, and from them arise the muscles of the eyeball. Associated with these three somites, and innervating the muscles which they form, are three small cranial nerves—III, IV, and VI of the numbered series (pp. 555, 556).

In the vast majority of vertebrates we find six typical straplike muscles developed from these somites (Fig. 156). They arise from the surface of the braincase and fan outward to attach to the eyeball somewhat internal to its "equator;" in varied combinations, their pull will rotate the eye in any desired direction. Four of them, the *rectus muscles,* arise posteriorly, often close to the eye stalk or optic nerve; the other two, the *oblique muscles,* spring from the anterior part of the eye socket. Typically, the superior rectus and the superior oblique attach to the upper margin of the eyeball, a pair of inferior muscles to its lower edge; the two remaining rectus muscles insert on the anterior (or internal) and posterior (or external) margins.

Four of the six muscles are innervated by nerve III, the superior oblique by nerve IV, the posterior rectus by nerve VI. As this would lead us to suspect, we find that embryologically four of these muscles usually arise from the first of the three somites concerned, and one each from the other two.

Variations are, of course, found on this basal pattern of six eye muscles. In hagfishes and other vertebrates with degenerate eyes the muscles are degenerate or absent. Accessory muscles are frequently present, particularly in land vertebrates. In a majority of tetrapods (birds and primates are exceptions) there is a *retractor bulbi* muscle which tends to pull the eyeball deeper in its socket; in most amniotes we find a *levator palpebrae superioris* derived from the eyeball muscles, but raising the upper lid, and rather variable slips which move the nictitating membrane of the eye.

LIMB MUSCLES

The musculature of the paired appendages is derived, historically, from the general myotomic musculature of the trunk and hence is part of the somatic system. The limb muscles, however, are so distinct in position and nature and so important in higher vertebrates that they deserve special treatment.

We have noted previously the decline of axial musculature in tetrapods; limb muscles, modest in size and simple in composition in fishes, grow, on the other hand, to relatively enormous bulk and to great complexity in land vertebrates, as they take over the major duties of body progression. As a homely example to emphasize this contrast we may compare our utilization as food, of a fish and of a domestic mammal. Fish as food is axial muscle, and the fins are discarded as not worth bothering with. In a steer, lamb, or hog, however, there is little axial muscle except small morsels which cling to the ribs or backbone. Our meat is almost entirely limb muscle. A chop with bone attached seems to testify to its being of axial origin. However, the main part of the chop is a limb muscle which takes origin from the trunk, and only the scrapings off the bone itself are really axial.

As derivatives of the somatic system, the limb muscles should, in theory at least, originate in the embryo from the myotomes. In some lower vertebrates—specifically, sharks—this origin appears to be demonstrable. Paired finger-like processes extend out from the ventral ends of a considerable number of myotomes developing near the base of the paired fins; these form masses of premuscular tissue above and below the fin skeleton (Fig. 153, p. 255). In tetrapods, however, the myotomic origin of appendicular muscles has not been demonstrated, and the muscle masses of the tetrapod limb appear to develop as condensations in the mesenchyme of the region of the budding limb. It is, of course, possible that this mesenchyme is derived from the myotome, but proof is lacking.

Paired Fins. In the fins of fishes the limb musculature is usually simple. Two opposed little masses of muscle are usually readily discernible, running outward from the girdle to the base of the fin (Fig. 157, *B*). A dorsal muscle mass serves primarily to elevate or extend the fin, a ventral muscle mass to depress or adduct it. There may be, in addition, special slips de-

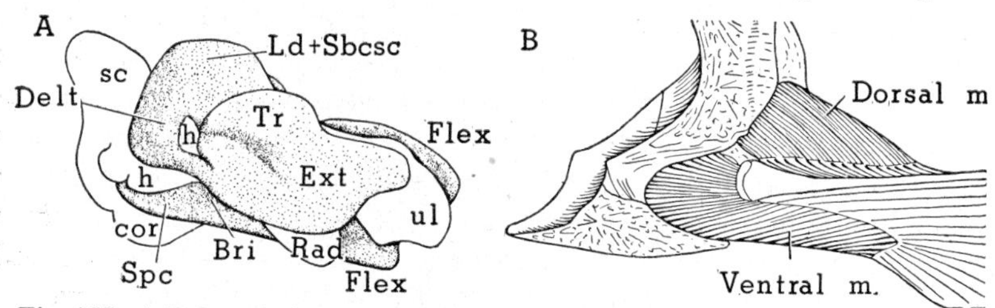

Fig. 157. *A*, External view of the left pectoral girdle and limb and its musculature in a lizard embryo (the skeleton unshaded, the muscle tissue stippled); *B*, the pectoral girdle and fin in a fish (sturgeon). In the fish fin the musculature consists simply of opposed dorsal and ventral muscle masses. In the adult land vertebrate the limb has a large number of discrete muscles, but in the embryo these are arranged in two opposed masses comparable to those of the fish fin. At the stage figured the two masses are barely beginning their differentiation into the muscles of the adult (cf. Fig. 158, *A*). The dorsal mass is well shown; the ventral mass is mostly concealed beneath the limb (in which the foot is not yet developed). *Bri*, brachialis inferior; *cor*, coracoid region; *Delt*, deltoid; *Ext*, extensor muscles; *Flex*, flexor muscles; *h*, humerus; *Ld*, latissimus dorsi; *Rad*, radius; *Sbcsc*, subcoracoscapularis; *Sc*, scapula; *Spc*, supracoracoid; *Tr*, triceps; *ul*, ulna.

veloped from either group which serve to give rotary or other special fin movements.

Tetrapod Limbs. In land vertebrates we meet with a different situation. Not only is the limb musculature much more bulky, but it is also much more complex. The mode of development affords, however, a clue to a natural classification.

Early in the ontogeny of a land limb, while the arm or leg is still a relatively short bud from the body, a mass of premuscular tissue (Fig. 157, *A*) is formed on both the upper and lower surfaces of the developing skeleton. It is clear that these two masses are equivalent to the opposed dorsal and ventral muscle masses which in fishes raise and lower the paired fins. From them arise, by growth and cleavage, all the complicated muscles of the mature limb. From the dorsal mass develop the muscles on the extensor surface of the limb and related muscles of the girdle region; from the ventral mass arise the flexor muscles of the opposite limb surface.

A distinction is frequently made between "extrinsic" and "intrinsic" limb muscles, the former name being given to muscles that arise from the body and run either to girdle or limb, the latter to muscles which do not leave the confines of the limb. This distinction is misleading. Embryological study shows that on the one hand muscles which arise from the body and run only to the limb girdle are not limb muscles at all, but true body muscles, which arise directly from somites. On the other hand, various true limb muscles may take origin from the trunk and run outward onto the limb. Their real nature is shown by the fact that they arise in the embryo as part of the limb muscle mass and only during the course of development spread inward on to the trunk. Where a muscle inserts, not where it arises, is the test as to whether it is a true limb muscle or not.

How to treat such a complex series of structures as limb muscles in a brief work of this sort is a problem, for there is an immense amount of variation from group to group in the arrangement of the musculature (and also, sad to say, in the nomenclature). On the one hand, one might describe, seriatim, the limb muscles of the various groups, in a fashion which would be not merely exhaustive, but exhausting. On the other hand, one might (as is sometimes done) make a few hasty generalizations about extensors, flexors, and so on, and pass hurriedly on. We shall here take a middle course (which probably includes the disadvantages of both extremes). The various muscle groups of the limbs will be considered and compared in fair detail in two types: (1) lizards, as representing a rather primitive and generalized tetrapod condition;* and (2) a primitive mammal, the opossum (Didelphis), as representing a type of specialization of particular interest. A few remarks will follow on the musculature of other forms, and a table of homologies presented.

Embryology shows that the musculature of each limb can be naturally divided into dorsal and ventral groups, proximally none too readily distinguishable in the adult, but distally clearly forming extensor and flexor series of the limb. We shall follow this logical order in considering successively the muscles of pectoral and then of pelvic limbs. Little will be said of the functions of individual muscles. It must be emphasized that rarely is a movement performed by the contraction of a single muscular element. In general, several muscles cooperate; the movement is a resultant of forces. The function of a given muscle varies according to the company it keeps.

Pectoral Limb. DORSAL MUSCLES OF SHOULDER AND UPPER ARM (Fig. 158). In lizards a number of muscles are present which attach to the humerus near its head and are responsible for much of the movement of that bone on the shoulder girdle. Superficially there are seen two prominent dorsal muscles of this sort, the *latissimus dorsi* and the *deltoideus.*

* Other reptiles (except Sphenodon) and the frog are more specialized, and the urodeles are degenerate in their musculature.

The former arises from the surface of the back and flank as a broad, thin sheet which narrows and thickens to insert on the back margin of the humerus near its head. The latter, a fan-shaped muscle, arises from the anterior rim of the girdle—from the front margin of the scapula and from the clavicle—to insert near the front edge of the humeral head. A deeper muscle associated with the latissimus at its insertion is the *subcoracoscapularis,* which arises from the inner surface of the scapula and coracoid; beneath the deltoid, arising from the surface of the scapula, is a small *scapulohumeralis anterior.*

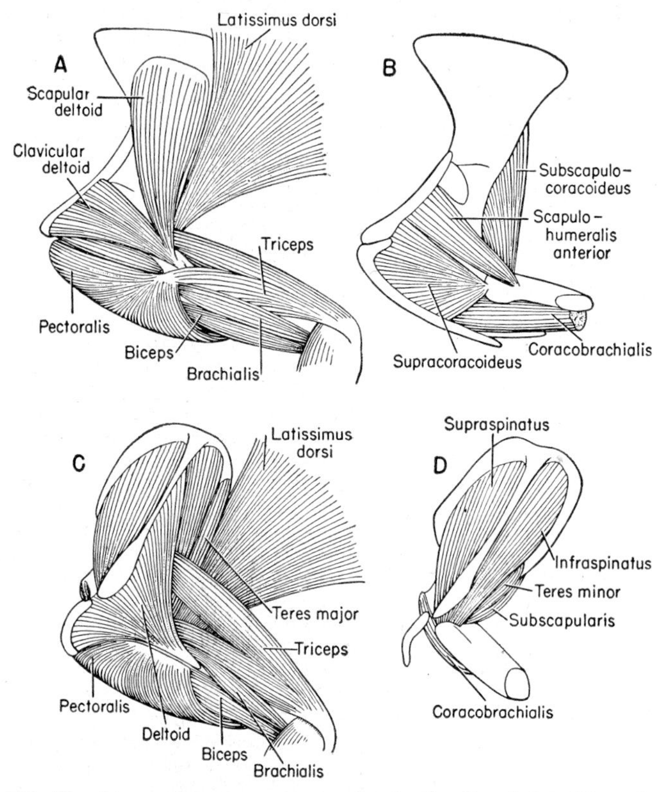

Fig. 158. Shoulder and upper arm muscles in the lizard (*A, B*) and opossum (*C, D*), lateral views; the right hand figures in each case are comparable deep dissections with latissimus, deltoid, pectoralis, and long muscles (triceps, biceps, brachialis) removed. Notable is the upward migration of the supracoracoideus to become the two "spinatus" muscles. (Subscapulocoracoideus = subcoracoscapularis of text.)

The dorsal surface of the humerus is covered by a stout muscle, the *triceps,* which arises from the humerus and by one or more heads from adjacent parts of the girdle; the muscle attaches distally to the olecranon of the ulna—this attachment is in fact the reason for existence of that process—and serves to extend the forearm.

In the opossum much the same elements are present, but there are important changes in distribution to be noted. Part of the original latissimus

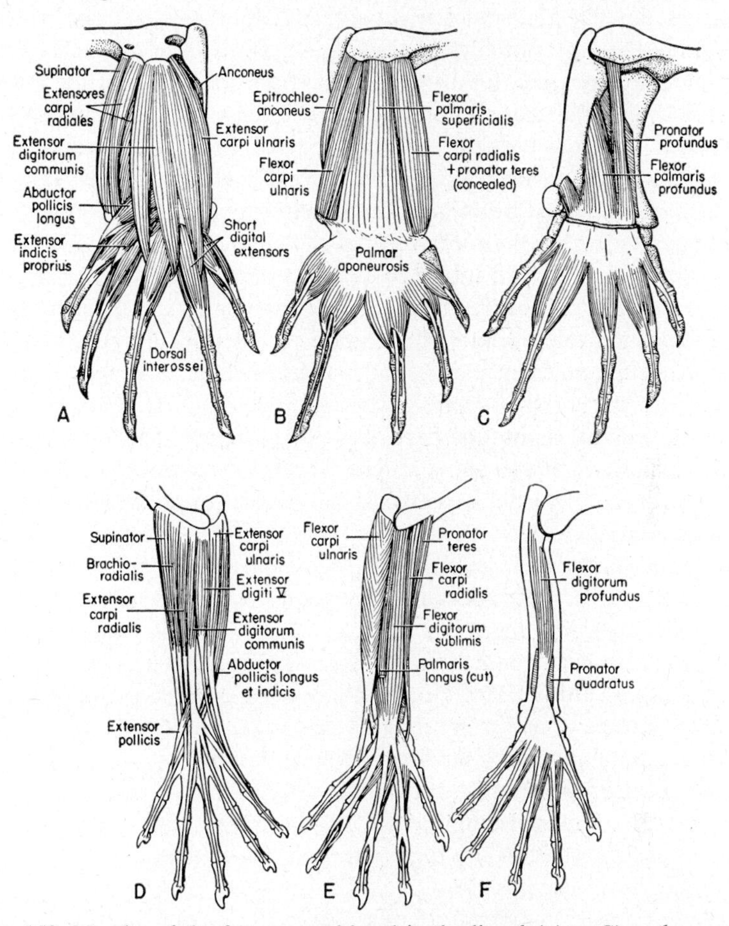

Fig. 159. Muscles of the forearm and hand in the lizard (*A* to *C*) and opossum (*D* to *F*), somewhat diagrammatic and simplified. *A* and *D* are views of the extensor surface; *B* and *E* superficial, *C* and *F* deeper dissections of the flexor aspect. On the extensor surface the most prominent change from reptile to mammal is the reduction of short muscles on the manus and the development from the common extensor of tendons to the toes. Long special muscles have developed for movement of the "thumb" and fifth digit. On the flexor aspect a prominent feature in reptiles is the presence of a stout and complex aponeurosis on the "palm" with which connect the long flexors proximally and a variety of tendons and short muscles to the toes. In mammals this is broken up; the palmaris longus inserts into a superficial aponeurosis over the wrist, which is cut away in the figure, and the two deeper flexors present here have each developed a broad palmar tendon. Deeply placed in the hand are various short muscles of the digits, which are not shown. (In *C*, flexor palmaris profundus = flexor digitorum profundus of text.)

has gained contact with the back margin of the scapula as the *teres major*. There is, in typical mammals, no developed coracoid, and the internal muscle is restricted to the scapula as a *subscapularis.* The scapular part of the deltoid does not arise from the surface of the bone, but from the spine of the scapula, and the little scapulohumeralis muscle has been pushed to the posterior edge of the scapula as the *teres minor.* These last two modifications are due to the invasion and crowding of the scapular surface by muscles present in reptiles on the ventral coracoid plate.

DORSAL MUSCLES OF FOREARM AND MANUS (Fig. 159). The dorsal arm musculature, in lizards, is continued beyond the elbow by muscles of the extensor series. Most prominent is a sheet of long muscles running down the upper surface of the forearm from the elbow region. Three parts can be distinguished, radial, intermediate, and ulnar; the two lateral groups are concerned with performing (or preventing) rotary movements, the central portion solely with digital extension. The radial group, running to radius and the adjacent part of the carpus, includes a *supinator* and *extensores carpi radiales.* Centrally, the *extensor digitorum communis* runs downward to insert onto the base of the metacarpals. The ulnar group includes a short *anconeus* and a longer *extensor carpi ulnaris.*

Beneath the long extensors distally is a series of short forearm extensors running diagonally downward and inward from the ulna and the adjacent part of the carpus to spread out onto the digits. These are the *extensores digitorum breves;* there are special long and strong muscles to the first two digits. On each digit there is a small *dorsal interosseous* muscle, running from the metacarpal to the base of the first phalanx.

In the opossum much of the general pattern is the same. However, the common extensor, which in reptiles stops at the base of the metacarpals, has in mammals acquired tendons running on to the digits. The pollex has a long extensor running down from the head of the ulna, and the fifth toe also has a long muscle running the length of the ulna to it. Distally, the short dorsal muscles of the digits are reduced or absent, and the dorsal interossei have disappeared.

VENTRAL MUSCLES OF SHOULDER AND UPPER ARM (Fig. 158). On the under side of the shoulder a superficial and important muscle in reptiles (and in all tetrapods) is the *pectoralis,* the chest muscle, which spreads far back over the sternum and ribs and inserts on a powerful process beneath the proximal end of the humerus. This muscle gives a strong pull on the humerus downward and backward, and thus is of great importance in locomotion. Beneath the front part of the chest muscle is the *supracoracoideus,* important in body support; it arises from a broad area on the coracoid and runs to the under surface of the humerus. From the back end of the coracoid the *coracobrachialis,* consisting of long and short elements, runs out on the under side of the humerus. Two muscles extend out to flex the radius and ulna: the *brachialis,* running down the outer and anterior margin of the humerus, and, ventrally, the *biceps,* originating from the coracoid.

Most of these muscles are present in mammals, but their arrangement is different. The coracoid plate of the shoulder girdle has disappeared; the biceps and coracobrachial muscles which originated from this plate in reptiles here take origin from the nubbin which still represents the coracoid in mammals. Further, the prominent supracoracoid muscle which arises from this plate in reptiles is seemingly absent.

It is, however, actually present in the form of the *infraspinatus* and *supraspinatus* muscles on the scapula; this major muscular migration is presumably responsible for the reduction of the coracoid region of the girdle which afforded it origin in reptiles. The reptilian muscle has pushed its way upward beneath the deltoid and has (*a*) taken over the old scapular blade as the infraspinatus (and restricted the scapular deltoid muscle to the old front margin, now the spine of the scapula), and (*b*) occupied a new shelf, the supraspinous fossa, built for its reception in front of the old anterior margin of the scapula. With the changed limb posture of mammals the supracoracoid muscle ceased to function in its old position. It has, however, retained its supporting function by the insertion of the two homologous mammalian muscles at the very tip of the humerus anterior to the glenoid; the resulting lever motion tends to swing the limb downward and forward or, conversely, to pull the body up and back on the arm.

VENTRAL MUSCLES OF FOREARM AND MANUS (Fig. 159). In the distal part of the limb the main propulsive effort is a backward push of the forearm and digits, accomplished by the muscles of the ventral, flexor surface. These muscles are in consequence more powerful than the extensors, and are more complicated as well.

Long flexors in reptiles arise from the strong entepicondyle of the humerus and can be divided (like the forearm extensors) into radial, intermediate, and ulnar portions, the two marginal groups being in part rotary in function, the central one concerned with toe flexion. The radial elements include a *pronator teres* and *flexor carpi radialis;* centrally is a *flexor palmaris superficialis;* on the ulnar side are the *epitrochleoanconeus* and *flexor carpi ulnaris*. Flexion of the digits by a muscle from the elbow is rendered difficult by the fact that it must pass round the curve on the under side of the wrist. This difficulty is avoided by the development of a *palmar aponeurosis,* a pad of connective tissue beneath the carpus. Into this run proximally the superficial palmar flexor already mentioned and a deeper *flexor digitorum profundus;* out from it distally run not only tendons to the digits, but a double series of short toe muscles. Still further, short toe muscles are present beneath the aponeurosis, and proximally a deep *pronator profundus* runs diagonally from ulna to radius.

In mammals major changes include the presence of three proximal flexors instead of the two palmar flexors of a lizard; there are the *palmaris longus, flexor digitorum sublimis,* and *flexor digitorum profundus*. Distally, instead of a single palmar aponeurosis, there are three superimposed sheets: an aponeurosis over the wrist into which the long palmar muscle inserts, and two broad tendons through which the two deeper muscles send tendons on to the toes.

In mammals the epitrochleoanconeus is small or absent, the pronator teres often reduced, and the pronator quadratus is confined to the distal part of the forearm.

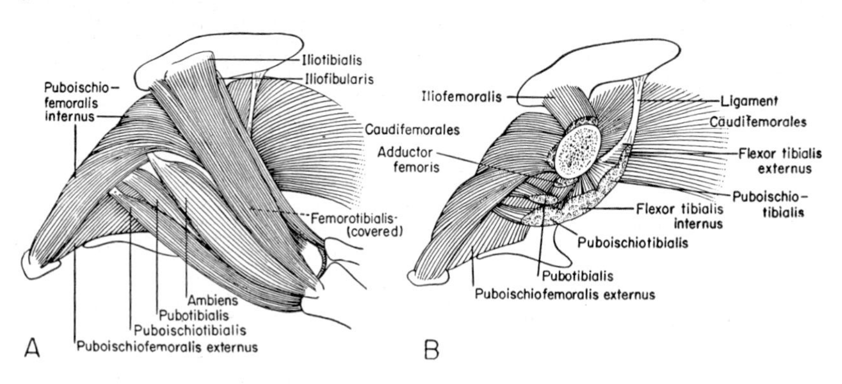

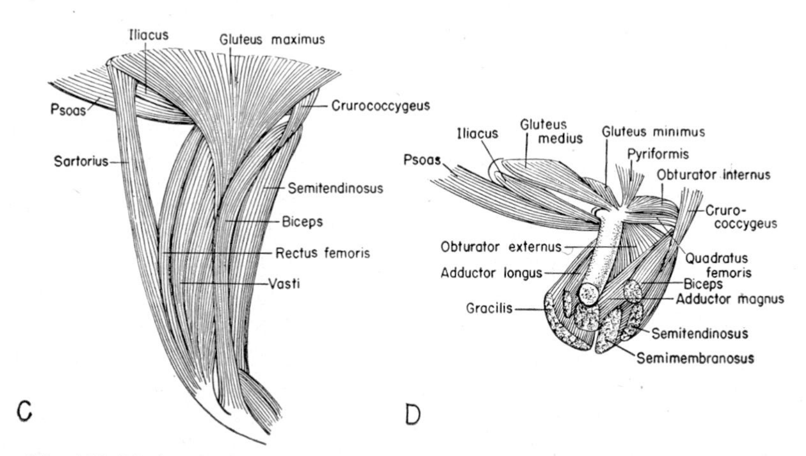

Fig. 160. Limb muscles of the pelvis and thigh in a lizard (*A, B*) and opossum (*C, D*), lateral views. *A, C,* Superficial views; *B, D,* dissections to show deeper layers of musculature.

Pelvic Limb. DORSAL MUSCLES OF HIP AND THIGH (Fig. 160). As in the pectoral appendage, the muscles of the distal part of the limb are readily separable into dorsal and ventral (flexor and extensor) groups, and the proximal muscles are equally divisible on an embryological basis. Dorsally, there are in lizards two short muscles inserting onto the femur and a series of long muscles extending to the lower leg. Of the short muscles, the *iliofemoralis* takes origin from the outer surface of the ilium and inserts onto the proximal upper surface of the femur. The *puboischiofemoralis internus* (what a name!) arises broadly from the inner surface of the pubis and ischium and, emerging anteriorly, inserts onto the femur near its head.

A major long muscle group extending by a common tendon to the tibia is that which corresponds in great measure to the *quadriceps femoris* in man and has as its major function extension of the knee joint. It includes in reptiles the *iliotibialis,* with an origin on the upper rim of the ilium, and powerful fleshy heads, from the femur, corresponding to the *vasti* muscles of mammals (usually called the *femorotibialis* in reptiles). Pertaining to the same group is the *ambiens,* which arises from the upper edge of the pubis and runs thence to the side of the tibia. More posteriorly an *iliofibularis* (quite distinct from the quadriceps group) runs from the ilium down the back or outer side of the thigh to the fibula.

In mammals the ilium is greatly changed in structure and rotated forward (Fig. 106, p. 188); the relative position of the muscles concerned has changed in like fashion. So great, however, have been the muscular changes that the homologies are none too certain in some cases. The iliofemoral muscle is homologous to part, at least, of the *gluteal muscles* running from the ilium over the buttocks to the region of the great trochanter of the femur. The puboischiofemoralis internus is represented by several deep muscles that have a similar insertion on the femur, but differ in origins. One, the *iliacus,* arises from the lower margin of the ilium; a second, the *psoas,* originates deep in the body in the lumbar region; the *pectineus* consists of fibers still remaining attached to the pubis. The quadriceps retains its femoral heads. An iliac head is here the *rectus femoris,* arising close to the acetabulum. The reptilian ambiens is presumably the *sartorius,* the "tailor's muscle" of man, with an origin that has migrated upward on to the ilium by way of an intervening ligament. The fate of the reptilian iliofibularis is uncertain. It may, however, be the *gluteus maximus,* running from the ilium down the posterior edge of the femur (in the forwardly rotated position of the mammalian limb, an iliac muscle cannot retain a straight course to the fibula).

DISTAL DORSAL MUSCLES OF THE HIND LEG (Fig. 161). Beyond the knee the extensor muscles of the pelvic limb in lizards show a pattern essentially comparable to that of the pectoral appendage. A series of long extensors arises from the knee region. The *tibialis anterior* runs down from the upper end of the tibia to the first metatarsal. Centrally, the *extensor digitorum communis* extends from the femur down to the base of a variable number of metatarsals. On the fibular side, *peroneal muscles,* long and short, take origin from much of the length of the fibula and run to the outer side of the tarsus. A shorter group of extensors fans out from the lower end of the fibula and the adjacent part of the tarsus to the various toes as the *extensores digitorum breves;* there is an especially long and

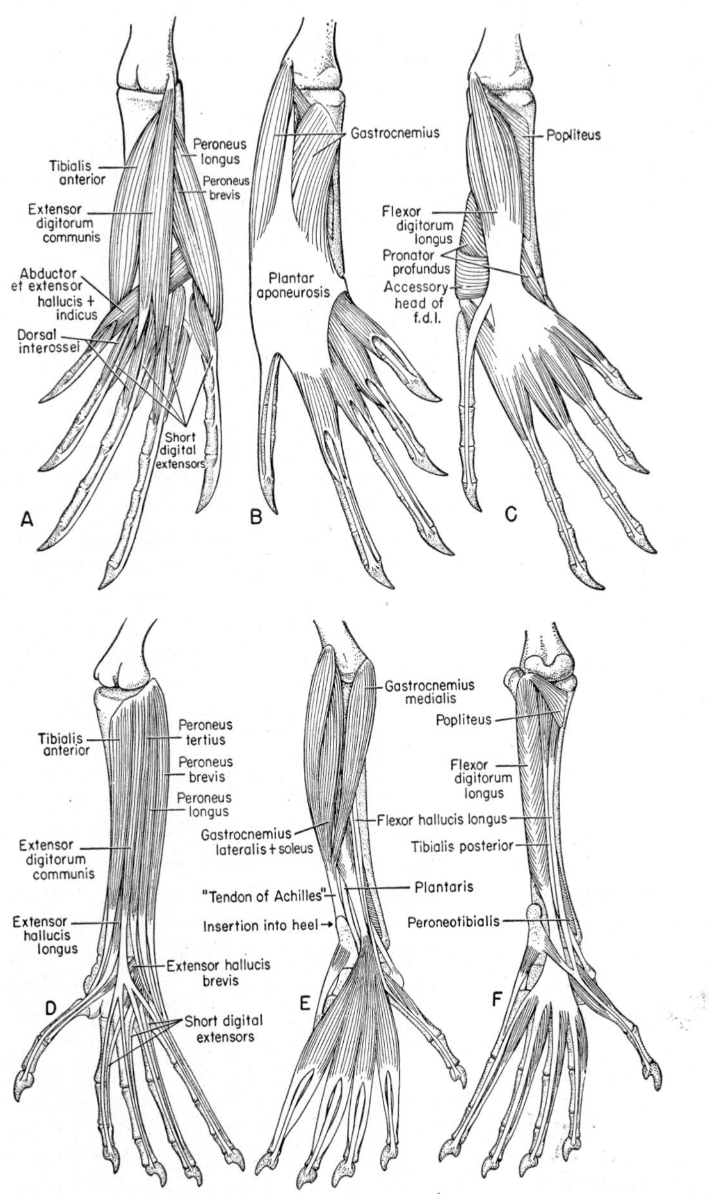

Fig. 161. Muscles of the lower leg and foot in a lizard (*A* to *C*) and opossum (*D* to *F*), somewhat diagrammatic and simplified. *A* and *D* are views of the extensor surface; *B* and *E* superficial, *C* and *F* deeper dissections of the flexor aspect. The extensor surface of the lizard hind leg is comparable to that of the forearm and manus except for a lesser development of individual muscles on the inner (tibial = radial) side. In the change to mammals the modifications are similar to those seen in the front leg, including the development of toe tendons from the common extensor and the development of long muscles working on the first and fifth toes. The flexor aspect of the lizard hind leg resembles that of the front in many regards, including the development in reptiles of a stout plantar (sole) aponeurosis, into which

strong division of this group running to the inner toes. In addition, there are *dorsal interosseous muscles* extending out over the proximal part of the toes.

In the mammals there have arisen changes somewhat comparable to those seen in the front foot extensors. As there, the dorsal interosseous muscles have disappeared, the common extensor has acquired tendons extending into the toes, and there is a long extensor for the hallux, and one running to the fifth toe.

VENTRAL MUSCLES OF HIP AND THIGH (Fig. 160). The muscles of this region mainly adduct the femur and flex the knee joint; in locomotion, that is, they raise the body off the ground and push it forward. They are hence important, large, and complex. They are disposed in three main groups:

1. The reptilian muscles from pelvis to femur are deeply buried. They include three members. A large muscle mainly corresponding to the *obturator externus* of mammals (puboischiofemoralis externus) arises fleshily from much of the outer surface of the puboischiadic plate (the fenestration of the plate is related to this muscle attachment) and attaches to the under side of the femur near the head; it acts powerfully in a down-pull of the femur. Homologous to the *obturator internus* of mammals is a small muscle (ischiotrochantericus) which emerges from the inner side of the ischium and runs to the femur head. The *adductor femoris,* superficial to the obturator externus in its origin from the puboischiadic plate, runs far down the under side of the femur.

2. Covering the ventral surface of the thigh is a large and complicated group of long muscles which flex the tibia. A superficial sheet is the *gracilis* (or *puboischiotibialis*); beneath and behind it a number of other rather variable bellies are termed collectively *flexor tibialis internus* and *flexor tibialis externus*. A separate deep slip is the *pubotibialis*.

3. Powerful ventral limb muscles, the two *caudifemorales* (long and short) arise in typical reptiles from the under side of the tail vertebrae. Running forward, these two muscles narrow to tendons which insert midway along the femur; they act powerfully in a backward pull of the femur and hence contribute greatly to forward locomotion in a reptile.

In mammals the obturator externus has much the position of its reptilian homologue (a small muscle, the *quadratus femoris,* has separated

---

much of the musculature attaches both proximally and distally. It differs, however, in the absence of the flexors running down either side and the development of a powerful two-headed "calf" muscle, the gastrocnemius. In mammals the calf musculature has (except for the plantaris) changed to a new attachment on the "heel bone." The long flexor of the digits, however, runs on to the reduced plantar aponeurosis. From this extend, as in reptiles, distal tendons and toe muscles (in *F* the superficial muscles of this set have been removed to show the deeper tendons and muscles). In both reptile and mammal there are, on the flexor surface, deep, short toe muscles not shown in the figures.

from it). Several mammalian adductors correspond in general to the single reptilian adductor; an *adductor magnus* may be the reptilian pubotibialis with a shortened point of insertion. The reptilian tibial flexors may have given rise to the mammalian *semimembranosus, semitendinosus,* and *biceps* (the homology of the last is doubtful), in addition to the gracilis. In mammals the tail has been reduced, and the caudal muscles of the limb with it; they are represented only by small and variable muscles such as the *pyriformis.*

VENTRAL MUSCLES OF LOWER LEG AND FOOT (Fig. 161). As in the pectoral limb, the ventral, flexor muscles are of more locomotor importance than the corresponding extensors, and in consequence are complicated in structure and large in bulk, the expansions of the major muscles producing a swollen "calf" region in reptiles and mammals alike. The construction of the calf musculature is, however, quite different in reptiles from that of the corresponding arm region. Notable is the entire absence of lateral members of the long flexor series. Instead, we find all the superficial musculature gathered into a large *gastrocnemius,* beneath which is a complex *flexor digitorum longus.* Comparable to the palmar aponeurosis, we find a stout *plantar aponeurosis* rounding the curve of the tarsal region. On to this the gastrocnemius and long flexor attach proximally; distally, a complex series of tendons and short muscles run outward to the toes. Proximal deep muscles include a *popliteus* and *pronator profundus;* distally there are complex, deep short muscles running from tarsus on to the toes.

A new type of foot-raising device has been evolved in mammals by the leverage action of calf muscles inserting in the heel tuber of the calcaneum. As a result, the two major gastrocnemius heads no longer extend to the sole of the foot, but insert by the "tendon of Achilles" on to this tubercle. From the outer (fibular) head of the gastrocnemius there have split off in mammals two smaller muscles, the *soleus* and *plantaris.* The former ties in also to the heel bone, but the latter persists in following the old course of the long muscles past the heel bone to end in the sole of the foot. The long digital flexor alone extends on to connect, by a constricted "neck," with the remains of the plantar aponeurosis, from which, as in reptiles, various toe muscles and tendons arise. The hallux has developed a separate long flexor.

Limb Muscles in Other Tetrapod Groups. We shall not attempt to note in detail all the variations from the lizard pattern found in other nonmammalian tetrapods. Among reptiles, Sphenodon is similar in almost every regard to lizards. In chelonians the presence of the shell has tended to make locomotion a specialized affair. Limb muscles as well as limb skeleton have been modified accordingly, but the basic pattern is comparable to that of lizards; the most notable differences are those caused by the peculiar orientation of the shoulder girdle noted in discussing the skeleton

(p. 182). In crocodiles most of the muscles are comparable to those of lizards, but certain modifications are suggestive of the avian relatives of that group. The modified pelvic structure is associated with marked changes in the arrangement of the ventral muscles of the thigh.

In birds the pattern of limb muscles is similar to that of reptiles, but there are numerous differences. In the wing the pectoral muscles which give the main down and back pull to the wing in flight are enormous; they constitute most of the "white meat" of the fowl. The dorsal wing muscles are relatively slender. Most of the short musculature of the reptilian "hand" has disappeared; on the other hand, there are developed special dorsal slips to the thick dermis of the wing membrane. Even more than in the crocodilians the pelvic muscles are specialized in relation to the forward rotation of the hind limbs and the associated modifications of pelvic girdle structure.

The urodeles show a simply-built type of limb musculature, but there is considerable evidence that this is more a degenerate than a truly primitive condition. In the urodele shoulder girdle the loss of the primitive dermal structures has resulted in specialization of the shoulder muscles. The clavicular deltoid arises from a long anterior process of the coracoid as a *procoracohumeralis* muscle. The short dorsal toe extensors are not developed. There is no biceps in the arm in amphibians. Probably this amniote muscle has arisen by a fusion of fibers of the supracoracoideus with those of the adjacent brachialis; in some urodeles and in frogs part of the former muscle separates as a *coracoradialis,* which sends out a long tendon to the radius and thus serves somewhat the functions of a biceps in flexing the elbow.

There are numerous differences from reptiles in the urodele hind leg. In the quadriceps femoris equivalent there is no development of the strong femoral heads present in all amniotes. On the ventral surface of the thigh, the long tibial flexors are much simplified, and the caudifemoralis longus does not reach the femur, but simply ties into the side of the gracilis. There is no development of a gastrocnemius of reptile or mammal type.

In the frog the musculature is highly developed, but highly modified in relation to the jumping habits of the animal and the peculiar structure of the girdle and limb skeleton. In Table II (p. 278) we have given the usual names accorded the muscles; the nomenclature is patterned on that of mammals, but, as may be seen, many of the muscles are not homologous to those whose names are applied to them.

In mammals, as might be expected from the wide variety of limb skeletal structures, there are many variations from the generalized muscle pattern; no attempt will be made to discuss them. The loss of the clavicle tends to bring about confusion in the deltoid and pectoral regions; the great modifications in the feet have, of course, resulted in wide differences in the build of the distal musculature of the limbs.

TABLE II

Muscle Homologies in Various Tetrapod Groups

No attempt has been made to include all the variants present in different members of the groups tabulated, and many details are omitted. For each group only one common name is given for each muscle; often there are synonyms. In many instances homologies are doubtful.

Pectoral Limb, Dorsal Musculature

Mammal	Reptile	Urodele	Frog	Bird
Latissimus dorsi Teres major	Latissimus dorsi	Latissimus dorsi	Latissimus dorsi	Latissimus dorsi
Subscapularis	Subcoracoscapularis Scapulohumeralis posterior	Subcoracoscapularis	———	Subcoracoscapularis Scapulohumeralis posterior
Deltoideus	Dorsalis scapulae Deltoides clavicularis	Deltoides scapularis Procoracohumeralis longus	Dorsalis scapulae Deltoideus	Deltoideus
Teres minor	Scapulohumeralis anterior	Procoracohumeralis brevis	Scapulohumeralis brevis	Scapulohumeralis anterior
Triceps	Triceps	Triceps	Anconeus	Triceps
Supinator	Supinator	Supinator longus	Extensor antibrachii radialis	Extensor antibrachii radialis
Brachioradialis Extensores carpi radiales	Extensores carpi radiales	Extensores carpi radiales	Extensor carpi radialis	Extensor carpi radialis
Extensor digitorum communis Extensor digiti quinti	Extensor digitorum communis	Extensor digitorum communis	Extensor digitorum communis	Extensor digitorum communis
Extensor carpi ulnaris Anconeus	Extensor carpi ulnaris Anconeus	Extensor carpi ulnaris	Extensor carpi ulnaris Epicondylocubitalis	Extensor carpi ulnaris Extensor antibrachii ulnaris
Abductores pollicis Extensores digitorum 1–3	Abductor pollicis longus Extensores digitorum breves	Supinator manus Extensor digitorum brevis	Abductor indicis longus Extensores digitorum breves	Abductor pollicis Extensores digitorum breves
———	Dorsal interossei	Dorsal interossei	Dorsal interossei	———

Pectoral Limb, Ventral Musculature

Mammal	Reptile	Urodele	Frog	Bird
Pectoralis	Pectoralis	Pectoralis	Pectoralis	Pectoralis
Supraspinatus Infraspinatus	Supracoracoideus	Supracoracoideus Coracoradialis	Coracohumeralis Coracoradialis	Supracoracoideus
Biceps brachii	Biceps brachii	(Not formed)	(Not formed)	Biceps brachii
Coracobrachiales	Coracobrachiales	Coracobrachiales	Coracobrachiales	Coracobrachiales
Brachialis	Brachialis inferior	Brachialis	———	Brachialis
Pronator teres Flexor carpi radialis	Pronator teres Flexor carpi radialis	Flexor carpi radialis	Flexor antibrachii medialis Flexores carpi radiales	Epicondyloradialis
Flexor digitorum sublimis Palmaris longus	Flexor palmaris superficialis	Flexor palmaris superficialis	Palmaris longus	Flexor digitorum sublimis
Epitrochleoanconeus	Epitrochleoanconeus	Flexor antibrachii ulnaris	Flexor antibrachii laterales	
Flexor carpi ulnaris	Flexor carpi ulnaris	Flexor carpi ulnaris	Flexor carpi ulnaris Epitrochleocubitalis	Flexor carpi ulnaris
———	———	Ulnocarpalis	Ulnocarpalis	———
Flexor digitorum profundus	Flexor digitorum profundus	Flexor palmaris profundus	Flexor palmaris profundus	Flexor accessorius
Pronator quadratus	Pronator profundus	Pronator profundus	Pronator profundus	Pronator profundus

Superficial short digital flexors: palmaris brevis, contrahentes, lumbricales, etc.

Deep short digital flexors, digital interossei, etc.

TABLE II (*Continued*)

Hind Leg, Dorsal Musculature				
Mammal	Reptile	Urodele	Frog	Bird
Sartorius	Ambiens	Iliotibialis	Tensor fascia latae	Ambiens
Rectus femoris	Iliotibialis	Ilioextensorius	Crureus glutaeus	{Iliotibialis Sartorius}
Vasti	Femorotibialis	———	———	Femorotibialis
Gluteus maximus	Iliofibularis	Iliofibularis	Iliofibularis	Iliofibularis
{Iliacus Psoas Pectineus}	Puboischiofemoralis internus	Puboischiofemoralis internus	{Iliacus Pectineus Adductor longus}	Iliofemoralis internus
Gluteus {medius minimus}	Iliofemoralis	Iliofemoralis	Iliofemoralis	{Iliofemoralis externus Iliotrochantericus}
Tibialis anterior	Tibialis anterior	Tibialis anterior	Extensor cruris brevis	Tibialis anterior
{Extensor digitorum longus Extensor hallucis longus Peroneus tertius}	Extensor digitorum communis	Extensor digitorum communis	Tibialis anticus longus	Extensor digitorum communis
Peroneus longus Peroneus brevis	Peroneus longus Peroneus brevis	{Peroneus longus Peroneus brevis}	Peroneus	{Peroneus longus Peroneus brevis}
Extensores digitorum breves	Extensores digitorum breves	Extensores digitorum breves	Tibialis anticus brevis	Extensores digitorum breves
———	Dorsal interossei	Dorsal interossei	———	———

Hind Leg, Ventral Musculature

Mammal	Reptile	Urodele	Frog	Bird
Obturator externus Quadratus femoris	Puboischiofemoralis externus	Puboischiofemoralis externus	Adductor magnus (pt.)	Obturator externus
Obturator internus Gemelli	Ischiotrochantericus	Ischiofemoralis	Gemelli Obturator internus	Ischiofemoralis
Adductores femoris brevis et longus	Adductor femoris	Adductor femoris	Obturator externus Quadratus femoris	Puboischiofemoralis
Adductor magnus	Pubotibialis	Pubotibialis	Adductor magnus (pt.)	———
Crurococcygeus	Caudifemoralis longus	Caudifemoralis longus	Caudalipuboischio-tibialis	Coccygeofemorales
Pyriformis	Caudifemoralis brevis	Caudifemoralis brevis	Pyriformis	———
Gracilis	Puboischiotibialis	Puboischiotibialis	Semitendinosus Sartorius	———
Semimembranosus Semitendinosus	Flexor tibialis internus	Ischioflexorius	Semimembranosus Gracilis	Ischioflexorius
Biceps	Flexor tibialis externus	———	———	Caudilioflexorius
Gastrocnemius medialis	Gastrocnemius internus	Flexor digitorum sublimis Flexor digitorum longus	Plantaris longus	Gastrocnemius internus Flexor hallucis longus Flexor profundus
Flexor hallucis longus Flexor digitorum longus	Flexor digitorum longus			
Tibialis posterior Popliteus	Pronator profundus Popliteus	Pronator profundus Popliteus Fibulotarsalis (?)	Tibialis posticus	Tibialis posticus Popliteus Gastrocnemius externus
Gastrocnemius lateralis Soleus Plantaris	Gastrocnemius externus			
Interosseus	Interosseus	Interosseus	———	———
	Superficial short digital flexors: flexor digitorum brevis, contrahentes, lumbricales, etc.			
	Deep short digital flexors, digital interossei, etc.			

BRANCHIAL MUSCULATURE

Markedly different from the striated musculature so far considered is the branchial musculature, highly developed in the gill region of the ancestral vertebrates, and persistently prominent, although in much modified form, in even the highest groups. We have noted earlier the distinctive nature of the skeleton of the pharyngeal region of primitive vertebrates and shall note later the equally distinctive nature of the nerves innervating the pharynx. Its musculature is also noteworthy. In contrast to all other striated musculature, it arises, not from the myotomes, but from mesenchyme derived from peritoneum of the lateral plate (Figs. 49, *C*, p. 99, 153, p. 255). The smooth musculature of the gut proper, posterior to the pharynx, arises in similar fashion, and branchial and gut muscles, striated or smooth, are but anterior and posterior parts of a single great visceral system of muscles whose primary locus is in the walls of the digestive tract.

The contrast between the striated condition of the anterior, branchial, part of the digestive tract and the presence of smooth muscle posteriorly may be associated with the contrast in functions between anterior and posterior regions. For food gathering and gill breathing, the primitive functions of mouth and pharyngeal regions, the vigorous movement characteristic of striated muscles is appropriate; for the slower movement fitting for digestive processes, smooth musculature suffices.

In higher vertebrates much of the striated visceral musculature has abandoned the gut lining and assumed a variety of natures as facial and jaw muscles and even a part of the shoulder musculature. Primitively, however, the pharyngeal walls and pharyngeal gill slits appear to have been its proper, truly visceral locale. In many instances the branchial muscles are associated with skeletal structures, primarily the gill bars, whereas the more posterior smooth visceral muscles lie exclusively in soft gut tissues. But in cyclostomes the skeletal connections are of little importance. And of great theoretical interest is the fact that the striated muscles of the gut are not confined to the pharyngeal region of the digestive tube, where these skeletal structures lie. In fishes generally the striated muscles may extend backward into the esophagus, and the same condition is repeated in mammals (although here their presence is perhaps secondary). There is thus no sharp, definite line of demarcation between branchial and smooth muscle regions, a fact which emphasizes their relationship.

From the gnathostome fishes up the most anterior set of branchial muscles have been much modified in function to serve the jaws, and in land vertebrates the surviving branchial elements serve a further series of aberrent functions, as facial, neck, and shoulder muscles, and even as elements accessory to the auditory apparatus.

Branchial musculature is well developed in the cyclostomes, both as sheets of muscles constricting the gill pouches and as specialized muscles operating the peculiar "tongue." The construction of the lamprey musculature is, however, quite unlike that of other vertebrate groups and will not be considered further here.

In the sharks (Fig. 163, *A*) we find the branchial muscles well developed, with a pattern that in many regards may be considered basal to that in all other gnathostomes. The jaw and hyoid arches have a highly developed series of visceral muscles. In the jawless ancestors of the gnathostomes the musculature of these two arches may have been originally generalized; it is, however, already specialized to a considerable degree in sharks. We shall therefore use an indirect method of attack, and consider first the simple and presumably more primitive arrangement seen in the typical gills farther back in the shark pharynx. We shall follow the fate of the muscles of these arches upward through the higher vertebrates, then return to consider the muscles farther forward in the region of the fish hyoid and mandibular arches.

The branchial musculature is, of course, primarily a pumping device to force water through the gills (pp. 314, 318). We may note, incidentally, that although the branchial muscles are those most intimately connected with the visceral skeletal elements, somatic trunk muscles may also become associated with the gill bars. We have already mentioned (p. 262) that the hypobranchial muscles of the floor of the throat may attach to them below; and in sharks the epibranchial muscles of the neck may also gain contact with the upper margins of the gill arches.

Behind the hyoid arch there are typically, in fishes, five gill slits with four intervening arches, each arch with its own proper musculature as well as its own skeletal bar. Even when, in land vertebrates, the gills themselves have disappeared as landmarks, muscles derived from various parts of the branchial system can be readily traced because of their innervation. The gill muscles are supplied by a special series of cranial nerves, numbers, V, VII, IX, and X (cf. pp. 552–555, and Fig. 336). The jaw arch is supplied by nerve V; the hyoid by nerve VII; the first typical gill bar is the territory of nerve IX; and the further gill bars are innervated by special branches of nerve X (which continues onward far down the gut).

Muscles of the Typical Gill Bars and Their Derivatives (Figs. 162, 163, *A*). Although there is frequent fusion of muscle tissues above and below the gill openings, each typical gill arch of a shark has a characteristic series of muscle slips proper to it, which are distinguishable by innervation from those belonging to neighbors in the series. The most prominent element in a shark gill is the *superficial constrictor;* this is a broad, if usually thin, sheet of muscle whose fibers generally run dorsoventrally in the flap of skin extending outward in the gill septum. Dorsally and ventrally most of the constrictor fibers terminate in sheets of fascia on the back and throat. The deeper fibers, however, may attach to the outer surface of the gill bar elements and may form separate *interbranchial* muscles.

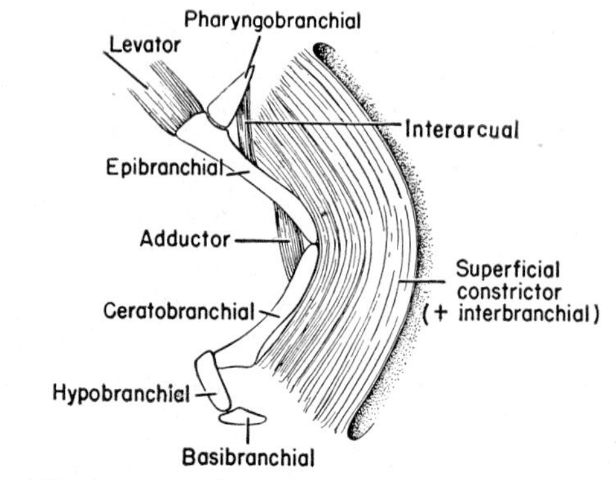

Fig. 162. A single gill arch of a shark and its musculature.

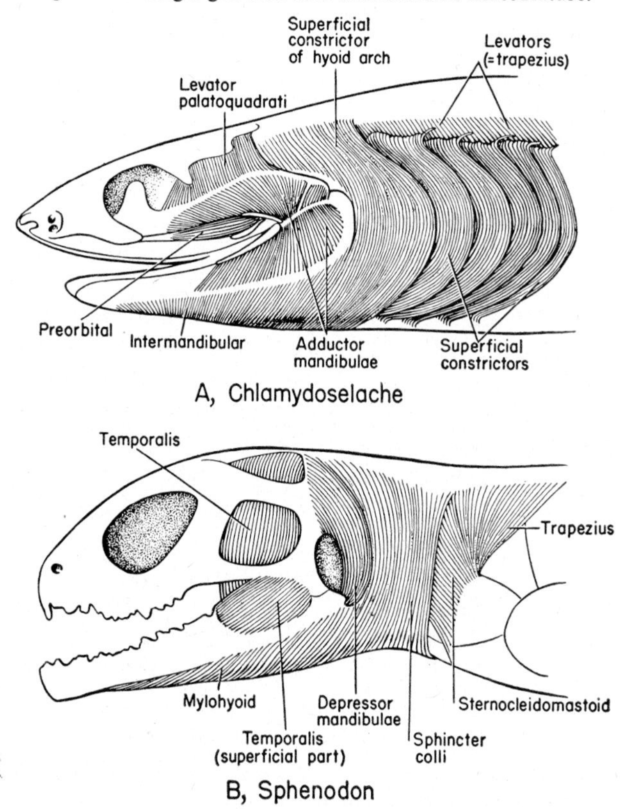

Fig. 163. Lateral views of the branchial arch musculature and its derivatives in a shark and the reptile, Sphenodon. (*A* after Allis; *B* after Adams and Fürbinger.)

In addition, there develop deeper muscles. The *adductors* of the visceral arches run from epibranchial to ceratobranchial and tend to bend the two together; dorsal *interarcual* muscles connect the epibranchials with the pharyngobranchials of the same or neighboring arches.

Dorsal to the constrictor, muscle fibers from the dorsal fascia run diagonally downward and backward and may insert on the dorsal region of successive gill bars as arch *levators*. In many sharks, however, most or all of these fibers slant back to attach to the shoulder girdle; this is characteristic of the tetrapod muscle derived from this muscle series, the trapezius, and this name is frequently applied to the shark muscle.

In bony fishes the development of the gill musculature is more restricted. Since the gill septa are reduced (in the presence of an operculum), the superficial constrictors are absent, but, ventrally, slips of this system persist as varied *subarcual muscles;* in teleosts the levators are lost, and the adductors and interbranchial muscles are reduced or absent.

In tetrapods, gill-breathing larval amphibians retain a series of gill muscles which resemble those of bony fishes; but in true land forms the muscles moving the typical gill bars have disappeared, except for small slips associated with the hyoid apparatus and larynx. There is, however, one conspicuous if aberrent group of muscles of this series which survives in land vertebrates—the *trapezius* musculature (Fig. 163, *B*), derived, as mentioned, from the arch levators of fishes. The trapezius is always present and frequently greatly expanded in land vertebrates, and its origin may extend far down the back. The thin trapezius sheet converges downward from the occiput and back fascia to insert along the anterior margin of the shoulder girdle. Primitively, the attachment was to the dermal elements—cleithrum and clavicle; but with their reduction or loss the insertion lies along the primitive front margin of the scapula (the scapular spine in mammals) and may reach the sternum ventrally. More anterior and ventral slips of the trapezius group may separate as individual muscles —*sternomastoid, cleidomastoid,* and so on—and with reduction of the clavicle may (in many mammals) fuse with slips from the deltoid to form long, slender compound muscles extending directly from head to front limb.

Muscles of the Hyoid Arch. Presumably in the ancestral jawless fishes the muscles of the hyoid arch were comparable to those of the typical gill arches. But in all living jawed vertebrates the hyoid arch (and its gill slit) has become highly modified, and the musculature is correspondingly modified. As we have noted, the muscles of this arch are always identifiable through their innervation by nerve VII, the facial nerve.

Related, presumably, to the loss of independence of the hyoid arch, all the deeper muscles are absent here in sharks, and only the superficial constrictor remains (Fig. 163, *A*). This muscle, however, may be variously subdivided in many fishes. Some deep slips may separate to connect hyoid arch elements with one another and with the jaw joint. Certain of these slips may persist in tetrapods in the hyoid and ear region; some are insignificant, but two of them, as noted later, achieve importance in land animals in connection with mouth opening. A ventral part of the constrictor may form, in fishes, much of the muscle (intermandibular) connecting the jaws ventrally, but in land vertebrates the hyoid contribution to this muscle sheet vanishes.

The most important part of the hyoid musculature is the dorsal part of the constrictor sheet developed in the gill septum of the hyoid arch. In bony fishes this septum is greatly expanded to form the bony operculum covering the gill chamber; this region of the hyoid constrictor is highly developed to control the movements of this gill cover.

In land vertebrates the operculum disappears; the constrictor of the hyoid, in the absence of the more posterior constrictors (they had disappeared in bony fishes), is the only superficial muscle remaining in the developing neck region. It expands over this territory as a thin sheet, the *sphincter colli,* circling the neck ventrally and laterally (Fig. 163, *B*). This muscle tends to adhere to the skin. In mammals it expands in spectacular fashion over the surface of the head to form the muscles of expression, discussed later (p. 289).

Mechanisms for mouth opening have apparently not been taken seriously (so to speak) by vertebrates (mouths tend, rather, to open by themselves), and various makeshift devices for this purpose are seen in different groups. Ventrally, the primitive series of hypobranchial axial muscles run forward from shoulder to jaw, and a backward pull of these muscles is sometimes utilized for jaw opening in fishes. In most tetrapods except mammals there is substituted a *depressor mandibulae* (Fig. 163, *B*), an anterior slip of the hyoid constrictor which arises laterally from the posterior part of the skull, and passes down behind the one-time spiracle (now the eardrum region) to attach to the back end of the lower jaw (there is frequently a retroarticular process of the jaw which gives the muscle good leverage).

In mammals we have seen that the jaw is refashioned; the elements about the former region of attachment of the depressor are either lost or taken into the ear. The depressor muscle, its function lost, vanishes.

Another slip of hyoid musculature, however, emerges to take the place of the depressor and contributes to the formation of the *digastric* (Fig. 164, *B*). This mammalian jaw-opener is, as its name implies, a compound muscle with two bellies. An anterior belly is formed by fibers of jaw-arch origin running back beneath the jaw ramus. Posteriorly, this connects with a slip of hyoid muscle which runs upward behind the jaw to attach to the skull in the ear region. The two elements may be at a sharp angle to one another, but between them they successfully perform the none too arduous task of depressing the jaw.

Jaw Muscles (Figs. 163, 164). With the modification in gnathostomes of the elements of an anterior gill arch to basic jaw structures, the muscles of this arch have become highly modified to serve special jaw functions. Since no typical gill opening adjoins the jaw, there is, of course, no development here of an expanded gill flap; in consequence there is no development of a typical constrictor sheet. The jaw musculature, as seen in sharks, consists of three parts:

1. The upper jaw is in sharks but loosely attached to the skull. A dorsal segment of the jaw musculature runs out and down from the braincase in front of the spiracle to attach to the upper jaw cartilage as an elevator of the upper jaw, a *levator palatoquadrati* rather comparable serially to the trapezius components which attach to the ordinary gill arches.

2. The major muscle mass of the jaw segment is the *adductor mandibulae,* which performs the most important function of that of any single

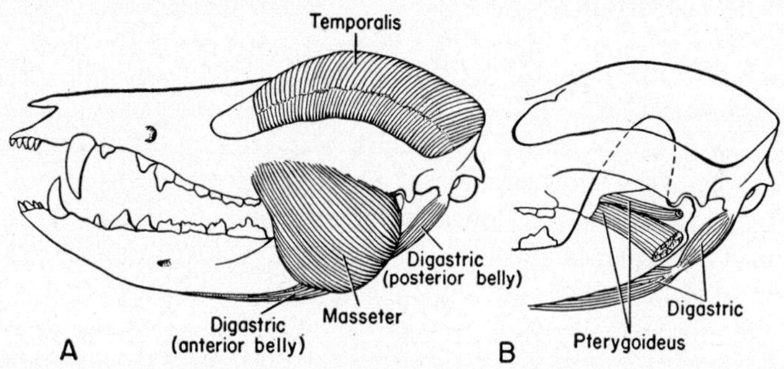

Fig. 164. Jaw musculature of the opossum. *A*, Superficial view; *B*, deeper dissection. The jaw is represented as transparent to show the pterygoid muscles, which attach to its inner surface.

branchial muscle in gnathostomes—the pressing together of the jaws in the biting or grinding motions essential for the ingestion of food. In sharks the main mass of the adductor muscle is arranged in a simple fashion, running from the palatoquadrate cartilage, which here forms the upper jaw, down to the mandibular cartilage. (A specialized portion, a *preorbital* muscle, runs forward from the main adductor to the front of the orbit, and, pulling the upper jaw forward, aids in anchoring it to the braincase.) The adductor of the mandible is comparable in position to the adductors of the normal branchial arches, but probably includes in its composition the material that farther back forms the main bulk of the constrictor, absent here.

3. Unimportant is a ventral element: fibers representing the most ventral part of the constrictor, which in fishes forms (with a hyoid derivative) an *intermandibular* muscle, a thin sheet of fibers between the two lower jaw rami.

In higher vertebrates the more ventral and dorsal parts of the jaw arch musculature are unimportant. In tetrapods the ventral fibers continue to form a thin sheet between the lower jaws as the *mylohyoid* muscle; a slip from this forms (as noted earlier) one belly of the digastric muscle which lowers the jaw in mammals. The upper jaw and palate continue to be movably articulated with the braincase in most bony fishes and a number of tetrapod groups. In such types the levator of the palatoquadrate persists, but in varied form. In bony fishes it may expand to connect the hyomandibular and even the operculum with the braincase. In land vertebrates it is represented by small muscles connecting the braincase with the quadrate (as in birds) or pterygoid (as in lizards and snakes). In vertebrate groups where the upper jaw is fused to the braincase—such as chimaeras, lungfishes, living amphibians, turtles, crocodiles, mammals—this type of musculature is useless. It has usually disappeared; in many reptiles and amphibians, however, a relic portion is found in the inner part of the orbit where its swelling or contraction tends to push the eyeball outward.

The adductor mandibulae is prominent throughout the gnathostomes. In bony fishes it is bulky. The "cheek" region in which it lies is covered over by dermal bone, and the attachments of the jaw muscles are in part on the under side of this dermal roof and may extend inward to the braincase as well. In tetrapods the adductor musculature is divided into two main groups: (1) temporalis and (2) pterygoideus. The *temporalis* muscle primitively arose from the under surface of the cheek region of the skull roof and from adjacent parts of the palatal complex and braincase, to insert in and about the fossa in the lower jaw. As discussed in connection with the reptilian skull, fenestration of the originally solid skull roof allows greater freedom of action for the temporal musculature. Associated with modification of the jaw structure and articulation in mammals, we find that the primitive temporal musculature has been divided into two parts. One, the temporalis in a narrow sense, has much of the original position of the muscle and inserts into the coronoid process of the mandible. A second muscle, the *masseter,* is more superficial in position. Its fibers run at a considerable angle to those of the temporalis, and pull the jaw forward and upward. In rodents, particularly, it may be highly developed and is often stronger than the temporalis. The *pterygoideus* muscles form a deep division of the adductor mass. They typically originate from the pterygoid region of the palate (but may extend on to the braincase) and insert on the inner or back surface of the jaw.

SKIN MUSCULATURE

In fishes the surface of the skeletal muscles is tied in closely to the dermis. In tetrapods, particularly the amniotes, this is no longer the case, and the skin lies relatively loosely over the underlying musculature. However, we find in many cases thin sheets or ribbons of muscle partly or completely embedded in the skin, and functioning in skin movements; these muscles are derived during development from the body muscles beneath them. In some cases the distinction between the major skeletal muscle system and its superficial dermal derivatives is clear; in other cases, however, this is not so, and the student dissector is all too often troubled to decide whether a given muscle sheet is a proper part of the trunk or limb segment which he is dissecting or a skin muscle which he may, for his immediate purpose, disregard.

In amphibians and reptiles the dermal musculature is in general little developed, except for the occasional presence of a dermal slip of the pectoral musculature—a condition repeated in many mammals. A notable exception lies in the Ophidia. The large scales of snakes each have attached to them a stout dermal muscle which can erect the scale so that it acts as a holdfast on the ground and prevents a backward slip of the body in undulatory motion; the scales thus have much of the functions of the absent feet. In birds, dermal elements, developed from a number of

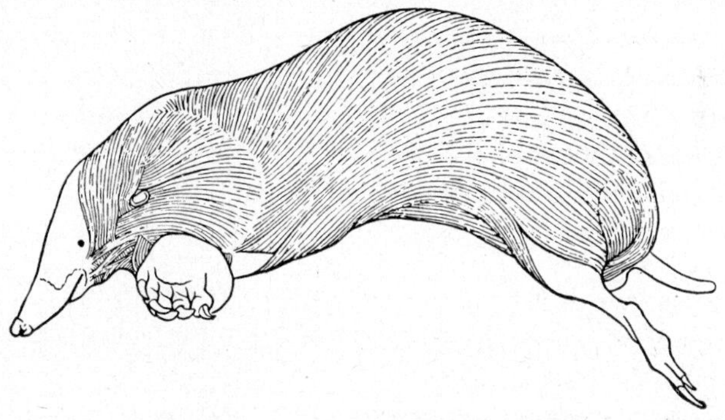

Fig. 165. Dermal muscles sheathing the body of a mole. (After Nishi.)

the shoulder and wing muscles, insert distally into the thick skin of the wing and aid in flight.

We find the greatest development of dermal musculature in mammals. In most mammals (higher primates are exceptional) almost the entire trunk and neck are wrapped about by a continuous sheath of skin muscle termed the *panniculus carnosus* (Fig. 165); the twitch of a horse's skin where a fly has settled is evidence of the presence and functioning of this dermal muscle sheet. The panniculus carnosus is, so to speak, indifferent as to the underlying source of its muscle fibers, for we find that while the trunk sheath is derived from the underlying axial muscles, the ring of superficial fibers about the neck, the sphincter colli, is, as we have noted, a part of the visceral musculature of that region and innervated by the facial nerve (VII). In mammals this dermal musculature undergoes a striking development (to which, indeed, this nerve owes its name). Slips grow forward from the neck over the skull and on to the cheeks to form the *facial muscles,* or muscles of expression (Fig. 166). Although widely distributed over the scalp and face, these muscle bands tend to be especially concentrated about the orbits, the outer ear, and lips (which last, hence, acquire a mobility and usefulness unknown in lower vertebrates). Lacking facial muscles, nonmammalian vertebrates can neither smile nor frown; lacking a mouth sphincter, they can neither kiss nor suck; lacking

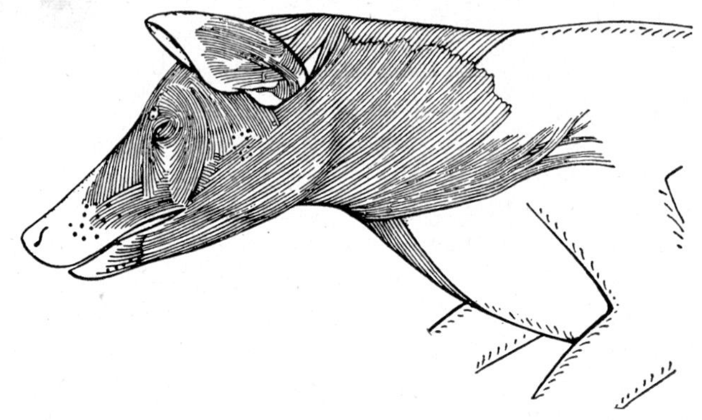

Fig. 166. Facial musculature; the head and neck of a dog. (After Huber.)

ear muscles, movement of the ear pinna (as seen in most mammals, including especially gifted human beings) would be impossible.

ELECTRIC ORGANS

In three types of fishes—rays of the genus Torpedo, the electric "eel" (Gymnotus), and the electric catfish (Malapterurus)—there is a develop-

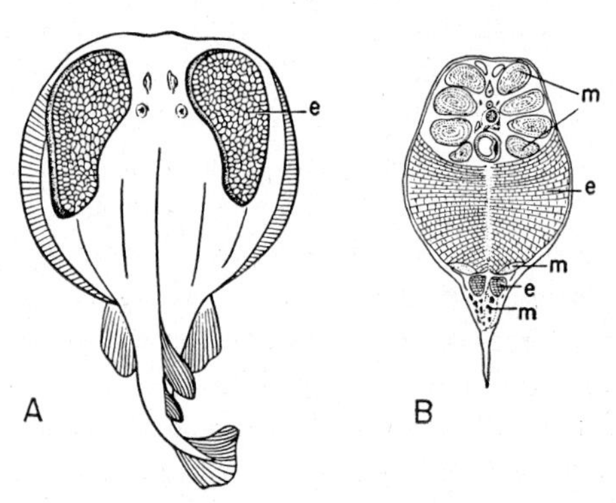

Fig. 167. Electric organs. *A,* The torpedo, a ray in which musculature of the expanded pectoral fins has been transformed into *electric cells* (*e*). The skin is dissected away to show the electric organs. *B,* Section of the tail of the electric "eel" of South America (Gymnotus). Typical axial musculature (*m*) is present above and below, but much of the tail muscle has been transformed into electricity-producing tissue (*e*). (*A* after Garten; *B* after du Bois-Reymond.)

ment of special organs capable of producing a heavy electric shock to any animal in contact with them, and in several other types of rays and teleosts similar if weaker phenomena are produced. The present chapter is the most appropriate place for their consideration, for these electric organs (Fig. 167) appear to be, in most if not all cases, modified muscular tissues. Muscle fibers are structures chemically adapted, as we have seen, for the rapid release of energy; in the present instance the energy is utilized, however, for the production of electricity rather than for muscular contraction.

The elements of which these electric organs are composed are generally flattened plates of multinucleated protoplasm, each innervated by a nerve fiber. In certain instances where their development is known they are seen to arise from embryonic striated muscle cells, but become much altered in shape and internal structure. In most cases these electric plates are arranged, one above another, in series of piles. This arrangement is immediately comparable to the old-fashioned voltaic pile, famous in the history of electrical discovery; these piles of electric cells form essentially an organic storage battery, made effective by plus and minus differences between the two surfaces of each plate. In the torpedo more than 200 volts and 2000 watts of current have been recorded. Obviously such a shock, or even one much milder, might be of considerable value to a fish either in warding off enemies or paralyzing its prey.

Despite the basic similarity of their construction, the electric organs vary greatly in position and appearance in the different types which have them; it seems probable that such organs developed independently a number of times. There is even evidence suggesting that similar structures were developed in certain of the more ancient of vertebrates, the Silurian and Devonian ostracoderms. In the torpedo they are present as two large groups of piled-up plates, one on either side of the head in the expanded pectoral fin of this flattened elasmobranch. There are hundreds of piles and some hundreds of plates in each pile; in one species of torpedo it has been calculated that over 200,000 plates are present. In the electric "eel" Gymnotus of South America, in which the tail makes up four-fifths of the body length, the massive electric organ is formed from the modified ventral half of the tail musculature. In the catfish Malapterurus from the Nile the electric tissue forms a sheath encircling the whole body just beneath the skin from the back of the head to the base of the tail. In this case the plates, although numerous, are not arranged in neat piles, and since the electric tissue is superficial to the body musculature, its origin from musculature is not so certain (the embryology is unknown).

CHAPTER X

MOUTH, PHARYNX, RESPIRATORY ORGANS

THE DIGESTIVE tract, with its various outgrowths and accessory structures, looms large in both bulk and importance in body organization. The major digestive organs, including stomach and intestine, form the more posterior segments of the tract. In the present chapter we shall consider the mouth and pharyngeal regions, "introductory" sections of the digestive tube. These play little role in alimentation beyond the reception of food, but are highly important in other regards—notably as the place of origin of respiratory organs and of important glandular structures.

THE MOUTH

From the mammalian or human point of view the mouth appears to be a well-defined structural unit, with fixed, uniform features such as the

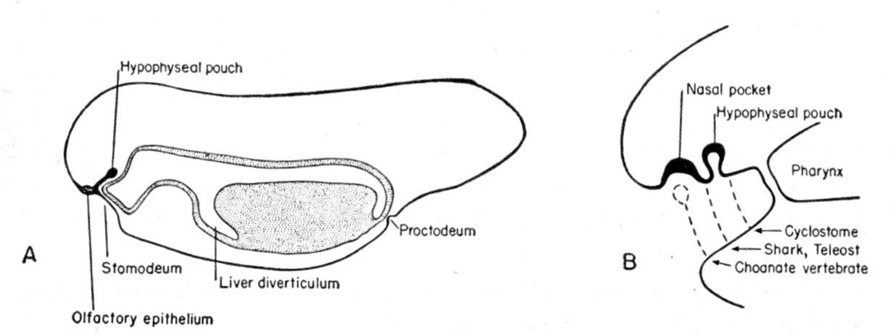

Fig. 168. *A,* Diagram of a larval amphibian (at about the stage of Fig. 53, *E*), in longitudinal section, to show the extent of the endoderm (stippled) and its relation to structures of the mouth region. *B,* Diagram to show the comparative position of the mouth margins in various types of vertebrates (cf. text).

lips, tooth row, tongue, and salivary glands. But a broad survey of the vertebrates shows that mouth structures vary widely; every one of the familiar landmarks may be absent in one group or another. Except that it is an inturned area leading to the pharynx, we can make few statements about the mouth that will hold true for all vertebrates.

Development. In the embryological story the opening from mouth to pharynx is late in appearance, suggesting that a mouth opening into the digestive tube may have been a relatively late phylogenetic development. In types of vertebrates with a mesolecithal egg, the early embryonic gut may open posteriorly at the blastopore, roughly homologous with the anus of the adult, but invariably the archenteron ends blindly at its anterior end. This blind end (Figs. 168, *A*, 207, p. 347) is the region of the future pharynx. Above and in front of the pharyngeal region there early appears an expanded brain and a consequently expanded head region. The head turns downward over the surface of the yolk-swollen body or yolk sac, producing beneath it an inturned fold or pocket of ectoderm, the *stomodeum*. This forms the primitive mouth cavity. At its inner end this fold lies close to the anterior, pharyngeal end of the gut tube. A membrane long persists at this point, separating the two structures. Eventually this membrane breaks down; mouth and pharynx are placed in continuity, and the digestive tube has acquired an anterior opening. The epithelia of the two regions concerned blend with each other and are difficult or impossible to distinguish at later stages. In a broad way, however, it is clear that the pharynx is lined with endoderm, but the mouth epithelium is ectoderm, essentially a continuation of the skin epidermis.

The embryonic gut cavity of vertebrates, the archenteron, was long ago recognized as comparable to the adult digestive cavity of coelenterates and a number of other simple invertebrate animal types. In these forms the adult gut has but a single opening, to which the vertebrate embryonic blastopore and adult anus are frequently compared. But in the more highly organized invertebrate phyla—annelids, molluscs, arthropods, echinoderms—a second opening, a mouth, has developed as a progressive feature; vertebrates presumably followed a similar evolutionary course, as suggested by the ontogenetic history.

Mouth Boundaries. The extent of the mouth cavity in the various vertebrate groups is highly variable (Fig. 168, *B*). One would, a priori, expect that the mouth of the adult would correspond roughly in extent with the embryonic stomodeum formed from inturned ectoderm. It is probable that in general the posterior boundary of the mouth in the adult is indeed approximately the boundary between the stomodeum and the embryonic gut. Are, however, the anterior and outer margins of the mouth fixed in position? Does the mouth always include the same area of infolded stomodeal tissue? Not at all.

Two good landmarks are always present in the roof of the stomodeum. Near the outer end of this funnel, beneath the swelling forebrain, is the embryonic nasal region, a pair of ectodermal pits or thickenings in most vertebrates, a single pit in cyclostomes. Farther back in the potential mouth roof is a median pit—the *hypophyseal pouch* (Rathke's pouch)—which later closes off and forms a major part of the pituitary body lodged beneath the brain. In an adult shark or ray-finned fish the nasal sacs lie external to the mouth, while the area from which developed the embryonic hypophyseal pouch lies well inside it; the jaw margins bounding the mouth cavity thus lie between these two landmarks. In the Choanichthyes and all land vertebrates the situation is different. The jaw margins form a bridge under the nasal pockets, so that both external and internal narial openings are present. Here, then, we have a more fully developed, more extensive, mouth cavity than in the shark or actinopterygians.

The opposite extreme is seen in the cyclostomes. A lamprey or hagfish has a well-defined mouth; but the embryonic development shows that this mouth corresponds only with the inner recesses of the mouth cavity of other vertebrates. In the larval cyclostome (Fig. 169) a nasal pit and a hypophyseal pouch are found in the stomodeal depression, much as in other vertebrates. But as the embryo increases in size, forces of differential growth cause a rotation of both nostril and pit (the two are closely connected) forward and upward on to the outer surface of the head (as in the hagfish, Fig. 13, p. 36) and, in the lamprey, to a position high on the dorsal surface, far removed from the adult mouth (Fig. 189, p. 320).

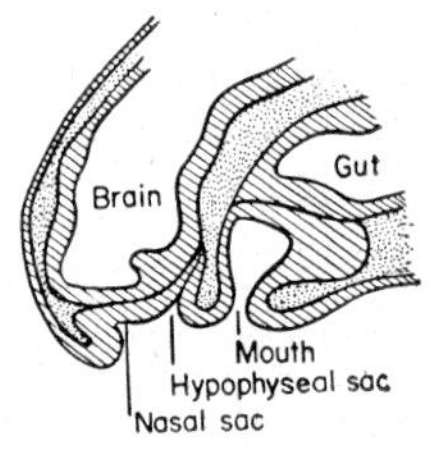

Fig. 169. Section of the head of a larval lamprey. At this stage nasal and hypophyseal sacs are still ventral in position (cf. Fig. 189, p. 320).

Most of the surface of the lamprey head is thus covered by ectoderm which in gnathostomes lies within the mouth.

In a majority of vertebrates the mouth margins are formed by *lips,* soft pliable structures of epidermis and connective tissue. In cyclostomes the mouth opening is rounded (a feature to which the group owes its name), and in the lampreys the circular margins constitute an effective sucker by which the animal attaches itself to its prey (Fig. 189). In jawed fishes, amphibians, and reptiles, the lips are in general small and unimportant skin folds, external to teeth and jaws. The lips have disappeared in forms in which a horny bill or beak is present—as in turtles, birds, and a few mammals. In typical mammals, on the other hand, lips are highly developed, separated by deep clefts from the jaw margins and rendered mobile by the presence of the facial musculature peculiar to this class. The mouth opening between upper and lower lips generally terminates posteriorly in mammals (in contrast to many lower vertebrates) well forward of the jaw articulation, thus creating a skin-covered *cheek* region. In rodents and Old World monkeys the potential storage spaces in the cheeks have been expanded into *cheek pouches,* useful in the carriage of food. (The pocket gophers—Geomys, and the like—have enormous pouches developed in this region, which, however, are hair-lined external pouches, opening to the surface, not into the mouth.)

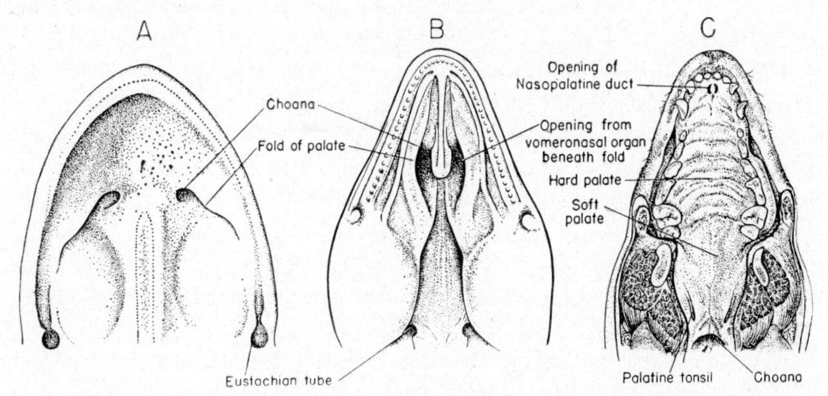

Fig. 170. The roof of the mouth in *A*, a urodele; *B*, a lizard; and *C*, a mammal (dog); to show particularly the position of the choanae.

Palate. In typical fishes the roof of the mouth is a rather flattened vault, unbroken by any opening (once the embryonic hypophyseal pouch has closed). In the Choanichthyes and amphibians, however, we find that

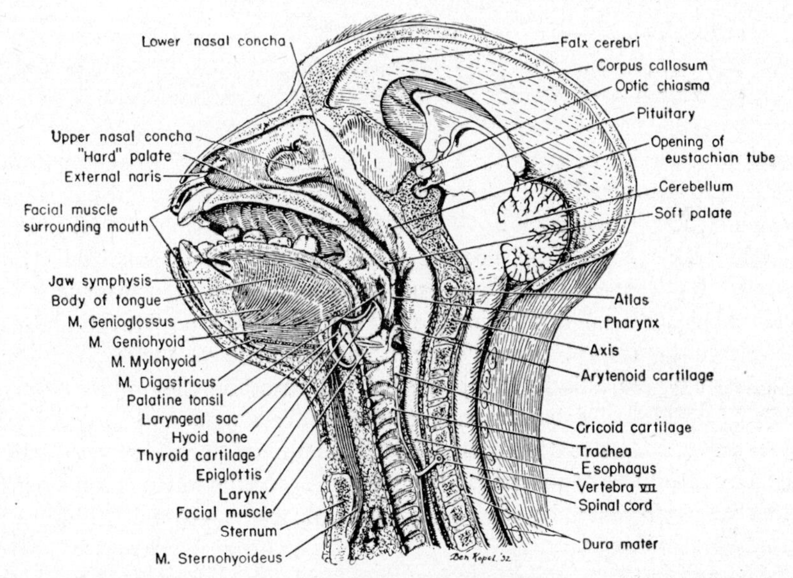

Fig. 171. Median section of the head and neck of a rhesus monkey. (After Geist.)

paired openings, the *choanae* or internal nares, are present near the anterolateral margins. In amphibians (Fig. 170, *A*) the mouth roof is nearly flat; in reptiles (Fig. 170, *B*) and birds, however, the roof is well vaulted, and a pair of longitudinal *palatal folds* aid in forming a passage from the choanae back through the mouth above the tongue to the pharynx for the freer passage of air. In crocodiles and mammals (Figs. 141, *C*, p. 235; 142, p. 236; 170, 171) this air channel has been completely shut off from the mouth cavity by the development of a bony secondary palate, noted in our discussion of the skull; this "hard palate" is extended backward in mammals by a thick membrane, the *soft palate* (Figs. 170, 171). The vomeronasal organs (p. 493) gain openings to the mouth roof independent of the choanae in reptiles; these organs persist in the majority of mammals, opening through the secondary palate into the mouth cavity through anterior palatine foramina.

The epithelium of the palate, and of the mouth cavity in general, is usually of a stratified squamous type. In living amphibians a rich capillary network underlies it, and it functions as an important breathing organ. In bony fishes, amphibians, and reptiles, teeth are usually present on the palate. In mammals there are often present on the palate transverse ridges of cornified epithelium which aid in the manipulation of food. They are highly developed in most ungulates and carnivores. In the toothless whalebone whales (Mysticeti) they have developed into long parallel plates of horny "whalebone" hanging down into the mouth cavity; fringes on their margins pick up small marine organisms which are licked off by the tongue and form the food supply of these giant mammals.

Tongue. In fishes the lower ends of the gill bars slant forward below and between the jaws into the floor of the mouth. In some teleosts these bars bear teeth, which work against those of the palate; in other fishes, however, the mouth floor is relatively smooth. Movements of the gill system may raise or depress it, but there is seldom any structure comparable to a tongue. In cyclostomes this name is given to an extrusible mouth structure bearing horny "teeth" wherewith to rasp the flesh of the animal's prey (Figs. 13, p. 36; 84, p. 154), but this apparatus is surely a parallel development rather than one homologous with a true tongue.

In the water the manipulation of food in the mouth is a relatively simple matter; on land, food materials are more difficult to deal with. With the reduction of the gills in tetrapods, the gill bars and their musculature became available for other uses, and the mobile tongue characteristic of most land vertebrates developed. The tongue musculature, we have noted, is derived from the hypobranchial musculature, and the tongue is anchored at its base by a skeletal system—the hyoid apparatus—formed of a series of modified gill bars.

Apart from its normal use in handling food, the tongue serves various minor functions: in mammals it is the sole bearer of taste buds; in man it is an aid to speech; cornified tongue papillae aid in the manipulation of food, and so forth. In a number of groups the tongue may elongate and take on an active role in the gathering of food, particularly insect food. Thus, while some anurans are tongueless, common frog and toad types are able to flip out an elongate tongue (attached anteriorly, free at the back) to pick up an insect with its sticky tip. The chameleon's tongue can be protruded and retracted with lightning rapidity for the same purpose. In the woodpeckers the tongue is extremely elongate, and in mammals several types of termite eaters and anteaters have similar adaptations. Rapid tongue extension may be at least partly due to muscular action, but in frogs protrusion is mainly accomplished by the sudden filling of a lymph sac at the tongue base, and in chameleons by a distention of the tongue through the filling of blood sinuses. In the woodpeckers extension is brought about through the protrusion by hyoid muscles of an extremely elongate but slender and flexible basihyal element; there is a special bony tube curving around the occiput and forward on the upper skull surface for its reception when withdrawn.

Mouth Glands. In fishes there are numerous mucus-secreting cells in the mouth epithelium. Seldom, however, is there any marked development of formed mouth glands. One of the few exceptions is a special gland in the lamprey, the secretion of which tends to prevent coagulation of the blood of its prey.

In water dwellers the water itself is a major aid in the moistening and swallowing of food. In land vertebrates, in the absence of a water medium, *salivary glands* make their appearance (they tend, however, to be secondarily reduced in groups which have returned to an aquatic existence). Such glands in general secrete mucin and a more watery material to produce the saliva.

In amphibians such glands are relatively poorly developed, but are usually to be found on the tongue, and there is normally a large *intermaxillary gland* in the anterior part of the palate. Reptiles have a good series of salivary glands along the lips, on the palate, and on and beneath the tongue. Modified salivary glands form the poison glands of the snakes and the Gila monster (the one poisonous lizard); the gland opening lies at the base of the poison fang, whence the venom passes into the wound via a duct or groove in the substance of the tooth. Many birds, particularly water-feeders, are almost devoid of mouth glands, but in others an abundant series may be present. Highly developed salivary glands are present in mammals; most prominent are the *parotid* and *submaxillary glands,* and the *sublingual,* lying in the cheek region and opening into the mouth above and below by long ducts. For the most part, salivary glands lack chemically active materials; but in some frogs, on the one hand, and in a number of mammals (man included), on the other, there is present in the saliva an enzyme—*ptyalin*—which initiates starch digestion. True digestion appears to have been confined exclusively to more posterior regions of the digestive tube in the lower vertebrates; in this salivary enzyme we find an advance guard of the enzyme army present at the portals of the alimentary canal.

DENTITION

Although the *teeth* are in reality modified parts of the dermal skeletal materials, they may be appropriately discussed here, since they are "inhabitants" of the oral cavity. Teeth are unknown in lower chordates and in jawless vertebrates, living or fossil. With the advent of jaws, teeth simultaneously appeared. Their presence made the jaws useful as biting structures; through the changed habits made possible by this new type of feeding device the way was paved for the rise of the gnathostomes to their present estate.

Tooth Structure (Fig. 172). In their simplest and presumably most primitive form, teeth are conical structures of a type such as may be seen in many fishes and reptiles and in the anterior part of the mammalian dental battery. Frequently, however, more complex forms appear, notably teeth in which the upper surface is more or less flattened to form a *crown*

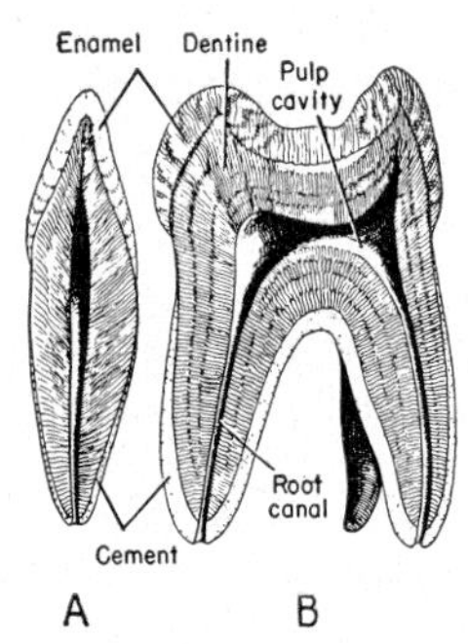

Fig. 172. Sections through a mammalian incisor, *A*, and molar, *B*. (After Weber.)

for crushing or chewing food materials. In the interior of a tooth is a *pulp cavity,* with soft materials including small blood vessels and nerves. At the base there may be present one or more *roots,* projecting processes by which the tooth may be firmly implanted in the jaw.

The major tooth materials are enamel and dentine. The *enamel,* exceptionally hard and shiny in appearance, forms a thin layer over the tooth surface. Enamel is almost entirely devoid of organic matter. Two-thirds of its substance consists of a carbonate apatite arranged in long prisms with axes at right angles to the surface; the remainder is principally a filling of calcium phosphate. *Dentine* forms the main bulk of the tooth. It is almost identical with bone in chemical composition, and half of its substance is a deposit of a form of calcium phosphate. Structurally, however, dentine differs notably from bone. The cell bodies are situated in the pulp cavity, and send out long, straight processes into the dentine through numerous fine parallel tubules—*canaliculi. Vitrodentine* is similar to dentine, but is, as the name implies, a harder enamel-like material which appears to take the place of enamel as a covering for shark teeth. In many vertebrates the teeth are tied into the jaw with spongy, bonelike materials. The specific substance of this sort found in mammals is *cement;* as well as surrounding the roots, the cement in some ungulates forms, before their eruption, a complete surface covering for the cheek teeth (Fig. 181, *B*, p. 306).

In sharks the teeth are attached to the jaw cartilages by means of fibrous connective tissue; in other groups they are more firmly attached to the underlying bony elements (Fig. 173). In some the tooth base itself may expand to form a supporting plate of bonelike material. In a large proportion of vertebrates the teeth are implanted in sockets—a *thecodont* condition; and in mammals, notably, the sockets accommodate deep roots (multiple in the cheek teeth). In some forms the tooth may be fused (as in Sphenodon and most teleosts) to the outer surface of the bone or the summit of the jaws—the *acrodont* type. A variation on this last is the *pleurodont* condition, common in lizards, where the tooth is attached by one side on to the inner surface of the jaw elements.

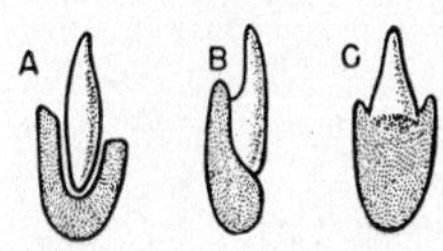

Fig. 173. Diagrammatic sections through reptilian lower jaws to show the distinction between thecodont, *A*, pleurodont, *B*, and acrodont, *C*, tooth attachments.

Tooth Position. At the top and bottom of the series of tooth-bearing vertebrates—in mammals on the one hand and sharks on the other—we find the functioning teeth limited to a single marginal series along each upper and lower jaw. Such marginal teeth are to be found in representatives of every class of gnathostomes, and are the most important element in the vertebrate dental equipment.

It must be noted, however, that in many groups teeth are by no means confined to the jaw margins. Wherever ectoderm is present, teeth or toothlike dermal structures may be found;* the mouth is lined with epidermis derived from the skin ectoderm; and in a wide variety of bony fishes, amphibians, and reptiles teeth are found in the mouth cavity internal to the marginal series. Teeth are present on the palatal elements of the mouth roof in many members of these groups, on the parasphenoid bone underlying the braincase, and on the inner surface of the lower jaw. In many fishes and amphibians certain of the teeth on the palate are large and may form a row paralleling the marginal teeth (cf. Figs. 130, *B*, p. 218; 131, *B*, p. 224). We have noted (p. 221) that the shark jaws are not identical with the jaw margins of bony fishes and tetrapods, but instead appear to

* Teeth are found outside the mouth in the sawfish (Pristis), a sharklike form with a long, slender, forward-projecting rostrum which bears a long row of teeth on each margin.

correspond to palatal structures. This suggests that the marginal teeth of sharks are not really identical with the marginal row of higher vertebrates, but are homologous with the palatal tooth rows of other groups.

We assumed that the ectodermal lining of the digestive tube gives way to endoderm, in the adult as in the embryo, at about the boundary between mouth and pharynx. Conditions in actinopterygians, however, more than suggest that there the ectoderm, with its tooth-forming potentialities, grows back into the pharyngeal region. In teleosts the maxilla is toothless, and in some the entire marginal dentition disappears. In compensation, there is a high degree of development of teeth in the inner parts of the mouth cavity, and also of *pharyngeal teeth* situated on the bony elements of the visceral arch system.

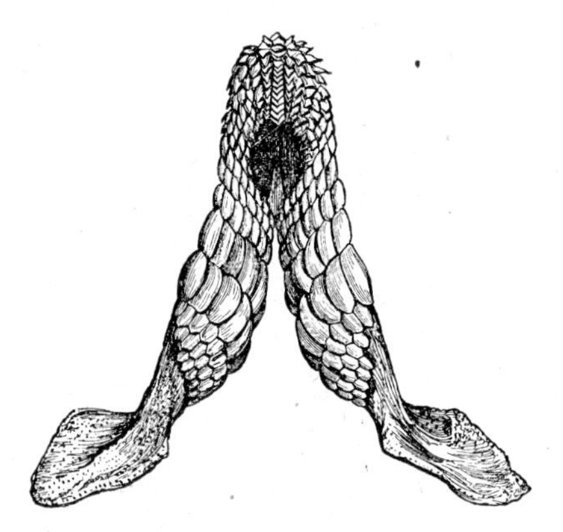

Fig. 174. The lower jaws of a primitive living shark, Heterodontus (Cestracion), the Port Jackson shark of the Pacific. The teeth differ greatly between front and back parts of the jaw. Rows of successional teeth are formed down within the inner surface of the jaws. (From Dean.)

Tooth Shapes in Lower Vertebrates. The vast majority of vertebrates have retained in part or all of their dental battery a primitive pattern of conical teeth or simple variants of the conical type. The cone is characteristic of some sharks, most bony fishes, crossopterygians, all amphibians, a majority of reptiles living and extinct, and the oldest birds. In sharks the cone may become compressed and triangular in outline, frequently with sharp, serrate edges or with additional cusps at the base. Another simple modification is the chisel-like cutting shape found in some bony fishes and repeated in the incisors of many mammals. The labyrinthodont structure (Fig. 175, *C, D*) is found in crossopterygians and repeated in the older fossil members of their descendants, the Amphibia. Longitudinal grooves on the surface of these pointed, conical teeth are indications of an infolding of the enamel into the substance of the tooth, giving in cross section a labyrinthine pattern.

From the lowest to the highest of jawed vertebrates we find numerous examples of the development of a flattened crown on some or all of the teeth, or of *tooth plates* for crushing or grinding purposes, sometimes formed by a fusion of originally discrete teeth. Tooth plates appear to have developed as an adaptation for dealing with molluscs or other hard invertebrates. Among the sharks, the Port Jackson shark (Heterodontus) exemplifies a type in which the posterior teeth are flattened, and rows of successional teeth come into play simultaneously to form a crushing apparatus (Fig. 174). Most skates and rays have a comparable battery of crushing teeth (about 1000 in a common ray), and some teleosts have plates of flat-crowned teeth (9000 of them in one catfish genus). In the chimaeras (Fig. 175, *A*) the dentition is reduced to a pair of crushing plates in both upper and lower jaws and an accessory pair of upper plates. An analogous development is present in dipnoans (Fig. 175, *B*); the mar-

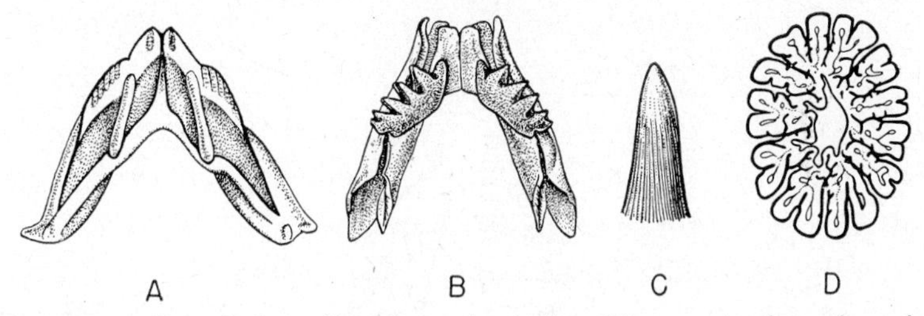

Fig. 175. *A,* Dorsal view of the lower jaws of a chimaera, to show the pair of large tooth plates. *B,* Similar view of the bony elements of the jaws of the lungfish Epiceratodus, to show the pair of fan-shaped toothplates. *C,* External view of a grooved labyrinthine tooth characteristic of crossopterygians and ancestral tetrapods. *D,* Section through such a tooth to show the complicated folding of the enamel layer (heavy black line). (*A* after Dean; *B* after Watson; *C* and *D* after Bystrow.)

ginal teeth are lost, and the dentition typically consists of four fan-shaped compound dental plates (plus an accessory upper pair). In the living Epiceratodus of Australia the main tooth plates exhibit a series of smooth radiating ridges; the embryological and phylogenetic stories agree in showing that each ridge is composed of a fused row of once discrete conical teeth. There are no examples of tooth plates or other pronounced crushing dentitions among amphibians or living reptiles except for some lizards, but flattened plates suggestive of a molluscivorous diet are present in a variety of extinct reptiles.

Tooth Differentiation in Mammals. In mammals the teeth are reduced to a short marginal series in each jaw half, above and below. In the more generalized members of the group, four tooth types can be distinguished in the adult in anteroposterior order (Fig. 176, *A*). Most anteriorly are the *incisors,* nipping teeth with a simple conical or chisel-like build. Next in each series is a single *canine,* primitively a long, stout tooth with a conical shape and sharp point, useful in attacking a carnivore's prey. Following the canine is a series of *premolars* (the "bicuspids" of the dentist), which frequently show a degree of grinding surface on the crown, and finally, a *molar* series, the members of which generally assume a chewing function and develop a complex crown pattern.

In early fossil mammals the number of teeth in these various categories appears to have been rather variable and the total number rather higher than in living forms. The primitive placentals, however, had settled down to a count in which there were present in each half of both upper and lower jaws three incisors, one canine, four premolars, and three molars. Few animals today retain precisely this number. But while there is often a reduction of teeth in one or several categories, there is seldom any increase over the primitive placental number (some whales and porpoises are conspicuous exceptions). There has been devised a simple "shorthand" for the nomenclature of placental teeth, and a formula to express the number of teeth present in any mammal. The letters *I, C, P,* and *M,*

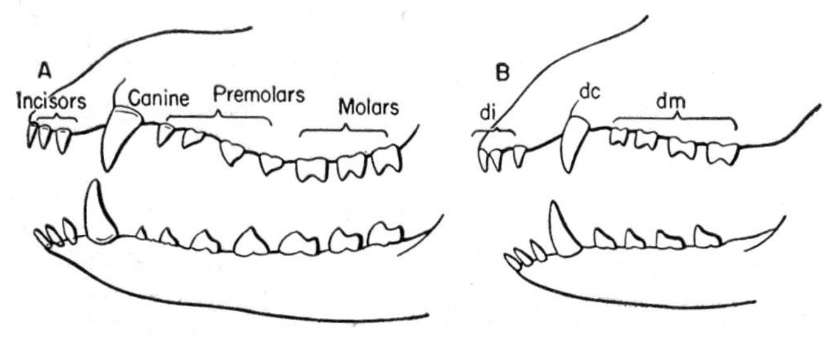

Fig. 176. Left side view of the dentition of a generalized placental mammal, showing permanent dentition in *A*; *B*, deciduous teeth. *dc,* Deciduous canine; *di,* deciduous incisors; *dm,* "milk molars" (= deciduous premolars).

followed by a number in an upper or lower position will define any tooth in terms of the original placental formula: I^1, for example, refers to the most anterior upper incisor; M_3, the last lower molar. The number of teeth of each type present, above and below, in the dental equipment of any animal can be formulated in succinct fashion. The formula $\frac{3.1.4.3}{3.1.4.3}$ indicates that in either side of both upper and lower jaws the primitive placental number of teeth is present (the four successive figures represent the number of incisors, canines, premolars, and molars present). The total number of teeth present in the mouth of such an animal is forty-four. The human formula is $\frac{2.1.2.3}{2.1.2.3}$; i.e., we have lost an incisor and two premolars of the primitive placental series from each half of each jaw and reduced our teeth to a total count of thirty-two.

Incisor teeth, useful in nearly any mode of life, are retained in most mammalian groups. Various herbivores, however, have evolved specialized cropping devices in which the incisors are modified or lost. Ruminants, such as cows, sheep, and deer, have, for example, lost the upper incisors, but retained the lower ones (Fig. 177, *B*); the elephant's tusks are greatly elongated upper incisors. Rodents have developed a pair of upper and of lower incisors as gnawing chisels which grow persistently throughout life at their roots as their tips are worn away.

Canines are primitively prominent biting and piercing weapons and as such are emphasized in carnivores; they reached their peak in the now-extinct saber-tooth "tigers" (Fig. 177, *A*). In nonpredaceous mammals the canines are sometimes prominent, and are used in defense or, in males, in mating battles. More often, however, they are reduced (as in man) and are insignificant or absent in most herbivores.

The cheek teeth—the premolars and molars—whose patterns are discussed below, have a varied history. In herbivorous forms, such as the

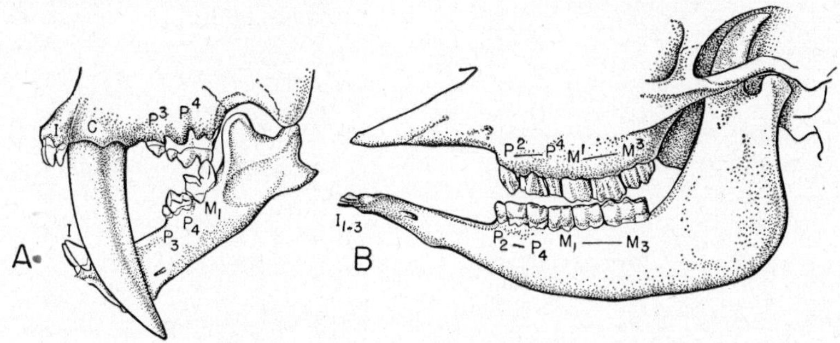

Fig. 177. Two specialized mammalian dental types. *A*, Felid; *B*, the cow. (For the felid, the extinct saber-tooth, with extreme specialization, is shown.) In the felids the dentition is much reduced; there are modest incisors and lower canine (concealed in the figure) and stabbing upper canine. In the cheek little remains except the carnassials $\frac{(P^4)}{(M_1)}$. In the ruminants the cheek teeth are expanded into a grinding battery; they are separated by a diastema from the cropping teeth—here consisting only of lower incisors, working against a horny pad on the upper jaw.

ungulate groups, these teeth are generally retained (except for the occasional loss of the first premolar). They are usually elongated into an efficient grinding battery, which (as may be seen in a horse or a cow jaw) becomes separated, spatially as well as functionally, from the cropping teeth by a gap in the tooth row—a *diastema*. In carnivores in which there is little or no chewing of food the cheek teeth are generally reduced in number—strongly so in the cat family.

Mammalian Molar Patterns. In mammals the cheek teeth, particularly the molars, generally develop an expanded crown with a varied and complicated pattern of projecting cusps—a structural arrangement which appears to be of major functional value in chewing and grinding food. The variations in this pattern are of diagnostic importance and are of great interest to systematists, paleontologists, and anthropologists, for a single molar tooth will often furnish precise generic or even specific identification of the animal to which it pertains. Such patterns have been much studied, but only the general features will be noted here (Figs. 178, 179).

The nomenclature used is basically simple. Each cusp is termed a *cone.* The names of specific cones are formed by adding the prefixes *proto-, para-, meta-, hypo-,* and *ento-;* and, where necessary, by suffixes: *-ul(e)*

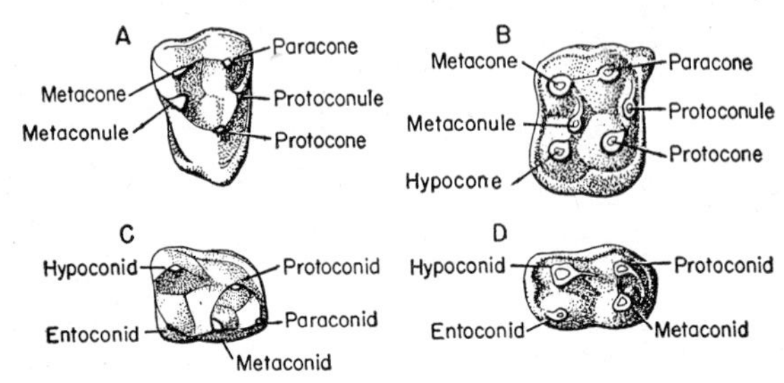

Fig. 178. Diagrams of molar tooth patterns of placental mammals. *A,* Right upper molar of a primitive form (outer edge of tooth above, front edge to right). *B,* The same of a type in which the tooth has been "squared up" by addition of a hypocone at the back inner corner. *C,* Left lower molar of a primitive form with five cusps (outer edge of tooth above, front end to right). *D,* The same of a type in which the tooth has been "squared up" by the loss of the paraconid.

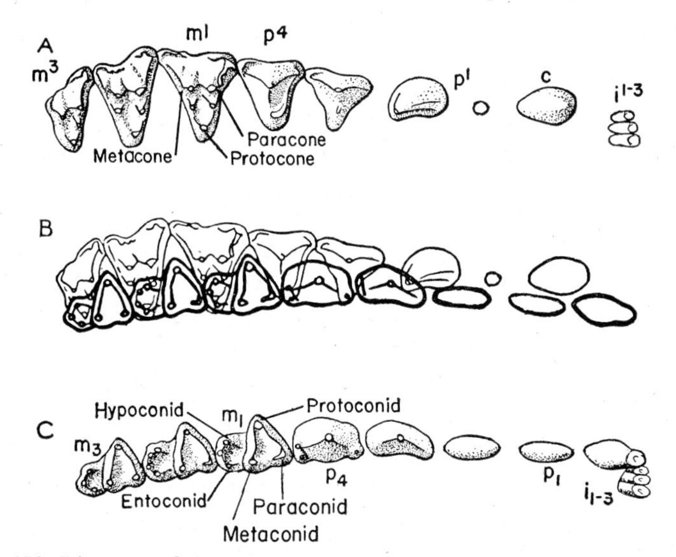

Fig. 179. Diagrams of the dentition of a primitive placental mammal (based on the Eocene insectivore Didelphodus). *A,* Crown view of right upper teeth; *C,* crown view of left lower teeth; between, in *B,* the teeth are placed in occlusion, the outlines of the lower teeth (heavy lines) superposed on those of the uppers (After Gregory.)

indicates a minor cusp, and *-id* a lower jaw element. The term *cingulum* is applied to accessory ridges of enamel around the margins of cheek teeth, the term *style* to vertical marginal ridges.

Neglecting some variations in pattern among early fossil mammals and among marsupials, we find that the earliest and most primitive of placentals have upper molar crowns in the general shape of a right-angled triangle, with two cusps along the outer edge of the tooth row and the right angle at the antero-external margin. There are three major cusps, the *protocone* at the apex, the *paracone* and *metacone* along the external border. Between the protocone and the external cusps is usually found a pair of conules (with which we shall not here concern ourselves further).

The primitive lower molar has as its major portion a high, well-developed triangular area—the *trigonid*—which has a pattern similar to that of the upper molar, but reversed; there is an external apex, the *protoconid,* and an internal base, where *paraconid* and *metaconid* are present. If upper and lower teeth are fitted properly together—that is, placed in *occlusion* (Fig. 179, *B*)—it will be seen that each lower tooth (which lies somewhat internal and to the front of its upper mate) would bite past its neighbors above without meeting them effectively, if it consisted only of a trigonid. This situation is prevented through the fact that each lower tooth has, posteriorly, a second crown portion, a *talonid,* primitively lower and less developed than the anterior part of the tooth, and bearing two additional cusps, *hypoconid* (externally) and *entoconid* (internally). The protocone of the upper tooth bites into the basin of the talonid, and proper occlusion is effected. We thus have a generalized primitive tooth pattern of cheek teeth with three main upper cusps and five lower ones. The full pattern is present in the molars. Passing forward in the premolar series, the pattern tends to become more simplified and approaches the essentially conical shape of the canine and incisors.

In primitive placental mammals the cusps were sharp-pointed structures, well adapted to deal with insects and similar food materials, and some living insectivores still have teeth of this sort.

Most modern mammals, however, have departed widely from this primitive pattern and have evolved a variety of tooth types which can be discussed only in summary fashion here. Carnivores which tend to reduce the number of cheek teeth generally preserve a simple and primitive pattern in those which remain. Notable, however, is the development in the carnivore cheek of a pair of sharp-crested, shearing teeth which deal with tendons and bones—the *carnassial teeth,* the last premolar in either upper jaw shearing against the first lower molar (Fig. 177, *A*).

Mammals with a mainly herbivorous diet usually retain the cheek teeth with little or no loss. The premolars generally remain simpler than the molars in cusp pattern. In certain herbivores, however, they are "molarized"—that is, assume a molar-like cusp pattern; this is exemplified in the horse. Generally there has been a "squaring up" of the molars by alterations in the cusps present (Fig. 178, *B, D*). In the originally triangular upper molars there is often added a fourth cusp, a *hypocone,* at the back inner corner, to turn the triangular tooth into a rectangular one.

Below, a four-square condition is attained, not by adding a cusp, but by dropping one—the paraconid at the anterior edge of the trigonid.

In animals with a mixed diet, such as men and swine, the cheek tooth cusps tend to become low and rounded, forming small "hillocks" on the

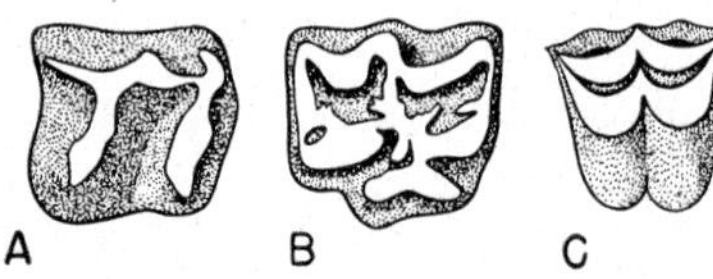

Fig. 180. Crown views of molar teeth of *A,* rhinoceros; *B,* horse; *C,* ox; to show types of molar patterns. The white areas are worn surfaces showing dentine; the black line surrounding the white is the worn edge of the enamel; stippled areas, cement covering or unworn surface. *A* is a simple lophodont type, with an external ridge (above) and two cross ridges; the development of this pattern by connecting up the cusps seen in Figure 178 can be readily followed. The horse (*B*) shows a development of the same pattern into a more complex form in which the primitive lophodont pattern is obscured. *C* shows a selenodont pattern characteristic of the ruminants; each of four main cusps takes on a crescentic shape.

crown—the *bunodont* type. In many ungulates which have taken up a herbivorous mode of life the cusp pattern has become complicated; individual cusps may, for example, assume a crescentic outline, a *selenodont* pattern; or cusps may connect to form ridges, the *lophodont* condition (Fig. 180).

Grazing habits present a serious "problem" to an ungulate, for grass is a hard, gritty material which rapidly wears down the surface of a grinding

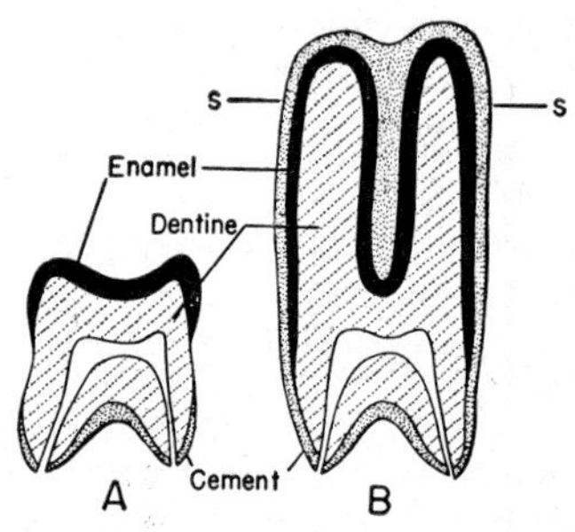

Fig. 181. Diagrams to show the development of a hypsodont tooth. *A,* Normal low-crowned tooth (cf. Fig. 172). *B,* Hypsodont tooth, with cusps elevated and the whole covered by cement. As a tooth of this sort wears down to any level, such as that indicated at *s-s,* it will be seen that no less than nine successive layers of contrasting materials are present in cross section.

tooth. If the low-crowned tooth of a primitive mammal were retained in a large grazing mammal, it would be worn to the roots in a short time. In relation to this we find in such hoofed mammals as the horses and cattle the development of a high-crowned and prism-shaped tooth—the *hypsodont* condition (Fig. 181). One could, in imagination, develop a high-crowned tooth by elongation of the dentine-filled bulk of the tooth body, with the cusps in their original shape on the grinding surface. This type of high-crowned tooth did form in some early fossil mammals, but has generally proved unsuccessful. Once the enamel surface is worn away, most of the wear comes on the relatively soft and easily abraded dentine; the resistant enamel remains only as a thin rim. Successful hypsodont forms have teeth built on quite another plan. The height is attained by a skyscraper-like growth of each cusp or ridge on the tooth; these slender peaks are fused together by a growth of cement over the entire tooth surface before eruption. As wear takes place, it grinds down through a resistant complex of layers of all the tooth materials—enamel, dentine, and cement.

Dental History. In the preceding sections we have noted many salient features of the dentition of the various types of vertebrates; a brief résumé by groups may now be given.

No teeth are present in lower chordates or in jawless vertebrates. The ostracoderms, like the lower chordates, presumably gained their food by straining particles from the water current passing through their gill systems; teeth were unnecessary and could not have functioned in the absence of jaws. The larval lamprey is likewise a food strainer. The adult cyclostomes are predaceous and have, in their rasping "tongue," a jaw substitute. This is armed (as is the mouth cavity as well in the lamprey) with conical, pointed, toothlike structures. These, however, are not true teeth, but consist of cornified epithelium.

In the extinct placoderms teeth as well as jaws made their appearance. In some acanthodians simple teeth comparable to those of sharks are found; in the arthrodires teeth are sometimes fused to the jaw plates, but in general these ancient fishes relied for biting purposes on sharp cutting edges of the jawbones themselves.

In sharks the functional dentition generally consists of a single row of teeth of simple structure, but numerous rows of replacement teeth are present along the inner jaw margins (Fig. 174, p. 300). In the skates and rays, most of which feed on molluscs and other hard invertebrates, a number of these successional rows may all come into play simultaneously for the more effective crushing of difficult food materials, and among the rays there may develop flattened crushing plates formed by a fusion of a series of teeth. In the chimaeras, also primarily mollusc feeders, a few crushing plates form the entire dentition (Fig. 175, *A*, p. 301).

In the bony fishes the primitive dentition included a marginal series of sharp teeth and a considerable palatal dentition. In the higher actinopterygians the maxilla becomes toothless, and in some teleosts the marginal dentition disappears entirely; the main reliance may be placed on a pharyngeal and palatal tooth battery.

The dipnoans have paralleled the chimaeras in the development of large tooth plates internal in position; marginal teeth disappeared early in the history of the group (Fig. 175, *B*). The early crossopterygians, on the other hand, had a powerful dentition of predaceous type, with both sharply pointed marginal teeth and numerous palatal teeth, some of them powerful fangs. Notable is the labyrinthine infolding of the enamel in this group (Fig. 175, *C, D*).

The ancestral amphibians, the labyrinthodonts, had a dentition exceedingly comparable to that of their crossopterygian ancestors. In modern amphibians, however, the dental equipment is much reduced. The teeth are small and simple in structure; in some toads they are entirely absent.

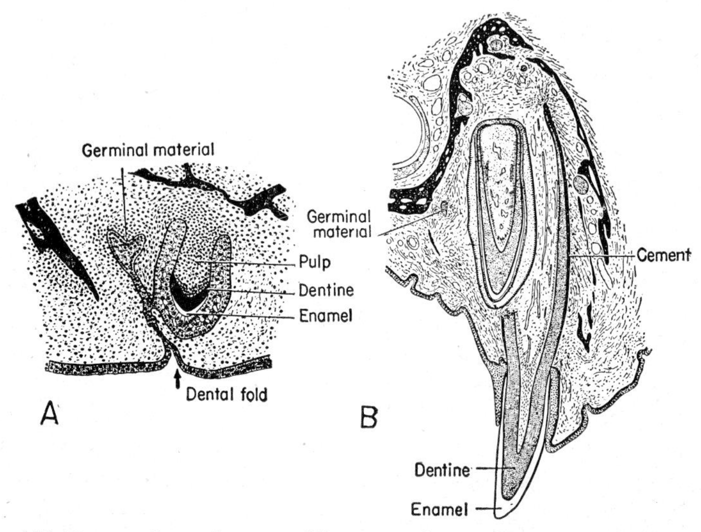

Fig. 182. Two sections of a crocodilian upper jaw to show tooth development in lower tetrapods. *A*, Embryo with a first tooth developing and germinal material in reserve at the bottom of the dental fold. *B*, Mature animal; a tooth is functioning, but is being resorbed at the inner side of the root, where the next successional tooth is already in process of formation. Germinal material for successional teeth persists. (After Röse.)

Among reptiles teeth have vanished entirely in the turtles. Marginal teeth, usually conical, are retained by the other modern orders. Palatal teeth are absent in crocodilians, but retained by the Squamata and Sphenodon. In the ancestral tetrapods the teeth were inserted in sockets; this is, today, the case in crocodilians but in Sphenodon and the Squamata acrodont and pleurodont conditions are developed. In ancestral Mesozoic birds marginal teeth were still present; these structures have given place to a bill in the living forms.

In the fossil mammal-like reptiles are seen stages in the initiation of the mammalian type of dentition. The most ancient members of this series, the pelycosaurs, show the development of large canines and the consequent separation of an incisor series from the cheek teeth (Fig. 139, *A*,

p. 232); various therapsids (Fig. 139, *B*, *C*) show further stages in the evolution of the mammalian type of dentition discussed above.

Tooth Development and Replacement. Embryologically, enamel is an ectodermal product, the remainder of the tooth structure mesodermal in origin. It is, however, the ectoderm which appears to be the active agent in tooth development. The first embryological indication of the appearance of teeth is an infolding of the epidermis. This infolding may take the form of discrete pockets or may, in marginal tooth rows, develop as a continuous furrow the length of the jaw, a *dentinal lamina* from which individual *tooth germs* develop (Fig. 182). If, as is the case in most lower vertebrates, there is to be a continuous succession of teeth during life, the germinal material is not used directly in tooth formation. Instead, the tooth germ remains, protected, well below the jaw surface, while successive teeth take their origin from bits of tissue which bud off from the germ and gradually move upward toward the surface. These tooth buds typically form as hollow cones which outline the surface of the future tooth. At their lower surfaces the cells appear to secrete their contribution to tooth structure—the enamel; from this fact the tooth bud is frequently termed the *enamel organ*. Within the crown of enamel-secreting ectodermal cells there gathers a mass of mesenchymal cells—the *odontoblasts*. These form an embryonic tooth pulp, a *dental papilla* with which small blood vessels and nerves become associated. The odontoblasts proceed with the task of dentine deposition, and come to form, at the end of growth, cells of the pulp cavity. As the tooth takes shape, it works toward the surface, erupts into functional position and, in bony vertebrates, becomes affixed to the jaw bones by cement or other spongy bone materials. In forms with continuous tooth succession each tooth eventually becomes subject to a process of resorption at its base and is shed from the jaw. Meanwhile, one or more successional buds have already formed from the germinal material and push outward as replacements.

The prominence of the ectoderm in tooth development is in agreement with our general ideas of the phylogenetic origin of teeth. It has long been recognized that teeth and the dermal denticles of the shark skin (cf. Fig. 78, p. 145) are essentially similar structures and hence presumably homologous. Mouth lining and body skin are comparable in nature, both being formed from ectoderm and connective tissues gathered beneath it. They hence may be expected to develop homologous structures.

It was once thought that teeth originated directly from dermal denticles which lay in the margins of the mouth overlying the forming jaws. Our current conceptions of the nature of the primitive fish as having a bony external skeleton demands a modification of this idea. The surfaces of the plates shielding the body of primitive fishes (Fig. 77, *A*, p. 144) are frequently ornamented with tubercles resembling both shark denticles and primitive teeth. Shark denticles, we noted previously, appear to represent the last surviving parts of these plates in the skin of modern sharks; the deeper layers have, so to speak, melted away in the course of evolutionary history. When jaws evolved in ancient fishes, dermal plates were present along the jaw margins. The denticles present on these jaw plates became the teeth. Teeth and dermal denticles are thus, in modern concepts, still regarded as homologous. Teeth are not, however, derived from shark denticles; instead, both are modified derivations of tubercles found in the dermal bony plates of armored ancestral vertebrates.

In many groups the mode and timing of tooth renewal is poorly known. There appears to be little replacement at all of the large tooth plates present in various fishes and reptiles. Such knowledge as we have of tooth replacement mainly concerns lateral tooth rows.

In sharks we see, on the inner side of upper and lower jaws, the formation of numerous rows of successional teeth long before they roll out-

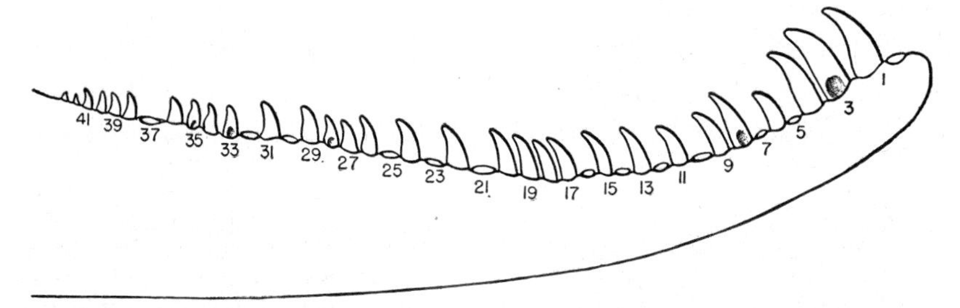

Fig. 183. Inner surface of the lower jaw of a primitive fossil reptile (Ophiacodon) to show the alternation in tooth replacement between odd and even tooth series. The gaps in the tooth row seem haphazard at first sight; however, if the two series be considered separately, it will be seen that each includes several waves of replacement arranged in a regular alternating manner. (Certain teeth — 3, 8, 28, 33, 35 — show resorption at the root and are at the point of being lost.)

ward into functioning position on the jaw margins. In early fossil amphibians and in many living and extinct reptiles tooth replacement takes place in an interesting way which guarantees a continuous function of the dentition despite frequent renewal of individual teeth (Fig. 183). The teeth and tooth germs are present in two series, the "odds" and "evens" in each tooth row. One may find a condition in which, for example, in a given area of the jaw, the odd-numbered teeth are functional; between them, in place of the even-numbered ones, are sockets beneath which new teeth are in process of formation. Later this region may show a condition in which both sets of teeth are in place at the same time, but the odd set shows evidence of age and wear. Still a bit later, the odd-numbered teeth will drop out, leaving the even series the functional set, and so on. This neat device guarantees that at least half the teeth in any region will be functioning at any time. The situation is complicated by the fact that replacement may take place in waves which travel along the jaw rami; at a given time the even set may be functioning in some areas of the jaw ramus, the odd set in others.

In mammals tooth renewal (to the regret of all except dentists) is a much reduced process; there is only one replacing set, and that is incomplete, for the most posterior—molar—teeth have no successors. We usually think of the mammalian dentition as comprised of a set of "milk" teeth—a *deciduous* or lacteal dentition—which has no molars included, and a complete permanent set (Fig. 176, p. 302). If, however, we consider the development of the teeth we gain a different concept (Fig. 184). The "milk" teeth, in a typical mammal (man is representative), usually develop successively in anteroposterior order from incisors to deciduous premolars or "milk molars" (the canines, large, or large-rooted, may lag in eruption). After the appearance of the last "milk molar" there begins the successive eruption of the three molar teeth. Meanwhile, without waiting for the completion of this first wave, the anterior teeth are replaced in

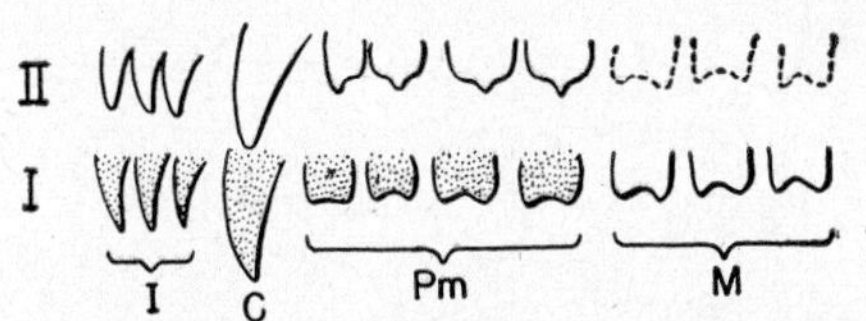

Fig. 184. Diagram to show the composition of the "permanent" dentition of a mammal; left upper teeth of a generalized placental mammal. One complete set of teeth (I) develops from incisors back to molars. All of these except the molars, however, are shed (as indicated by stipple). A second set of teeth develops (II), but never produces molars. Hence the "permanent" dentition includes portions of two tooth series (cf. Fig. 176).

order back to the premolars. But the succession proceeds no further, and the molars are not replaced. Obviously, we have one complete set of teeth and a partial second set; the molars, although permanent, belong in series to the set which, farther forward, is deciduous in nature. An adult mammal thus has in its mouth parts of two dental series; the molars are part of the first set, the more anterior teeth belong to the second set. This concept explains various peculiarities such as the fact (well known to paleontologists and anthropologists) that the molars in many cases resemble much more closely the short-lived "milk molars" than they do the permanent premolars which they adjoin for most of the span of life.

GILLS

The *pharynx* is a short and unimportant segment of the digestive tract in the higher vertebrate classes—a minor connecting piece between mouth and esophagus where ventrally the *glottis* opens to the lung apparatus, and dorsally the paired *eustachian tubes* lead to the middle ear cavities (Fig. 171, p. 295). In mammals the pharynx is merely the place where air and food channels cross one another in awkward fashion, and where there accumulate masses of lymphoid tissues, the *tonsils*. But from both phylogenetic and ontogenetic viewpoints the pharynx is an area of the utmost importance; it is the region from which are developed the gill pouches, basic in the construction of respiratory devices in lower vertebrate classes, and persistently important in the developmental story in higher groups.

The body cells constantly demand oxygen for their metabolic activities and as constantly must be rid of the carbon dioxide, which (with metabolic water) is the major waste product resulting from oxygen utilization. Internal carriage of these gases to and from the cells is by means of the circulatory system. The oxygen source and the carbon dioxide destination is the surrounding medium—water or, in higher groups, air. Exchange of gases—the process of respiration—occurs by diffusion through moist vascularized membranes.

In many small water-dwelling invertebrate organisms this exchange takes place through the general body surface. Such surface exchange may persist among vertebrates, and, we note, certain salamanders respire entirely through the skin, and have neither gills nor lungs. In general, however, all vertebrates and most multicellular invertebrates have developed special respiratory organs. Skins are often too thick or hard, and too little vascularized to function effectively in respiration; and surface-volume relations generally come into play even when the body surface is competent as an exchange medium. In water dwellers of various phyla, *gills* of one sort or another are developed—structures with a moist, thin surface membrane, richly supplied with blood vessels, and folded in complicated fashion to give a maximum amount of surface for gas exchange.

Among invertebrate groups such gills are usually situated outside the main mass of the body as external gills. In a limited number of cases analogous developments are seen among vertebrates—the external gills of larval fishes and amphibians (described later), projecting gill filaments of other larval fishes, hairlike respiratory growths on the slender pelvic fins of the South American lungfish and on the pelvic limbs of the male "hairy" frog of Africa (Astylosternus), for example. But the characteristic structures of the lower vertebrates—indeed, as we have earlier suggested, the most fundamental trade-mark of the chordates as a whole—are *internal gills:* respiratory structures located in a series of slits or pockets leading from the pharyngeal region to the outer surface of the body. As the locus of origin of these slits the pharynx is a highly developed and highly important part of the digestive tube in the lower classes of vertebrates.

The Gill System in Sharks. The characteristic development of the vertebrate gill system is that seen in the sharklike fishes, the Chondrichthyes, on the one hand, and the higher bony fishes on the other. Gill construction in the two groups differs in certain important respects, but the basic pattern is the same.

In the sharks (Fig. 185, *C*) the pharynx is a long, broad duct, continuing backward from the mouth region without break and terminating posteriorly at the constriction whence the short and narrow esophagus leads on to the stomach and intestine. On either side a series of openings runs outward to the surface of the body. Anteriorly, there is generally present a small and specialized paired opening, the spiracle, described later. Back of this point, on either side, are found the typical *gill slits,* five in most sharks, but six or seven in a few special cases. In these slits are

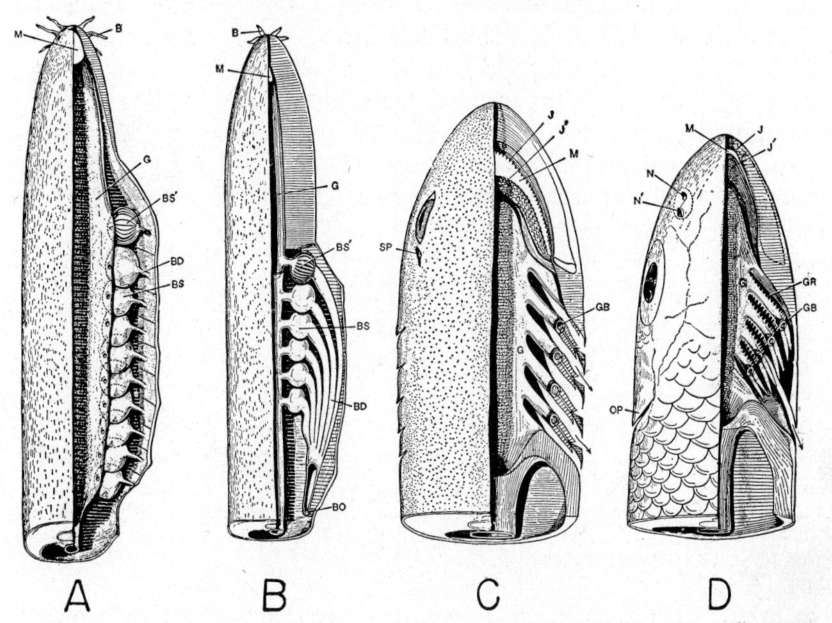

Fig. 185. Heads of various fishes to show gill arrangement. *A,* The slime-hag Bdellostoma; *B,* the hagfish Myxine; *C,* a shark; *D,* a teleost. The right half of each is sectioned horizontally through the pharynx. Abbreviations: *B,* barbels around the mouth; *BD,* ducts from gill pouches; *BO,* common outer openings of gill pouches; *BS,* gill sacs; *BS',* sacs sectioned to show internal folds of gill; *G,* gut (pharynx); *GB,* cut gill arch; *GR,* gill rakers; *J, J',* upper and lower jaws; *M,* mouth; *N, N',* anterior and posterior openings of nasal chamber; *OP,* operculum; *SP,* spiracle. (From Dean.)

found the respiratory organs, the gills. Water is typically brought into the pharynx through the mouth and flows outward through the gill slits; in its passage past the gill surfaces, the respiratory exchange takes place. The term "gill arch"* is applied to the tissues lying between successive openings. The region from mouth to spiracle is the *mandibular arch;* that

* There is a source of confusion here, for the term "arch" is often used in three different senses in connection with the gill region. As noted in Chapter VII, it may refer to the series of bars which form the skeleton of each gill segment. The word is also used to describe the arterial blood vessels—aortic arches—which traverse each gill (cf. Chap. XIII). In the present chapter it is used in a broader, inclusive sense, to describe the total gill structure lying between two successive gill slits.

between the spiracle and the first normal gill slit, the *hyoid arch;* more posterior gill arches are generally referred to by number.

Typical shark respiratory movements are a combination of suction and force pump effects, produced by the action of the branchial muscles described in the preceding chapter. With the mouth open and the external gill openings closed, the pharynx is expanded and water drawn in via the mouth; the mouth is then closed, the pharynx constricted, and the water forced out through the slits.

Each of the series of gill arches includes a number of characteristic structures (Fig. 186, *A*). Running dorsoventrally, close to the pharyngeal wall, are the skeletal supports, the gill bar elements of the arch; skeletal *gill rakers* may extend inward from them into the cavity of the pharynx, screening food particles from the gills and diverting them backward toward the esophagus; *gill rays* run outward toward the surface, stiffening the gill structure. In each gill arch there is in the embryo a finger-like up-

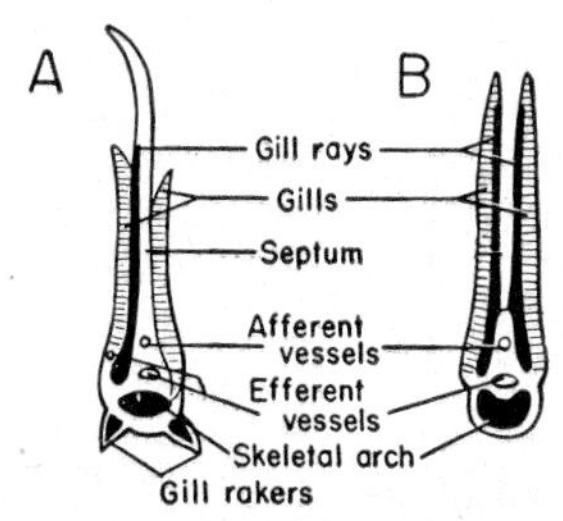

Fig. 186. Horizontal sections of a gill arch of *A*, a shark; *B*, a teleost. External surface above, pharynx margin below; cartilage or bone in black. In the shark there is a projecting septum, reduced in *B*.

ward extension of the celomic cavity (Fig. 153, p. 255); this disappears in the adult, but from its mesodermal walls is derived the branchial musculature of the arch (Fig. 162, p. 284). Each gill arch has a set of aortic arch blood vessels which bring blood from heart to gill and carry it thence, after aeration, to the body (pp. 438–442). Down each gill arch passes a special cranial nerve trunk or branch (cf. Fig. 336, p. 553). Extending outward in the center of each arch is a *gill septum,* a vertical sheet of connective tissue.

The gill passages of elasmobranchs are slits, narrow anteroposteriorly, deep in a vertical plane; primitively they appear to have opened widely both to the pharynx and to the exterior, but in most forms these entrances and exits are narrowed down to relatively short slits. In sharks the external openings are lateral in position; in skates and rays, with flattened bodies, the typical gill openings are on the ventral surface. In elasmobranchs the gill septum and its epithelial covering extend out to the surface of the body as a fold of skin which overlaps and protects the gill next posterior to it.

The gill itself is an elaborately folded and richly vascularized structure, covered with a thin epithelium. The gill surface takes the form of numerous parallel *gill lamellae*. Through them course capillaries, connected at one end with the afferent vessels from the heart and at the other with the efferent vessels leading to the dorsal aorta and thence to head and body. For the most part the gill is, in elasmobranchs, closely attached to the gill septum; at its outer tip, however, it may extend freely for a short distance.

Such a gill may develop as a dorsoventral sheet on either anterior or posterior side of any gill slit, or, stated in another fashion, it may develop

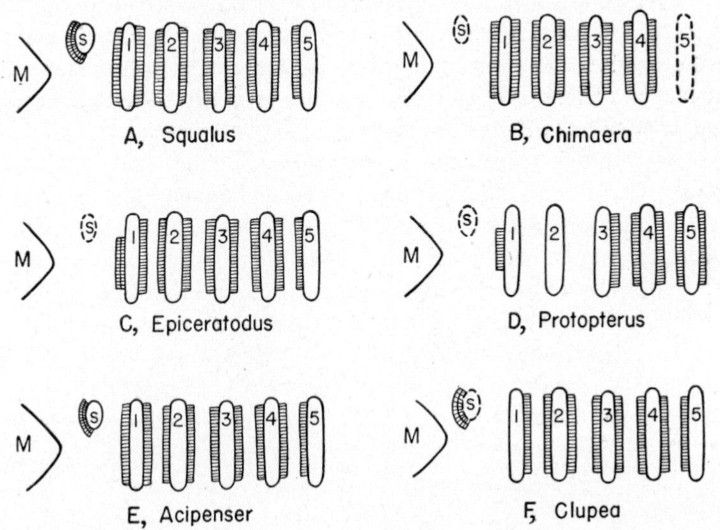

Fig. 187. Diagrams to show the gill arrangement. *A*, Shark; *B*, a chimaera; *C*, the Australian lungfish; *D*, the African lungfish; *E*, sturgeon; *F*, a teleost (herring). Broken line indicates a closed slit. Hatched area adjoining slit indicates gill surface; vertical line through hatching indicates pseudobranch. The presence in Protopterus of a gill surface on the posterior side of the last slit is unique. *M*, mouth; *s*, spiracle; postspiracular slits numbered.

on either surface of a gill arch (Fig. 187, *A*). In most cases the gill arch carries a gill on both surfaces of the septum, and such a gill arch is considered to be a complete gill, a *holobranch*. More exceptionally, a gill may be developed on only one surface of a gill arch; in this case the term *hemibranch* is applied. No vascular aortic arch is present behind the last gill slit; and, in consequence, there is almost never any gill development on the posterior surface of the last slit. All other typical slits in sharks, however, bear gills on both sides; in terms of gills rather than slits, there are four holobranchs. There is no gill development on the posterior side of the spiracle; hence the hyoid arch behind it bears merely the hemibranch on its posterior surface.

The *spiracle* is a gill slit lying between mandibular and hyoid arches which appears to have gained its specialized nature early in the history of jawed vertebrates. We have noted that in typical jawed fishes the hyomandibular element of the hyoid arch has become tied into the jaw joint as an effective prop for the jaws. Presumably the ancestral vertebrates had a fully developed gill slit between jaws and hyoid arch. When, however, the hyomandibular extended across this area, the slit was necessarily interrupted, and has become restricted to a dorsal opening from throat to surface (Fig. 127, *C*, p. 212).

But though we may account for the special nature of the spiracle as due to skeletal developments, this opening has a positive functional history. If we survey the sharklike fishes, we find that among the sharks themselves the spiracle is small in many forms and is absent in certain of the more active, fast-swimming genera. On the other hand, some of the more sluggish sharks, which spend much time on the bottom, show a considerable enlargement of the spiracle, and in the depressed, bottom-dwelling skates and rays it is an enormous opening on the dorsal surface behind the eye (Fig. 19, p. 41). This correlation of spiracular development with habits is indicative of the use to which the spiracle is put. In bottom dwellers the mouth, the usual point of entrance of the water current, is more or less buried in the mud or sand; the spiracle, dorsally placed, then affords a channel for the inflow of clear water to the gills.

Presumably the spiracular slit, when fully developed, originally carried a typical pair of gills. Even in its reduced state a small gill structure is generally to be found on the anterior side of the spiracle—that is, on the posterior surface of the jaw arch. This gill, however, is of a peculiar nature. Every proper gill receives blood for aeration directly from the heart by way of one of the aortic arches. But though such an aortic vessel runs to the mandibular gill arch in the shark embryo, it disappears before the adult stage is reached, and this little spiracular gill is supplied by blood that has already been purified by passage through the gills behind it (Fig. 264, *B*, p. 441). It is, thus, a deceptive "imitation" gill—a *pseudobranch*. It may, however, serve some useful purpose, for the blood from this gill passes, in the shark, to eye and brain, where a supply of "superpure," doubly oxygenated blood is presumably of especial importance.

The chimaeras, although related to the sharks and skates, show certain notable differences (Fig. 14, p. 37; 18, *C*, p. 40; 187, *B*). The spiracle has been lost, and the last gill slit is closed. A major contrast lies in the system of gill protection. In sharks a skin fold is developed from each gill arch which protects the next posterior gill from internal injury. In chimaeras, however, there is but a single large skin flap, which grows backward from the hyoid arch and covers the entire gill series. Skin folds and supporting rays have been lost on the more posterior arches. This protective device is an *operculum;* with its development we find but a single outer gill opening on either side of the body at the back end of the gill region. Comparable developments are found in both higher and lower fish types.

Gill Development (Fig. 188). The gill slits of elasmobranchs arise embryologically in a fashion characteristic of vertebrates generally. Early in development paired pouches push out from the endodermal lining of the anterior end of the embryonic gut. Extending toward the surface, they interrupt the continuity of the mesodermal plates of this region and come in contact with infoldings of the surface ectoderm. Presently the intervening membranes break down, and ectoderm and endoderm join to form a continuous lining for the gill slits. After this occurs the development of gill structures along the lining of the slits.

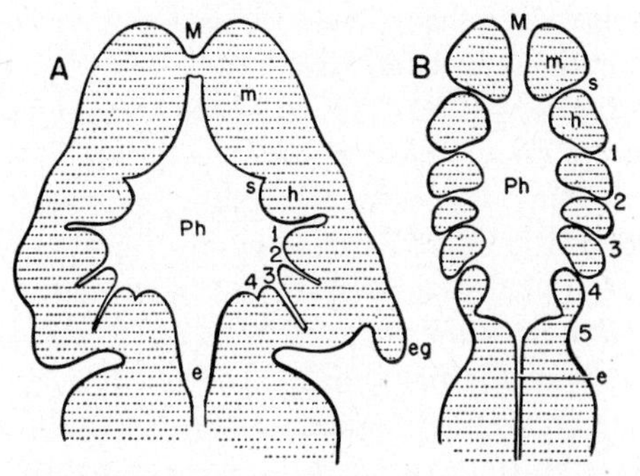

Fig. 188. Horizontal sections through the head and pharynx regions of *A*, a frog (Rana); and *B*, an elasmobranch; to show development of gill pouches. In *A* the pouches are developing as narrow slits, which have not as yet opened to the surface. (The mouth is also closed.) In *B* all the slits except the last one are open. *e*, Esophagus; *eg*, external gill developing; *h*, hyoid arch; *M*, mouth; *m*, mandibular arch; *Ph*, pharynx; *s*, spiracular gill slit; *1* to *5*, postspiracular slits.

In both Chondrichthyes and bony fishes the gills appear to be formed from the ectodermal part of the epithelium of the slits. In cyclostomes, exceptionally, the evidence suggests that gill formation is endodermal (but the evidence for this has been questioned). Some have argued that if cyclostome gills are of endodermal derivation, they cannot be truly homologous with those of other fishes. However, this point of view, which regards the germ layers as sacrosanct, should not be taken too seriously. The lining of the gill passages in any fish includes derivatives of both ectoderm and endoderm, indistinguishably fused and apparently similar in nature. The development of a gill is presumably due to the action of morphogenetic forces upon this epithelium. It seems reasonable to assume that either ectoderm or endoderm, if lying in the area of potential gill development, might respond.

Gill pouches are formed in paired longitudinal series, and it is only natural that many workers have tried to correlate the gill arches with the segmental arrangement seen in the myotomes and the segmentation of skeletal structures and nerves associated with this somite musculature. There is, however, no conclusive evidence of any real relation between these two systems (cf. Fig. 153, p. 255). The gill "segmentation" is based on pouches of endoderm; that of the trunk muscles and associated structures is based on somites of mesodermal origin; the two seem to be quite independent developments.

Gills in Bony Fishes. The gills of the Osteichthyes (Fig. 185, *D*, p. 313) are built on the same basic plan as those of sharklike forms; there are, however, important differences due to the universal presence of an operculum. This structure is basically comparable to that just mentioned for the chimaeras, developing as a fold from the hyoid arch. The operculum of Osteichthyes, however, has developed as a bony structure, forming a part of the armor between head and shoulder structures on either side (Figs. 99, p. 172; 131, p. 224; 132, p. 225). A major skeletal element is a large platelike *opercular* bone; *preopercular, subopercular,* and *interopercular* elements may be present in addition, and more ventrally, where the gill cavity below the opercular fold curves forward beneath the jaws, there may be ventral plates—*gulars,* or, in teleosts, slender, raylike bones, the *branchiostegal rays.* A considerable *branchial chamber* may be present on either side between the opercular cover and the gills. The opercular opening is primitively a great crescentic slit, convex in contour posteriorly, lying just in front of the dermal shoulder girdle. In some specialized teleosts, however, the opening may be reduced to a small hole.

Breathing methods in bony fishes frequently are comparable to those of sharks, except that opercular movements in great measure replace the shark's movements of individual gill openings. Water enters through the mouth, with a closing of the external gill opening behind the operculum and an expansion of the branchial chamber. Valvular skin folds about the mouth help to close that orifice during the reverse movement of constricting the pharynx and expelling water through the external gill aperture. Some stiff-bodied bony fishes utilize the water current through the gills as a locomotor device, after the fashion of a jet plane; on the other hand, such forms as the mackerel rely on their fast-swimming movements for the production of a stream of water through the gills, simply holding the mouth open as they go. Individual gill flaps are unnecessary in bony fishes as protective devices, and we find that the gill septum does not extend (as in sharks) outward beyond the gill surface itself. In teleosts, in fact, reduction has gone so far that the gill septum has shrunk away from the area between the pair of gills on the arch, leaving these elements projecting outward without any extraneous support as two branches of a V-shaped holobranch (Fig. 186, *B*, p. 314). Gill rakers are frequently found, as in sharks; gill rays are present, but, in contrast to elasmobranchs, form in two rows, rather than one, on each gill arch, stiffening the delicate pair of gill structures.

Normal gill slits are, almost universally, five in number in bony fishes as in typical elasmobranchs (Fig. 187, *C-E*, p. 315). Polypterus, however, has (like the chimaeras) lost the last pair of gill slits, and aberrent conditions are found in some teleosts. The pattern of distribution of gills along the slits is subject to reduction in various types. In actinopterygians the hemibranch on the anterior surface of the first slit is lost except in a few primitive types (sturgeons, paddlefish, gar pikes); this is understandable, since the first slit, covered closely by the operculum, is in an unfavorable position to gain a good flow of water. The lungfish Epiceratodus of Australia has a full complement of gills; but that on the anterior side of the first slit is without a direct blood supply and hence is a pseudobranch. The other lungfishes exhibit further gill reduction; these forms tend to rely in considerable measure on lungs rather than gills for respiration. In Lepidosiren the first gill slit is closed, but three holobranchs remain. In Protopterus gills are reduced in both size and number. All the normal gill slits are open, and there is a true if small gill on the anterior surface of the first slit. The two branchial arches behind this slit have no gills at all, but there then follow two holobranchs.

That the spiracle was primitively present in bony fishes is attested by its retention, with an accompanying pseudobranch, in the three living types of chondrosteans. In all other living bony fishes it is lost, although a few additional actinopterygians show a blind internal pouch in its former position. Curiously, however, we find that a majority of actinopterygians have, despite its loss, retained the pseudobranch which was associated with it. Tucked beneath the edge of the operculum, or beneath the base of the skull, it sometimes retains its gill-like structure, sometimes forms a body sunken beneath the surface and containing secretory gland cells. Probably this former gill has become a gland of internal secretion, but its nature is unknown.

An important function in addition to respiration is served by the gills in marine teleosts. Such animals are living in an external environment containing a greater concentration of salt than is present in their own proper "internal environment." Osmotic processes constantly tend to increase the body salt content to a lethal condition, and the kidneys are unable to counteract this entirely. It is found that in at least some marine teleosts the gill epithelium has the ability to take excess salt from the blood and excrete it into the water current flowing through the gills. Still further, the gills of teleosts have an additional functional resemblance to the kidneys in that they appear to be able to excrete nitrogenous wastes such as ammonia.

Although ancestral bony fishes presumably had lungs as accessory breathing organs, these structures were later transformed by almost all actinopterygians into a hydrostatic air sac, described later, of little or no value in respiration. A variety of modern teleosts, however, live today under conditions in which air breathing may become a necessity. Some are more or less amphibious, spending part of their lives on land; others live in a muddy environment, where the water often forms too thick a gruel for satisfactory gill respiration. For a number of teleosts, air gulped into the digestive tract has some respiratory value, with the air bladder lining or the gut epithelium functioning in gas exchange. More frequently, however, we find lung substitutes in the form of structures developed in or from the branchial chambers, kept moist beneath the oper-

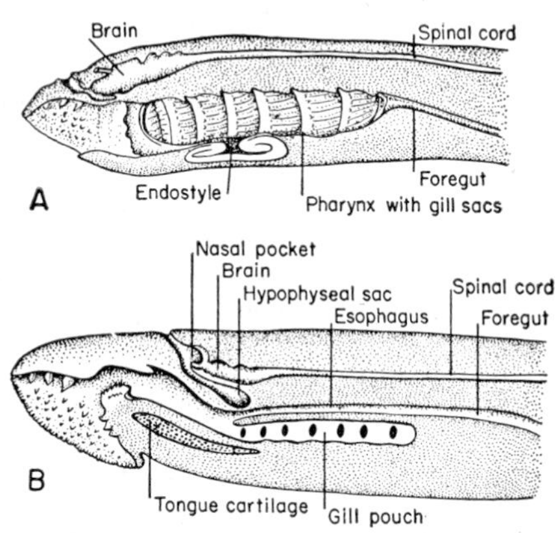

Fig. 189. Sagittal sections of *A*, the ammocoete larva of a lamprey; *B*, adult lamprey; to show especially the division of the adult pharynx into two parts. The endostyle of the larva becomes the thyroid gland of the adult (not shown). (After Goodrich.)

culum. In various catfishes, for example, and the "tree-climbing perch" (Anabas) of the East Indies, familiar in popular lore, rosettes or branching filaments of vascularized epithelium are developed behind or above the gills in the branchial chambers. In several cases these chambers themselves are expanded into lunglike shapes, extending backward into the trunk as a pair of elongated sacs lined with a respiratory epithelium.

Gills in Jawless Fishes. Descending the scale from the gnathostomes, we find that in the lampreys and hagfishes (Figs. 185, *A*, *B*, p. 313) gills are equally well developed, but in a rather different fashion. The gill passages are not in the form of slits, but of spherical pouches, connected by narrow openings with the pharynx on one hand and the exterior on the other. In the lamprey larva the gills are arranged as hemibranchs on anterior and posterior surfaces of each pouch; in the adult cyclostome, however, the gill lamellae have come to form a ring around the margins of the gill sac. The branchial skeleton forms a continuous but flexible latticework about the pouches (Figs. 84, p. 154; 125, *A*, p. 210). There is a well-developed musculature which, working upon these structures, constricts and expands the sacs. In this fashion a pumping effect can be obtained; water can be pulled in through the external openings and forced out again through them, thus permitting respiration to continue when the adhesive lamprey mouth is attached to its prey or the hagfish "nose" buried in its victim's flesh, with the blocking of any anterior entrance of water to the pharynx.

Such anterior entrances vary within the group. In the larval lamprey, as in the hagfish, the mouth leads to a pharynx of normal construction which in turn connects posteriorly with the esophagus. In the metamorphosis to the adult, however, a radical change takes place in the pharynx. This splits into two separate tubes (Fig. 189). A dorsal duct leads back without interruption from mouth to esophagus. Below is a large tube with which the gills connect; this ends blindly at its posterior end, and is a purely respiratory part of the pharynx.

In the hagfishes no such subdivision of the gut is found, but an equally peculiar device is present in that the combined nostril and hypophyseal sac breaks through posteriorly into the mouth roof (Fig. 13, p. 36); water can thus flow back into the gills with the mouth "otherwise engaged," as long as the nostril opening is free.

In contrast to the usual gnathostome condition of five normal slits plus a spiracle, we find in most cyclostomes a higher count of gill pouches. In the lamprey Petromyzon seven pairs of pouches are present; in the hagfishes the number varies from six to fourteen. In addition, all members of the hagfish group have an extra unpaired gill-less duct on the left side, usually at the back of the pouch series, leading from pharynx to exterior. The pouch count in cyclostomes suggests that in primitive vertebrates the number of gills was higher, but not much higher, than in modern jawed fishes.

Variations are seen in the arrangement of the porelike external openings of the gills of cyclostomes. In lampreys and the slime hag Bdellostoma each gill pouch opens by a separate orifice. But in another hagfish (Paramyxine) the tubes of each side, although separate, are crowded close together at the surface, and in the common hag, Myxine, the whole series of tubes on either side fuses externally to form a single outer opening. There is thus developed here a condition somewhat similar to that of operculate jawed fishes. Whereas in most vertebrates, and even in lampreys, the gills are situated close up to the head structures, they are removed some distance back in the hagfish group. This appears to be an adaptation to the semiparasitic life of these forms; the gill openings may remain free even if the head region of the hagfish is completely buried in the flesh of its unwilling "host."

The Ammocoetes larva of Petromyzon spends its existence half-buried in the bottoms of streams and ponds. Lacking the "tongue" device which develops in the adult for a carnivorous mode of life, it subsists on microscopic food particles. The gill system forms the food-collecting mechanism. The particles contained in the water current entering the mouth are strained in passage through the pharynx and are carried on down the digestive tract by a series of ciliated channels similar to those described for Amphioxus in an earlier chapter (p. 19).

It is apparent that this food-collecting function of the gill system in the Ammocoetes is not merely an accessory one, but a function even more important here than that of respiration—which, one would think, could have been largely carried out in the skin of these small, soft-bodied ani-

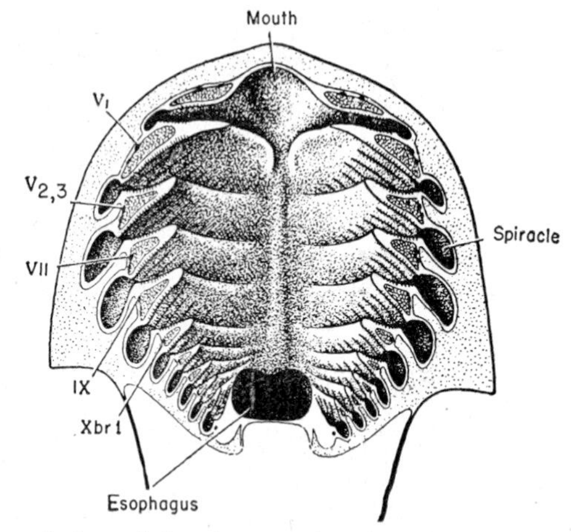

Fig. 190. Restored view of the pharyngeal region of a late Silurian ostracoderm of the cephalaspid type; the "head" viewed from below after the removal of the bones of the "throat" (cf. Fig. 16, *B*, p. 38). The mouth is tiny, the pharynx an expanded food-straining device. V_1, $V_{2,3}$, VII, IX, *Xbr1*, section of branchial nerves associated with successive gills (cf. Fig. 336, p. 553). Two gill pouches are present anterior to that which appears to represent the spiracle. (After Stensiö.)

mals. This situation leads one to the concept that in the vertebrate ancestors, food collecting was the primary raison d'être of the internal gill system, and that originally respiration may have been an accessory function only. The development of actual gill structures along the gill tubes may have been an afterthought. The fact that, as we have seen (p. 35), lower chordates likewise make their living by straining water through a highly developed system of pharyngeal gill slits adds emphasis to this concept.

Further evidence can be adduced from a consideration of the most ancient jawless vertebrates, the fossil ostracoderms (Figs. 15, 16, p. 38). A considerable amount of data is available regarding the gill system of these Paleozoic forms. The mouth was generally small; the gill region and pharynx, on the other hand, are highly developed. In some the gills opened to the exterior, somewhat as in hagfishes, by a common opening on either side; in others a series of round holes is present on the lateral or under body surface, showing a condition similar to that of lampreys. The number of gills is somewhat variable, but no ostracoderm yet investigated shows more than ten pairs—a number modestly above that of jawed fishes, but within the general range of modern cyclostomes. The internal structure is often unknown, or poorly known, but in some cases is so well preserved that a fairly accurate reconstruction of the gill region can be made (Fig. 190). The gills here were certainly in part respiratory in function, but the relatively enormous size of the gill aparatus in such a form as that figured suggests that there were other uses as well. The mouth was small, and there is no possibility that any other feeding method than that of screening food particles could have obtained.

The history of the pharynx and slits, it would thus seem, began with their development as a feeding apparatus—a function proper to the inclusion of these structures in an introductory part of the digestive tract. The passage through this food-screening system of a constant current of water from which oxygen could be readily obtained presumably led to the development here of gill structures; in ancestral vertebrates this pharyngeal system served a double function, as it does today in the larval lamprey. With the development of jaws (or cyclostome substitutes for them), the feeding functions of the gill system were, for the most part, lost; but in typical fishes the pharynx and gill apparatus are structures which, though relatively somewhat reduced in size, are still highly developed and generally indispensable for respiratory purposes.

Pharynx and Gills in Tetrapods. With the appearance of lungs and the subsequent shift of tetrapods to a terrestrial life, there came about a further decrease in the size and importance of the pharynx and gill system. In amphibians, however, we see a transitional stage. In water-dwelling larval forms of this group pharyngeal gill slits are still present. They are, however, much reduced even here. The bony operculum of the fish ancestors is absent; the anurans develop a saclike opercular structure, but this is probably a new, larval adaptation. The erstwhile spiracle seldom opens and is generally incorporated into the auditory apparatus. Four of the five typical gill slits open up in the larvae of most amphibians. But never are typical internal gills developed in the walls of these slits (although the peculiar gills of the tadpole, noted later, may be in part derived from internal gill structures). Further, the gill openings disappear with the end of the larval period; only in such urodeles as fail to metamorphose completely do certain of the slits (from one to three of them) persist. The function of the slits seems to be simply the insurance that a water current is set up near the external gills (described below), upon which larval amphibians mainly depend for blood aeration.

In amniotes, gill pouches always push outward from the pharynx in every embryo, and surface furrows push inward to meet them; but gill slits open only in a transitory manner in the embryo of reptiles and birds, and open seldom if at all in mammals. In no case is there, in amniotes, the slightest development of respiratory functions in the gills. The amniote embryo, in conservative fashion, repeats the embryological processes of countless ancestral generations; but never is there any true ontogenetic repetition of the development of the full-blown gill system found in the adult fish ancestor.

In the adult amniote, as was said in the introduction to this section, the pharynx retains hardly a vestige of its former importance. It has greatly shortened, permitting the development of the constricted neck region, traversed by an elongated esophagus. The lungs developed from the pharyngeal region in fishes, and the lung entrance is still to be found in the pharyngeal floor. Of the gill pouches, the spiracle persists in modi-

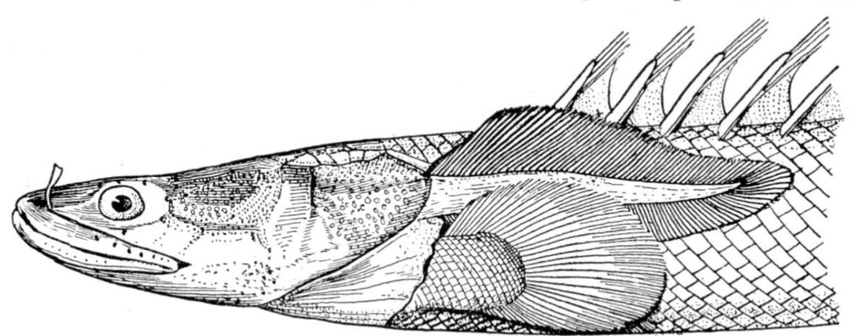

Fig. 191. Larval form of the primitive African ray-finned fish, Polypterus; the large external gill extends back above the pectoral fin. (From Dean.)

fied form in connection with the ear; the others have disappeared completely in the adult.

External Gills. In the larvae of a few bony fishes and of amphibians, accessory respiratory structures are present in the form of external gills. Among fishes they are found only in Polypterus and in the African and South American lungfishes. As seen in the lungfishes, they consist of four feathery processes growing out on either side of the "neck" region above the gill openings. In life they have a reddish color, due to the presence of an abundant supply of blood—the blood coming from the vascular arches which traverse the gill arches beneath. Polypterus (Fig. 191) shows similar outgrowths, but they are limited to a single pair.

The climatic conditions which (we have noted elsewhere) tend to favor the presence of lungs in these same animals obviously are significant in relation to the presence of external gills. Drought and a consequent paucity of oxygen in stagnating waters give a situation especially unfavorable for young and growing fishes, which have a high rate of metabolism. External gills increase considerably the respiratory surfaces and enable the larva to make better use of such poor oxygen supplies as are available. As is the case with lungs, external gills are of little importance in the ray-finned fishes. Since, however, we find them present in some members of both major groups of Osteichthyes, we can reasonably assume that (again like lungs) they were present in the ancestral bony fishes, to be lost later among the more advanced actinopterygians.

Fish external gills appear at an early embryonic stage, before the operculum is formed, and arise from the dorsal ends of successive gill bars. Possibly they are ancient specializations from the internal gill system.

To be distinguished from true external gills are filaments protruding from the internal gills in various fish types. Such external filaments are found in the early stages of all elasmobranchs. They float in the albuminous fluid lying in the egg case and appear to serve for the absorption of food material as well as for respiration; they disappear at the time of hatching. Somewhat similar, but purely respiratory, filaments are found in sturgeons and paddlefishes and a limited number of teleosts.

Presumably, external gills were present in the crossopterygian ancestors of the tetrapods, and we find them present in the urodele amphibians in a form similar to that seen in their lungfish cousins—feathery, coral-colored outgrowths, usually three pairs of them, above the gill slit region of larval forms (Fig. 53, *E*, *F*, p. 104). Throughout life certain urodeles tend to remain water-dwelling larvae; retention of external gills is a diagnostic feature of these "perennibranchiate" forms. In the Apoda there is a development of external gills similar to that of fishes and urodeles. In the Anura there is a specialized larval gill system. Four gill slits break through in the early tadpole stage of typical frogs and toads. Their external openings, however, are sheathed by a peculiar "opercular" fold so extensive that it envelops the under surface of the throat region and even the developing fore limbs; in most anurans it has but a single opening to the exterior, usually on the left side. The operculum encloses a considerable chamber filled by a mass of vascularized filaments. Part, at least, of this gill tissue is derived from external gills which bud out above the gill slits at an early stage. There are, however, additional filaments, developed later, which arise more ventrally in the gill slit region and are sometimes thought to be some sort of modified internal gill structures.

Many amphibians, including representatives of all three orders, develop in a variety of situations away from streams and ponds, such that they must breathe air from an early developmental stage. A number of different types of respiratory devices have been evolved in such cases; frequently, elaborate external gills function as embryonic or larval lungs.

Like the internal gills, these external ones have disappeared, without trace, in amniotes.

THE SWIM BLADDER

Characteristic of most actinopterygian fishes is the presence of a swim bladder, an elongate sac arising as a dorsal outgrowth from the anterior part of the digestive tube (Figs. 192, 193, *A, B*). The swim bladder is usually distensible and is filled with air or other gases; its major function

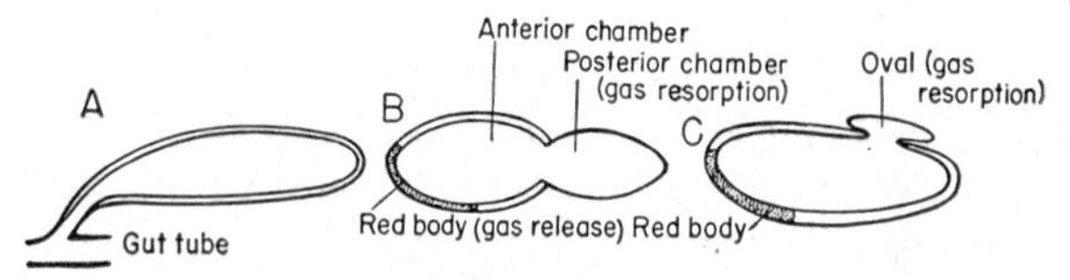

Fig. 192. Diagrammatic longitudinal sections of the air bladder in various teleosts. *A*, Primitive type, open to the gut; *B, C*, closed types, with gas-producing red body and other areas for gas resorption.

appears to be that of a hydrostatic organ. The specific gravity of a fish (or any vertebrate) is slightly greater than that of sea water. Filling or

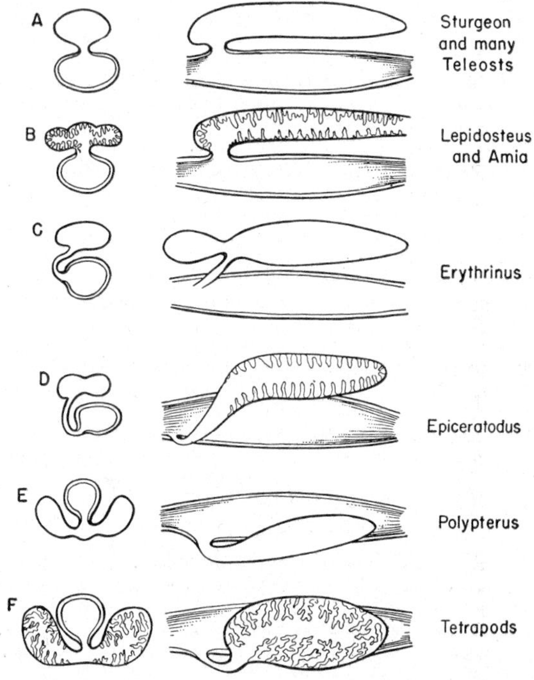

Fig. 193. Diagrammatic cross sections and longitudinal sections of the air bladder or lung in various types. *A*, Typical dorsal air bladder of actinopterygians; *B*, a more primitive bladder type, found in holosteans, with a folded inner surface capable of some breathing function; *C*, an unusual teleost type with a lateral opening suggestive of a transition from lung to air bladder; *D*, the Australian lungfish, with the bilobed lung rotated dorsally, although the opening remains ventral in position; *E*, the archaic actinopterygian, Polypterus, with a ventral lung, probably primitive and antecedent to all other lung or air bladder types; *F*, the type of lung developed in land vertebrates, with complex internal structure. (After Dean.)

emptying the swim bladder results in a change in the specific gravity of the fish as a whole, and facilitates its keeping to a depth, high or low in the water, proper to its habits. It is thus obvious that this structure is of greatest use to a form living in the ocean or other deep-water bodies.

Among actinopterygians the most primitive type, Polypterus, has no air bladder as such, but has, instead, paired ventral lungs. A typical dorsal air bladder is, however, present in all other living ray-finned types below the teleost level, and is retained in most teleosts. It is absent, for example, in the bottom-dwelling flounders, in which a hydrostatic organ would be functionally useless, but even there it is present in the embryo. It seems certain that the dorsal swim bladder was developed by the actinopterygians early in their history; it is not found in any other fish group whatsoever.

A *pneumatic duct,* leading from the digestive tract upward along the dorsal mesentery to the bladder, is present in all the more primitive teleosts and in lower ray-finned forms. Primitively, the duct is short and broad and may open from the back end of the pharynx in lower actinopterygians. In teleosts, however, it becomes a narrow and often elongate tube, opening frequently from the esophagus, but in some cases from a point farther back in the gut. Finally, in a large series of advanced teleosts the connection with the digestive tract is entirely lost.

In its more primitive form the swim bladder is an elongate oval sac, lying above the celomic lining of the abdominal cavity, below the dorsal aorta and vertebral column. Its shape is variable; frequently it is constricted into anterior and posterior subdivisions. An elastic connective tissue surrounding it allows for expansion, and both smooth and striated muscle fibers function in contraction.

In some primitive surface-dwelling forms air is taken into the sac through the pneumatic duct. But even in many forms in which the duct is present it appears to serve merely as a release valve, and in most cases the swim bladder depends upon its own internal mechanisms for supply and absorption of its gaseous content. In the anterior part of the bladder of most teleosts is found the *red body,* which owes its name to the marvelous network of capillaries—a rete mirabile—contained in it. This is the gas-producing organ; the gas includes, as in air, nitrogen, oxygen, and carbon dioxide, but in varying proportions. A gas-absorbing tissue is present in the posterior part of the bladder—in the posterior chamber when two are present. Primitively, this is a broad area of thin epithelium lying in the general surface of the bladder. In advanced forms, however, gas absorption is restricted to the *oval,* a sunken pocket which can be closed off by a sphincter.

Although the major use of the teleost air bladder is as a hydrostatic organ—a function of considerable value in such a predominantly marine group—other attributes can be ascribed in some cases. As noted in a later chapter, it is a hearing aid in some teleost groups (p. 518). Still further, it is sometimes an accessory air-breathing organ. This is particularly noticeable, and significant from an evolutionary viewpoint, in the gar pike and Amia, primitive fresh-water actinopterygians. In these fishes the inner surface of the bladder is rather cellular and lunglike in texture, quite in contrast to the smooth lining normally found in teleosts. This condition gives a strong suggestion that air breathing was the original bladder function.

In its dorsal position and single rather than paired nature the swim bladder contrasts strongly with the lungs. However, most anatomists believe that the two are in some fashion and to some degree homologous structures. We shall discuss this further after a consideration of the early history of the lungs.

LUNGS

Structurally distinct from the gills, although similar in basic function, and, like them, pharyngeal derivatives, are the lungs, which in typical

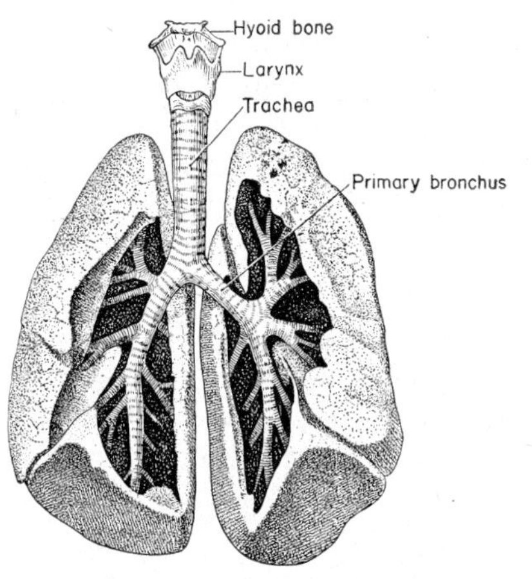

Fig. 194. Ventral view of the respiratory system of man. The lungs dissected to show the bronchi. (After Toldt.)

air-breathing vertebrates replace the gills as the medium through which oxygen is brought to the blood and tissues. In most tetrapods the air-breathing apparatus has a characteristic pattern which may be briefly sketched at this point (Fig. 194). Entrance to the air duct is gained by a median ventral opening in the pharynx, the *glottis*. Immediately beyond the glottis the duct enlarges to a chamber, the *larynx;* beyond this a ventral median tube, the *trachea,* extends backward to divide into primary *bronchi,* one leading to each lung. The paired *lungs,* primitively ventral structures, may expand to occupy a lateral or even dorsal position in the anterior part of the celomic cavities.

Embryologically, the lung structures first appear in the form of a ventral outpocketing of the floor of the throat, median in position, but in many cases distinctly bilobed at an early stage. Growing backward, this endodermal pouch structure divides into paired outgrowths from which the lungs develop; associated with it are tissues of mesenchyme origin which strengthen its walls and produce skeletal elements related to larynx, trachea, and bronchi.

Lung Structure. The inner surface of the lung and its compartments is, of course, lined by an epithelium of endodermal origin; the external surface, usually lying in the celom or a subdivision of it, is covered by the epithelium of that cavity. Within the substance of the lung are variable amounts of connective tissue, smooth muscle fibers, and an abundance of blood vessels.

The lungs function in the body economy in a fashion basically similar to that of the gills. In many reptiles and in mammals a lung is a spongy structure with air passages, *bronchioles,* radiating from the bronchi, and terminating in small pouches, the *alveoli,* through the walls of which the exchange of oxygen for carbon dioxide takes place (Figs. 194, 196, pp. 328, 332). The lung is richly supplied with blood capillaries, which branch from a special pulmonary artery coming from the aortic arch of the last gill or, in higher tetrapods, from the heart itself (cf. p. 444). From the lung the blood returns by a pulmonary vein to the heart. Between capillaries and lung alveoli there lie but thin membranes, through which diffusion of gases can readily take place.

In such fishes as bear lungs, and in amphibians, the lung sacs have a relatively simple internal construction. The efficiency of a lung depends, of course, in the main upon the amount of internal membrane surface present for gas exchange. Advanced tetrapods, with greater activity and greater needs for oxygen, increase the exchange area, not merely by increasing lung size, but also by increasing the complexity of internal subdivision. Here, as usual, we must keep in mind the problem of surface-volume relationships; in large animals the lungs must be disproportionately increased in size, or attain greater complexity of subdivision, so that the exchange surface may keep pace with the volumetric growth of tissue demanding oxygen.

Although lungs are the structures normally present in tetrapods for breathing functions, they are not universally present in amphibians, and in many air-breathing members of that class much of the oxygen intake is gained by other means. Exchange of oxygen and carbon dioxide with the air can take place at the surface of any moist mucous membrane properly supplied with blood vessels. Even though lungs are present in frogs and toads, they are of relatively little importance. If a frog is observed it will be noticed that the floor of the mouth is periodically lowered and raised. This movement expands and contracts the mouth cavity, drawing in and expelling air through the nostrils; the moist mouth epithelium acts as a substitute lung. Only relatively rarely is there a convulsive movement of the throat indicating inspiration into the lungs. Even more important than the mouth as a breathing organ is the moist skin of modern amphibians. This is richly vascular and appears to be so effective that in some salamanders "breathing" is accomplished by the skin in satisfactory fashion despite a total absence of lungs.

Lungs in Fishes. Although lungs are most characteristically developed in tetrapods, they are phylogenetically older structures. Lungs are present in the dipnoans, who gain their popular name of lungfishes because of this fact, and presumably were present in their close relatives, the ancient crossopterygians, from which land animals are descended. Still further,

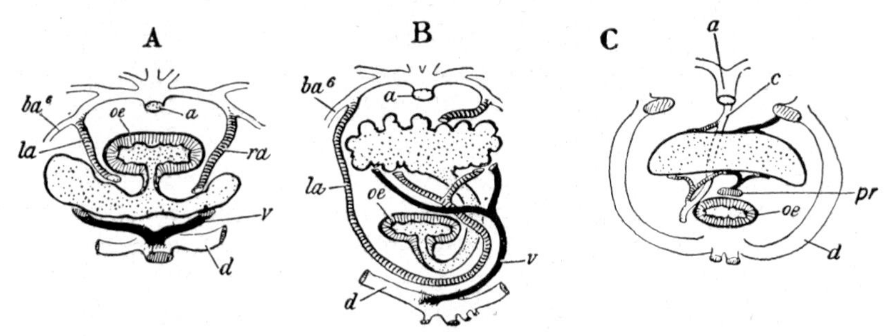

Fig. 195. Diagrammatic cross sections of bony fishes (seen from behind) to show the position of the lung or swim bladder and the blood vessels connected with it. *A*, Polypterus, a primitive actinopterygian, with a ventral paired lung; *B*, Epiceratodus, a lungfish, with a single lung dorsal in position, but with a ventral opening from the gut; *C*, a teleost with air bladder (duct lost). In the lung, the arterial supply is from the last aortic arch; in *B* the curve of the vessel from the left arch below the gut indicates the path of dorsal migration of the lung. In the teleost the arterial supply to the air bladder is from the dorsal aorta (by way of the celiac artery). In the lung the venous return is directly to the heart region. The Epiceratodus veins show an asymmetrical condition comparable to that of the arteries. In the air bladder the blood returns to the heart via the normal venous system. *a*, Dorsal aorta; ba^6, sixth aortic arch; *c*, celiac artery, whence branches run to teleost air bladder; *d*, common cardinal vein (duct of Cuvier); *la*, left pulmonary artery; *oe*, esophagus; *pr*, portal vein, draining part of teleost air bladder; *ra*, right pulmonary artery; *v*, pulmonary veins. (After Goodrich, 1909.)

lungs are present in Polypterus, the most primitive of all members of the ray-finned group of bony fishes.

Fish lungs are simple in construction. In Polypterus (Fig. 193, *E*) an opening (with a sphincter) in the floor of the pharynx leads to a bilobed sac; the lobes pass back on either side of the esophagus, and the right lobe, much the longer, continues backward and upward into the mesentery above the gut. The lung sacs are of simple construction, with but a few internal furrows. Except for the elongation of the right lobe, this would appear to be a primitive lung type, resembling the simple lung of amphibians (Fig. 193, *F*).

In the Dipnoi this generalized pattern has been modified (Fig. 193, *D*). The entrance, as always, lies in the floor of the pharynx. Instead, however, of a symmetrical development, we find that the duct to the lungs, long and unbranched, curves upward around the right side of the esophagus to a lung or lungs dorsal in position. The lung structure is more efficient than that of Polypterus, for there is some degree of internal subdivision into pockets and alveoli. In the Australian lungfish the lung is a single structure, but in the African and South American genera the lungs are definitely double. The course of the duct suggests that the dorsal position of the lungs is a secondary condition, the entire paired lung structure having twisted upward around the right side of the body. Strong proof of this assumption is afforded by the blood supply (Fig. 195). If the dorsal position were the original one, we would expect the arterial blood supply to come from adjacent vessels in some simple fashion. We find, however, that an artery from the left side of the body reaches the lungs by looping ventrally around the esophagus and then up on the right side, showing by its course the route of phylogenetic migration of the lung system.

Origin of Lungs. Although gills are present in lung-bearing fishes, the lungs are important breathing structures; indeed, the African lungfish will drown if air breathing is prevented. We have earlier commented on the climatic conditions of seasonal drought which appear to give lungs a considerable survival value in those few modern tropical fishes that have them, and on the fact that such conditions were presumably widespread in the Devonian period when the bony fishes arose. That lungs were probably present as adjunct breathing organs in early bony fishes of all types is indicated by the fact that they are a feature, not only of the Choanichthyes, but also of the most primitive of living actinopterygians. Lungs may perhaps be still more ancient in vertebrate history. There is no trace of them in the modern Chondrichthyes. But fossil evidence indicates the presence of a pair of ventral pharyngeal pouches which seem surely to be primitive lungs in a Devonian member of the Placodermi—a group at the very bottom of the scale of jawed vertebrates.

The mode of origin of lungs, phylogenetically, is none too certain. However, as we have noted, air breathing can be carried on through any moist and permeable membrane, and it may be that in some ancient fishes, much as in amphibians today, the lining of the mouth and pharynx functioned in this fashion. Any pocket-like outgrowth of the pharynx increasing the respiratory surface would, of course, be advantageous. In amphibians the lungs sometimes appear in early embryonic stages as pockets which have the appearance of an extra pair of gill pouches. Perhaps the oldest lungs were a modified pair of posterior gill pouch structures.

Lung Versus Swim Bladder. If we assume, as most students of the subject have done, that lungs and swim bladder are homologous structures, which is the ancestral type? Early writers took it for granted that, since the swim bladder is a fish structure and the lung a characteristic tetrapod feature, the swim bladder was the progenitor of the lung. But consideration of the evidence now available strongly indicates that, despite the seeming reasonableness of the older theory, the reverse is the case: the lung is the more primitive of the two. The swim bladder is found only in one subdivision of the bony fishes, the actinopterygians, whereas the most primitive member of that group, and members of the Choanichthyes as well, have a lung. The air bladder, thus, would appear to have evolved from a more ancient and widespread lung structure during the course of actinopterygian history.

The shift from lung to swim bladder in actinopterygians can be correlated with known facts in their history. The lung was an adaptation for survival of the individual among the early, Devonian bony fishes during seasons of drought. It is still of advantage to the few fishes which have lungs, living as they do in tropical areas in which such conditions prevail today. For actinopterygians generally the lung became of little advantage. Most members of this group lay thousands or even millions of eggs; individual

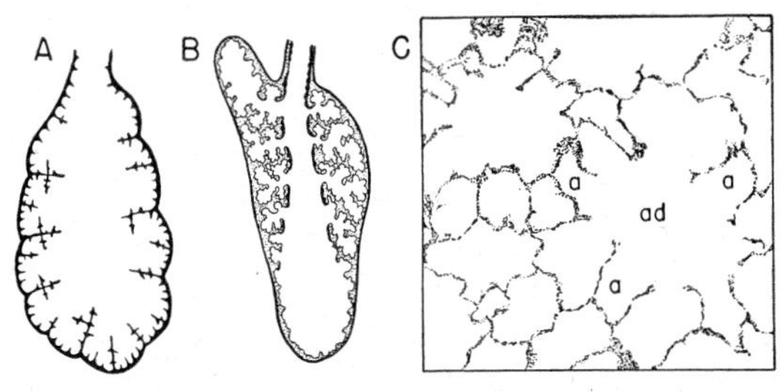

Fig. 196. Diagrammatic sections through the lung of *A*, a frog; and *B*, a lizard. *C*, Section of a small area of a human lung, × about 50, showing its complex construction. *ad*, An alveolar duct, smallest component of the duct system; *a*, one of the alveoli to which it leads. (*A* after Vialleton; *B* after Goodrich.)

survival is of little importance, for a few survivors can soon remake a population. Further, as a ventral structure, a lung makes a fish top-heavy; if a dorsal shift of the air sacs to the swim bladder position could be accomplished, greater stability would be gained. A final point to be noted is that later ray-finned fishes migrated in great numbers to the seas. In the oceans there is no drought and no need for air breathing. And on the other hand an air bladder, if present, would prove highly useful to dwellers in the deep as a hydrostatic organ.

We have no satisfactory structural intermediates in living actinopterygians between the lung of Polypterus and the swim bladder of more progressive forms. However, the lungfishes, although of course not belonging to this series, suggest the method of transition. The dipnoan lungs open by a ventral duct, but are actually in the dorsal position of a swim bladder, and in Epiceratodus only a single symmetrical dorsal lung is present. The further changes needed to parallel the presumed development of a swim bladder are shifts in the duct opening and in the blood supply, and one existing teleost shows a transition in the first regard (Fig. 193, *C*, p. 326).

Lungs in Lower Tetrapods. In amphibians the lungs show little advance over the fish condition, and are relatively inefficient structures. In some urodeles the inner walls remain smooth; generally, however, there is, in amphibians, a development of septa—ridges containing connective tissue which somewhat increase the respiratory surface and give the internal surface of the lung a honeycombed appearance (Fig. 196, *A*).

Among reptiles, the lungs of Sphenodon and many lizards are not much further advanced (Fig. 196, *B*). They are still saclike structures, with a modest internal subdivision of the walls into alveolar pockets. (Parenthetically we may note that in many limbless lizards, many snakes, and in apodous amphibians as well, only a single lung—usually the right—attains any degree of development in the slender body.) In some lizards and in turtles and crocodiles the lungs are built in more complicated fashion. The septa, originally confined to the outer walls of the sac, have grown inward and subdivided in complicated fashion to give the lung as a whole a rather solid if spongy texture. A main central tube still runs down the length of the lung as a continuation of the bronchus leading to it from the trachea. From this, branches lead outward between the major septa to subdivide and finally reach the alveoli.

In amphibians, in the absence of ribs, the lungs are filled by "swallowing" air, a force pump mechanism; in reptiles the action is that of a suction pump. In most reptiles the elongate lungs lie in the general body cavity. The action of trunk muscles raises the ribs and expands the capacity of the trunk as a whole; air is sucked into the lungs, which expand to fill the additional space. Turtles, of course, cannot move their ribs, and rely mainly on movements of the limbs, and particularly of the pectoral girdle, for expansion and contraction of the lungs.

The Lung in Birds. In birds, with their demand for great amounts of oxygen, the lung apparatus has developed into a complex series of structures. The lungs themselves are relatively small and compact. Out beyond them, however, grow numerous extensions as paired *air sacs* which invade every major part of the body (Figs. 197, *A*, 198). A pair of cervical sacs grow upward along the neck; clavicular sacs (usually fused) are found beneath the "wishbone;" two pairs of lateral sacs are present in the thoracic region; a pair of abdominal sacs extends far back among the viscera. With sac development is associated a complicated subdivision of the body cavity (cf. p. 481; Fig. 293, p. 482). From the sacs further air spaces invade the bones of the skeleton in most birds and partially replace the bone marrow. The lining of the air sacs is in general smooth; they are thus of little value as respiratory surfaces, although they may aid, as cooling devices, in the regulation of body temperature. They nevertheless play a major role in respiration. In the birds, as in other amniotes, the breathing mechanism is that of a suction pump, but the lung itself does little in the way of expansion or contraction. When the bird is at rest, air is drawn in by a forward and upward movement of the ribs which expands the trunk capacity; during flight, the action of the wing muscles causes a similar movement. Most of the inspired air passes through the lungs, by channels continuous with the bronchi, directly to

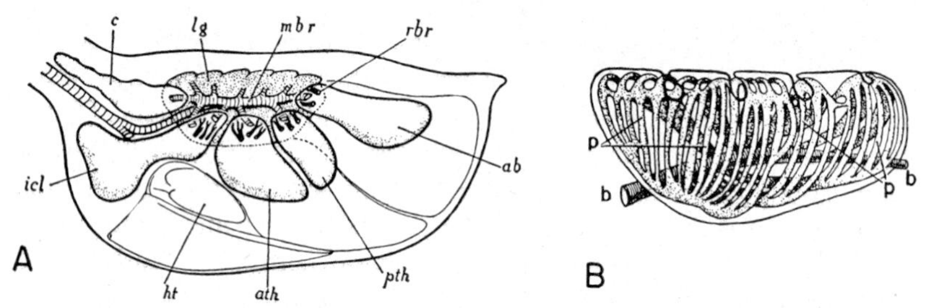

Fig. 197. *A,* Diagram of the trunk of a bird, left side view, to show the position of the respiratory sacs. *ab,* Abdominal air sac; *ath,* anterior thoracic sac; *c,* cervical air sac; *ht,* heart; *icl,* interclavicular air sac; *lg,* left lung; *mbr,* bronchus running through the lung and leading from trachea to various air sacs; *pth,* posterior thoracic air sac; *rbr,* recurrent bronchi which return air from sacs to respiratory areas of lung. *B,* Lateral view of the left lung of a bird. The primary bronchus, *b,* traverses the length of the lung to the abdominal sac. Branches are given off, leading to other air sacs and to parabronchi (*p*). These connect at both ends with the bronchi. Notches in the dorsal outline of the lung are rib impressions. (*A* after Goodrich; *B* after Locy and Larsell.)

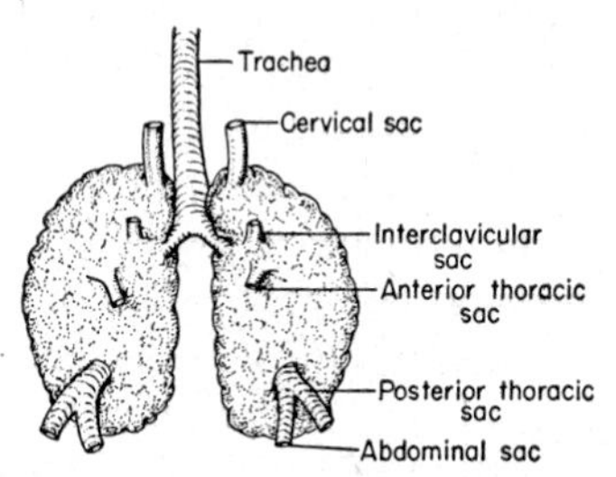

Fig. 198. Ventral view of the lungs of a bird to show the connections with the various air sacs.

the air sacs without much respiratory exchange taking place on the way. The sacs, however, are connected with the lung proper by smaller tubes in addition to the continuations of the bronchi through which the air enters them; when, on expiration, the air is forced out of the sacs, it passes by these openings to the lung tubes and respiratory surfaces.

The structure of the bird lung itself is unique among vertebrates. In other amniotes the respiratory membranes are in "dead-end" alveoli. Nothing of the sort is found in bird lungs; every passage, large and small, is open at both ends so that there is a true circulation of air (Fig. 197, *B*). Within the lung a series of tubes, the *parabronchi,* loop from the bronchi through the lung tissue and back to the bronchi; the respiratory exchange takes place in the walls of countless tiny *air capillaries* which loop outward from the walls of the parabronchi and return to them again.

The Mammalian Lung. In mammals there is seen a rather simpler lung pattern than that found in birds. The lungs are short, anteroposteriorly, but expanded, frequently with a varied subdivision into lobes; together with the heart, lying between them, they occupy most of the thorax. The lungs are finely subdivided into tiny but exceedingly numerous alveoli, to which the air passes by means of a branching series of larger and smaller bronchi and, finally, bronchioles; the alveoli cluster about the terminal

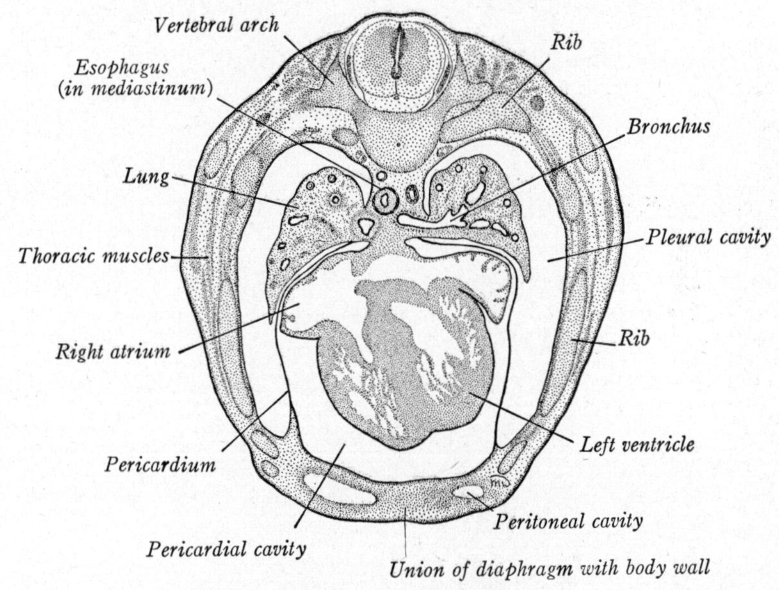

Fig. 199. Cross section of a human embryo (at eight weeks), to show the development of the lungs and of pleural and pericardial cavities. (From Arey.)

alveolar ducts like grapes on the stem. Phylogenetically, as indicated by conditions in amphibians and more primitive reptilian types, increasing complexity of lung structure was attained by increasing development of septa, large and small, which brought about finer subdivisions of the cavities of the sac. By the time the mammalian condition is reached, however, it is, so to speak, the alveolar "hole" rather than the septal "doughnut" which is prominent. The tubes and terminal alveoli form the lung pattern; the mesodermal tissues about them appear to play a relatively passive part in lung ontogeny. In the embryo the primary lung bud invades a compact mass of mesenchyme (Fig. 199), within which by repeated budding and subdivision, the adult lung structure finally emerges.

Breathing is accomplished by a highly efficient suction mechanism. In contrast with the condition in primitive amniotes, the lungs are not in the general celomic cavity; instead, a tendinous and muscular partition, the diaphragm (cf. p. 482), separates the thorax from the abdominal region, and each lung lies anterior to the diaphragm within an individual *pleural cavity,* a subdivision of the embryonic celom (Fig. 200). Expan-

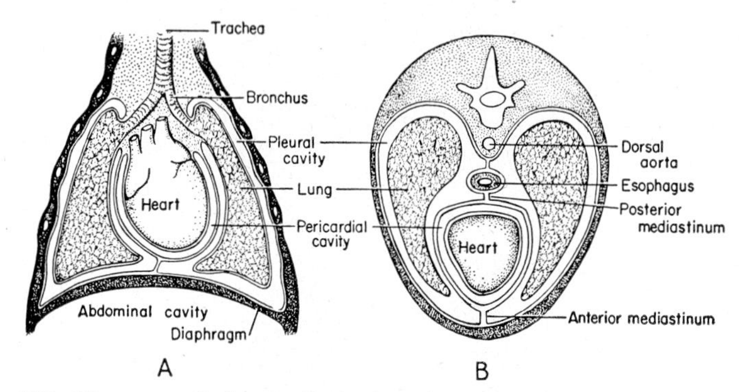

Fig. 200. Diagrammatic longitudinal, *A,* and cross sections, *B,* of the thorax in mammals to show the position of heart and lungs and of pleural and pericardial cavities.

sion of the thorax is accomplished in part by upward and outward movement of the ribs, as in reptiles and birds. More important, however, is a posterior or downward movement of the diaphragm. Each lung lies free within its pleural cavity, attached only at the anterior end, where the

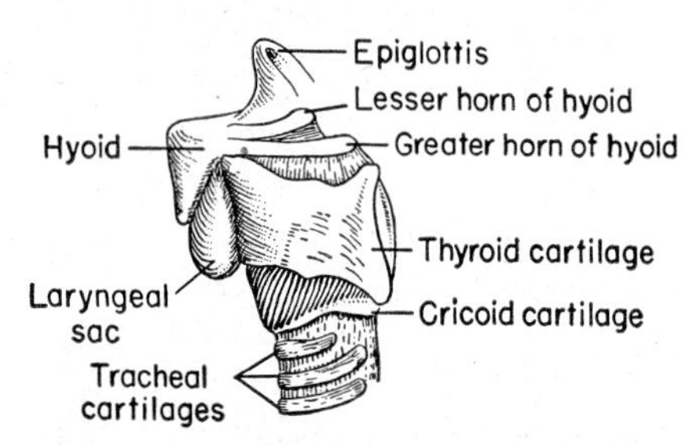

Fig. 201. The larynx of the rhesus monkey, in side view. The hyoid apparatus, with its major and minor horns, is seated above and anterior to the larynx. The rhesus monkey has a resonating chamber in the laryngeal sac. Shown are the thyroid and cricoid cartilages, tracheal cartilages, and a membrane connecting the hyoid and the thyroid cartilages. A muscle runs from thyroid to cricoid and covers most of the latter; there are several other small muscles proper to the larynx more deeply situated and not shown. (After Hartman and Straus: Anatomy of the Rhesus Monkey. Williams and Wilkins Co.)

bronchus and blood vessels enter. Rib and diaphragm movements expand the chest and consequently the pleural cavities; air is pulled into the lungs as they expand to fill the vacuum which would otherwise exist.

Larynx. (Figs. 171, p. 295; 201). With the increased importance of the lungs in land life, there develops in tetrapods a differentiation of the entrance to the lung passage, in the form of the larynx, an enlarged vestibule antecedent to the trachea. The laryngeal cavity is surrounded by a complex of cartilages or bones (Fig. 126, p. 211). Associated muscles adjust the laryngeal cartilages and open and close the glottis. This slit-like opening from the pharynx is protected in many forms by a flap of skin folding over it from the front, and in mammals develops as the *epiglottis,* containing a special cartilage.

Many vertebrates are voiceless in any true sense. This is particularly true of the salamanders and Apoda among the amphibians, and of the great majority of reptiles, although certain of these lower tetrapods can make hissing or roaring noises by a violent expulsion of air through the glottis. In frogs and toads, a few lizards, and, notably, most mammals, the larynx is a vocal organ. Voice production is accomplished through

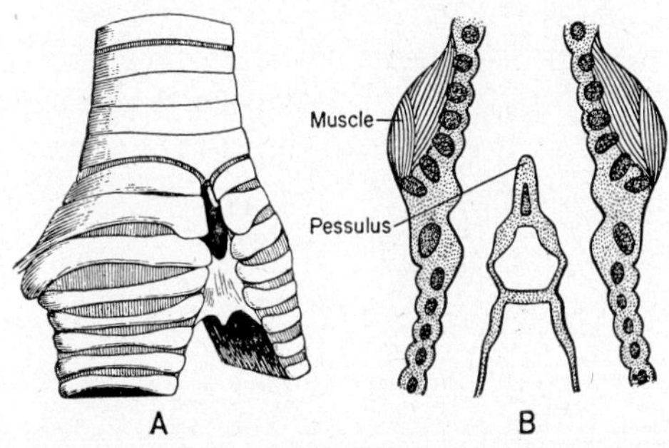

Fig. 202. The syrinx of a songbird (magpie). *A*, External view; *B*, in section. Vibratory membranes on the inner aspect of the two bronchi meet at the base of the trachea to form the median pessulus; further membranes may develop between the expanded rings at the tracheal forks. The syrinx is controlled by musculature of the hypobranchial group. (After Haecker.)

the presence of a pair of *vocal cords,* ridges containing an elastic tissue, which are stretched across the larynx. The two cords can be closely appressed to one another, and set in vibration by the passage of a current of expired air between them.

In birds the larynx is present, but lacks vocal cords, and voice production takes place in a special organ, the *syrinx* (Fig. 202). This is a structure somewhat comparable to the larynx, but situated farther down the air passage, typically at the point at which the trachea divides into the two major bronchi.

Trachea and Bronchi. In Polypterus the lung sacs expand directly from the pharyngeal opening, but in other lung bearers there is usually, for a greater or lesser distance, a median air passage, usually with a ciliated epithelium, termed the trachea; this leads in tetrapods from the larynx to the point where the two lung structures diverge (Fig. 194, p. 328). In lungfishes the duct leading upward around the esophagus to the lungs is a trachea of sorts, but one which has developed independently of that seen in land animals. In anurans and typical salamanders, in the absence of a developed neck region, the trachea is of negligible length; a formed tracheal tube is, however, present in some long-bodied salamanders and in the Apoda. In amniotes, where head and trunk are separated by a distinct neck region, a trachea is always present, its length, naturally, varying with the degree of elongation of the neck. Some birds (the swan, for example) have an extra length of tracheal tube coiled, for some purpose unknown, beneath the sternum. In almost every case the trachea is stiffened by cartilaginous structures within its walls. In amphibians these are rather irregular nodules; in amniotes they are incomplete rings (Figs. 126, 171, pp. 211, 295; 200–202).

In amphibians and some reptiles the lungs spring almost directly from the end of the trachea; in most amniotes, however, the air duct itself bifurcates before reaching the lungs, giving rise to a pair of primary bronchi which may also be stiffened by cartilaginous rings. When the internal structure of the lung attains the complicated pattern found in some reptiles and in all birds and mammals, the primary bronchi run on into the lung substance before dividing into smaller bronchi and these in turn into bronchioles and subsidiary ducts.

PHARYNGEAL GLANDS

Despite its reduction in the higher vertebrate classes, the pharynx is important in every group as the embryonic source of glands of internal secretion. In fishes certain glands of this type arise from the endodermal pharyngeal epithelium, and such structures are even more important in tetrapods. Nature is a tidy housewife, seldom wasteful, and the pharyngeal pouches appear to be areas of epithelium which are easily adaptable to glandular development. We have noted that the "normal" gill epithelium in teleosts may take on glandular activity in the excretion of salts. Other gill-derived glands are endocrine in function. As was said in an earlier section, the spiracular gill appears to have been transformed into a gland of this type in many teleosts. Other glands are derived from the endodermal epithelium of the gill pouches in all gnathostome classes. These glands are formed by thickenings of gill pouch epithelium at either dorsal or ventral borders of the pouches—thickenings which later separate from the epithelium to become varied masses of tissue buried in the throat region of the adult (Figs. 203, 204).

Thymus Gland. Almost universally present, often prominent and seemingly important, but of uncertain function, is the thymus gland. This consists of clusters of soft glandlike material found in variable arrangement in the gill or throat region. When well developed, the thymus tissue is arranged in lobules, each containing a cortex and medulla (Fig. 205). Most characteristic are *reticular cells* which form a network throughout the gland and constitute most of the bulk of the medulla. The cortex mainly consists of cells identical in all regards with lymphocytes (a common type of white blood corpuscle). In addition, there may be other histological features—elongate muscle-like cells, spherical corpuscles

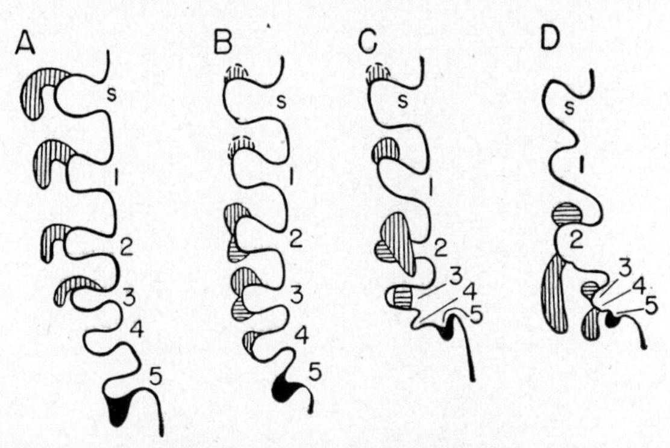

Fig. 203. Diagrams of the gill pouches of the left side of the pharynx in *A*, a shark; *B*, a urodele; *C*, a lizard; *D*, typical mammals; to show the derivation of thymus, parathyroid, and ultimobranchial bodies. The dorsal part of each gill pouch is, for purposes of the diagram, at the upper side. Broken outline, variable thymus derivatives; vertical hatching, thymus; horizontal hatching, parathyroid; solid black, ultimobranchial body. *s*, Spiracular pouch. (Mainly after Maurer.)

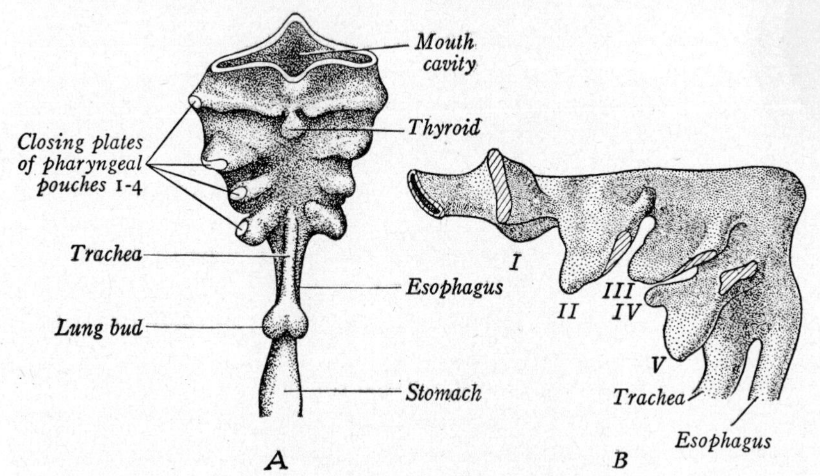

Fig. 204. *A*, Ventral view of a model of the anterior part of the gut tube and its outgrowths in an embryo mammal (Homo). *B*, Lateral view of the pharynx in a slightly later embryo. Closing plates separating tips of gill pouches from the surface are hatched; pouches, inc'uding the spiracular pouch (middle ear), indicated by Roman numerals. (From Arey.)

which perhaps are masses of degenerating cells in the medulla, and so on. A thymus has not been identified in hagfishes, and there is some doubt as to its presence in lampreys; it is, however, found in all higher fishes, usually as a paired series of soft, irregular masses of tissue situated deep to the surface above most or all of the gill slits. In amphibians the thymus is found deep within the lateral surface of the neck region, behind the pharynx. In frogs the thymus consists of a single pair of compact oval bodies; in addition, a pair of *jugular bodies* are thymus-like lymphoid structures which are gill pouch derivatives. In the other two amphibian orders the thymus is generally a rather diffuse, lobulated mass of tissue. The reptilian thymus is again found in the lateral part of the neck. In young reptiles it is frequently in the form of a strand of tissue running the length of the neck on either side. It retains this form in adult crocodilians, but usually forms more compact bodies in reptiles—two pairs of them in Sphenodon, lizards, and snakes, a single pair in turtles. In birds, as in their crocodilian relatives, the thymus is generally a long strand of tissue extending the length of the neck; it is somewhat irregularly and variably

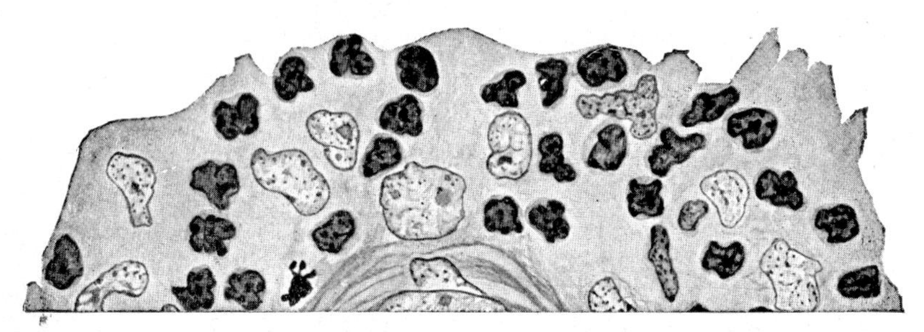

Fig. 205. A small part of the cortex of a mammalian thymus. The lightly stained cells are reticular cells derived from the gill pouch epithelium; the darker cells are lymphocytes. (After Maximow and Bloom.)

lobate. Among mammals the most common condition of the thymus is that of a pair of glands which have migrated from the throat back to a position in the anterior ventral part of the thorax, deep to the sternum and near the point at which the major vessels leave the heart. However, the thymus is sometimes cervical in position, and in some forms both cervical and thoracic bodies may be present.

The thymus is, in part at least, derived from the endodermal lining of the embryonic gill pouches, generally from epithelial thickenings formed at the dorsal side of such pouches. There is, however, much variation in the specific pouches concerned (Fig. 203). In fishes every typical gill pouch may produce a thymus bud, and even the spiracle is implicated in many sharks. Among the amphibians the Apoda produce thymic material from every embryonic pouch; but in salamanders the number concerned is usually reduced, and in frogs and toads only the first posthyoid cleft contributes. Reptiles also show a variable condition, but usually only two pairs of pouches contribute to the adult thymus. In birds the thymus usually comes from the second and third posthyoid pouches. In mammals the second postspiracular gill cleft is the usual seat of thymus origin. The mammalian thymus buds are, in contrast to the usual situation, derived from the ventral rather than the dorsal parts of the gill pouches. Another odd feature of mammals is that the cervical part of the thymus, which we have mentioned as present in various forms, does not come from the gill endoderm at all, but from an inpouching of skin in the neck of the embryo; it is, seemingly, a new development, comparable in structure to a true thymus, but not homologous with it. The reticular cells of the thymus appear definitely to be derived from the gill pouch epithelium. The lymphoid cells, however, do not appear to have this origin, but develop directly (as do other blood cells) from mesenchyme.

The thymus grows rapidly during embryonic life and is frequently a massive structure of the throat region of young or larval vertebrates (the calf thymus forms a good proportion of commercial sweetbreads). By the time adult life is reached, however, the thymus has ceased to grow and often shrinks; it thus becomes relatively, and often absolutely, smaller than in earlier stages. Its tissues tend to degenerate, and in many adult mammals the thymus has disappeared entirely.

Despite much work and study, little is known of the functions of the thymus. One is tempted to believe, nevertheless, that it has some important relation to the growth and welfare of the young vertebrate. The wealth of lymph cells associated with it strongly indicates that it may be a seat of origin of white blood cells in the young. Beyond this, we have little idea of the nature of the thymus.

Parathyroid Gland. In the different vertebrate classes we find that a variety of other glandular organs, usually of small size, may be budded off from the embryonic gill pouches and may be found in the throat region of the adult. In contrast to the thymus, these little structures retain an epithelial nature, and in section generally appear as clusters of follicles, tubules, or cell cords. In consequence, the term *epithelial bodies* is sometimes applied to all pouch derivatives other than the thymus.

The most important member of this epithelial series is the parathyroid gland, found in all tetrapod groups. The parathyroid usually consists of two (occasionally three or one) pairs of small rounded bodies found in the floor of the throat region. Although there is some variation, they usually arise in the embryo as buds from the third and fourth pairs of gill pouches. Under the microscope the mature parathyroid exhibits densely packed masses of cells, usually arranged in cords (Fig. 206). The greater part of them are *principal cells,* with a pale, clear cytoplasm, which appear to be the essential elements of the gland; in addition there is, in mammals, at least, a small percentage of *oxyphil cells,* filled with granules which stain with acid dyes.

The name "parathyroid" is an unfortunate one; it is due to the fact that in many mammals (including man) the parathyroids are embedded in the thyroid gland, although in most other vertebrates they are independent of neighboring structures. Functionally they are quite distinct from the thyroids. Attention was called to them when it was discovered that extirpation of parts of the human thyroid containing these little bodies resulted in death of the patient because of disturbance of calcium metabolism. They elaborate a powerful hormone which functions in keeping up the concentration of calcium in the blood plasma.

No specific predecessors of the parathyroids are known in fishes. Why, with the shift from aquatic to terrestrial life, there became necessary a

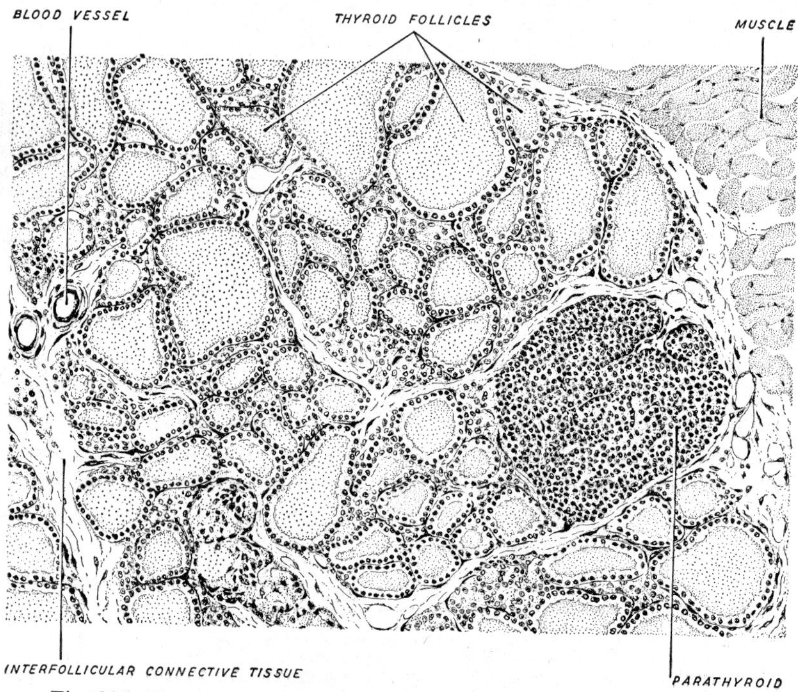

Fig. 206. Thyroid and parathyroid tissues of the rat. (From Turner.)

mechanism for calcium regulation for which there was no need in fishes, is an interesting but unsolved problem.

Ultimobranchial Bodies. In all groups of jawed vertebrates (apart from teleosts) there are present small obscure types of glandular epithelial bodies which arise in the neighborhood of the last pair of gill pouches and are hence termed ultimobranchial bodies. They arise as small, pouchlike structures in the embryo in a fashion rather comparable to that of the glands already described (Fig. 203). In some cases they are seen clearly to be derivatives of the last pouch; in others they are more independent and more posteriorly placed in the embryonic pharynx, and are sometimes regarded as vestiges of lost posterior gill slits. The fate of these outgrowths is variable. They usually form small clusters of glandular follicles, tubes, or cell strands. These frequently lie in the neighborhood of the heart cavity in amphibians; in reptiles they lie in the neck; and in mammals they are incorporated in the mass of thyroid tissue. Little is known with certainty of the function of these small structures, probably endocrine in nature.

Thyroid Gland. In contrast to the glands already mentioned, the thyroid, although a pharyngeal derivative, does not arise from the gill pouches, but is formed by a median ventral outgrowth from a point well forward in the pharyngeal floor (Fig. 204). Found in every group of vertebrates, the thyroid is situated in fishes beneath the gill chamber, in tetrapods generally ventral to the windpipe at some point along the length of the neck. The gland remains a single structure in many fish groups, in reptiles, and in mammals, although distinctly bilobed in many cases; its name is due to its bilobed, shield-shaped appearance in man. In teleosts it has separated into right and left halves, and these in turn may be broken up into smaller units in some forms. In amphibians, and likewise in birds, the thyroid is paired. Small accessory masses of thyroid tissue are present in a variety of amniotes; these are sometimes termed parathyroids, making for regrettable confusion with the different gland described above.

The thyroid is a major gland of internal secretion, itself regulated in great measure by the pituitary body at the base of the brain. Its secretions are of great importance in the regulation of body metabolism and growth; their effect upon the metamorphosis of amphibian larvae is particularly striking, and in man the abnormal conditions caused by underactivity or overactivity of the thyroid are familiar. The hormone extracted from the gland has as its active principle *thyroxine,* a simple organic compound containing iodine. A small but steady supply of iodine for the manufacture of this hormone is in consequence necessary in the food supply of any vertebrate.

The gland, rich in blood vessels, is composed of numerous small, spherical *thyroid follicles,* bound together by connective tissue (Fig. 206). Each follicle consists of a cuboidal epithelium of secretory cells, surrounding and discharging into a central cavity filled with a "gluelike" *colloid substance.*

The thyroid has a pedigree which stretches far back in chordate history. None of the lower chordates has a thyroid gland or any organ producing a thyroid hormone. Its antecedents, however, are recognizable in lower chordates in the structures to which the term *endostyle* is applied.

In both Amphioxus and tunicates there are present ciliated and sometimes glandular channels along which food particles strained out of the water current are carried back to the intestinal region. Such a channel in the floor of the pharynx of Amphioxus (Fig. 5, p. 18) is frequently termed the endostyle. In tunicates the endostyle is a similar prominent groove along one margin of the pharynx—that margin which, as determined by the build of the embryo, is morphologically ventral (Fig. 6, p. 20).

In the Ammocoetes larva of lampreys, alone among living vertebrates, we find a similar feeding habit and a series of ciliated grooves dorsal and ventral, similar to those of Amphioxus. The ventral groove terminates posteriorly, however, in a deep pouch in the pharyngeal floor (Fig. 189, *A*, p. 320). This pouch is subdivided in complicated fashion, and contains a number of epithelial cell types, some of which are glandular; their action is unknown. The term endostyle is applied to this pouch (although, of course, the homology of this gland with the groove from which it arises is not an exact one).

The endostyle, including the Ammocoetes gland, is thus a median ventral pharyngeal structure, comparable in position to the thyroid gland. Is it truly homologous? The lamprey gives us a positive, conclusive answer. At metamorphosis the larval endostyle pouch closes off from the gut and breaks up into a series of follicles indisputably thyroid in structure and producing a true thyroid hormone.

It would appear that the history of the thyroid gland has been one in the course of which a feeding structure has been transformed into a gland of internal secretion. It is of great interest that clinically the thyroid hormone can be taken by mouth, whereas other hormones, when introduced into the digestive tract in an unaltered state, are destroyed by gastric and intestinal juices. Presumably at some stage in chordate evolution, secretion began in the endostyle while this was still a part of the feeding mechanism, the thyroxine produced being taken on into the gut with food materials. With the abandonment of filter feeding the endostyle lost its original function as a constituent of the pharynx; the thyroid gland has, as an endocrine structure, continued its glandular activity.

CHAPTER XI

DIGESTIVE SYSTEM

THE MOUTH AND PHARYNX, described in the last chapter, are part of the digestive tube. They constitute, however, only its forward outposts, with gathering of food materials as their original major duty. The business of digestion is the function of the remainder of the digestive tract, to which the ancient and simple Anglo-Saxon term "gut" may be properly applied. The gut, so limited, together with its outgrowths—liver and pancreas—will be considered in the present chapter.

Gut Functions. The functions of the gut may be considered under four heads: (1) *Transportation.* Once food materials are gathered, they must be carried along the "dis-assembly line" of the successive gut sectors, and wastes must be ultimately disposed of posteriorly as the feces. (2) *Physical treatment.* Although food may be taken in as small particles or brought down to small size in the mouth by chewing, food materials frequently pass back into the gut in large masses, the size of which must be reduced before an efficient chemical attack on them can be made. Further, fluid materials must be added to bring the food to a pasty condition and thus facilitate digestive activity. (3) *Chemical treatment.* This is digestion in the technical sense, the breakdown of potentially useful "raw materials" in the food to relatively simple substances which can be taken into the body. (4) *Absorption.* When this chemical breakdown has been accomplished, the products are absorbed by the intestinal wall for circulation to body cells and storage areas.

Transportation is the function of the visceral musculature which surrounds the digestive tube for its entire length. At the anterior end of the gut proper (the region of the esophagus), striated musculature, an extension of that of the pharynx, may be present to a variable degree. Along the remainder of the tube the musculature consists of smooth muscle cells, most typically including layers of both longitudinally and circularly arranged fibers (Fig. 209, p. 350). These gut muscles are innervated by nerves of the autonomic (involuntary) nervous system which may either stimulate or inhibit their movement. To a considerable extent, however, they operate quite independently of any control by the central nervous system, being stimulated by the contraction of neighboring fibers or through a local nerve net, which is in turn stimulated by the chemical and physical conditions present locally in the gut. The major muscular activity causing movement of food materials is *peristalsis*—successive waves of muscular contraction causing constrictions of the gut which travel backward and push the food before them.

Food taken into the digestive tract must be reduced to a soft pulp, the *chyme,* before effective chemical attack can be made. This reduction is mainly accomplished by muscular action. Most effective here are (in contrast to peristalsis) rhythmic contractions of the gut musculature in food-containing areas. The gut epithelium includes throughout its length mucus cells or mucus-producing glands; the slime they produce facilitates the passage of food and, further, gives an appropriate consistency.

In Chapter IV we reviewed the food requirements of the cells of the body. Needed are oxygen, water, a variety of inorganic salts, glucose, fats, amino acids, and small amounts of other organic compounds—the vitamins—which the cells themselves cannot manufacture. Oxygen is obtained via the gills or lungs; for all other materials reliance is had on the food materials brought to the digestive tract and after absorption carried on to the cells by the circulatory system. Certain of the requirements—water, simple salts—can be obtained from nonliving sources: sea water, for example. The organic materials, however, can be obtained only from other living organisms—from the plants by which alone most needed substances are manufactured, or from animals which have already obtained these substances from plants.

Water, necessary salts, and vitamins as well, are readily absorbed into the circulation by the lining of the intestine. But the situation is different as regards the glucose, fat materials, and amino acids needed in the body. The body cells utilize these in general as building stones to form more complex materials, and in the animal food eaten they are present, naturally, in similar complex form. As such they would not, of course, be immediately useful to the body cells if they were taken into the circulation; but more than that, they are incapable of being absorbed into the gut lining while in this state. A chemical breakdown is requisite; this is accomplished by the action of the digestive enzymes, together with accessory substances which activate these enzymes or facilitate their work.

We have previously mentioned the nature of enzymes, organic catalytic agents. Varied enzymes are present in every cell to bring about the internal chemical changes associated with life processes. In the gut we have a special situation, for in the digestive tube and its glands there is formed a variety of enzymes which act outside the confines of the individual cell. Such enzymes are discharged into the gut cavity and attack the organic compounds in the food materials, reducing the complex carbohydrates to simple sugars, the fats to glycerine and fatty acids, and the proteins to amino acids. When the work of the enzymes has been accomplished, the organic materials are reduced to molecules of a size and nature which enables them to pass through the cells lining the intestine, and to enter the circulation through the blood vessels and lymphatics surrounding that organ. Sugars and amino acids pass via veins to the liver and thence to the heart and body; part of the fats, reconstituted once the intestinal lining is passed, reaches the general circulation by way of lymphatic vessels.

Development. In Chapter V we described the earliest stage in the development of the digestive tract: the formation, in one fashion or another, of an archenteron, or primitive gut.

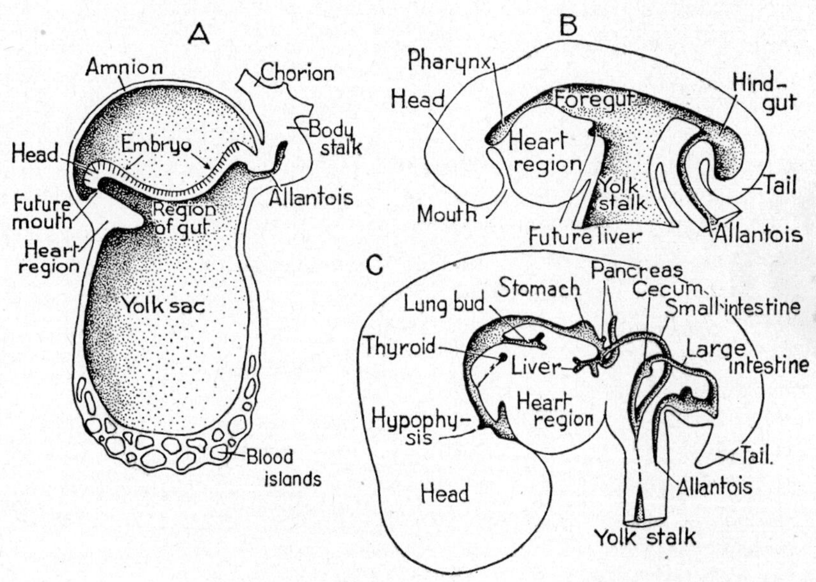

Fig. 207. Diagrams to show the development of the digestive tract in a mammal (Homo). *A*, Stage somewhat later than that of Figure 45, *C; B*, stage slightly later than that of Figure 45, *D; C*, an embryo of about the age of that in Figure 57. (After Arey.)

In Amphioxus (Fig. 47, p. 97) this is a simple cylindrical pouch, lined by a single layer of endodermal cells which contain a modest amount of yolk. In mesolecithal egg types, such as amphibians, the early gut is similarly constructed, except that the floor of the gut is not a simple epithelium, but a thick mass of endodermal cells rich in yolk. The embryo's belly is in consequence a paunchy, distended structure (Fig. 53, p. 104) until, at a later stage, this yolk material has been absorbed into the animal's circulation and has contributed to its growth.

In telolecithal eggs the great amount of yolk material has, we have seen, caused a radical change in the nature of the embryonic gut. Instead of forming a mass of yolky endodermal cells, the yolk fails to cleave at all, and the gut lining is at first simply a flat plate lying over the inert yolk. As the embryo develops, the endoderm grows down around the yolk, to enclose it eventually in a yolk sac which is an extension of the gut cavity (Figs. 45, p. 95; 54, *D*, *E*, p. 105; 57, p. 108). The yolk is gradually digested, the sac being gradually reduced in size as absorption proceeds. Above the yolk sac the embryo proper has meanwhile begun to take form; the endoderm underlying its other structures gradually separates from its yolk sac extension, although the two regions remain connected by a yolk stalk. The gut proper eventually becomes tubular, comparable at last to that of mesolecithal types. Mammals, despite the secondary absence of yolk, follow the telolecithal pattern of gut development and produce a yolk sac, although an empty one (Figs. 45, *D*, 57, 207). In amniotes we noted (Fig. 45, *B*, *D*) the presence of a second diverticulum from the embryonic gut, the allantois.

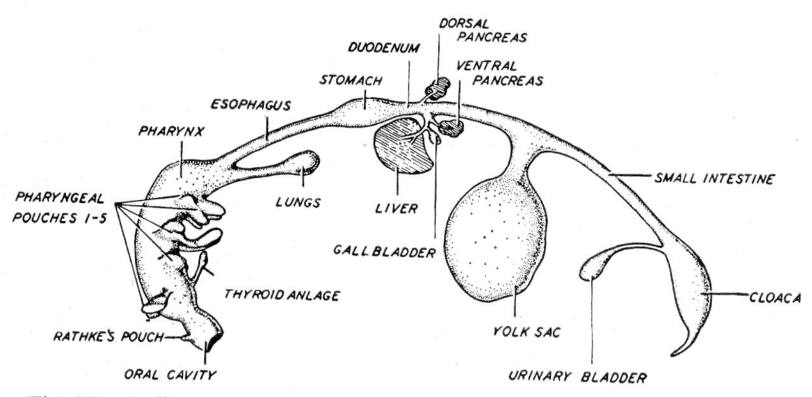

Fig. 208. A diagram of the digestive tract and its outgrowths in an embryonic amniote similar to that of Figure 207, *C*, but with the structures concerned shown as solid objects rather than in section. (After Turner.)

For part of its development the embryonic gut, whether in mesolecithal or telolecithal types, usually has the essential form of a tube closed at both ends. We noted in the last chapter the development anteriorly of a stomodeal depression of the ectoderm which forms the mouth cavity and eventually breaks through to connect with the pharyngeal region of the gut (Figs. 168, p. 292; 207). Posteriorly, mesolecithal types have in the gastrula stage a posterior opening, the blastopore, in approximately the region of the anal or cloacal aperture of the adult. But even in most mesolecithal types the blastopore closes over in an early postgastrula stage; in telolecithal eggs the blastopore is at the best transitory and atypical in its appearance, and, posteriorly as anteriorly, the gut tube ends blindly for much of embryonic life. Analogous to the development of the stomodeum anteriorly is the appearance posteriorly of the *proctodeum,* an indipping pit of ectoderm beneath the tail, which comes to be closely apposed to the distal terminus of the embryonic gut, separated from it by a membrane which eventually disappears.

Within the gut tube, postgastrular phenomena consist primarily of a lengthening of the tube, a serial differentiation of successive regions into adult organs, and the building up of the complex gut wall structure (Figs. 207, 208). The most anterior gut compartment, the pharynx, has been described previously. The most posterior subdivision is the cloaca; this becomes associated with urinary and genital as well as digestive systems, and its consideration is best postponed until these other systems have been described. Subtracting these specialized terminal areas, there remain for consideration here the major segments of the endodermal tube to which the term gut is for present purposes restricted.

Gut Regions. A study of the higher vertebrates gives one the impression that the succession of structures along the gut tube is consistent and uniform: esophagus, stomach, small intestine, large intestine, and on to rectum and anus. But when we extend our view to the lower groups, the picture becomes confused (Fig. 210). As inspection of a shark or cyclostome gut makes evident, the distinction between large and small intestine breaks down completely in primitive vertebrates; the esophagus may be absent as a distinct region; and in some fishes the stomach does not exist. In Amphioxus and cyclostomes the postpharyngeal gut is essentially a single tubular unit; most familiar landmarks are absent.

It is, however, possible to establish in most vertebrates one constant and definite point of division along the length of the gut. Between stomach and intestine there is generally found a distinct *pylorus,* a constriction in the gut tube which guards the gateway into the intestine and is usually furnished with an efficient band of sphincter muscle fibers. In some fish types the pylorus may be poorly marked or absent, but the fact that the bile duct from the liver always enters the intestine a short distance beyond the pylorus enables us, even so, to establish a line of demarcation.

The region anterior to the pylorus may be termed the *foregut;* the intestinal area beyond, the *hindgut.* In most vertebrates there is some overlapping of function between the two areas, but primitively, it would seem, the two served different purposes: the foregut was merely a short and unimportant connecting link with the pharynx; the intestine, forming the hindgut, was solely responsible for chemical treatment of food as well as its absorption. In the course of vertebrate evolution, however, the foregut has assumed a greater importance; stomach and, in land forms, esophagus have grown to be gut segments of prominence, and the former has assumed a true digestive role.

Gut Tube Structure. Although highly variable from region to region and from form to form, the gut tube is generally formed, from the inside outward, of four successive layers, or "tunics" (Fig. 209). (1) The *tunica mucosa* includes the internal epithelium and a connective tissue layer beneath it; this latter often contains smooth muscle fibers. The epithelium is the sole part of the tube formed from the endoderm. Anterior and posterior boundary zones (esophagus, rectum) may develop a stratified epithelium; over the greater length of the gut, however, the lining is of a simple columnar type. The cells of this epithelium may have varied functions, as secretory cells or (posteriorly) absorptive areas. The epithelium may produce a variety of glandular structures, some the seat of formation of enzymes or other chemically active materials, others producing (as do

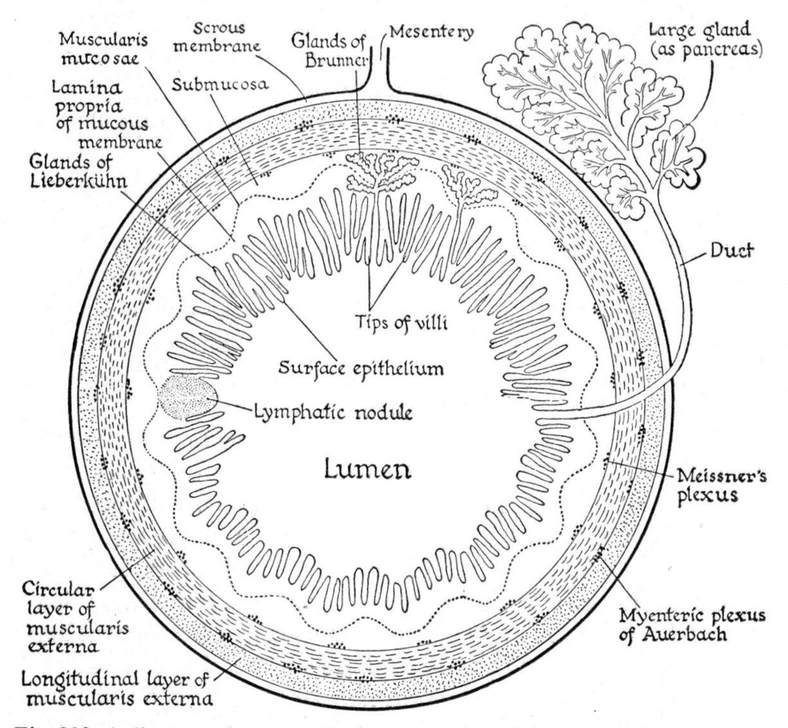

Fig. 209. A diagram of a generalized cross section of the gut. In the upper half of the drawing the mucous membrane is provided with glands and villi; in the lower half it contains only villi. Meissner's plexus and the myenteric plexus are part of the autonomic nervous system. The glands of Brunner and of Lieberkühn are two types of glands characteristic of the mammalian small intestine. (From Maximow and Bloom.)

some, at least, of the superficial cells in every gut region) a mucous material useful in softening and transporting food. (2) The *submucous tissue* is usually a thick layer, mainly connective tissue, but containing numerous small blood vessels and nerve cells and fibers. (3) The *muscular tunic* generally includes two prominent layers of smooth muscle: internally a circular layer capable of constricting the gut, externally a longitudinal layer capable of locally shortening its length. Between and adjacent to the two layers there may be diffuse plexuses of cells and fibers of the autonomic nervous system (cf. p. 545). (4) For most of its length the gut, lying in the abdominal cavity, is surrounded externally by a *serous tunic* consisting of celomic epithelium underlain by connective tissue.

Esophagus. Not only in Amphioxus and cyclostomes, but in a fair number of jawed fishes as well—chimaeras, lungfishes, certain teleosts—

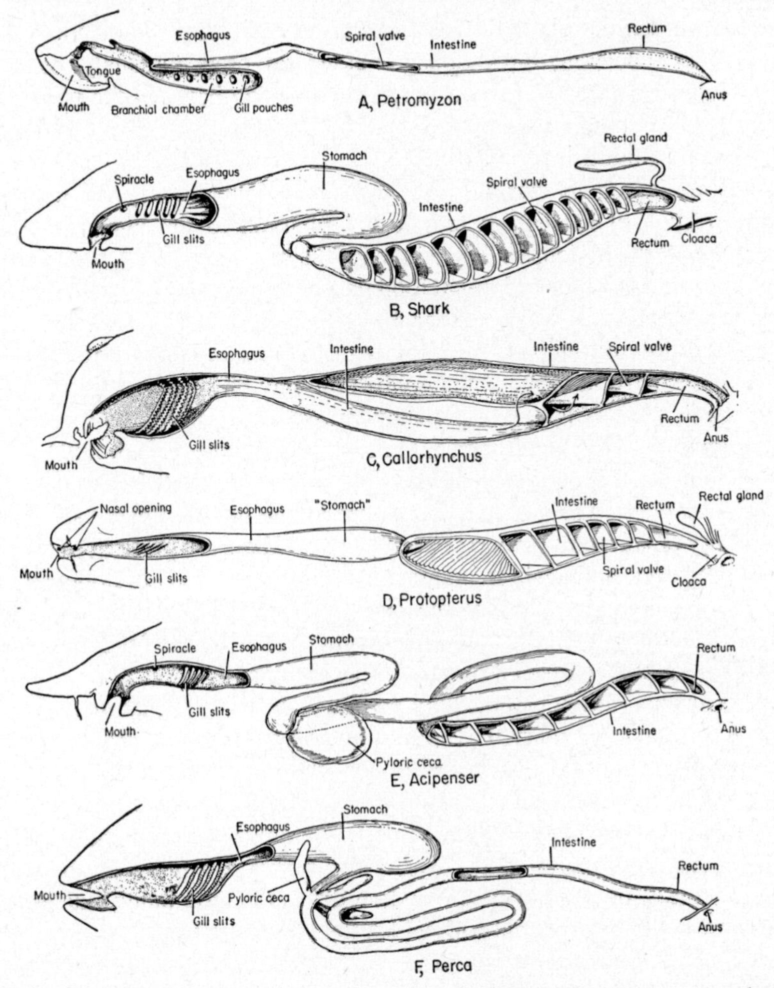

Fig. 210. Digestive tracts of *A*, a lamprey; *B*, a shark; *C*, a chimaera; *D*, a lungfish; *E*, a sturgeon; *F*, a teleost (perch). The "stomach" of the lungfish is nonglandular and is simply a somewhat enlarged section of the esophagus. (From Dean.)

the entire foregut consists merely of a simply constructed piece of tube interjected between pharynx and intestine (Fig. 210). Although in more progressive vertebrates stomach as well as esophagus develop from this region, no stomach is present in the forms listed; the term "esophagus" may perhaps be applied to this simple gut segment. In all other living fishes

—elasmobranchs and most ray-finned forms—a stomach is present in the foregut, leaving a short and ill-defined area anterior to it to be considered an esophagus.

It is not until we reach land vertebrates that the esophagus assumes prominence. Here, with the loss of gill breathing, the pharynx becomes restricted in length to a short area within the confines of the head. With pharyngeal reduction, the esophagus elongates proportionately, as the neck region develops, to form a tubular connection between the pharynx and the remaining gut segments, concentrated posteriorly in the abdominal region of the trunk.

Only exceptionally has the esophagus any function beyond that of transportation of food from mouth and pharynx to the important gut structures of the abdomen. The epithelium is in general of a tough stratified type, able to withstand rough food particles; there are generally longitudinal internal folds which render the esophagus distensible. Glands are in general little developed. In a number of elasmobranchs, amphibians, and reptiles part, at least, of the epithelium may be ciliated; but, in general, food transport is due to peristaltic movements, and there is a stout muscular coat. This is, of course, visceral musculature (cf. p. 282), but the esophagus lies at the boundary zone between striated and smooth muscle regions of the gut, and the nature of the muscle fibers varies. In a majority of vertebrate groups smooth muscle fibers are present; in many fishes, however, and again in mammals there is a notable development of striated musculature. Most of the digestive tract lies more or less free within the abdominal cavity; the esophagus, however, lies for most of its length within the compact structures of the neck and "chest."

Two unusual developments may be noted. In the adult lamprey (Fig. 189, *B*, p. 320) the pharynx is split into two parts. As we have seen, the lower segment is a blindly ending pouch, the upper part a tube leading directly from mouth to gut and hence giving the appearance of an elongate esophagus. In birds there develops part way down the esophagus a distensible sac, the *crop* (Fig. 211, C), which serves as a place for the temporary storage of grain or other food; in doves the crop lining exudes, in both sexes, a milky material with which the young are fed.

Stomach. The stomach, found in most (but not all) vertebrates, is a muscular, pouchlike expansion of the foregut, lying in the anterior part of the abdominal cavity. Its epithelial walls are generally highly folded and studded with mucus cells; in specific areas, deep-seated glands produce digestive secretions as well. Accustomed as we are to thinking of the stomach as a normal and useful part of the gut, its absence in a variety of types seems at first an anomalous situation. It is absent in Amphioxus; in true vertebrates it is absent not only in the cyclostomes, but also in chimaeras and lungfishes. Except for a fair number of teleosts (in which its loss may be secondary), all other vertebrates have a stomach which, despite great variations in form, has much the same functions in all cases.

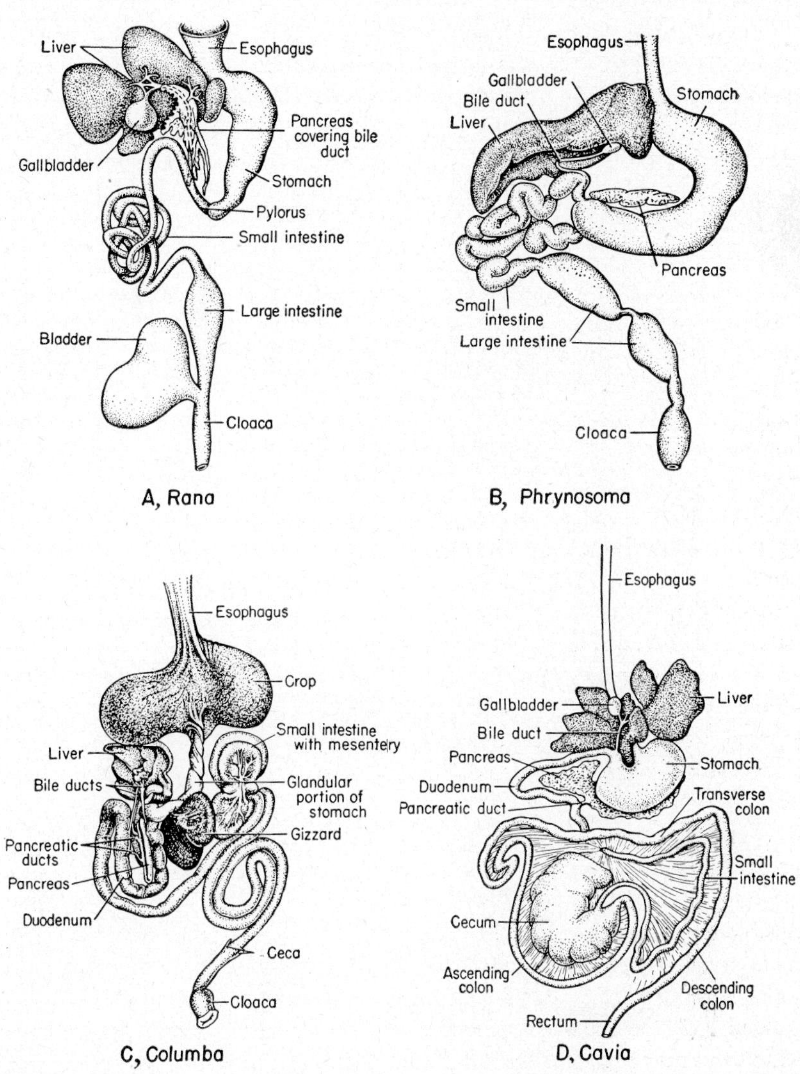

Fig. 211. Diagrams of digestive tract and appendages, seen in ventral view in *A*, a frog; *B*, a reptile; *C*, a bird (pigeon); *D*, a mammal (guinea pig). (*A* after Gaupp; *B* after Potter; *C* after Schimkewitsch.)

The stomach serves as a place for storage of food awaiting reception into the intestine; for the physical treatment of this food; and for initial chemical treatment of protein food materials. These functions give some suggestion of the history of the stomach. In "food strainers" such as

Amphioxus (and, we believe, the ancient jawless ostracoderms) food particles were small and collected more or less continuously; they could be passed on directly to the intestine without the necessity of storage or preliminary treatment. Quite in contrast are the food habits of such a predaceous jawed fish as a shark. The food intake is irregular; a large quantity may be taken in at one time and must be stored until the intestine

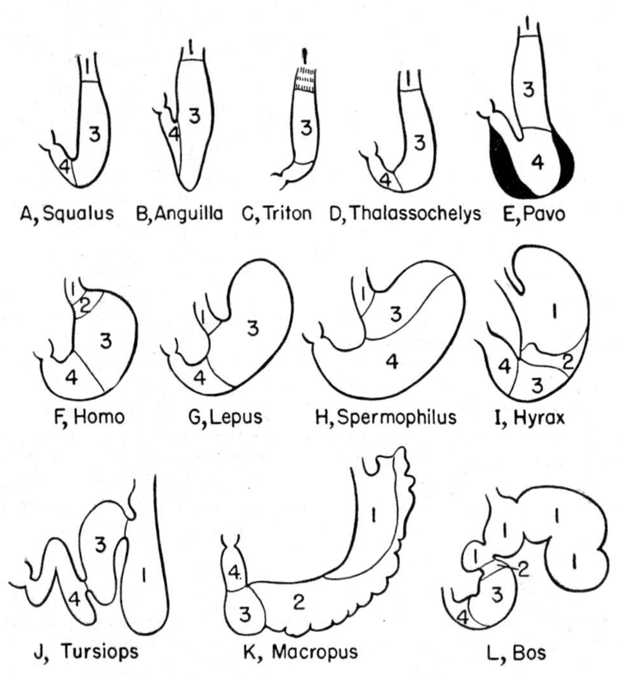

Fig. 212. Diagrams to show stomach form and nature of internal lining in *A*, a shark; *B*, a teleost (eel); *C*, a salamander; *D*, a turtle; *E*, a bird (peacock—the thickened wall of the gizzard is indicated); *F*, man; *G*, a hare; *H*, a ground squirrel; *I*, the coney of Africa; *J*, a whale; *K*, a kangaroo; *L*, a cow. *1*, Epithelium of esophageal type (ciliated in *C*) which may penetrate into stomach, particularly in mammals; *2*, cardiac epithelium (found only in some mammals); *3*, fundus epithelium; *4*, pyloric epithelium. (After Pernkopf.)

can take care of it. Further, food may be bolted in large chunks upon which the intestinal juices cannot act effectively without proper preparation. We may reasonably conclude that the primary functions of the stomach, when developed in the course of early jawed fish history, were, first, food storage and, secondly, physical treatment of the stored materials. The introduction of digestive ferments was presumably a phylogenetic "afterthought."

As regards external shape (Figs. 211, 212), the central type of stomach structure is that seen in forms as far apart phylogenetically as sharks and men. At the end of the esophagus the digestive tube curves to the left, "descends" and expands into a major sac, and then "ascends" at the right to end at the pylorus. Topographically, the proximal end of the stomach, closest to esophagus and heart, is termed the *cardiac region;* the expanded middle part, the *fundus;* the distal limb, the *pyloric region.* Stomach shapes may vary greatly from group to group. It should be noted that, as found from studies on man, the shape of the stomach in a dead, dissected animal may be quite different from that in life, and that violent motions with marked changes of shape may occur from moment to moment in an active stomach (Fig. 213).

Highly important is the nature of the internal epithelium lining the various regions of the stomach. Four types may be distinguished. (*a*) *Esophageal.* The proximal end of the stomach in many mammals has walls similar to the nonglandular, stratified epithelium of the esophagus; such a stomach region is essentially an expanded part of the esophagus. (*b*) *Cardiac.* In mammals alone there is usually distinguished a transitional

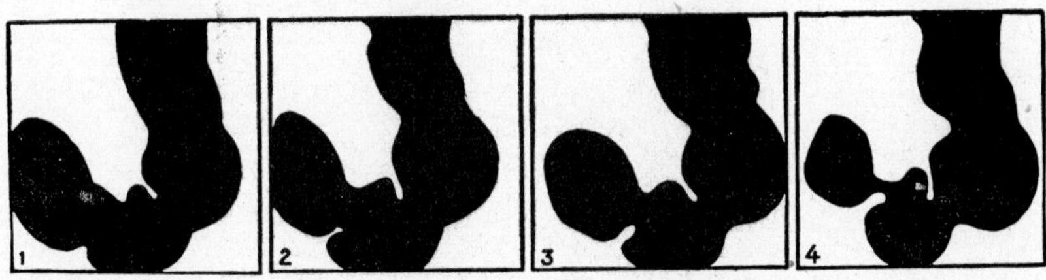

Fig. 213. Skiagrams of a human stomach taken at intervals after food, showing its variable shape when active. (After Cole, from Fulton.)

region at the proximal end of the stomach in which the epithelium contains glands, but glands of no chemical importance. In this and the following stomach regions the endodermal gut lining is a simple columnar epithelium, although often folded in complicated fashion. (*c*) *Fundus* (Fig. 214). Although the intestine is entirely competent to deal by itself with chemical digestion, the stomach, when present, produces in the fundus region digestive enzymes, prominently *pepsin,* which aids in the breakdown of proteins into a simpler and more soluble form preparatory to intestinal digestion. The fundus region is characterized by the presence of numerous tubular glands in which two kinds of cells, readily distinguished histologically, occur. Of these, the so-called *chief cells* produce pepsin; *parietal cells* yield hydrochloric acid and thus produce in the gastric juice an acid condition favorable for the work of pepsin. (*d*) *Pyloric.* This epithelial type, occurring in the distal part of the stomach, is glandular, but, like the cardiac region, produces a fluid which has little or no chemically active materials.

Although three of the four epithelial types just described are given names used in stomach topography, their distribution is not necessarily correlated with the regions so named. External appearance may be misleading as regards histological conditions. Thus, for example, the stomach of a mouse is shaped much like that of a man, but, in contrast to human conditions, the greater part of the stomach lining of a mouse is of esophageal type.

There are numerous variations among vertebrates in stomach form and epithelial linings (Fig. 212). As regards general proportions, shape may vary greatly even in forms with a simple type of stomach. Teleosts, for example, often develop a deep V-shaped fundus, while a variety of fishes,

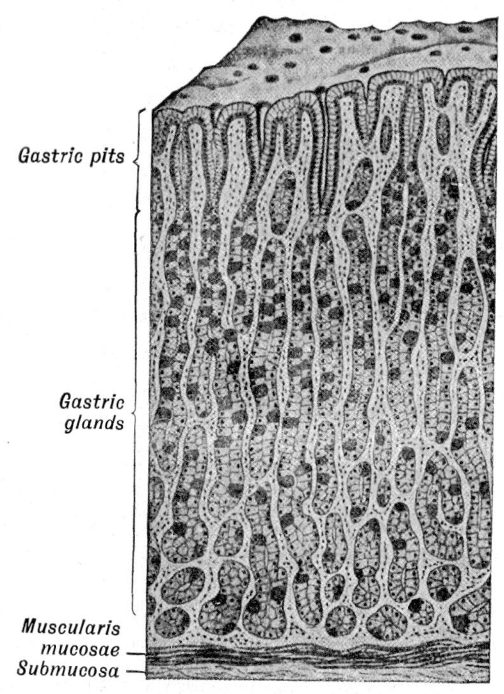

Fig. 214. Semidiagrammatic view of a portion of the fundus region of a mammal. In the glands the dark stained cells are the parietal cells, the lighter ones the chief cells. (From Maximow and Bloom, after Stöhr-v.Möllendorff.)

long-bodied amphibians, and snakes have a perfectly straight cigar-shaped stomach. The stomach has stout muscular walls which through periodic regional contraction may squeeze and soften the food and aid in preparing it for suitable intestinal treatment. Usually this muscular tissue is more or less evenly distributed; in birds, however, it is concentrated in a distal stomach compartment, the *gizzard* (Figs. 211, *C*, 212, *E*). In grain-eating birds, small stones are swallowed to lodge in the gizzard and aid the rough walls of this chamber in its function as a grinding mill which replaces the lost teeth.

Among many aberrant mammalian stomach types, that seen in the cud-chewing artiodactyl ungulates—the ruminants, such as cow, sheep, deer, and so forth—is notable (Fig. 215). In these forms four stomach chambers are present. Food when eaten descends to engage the attention of the first two of these chambers, the *rumen* and the *reticulum.* The for-

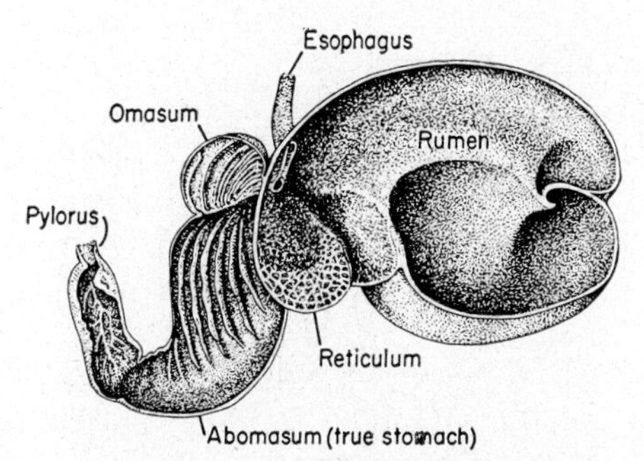

Fig. 215. The stomach of a sheep, sectioned to show the four compartments characteristic of higher ruminants. (After Pernkopf.)

mer is a large storage pouch where the food, reduced to a more workable pulp by the addition of liquid, is kneaded through the action of the muscular walls and subjected to bacterial fermentation. The reticulum is an accessory structure whose interior surface consists of a criss cross series

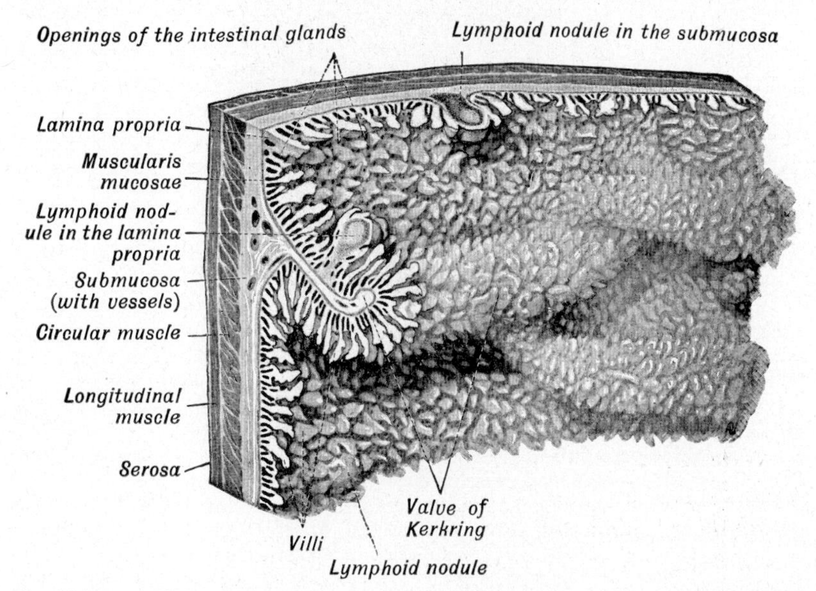

Fig. 216. Part of the wall of the mammalian small intestine. (From Maximow and Bloom, after Braus.)

of ridges with deep intervening pits ("honeycomb tripe"). At leisure the animal regurgitates the "cud" for further chewing—or rumination, if you will. On its second descent the food by-passes both the rumen and reticulum, by way of a deep fold in the anterior wall of the latter chamber, and enters the *omasum* or psalterium, whose parallel internal ridges reminded early workers of the leaves of a hymn book. After further physical re-

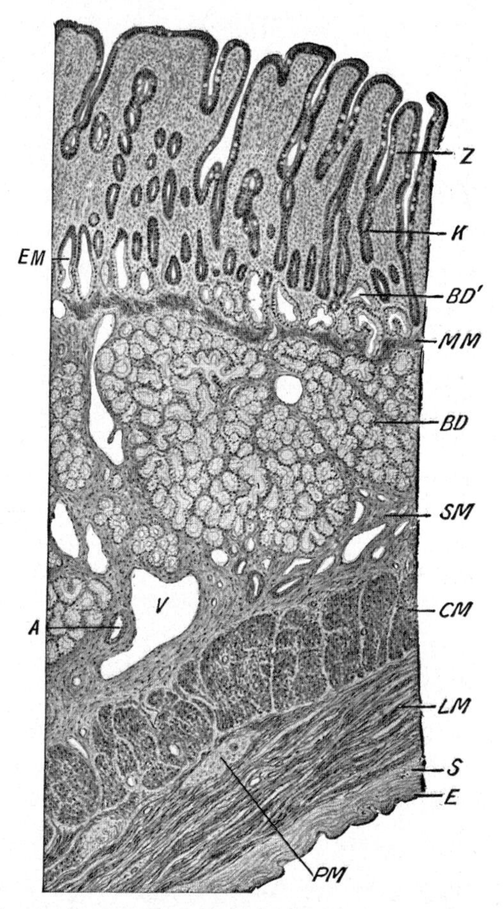

Fig. 217. A longitudinal section through the duodenum of man. Two types of glands are shown, the crypts of Lieberkühn (*K*) and the Brunner's glands (*BD, BD′*), which empty into them (cf. Fig. 209, p. 350). *A,* Artery; *CM,* circular muscle cut across; *E,* epithelium of serous tunic; *EM,* emptying of a Brunner gland into a crypt; *LM,* longitudinal muscle layer; *MM,* muscularis mucosae, a thin layer of smooth muscle underlying the mucous tunic; *PM,* myenteric plexus of autonomic nervous system, with a ganglion cell in section; *S,* serous tunic; *SM,* submucosa; *V,* vein; *Z,* villus. (From Maximow and Bloom, after Schaffer.)

working here the food at long last enters a final compartment, the *abomasum*. Here are present all three types of epithelium proper to the mammalian stomach—cardiac, fundus, pyloric. It becomes obvious that this compartment alone is the true stomach; the other three in advance of it are essentially elaborations of the lower end of the esophagus.

The Intestine (Figs. 210, p. 351; 211, p. 353; 216, 217). The major stages in the true digestive process are those which occur in the hindgut, the variably built intestine. More anterior segments receive, transport, store, and prepare food materials. In the intestine, however, occur most —and primitively, all—of the chemical processes which constitute digestion in a narrow sense. And here alone occurs the final crucial step, the absorption of food materials for body use.

Although the production of digestive enzymes may well have been in the vertebrate ancestors an exclusive function of the intestinal epithelium itself, that is no longer the case in living vertebrates. A special glandular outgrowth of the intestine, the pancreas, has evolved to become the place of origin of a considerable part of these enzymes, poured from it into the anterior end of the intestine; the enzyme pepsin has come to be formed in the stomach; even, as we have seen, an enzyme may be produced by the salivary glands of the mouth region. Despite all this, the intestine itself is still important as a seat of enzyme production. It secretes a series of these substances, forming an "intestinal juice" instrumental in the final breakdown of protein and carbohydrate materials.

Most intestinal enzyme production takes place in a series of small glandular outgrowths in the intestinal walls. The intestinal epithelium (Fig. 217) is of a simple columnar type. Occasional goblet cells supply mucus to lubricate food passage, and, despite the development of glands, a certain amount of enzyme formation may occur in the surface cells. In general, however, the intestinal epithelium has as its dominant function the absorption of digested food. Into the substance of its cells pass the water, salts, simple sugars, fats as glycerin and fatty acids, amino acids, and vitamins needed by the cells of the body, to be discharged on the opposite surface of the epithelium into the capillaries and lymphatics with which the intestine is richly supplied.

To afford efficient absorption, a large area of intestinal epithelium is necessary; a straight, smoothly lined tube will not suffice. Methods of increasing this internal surface are demanded by every vertebrate. Such increases have been brought about on three levels of magnitude. (1) Countless small internal foldings of the gut lining, essentially of microscopic size, which involve only the innermost, mucous tunic of the gut, are found in all vertebrates. Primitively these mucous layer folds appear to have been in the form of a network of tiny ridges; a more specialized condition, more prominent in higher vertebrate groups, is the presence of countless *villi,* small finger-like projections from the gut wall (Fig. 216). (2) In numerous instances somewhat larger folds may be present, often as a series of rings, in which the submucous layer is involved (as in the mammalian valves of Kerkring; Fig. 216). (3) Major structural developments occur which greatly increase the area and length of the intestinal surface. Two are most notable: the spiral valve intestine, short, but of complicated internal structure, characteristic of primitive vertebrates; and the slender tubular intestine capable of length increase by coiling, developed by the teleosts and by tetrapods.

The Spiral Intestine (Figs. 14, p. 37; 20, p. 42; 21, p. 43; 210, *A-E*, p. 351). In members of every major group of fishes we meet with a type of intestine seemingly primitive for vertebrates—the spiral intestine. This is characteristically developed in elasmobranchs, in which it is typically a large cigar-shaped body, extending nearly straight anteroposteriorly and occupying most of the length of the abdomen. Internally, it is seen to have a complex structure. In addition to minor epithelial folds, the surface area and the actual length of the food channel are greatly increased by the presence of a *spiral valve*. This is a fold of epithelium extending from one end of the intestine to the other, which in typical forms twists numerous times in spiral fashion around the walls of the gut. A similar structure would be formed if we took a carpenter's auger, or bit, and enclosed it in a tube. In some sharks the spiral valve may become exceedingly complex, with dozens of turns along its course. The development of the spiral fold greatly increases the surface area inside the gut, and also its functional length, for the food material must follow a long course down around the twists of the spiral staircase to reach the end of the seemingly short intestine. In a minority of sharks the spiral valve has a different form, but is equally effective. The base of the valve twists but little and hence is relatively short; the valve fold, however, is highly developed and rolled up into a great scroll running lengthwise of the intestine.

The primitive nature of the spiral structure is demonstrated by its widespread presence in other fish groups, although the number of turns is never high. The spiral intestine is present in chimaeras and cyclostomes, and fossils even demonstrate its presence in the ancient class of placoderms. Further, it was present in the early bony fishes, for living lungfishes and all lower actinopterygians retain a spiral valve, and even a few teleosts have a vestige of this structure.

Many workers on the digestive tract consider the intestinal region, here termed as a whole the hindgut, to be basically composed of two parts, which in mammals form the small and large intestines. Conditions in fishes show that this belief, based on human conditions, is unsound. In sharks the spiral intestine alone includes almost the entire length of the digestive tube between stomach and cloaca. Anteriorly, there is merely a short connecting piece, at the most, between stomach and spiral valve region. Posteriorly, there is only a short terminal segment, somewhat comparable to the rectum of higher types, leading to the cloaca. Sharks have in the "rectal" region a finger-shaped *rectal gland*, which has much the appearance of the cecum of land vertebrates. It is, however, not homologous with that structure, but is a mucous gland which presumably aids, by lubrication, voiding of the feces collected in this terminal gut segment.

The Teleost Intestine. The teleosts, and land vertebrates as well, have abandoned the old-fashioned spiral intestine for a new type of hindgut structure. In these forms (Fig. 210, *F*, p. 351) the intestine is a slender tube, without major internal folds. In compensation, however, it is generally greatly elongated; the intestinal length is, on the average, half again that of the whole body. A long, coiled part of the teleosts' hindgut appears to be a highly active digestive region, functionally comparable to the spiral intestine of their ancestors, on the one hand, and on the other roughly comparable to the small intestine of their tetrapod cousins. Distal to this we find a relatively short and straight end gut segment somewhat similar to the colon of land vertebrates, leading to the anus. Apart from gut coiling, teleosts and their lower actinopterygian relatives add to their intestinal surface by the development of distinctive *pyloric ceca*—pouches in the proximal end of the intestine into which food materials may enter and be absorbed.

In the vertebrate embryo the intestine is supported dorsally by a straight and simple dorsal mesentery. When, in teleost or tetrapod, intestinal coiling occurs, this mesentery becomes a complex, tangled structure, fanning out here, crumpled there; frequently the sheet supporting one intestinal coil may fuse with an adjacent one, or an intestinal loop may fuse with the dorsal celomic wall (cf. Figs. 291, 294, pp. 480, 483).

In both teleosts and land vertebrates there are enormous differences from form to form in the amount of convolution of the intestine and consequently in the amount of absorptive surface as well as in the length of the food channel. Two factors may be correlated with intestinal length: food habits and absolute size.

Plant food, often in the shape of compact masses of complex carbohydrates, particularly cellulose, appears to be difficult to digest, and in general the intestine is longer in herbivores than in flesh eaters. As regards the differences due to size of the animal, we encounter here again the question of surface-volume relationships. With increasing size, the volume of the body requiring nutriment increases faster than does the intestinal surface; a disproportionate elongation of the intestine is necessary to keep the area of absorptive surface in line with the demands made on it. From teleosts to mammals (although with many variants), small flesh eaters tend to have the shortest intestines, large feeders on plant materials the longest.

The Intestine in Tetrapods. Although a spiral intestine was presumably present in the crossopterygian ancestors of the land vertebrates, no living tetrapod has the slightest trace of this organ. Instead, land forms have paralleled the teleosts in the development of a rather simple tubular intestine, coiled to a variable degree. A seemingly generalized tetrapod type is that found in a variety of amphibians and reptiles (Fig. 211, *A*, *B*, p. 353). The hindgut is definitely divided into two distinct segments. The proximal, and longer, part is a *small intestine,* of narrow calibre, usually considerably coiled and often much longer than the entire body of the animal. At the end of the small intestine there is frequently a valvular constriction beyond which lies a well-defined end gut broadly homologous with the *colon,* or large intestine, of mammals; this is a relatively short, straight segment, expanded proximally. There is frequently a small outpocketing, termed a *cecum,* at the proximal end of the colon, but it is far from certain that it is identical with the structure of that name in mammals. Materials which have failed to be absorbed in the small intestine pass into the colon to be collected into the feces. Digestive juices carried down from the small intestine may continue their work while food materials are present in the colon. This region is, further, populated with bacteria which putrefy the food residues and break down proteins into a variety of substances which are of dubious value to the organism or are actually toxic. Through the walls of the colon is absorbed much of the water remaining in the prospective feces as well as some further food substance and part of the products resulting from putrefaction.

In birds (Fig. 211, *C*) the small intestine is similar to that of lower tetrapods in build, but more highly convoluted; on the average, the length of its coils is more than eight times that of the whole body. This great elongation is presumably correlated with the great activity and high metabolic rate of birds, which results in an increased food demand and the consequent requirement of additional absorptive surface in the intestine. The end gut is relatively short and not expanded proximally. Distally, it is expanded and not sharply marked off from the cloaca, so that, since birds have no functional urinary bladder, urine and feces are mixed, to be discharged as a "slushy" material. Paired ceca are present.

The Mammalian Intestine. In mammals (Fig. 211, *D*) the small intestine has undergone a development comparable to that of birds. In relation, again, to metabolic needs the intestine is complexly coiled, with a length which averages seven to eight times that of the body, and is several times greater than this figure in some large herbivores. In man it is customary to distinguish three sucessive regions of the small intestine—*duodenum, jejunum, ileum*—but the differences between them are not marked, and in many mammals these terms can be applied only in a rather arbitrary manner.

In some small mammals the large intestine, or colon, usually, but not always, separated by a valve from the small intestine, may be a short, straight structure; but in most mammals this gut segment is highly developed—much more so than is usual in other tetrapod classes. It is a tube of considerable diameter, but of irregular shape, usually with a series of outpocketings and longitudinal muscle bands (absent, however, in carnivores and ruminants). Its characteristic shape, as seen from the ventral surface, is that of an inverted U, or of a question mark. The morphologically proximal segment is the "ascending" colon, running anteriorly along the right side of the body; the small intestine enters this segment at a sharp angle. After a flexure, the "descending" colon runs posteriorly along the left wall of the body cavity. In primates a transverse segment is present between ascending and descending portions, and in most herbivores further coiling occurs.

Almost every mammal has a cecum, analogous to, although probably not homologous with, that of other tetrapods, and furnishing an additional area for colonic functions and a reservoir of intestinal bacteria. As its development shows, the cecum is a ventral side pocket of the gut tube; but in most mammals the adult condition gives it the appearance of being a part of the ascending limb of the colon, with the small intestine opening into its side. Frequently its diameter is as great as that of the colon itself, or greater. Its length is variable; it may be long and even much coiled in herbivorous animals. In man it terminates in the narrow *vermiform appendix*. This is frequently cited as a vestigial organ supposedly proving something or other about evolution. This is not the case; a terminal appendix is a fairly common feature in the cecum of mammals, and is present in a host of primates and a number of rodents. Its major importance would appear to lie in the financial support of the surgical profession.

The mammalian rectum, endowed with muscular sphincters of both smooth and striated types, is an end piece of the digestive tract, leading to the anus. As will be seen (p. 416), it is a part of the embryonic cloaca and hence not strictly comparable to the terminal gut segment to which this name is sometimes given in groups in which a cloaca is retained in the adult.

The Liver. In concluding this chapter we treat of two organs, liver and pancreas (derived embryologically from the gut endoderm), which are important both for secretions which they furnish to the intestine and for their functions in the metabolism of food already digested.

The liver, most massive of the abdominal viscera, always occupies a considerable area in the more ventral and anterior part of the abdomen. As shown by its embryonic history, the liver develops in the ventral mesentery below the anterior part of the intestine (Fig. 2, *D*, p. 6) and is, hence, covered by celomic epithelium. It remains connected to the stomach by a part of the ventral mesentery, termed in mammals the *lesser omentum*. Below the liver a persistent part of the ventral mesentery, the *falciform ligament,* ties it to the ventral body wall. Its major attachment, however, is an anterior one. In lower vertebrates a transverse septum develops separating the heart chamber (the pericardial cavity) from the remainder of the celom (cf. p. 477); in mammals this septum is incorporated in the diaphragm separating thorax from abdomen. The liver is attached to septum or diaphragm—broadly, or by a narrowed *coronary ligament*—and from this point extends backward to expand into the abdominal cavity.

From the point of view of gross morphology the liver is of little interest. It has no constant form, and needs none; sufficient bulk and an appropriate internal arrangement, on a microscopic level, of its cells and vessels are all that are necessary for its proper functioning. In various types more or less distinct right and left lobes are developed; but details of arrangement and lobation vary greatly from form to form and even between individuals. Ancient priests divined the future from the liver patterns of animal sacrifices; there was plenty of scope for interpretive imagination. Essentially one may say that the liver accommodates its form to that of the other viscera and expands into any part of the abdominal area not needed by other structures.

In general, as we have noted, the endoderm forms only the thin lining of the gut tube. Only in the liver (and to a lesser extent in the pancreas) does endodermal material bulk large in body composition. Except for a

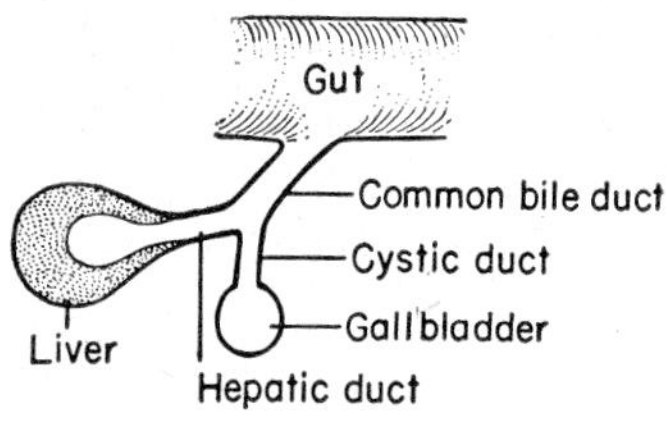

Fig. 218. Diagram of an early stage of liver development to show the relationship of the ducts.

framework of connective tissue and the varied vessels which weave through its substance, the liver is composed entirely of the countless *liver cells,* of endodermal derivation. These are polyhedral structures, morphologically similar in all parts of the liver. Despite a deceptive simplicity in microscopic appearance, these cells play a complex series of varied roles in the life of the body.

One function of the liver is that of a gland tributary to the intestine. In the adult vertebrate this function is a minor one, but historically the liver presumably arose as a glandular structure and embryologically arises as a glandlike outgrowth from the gut tube (Figs. 168, *A*, p. 292; 207, *C*, p. 347; 208, p. 348; 289, p. 478). The liver appears at an early stage in the development of the primitive gut as a pair of thickened masses of tissue which push out ventrally and anteriorly from the gut wall on either side of a median tubular outgrowth. The main mass of liver tissue extends forward in the developing ventral mesentery to the region of the transverse septum, leaving behind a side pocket of the tube which forms the gallbladder. From its anterior base in the septum or diaphragm the liver tissue rapidly expands posteriorly into the celom.

In the adult the primary glandular structure of the liver is for the most part lost. Throughout the liver there remains, however, a network of tiny *bile tubules* with which every cell is in contact. These tubules gather the cell secretions, the *bile,* into a *hepatic duct* draining the liver (Fig. 218). Part way toward the intestine this main duct is joined by the *cystic duct,* connected with the *gallbladder,* in which bile may be stored. Beyond this junction a *common bile duct* (ductus choledochus) empties into the small intestine, sometimes in association with a duct from the pancreas. This normal arrangement is subject to great variation. The gallbladder may be absent; several separate bile ducts may open into the intestine. Of the bile contents, most are purely excretory, including products of decomposition of protein materials and bile pigments derived from the hemoglobin of worn-out red blood cells. The only actively useful materials are the bile salts, which, discharged into the intestine, aid a pancreatic enzyme in the splitting and absorbing of fats.

The major duties of the liver cells, however, are not those directly connected with digestion, but with the treatment, within the confines of the liver, of food materials after their digestion and absorption into the body. In part this activity is that of a storage depot. All classes of food materials may be stored in the liver, but particularly notable from a quantitative point of view is carbohydrate storage. Large quantities of the simple sugars received into the body are stored in the liver as glycogen, and released again into the blood stream when needed. Every cell of the body is capable of producing, for its own needs, various changes in the food materials it receives; the liver cells act as complex chemical retorts whose activities benefit the body as a whole. Proteins may be synthesized, fats altered in composition, proteins and fats made into carbohydrates, nitrogenous cell wastes such as ammonia transformed into uric acid or urea to be eventually excreted by the kidneys.

For its major functions as storage depot and manufacturing plant the liver is strategically situated along a "main line" of the transportation system formed by the blood vessels. The liver receives by way of an arterial blood supply the oxygen necessary for its activities. Bulking much larger is the venous blood flow through the liver. As will be described later (pp. 451–453), all the venous vessels from the intestine carrying absorbed food collect as a hepatic portal system of veins which carry blood to the liver; after passing through a series of capillaries and sinuses in that organ, this venous blood finally reaches a hepatic vein which runs on toward the heart. From the capillaries which they border, the liver cells thus have the first opportunity to select, for storage or chemical modification, food materials newly arrived in the body.

Because of the overwhelming dominance of its metabolic rather than its secretory functions, the liver has tended to abandon its original glandular structure in favor of one in which a pattern of proper arrangement of cells in relation to the circulatory vessels is dominant. The pattern varies from group to group and, owing to its complexity, is not too well understood in many cases. In mammals much of the liver appears to consist of *lobules* of microscopic size (Figs. 219, 220). These are, crudely, polyhedral blocks of cells arranged in what one may term a three-dimensional honeycomb. Surrounding each lobule are connective tissue sheets in which are located branches of the bile duct and of both the portal vein and the

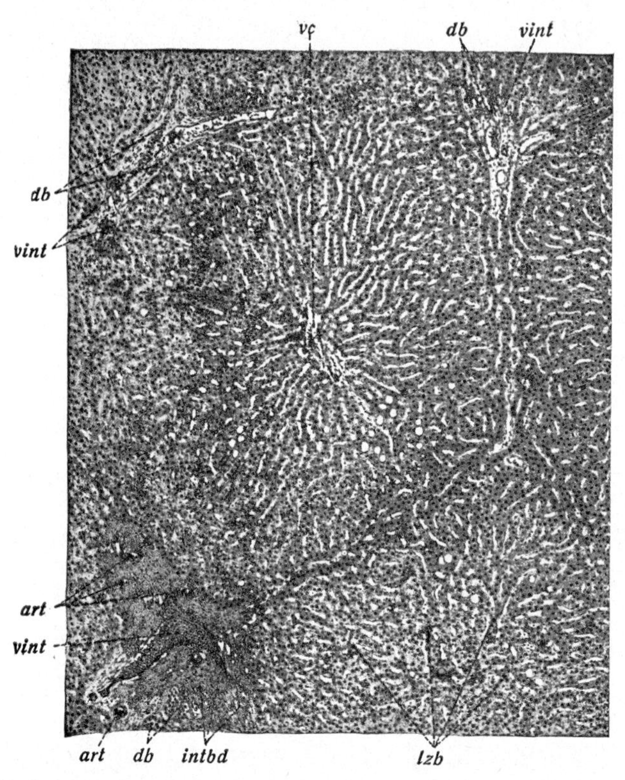

Fig. 219. Section of a mammalian liver, showing a lobule and portions of others. *art,* Branch of hepatic artery; *db,* a bile duct; *intbd,* connective tissue between lobules; *lzb,* cords of liver cells in lobule; *vc,* central vein; *vint,* interlobular vein (from portal system). (From Maximow and Bloom, after Sobotta.)

hepatic artery bringing blood inward. Bile tubules and capillaries from both vessels penetrate into the lobule from outside. Inside each lobule a central cavity contains a *central vein* which carries the blood from the capillaries onward to the hepatic vein.

The Pancreas. Common to all vertebrates, this organ is generally found as a mass of soft tissue lying in the mesentery not far distal to the stomach, and opening by one or more ducts into the proximal part of the intestine near the point of entrance of the bile duct. It is the major item among the glandular materials which appear on the table as "sweetbreads." In most vertebrates it is a moderately compact if somewhat amorphous body; in various cases, however, it may be diffuse, spreading out thinly along the mesenteries and even invading the substance of the adjacent liver or spleen. In many instances there is but a single pancreatic duct. The opening of this duct may be variable in position; it may enter either morphologically dorsal or ventral surfaces of the intestine, and, if ventral, may be combined with that of the common bile duct. Further, two or three (or, exceptionally, a higher number) outlets to the intestine may be present.

Histologically, the greater part of the pancreas is a typically exocrine glandular structure, with fine branching tubules collecting secretions pro-

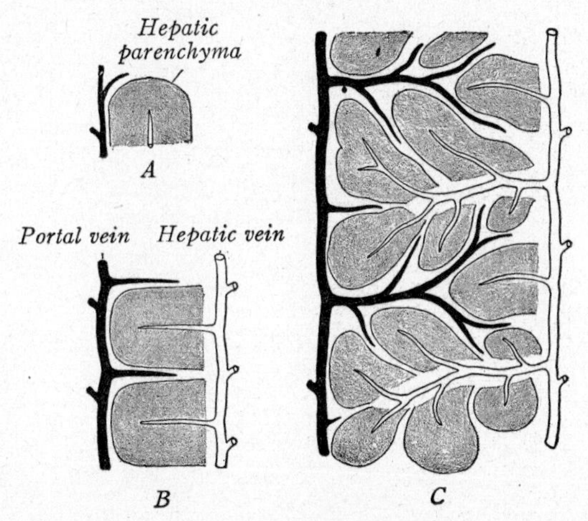

Fig. 220. Diagrams to show the development of liver lobules. Masses of liver tissue (parenchyma) become clustered about branches of the excurrent hepatic vein; the portal vein branches (and those of the hepatic artery and bile duct, not shown) ramify external to the developing lobules. (From Arey, after Mall.)

duced by the surrounding cells and passing them on to the duct system (Fig. 221). In nearly all forms studied, however, areas of tissue of another type can be seen distributed through the gland as isolated islands. These *pancreatic islands* (islets of Langerhans) consist of cells glandular in nature, but not furnished with ducts; they thus obviously form an endocrine organ, sending secretions out into the blood stream. Although usually diffused among the ordinary tissues of the pancreas, the insular material is found in a few forms grouped to form a small special organ of its own.

The general functions of the pancreas are well known. Its major exocrine portion secretes a number of enzymes which are discharged into the upper end of the intestine and are responsible for a great part of the digestive activities of the gut. The pancreatic enzymes are important for the digestion of all three major types of organic food materials. It was once thought that the pancreas produced a single enzyme for each of the three. Modern work has, however, shown the presence of more than half a dozen enzymes in the pancreas. Only one is known for fat digestion, but there are two or more for both carbohydrates and proteins, and it is not unlikely that still others may be discovered.

A function quite different (although also associated with metabolism) is that of the islands. It has long been known that disease or injury to the

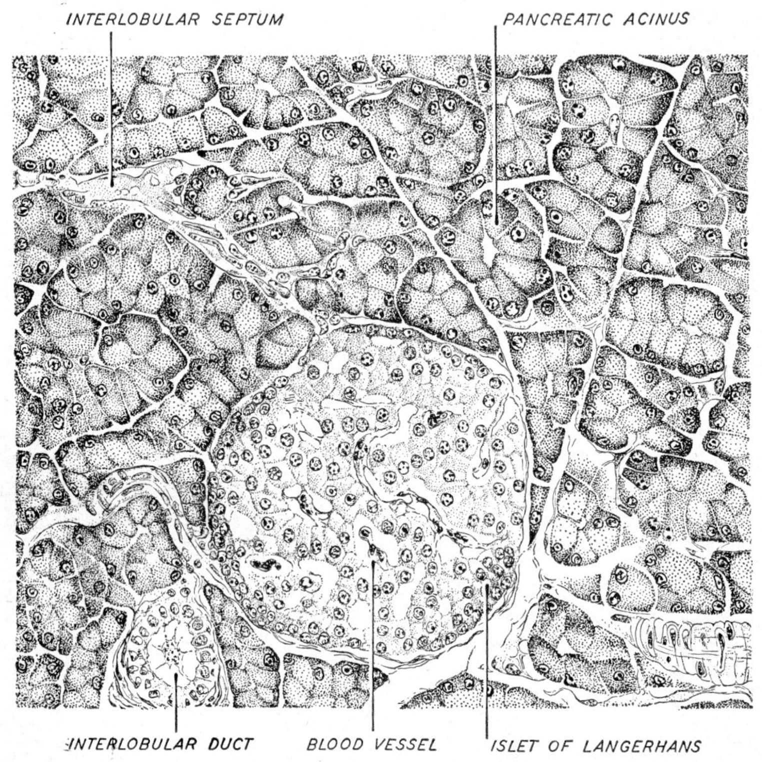

Fig. 221. A section through the pancreas of a rat, to show both exocrine tissue (pancreatic acini) and a pancreatic island. A duct of the exocrine part of the gland and a connective tissue septum between lobes are also shown. (From Turner.)

pancreas interferes with carbohydrate metabolism and produces the condition known as diabetes. We are now familiar with the fact that the islands produce a specific material, *insulin,* vital in carbohydrate metabolism. In the absence of this hormone, liver storage of glycogen decreases; the amount of glucose in the blood increases and is lost to the body through kidney action.

The functional "reason" for the evolution of the pancreas is readily deduced. Originally, we may well believe, the intestine itself was the place of production of all the enzymes which it utilizes. Some, as we have seen, are still produced in the gut; but the segregation of much of this type of activity in a distinct organ would appear to have been an advantageous development. Conditions in Amphioxus and the lamprey render such a story reasonable. In the former there is no separate pancreas, but groups of cells lying in the wall of the anterior part of the intestine show the characteristics of pancreas cells. The lamprey shows a transitional condition. There is no formed pancreas, but these cells have formed a cluster of tiny separate glands around the gut near the opening of the bile duct.

The embryological picture also agrees with this interpretation and, further, throws light on the varied nature of pancreatic ducts (Fig. 222). The pancreas is not a single structure, but a compound one. In the early stages of representatives of a number of vertebrate groups clusters of

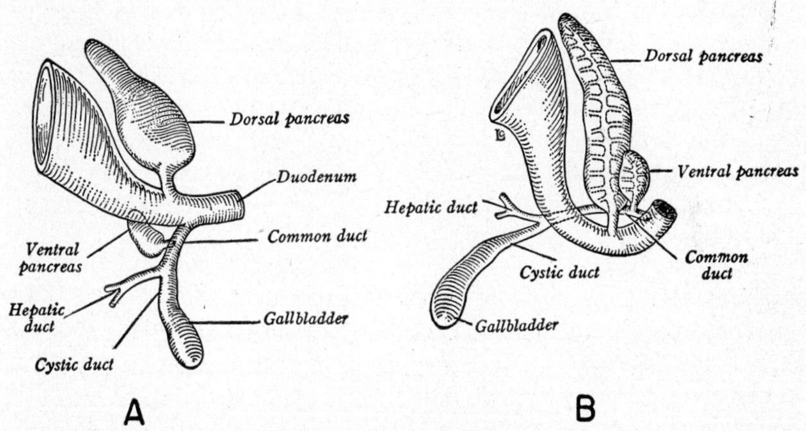

Fig. 222. Diagrams showing two stages in the development of the human pancreas. *A*, An early stage, with both dorsal and (smaller) ventral pancreas developing. *B*, Later stage in which dorsal and ventral portions are beginning to fuse. (After Arey.)

pancreatic cells are found in the gut wall near the area in which the liver diverticulum is developing. A major outpocketing of pancreas tissue takes place at the dorsal edge of the gut tube. Part, however, of the pancreatic tissue becomes involved in the mouth of the future bile duct, and one or a pair of pancreatic outgrowths forms in the walls of that tube. These tend to grow upward and fuse with the dorsal pancreas. Not always do all three structures persist in the adult, but usually the dorsal outgrowth and one, at least, of the ventral pair contribute to the adult pancreas. Each part may retain a separate duct; on the other hand, the duct systems may fuse, and a single outlet, dorsal or ventral, may serve the entire gland. As still another variant, the dorsal pancreas may fail to separate as distinctly as is usual from the gut wall, and its various parts may all drain separately into the intestine.

CHAPTER XII

EXCRETORY AND REPRODUCTIVE SYSTEMS

FROM A FUNCTIONAL point of view, a joint consideration of urinary and genital systems seems absurd, for excretion and reproduction have nothing in common. Morphologically, however, the two systems are closely associated in their mode of development and their use of common ducts, and it is impossible to describe one without numerous cross references to the other. This association appears to be due mainly to embryonic propinquity; in the embryo the major organs of the two systems arise in areas of the mesoderm which lie close to one another in the lateral walls of the trunk near the upper rim of the celomic cavity (Fig. 239, p. 398).

URINARY ORGANS

Kidney Tubule Structure and Function. Paired kidneys (Greek *nephros,* Latin *ren*), developed in varied form in all vertebrates, are the major organs of the urinary system. The basic kidney structure is a *kidney tubule,* or *nephron;* the numerous tubules connect with a varied duct system which eventually leads to the body surface.

Among vertebrates as a whole the most generalized type of tubule is that shown diagrammatically in Figure 223, *A;* tubules of this type are present in such diverse forms as sharks, fresh-water teleosts, and amphibians. The proximal part of the unit is the spherical *renal corpuscle* (malpighian corpuscle). If sectioned, this is seen to consist of two parts, a *glomerulus* and a *capsule* (Bowman's capsule). The glomerulus is formed by the circulatory system; it is a compact little cluster of small blood vessels the size of capillaries, which coil through the mass in intricate loops (Fig. 224). The capsule is a proximal part of the tubule; its shape is that of a double-layered hemisphere. The inner layer is closely adherent to the lining of the vessels of the glomerulus, so that only thin membranes separate the blood stream from the cavity between the two layers of the capsule. This cavity is continuous with that of the *convoluted tubule.** The tubule may attain considerable (relative) length. Along its course it is closely bordered by a network of capillary vessels. Distally,

* Note that the term "tubule" is generally used in two ways: (1) as a synonym of nephron, i.e., a name for the entire nephric unit of renal corpuscle and convoluted tubule; and (2) for the latter structure only.

each tubule connects, in a fashion which varies in different groups, to ducts leading to the exterior of the body.

The result of the activity of the tubule and its corpuscle is the production of urine, destined for excretion and derived, obviously, from the associated circulatory vessels. Urine is composed mainly of water, but contains to a variable degree other substances in solution, including salt ions, particularly sodium, potassium, chloride, and sulfate; bulking much larger, however, are simple nitrogen compounds, mostly in the form of

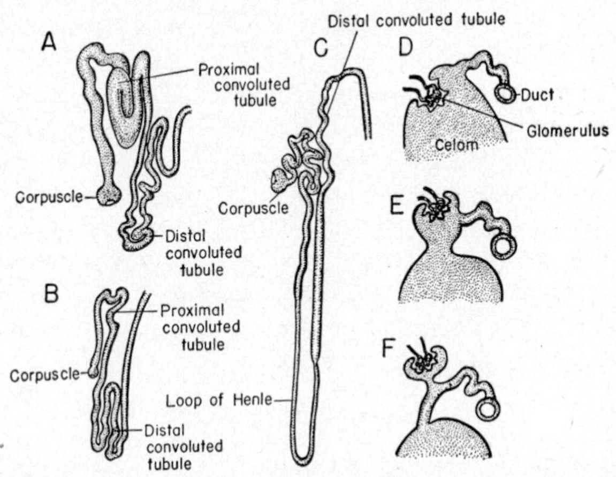

Fig. 223. Tubule types. *A* to *C,* The three major types common in adult vertebrates. *A,* Presumably the most primitive, with corpuscle of good size, found in elasmobranchs, fresh-water bony fishes, amphibians. *B,* Corpuscle reduced or absent, characteristic of salt water teleosts, reptiles. *C,* Corpuscle large; a loop of Henle inserted; found in mammals, birds. *D* to *F,* primitive tubule types found in lower vertebrates, principally in the embryo, and perhaps illustrating the early evolution of kidney tubules. *D,* Tubule opens out of celom; glomerulus, if present in celom, not associated with tubule. *E,* Special small celomic chamber formed for glomerulus. *F,* This chamber has become the capsule of a renal corpuscle; tubule still connects with celom; closing this opening leads to more progressive tubule type.

urea or uric acid. The major functions of the kidney tubules are twofold: (1) regulation of the internal environment and (2) elimination of waste.

In Chapter IV we pointed out that for the health of the body cells it is necessary that they live in a proper environment, and that a most important feature of this is the continued presence, in the interstitial fluids which surround them, of appropriate amounts and proportions of specific simple salts. These salt ions are present in the same proportions in the blood stream and in the interstitial fluid with which it readily interchanges through capillary walls. The maintenance of proper salt content in the body liquids as a whole demands a proper balance between body intake, mainly via the intestine, and output. Deficiencies in salt intake are the concern of the digestive apparatus; excesses in intake, which bring the salt concentration above the proper level, require excretory devices for their removal. Such structures as the salt-excreting glands of the teleost gills or the sweat glands of the mammalian skin are of use here. The kidney tubules, however, furnish the main means for the elimination of excess salts on the one hand and, on the other, for the removal of excess amounts of liquid which tend to dilute too greatly the body fluid.

Apart from regulation of salt and liquid content, however, there is a further important kidney tubule function: the elimination of wastes and harmful materials from the body. Digestion may bring into the body fluid compounds which are, for cellular utilization, inert, or may be actually destructive or poisonous; the kidneys are in most cases effective in removing them. Most important of materials needing removal are the broken-down products of protein metabolism. From carbohydrates and fats the main end products put out by the body cells are, as we have seen (p. 71), carbon dioxide and excess water, both of which are readily eliminated. Protein metabolism, however, produces other substances, primarily simple compounds containing nitrogen. The greater bulk of the nitrogenous waste as it leaves the cell consists of ammonia, which is highly toxic. In some vertebrates—teleosts, crocodiles—it remains in this form until excreted. In most, however, the ammonia is transformed in the liver, almost as fast as it appears in the blood, into harmless ammonia compounds—urea and uric acid. Urea includes the greater part of the excreted nitrogenous waste in sharks, amphibians, turtles and mammals; uric acid (an almost insoluble material) is the bulk of the excreted material in lizards and birds. Nitrogenous materials may be eliminated by the gills in teleosts, and a minor amount by the sweat glands in mammals. For the most part, however, the kidneys bear the responsibility for the excretion of nitrogen wastes.

The way in which a kidney functions is today fairly clear. Two distinct operations are present, one having to do with the renal corpuscle, the other with the tubule proper. The structure of the corpuscle suggests that we are here dealing with a relatively simple filtering device; that through the membranes separating the glomerulus from the cavity of the capsule there passes a filtrate of the blood plasma. Studies on amphibians show that this is indeed the case. Liquid drawn from the capsule by micropipette proves on analysis to have in it almost all of the materials found in the blood, and in the same proportions. Blood corpuscles, of course, do not pass through this filter, nor do large molecules, such as the proteins specific to the blood stream. All other materials, however, including not only wastes, but also valuable food materials, particularly glucose, pass through readily. Still further, the amount of liquid filtered is excessive. It has been calculated that if all the liquid passed through the glomeruli of a frog were to be eliminated, nearly a pint of urine would be produced daily by this small animal, and that a man would produce about 50 gallons of urine daily!

Obviously, nothing of the sort happens. If the only kidney action were simple filtration from the glomerulus, the body would be dehydrated in short order. Further, glomerular filtration would do little to alter the proportions of materials found in the body fluid; wastes and harmful or excess materials would still be present in the remaining fluids in the same relative amounts, and in addition valuable food elements would be sacrificed.

The corrective to this situation lies in the work of the convoluted tubule. From the nephric corpuscles the filtrate traverses the length of these tubules to kidney ducts and, in many groups, to a storage reservoir, the bladder. After this passage, analysis of the resulting urine tells a different story: we are dealing with a different substance. For one thing, the volume

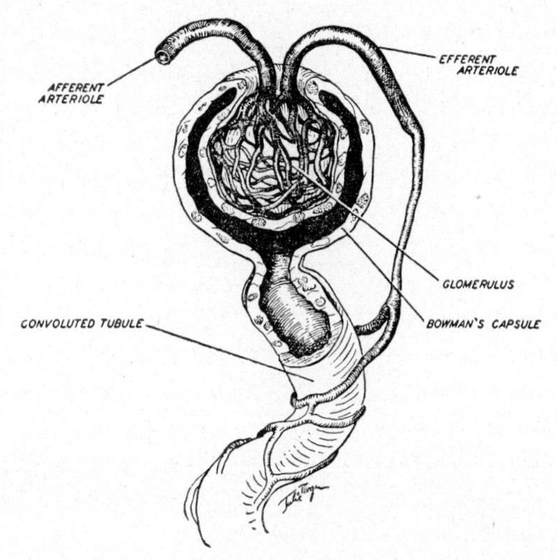

Fig. 224. A renal corpuscle. (After Turner.)

is greatly reduced. In a frog, not more than 5 per cent of the liquid filtered from the glomeruli reaches the bladder, and in a man hardly 1 per cent. Further, the composition is much changed. Useful food materials such as glucose, although readily filtered into the tubule, do not ordinarily reach the bladder, and of other substances, the relative proportions are radically altered.

It is thus obvious that, in addition to the glomeruli, the kidney tubules themselves play a major role in the process of excretion; they are not morphological curiosities, but highly important functionally. They are lined with epithelia which change from one part of the tubule to another in patterns which are quite variable in different vertebrate types, and the structure of the kidney is so fine and complex that in but few cases has the specific function of different tubule segments been determined. As a whole the tubule cells have two activities. A major one is that of reabsorption of most of the filtrate. Much of the water is obviously resorbed, and food materials are withdrawn from the fluid as it passes by. On the other hand, there appears definitely to be a certain amount of further excretion of waste substances from tubule cells into the urine as a supplement to the filtrate.

In a nephron the crude work is, so to speak, done by filtration from the glomerulus; the tubule proper adds the necessary refinement to the process.

Tubule Types and Vertebrate History. Whether fresh or salt water formed the original vertebrate home has been a problem of interest to

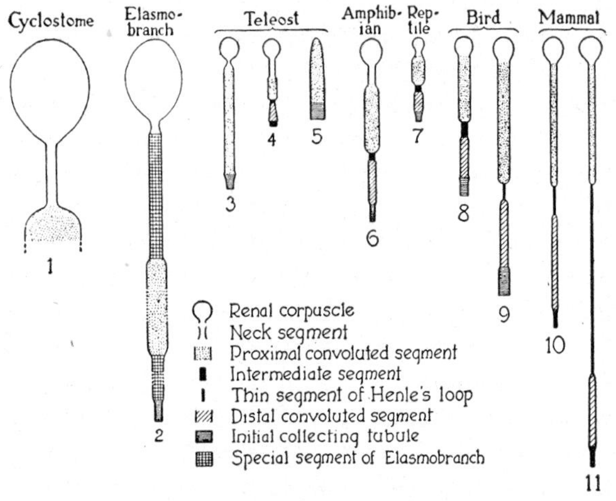

Fig. 225. Diagram of kidney tubules of various vertebrates, all reduced on the same scale, to show the relative size of the components in the different groups. The glomeruli are at the upper end in each case, and the tubules are represented as if straightened out. The glomeruli are well developed in most groups and of enormous size in cyclostomes and elasmobranchs, but of reduced dimensions in reptiles, and are done away with in some marine teleosts (*5*). All have a proximal convoluted segment of the tubule; an intermediate segment, followed by a distal convoluted tubule, appears in some fishes and is present in all land forms. The intermediate segment becomes the loop of Henle in birds (in part) and mammals; this loop may be much elongated in the latter group. The special segment of elasmobranchs may be related to resorption of urea. 1, Hagfish; 2, skate; 3, sculpin; 4, catfish; 5, toadfish; 6, frog; 7, painted turtle; 8 and 9, chicken; 10 and 11, rabbit. (From Marshall.)

students of function as well as students of classification and phylogeny. The paleontological record indicates that the earliest fishes dwelt in fresh waters and only later invaded the seas. A study of kidney tubule structure and function leads to the same conclusion.

Three main types of nephric units may be found in the functional adult vertebrate kidney (Figs. 223, 225). All have a well-convoluted tubular region. (*a*) In one type (Fig. 223, *A*), found characteristically developed in such varied forms as amphibians, fresh-water bony fishes, and elasmobranchs, there is a renal corpuscle of good size, and the amount of filtrate (and consequently of water output) is high. (*b*) A second type (Fig. 223, *B*) is that found, with variations, in many salt-water teleosts and in reptiles. Here the corpuscle is small or absent; obviously water output is low. (*c*) A third type (Fig. 223, *C*) is that seen in mammals and, in less extreme form, in birds. Here we see the interjection, into the middle of the convoluted tubule, of a long, slim extra loop, the *loop of Henle*. Although, with a large corpuscle, a large amount of fluid is produced, little reaches the bladder through this type of tubule.

From the distribution of these tubule types among the vertebrates a consistent story can be constructed regarding the environmental history of vertebrates. It is assumed that (*a*) is the primitive tubule type, possessed by early fresh-water vertebrates and retained by forms which still inhabit such waters. Such an animal lives in a medium more dilute than its own body fluids; in consequence, water tends to invade the body by osmosis at any membranous surface. To prevent overdilution of the body content and death, large water output is necessary, and this is afforded by the large corpuscle and its large output of water in the dilute urine.

In a marine fish, on the other hand, the sea water has a higher salt concentration than the body fluids, and the result tends to be loss of water into the surrounding "brine," dehydration, and an overconcentration of salts in the body. In marine teleosts salt excretion from the gills (cf. p. 319) partially remedies the situation; in addition, however, reduction or loss of the renal corpuscle from the tubule system conserves a vast amount of water.

Land vertebrates have basically much the same problem as a marine fish. They live in a dry environment, where water is continually lost through the body surface, with danger of dehydration and consequent increase in salt concentration in body liquids. Conservation of water is necessary. It is probable that the ancestral reptiles had kidney units of type (*a*), with a large corpuscle and normal tubule, and a free flow of watery urine. In living reptiles reduction of water output has been attained by reduction of the size of the corpuscles and a consequent reduction of the amount of water filtered through them. Birds and mammals, likewise descended from the ancestral reptiles, have developed a different method of conserving water. The normal corpuscle is present, and a large quantity of urine is poured into the tubule. The tubule, however, is an "Indian giver." There is a considerable resorption of water in any type of tubule; in these forms there is inserted between proximal and distal convoluted tubules the long, thin loop of Henle, which is believed to be especially potent in water resorption. As a consequence, the urine is highly concentrated and water is efficiently conserved.

The argument is essentially reasonable, leading to the conclusion that the presence of a large glomerulus is a primitive character; that this is due to its efficiency as a pump for the discharge of quantities of water; and that the desirability of such discharge is due to the life of early vertebrates in a fresh-water environment.

But one could still, as far as we have here carried the argument, advocate a marine environment as primitive and suggest that the development of a large glomerulus came later, when vertebrates moved into fresh water. The thesis that a large glomerulus is associated with fresh-water existence seems to be negatived by the sharks, for here we have fishes which have a large glomerulus and a tubule of type (*a*) and yet are marine!

It is the sharks, however, which clinch the argument for fresh water, when the situation is studied. The shark lives in a medium in which the salt concentration is greater than that of his own body fluids; yet he does not lose water to his environment and can pump water out of his system through large glomeruli with as little concern as a fresh-water fish. For although his salt concentration is that of any other animal, and low compared to sea water, his total osmotic pressure (and that is what counts) is comparable to that of the sea. The shark has adopted a unique mechanism; it retains much of its nitrogenous waste, in the form of urea, in the blood stream (without apparent harm) and thus raises its total concentration of materials in solution without increasing its salts.

We have thus in sharks and salt-water teleosts two radically different kidney adaptations to salt water. If the sea had been the original home of fishes, we would hardly expect such differences to exist. It is more reasonable to assume that the ancestors of both sharks and marine teleosts lived in inland waters; that (as the fossil record indeed shows) they invaded the seas independently; and that in the two cases quite different methods were evolved for combating the dangers of too great salinity.

Primitive Tubule Structures. The types of nephric units described in the foregoing discussion are those most characteristic of adult vertebrate kidneys. Still other types, however, may be noted; they are more often found in embryos, particularly in the first units developed, than in adults, and more common in lower than in higher vertebrate types; there is hence considerable reason to consider them primitive in nature.

These types (Fig. 223, *D-F*) have in common the fact that they have *nephrostomes*—openings into the celomic cavity which are commonly ciliated funnels. In some cases there is, in addition to the celomic opening, a typical renal corpuscle connected with the tubule. In a few instances we find a glomerulus which protrudes into the celomic cavity near the funnel instead of having a position in the tubule structure. And in still other cases no corpuscle or glomerulus of any sort is present.

One theory of the origin of the vertebrate nephric structures suggests that the last-described condition was the truly ancestral one; that the tubules first functioned in draining waste products and excess liquid from the celomic cavities. A subsequent development of blood vessels, as glomeruli, in the walls of the celom afforded a greater supply of liquid for the tubules to work upon. A third stage, the theory suggests, was the incorporation of the glomerulus in a corpuscle forming an integral part of the nephric unit; the final stage, a loss of the primitive celomic connection, and dependence upon the glomerulus alone for a supply of filtrate.

Most major invertebrate groups have excretory tubules of some type with functions similar to those of vertebrates. Renal corpuscles are never developed, but in some annelid worms there are tubules which have celomic openings rather like those of vertebrates. Since, however, vertebrates appear to have no phylogenetic connections with annelids, this similarity is presumably due to parallelism. Echinoderms, the closest relatives to the chordates, have no kidney tubules of any sort, but acorn worms, among the lower chordate types, have developed excretory structures with glomeruli rather like those of the vertebrates. We may assume thus that the chordate kidney organs have evolved independently from those of invertebrate phyla.

Here, however, Amphioxus strikes a discordant and disturbing note. Amphioxus has highly developed excretory organs, but these have neither nephrostomes nor glomeruli; instead, the urine is collected by peculiar cells termed *solenocytes*. These are beaker-shaped elements with a long "neck;" within is an active flagellum, extending into the nephric cavity. Solenocytes are otherwise unknown except in a few annelids. So similar is Amphioxus to the vertebrates in many basic features that its close relationship is undoubted. Why, then, does it have so radically different a kidney structure? To this question there is, at present, no answer.

Organization of the Kidney System. So far we have discussed merely the nature of the vital microscopic kidney elements. Now to be considered is the organization of these secretory units and the duct systems leading from them to form the gross structure of the urinary system.

As seen in the higher, amniote classes of vertebrates, the pattern of the kidney system is fairly uniform and apparently simple, with paired, compact kidneys projecting into the abdominal cavity from its dorsal walls; paired ureters carrying urine from the kidneys and often emptying into a median bladder. Rather comparable structures are seen in various lower types. Study, however, shows that there are great differences from group to group in the construction of the urinary system. The kidneys themselves are varied in structure; the ducts vary; the bladder is variable.

These variations have two major causes. (1) Unlike most body organs, the kidney must begin its operations at a rather early stage of development, to take care of the metabolic wastes of the rapidly growing embryo. In consequence there is no possibility of the peaceful and undisturbed development of a kidney structure suitable for postnatal life alone; there must be rapidly constructed a functioning embryonic kidney which, however, is subject to modification or replacement in later developmental stages and adult existence. (2) The sex glands, testis and ovary, lie adjacent to the kidney structure; these glands—the testis particularly—tend to "invade" the urinary system and take over part of its tubes and tubules as a conducting system for their products. The urinary organs have been markedly modified in most vertebrate groups as a result of this invasion.

Holonephros and Opisthonephros. We may begin our discussion of types of urinary systems by the description of the structure and embryonic development of an "idealized" primitive kidney which may be termed a

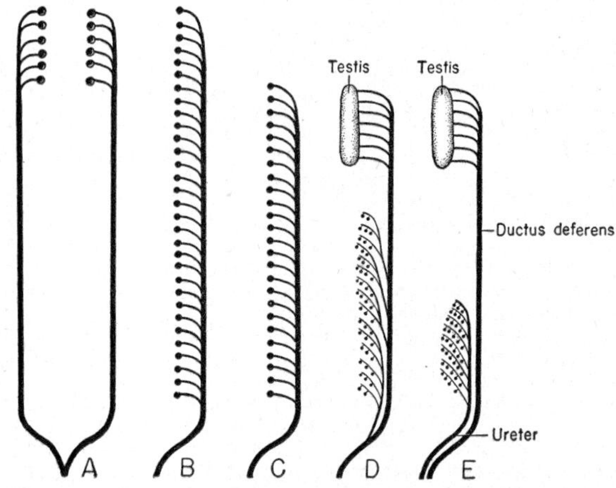

Fig. 226. Diagrams of kidney types. *A*, Pronephros (embryonic); *B*, theoretical holonephros (each trunk segment with a single tubule); *C*, primitive opisthonephros: pronephros reduced or specialized, tubules segmentally arranged, as in hagfish. *D*, Typical opisthonephros: multiplication of tubules in posterior segments, testis usually taking over anterior part of system, trend for development of additional kidney ducts (most anamniotes). *E*, Metanephros of amniotes: an opisthonephros with a single additional duct, the ureter, draining all tubules. In *A*, both sides of the body are included; in *B* to *E*, one side only.

holonephros. This ideal is, sad to say, not realized in any living adult vertebrate, but a larval hagfish kidney approaches it so closely as to be capable of inclusion in this category.

In our embryological story we noted that in every trunk segment the mesoderm includes, on either side, a nephric region, often segmentally distinct as a *nephrotome*, a small discrete block of tissue interposed between somite and lateral plate (Figs. 43, *D*, p. 93; 49, *C*, p. 99; 227). In the ancestral vertebrate, we believe, each nephrotome gave rise to a single nephric unit; the small celomic cavity of the nephrotome became the cavity of a renal corpuscle, and out from this grew a kidney tubule. There would thus be established a row of nephric units, segmentally arranged, down either side of the body. As with the somites, the differentiation of nephrotomes takes place in embryos from before backwards; in consequence the oldest members of the tubule series in our ideal ancestral type would be those at the anterior end of the series, the last-formed those at the posterior end of the trunk. With the downgrowth of the myotomes, the kidney units would be separated from the surface of the body by the trunk musculature and associated skeletal elements. The nephrotome lay above the celom, and the developing kidney units would in later stages lie dorsal to the celomic region, well to either side of the midline (Fig. 237, p. 396). As noted earlier, conditions in embryos of lower vertebrates suggest that the nephric units in an ancestral vertebrate retained the celomic connection now seen in early embryonic stages and occasional adults in various forms.

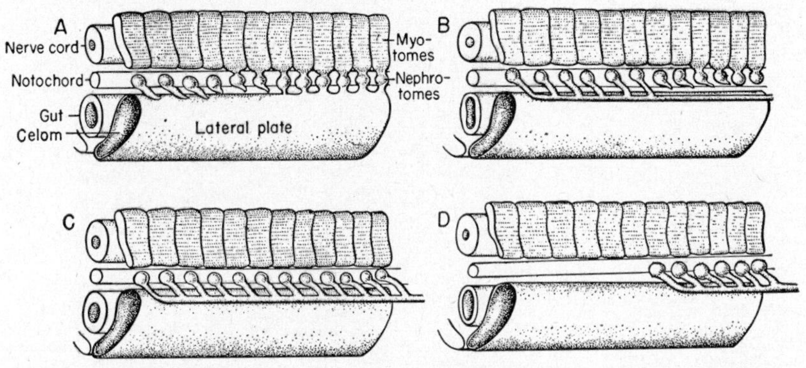

Fig. 227. Diagrams of the anterior part of the trunk of an embryo (skin removed) to show the development of the archinephric duct. *A,* Most anterior—pronephric—nephrotomes are budding out tubules which tend to fuse posteriorly. *B,* The pronephric tubules have formed the duct; some of the nephrotomes farther posteriorly are forming tubules which are to enter the duct. *C,* The more posterior tubules have joined the duct. *D,* The pronephros lost, but the archinephric duct formed by it persists to drain the more posterior part of the kidney.

In many invertebrates the kidney elements open individually to the exterior. In the vertebrate embryo, however, a longitudinal duct develops on either side, gathering the urine from the series of segmentally arranged units; the two ducts often unite before emptying to the exterior in the region of the cloaca. The primitive kidney duct has been variously named; it will here be termed the *archinephric duct.** This duct is, like the tubules, of mesodermal origin. In embryos it typically forms by a fusion of the tips of the tubules of the most anterior and first formed nephric units (Figs. 226, *A*, 227, *A*, *B*). It grows backward along the lateral surface of the nephrotomes, sometimes using further nephrogenic material in its formation. As tubules develop farther back, they grow outward to con-

* It is often called the *wolffian duct,* after its discoverer. *Pronephric duct* and *mesonephric duct* are terms applied to it on embryological grounds. When serving as a kidney duct in the adult of lower vertebrates, it is sometimes incorrectly called a ureter. If taken over by the genital system, it becomes the duct of the epididymis and the ductus deferens (pp. 411–414).

nect with this previously formed structure. The end result (Figs. 226, *B,* 227, *C*), ideally, is a *holonephros:* a kidney with a single nephric tubule in each trunk segment on either side of the body, the series draining through a pair of archinephric ducts. Such an ideal is found only in the larval hagfish.

The adult hagfish has a kidney which does not differ greatly from this type, but even here the most anterior and first-formed part of the kidney tubule series is specialized or degenerate; and the same is true of all higher vertebrates. These anterior tubules are termed the *pronephros*

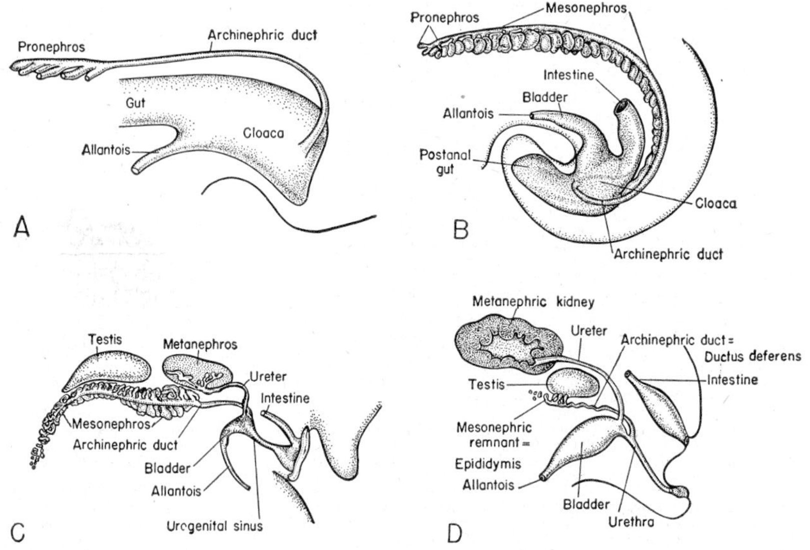

Fig. 228. Diagrams to show formation of the metanephros of an amniote (male) embryo as seen from the left side. *A,* Pronephros and duct formed; *B,* mesonephros partly formed; *C,* pronephros reduced, posterior part of mesonephros functional, ureter formed, and metanephros beginning to differentiate; *D,* definitive stage; mesonephros reduced, and tubules and duct utilized only for sperm transport; metanephros the functional kidney.

(Figs. 226, *A*, 227, *B*, 228, *A*). The remaining, major part of the kidney system, from which is formed, in one fashion or another, the kidney of adult living vertebrates, may be termed as a whole the *opisthonephros,* the "back kidney" (Fig. 226, *C, D*). This opisthonephros generally differs from the theoretical holonephros in three main particulars: (1) The anterior tubules, the pronephros, do not form part of this structure. (2) The simple segmental arrangement is no longer present (above the hagfish level). For most, at least, of the length of the adult vertebrate kidney numerous tubules may develop in each segment; in fact, separate nephrotomes may fail to develop, and, particularly toward the back end of the trunk, there may be a development of a mass of tubules from a longitudinal band of unsegmented nephric tissue. (3) In most vertebrate groups the archinephric duct is utilized by the male for sperm transport, and part or all of the business of urine transmission may be performed by a newer type of duct, a ureter or some equivalent.

Amniote Kidney Development. To show an extreme contrast with the ideal holonephros, we may describe the development of the kidney in mammals (Fig. 228); the story is similar in the other amniote groups—reptiles and birds. Our departure here from a logical sequence is due to the fact that study of the mammalian kidney has been influential in the development of concepts of kidney types and their nomenclature.

As the differentiation of mesodermal somites progresses, simple, segmental, rounded nephric vessels, or nephrotomes, appear below them in a limited number—not over a dozen—of the anterior body segments, at the back of the head and along the future neck region. These vesicles form the pronephros. From them short tubules extend outward and turn back, fusing at their tips, to form an archinephric duct which grows rapidly backward to the cloacal region. In a mammal the tubules constituting the pronephros disappear at an early stage; indeed, the first ones have begun to degenerate before the last members of this short series are formed.

Segmental growth of the kidney structure continues backward without interruption beyond the end of the pronephric tubule series. The tubules now formed belong to a second embryonic nephric structure, the *mesonephros,* which is the functional kidney for a considerable period of embryonic life in mammals and may persist until after birth in reptiles. Still segmental, to begin with, mesonephric tubules continue to form along the trunk as far as the lumbar region. Pronephric units are in mammals simple structures, without glomeruli and with but a short tubule; the mesonephric elements, on the other hand, have characteristic glomeruli and well-coiled tubules. At first there is but one tubule to a segment in the mesonephric region; later, however, additional tubules develop, and the segmental condition is lost.

As the back portions of the mesonephros are developed, the more anterior tubules are degenerating. Long before birth in many mammals the entire mesonephric series has lost its original urinary function, as has the original urine tube, the archinephric duct. As will be seen, mesonephric tubules and duct persist in modified form as part of the reproductive system of the adult male (pp. 411–414; Figs. 228, 235, p. 389; 245, p. 409).

The most posterior part of the nephrogenic tissue is not utilized, however, for the mesonephros. It does not become segmented, but instead forms into a spherical mass in the roof of the lumbar region of the body cavity. In this mass of mesoderm there presently appear large numbers of kidney tubules. These tubules are to form the *metanephros,* the functional kidney of the late embryo and the mature mammal. Unlike the mesonephric units, these tubules do not empty into the archinephric duct; there develops a new tube, the *ureter,* to drain them. This is an outgrowth of the archinephric duct. From near the point where each archinephric tube enters the cloaca, this new duct buds off and grows forward and upward to enter the metanephric tissues. Here it forks repeatedly into smaller tubules. At the tips of the smallest branches of the ureter, the *collecting tubules,* clusters of nephric tubules come to open. Further growth and differentiation of this metanephric mass produce, on either side of the body, the adult kidney, which is thus in form, position, and mode of drainage quite distinct from pronephros and mesonephros.

Pronephros, Mesonephros, and Metanephros. We see in this story the development in the amniote embryo of three successive kidney structures —pronephros, mesonephros, metanephros. It is often stated or implied that these three are distinct kidneys which have succeeded one another phylogenetically as they do embryologically. Upon consideration, however, it will be seen that there is no strong reason to believe this. The differences are readily explainable on functional grounds; the three appear to be regionally specialized parts of the original holonephros which serve different functions.

In many vertebrates the pronephric tubules differ in structural features from those of the mesonephros; but these differences are not constant, and sometimes part or all of the pronephric units are identical in nature with those behind them. The one really distinctive feature of the pronephros is that its tubules form the archinephric duct. There is, however, nothing really significant or mysterious about this; it is a practical matter. An actively growing embryo has wastes to excrete. When kidney tubules are formed anteriorly, they begin the process of urine formation. The formation of a urinary tube for drainage cannot be delayed until the entire kidney is formed; the anterior tubules just cannot wait that long.

In amniotes the metanephros is readily distinguished from the mesonephros. It is a discrete formation; it attains a vastly greater size; its drainage by the ureter rather than by the archinephric duct is definitive. But although the metanephros assumes a distinctive nature, it is, to begin with, simply a greatly enlarged posterior termination of the band of nephrogenic tissue from which mesonephros and pronephros were also formed. Its large size is, of course, due to the presence of large numbers of tubules; the need for large numbers of tubules in the amniotes is correlated in part with the relatively large size of the late embryo and adult, in part with the high metabolic activity of birds and mammals. The efficiency of a kidney depends upon the amount of surface present in its glomerular filters and tubule walls, and the same surface-volume relationship confronts us here as in the case of many other organs. The presence of a special duct system is obviously related to the impossibility of attaching the system directly to a single unbranched archinephric duct. Further, the utilization of the archinephric duct for sperm transportation in the mature male makes a separation of a urinary transporting agency desirable.

The adult kidney of fishes and amphibians is frequently termed a mesonephros and assumed to be homologous with the embryonic kidney alone in amniotes. This position appears to be untenable; it implies that there is nothing in the kidney of a shark or frog comparable to the metanephros of amniotes. But the embryological story shows that the entire length of the band of nephrogenic tissue is used in the development of the kidney of lower vertebrates, and hence the back portion of their kidney is formed from the same materials as the metanephros. Never, it is true, is the adult kidney of a lower vertebrate so completely concentrated in one short region as is the case with the metanephric type, but somewhat similar conditions may be observed. It is often stated flatly that the anamniote kidney is drained by the archinephric duct, as in the case of the embryonic mesonephros; but, as we shall see (Fig. 235, p. 389), various fishes and amphibians have structures comparable to ureters. All in all, it is best to confine the term mesonephros solely to the kidney of embryonic amniotes. The adult anamniote kidney should be termed the opisthonephros, with the understanding that the metanephros is a special variety of this structure.

Pronephros and Head Kidney. We have described the pronephros as the most anterior and first developed part of the embryonic kidney—essentially that set of tubules which are concerned in the formation of the archinephric duct. In amniotes and Chondrichthyes the pronephros is an exceedingly short-lived structure and never survives in the adult. In contrast are conditions in other fishes and amphibians, with small-yolked eggs. Here the embryo generally becomes an active food-seeking larva at an early stage. The pronephros persists in these larvae to satisfy excretory needs; it is, however, highly specialized and is better termed a *head kidney,* in reference to its anterior position. In most cases a fair number of tubules from the more anterior body segments are involved; however, the number is usually reduced during embryonic development, so that the larval head kidney has but from one to three large convoluted tubules, and in frogs and salamanders only two or three units are involved at all.

In the head kidney the tubules generally have ciliated funnels opening from a celomic space containing a single large glomerulus. This nephric part of the celom may be closed off from the remainder of the celomic cavity as a closed pocket—a sort of large renal corpuscle. In the cyclostomes the portion of the celom drained by the head kidney is the pericardial cavity; as a further peculiarity in this group, we find that the liquid drained from the pericardial cavity does not pass into the kidney duct, but, instead, into an adjacent vein. The larval head kidney disappears

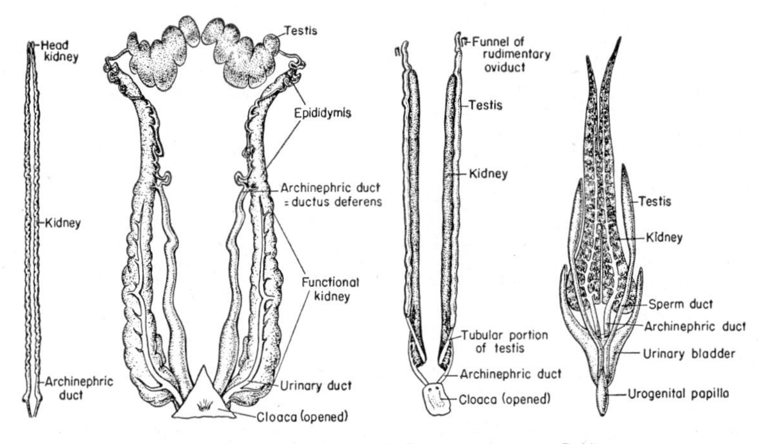

Fig. 229. Urogenital systems in ventral view of males of *A*, the slime hag, Bdellostoma; *B*, the elasmobranch Torpedo; *C*, the lungfish Protopterus; *D*, a teleost, the seahorse Hippocampus. In *A* the testis, not shown, is pendent from a mesentery lying between the two kidneys and has no connection with them. In *B* the testis has appropriated the anterior part of the kidney as an epididymis, much as in most land vertebrates, and utilizes the entire length of the archinephric duct as a sperm duct. In *C* the testis ducts drain, on the contrary, only into the posterior part of the kidney and thence to the archinephric duct. In *D* the sperm duct is entirely independent of the kidney system. (*A* after Conel; *B* after Borcea; *C* after Kerr, Parker; *D* after Edwards.)

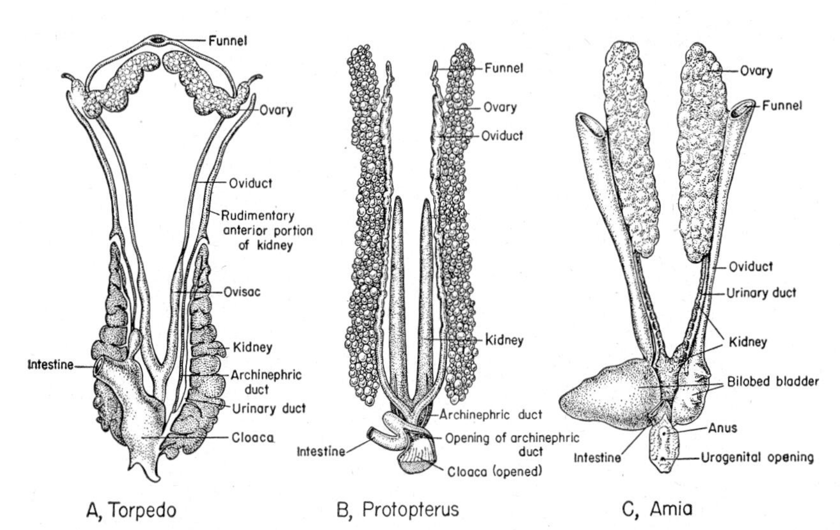

Fig. 230. Urogenital systems in ventral view of females of *A*, the elasmobranch Torpedo; *B*, the lungfish Protopterus; and *C*, the primitive actinopterygian Amia. (In Torpedo the shell gland is not developed.) (*A* after Borcea; *B* after Parker, Kerr; *C* after Hyrtl, Goodrich.)

as a functional organ in the adult amphibian and in many fishes; in teleosts it is commonly modified into a mass of lymphoid tissue. It persists, however, throughout life in many teleosts and in hagfishes.

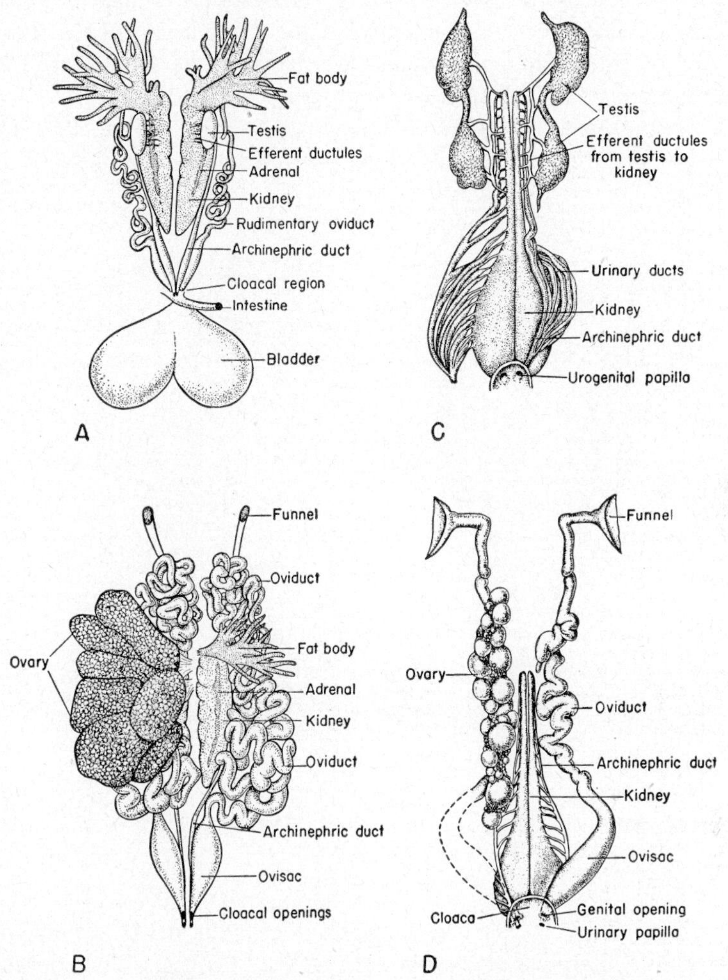

Fig. 231. Urogenital system of amphibians. *A, B,* Male and female organs of a frog (Rana); in *B* the ovary (shown only on the right side of the body) is in a condition close to breeding maturity. The bladder and intestine are not shown in *B*. *C, D,* Male and female organs of the urodele Salamandra. In *C* the urinary ducts of the right side are detached and spread out to show their connections with the kidney. In *D* the ovary is shown only on the right side; the oviduct of the same side is partly removed to show the more posterior urinary ducts. (*A* and *B* after McEwen.)

The Opisthonephros of Anamniotes (Figs. 229, 230, 231). We shall here survey the varied form which the opisthonephros assumes as the functional adult kidney in lower vertebrates. In larval hagfishes, as we have seen, the kidney is essentially a holonephros; in adult cyclostomes this is not the case, for the pronephric part has disappeared from the kidney system. The structure is nevertheless simple. In an adult hagfish the opisthonephros consists simply of a long series of tubules arranged in segmental fashion most of the length of the trunk, each tubule draining directly into the archinephric duct. In a lamprey the tubules are rather more numerous, but the structure is otherwise similar.

In gnathostomes the kidney is universally more complex. Two major developments are seen in the various jawed fishes and amphibians: (1)

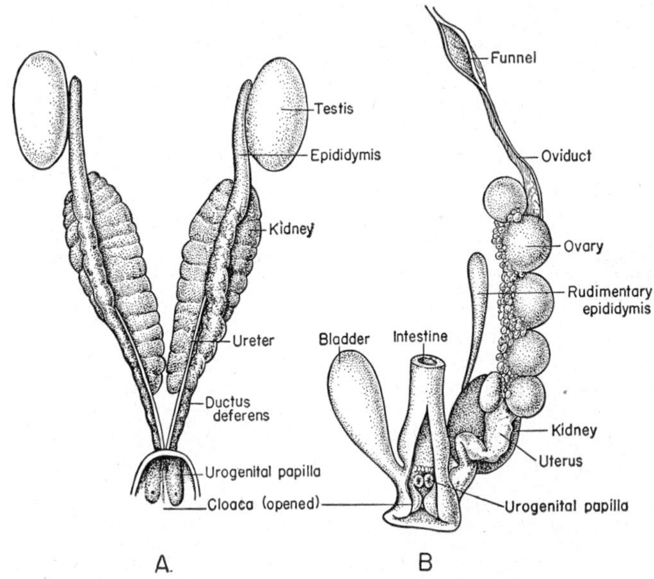

Fig. 232. Urogenital organs in reptiles. *A,* Male organs of the lizard Varanus. *B,* Female organs of Sphenodon. In *A* the bladder is omitted; in *B* it is shown turned to one side. In *B* the organs of the left side only are figured. (*A* after Vandebroek; *B* after Osawa.)

There is a notable increase in tubule number, with loss of segmental arrangement and growth of kidney bulk, often associated with a concentration of the functional kidney into a small portion of its original anteroposterior extent; (2) part of the opisthonephric kidney system in the male is utilized for the passage of sperm. These two features are correlated, for in many groups the testis tends to take over some one specific region of the kidney—usually at its anterior end—for its special use; this region may lose all or part of its proper renal function, and the remaining portion becomes in compensation greatly hypertrophied.

The anamniote kidney is generally elongate; in its length are included most of the trunk segments behind the pronephric region. Seldom, however, is the entire length of the kidney well developed and functional as a urinary organ. In sharks and chimaeras the anterior part is slender, and the great bulk of the kidney substance is concentrated posteriorly in a dozen body segments or less, a condition obviously antecedent to the development of the metanephric kidney type. The reason for this concentration appears to be the fact that the testis has become associated with the anterior part of the kidney and has come to use its tubules for sperm transport to the archinephric duct. This portion of the kidney loses part or all of its urinary function; it may become in the male essentially an accessory to the testis (comparable to the mammalian epididymis) and may degenerate in the female. The salamanders and Apoda among the amphibians show a kidney structure comparable to that of the cartilaginous fishes. The anurans, in which only a few segments are present in the entire trunk, have, as one would expect, a short, compact kidney.

The general trend in kidney evolution in anamniotes has been toward posterior concentration of urinary function in the opisthonephros and an approach to the metanephric kidney type of amniotes. The actinopterygians, however, have struck out along a line of their own in kidney evolution as in other regards. In teleosts (and Polypterus as well) the testis has acquired a duct quite independent of any connections with the urinary system, and an elongate opisthonephros persists.

The Amniote Kidney. As indicated by our story of mammalian development, the amniote kidney is a specialized end type in which the trends seen in lower classes toward compact structure and separation from the genital system have attained a peak development. There remains no connection between the formed kidney and the male genital organs, for the old archinephric duct has been completely abandoned to testicular use in the male, and the kidney is drained by the ureter and its branches. The anterior part of the opisthonephros is functional in the embryo as the mesonephros, but has disappeared in the adult except for vestiges attached to the testis. The mature kidneys, the metanephroi, are formed by a great expansion of nephric material in the region of the most posterior body segments and are relatively short and stout structures, bulging dorsally into the abdominal cavity in the lumbar region.

The reptilian kidney (Fig. 232) has a crenulated appearance due to the fact that it is built of numerous small lobules. The lobulation, in turn, is due to the arrangement of the urine-collecting system. The ureter has, typically, numerous long branches into which series of collecting tubules empty; each branch of the ureter is the center of formation of a lobule. Compared to other amniotes, the number of tubules is low; estimates of 3000 to 30,000 are recorded for various lizards. This may perhaps be associated with the generally low rate of metabolic activity in reptiles as compared with birds and mammals.

Embryology indicates that the adult bird kidneys (Fig. 233) arise from nephrogenic tissue belonging to but a single segment at the end of the trunk, but these organs are of relatively enormous size, typically filling large paired cavities in the roof of the abdomen, shielded by the sacrum

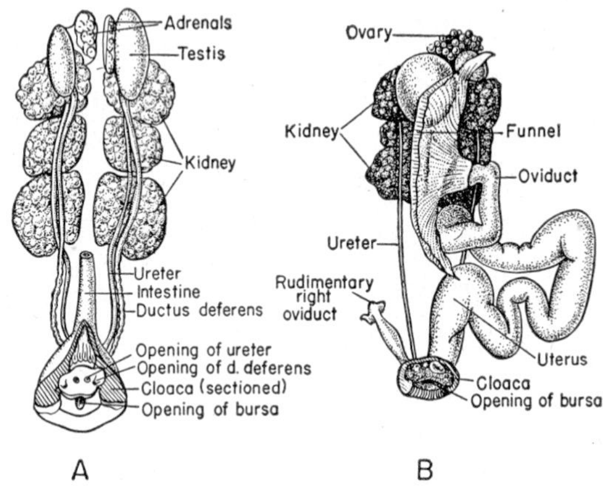

Fig. 233. Urogenital organs of pigeon: *A,* male; *B,* female. The bursa (of Fabricius) is a pouch, of uncertain function, opening dorsally into the cloaca of birds. (*A* after Röseler and Lamprecht; *B* after Parker.)

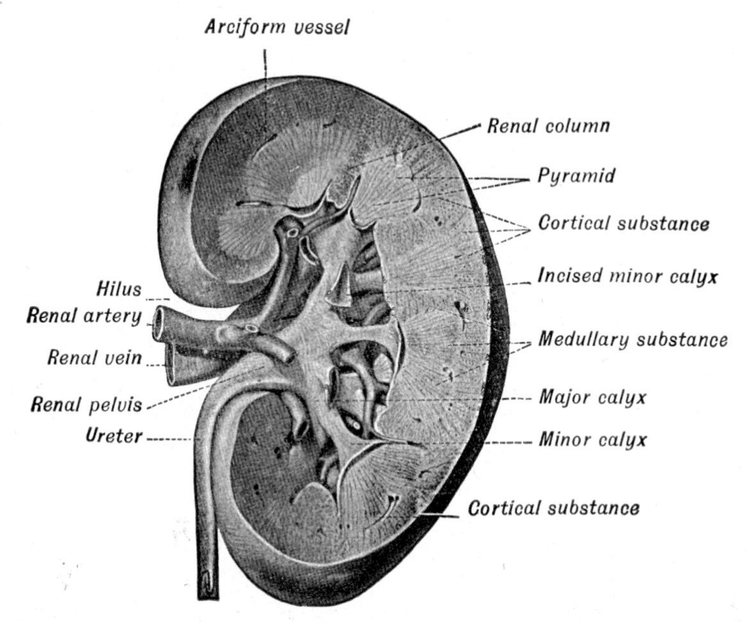

Fig. 234. Section of a human kidney, and its blood vessels (arciform arteries and veins, supplying the tubules, lie between medulla and cortex). (After Braus, from Maximow and Bloom.)

and ilia. Each of the rather elongate kidneys is usually divided into three or more irregular lobes, and these into a large number of small lobules, each containing a branch of the ureter. The glomeruli are small, and tubules are exceedingly numerous; a fowl, for example, has been calculated to have some 200,000 of them, or twice as many as a mammal of comparable size.

The mammalian kidney (Fig. 234) has generally a smooth external surface. Internally, it shows a markedly different build from that of other amniotes. Within the substance of the kidney the ureter expands into a

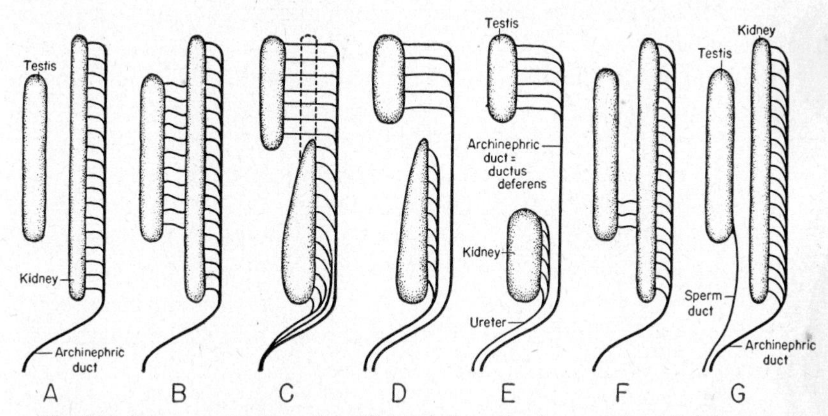

Fig. 235. Diagrams to show the differentiation of urinary and genital ducts in the males of various vertebrates. Ventral views, left side only shown. *A,* Condition seen in cyclostomes; archinephric duct solely for urinary system; gonad not involved. *B,* Condition presumably primitive for gnathostomes, preserved in the sturgeon and gar pike. Testis connects at various points with kidney and thence with archinephric duct. *C,* Stage beyond *B* toward amniote condition as seen in many sharks and urodeles. The testis has taken over the anterior part of the original kidney; the functional posterior part of the opisthonephros tends to drain by a series of ureter-like ducts. *D,* Stage more advanced toward the amniote condition, found in some sharks and urodeles; the kidney drainage is by a single ureter-like duct. In females (not shown) of the types represented by males in *C* and *D,* the tendency toward development of new ducts for kidney drainage is generally not so marked as in the males; the condition shown in *D* is reached by few female forms below the amniote level. *E,* Amniote condition; a definite single ureter formed in both sexes. *F* and *G,* Lungfish and a teleost, showing a type of specialization peculiar to modern bony fishes. The testis tends to concentrate its connections toward the back end of the kidney (as in *F*), and in teleosts evolves a separate sperm duct, releasing the archinephric duct for its original urinary functions. Note that the two ducts for sperm and urine in *G* are *not* respectively homologous with ductus deferens and ureter, as might be thought at first sight.

cavity, the *renal pelvis,* into which the collecting tubules drain. Frequently the pelvis is partially subdivided to form small cup-shaped cavities, *calyces.* The kidney is most commonly a bean-shaped structure, and these cavities lie along its inner curve (the *hilus*). Into each calyx converge collecting tubules in a radiating structure termed a renal *pyramid.*

We have earlier (Fig. 223, *C*, p. 371) noted the structure of the mammalian nephric unit; characteristic are large glomeruli and the presence of the long straight loop of Henle between proximal and distal convoluted regions. In section, the mammalian kidney usually shows distinct "rind" and marrow, *cortical* and *medullary* portions. The former contains principally the renal corpuscles and the convoluted portions of the tubules. The medulla is mainly composed of long, straight collecting tubules and loops of Henle; it is these which give the striated appearance to the pyramids. Although the number of nephrons is not so great as in birds of similar size, it is nevertheless high; even a mouse appears to have about 20,000, and in such large mammals as a man or cow the number may run into the millions.

Blood Supply to the Kidney. The functions of the kidney demand a profuse blood supply for the filtration activities of the glomeruli and for the activity of the tubules. The nature of the kidney circulation will be touched upon in the next chapter, but may be briefly noted here.

In cyclostomes and in mammals the entire blood supply to the kidney is arterial. In all intermediate groups, however, from jawed fishes to reptiles and birds, we find an accessory blood supply in the renal portal system (cf. pp. 455, 456). Venous blood from the tail or hind legs, or both, in its course to the heart is forced to pass through a venous capillary system within the kidney.

In vertebrates with only an arterial blood supply the arteries furnish blood both to the glomeruli and to capillaries which cluster about the tubules. In forms with a renal portal system the functions of the two blood supplies are distinct. The glomeruli are always supplied by arterial blood, and there may be some arterial capillaries at the proximal end of the tubules. Venous blood, on the other hand, is never concerned with the glomeruli, but forms most of the network around the convoluted tubules.

No fully satisfactory reasons have been advanced for the development of the renal portal system and its later abandonment in mammals, nor for the curious division of labor between arterial and venous systems when both are present. However, a partial explanation, at least, may be found in the blood pressure contrast between arteries and veins and in the pressure differences in the system as a whole between higher and lower vertebrates.

The glomerular filter is in some measure to be compared to a pump. For the passage of filtrate through the glomerular walls, blood under the highest possible pressure is desirable; this is, of course, the arterial blood stream.

For the work of the tubules, high pressure is not necessary; what is needed is the presence, adjacent to the tubules, of a large amount of fluid to take up materials resorbed by the tubule cells. To supply enough rapidly circulating arterial blood so that the voluminous kidney tubule capillaries would be constantly filled is a task apparently beyond the power of the circulatory system in typical lower vertebrates. Since venous blood travels relatively slowly and hence would be required in much smaller quantities to fill a capillary system, the development of a venous portal system appears to have been advantageous.

Once true land vertebrates evolved and the gill circulation was done away with, the arterial system underwent a major improvement. The pressure drop between heart and body formed by the gill capillaries was done away with, and arterial blood under high pressure could be supplied to the body organs, including the kidney. Much of the raison d'être of the renal portal system has therewith disappeared. A majority of tetrapods have conservatively retained the old renal portal system, but in some reptiles and in birds part of the renal portal blood by-passes the kidney, and mammals, finally, have abandoned the renal portal to return, full circle, to a purely arterial kidney supply.

Evolution of Urinary Ducts (Fig. 235). In the primitive vertebrate kidney, as seen in cyclostomes, a pair of simple archinephric ducts sufficed for the drainage of urine. In gnathostomes, however, a complicating factor enters the situation: the archinephric duct is invaded in the male by the genital system.

In cyclostomes the sperm are shed into the celomic cavity and reach the exterior through a posteriorly placed abdominal pore. In primitive gnathostomes, it appears, there arose connections between the sperm tubules and kidney tubules; the path of sperm discharge came to be via the archinephric duct. As a result, this duct came to serve a dual purpose. Among modern vertebrates it functions fully for both sperm and urine conduction in a few scattered forms—the Australian lungfish, lower actinopterygians such as sturgeons and gar pikes, the common frog, and a few apodans and urodeles such as Necturus (the mud puppy).

A condition of this sort is, however, unsatisfactory. It is clear that in the course of gnathostome history there has been a struggle between urinary and genital systems for possession of the archinephric duct.

Among bony fishes the urinary system has been the winner. In African and South American lungfishes and Polypterus (Fig. 235, *F*) sperm enter this kidney duct, but only near its posterior end; for most of its course it is a purely urinary tube. In teleosts this development of a discrete sperm tube has been completed; there is a separate route for sperm conduction back from the testis, and the archinephric duct is restored to its original function of conduction of urine alone (Fig. 235, G).

In all other gnathostomes the fight has gone the other way; in most the archinephric duct has been completely taken over in the male sex as a sperm duct. In both sexes, new tubes have appeared, to care for drainage of urine from the kidney.

Among the amphibians are seen a number of stages in the development of functional replacements for the archinephric duct (Fig. 235, *C, D*). In some members of all three orders the archinephric duct still carries both products; a series of short collecting ducts empty into it along the length of the kidney. In other members of both anuran and urodele orders, however, the more posterior collecting ducts make their own way to the cloaca independently of the archinephric duct, and in males of such anurans as Alytes, the "obstetrical toad," and of the common newt Triturus, these tubules have joined to form a single duct draining the entire kidney quite independently of the archinephric duct. Such a duct appears to be in every way the equivalent of the ureter of amniotes. In female amphibians a similar but rather more retarded development of a ureter can be seen, despite the lack of other use of the archinephric duct.

In the Chondrichthyes a parallel development has occurred. In no male shark or skate is the archinephric duct concerned with urine transport. A fan of tubes collect urine from the opisthonephros and unite posteriorly into a common stem—a ureter. In sharks and skates, as in amphibians, ureter development is not so marked in the female, where there is no conflict with sex organs, as in the male. In some female elasmobranchs the archinephric duct still drains the anterior part of the kidney, while more posteriorly the drainage is by a fanwork of ureter-like tubes emptying posteriorly into the end of the original duct. In other elasmobranchs, however, the most anterior part of the kidney is reduced or absent, and the ureter type of drainage is complete.

In all amniotes, as we have seen (p. 382), a definite ureter has been developed; and the archinephric duct has been abandoned to the use of the genital system as a ductus deferens for sperm (Fig. 235, *E*).

Urinary Bladder in Fishes. In a majority of vertebrates there is developed a bladder of some sort, a distensible sac in which urine may be stored before being voided. In elasmobranchs and primitive bony fishes such structures are developed from the urinary ducts themselves. In female elasmobranchs the posterior end of each archinephric duct may become enlarged for urine storage; since the two fuse near their external opening, there may be a median portion to this simple type of bladder. In the male elasmobranch these primary ducts are utilized for the passage of sperm, and no bladder develops along their course. But, as we have just seen, the sharks and rays tend to develop on either side new tubes distinct from the primary duct, for urine passage. Much like the primary ducts in the female, these urinary ducts may expand in bladder-like fashion.

In primitive bony fishes generally—lungfishes and lower actinopterygians—a small urinary bladder develops out of the conjoined ends of the archinephric ducts and hence seems essentially similar to that of female sharks. In teleosts there may be similarly a bladder just inside the urinary opening; here, however, a pocket of the cloaca—present in the embryo, but not in the adult—takes part in the formation of the bladder walls. In the lamprey there is a comparable small bladder or sinus formed from a pinched-off portion of the cloaca into which the urinary duct opened in the embryo (Fig. 249, *A*, p. 415).

The Bladder in Tetrapods. In tetrapods a bladder is useful as a rudimentary sanitary measure; it is, further, useful in water conservation, and in anurans and turtles, for example, resorption of bladder water is important in preventing desiccation under terrestrial conditions. The tetrapod type of bladder appears among the amphibians. It is a new type of structure, not seen in fishes. It appears as a pocket in the floor of the cloaca. There is no direct connection between the bladder and the kidney ducts; to reach it, urine must first be passed into the cloaca (Fig. 250, *A*, p. 416). Excretion of urine from the bladder takes place through the common cloacal outlet.

The reptilian bladder is of similar construction and position. It is found in Sphenodon, in turtles, and in most lizards. It is lost, however, in other

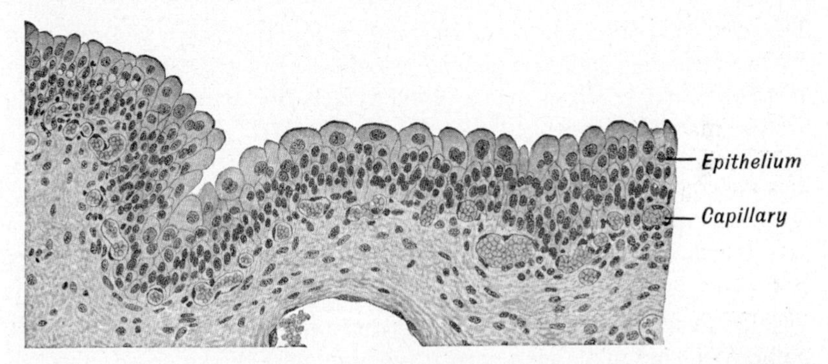

Fig. 236. Section of the wall of a mammalian bladder, showing "transitional epithelium," × 50. (From Maximow and Bloom.)

lizards, in snakes, and in crocodilians; and, further, the bladder has disappeared in birds except in the ostrich. In the absence of a bladder the urine is poured directly into the cloaca to be absorbed by the feces.

We have noted in an earlier chapter (p. 107) that the tetrapod bladder plays an important part in amniote embryology. In all forms with this developmental pattern an important embryonic membrane is the allantois, a pouchlike structure arising from the floor of the most posterior part of the primitive gut. This outgrowth is obviously a much expanded urinary bladder. At birth the allantois itself disappears, but the adult bladder develops as an expansion of the allantoic stalk.

In mammals the tetrapod urinary bladder at long last becomes anatomically incorporated into the urinary system. As noted later (p. 416), the ureters, with modification of the cloaca during development, come to enter directly into the wall of the bladder, and no passage of urine through a cloacal cavity is necessary to reach it.

The tetrapod bladder is a highly distensible organ with stout walls. It is, particularly in mammals, endowed with thick coats of smooth musculature. The inner lining is of a peculiar type known (inappropriately) as *transitional epithelium* (Fig. 236). When the bladder is empty it appears to be of thick, stratified nature. When the bladder is filled and the epithelium expanded in area, it becomes thinned down to a layer or two of flat squamous cells. An epithelium of such a readily distensible type is especially necessary in bladder construction, but is often found in other parts of the urinary system as well.

The Adrenal Cortex. In most tetrapods there is found, adjacent to the kidneys and capping them, a pair of endocrine structures termed the *adrenal glands*. Microscopic examination shows the presence of two different types of tissues, intermingled in lower tetrapods, but "sorted out" in mammals to form a distinct cortex and medulla (Fig. 332). The medullary tissue is a modified part of the nervous system, and will be discussed later (p. 546). The cortical substance is of quite another nature. In fishes there are no formed adrenals, but investigation has shown that cell clusters representing both components can be found between the kidneys, along the course of the major blood vessels dorsal to the celom. The cortical materials form small *interrenal bodies*.

That the cortical material is vital for the maintenance of life was recognized a century ago, for human deaths from an ailment known as Addison's disease were invariably associated with deterioration of the adrenal cortex. The cortex cells have since been discovered to secrete a whole series of hormones (simple organic substances, sterols, similar in composition to sex hormones) which influence the body in a variety of ways, notably (1) in maintaining normal functioning of the kidneys, (2) regulating the distribution of water and of salts in solution among blood, intestinal fluid, and cells, and (3) affecting in some way the production of glycogen and its utilization by muscles.

The first two of these categories suggest a relationship of some sort between the cortical materials and the kidneys. That this physiological association and the topographic propinquity of kidney and adrenal are not accidental, but have historical significance, is indicated by the embryological origin of the cortical substance. This appears as strands of cells which bud off from the epithelium of the roof of the celom medial to the developing kidney tubules. Kidney and adrenal cortex are thus derived from adjacent regions of the embryonic mesoderm.

GENITAL ORGANS

Sexual reproduction is universal in vertebrates as in multicellular plants and animals generally. Among certain groups of invertebrates the hermaphrodite condition, in which the same individual may function as both male and female, is commonly found; in the vertebrates the sexes are functionally separate in every case.

The sex of the individual depends basically upon the nature of its chromosomal inheritance, a balance between male and female potentialities received from the two parents. This balance appears frequently to be a delicate one, easily influenced by environmental factors. Further, the differentiation and maintenance of sex organs is subject to a complex series of hormonal controls, likewise subject to disturbance. It is, in consequence, not unexpected that in many cases, particularly in lower groups, hermaphroditic trends are to be seen; "intersexes" which are not strongly either male or female are met with, and even changes from one sex to another during the lifetime are recorded.

The basic reproductive structures are the *gonads*—ovary or testis. In these organs are produced the *gametes*—eggs (ova) or sperm—by the union of which, in the process of fertilization, the new generation is produced. In all gnathostomes there are associated with the gonads tubes or ducts (with accessory structures) for the transport of gametes and in certain cases for the protection and nourishment of the growing young within the female body. In addition, copulatory organs, which aid in internal fertilization of the eggs, may develop in the region of the external opening of the genital tract. And, finally, secondary sexual characters are frequently found in other bodily structures; sex may affect general body size or proportions, and the development of such features as the plumage in birds, mammary glands in mammals, antlers and horns in ruminants.

Sex Hormones. Hormonal control over the reproductive organs is exercised in great measure by that "master mind" of the endocrine system, the pituitary gland at the base of the brain. This produces, among other products, *gonadotrophic hormones,* identical in the two sexes. Two such hormones are known to be present in the pituitary in every group from elasmobranchs to mammals, and a third substance of this type appears to be present in higher vertebrate groups.

In most vertebrates reproductive activity is a cyclical affair, recurring with the natural seasons and often annual in nature.* Seasonal reproduction is reasonably interpreted in general as a device to bring about the birth of the young at a time when they will have the best opportunity for survival. Sex activity at the mating season appears to be due to increases in the supply of gonadotrophic hormones from the pituitary; the pituitary, in turn, is influenced by environmental climatic conditions—notably increase or decrease in light. Changes in production of pituitary hormones affect the size and functions of the gonads; specifically, an increase promotes maturation of sperm in the male and ripening of eggs in the female. Much of sexual activity, however, is brought about, not directly by the pituitary hormones, but indirectly, through their stimulation of the gonads to produce sex hormones of their own.

Male and female hormones of gonad origin are distinct but related

* Certain domestic animals and the higher primates are exceptional in this regard.

simple organic compounds (sterols). The male hormone—termed an *androgen*—is produced by the interstitial cells of the testis. This hormone is apparently a known specific substance, *testosterone*. The similar female principle, *estrogen,* produced by follicle cells or interstitial tissue of the ovary, is *estradiol*. These substances appear to be responsible for the maintenance in appropriate condition of the accessory sexual organs, and for the proper development of secondary sexual characteristics. The

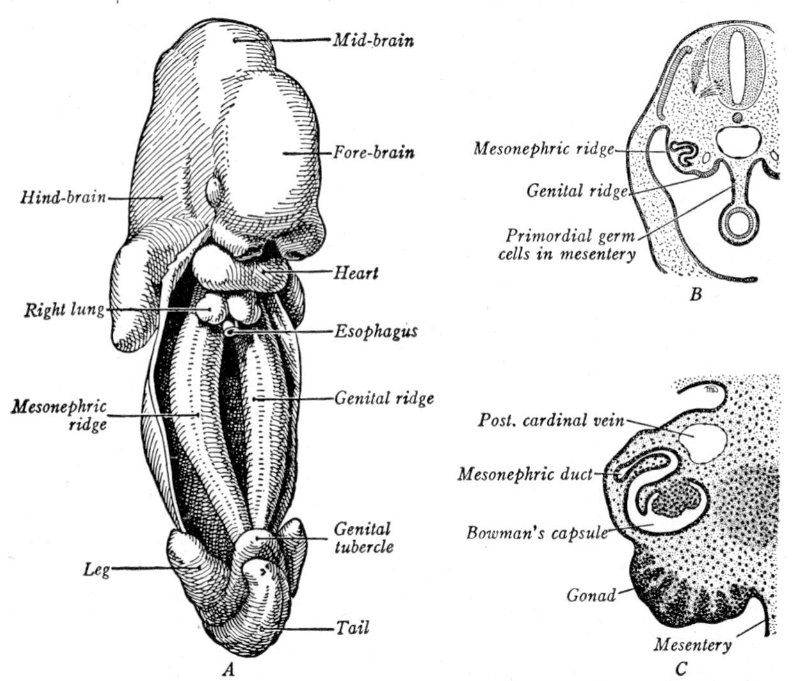

Fig. 237. *A,* Ventral dissection of a human embryo of 9 mm., digestive tract removed, to show genital ridge and kidney (mesonephric ridge) projecting downward into the (opened) celomic cavity. *B,* Cross section of an embryo at an earlier stage (7 mm.); and *C,* a slightly later one (10 mm.). In the latter the primary sex cords are forming in the still "indifferent" gonad, and capsule and glomerulus are forming in the kidney tubules. (From Arey.)

gonadal hormones are responsible for the stimulation of mating activity at an appropriate time in each cycle, the period termed in the female of higher vertebrates that of *estrus,* or "heat."*

Sex Development. Although in gnathostomes the genital systems differ greatly in the two sexes, there are frequently found in the one sex rudiments of the structures characteristic of the other. This is due to the fact that for some time during embryonic life there is an *indifferent stage,* during which gonads and ducts proceed far in their development without any

* The menstrual cycle in the human female is a phenomenon partly, but not wholly, comparable to the estrous cycle of other mammals.

indication of which sex the individual is to become (Figs. 237, *C*, 238, *A*, *B*). The gonads may attain considerable size without showing specific features of either ovary or testis; and both male and female duct systems may differentiate to a considerable degree in potential members of both sexes. Eventually, however, there appears a definite sexual stage; the gonads become definitely testes or ovaries, and only the ducts and other accessory structures appropriate to one sex or the other continue their development. The nonpertinent structures of the opposite sex cease to develop and may be resorbed, but are sometimes merely arrested in their growth to persist as rudiments in the adult (Figs. 239, 240).

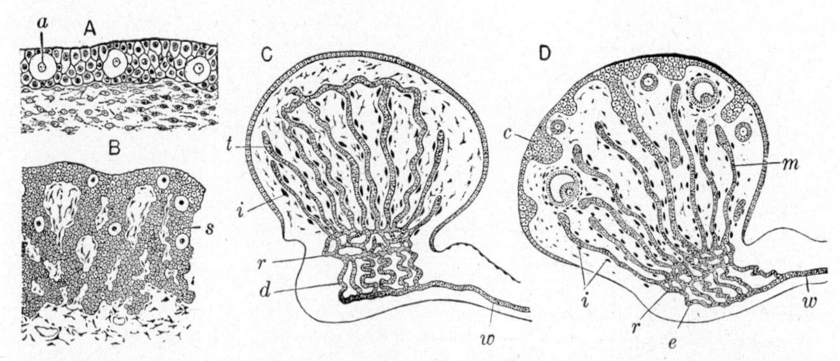

Fig. 238. Development of the gonads. *A*, Section through the primordium of an early "indifferent" gonad in a stage such as that of Figure 237, *B*. *B*, Section through a later "indifferent" gonad of a stage such as that of Figure 237, *C*. Primary sex cords are forming. *C*, Later stage in the formation of a testis. Seminiferous tubules are forming from sex cords, and at the base a junction is formed with the urinary system through the rete testis. *D*, Later stage in the development of the ovary. Structures similar to those of the testis have formed, but are destined to degenerate; secondary cords are forming from the cortex; these give rise to eggs and follicles. *a*, Primordial germ cell; *c*, follicle forming in ovarian cortex; *d*, ductuli efferentes; *e*, epoophoron (= female equivalent of epididymis); *i*, interstitial cells; *m*, medullary cords of ovary (= seminiferous tubules of male); *r*, rete testis in male, rete ovarii (rudimentary in female); *s*, sex cords; *t*, seminiferous tubules; *w*, archinephric (wolffian) duct. (From Maximow and Bloom, after Kohn.)

The presence of the indifferent stage in sexual development is correlated with the nature of vertebrate sex determination. As discussed in Chapter V, the early development of the embryo is mainly due to the organization already present in the unfertilized egg; the influence of the sperm and the hereditary characters which it introduces are not appreciable until a relatively late stage. Whether an individual is to become a male or female depends for the most part upon the chromosome complement of the fertilized egg. The egg has, so to speak, no knowledge of which sex it will become, and the early embryo is prepared for both possibilities. The influence of the sex chromosome mechanism is eventually manifested in the gonad. Once this has become definitely either testis or ovary, male or female patterns in other sexual organs appear in the later stages of embryonic life. Evidence indicates that these events are affected by hormones produced by the gonad, although it is thought that these hormones are not identical with those produced by ovary or testis in later life.

There is in vertebrates a great difference in the rapidity with which different organ systems develop. The nervous system, for example, grows

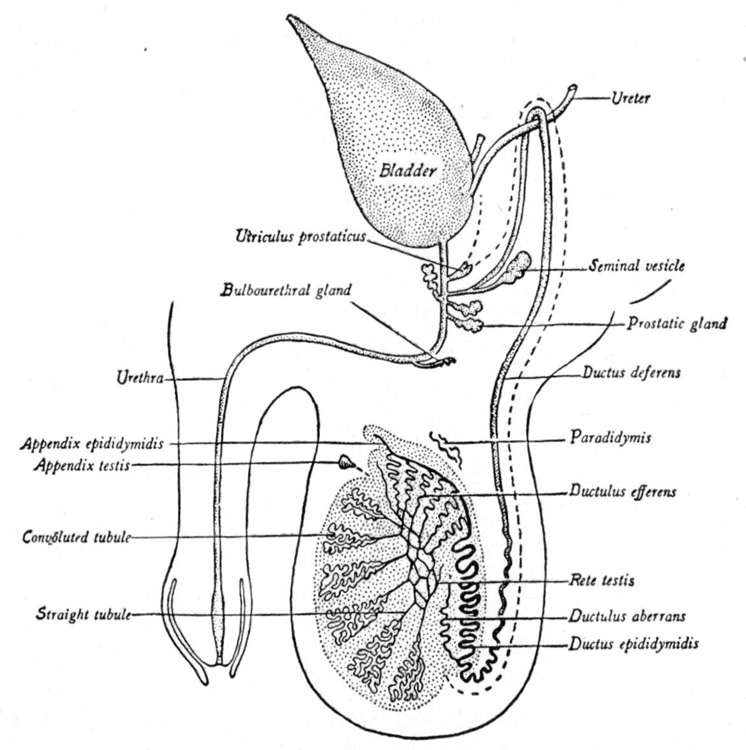

Fig. 239. Diagram of the male human sexual organs. The paradidymis, ductulus aberrans and appendix epididymidis are remnants of the mesonephros not utilized in construction of the epididymis; the appendix testis is a vestige of the upper end of the oviduct; the utriculus prostaticus, a vestige of the lower end of that tube, of which the former course is shown in broken line. (After Eberth, from Bremer and Weatherford, "Textbook of Histology." The Blakiston Co.)

exceedingly rapidly in early stages; the genital organs, primary or secondary, are, on the other hand, the slowest of any to develop. This is to be correlated with the fact that whereas most of the bodily structures must be put to use at birth or even during embryonic existence, the sex organs do not function until maturity of the individual has been attained.

Gonads. The gonads make their appearance only at a stage in embryonic history when most of the main features of other organ systems have been blocked out, and when the celomic cavities are well developed. At this time paired longitudinal swellings, the *genital ridges,* are formed along the roof of the celom, lying lateral to the root of the mesentery and medial to the embryonic kidney (Fig. 237). Elongate to begin with, the gonads developed from such ridges often become relatively short and compact in later stages. Usually the definitive gonad region is toward the anterior end of the abdominal cavity; fat bodies or other structures may arise from abandoned portions of the genital ridge. The *germinal epithelium* of the

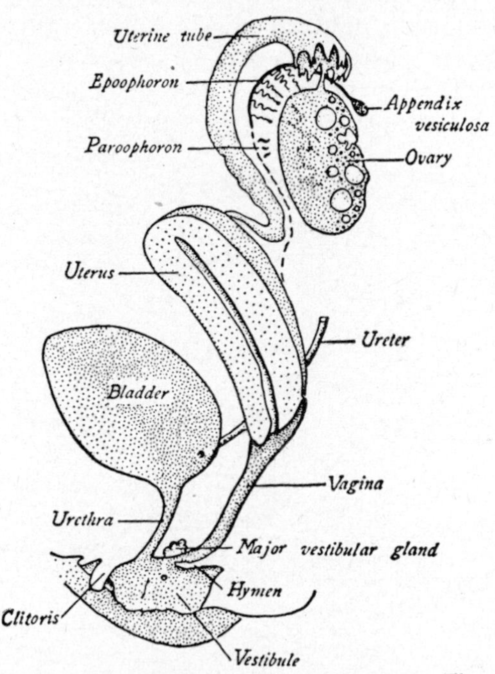

Fig. 240. Diagram of the female human sexual organs. The epoophoron and appendix vesiculosa are remnants of the epididymis; the paroophoron (like the paradidymis) is a vestige of the embryonic mesonephros. The former course of the archinephric duct is shown in broken line. (From Bremer and Weatherford, "Textbook of Histology," The Blakiston Co.)

ridge, continuous with the mesodermal lining of the rest of the celom, forms the more important structural elements of the gonad (Figs. 237, *B, C,* 238, *A*); mesenchyme lying beneath the epithelium forms connective tissues and in higher vertebrates gives rise to special *interstitial tissues.* It is believed that the interstitial cells are a source of the gonad hormones, androgen or estrogen. There is, however, little evidence of the presence of such specialized cells in fishes in either sex, and in the ovary interstitial cells have been identified only in mammals.

Before the end of the indifferent stage the gonad develops in many cases into a compact, swollen structure extending out into the celomic cavity from its dorsal wall in the neighborhood of the developing kidney and often supported by a special mesentery—the *mesovarium* in the female, the *mesorchium* in the male. From the germinal epithelium covering its surface, finger-like structures, the *primary sex cords* (Figs. 237, *C*, 238, *B*, *C*), grow inward into the substance of the gonad. These cords contain germ cells as well as less specialized supporting elements.

Origin of Germ Cells. Germ cells from which eggs or sperm develop—large cells with a characteristic clear cytoplasm—may be identified at a relatively early stage in the germinal epithelium of the gonad, and it seems logical to assume that the germ cells are of local, mesodermal origin.

There is, however, a peculiar quirk in the story. It may well be that the functional eggs and sperm of the adult do develop in this fashion. But a

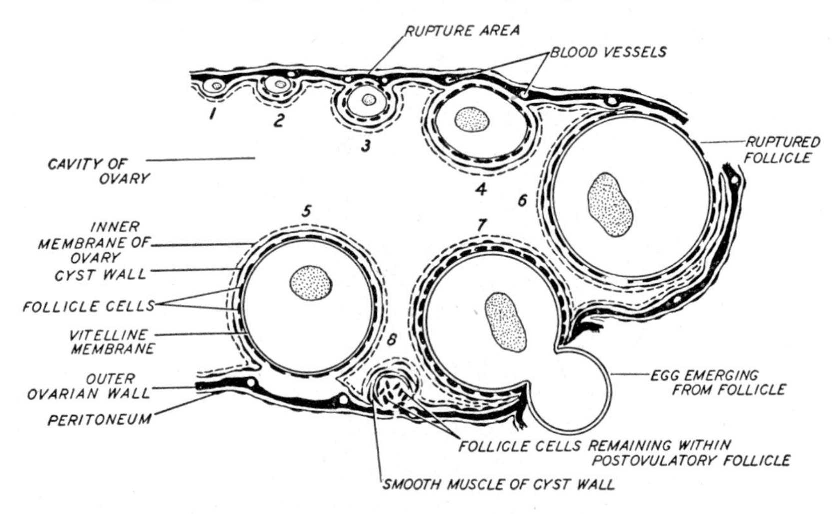

Fig. 241. Diagrammatic section through a lobe of the frog's ovary. *1* to *5* illustrate stages in the growth of the follicle; *6* and *7* rupture of a follicle and emergence of egg; *8*, postovulatory follicle. (From Turner.)

considerable body of evidence indicates that the first germ cells to appear in ovary or testis are actually derived from the endoderm. In embryos of every major vertebrate group from cyclostomes to mammals there have been observed in the gut lining, at an early stage, cells which histologically are quite distinct from the ordinary cells of the digestive tract. If a series of stages is followed through, these distinctive cells can be seen to leave the gut walls and migrate—either through the intervening tissues (Fig. 237, *B*) or by way of the blood stream—into the celomic epithelium from which the genital ridges form. These cells are primary germ cells, from which eggs or sperm later develop.

Biologists have often emphasized the fact that in animals in general the germ cells form an exceedingly independent tissue; the rest of the body, the "soma," is, on this point of view, merely a temporary structure shielding and conserving the potentially immortal germ plasm. In many invertebrates the future germ cells become distinct from the rest of the embryo at an early cleavage; this migration of vertebrate germ cells is, possibly, a demonstration of their equally distinctive nature.

But there are puzzling features in the vertebrate story. Most workers believe that the primordial germ cells, once arrived at the gonads, give rise to eggs or sperm. Some, however, believe that this is not the case, and that these cells degenerate after a fruitless migration. Many hold an intermediate viewpoint that these first germ cells give rise to a first-developed series of gametes, but that the functional eggs and sperm of the adult are derived from other germ cells, which arise in situ in the mesodermal epithelium of the gonad. Under this view the function of the immigrant primary germ cells may be merely the initiation and stimulation of the process of gamete formation.

Intersexes. Beyond the embryonic indifferent stage there usually occurs a clear-cut divergence of developmental paths leading to the formation of a definite ovary or testis. In many vertebrates, however, the mechanisms of sex determination are so delicately balanced that the gonads hesitate (so to speak) between these two alternatives, and both eggs and sperm may tend to develop. But never, as far as known, do both types of gametes come to maturity in the same individual at the same stage of life; there is never a functional hermaphrodite, but intersexual conditions of one sort or another may be present.

In cyclostomes both eggs and sperm begin development in the gonad of young individuals. For the most part one sex or the other dominates in maturity, but a percentage of hagfishes remain sterile intersexes throughout life. Among the teleosts, the sea bass (Serranus) and its relatives exhibit in the young a development of both ovary and testis, although only one comes to maturity. Amphibians show the most delicate balance between sexes of any vertebrate group, and variants of one sort or another are numerous. For example, in the common toad (Bufo) the males have an ovary-like structure—*Bidder's organ*—in addition to a testis. In the common frog genus Rana, young males show female tendencies, and genetic females may in old age shift to the male side of the balance and produce sperm.

Ovary: Egg Production (Figs. 241, 242). In the development of an ovary beyond the indifferent stage there is generally found as a diagnostic feature the proliferation inward from the germinal epithelium of *secondary sex cords* (Fig. 238, *D*, p. 397). The set first formed degenerates, and it is from this succeeding series that the ova of the adult arise. In the developing ovary the germ cells contained in these cords divide repeatedly and develop eventually, by a complicated maturation process —*oogenesis*—into definitive egg cells. Each egg cell becomes surrounded by a cluster of other cells derived from the sex cords to form a *follicle;* connective tissue cells may form a further external sheath. The follicular

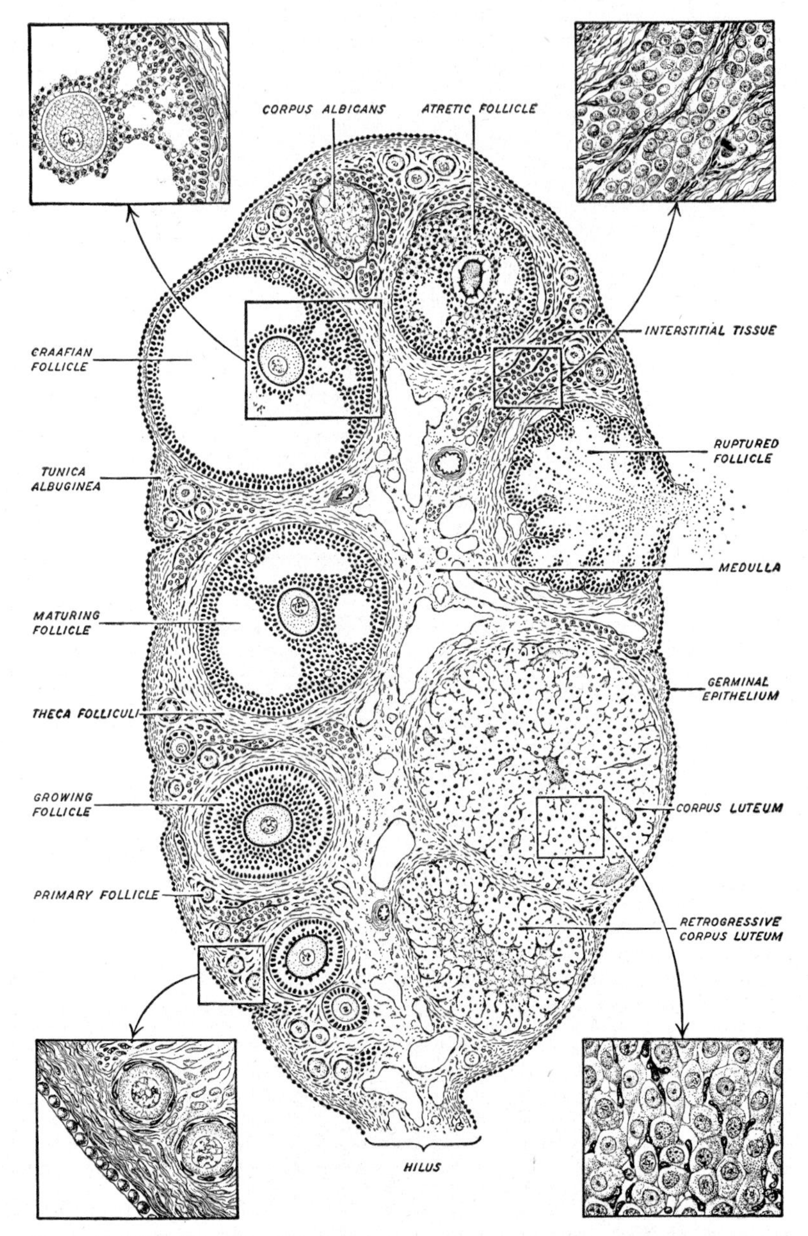

Fig. 242. Diagram of a composite mammalian ovary. Progressive stages in the differentiation of a follicle are indicated on the left. The mature follicle may ovulate and form a corpus luteum (*right*) or degenerate without ovulation (atretic follicle). (From Turner.)

cells aid in the sustenance of the growing egg; in addition, the follicle is a seat of estrogen formation. As maturity is approached, egg and follicle increase in size. In forms with large-yolked eggs the follicle comes to be relatively enormous, and to form a major expansion on the ovarial surface. In viviparous mammals the egg is small, but the follicle surrounding it nevertheless enlarges, causing the development of a liquid-filled space in its interior.

At seasons of reproductive activity follicles ripen at the surface of the ovary and burst, releasing the egg into the surrounding celomic space—

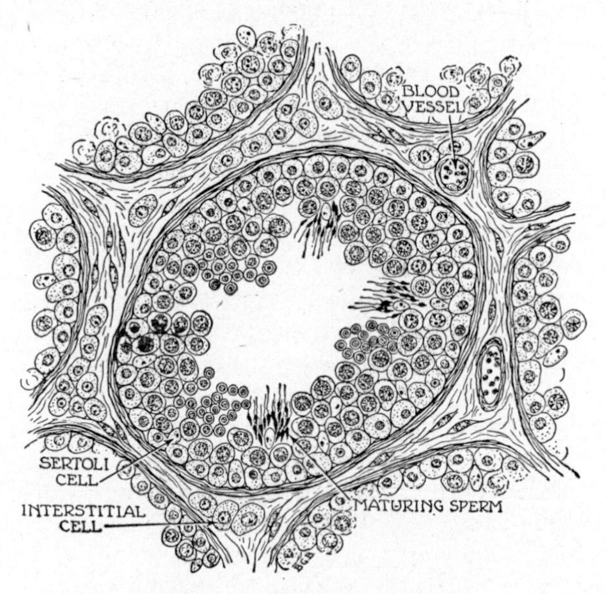

Fig. 243. A small area of a mammalian testis, showing one tubule and parts of several others in cross section and the intertubular tissue. (Hooker in Fulton-Howell.)

the process of *ovulation*. In most lower vertebrates the follicle material is quickly resorbed. In amniotes, however, and in some elasmobranchs the follicle persists for some time after the egg has been released, its cavity filled with a mass of yellow material from which its name of *corpus luteum* is derived. This "yellow body" secretes a hormone, *progesterone,* which prepares the uterus for the implantation of the ovum in its walls and, if fertilization and implantation are accomplished, stimulates the development of a placenta. (The placenta "aids" the ovary in many mammals by becoming itself a seat of hormone production.)

The number of eggs in a mature state in the ovary at any one time is, in most groups of vertebrates, small—from two to a dozen in many cases. In amphibians, however, hundreds or even thousands of ripe eggs may be present at breeding time, and in actinopterygians there may be hundreds of thousands or even millions (the codfish is estimated to lay 4,000,000 eggs in one season). At an early stage of ovarian development a large number of tiny eggs are produced as a stock from which those matured at successive seasons may be drawn. It was once believed that there is no further production of egg cells beyond this early stage; but the large number of eggs produced by many fishes and amphibians indicates that there must be later proliferations of germ cells from the germinal epithelium on the surface of the ovary. On the other hand, forms in which the number of eggs which mature is small, obviously do not use up the abundant supply of egg cells; in such types investigation shows that a large proportion of the eggs degenerate, even if far advanced in follicle construction, without reaching maturity.

The ovary is usually a paired structure, often with an oval shape when in a resting phase, but frequently distended and irregular in outline at the breeding period. In cyclostomes the pair of gonads fuse into a single median structure, and the two ovaries fuse in many teleosts as well. In many elasmobranchs the left ovary remains undeveloped; in birds and in the primitive mammal Ornithorhynchus (the duckbill) the left alone matures. In amphibians and reptiles the ovary is hollow, containing central lymph-filled cavities. In other cases the central part—the medulla—is mainly a connective tissue structure; the eggs and their follicles, together with the germinal epithelium on the surface, form the cortex.

Testis. The early history of the testis is, through the indifferent stage, identical with that of the ovary in its development from a portion—usually an anterior part—of the genital ridge, in the development of a germinal epithelium, and in the growth inward from this surface of primary sex cords containing germ cells (Fig. 238, *C*, p. 397). From this point onward, however, the history of the two organs diverges. In contrast to the ovary, no secondary set of sex cords is developed; the primary cords form the sole source of male gametes. No germinal epithelium remains at the surface of the testis, which, beneath a normal celomic epithelium, is surrounded by a connective tissue sheath.

The testis tends to be, on the whole, a more compact and regularly shaped structure than the ovary; it is subject to seasonal change, but generally does not exhibit marked differences between breeding and resting stages as does the female gonad. In cyclostomes the testis, like the ovary, is a median rather than a paired structure. In sharks the two testes may fuse posteriorly, and the right member of the pair may be the larger. In various birds and mammals the left testis tends to be moderately larger than the right.

The testis includes in its structure a considerable amount of connective tissue and, in amniotes, interstitial cells producing the male hormone. Its reproductive tissues are produced from the primary sex cords. The material of the cords breaks up in the substance of the testis into numerous hollow structures. In fishes and amphibians these are usually more or less spherical sperm *ampullae;* in amniotes and some teleosts there are formed elongate *seminiferous tubules* (Fig. 243). These structures are lined by epithelia in which countless numbers of sperm are produced at each breeding season. The epithelium of tube or ampulla is stratified and contains both reproductive cells and relatively rare supporting cells—*Sertoli cells.* At the base of the epithelium there are generally to be found germ cells which are little differentiated, the *spermatogonia.* From these primary cells sperm are produced in the process of *spermatogenesis.* The spermatogonia by division give rise to *spermatocytes.* Each of these, by two further divisions, forms cells which are released at the surface of the epithelium, as *spermatozoa.* They include in their structure little except a head, containing the nuclear material, and a long, motile tail. Spermatozoa

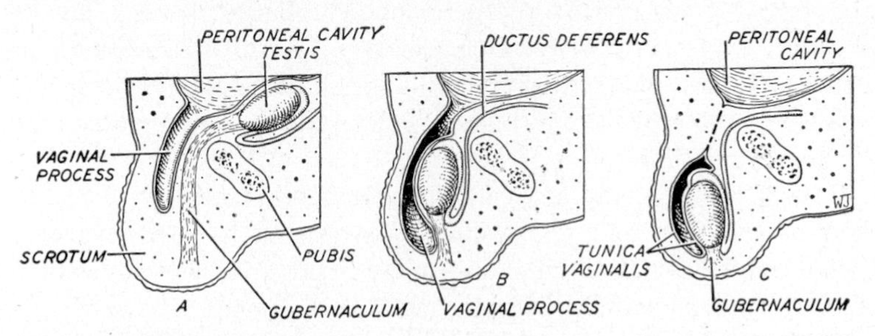

Fig. 244. Descent of the testis in mammals; ventral surface of body at left. A vaginal process develops from the body cavity; its peritoneum forms the vaginal tunic of the scrotal sac. Broken line in *C*, position of inguinal canal in mammals in which sac is not completely closed, and site of inguinal hernia. (From Turner.)

are exceedingly small but exceedingly numerous, and even in small animals the total production may be measured in billions.

In most vertebrates sperm production, like that of eggs, is a cyclic phenomenon. The sperm ampullae of lower vertebrates are expendable and are thus rather comparable to egg follicles. All the germinal materials of a given number of ampullae ripen at a specific breeding season, the sperm are discharged, and the ampullae resorbed, to be replaced by others which have meanwhile developed more slowly. The seminiferous tubules of higher vertebrates, on the other hand, are characteristically permanent structures, producing successive crops of sperm and persisting little changed throughout the breeding life of the individual.

Descent of the Testis (Fig. 244). In most vertebrates the gonads of the adult retain their primitive position in the upper part of the celomic cavity. In a majority of mammals, however, a radical change in the position of the testes occurs during embryonic development. Paired pouches form in the floor of the abdominal cavity and cause external swellings on the ventral surface of the body just anterior to the pelvis. These are the *scrotal sacs,* lined internally with celomic epithelium, and constituting separate portions of the celomic cavities. Each testis moves posteriorly and ventrally from its original position, and descends into a sac, accompanied, of course, by the ductus deferens for sperm transport, and a fold of its proper mesentery—the *gubernaculum.* In some cases the scrotal sacs remain in open connection with the abdominal cavity through an *inguinal canal,* and the testes may withdraw from the sacs into the body between breeding seasons. In other mammals the sacs may be closed off from the abdominal cavity by a fold of tissue. Rupture of this fold—a weak spot in the abdominal wall—may lead to the condition known in man as inguinal hernia. Most commonly the two scrotal sacs may push downward and backward to fuse externally below and behind the penis; more rarely, as in many marsupials, they may be anterior to that organ.

Experiments and the study of abnormalities have revealed the functional basis for this unusual phenomenon of testicular descent. The delicate process of sperm maturation cannot go on at high temperatures. The temperatures maintained internally in the bodies of mammals are in general too great for sperm formation; temperatures in the exposed scrotal sacs are several degrees lower.

It may be asked why a parallel phenomenon is not found in birds, where even higher body temperatures are present. It may, however, be pointed out that a pair of air sacs lie close beside the testes; it is possible that they cause some reduction in avian testicular temperature.

The Oviduct and Its Derivatives in Lower Vertebrates (Figs. 230, 231, 232, pp. 384, 385, 386). In cyclostomes both eggs and sperm are shed into the celom, and must find their own way to the posterior end of this cavity, whence a pair of abdominal pores afford them passage to the exterior. In all gnathostomes the sperm are conducted to the outside by way of closed tubes, but the eggs are still shed in most cases into the celom. They are not, however, really freed into this cavity, for they are (except by accident) received, through a funnel-like structure close beside the ovary, into a tube which cares for them on the remainder of their journey. This is the primitive *oviduct* (or müllerian duct), present, with various regional modifications, in all jawed vertebrates. Along the course of the oviduct there may be formed specialized enlargements for various purposes: storage of eggs before laying, the deposition of a shell, or the retention of the egg during embryonic development and a subsequent "live" birth—the *viviparous* condition.*

The oviduct and specialized structures derived from it arise in the dorsolateral wall of the celomic cavity, and generally come to be suspended in the celom by special mesenteries; that for the oviduct itself

* In contrast to the primitive *oviparous* method, in which the egg is laid and development takes place externally. Some distinguish from viviparity an intermediate *ovoviviparous* type, in which development is internal, but takes place without the formation of a placenta or other means of obtaining nourishment from the mother.

is termed the *mesosalpinx*. The embryonic course of the oviduct parallels that of the archinephric duct. In elasmobranchs it is formed by a splitting in two of this duct. In some tetrapod groups there are indications of a similar origin, and in some lower actinopterygians—sturgeons, paddlefishes—the oviduct remains, even in the adult, as a side branch of the archinephric duct. On the other hand, in a majority of land vertebrates the oviduct arises as a distinct structure, formed by an infolding of the celomic epithelium. The evidence on the whole thus indicates that the female genital duct (like that of the male, described below) is derived from the urinary system, but has become so specialized that embryonic evidence of its origin may be lost.

The oviduct is seen in its simplest form in lungfishes and amphibians—groups with reproductive habits which apparently are primitive, the female laying at each season a moderate number of shell-less eggs of modest size. It is here for the most part a simple ciliated tube, with glands or gland cells which secrete mucus or a gelatinous covering for the eggs. In the resting phase the egg tube is usually small in diameter and relatively straight; at the breeding season it may elongate to become highly convoluted, and its diameter may increase greatly.

There is no direct connection between ovary and oviduct in primitive forms (nor, for that matter, in most advanced types). The anterior end of each oviduct, however, lies close beside the ovary; it consists of a ciliated funnel, the *infundibulum,* with a frilled margin. (The funnels are normally separate; in the Chondrichthyes, exceptionally, the two are fused.) Eggs released from the ovary are normally caught up by the funnel to begin their journey down the tube. We have noted that anterior kidney tubules in embryos and in lower vertebrates frequently connect with the celomic cavity by funnel-shaped nephrostomes; if the oviduct is derived from kidney structures historically, its funnel may be compared to a greatly enlarged nephrostome.

Posteriorly, there is little specialization of the oviduct in the more primitive condition. There is generally an expanded distal part of the tube, an *ovisac,* which serves for the storage of eggs. A few salamanders and toads are viviparous; fertilized eggs are retained within this sac until embryonic development has been completed and the young are "hatched."

The oviducts open posteriorly into the cloaca, when present, or to the surface of the body. In variable fashion the two tubes may remain separate over their entire length, as in salamanders and some anurans, or fuse into a common tube, as in other anurans and in lungfishes.

An exceptional situation is that which must be dealt with by the actinopterygians, and more especially the teleosts, in which countless thousands or even millions of eggs are released during a short breeding season.*

* A few teleosts have become viviparous, the eggs developing within the ovarian cavity. In such forms the number of eggs is secondarily reduced.

It is obvious that the normal system of egg reception through an open funnel would be inadequate; there would be the danger, or rather the certainty, that the entire body cavity would be hopelessly choked with eggs. In the teleosts (and in the related gar pikes as well) the difficulty has been met with by a complete enclosure of the ovary and tube system so that no escape of eggs into the general body cavity is possible. The leaf-shaped ovary is so folded that it encloses a cavity within its substance, or separates off next to it a part of the celom. Into this cavity the eggs are shed, and directly from it a funnel-like duct leads backward to the exterior; there is no possible opportunity for the escape of eggs into the interior of the body. So different is this apparatus from that of other vertebrates that many workers believe that the duct here is a special teleost structure not homologous with the oviduct of other vertebrates.

In the remaining vertebrate classes—cartilaginous fishes and amniotes—a large, shelled egg has been evolved, and special regions of the tube are developed which produce the egg white and shell. In elasmobranchs and chimaeras there is, part way down the tube, an enlarged and specialized region, the *shell gland* or nidamental gland. Although this usually appears to be a uniform structure externally, there are internally two types of epithelia present when the gland is well developed. In the upper part of the nidamental region of the tube albuminous material—"egg white"—is secreted about the egg; in the lower part a hard, horny egg case is formed. Fertilization must, of course, occur before the shell is added to the egg, and in the Chondrichthyes internal fertilization takes place. With the aid of claspers on the pelvic fins of the male the sperm are injected into the cloaca. Thence they travel "upstream" in the oviduct, and fertilize the egg before it descends to the shell gland.

We have noted that the young are born alive in a few amphibians and teleosts. Viviparity requires internal fertilization of the egg, and is hence the exception in vertebrates with primitive reproductive habits; for external fertilization appears to have been the original method of insemination. With the development of shelled eggs and consequent internal fertilization in cartilaginous fishes, opportunities for the development of viviparous conditions are increased.*

Distally, there is present in the elasmobranch oviduct an ovisac, which in many forms is simply a repository for eggs before discharge. In a variety of sharks and skates, however, viviparity is present, and this sac serves as an organ in which the embryo is sheltered and nourished until its birth as a "pup."The ovisac in such forms supplies food materials in one fashion or another to the young. Most common is the development from the inner wall of the ovisac of highly vascular leaflike or filamentous

* Hagfishes have a thin horny shell and yet have external fertilization; a soft spot (micropyle) in one end of the shell allows the sperm to enter. In the absence of any oviduct system the capsule is formed about the egg before it leaves the ovary.

outgrowths which gain contact with the embryo's yolk sac; a connection is formed somewhat comparable to the placenta of a mammal.

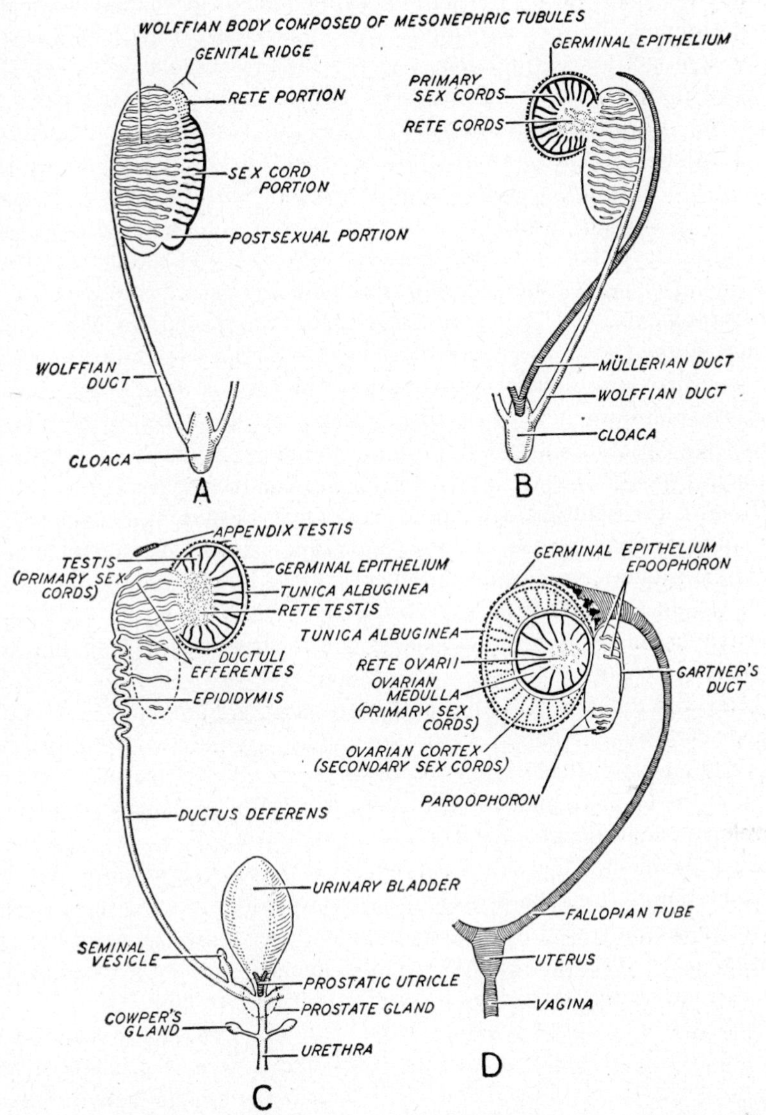

Fig. 245. Embryonic development of the genital system of amniotes. *A*, Early indifferent stage, showing sex cords, developing rete testis, mesonephric kidney and archinephric duct (wolffian body, wolffian duct). *B*, Somewhat later stage, in which the embryonic oviduct (müllerian duct) has appeared. *C*, The adult male (cf. Fig. 239). *D*, Adult female (cf. Fig. 240). (From Turner.)

The Oviduct in Amniotes—Uterus, Vagina (Figs. 232, *B*, p. 386; 233, *B*, p. 388; 240, p. 399; 245). In amniotes, as in sharks, a shelled egg has brought about specializations of the tube for the production of albumen and shell. In reptiles and birds the successive divisions of the oviduct are given names comparable to those used in mammalian anatomy, although the functions differ. The greater part of its length is the oviduct proper, or *uterine tube*. This is a broad, muscular tube, capable of further distention at the breeding season. The *uterus** lies close to the distal, cloacal end of the tube; it is in the resting phase somewhat greater than the tube in internal diameter and may be greatly expanded during the breeding season. Its walls contain a heavy coat of smooth muscle. Glands are present in the epithelia of both tube and uterus; that of the tube secretes the albumen or egg white, that of the uterus forms the shell. Beyond the uterus a short terminal segment leads to the cloaca; this is often termed a vagina, although it lacks the special function of that region in mammals. The two vaginae of reptiles open separately into the cloaca. Birds are universally oviparous, but various lizards and snakes (like the elasmobranchs) bear their young alive, and in some cases have paralleled elasmobranchs and mammals in the development of structures through which they may receive nutriment from the mother.

In birds the right oviduct, like the right ovary, is absent, but the structure of that remaining has retained the reptilian pattern. Among the mammals the monotremes, which produce a shelled egg, although of small size, have a structure essentially similar to that of reptiles, with paired uteri containing shell-forming glands, and separate openings of the two ducts into the cloaca.

Markedly different from those of more primitive amniotes are conditions in marsupials and placental mammals, for here, with viviparity, shell formation has ceased; the uterus becomes the site for the development of the embryo; and the ends of the two oviducts, usually conjoined, develop as the vagina for reception of the male organ.

Since the mammalian ova are of small size, the upper part of the original oviduct, the uterine tube, is a slender structure, and lacks, of course, the albumen-secreting glands found in other amniotes. Fertilization of the egg usually takes place at the upper end of the tube, and the trophoblast (p. 86) has already developed before the uterus is reached.

The uterus, or womb, is a thick-walled muscular structure, within which the embryo is destined to pass its embryonic existence. It has a richly vascular epithelium, the *endometrium,* which may be relatively thin during resting phases, but is greatly thickened at seasons of reproductive activity; its growth rhythms are controlled by hormones of ovarian origin. The placenta, affording maternal nourishment of the developing embryo,

* The distal expansion of the oviduct in elasmobranchs and amphibians, here termed the ovisac, is frequently called a uterus. Since its homology with the mammalian uterus is none too certain, it is probably better not to extend the use of this term to these lower vertebrates.

is formed by a union of the uterine epithelium with the external membranes of the embryo. When the placenta is once formed, this structure itself may function as a source of hormones which aid in its own maintenance during the later stages of development.

In the most primitive mammalian condition the two uteri are quite separate, a *duplex* condition; this is found, for example, in marsupials and in many rodents and bats (Fig. 246). In most mammals, however, the distal parts of the two uteri are fused together to give a *bipartite* or *bicornuate uterus;* and in higher primates there is a complete fusion to a *uterus simplex.*

Although the uteri may remain distinct, the terminal portions of the two egg tubes are invariably fused in placental mammals to form a vagina. The cloaca, into which the genital tubes opened in ancestral types, is absent in higher mammals, but is represented by a urogenital sinus (cf. Fig. 251, *F*, p. 417). This sinus is frequently shallow; when of some

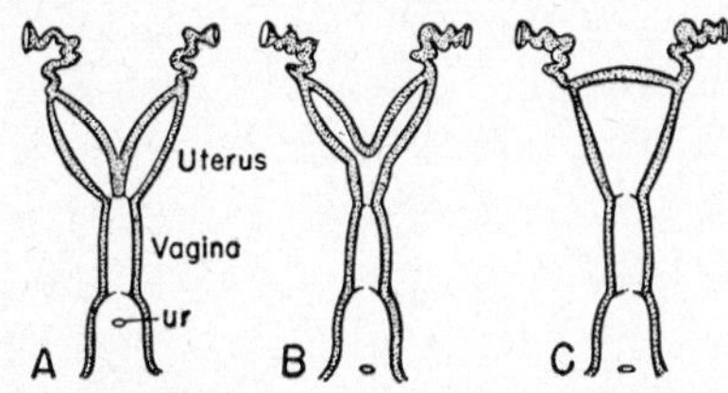

Fig. 246. Diagrams of common types of uterus in mammals. *A,* Primitive duplex uterus; *B,* uterus bicornis, with partial fusion of the two; *C,* complete fusion, forming uterus simplex. *ur,* Urinary opening in urinogenital sinus (vestibule). (After Weber, Portmann.)

depth, as in a considerable number of mammalian types, it has the appearance of a further continuation of the vagina itself.

The marsupials are structurally transitional between primitive reptilian and placental types in their reproductive system, but vaginal construction is of a specialized sort (Fig. 247). The vaginal tubes are paired, but are fused at proximal and distal ends, and a vaginal sinus between them sometimes extends outward to form a third, median exit from the uterus to the exterior.

Sperm Transport: Epididymis, Ductus Deferens. In the male cyclostome, as in the female, there is no duct system for sex products. Ripe sperm ampullae open to the celomic surface of the testis; the spermatozoa are discharged into the celom and must make their own way to the outer world through the abdominal pores.

In all higher vertebrates this inefficient mode of transport has been abandoned; the male has developed a system of ejaculatory ducts. This, however, contrasts strongly with the female duct system. It is always closed throughout its course. Further, though the female duct may be of somewhat questionable origin, the male transport system is demonstrably a modified part of the urinary structures. As the testis develops in any vertebrate, it lies close beside (and medial to) the embryonic kidney (Fig. 237, p. 396). A short distance away from the ampullae or seminiferous tubules are the tubules of the kidney, which empty into the archinephric duct leading to the cloaca. If the short gap between testis and kidney is bridged, there is available a closed conduction system leading to the exterior which could be followed by the sperm in safety, avoiding the vicissitudes of travel through the celomic wilderness. This gap was bridged by ancestral gnathostomes, and reproductive functions were superposed on the urinary system.

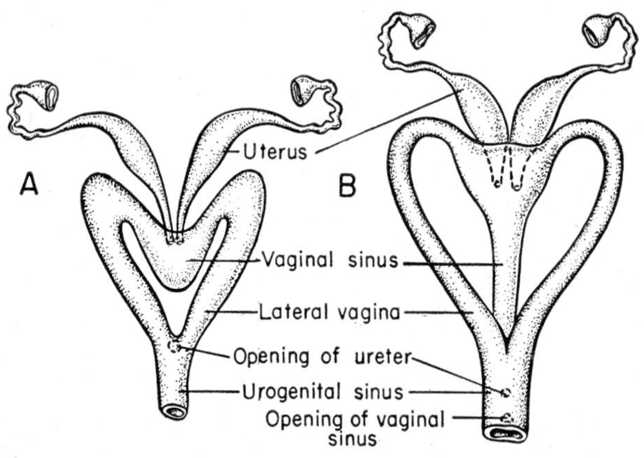

Fig. 247. Female genital organs of marsupials. *A*, Opossum; *B*, kangaroo. From a median vaginal sinus there develop, in the opossum, a pair of lateral vaginae which unite distally in the urinogenital sinus. In the kangaroo figured the vaginal sinus has developed into a median vaginal tube. (After Vandebroek.)

Although there are variations of one sort or another, the connections between the seminiferous structures of the testis and the archinephric duct follow a basically similar pattern in many vertebrates (Fig. 248). Ripe ampullae or the tips of the seminiferous tubules may be connected with one another by a longitudinal *central canal* of the testis or, in birds and mammals, by a network of small canals, the *rete testis*. From this a number of parallel tubules develop to run across the short distance to the edge of the kidney; these are the *ductuli efferentes*. In many vertebrates these ductules, when the kidney is reached, enter a *lateral kidney canal* which connects with adjacent kidney tubules. This canal does not develop in mammals; in any event, however, the sperm pass through the erstwhile kidney tubules, now termed *ductuli epididymidis,* and finally reach the archinephric duct.

Presumably the most primitive condition in gnathostomes is one in which the archinephric duct continues to serve impartially both urinary and genital functions, as in certain of the more primitive bony fishes and a number of amphibian types. As we have noted, however, in our discussion of the urinary system, most vertebrates have developed new ducts so that sperm and urine travel separate courses. The events in this history have been noted in discussing the urinary organs; they may be reviewed here from (shall we say) the point of view of the testis (Fig. 237, p. 396).

In actinopterygians the kidney has retrieved its original duct, and the spermatozoa pass back to an external opening behind the anus by way of a new sperm duct of simple construction which by-passes the kidney. In all other groups of vertebrates, we have noted, sperm transportation has triumphed in the fate of the archinephric duct. Parallel developments are witnessed in cartilaginous fishes and in land vertebrates. Urine conduction is in elasmobranchs and many amphibians restricted to the posterior portion only of the archinephric duct, and in all amniotes has been completely excluded. With such partial or total exclusion the archinephric duct has become a *ductus deferens,* a tube specific for sperm transport; the testicular connections are universally with the proximal, or anterior end of the duct.

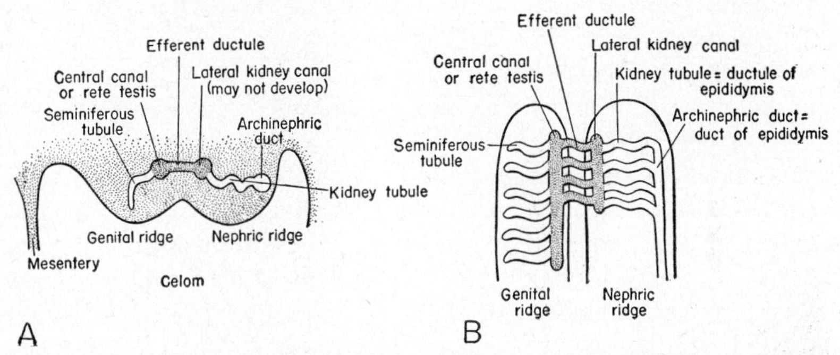

Fig. 248. *A,* Cross section of an amniote embryo, to illustrate the fact that testis and kidney are adjacent (cf. Fig. 239) and may thus readily come into connection by the development of efferent ductules bridging the gap between sperm tubules and kidney tubules. *B,* Diagrammatic ventral view of a section of nephric and genital ridges to show mode of connection, usually with a central testis canal or rete testis connecting the sperm tubules with efferent ductules and frequently with a lateral kidney canal at the nephric end of the efferent ductules.

For much of its length the ductus deferens remains a simple structure, but it may undergo major modifications in various regions and in various regards. Most notable are the modifications in the anterior portion of its length, where in both Chondrichthyes and amniotes it takes part in the formation of an appendix to the testis termed the *epididymis.* In sharks, skates, and chimaeras a limited number of efferent ductules from the testis enter the anterior end of the embryonic opisthonephros and reach the former archinephric duct by way of ductules of the epididymis (Fig. 228, *D,* p. 380; 235, *C, D,* p. 389). The anterior end of the archinephric duct is much convoluted and may be termed the *ductus epididymidis.* As it passes backward, before reaching the functional kidney, it is joined by a series of nephric tubules which have been modified into glandular structures, the *glands of Leydig;* these secrete a fluid medium which is believed to stimulate the sperm.

A parallel development has occurred in amniotes; in them the epididymis forms, as its name implies, a compact body resting close beside or upon the testis. The efferent ductules which cross from the testis to this former nephric structure enter, in reptiles, a longitudinal canal whence deferent ductules continue on to the head of the former archinephric duct; in birds and mammals the longitudinal canal is absent, and the sperm continue directly via the much convoluted ductules of the epididymis to the archinephric duct (Figs. 238, p. 397; 245, p. 409; 248). This is here, as in elasmobranchs, highly convoluted, and, as the ductus epididymidis, it fills the greater part of the bulk of the epididymis before leaving it to become the ductus deferens.

Distally, the ductus deferens may expand into an *ampulla* for sperm storage before ejaculation; such a structure is common, but not universal, in Chondrichthyes, amphibians, and the amniote groups. In many mammals (Fig. 237, p. 396) there are further present *vesicular glands* (seminal vesicles) which do not store semen (as the alternative name implies), but secrete a thick liquid as part of the seminal fluid. In mammals the *prostate gland,* secreting a thinner fluid, and the *bulbo-urethral glands* (Cowper's glands) are further structures along the path of the sperm duct. These, however, are derived from the walls of the embryonic cloaca rather than from the ductus deferens itself.

THE CLOACA AND ITS DERIVATIVES

In a great variety of fishes and tetrapods there is present at the back end of the trunk region a ventral pocket, communicating with the exterior, into which open the orifices of the digestive, genital and urinary systems. This structure, appropriately termed the *cloaca* (the Roman name for a sewer) appears to have been a primitive vertebrate feature. We shall in this section follow, through the vertebrate groups, the history of the cloacal region and the varied disposition of the outlets of the systems concerned. Inseparable from this general story is that of the urinary bladder of the land vertebrates and of the external genitalia of the amniotes.

The Cloaca in Fishes. The cloaca is typically developed in elasmobranchs (Fig. 249, *B*), where it forms a depression, which may be closed by a sphincter muscle, on the ventral surface of the body behind the pelvic girdle. The major opening into it is that of the posterior end of the intestine. Into it further open the urinary ducts and the vasa deferentia of the male or the paired oviducts of the female.

Embryologically, the cloaca has a twofold origin (cf. Fig. 168, *A*, p. 292). The major part in the embryo of any vertebrate consists of an expansion of the posterior end of the endodermal gut tube. This cloacal cavity is, until a relatively late stage of development, closed off from the exterior by a membrane, external to which lies a depression of the ectoderm, the *proctodeum*. When the cloacal membrane disappears, this ectodermal area is incorporated into the cloaca. It is generally assumed that the ectoderm plays little part in cloacal formation. But once the membrane breaks, it is difficult to tell which part of the cloacal lining comes from ectoderm, which from endoderm; it is known in some cases, and suspected in others, that the ectodermal contribution to the cloaca and its derivatives may be of considerable magnitude.

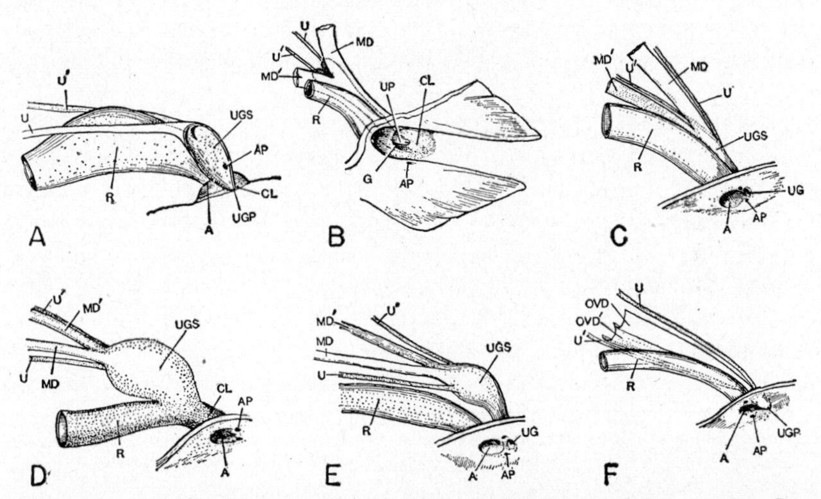

Fig. 249. Cloacal and anal region in fishes. *A*, The lamprey, Petromyzon; *B*, a female shark; *C*, a young female chimaera; *D*, the Australian lungfish, Epiceratodus; *E*, a female sturgeon; *F*, a female salmon. Abbreviations: *A*, anus; *AP*, abdominal pore; *CL*, cloaca; *G*, genital opening; *MD*, *MD'*, left and right oviducts (müllerian ducts); *OVD*, *OVD'*, left and right oviducts of teleost; *R*, rectal region of intestine; *U*, *U'*, left and right urinary ducts; *UG*, urinogenital opening; *UGP*, urinogenital papilla; *UGS*, urinogenital sinus; *UP*, urinary papilla. (From Dean.)

Among other fishes we find a well-developed cloaca only in the lungfishes (Fig. 249, *D*). In the chimaeras (Fig. 249, *C*) the cloaca has been entirely eliminated, anal, urinary and genital openings all reaching the surface separately. Among cyclostomes, hagfishes have a cloacal pocket into which anus, urinary tube, and abdominal pore all open, but in lampreys the anus has an opening separate from that of the conjoined urinary and genital tubes (Fig. 249, *A*).

In actinopterygians (Fig. 249, *E, F*) the cloaca has likewise disappeared. In lower ray-finned fish types and some teleosts urinary and reproductive tubes empty by a common sinus representing part of the cloaca, but the anus has a separate opening. In other teleosts we find that three distinct openings are present—anal, genital, and urinary.

The Cloaca in Lower Tetrapods (Fig. 250, *A*). In the amphibians a cloaca is present, presumably a direct inheritance from piscine ancestors. The wolffian ducts and, in the female, the oviducts open as paired structures dorsally into the inner end of the cloacal pouch. Ventrally, there develops from the cloaca a urinary bladder of varied shape, often distensible to large size (p. 393). In reptiles the cloacal structure is similar; the paired openings of the ureters are, except in turtles, typically on the dorsal wall of the cloaca, above the anal entrance; the urinary bladder, when present, is a distinct ventral pocket not connected with the ureters. In birds, entrances into the cloaca are similarly placed, but no urinary bladder is present.

Fate of the Cloaca in Mammals. In mammals the lowly monotremes still have a cloaca. Higher types have done away with this structure and

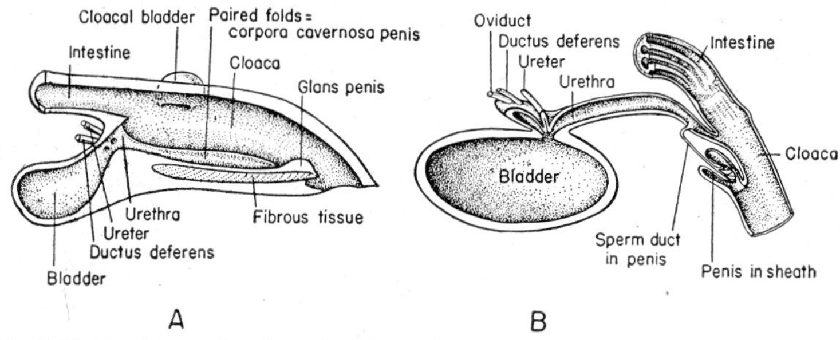

Fig. 250. Section of the cloacal region of *A*, male tortoise, *B*, a monotreme mammal (Echidna). In *A* a penis-like structure is contained in the floor of the cloaca; paired folds may meet to form a tube at the time of sperm emission. In the monotreme a formed penis is present within the cloaca; it contains a tube, divided into several branches, for sperm transport, but urine passes out via the cloaca. A turtle specialization is the presence of a pair of small bladder-like structures in the side walls of the cloaca. (*A* partly after Moens; *B* after Keibel.)

have a separate anal outlet for the rectum. The monotreme cloaca shows the initiation of this subdivision (Fig. 250, *B*). The cloaca as such includes only the distal part, roughly comparable to the proctodeum. The more proximal part is divided into (1) a large dorsal compartment into which the rectum opens, the *coprodeum,* and (2) a ventral portion, the *urodeum,* or *urogenital sinus,* with which the bladder connects. Ureters open into the proximal end of the urogenital sinus, which is partially equivalent to the placental urethra (cf. Fig. 251) and the urine may thus enter the bladder without being associated with fecal material.

In marsupials the cloaca is persistently represented by a slight proctodeal depression, with a sphincter muscle, into which rectum and urogenital sinus open, but in most higher placental mammals the cloaca has disappeared completely in the adult; coprodeum and urodeum have distinct orifices. The coprodeum simply becomes the end portion of the rectum, opening at the anus. The urodeum, however, has a more complex history and differs much in the two sexes. Conditions here are best understood by considering the developmental story; the development of the placental mammal recapitulates in many respects the phylogenetic story (Fig. 251).

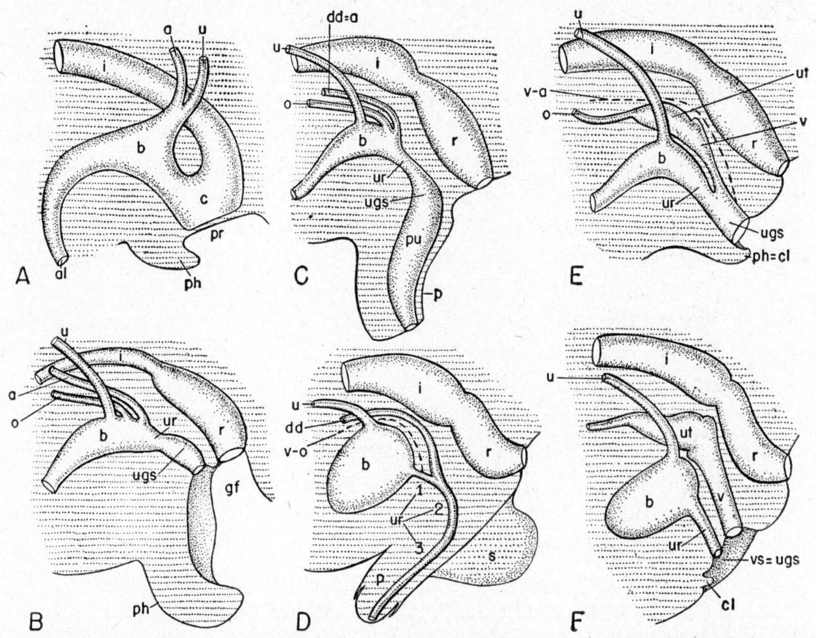

Fig. 251. Embryology of the cloacal region of mammals. *A,* Sexually indifferent stage, with intestine and allantois opening into undivided cloaca, archinephric duct and ureter opening together into base of allantois. *B,* Later indifferent stage; embryonic oviducts developed, ureter and archinephric duct separated, cloaca divided into rectum and urogenital sinus; phallic organ has begun development. *C,* Early stage of development of male, and *D,* diagram of adult male structure. In contrast to female, the oviduct disappears (broken line in *D*), the archinephric duct becomes the ductus deferens, and an extensive urethra includes, in addition, (1) the base of the allantois, (2) the urogenital sinus, and (3) a duct traversing the penis. *E,* Early stage of development of female, and *F,* diagram of adult female structure; disappearance of archinephric duct (broken line in *E*), development of bladder from base of allantois with short urethra distal to it, differentiation of uterus and vagina from embryonic oviduct, development of phallic organ as clitoris. *a,* Archinephric duct; *al,* stalk of allantois; *b,* bladder; *c,* cloaca; *cl,* clitoris; *dd,* ductus deferens; *gf,* genital fold; *i,* intestine; *o,* oviduct; *p,* penis; *ph,* phallic organ; *pr,* proctodeum; *pu,* penile urethra; *r,* rectum; *s,* scrotum; *u,* ureter; *ugs,* urogenital sinus; *ur,* urethra; *ut,* uterus; *v,* vagina; *v-a,* vestige of archinephric duct; *v-o,* vestige of oviduct; *vs,* vestibule.

In the sexually "indifferent" stage of a placental mammal there is a cloaca, formed as a simple, distal endodermal expansion of the gut. A membrane is present to separate the endodermal cloaca from the indipping ectodermal area which in ancestral types forms the proctodeal component of the cloaca. Into the walls of the endodermal cavity open the paired wolffian ducts (paralleled later by the oviducts); out from its floor extends the stalk of the allantois.

While the indifferent stage still persists, a septum develops, and extends out to the closing membrane. This divides the cloaca into two chambers, comparable to those seen in the monotreme cloaca: a coprodeum continuous with the gut above, and a urodeum or *urogenital sinus* below. The archinephric ducts—that is, the ducti deferentes of the male—enter the lower division, as do the female genital ducts. As development proceeds, part of the allantoic stalk region enlarges as the beginning of the typical amniote bladder, and a narrower tubular region develops between the urogenital sinus proper and the bladder. The sex tubes remain connected with the sinus. As the ureters develop, however, they separate from the parent archinephric ducts and gain separate distal openings. These openings migrate to the region of the bladder, their definitive position in placental mammals; urine in mammals (in contrast with lower tetrapods) enters the bladder directly, without having to traverse any part of the cloaca or its subdivisions.

From this stage onward the course of development of the two sexes diverges as regards the structures associated with the urogenital sinus. In the female, when the cloacal membrane disappears, this ventral part of the cloaca becomes the *vestibule* of the urinary and genital organs. This may attain a considerable depth (as in typical carnivores) or be a relatively shallow depression (as in primates). Into this vestibule open the conjoined distal ends of the oviducts, the vagina (the archinephric ducts degenerate); into the vestibule also opens a relatively short tube from the bladder, the *urethra.*

In the male the sinus has a different history. Instead of becoming shallow and broad, it becomes an elongate tube which continues into the penis. Into its proximal end open the ducti deferentes (the oviducts degenerate) and the tube from the bladder. In the female this short tube alone constitutes the entire urethra. In the male, however, the entire extent from bladder to end of penis is termed urethra. Female and male urethrae are, thus, not comparable structures; that of the male includes the homologues of both female urethra and vestibule, and also, more distally, an ectodermal component within the substance of the penis.

External Genitalia. In primitive water-dwelling vertebrates with shell-less eggs external fertilization is the rule; sperm and eggs are shed into the water, and fertilization takes place in that medium. But in forms with shelled eggs or with viviparous habits—such types include the Chondrichthyes and all amniotes—internal fertilization takes place, for the sperm must reach the egg before it has descended the oviduct to shell gland or uterus. Special structures are usually formed to facilitate the entrance of the sperm into the female genital tubes.

In sharks, rays, and chimaeras—fishes with heavily shelled eggs—

claspers are developed in the males to aid in fertilization. These are posterior extensions of the pelvic fins, which are stiffened by rods from the fin skeleton (Fig. 108, *C*, p. 191). The claspers are inserted into the cloaca of the female; rolls of skin folded into tubes along the clasper form a channel for the sperm. In a limited number of teleosts internal fertilization (and viviparity) is present; in the minnow family Cyprinodontidae the anal fin develops a clasper-like intromittent organ with a groove or tube to carry the sperm.

In the amniotes internal fertilization is rendered necessary, apart from the presence of the shelled egg or of viviparous type of development, by terrestrial life and the consequent absence of a watery environment for sperm and eggs at the time of breeding. In the ancestral amniotes direct contact of male and female cloacae would seem to have been sufficient for the transfer of sperm, for Sphenodon lacks copulatory organs, which are also absent in most birds (here their absence is perhaps secondary).

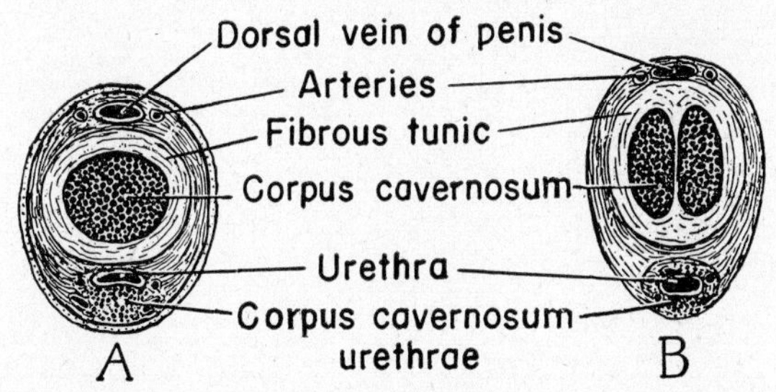

Fig. 252. Sections through the penis of a rhesus monkey, *A*, proximally and, *B*, distally. (From Wislocki.)

In many reptiles, however, the male has some type of accessory organ to aid in sperm transfer. In snakes and lizards there are unique structures, the *hemipenes*. These are a pair of pockets, often containing thorn-like spines, which lie in the skin adjacent to the cloacal opening; at the time of copulation the hemipenes may be turned inside out, extruded, and inserted in the female cloaca to guide the sperm. In turtles and crocodilians there are structures which may be morphological forerunners of the penis of mammals (Fig. 250, *A*). In the ventral wall of the cloaca are a pair of longitudinal ridges, the *corpora cavernosa penis,* with a groove between them; they are composed (as the name implies) of fibrous spongy tissue. A further spongy structure, the *glans penis,* lies at the outer end of the groove. At copulation these structures are filled with blood and thus distended; the glans is protruded and inserted into the female cloaca; the cavernous bodies expand to meet one another and turn the groove between them into a tube for the passage of the sperm. These structures appear in the embryo in the sexually indifferent stage, and persist in the adult female in a relatively undeveloped and functionless state as the *clitoris.*

The phallic organs of typical mammals are more highly developed (Figs. 239, p. 398; 252); the monotreme penis, however, is transitional in some regards, and the mammalian clitoris is simpler than the penis in structure. Glans and corpora cavernosa are present and form the main erectile structures; they are, however, permanently fused in the adult male penis, together with an extension of the urethra, carrying both sperm and urine. An additional erectile body continuous with the glans penis, the *corpus cavernosum urethrae,* surrounds the contained tube. The corpora cavernosa are in reptiles embedded in the walls of the cloaca; in typical mammals, however, the penis as a whole extends outside the body, and the glans is surrounded with a skin fold, the *prepuce*.

CHAPTER XIII

CIRCULATORY SYSTEM

IN MANY SMALL invertebrates a circulatory system is not necessary. Distances are short, structures are simple, and transportation of materials within the body may be effected by diffusion and by such flow of internal fluids as may result from body movement. With greater size and complexity a definite circulatory system is required. A comparison with human communities is a fair analogy. In a village a transportation system is unnecessary; stores, school, church are all close to the homes they serve. With growth of the community this is no longer the case, and an organized transportation system is a public necessity.

Among invertebrates we see the development of two types of circulation—open and closed systems. In the former a heart forms a pump whence blood is forced out through a series of vessels to various parts of the body. From the point where the vessels terminate, however, the blood oozes back to the heart through tissue spaces; return vessels are absent. In the more advanced closed system—the type present in the vertebrates—the blood never leaves the vessels and thus lacks direct contact with the body cells which lie in the separate interstitial fluid. The blood serves the tissues by diffusion of materials through small, thin-walled terminal vessels and returns to the heart again through a second series of closed channels.

Functions. Foremost of functions served by the circulatory system are those of transportation to the cells (via the interstitial fluid) of the materials needed for their sustenance and activity, and of the removal of wastes. The tissues must be constantly supplied with oxygen from gills or lungs, and fed with a smaller but steady stream of food materials from intestine or from storage and manufacturing centers, notably the liver; these materials include glucose, fatty substances, amino acids, minor amounts of other chemicals, minute amounts of vitamins. Wastes must be removed: carbon dioxide, destined for gills or lungs; nitrogenous wastes, to be excreted in the kidneys; excess metabolic water.

A second important function of the circulation lies in the influence which it exerts over the maintenance of the stable and narrowly defined internal environment necessary for cellular welfare. The constant circulation of liquid throughout the body in the blood stream makes for uniformity of composition in the interstitial fluids of every region, and aids in the maintenance of relatively uniform temperatures.

The vascular system has still further functions. It aids in the struggle against disease and in the repair of injuries; through the circulation of hormones, the blood acts as an accessory nervous system.

Components. The important gross structural features of the vertebrate circulatory system are familiar and readily demonstrated. A muscular pump, the heart, forces the blood stream by way of arteries to gills or lungs and to the varied body tissues. The tiniest circulatory units are networks of capillaries, through whose walls exchange of materials with the interstitial fluid surrounding body cells takes place. From the capillaries blood returns to the heart via veins; lymphatic vessels, originating in the tissues and unconnected with the arteries, form an alternative return route. In certain instances blood from body tissue capillaries does not return directly to the heart, but is forced through an intervening second capillary "portal" on the way. The veins between the two capillary systems are termed a *portal system.*

BLOOD

The blood, filling the vessels of the circulatory system, may be regarded as a tissue, in certain aspects comparable to the connective and skeletal tissues, to which, ontogenetically, the blood system is closely related. Each of these tissues consists of cellular elements lying in a "matrix." In bone the matrix is a hard substance; in connective tissue it is gelatinous in consistency; in blood, it is a liquid in which the cellular components float in free fashion.

Blood Plasma. The liquid part of the blood, plasma,* is a watery fluid of complex composition, of which—in mammals, for example—some 7 to 10 per cent of the total volume consists of materials in solution. We have previously (p. 74) noted the composition of the interstitial fluid surrounding the body cells, notably the complex and stable series of salts it contains in solution. The blood is to be regarded as essentially a part of this tissue juice enclosed within the walls of the blood vessels, and it has a practically identical salt content.

In addition, however, the blood contains materials peculiar to itself in the form of special *blood proteins*—albumin, globulin, and fibrinogen, formed, it is believed, by the liver cells. These molecules are so large that they are normally unable to pass through capillary walls and thus leave the blood stream. Since the salt concentration of the blood is comparable to that of the interstitial fluids lying outside the blood vessels, the presence of these blood proteins raises the osmotic pressure of the blood above that of the interstitial fluids—a point of importance in capillary function. The *albumin,* which in mammalian blood constitutes roughly half of the volume of the plasma proteins, has no further function. *Fibrinogen* makes up but a small percentage of the total, but is important as the material

* The term "serum" refers to the liquid remaining after the protein clot material (fibrin) has been removed from clotted blood plasma.

which, in modified form as fibrin, causes clot formation when a blood vessel is cut. Highly important are the abundant and varied *globulins*. One globulin type takes part in the formation of the blood clot from the fibrinogen proteins. Other globulins of a highly specific nature are sensitive to protein materials of "foreign" origin and will cause the clumping (agglutination) of red blood cells of another species or even (in man) of persons of another blood type. Still other globulins are active agents in disease protection and appear to be the antibodies effective against invading viruses.

The salts and blood proteins are stable, permanent, plasma constituents. In addition, however, blood analysis reveals various materials in transit. A quantity of glucose, on its way to be used as cell fuel, is always present, as well as smaller amounts of fats and of amino acids for intracellular formation of proteins. Waste materials, on the way to kidney or to gills or lungs, are also present, particularly carbon dioxide, mainly as a bicarbonate, and nitrogenous waste, usually in the form of urea or uric acid.

In more minute amounts enzymes may be identified in the plasma, and endocrine secretions—the hormones.

Hormones. In the embryonic development of vertebrates one finds numerous instances in which chemical materials produced by some one organ or tissue stimulate activity or change in other organs or tissues, and there is good reason to believe that chemical stimulation has been an important factor in body activities throughout the history of metazoan animals. With the development of the circulatory system there is established a medium through which such chemical influences can be spread widely and efficiently. In the vertebrates we find a series of endocrine glands which pour their secretions into the blood as hormones, "messengers," which may influence cells and tissues in any or every part of the body. There is considerable evidence that the stimuli produced upon muscular and glandular tissues by the nervous system are chemical; the hormone system is in this regard somewhat similar to the nervous system. An efferent nerve stimulus, however, can affect only a limited number of cells which lie within the influence of its peripheral fibrils; the messages poured into the blood stream by hormone secretion are "broadcast" to all parts of the body.

A hormone-producing endocrine gland need have no specific position adjacent to the structures upon which major effects will be produced. For blood vessels are all-pervasive, and the hormone message will be received as fully and promptly by a structure in the farthest removed corner of the body as by one close at hand. Endocrine structures need not, and do not, form an anatomically coordinated organ system; they are to be found in the most varied situations and associated with a variety of organs. Some have been noted in previous chapters; still others will be described in the chapter on the Nervous System. Since all their products are transported by the blood stream, the endocrine organs and their hormones may be reviewed here in brief form. It must be remembered that much of hormone study has grown out of medical work, and is confined in great measure to man and a few laboratory animals. In consequence there are, without doubt, many hormones and hormone functions in lower vertebrates of which we have as yet no knowledge.

1. *Pituitary.* This structure, lying at the base of the brain, is the greatest element in the endocrine system, for its hormones not merely act upon a variety of body tissues, but most especially are seen to regulate the activities of other hormone-producing glands. More than half a dozen pituitary hormones are well known; a number of others are less adequately known or reasonably suspected. The anatomical structure and subdivisions of the pituitary are noted elsewhere (p. 578). Most of the known hormones are produced by the pars distalis. Among other effects, these stimulate body growth directly; stimulate growth and metabolism by influencing the thyroid; stimulate the gonads and their hormone secretion; stimulate the mammary glands of mammals; stimulate secretion by the cortex of the adrenals. The function of the pars intermedia is unknown, except that it influences the chromatophores of the skin in lower vertebrates. The posterior lobe appears to be relatively much less important in hormone secretion than the anterior. Its products tend to cause contraction of smooth musculature in blood vessels and thus increase blood pressure, to stimulate contraction of smooth musculature in the mammalian uterus, and in some fashion tend to keep down the water content of mammalian urine.

2. *Adrenal Gland* (pp. 394, 546). Its medullary portion secretes adrenalin, which produces a general stimulating effect on body activity, but depresses digestive functions; the cortical part produces hormones with marked influence on carbohydrate metabolism, kidney function, and so forth.

3. *Thyroid* (p. 343). Thyroxin increases the metabolic rate of the body.

4. *Parathyroid* (p. 341). Its hormone conditions calcium metabolism.

5. *Pancreas* (p. 366). Insulin, produced by the pancreatic islets, regulates carbohydrate metabolism.

6. *Small Intestine.* Several hormones produced here regulate the flow of bile and pancreatic juice and influence stomach activity.

7. *Gonads* (p. 395). Interstitial cells and, in the female, follicle cells, produce hormones—testosterone (male), estrogen (female)—which affect the other sexual organs; in mammals the corpus luteum secretes progesterone, important in the initiation of pregnancy.

Still other body structures are suspected of being hormone sources, but proof is lacking or incomplete. The list here includes the *pineal body, thymus, spleen, stomach, liver* and *kidney.*

Blood Cells (Fig. 253). The normal cellular elements of the blood stream include (1) red blood corpuscles or erythrocytes, (2) white blood corpuscles or leukocytes, (3) thrombocytes.

Constant transportation of oxygen is, on a percentage basis, the most important function of the circulatory system. Various metallic compounds, particularly of iron or copper, are found in the blood of almost every form with a developed circulatory system; these compounds have the property of attaching to themselves large amounts of oxygen, and thus vastly increase the oxygen load that the blood can carry. The vertebrate compound is *hemoglobin,* a protein molecule which includes in its complicated structure four iron atoms, each capable of carrying an oxygen ion. (In addition to oxygen carriage, hemoglobin is a factor in the return of carbon dioxide from the tissues.)

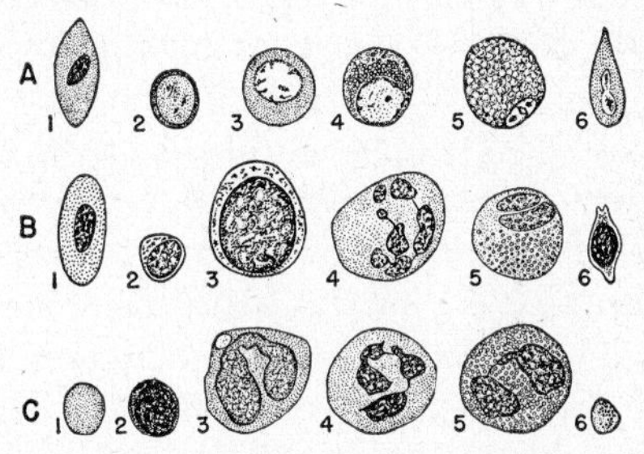

Fig. 253. Blood cells of *A*, a teleost (Labrax); *B*, a frog (Rana); *C*, a mammal (man). *A1, B1, C1,* erythrocytes; *A2, B2, C2,* lymphocytes; *A3, B4, C4,* neutrophile granulocytes; *A4,* fine-grained acidophile granulocyte; *A5,* coarse-grained acidophile granulocyte; *A6, B6,* thrombocytes; *B3, C3,* monocytes; *B5, C5,* acidophile granulocytes; *C6,* blood platelet. *A,* × about 1800; *B,C,* × about 1200 (*A* after Duthrie; *B* after Jordan; *C* after Maximow and Bloom.)

In some invertebrates the oxygen-carrying substances are loose in the blood stream; in others, however, they are carried by cells as blood corpuscles. This is the case in vertebrates, and necessarily so, for the presence of a large quantity of free hemoglobin in the blood would cause serious interference with other plasma functions. The hemoglobin carriers are the red blood corpuscles, the *erythrocytes.* In most vertebrates these corpuscles are properly nucleated cellular structures, flattened ovals in form, with their cytoplasm tightly packed with hemoglobin. The mammalian erythrocyte is unusual in structure. Except for camels and llamas, the shape is round rather than oval. Further, the mature mammalian erythrocyte extrudes its nucleus, thereby, it would seem, leading to a greater efficiency in oxygen carriage in the total mass of the red blood cells. Apart from mammals, the enucleated condition is present in some of the red corpuscles of various amphibians.

There is great variation in the size of red blood corpuscles. In man the diameter of the erythrocyte is but 7 or 8 microns, and in a goat but half this figure; some amphibians have corpuscles with a volume 100 times or more as great as that of man, and the erythrocytes are larger in all lower classes than in mammals. The amount of hemoglobin present in the blood is fairly constant, for the number of corpuscles present in the blood varies inversely with their size.

The white corpuscles characteristic of the blood and lymph streams are the *leukocytes*. They are much fewer in number than erythrocytes. In amniotes they never compose more than 1 per cent of the blood corpuscles; in fishes, however, the figure may rise to 10 per cent or over. Leukocytes vary greatly from one group of vertebrates to another, and it is difficult to establish any classification of them which will hold throughout. White blood corpuscles are basically derived from mesenchyme cells, as are connective tissue cells. In the body there appears to be a considerable interplay between the connective tissues and those of the circulatory system; white blood corpuscles when they leave the blood stream may play an active role in connective tissues, with a radical alteration of appearance.

In general we may distinguish two major groups of white cells—lymphoid types, with simple, single nuclei and a nongranular cytoplasm, and granulocytes, in which the nuclear materials are irregularly arranged and often subdivided, and the cytoplasm is granular in appearance.

Among *lymphoid leukocytes* the common forms are the *lymphocytes,* small, rounded cells with a large nucleus and a relatively small amount of clear cytoplasm. These are present in all classes of vertebrates and are especially abundant in the lower groups. They derive their name from the fact that in mammals they abound in the lymph nodes, but they are equally common in other classes of vertebrates where such nodes are few in number or absent. *Monocytes,* sometimes regarded as a separate category of white cells, appear to be merely large lymphocytes, with a larger amount of cytoplasm and a nucleus that may become somewhat irregular or kidney-shaped.

The second characteristic group of leukocytes is that of the *granulocytes* or *polymorphonuclear leukocytes*. These, like the monocytes, are large cells. The nucleus, as the alternative name suggests, is irregular in structure and more or less subdivided into lobes. Most characteristic is the fact that the cytoplasm is highly granular in appearance. In some cases the granules are stainable with acid dyes; in others basic dyes are effective; in still others both types of dyes take hold. These three types of "polys" are hence reasonably termed, respectively, *acidophilic* (eosinophilic), *basophilic,* and *heterophilic* (neutrophilic) *granulocytes*. Heterophilic elements of varied appearance are the common granulocytes in all vertebrate groups; acidophils are rare in the individual, but widespread among vertebrate groups; basophils are generally still fewer in numbers and seldom identified in fishes.

Leukocytes appear to be essentially inert while in the blood stream. They are, however, capable of ameboid motion and seem to be able to wriggle in and out of capillaries between endothelial cells. Their functions are incompletely known. They tend, however, to accumulate rapidly in injured, inflamed, or infected tissues. The heterophile granulocytes are capable phagocytes, "cell eaters," and a major means of reducing bacterial infection. The mononuclear leukocytes and lymphocytes may become transformed into connective tissue elements of the repaired region.

Thrombocytes are blood structures associated with the process of blood clotting. In mammals these structures take the form of *blood platelets,* tiny bodies that look like broken cell fragments and are believed to be formed by the fragmentation of giant cells present in the blood-forming tissues. In most nonmammalian vertebrates the thrombocytes take the form of *spindle cells*—small, oval, pointed structures with a central nucleus, which appear to be related in origin to other white blood cells. When thrombocytes leave the environment of the blood stream (as in a cut vessel), they tend to disintegrate, it is thought, and release a material which (through a series of complicated chemical steps) causes the blood protein fibrinogen to change into fibers of fibrin, masses of which form the blood clot.

HEMOPOIETIC TISSUES

In most body tissues cellular differentiation takes place in embryonic stages once and for all; in blood elements, proliferation and differentiation of cells continues throughout life. The life of an individual blood cell is short; human erythrocytes, for example, appear to live, at the most, but a few months, and lymphocytes but a few hours. In consequence, an animal must retain throughout its existence hemopoietic tissues in which blood cells are constantly formed, and must in addition provide for the removal of worn-out cells.

Students of the blood have devoted much study to blood cell formation, but many features of the process are in doubt, owing to the peculiar nature of the vascular system. Most other differentiated cells are parts of tissues which are fixed in position; their development takes place on the spot. But blood cells are free agents, and it is immaterial where they are formed, as long as the area is in communication with the blood vessels. A loose cell in the blood stream carries no clue in itself as to where it arose in the body, and blood-forming tissues are found in the most varied places in various vertebrates. Unlike most organ systems, no part of the circulatory system is derived from any of the epithelial sheets laid down in the early embryo. In its entirety it is derived from mesenchyme cells—vessels and blood corpuscles alike. The blood cells are thus formed from the same elements which give rise to the connective tissues; they frequently show comparable characteristics, and elements of one system can be demonstrated to become transformed into cells of the other category.

Except for certain sites utilized in early embryonic stages, areas in which blood cells arise have, in general, common structural features (Fig. 254). All are spaces which are enlargements of circulatory vessels or lie adjacent to such vessels. A network of fibers forms the "skeleton" of the tissue. Along the course of the fibers lie fixed *reticular cells* and *macrophages*. The reticular cells are responsible for the formation of the fiber

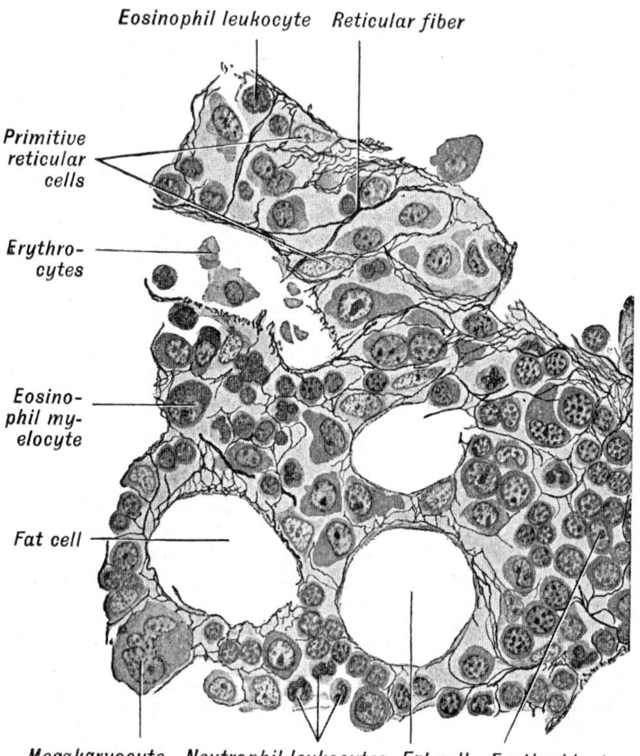

Fig. 254. A hemopoietic tissue—bone marrow from a mammalian femur. Much of the reticular framework and two reticular cells are seen, as well as two types of granular leukocytes in process of differentiation, and erythroblasts from which erythrocytes are formed. The megakaryocyte is a giant cell type from which it is believed blood platelets may be formed. (From Maximow and Bloom.)

network. The macrophages are large cells which have the power of engulfing waste materials, notably damaged blood cells. Such cells are not restricted to blood-forming tissues, it may be noted, but may be found along the course of blood vessels, notably in the liver, and occur as free, ameboid scavenging elements, in the connective tissues.

Enmeshed within the reticular framework of a hemopoietic tissue are found masses of blood cells in process of multiplication and differentiation. A primary unspecialized type, from which erythrocytes or leukocytes of any sort are thought to be derived, is the *hemocytoblast*. This is a cell of modest size, with a relatively small amount of clear cytoplasm. Such a cell is similar to a lymphocyte in structure, and it is not unreasonable to believe that the lymphocytes are essentially primitive blood cells released into the blood stream.

Embryonic Blood-Forming Sites. The first blood vessels which form in the embryo are those engaged in bringing food materials into circulation, and the first blood cells are erythrocytes formed in connection with them. In forms with a mesolecithal egg type, the food lies in the yolk-filled gut, and the first vessels and blood cells form in the mesenchyme of the belly floor. In large-yolked eggs, vessels and cells form in the yolk sac, typically as clusters of cells termed *blood islands* (Fig. 207, *A*, p. 347), and in mammals vessels and cells are early formed in the maternal tissues of the placenta as well. At a somewhat later stage, as vessels extend through the body, blood cells may form locally in the mesenchyme of various regions, and may long continue to arise from blood vessel walls. In later phases of embryonic development blood cell formation becomes more localized, but the hemopoietic tissues are often located in areas quite different from those found in adult life. The embryonic kidney is an important seat of this activity in many forms from sharks to reptiles and birds. The tissues of the throat region and the thymus in particular are an important source of blood cells in many embryos. In mammals the liver and spleen are no longer important as hemopoietic organs in the adult, but play a prominent role in the embryo.

Adult Blood-Forming Tissues. A great variety of organs are seats of such tissues in adult vertebrates. In many teleosts the kidney continues throughout life as a great source of blood cells. In sharks and lungfishes the gonads are areas in which white blood corpuscles are formed. In teleosts, amphibians, and turtles hemopoietic tissue is associated with the liver, and in urodeles this is a large source of leukocytes. In the sturgeon and paddlefish blood-forming tissue surrounds the heart. As a persistence of embryonic conditions, many vertebrates show in adult life blood cell formation in the throat region. Masses of lymphocyte tissue are present in the walls of the mouth and pharynx in amphibians, and are present in mammals as *tonsils,* palatine and pharyngeal. The esophagus of sharks shows a large mass of lymphocyte-forming tissue, and similar esophageal "tonsils" may be found in a number of amniotes. Nodules which are centers of lymphocyte formation are present in the intestinal wall in a great variety of vertebrates.

The *bone marrow* is a great center of blood cell formation in higher vertebrates. The hollowing out of large limb bones appears to be primarily an adaptation for greater efficiency in the use of bone tissue. The resulting "waste" space is, however, available for other purposes; fat storage and, secondarily, blood cell formation are the uses to which the marrow cavities are put. In frogs some hemopoiesis is found here in the males at the breeding season; in birds and mammals this site is universally used and highly important as a source of blood cells. In reptiles and birds

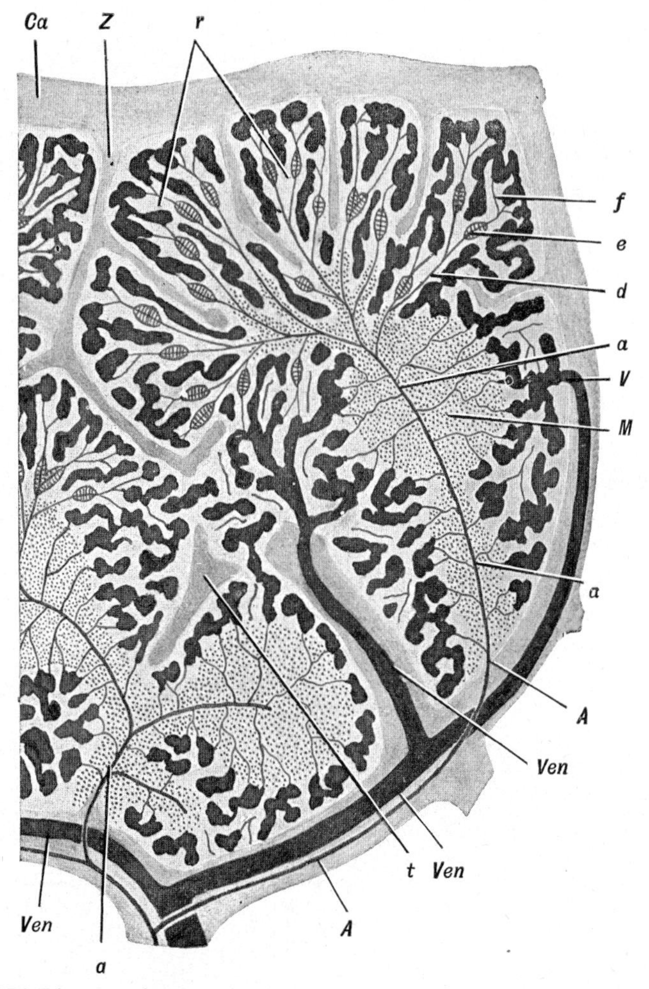

Fig. 255. Diagram of a part of the mammalian spleen: *A*, artery; *a*, artery covered by a sheath of white pulp; *Ca*, capsule; *d*, arterioles of pulp; *e*, a "sheathed artery" peculiar to the spleen; *f*, arterial capillary; *M*, white pulp; *r*, red pulp; *t*, connective tissue trabecula; *V*, *Ven*, veins; *Z*, base of a trabecula. (From Maximow and Bloom.)

the bone marrow produces blood cells of all sorts; in mammals, however, lymphoid cell formation is practically absent from the bone marrow, and production there is confined to erythrocytes and granular leukocytes. This mammalian specialization has led some histologists without a comparative background to the belief, apparently erroneous, that the lymphoid corpuscles are different from other blood cells.

Spleen (Fig. 255). It is only in the spleen and the lymph nodes characteristic of mammals that hemopoietic tissues assume the shape of discrete organs. The spleen is the largest mass of reticular tissue in the vertebrate body, and is an important organ in the formation, storage, and destruction of blood corpuscles; it is, further, of importance in defense against disease. A reddish body situated in the dorsal mesentery, it is surrounded by a connective tissue capsule. Extensions inward of connective tissue partitions form the large framework of the spleen. It contains an extensive reticular network, in the spaces of which are contained tissues of two types, a *white pulp* and a *red pulp*. As may be inferred, the former contains clusters of white corpuscles; the red pulp consists of thick masses of all the elements of the blood, including, of course, a large percentage of erythrocytes. In fishes and some amphibians the pulp is dominantly red; in higher tetrapods white pulp is often more common. The spleen is fed by an artery, and is drained by a vein; lymphatic vessels are never present. The internal circulation of the spleen is obviously complicated, and its nature much disputed. The white pulp clusters about the terminal arterioles; the red pulp lies near the venous outlets.

The spleen is always an important center of blood cell production. In the embryo of all vertebrates, erythrocytes as well as leukocytes are formed there, and this function persists in the adult in every group except mammals. In this last class, in which bone marrow has become the important seat of red blood cell formation, hemopoiesis in the spleen is confined to lymphocytes alone. Red corpuscles are stored in great quantities in the spleen, and destruction of such cells occurs in this organ, principally through the agency of numerous macrophages found there. The macrophages of the spleen are of great importance in destroying infectious materials in the blood stream.

The spleen was developed early in the course of vertebrate history, for it is present in most fishes and universal among tetrapods. Historically, it appears to be connected with the primary seat of blood cell formation in the embryo, for it arises in connection with the gut tube. Cyclostomes and lungfishes lack a formed spleen. In the former it is represented by a layer of reticular blood-forming tissue surrounding much of the gut; in the latter by a mass of tissue of a somewhat more compact nature, but still ensheathed within the outer lining of the gut (as is the pancreas in these forms). In the remaining fishes—Chondrichthyes, actinopterygians—this tissue has separated to form a definitive spleen, lying above the gut in the dorsal mesentery. The shape of the spleen is variable, and it appears to modify its outlines readily to those of adjacent organs. An elongate shape, paralleling much of the gut length, is presumably primitive, and is retained in a majority of fishes, urodeles, and reptiles. A more compact structure is, however, present in some fish, some urodeles, anurans, many birds, and all mammals above the monotreme level.

In many mammals there are present occasional small red structures termed *hemal nodes,* interposed between arterioles and venules. Their structure is that of a spleen in miniature, and they may be accessory organs of a similar nature.

Lymph Nodes (Fig. 256). For the most part, lymph nodes are a specialized mammalian type of structure. Although lymphatic vessels are present in all vertebrates from the bony fishes upward, we do not find any notable aggregation of blood cell tissues connected with them until we reach birds and mammals. In birds a few pairs of nodular lymphatic developments may be present; in mammals there are numerous and variable lymph nodes, both in the subcutaneous tissues and in the interior of the body.

These are small, rounded areas of reticular tissue, comparable to a small spleen in general structure, but to be sharply distinguished from that organ by being situated along the course of lymphatic rather than

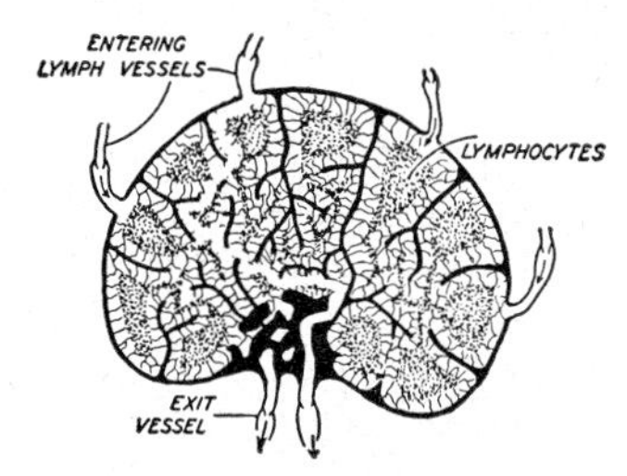

Fig. 256. Diagram to show the structure of a mammalian lymph node and associated lymphatics. Connective tissue capsule and trabeculae in black; reticular nature of tissue indicated. (After Carlson and Johnson.)

blood vessels. Their cellular contents consist almost entirely of lymphocytes of various sizes. These nodes are, in mammals, a special seat of lymphocyte development; in lower vertebrates the lymphocytes are developed in common with other types of corpuscles. Macrophages in the node reticulum, as well as, apparently, the lymphocytes themselves, deal with "germs" which have entered the body and reached the lymphatics; under such conditions, enlargement of the node may occur, owing to rapid proliferation of lymphocytes.

CIRCULATORY VESSELS

The vessels of the circulatory system, like the blood cells, are derived from the embryonic mesenchyme. As food-containing liquid begins to flow through the body of the early embryo, adjacent mesenchyme cells gather about such channels and surround them with a thin but continuous wall. All early formation of major blood vessels takes place in this fashion; later in development, minor vessels may arise by outgrowths from the cells lining the blood channels already established.

This inner lining of circulatory vessels is a special type of epithelium which has arisen *de novo,* and is not derived directly from any epithelial formation originally present in the embryo; it is termed the *endothelium.* It consists of a single layer of thin, leaf-shaped cells, continuous with one another at their margins. The absolutely continuous nature of the endothelial lining cannot be too strongly emphasized; nowhere, under normal conditions and circumstances, is the circulating fluid in continuity with the interstitial liquid and cells outside its vessels, although most of the plasma contents can freely interchange through the thin endothelial membrane.

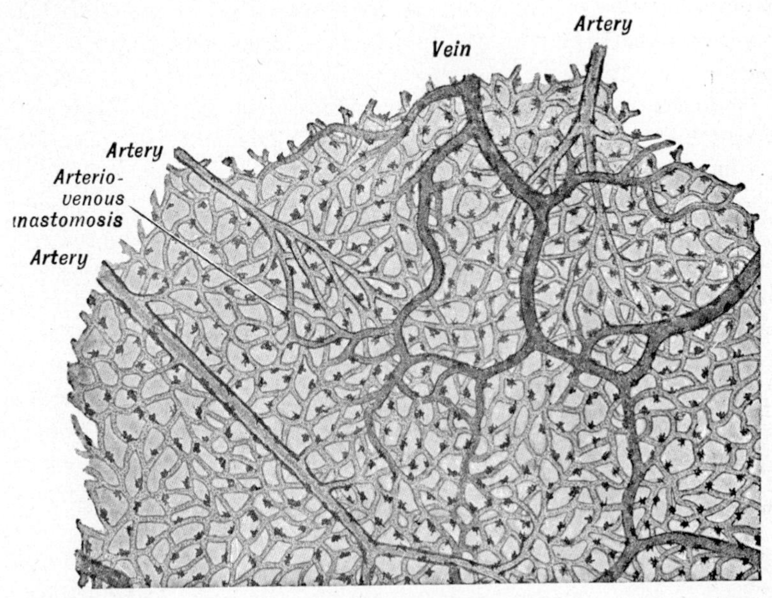

Fig. 257. Network of blood vessels in the web of a frog foot, showing the connection of arteries and veins with the capillary network and a direct connection between arterioles and venules. (From Maximow and Bloom.)

The vessels of the circulatory system include (1) the heart; (2) arteries, by definition vessels carrying blood from heart to body tissues; (3) capillaries and comparable structures, typically small vessels, connecting arteries and veins; (4) veins, returning blood to the heart; and (5) lymphatics, auxiliary vessels, prominent in higher vertebrates, which aid in the return of fluid from the tissues.

Capillaries (Fig. 257). These smallest of vessels, typically with only the "bore" necessary to allow an erythrocyte to pass, are simple in structure. Deployed from the ends of the arterioles or the veins of portal systems, they twine in among the cells of most of the body tissues so that no cell is far from a capillary. At their distal ends the capillaries are recollected into larger vessels—usually veins—in which the circulatory path of blood flow continues onward. The capillary walls consist simply of a thin if absolutely continuous layer of endothelial cells, although connective tissue cells or other cells of mesenchyme type may adhere to them.

Capillaries are too small to be dissected by ordinary means and hence are neglected from the point of view of gross anatomy. But it must never be forgotten that functionally the capillaries are the most important part of the circulatory system. Elsewhere, blood is merely in transit; here it is at work. The endothelium is in general a barrier to cells and large molecules; smaller molecules and ions, however, pass through readily, and in the walls of the capillaries an active exchange of materials takes place. In typical body tissues the capillary in the proximal part of its course gives out oxygen and food materials; in its distal part it takes on waste and carbon dioxide from the tissues. This shift of direction of transit, vitally necessary, is made possible by a change of balance between physical and osmotic pressures in the blood as it passes through the capillaries. The presence of its specific proteins makes the blood osmotically superior to the interstitial fluids and creates a tendency for materials to flow into the blood stream. At the proximal end of a capillary the physical pressure put on the blood by the heart pump more than balances this tendency for inflow; food materials and oxygen are given off. Distally, in the capillary network the physical pressure has been decreased by the friction of the capillary walls; osmotic forces dominate over physical pressure, and waste materials flow inward to the blood vessels from the surrounding tissue fluid.

Capillary networks in most areas of the body tissues are interposed between arterial and venous vessels. But the breaking up of a blood stream into capillaries and its reconstitution into larger vessels may take place also along the path of either the arteries or the veins in special regions. In gill-bearing vertebrates the course of the arterial flow is interrupted by a capillary system for aeration of the blood in the gills.* The return of the venous blood from the tissues may be interrupted also by its forced passage through capillary networks. In all vertebrates blood from the gut passes by a venous portal system to a capillary system in the liver; a majority of vertebrates (the highest and the lowest are exceptions) route part of the returning venous blood through a portal system to capillary networks in the kidneys.

Although capillaries are the major type of arteriovenous connection, there are other modes. Microscopic study sometimes reveals direct connections of larger caliber—anastomoses—between small arteries and veins. Sometimes the exchange of materials between tissues and blood

* We are accustomed to think, in visual images, of arterial blood as "oxygenated" and "red;" but arterial blood between heart and gills is, of course, of the "blue" venous type; the arteries and veins leading to and from the tetrapod lung likewise have a reversal of "blue" and "red" blood types.

may take place in small, thin-walled "ponds" of blood termed *sinusoids*. In some instances, usually connected with special organ functions, a blood vessel breaks up into a complicated, coiled mass of tiny blood vessels reasonably termed "a marvellous network"—*rete mirabile*; a kidney glomerulus is such a structure, and others are found, for example, in the red body of the teleost air bladder and the heat-sensing organ of pit vipers. The tails of many mammals (such as dog, cat, rat) show small tissue masses, termed *caudal glomeruli,* which contain retia mirabilia; man retains one such structure as a *coccygeal glomus*.

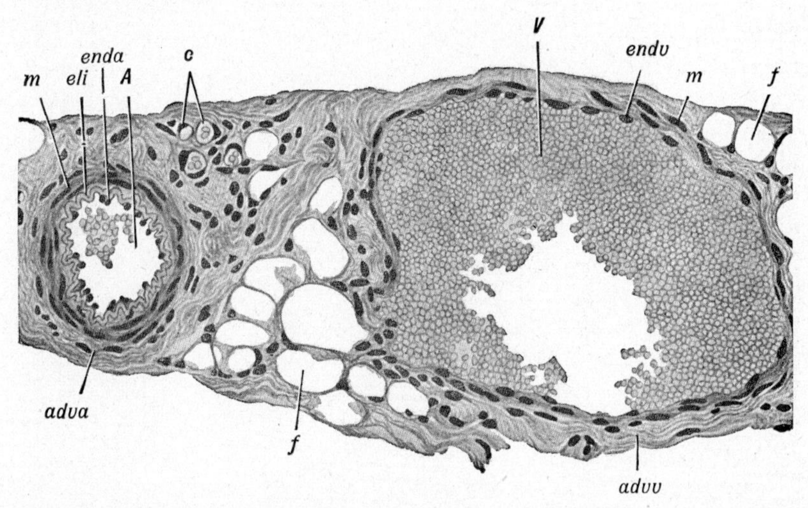

Fig. 258. Section through a small artery (*A*) and its accompanying vein (*V*), showing contrast in size and thickness of walls: *adva, advv,* adventitia; *c,* capillaries cut in section; *eli,* elastica interna; *enda, endv,* endothelium; *f,* fat cells in the loose connective tissue; *m,* muscle cells of the media. × 187. (From Maximow and Bloom.)

Arteries and Veins (Figs. 258, 259). The larger vessels of the body—*arteries* and *veins*—their smaller branches, the *arterioles* and *venules,* and likewise the major lymphatic vessels have sheathing materials in addition to the ubiquitous endothelium. Reinforcing elements include connective tissue fibers, elastic fibers, and smooth muscle cells in variable amounts and arrangements; nerve fibers for the smooth muscles are also present, and small nutrient blood vessels may be present in the walls of large arteries or veins. Although vessel sheaths are developed in all vertebrates, they are stouter in amniotes than in lower vertebrates, in which the blood, owing to the interposition of gill capillaries, flows at a lower pressure (p. 469).

The walls of blood vessels are customarily described as consisting of three layers or "tunics," the *tunica intima, tunica media,* and *tunica externa* or *adventitia*. As seen in simplest form in small arteries or veins, the intima may include only the endothelium, the media a few muscle fibers, the externa a bit of connective tissue. In larger vessels, however, thick sheaths of complicated and variable arrangement are present.

In a typical large artery the intima includes, besides the endothelium, a thin sheet of fibrous connective tissue just beneath it, and a sheath of elastic tissue which completely surrounds the vessel as an *internal elastic membrane*. The tunica media contains some connective tissue, but is dominantly the muscular layer, although smooth muscle fibers are sometimes found in the other two layers. When well developed, the muscle may be

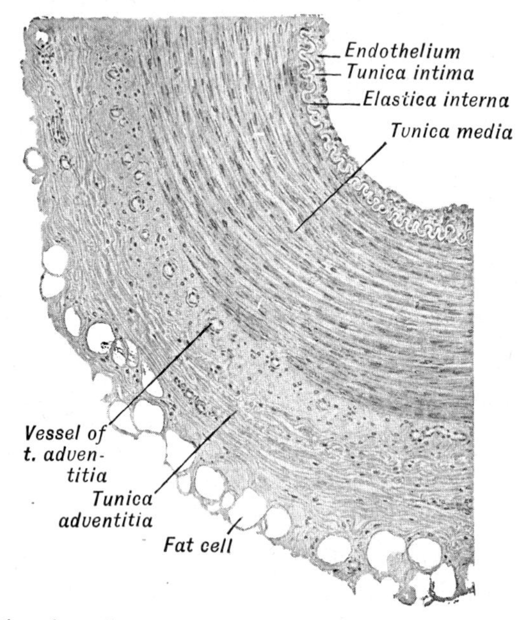

Fig. 259. Section through an artery of a mammal. (From Maximow and Bloom, after Schaffer.)

arranged in two layers, of circular and longitudinal fibers. Contraction of circular fibers, plus the action of elastic sheaths, constricts a vessel; the longitudinal fibers tend to expand its caliber. The tunica media is often bounded externally by a second elastic tissue structure, an *external elastic membrane*. Beyond this is the adventitia—connective tissue, often rather loose, which binds the vessel to adjacent structures.

Veins have a somewhat similar but simpler build; in general there is much less muscle and elastic tissue in proportion to connective tissue. There are many further contrasts between arteries and veins, as regards both gross and minute structure; much of the contrast is due fundamentally to the fact that the blood flows under higher pressure in the arteries. That pressure may be better withstood, arterial walls are thicker and more complex than those of veins. Again, the lumen of an artery is much smaller than that of a vein of comparable importance; this is correlated with the faster flow of arterial blood (just as, for example, a high-speed through highway needs fewer lanes than a street where, with the same amount of traffic, the speed is slower). Frequently, in histological sections, twin vessels can be seen beside one another (cf. Fig. 258), and artery and vein can be clearly distinguished by these criteria. The "bore" of an artery tends to remain constant, diminishing only as branches are given off; a vein sometimes expands along its course to form a large sac, or *sinus*. Again associated with speed and pressure differences is the fact that the arterial system presents few individual anomalies, but veins are highly variable. A fast-flowing mountain stream tends to take a direct and

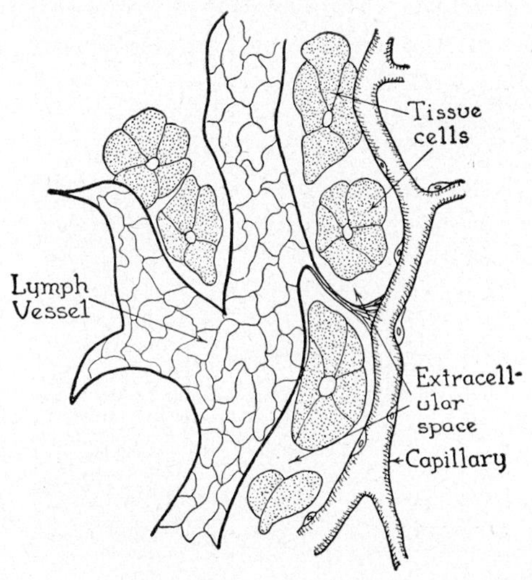

Fig. 260. A small lymphatic and a neighboring capillary. (Ponder in Fulton-Howell.)

undeviating course, whereas a sluggish stream meanders, branches and re-forms, produces islands; so in the embryo a vein often appears as a network of variable channels (cf. Fig. 283, p. 470). Which channels of the network are to form the definitive vein is not fixed; hence the frequent anomalies.

Lymphatic Vessels. The lymphatic system, well developed in the higher vertebrates as a return route from tissues to heart paralleling the veins, has a series of vessels comparable to that of the capillaries and veins of the blood system. Lymphatic capillaries, however, tend to be larger in diameter and more irregular than blood capillaries (Fig. 260). Larger lymphatic vessels, in which the lymph fluid flows but sluggishly toward the heart, have walls still thinner than those of veins, and may expand into thin-walled pockets, *lymphatic cisterns*.

Valves and Sphincters. Expansion and contraction of blood vessels is controlled by the action of the muscular and elastic tissues present along their walls; more positive control of blood and lymphatic flow is frequently exercised by the development of sphincters and valves. In small arteries and veins, circular muscle fibers may be greatly developed in a sharply localized region to form a sphincter capable of closing off the blood flow in the vessel concerned. A backwash of blood or lymph through a vessel may be prevented by the presence of valves. These are folds of endothelium and connective tissue, behind which lies a pocket-like depression. A backward flow of liquid distends the pocket and pushes the valve flap out into the lumen of the vessel; two such valves generally suffice to close a vessel (three or four, however, are occasionally found). Valves are common in the veins and in the lymph vessels of birds and mammals (and are highly developed in the heart). Almost never is a valve found in an artery, since arterial blood cannot ordinarily work back against the strong pressure under which it flows.

ARTERIAL SYSTEM

Aortic Arch System in Fishes (Figs. 261, 264). In primitive, gill-breathing vertebrates all the blood from the heart courses forward in a

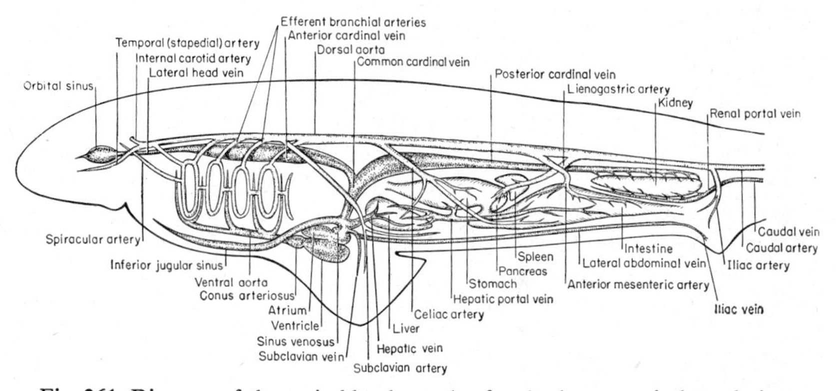

Fig. 261. Diagram of the main blood vessels of a shark as seen in lateral view.

ventral aorta. This median trunk, lying in the floor of the throat, gives off a series of paired vessels which arch upward on either side of the pharynx between successive gill clefts. Each typical aortic arch breaks up into a series of capillary vessels along the walls of the gill cleft; thence the blood is re-collected dorsally into further arterial vessels which (mainly via the dorsal aorta) pass to the various tissues of the body. In land vertebrates the gills are lost, but the gill arch vessels are found in every embryo. The history of these aortic arches is one of the most interesting chapters in the structural evolution of the vertebrates.

In a diagrammatic primitive vertebrate (Fig. 264, *A*) we may picture the aortic circulation as including, beside a ventral aorta running forward from the heart, paired aortic vessels passing upward laterally in front of and between each gill slit or pouch. In its upward passage each vessel breaks up into a network of capillaries on the gill membranes for aeration of the blood. Dorsally, the blood is re-collected into vessels which pass into the system of the *dorsal aorta*. Posteriorly, the aorta is a single median trunk; anteriorly, it consists of a pair of vessels passing along either side

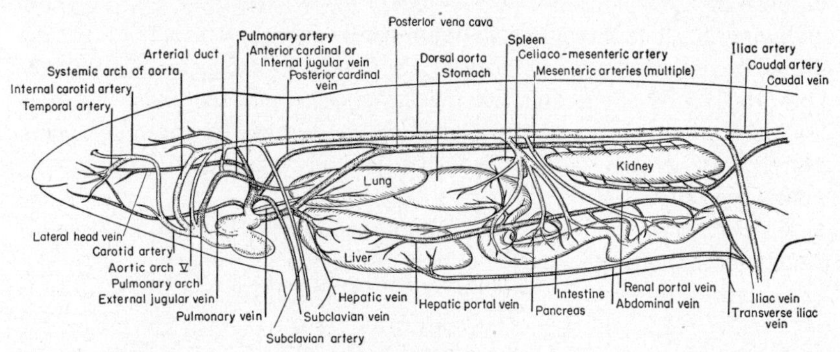

Fig. 262. Diagram of the main blood vessels of a urodele amphibian as seen in lateral view.

of the head. The number of aortic arches was presumably high and variable in the ancestral vertebrates; in most living jawed forms, however, there are but five normal gill slits plus a spiracle, and hence potentially six pairs of aortic arches, usually designated by Roman numerals. In the vertebrate embryo these arches are, to begin with, continuous vessels

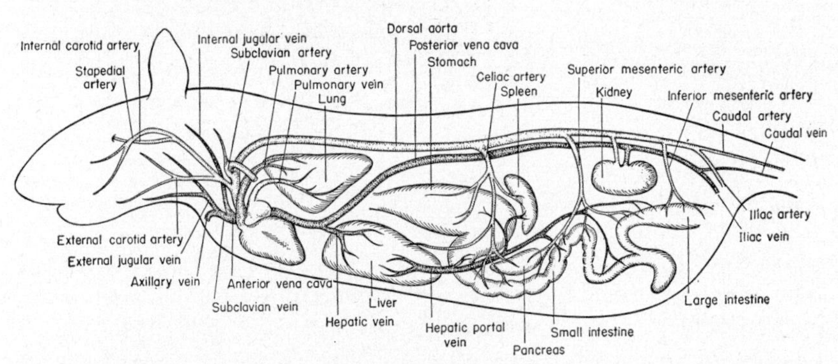

Fig. 263. Diagram of the main blood vessels of a mammal (rat) as seen in lateral view.

passing without interruption from ventral to dorsal aorta; only as the gill slits open and become functional does the gill capillary system develop, with consequent interruption of the loop of the arch. In all vertebrates the aortic arches develop in order, in the embryo, from front to back. The first, or mandibular, arch is, in the early embryo, the only passage from ventral to dorsal aorta (Fig. 268, *A*, p. 446).

The circulation in the gills may be noted briefly. In the cyclostomes the gills are subspherical, pouchlike structures; the arches break up into smaller afferent vessels and capillaries beneath each pouch and collect into a series of efferent vessels above. In all other living fishes the gill openings are slit-shaped, and the afferent vessel of an aortic arch typically runs up between each successive pair of slits (Fig. 265, p. 442). In selachians and lungfishes the afferent vessel continues as a single structure far up the gill, giving off capillaries as it goes. The efferent vessels of each gill are a pair, anteriorly and posteriorly placed; at the top of the gill slit the two members of the pair fuse to a single efferent structure. In actinopterygians there is a different pattern of arch vessels. Here both efferent and afferent vessels are single throughout.

Vertebrates invariably refuse to adapt themselves to a man-made structural diagram, and in the aortic arches, as elsewhere, departures from the "idealized" arrangement must be noted. The efferent branchial arteries are usually represented, diagrammatically, as leaving the gill region for the dorsal aorta at a point opposite the middle of each gill bar. This is usually the case; but in selachians (Fig. 264, *B*) there is frequently a shift in position so that each efferent branchial artery lies opposite a gill slit, not a gill bar. Again, the efferent vessels in successive arches are in theory discrete; but actually in many forms—including all cartilaginous fish and many of the actinopterygians—there may be connections at either side between the efferents of successive arches; these connections are more common dorsally, but may be present below the gill slits as well. In all gnathostomes there is commonly a ventral *hypobranchial artery* developed from the efferent system, which runs backward along the throat to supply that region and the heart with aerated blood. Characteristically this takes its root from the fourth arch. Still further, we frequently find in fishes an artery running forward ventrally from the efferent gill arteries to the mandibular region.

A condition of the aortic arches which may well be rather primitive is that found in the lampreys and hagfishes. There are usually, we have noted, numerous gill pouches, and hence six to fourteen pairs of arches stemming from a long, ventral aortic trunk. All are equally well developed; an exceptional feature, however, is that each arch supplies a gill pouch rather than adjacent parts of two successive pouches, as would be expected by analogy with gnathostomes.

In jawed fishes, with the exception of a few sharks, the gill pouches are six in number (including the spiracle). Six pairs of aortic arches develop, in the embryo at least, the first lying between mouth and spiracle, the last in front of the last cleft. For the most part the arches reach a characteristic development in the adult, but there are numerous variations; most prominent are modifications of the first two arches (Fig. 264, *B-D*).

In the fossil acanthodians there appears to have been a full development of gills about the first (spiracular) cleft; hence a well-developed arch I may have been present, running up the jaw region, in that group. In living gnathostomes a first aortic arch is found in the embryo, but never

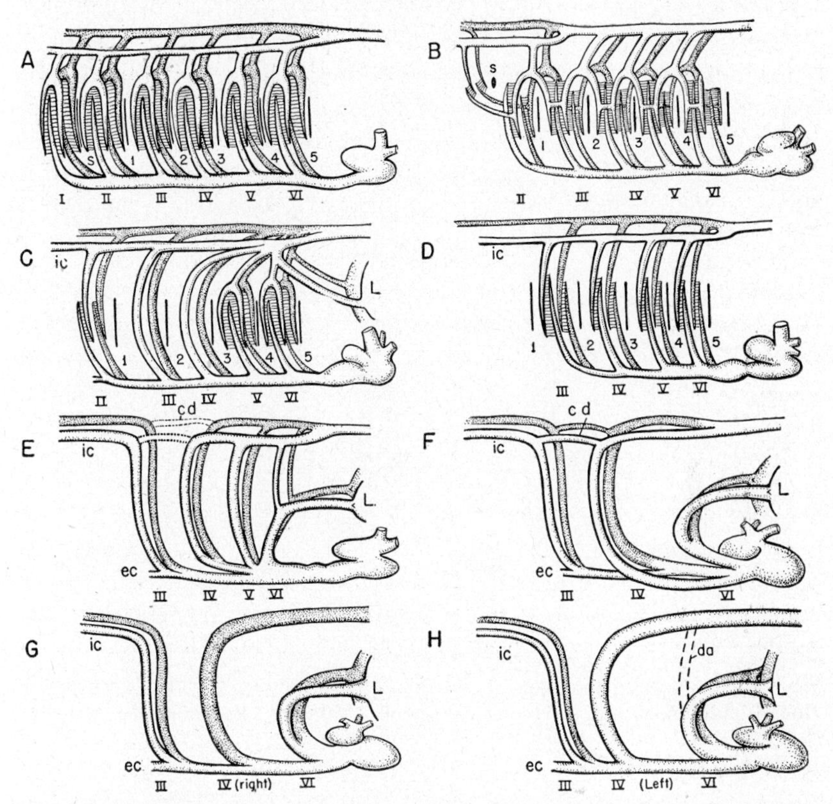

Fig. 264. Diagrams of the aortic arches and derived vessels in various vertebrate types. *A,* Theoretical ancestor of the jawed vertebrates with six unspecialized aortic arches; *B,* Typical fish condition as seen in a shark; *C,* the lungfish Protopterus; *D,* a teleost; *E,* a terrestrial salamander; *F,* a lizard; *G,* a bird; *H,* a mammal. Various accessory vessels are omitted. The vessels of the right side are heavily shaded. In terrestrial forms the position of the vessels is made (for purposes of the diagram) to correspond more or less to that of the arches from which they are derived. Aortic arches in Roman numerals; *s,* spiracular slit; following gill slits in Arabic numerals. *cd,* Carotid duct; *da,* embryonic ductus arteriosus; *ec,* external carotid artery; *ic,* internal carotid artery; *L,* lung. The carotid duct shown in the lizard is absent in other reptiles; in turtles the carotids arise by a separate stem directly from the heart. In *H* the embryonic arterial duct by-passing the lungs is shown in broken lines.

is present in typical form in the adult. Its afferent portion almost universally disappears. There is usually in fishes a gill-like structure on the anterior margin of the spiracle opening. In theory such a gill should be supplied by arch I. But its blood supply comes, in the adult, from the efferent system of the second arch or from a branch of the dorsal aorta.

The blood which reaches it has, thus, already been aerated. The spiracular gill is therefore not a true gill, but a "false gill" or pseudobranch (cf. p. 316). The efferent portion of arch I persists in fishes and runs up from the pseudobranch into the orbital region, supplies the eye with blood, and enters the braincase to connect with the arteries beneath the brain.

The second aortic arch is somewhat variable but more persistent. It is generally present in typical fashion in Chondrichthyes and is also found in some bony fishes. But in most actinopterygians it is lost, and the same is true of the lungfish Epiceratodus.

Arches III to VI are in general both normally and fully developed in fishes. In the lungfish Protopterus an interesting exception is that arches III and IV run without break past the gill region. This feature is associ-

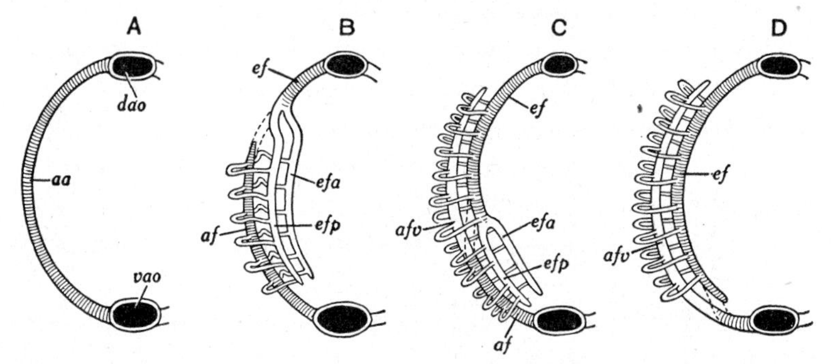

Fig. 265. Diagram of circulation in a fish gill, left side, from behind. *A,* Embryonic condition, with continuous aortic arch, *aa,* from ventral aorta, *vao,* to dorsal aorta, *dao. B,* Shark condition; the afferent gill vessel, *af,* is formed from the aortic arch; the paired efferent vessels, *efa, efp,* are new formations. *C,* Transitional condition, seen in sturgeons. *D,* Teleost condition. The embryonic arch gives rise to the efferent vessel, *ef;* the afferent vessel, *afv,* is a new formation. (After Sewertzoff, Goodrich.)

ated with the reduced gills and developed lung of that genus and shows a partial parallel to developments in the amphibian relatives of the dipnoans.

The introduction of a lung into the circulatory system in certain bony fishes makes, for most forms still in the fish stage, no great change in the gill arch system. The lung in Polypterus and lungfishes (Fig. 264, *C*) is supplied by an artery from the efferent portion of the sixth aortic arch; hence the blood reaching it has already passed through the normal aerating device of the gill capillaries.

Aortic Arches in Amphibians (Figs. 264, *E*, 266, *A*). In the amphibian stage marked changes are present, even in the larval forms. Arches I and II disappear early in development, although a lingual (or external carotid) artery may continue forward ventrally to the jaw region (this vessel may be related to a small artery found in this region in fishes). The remaining four aortic arches persist in the adult as continuous tubes in urodeles, since internal gills fail to develop (the external gills of the larva are supplied by accessory capillary loops). In frogs the extensive development of the peculiar gills leads to a larval interruption of continuity of the arches, which, however, is restored for the most part at metamorphosis. In the adult amphibian we generally find, in consequence, a system of complete tubular aortic arches, which always include the

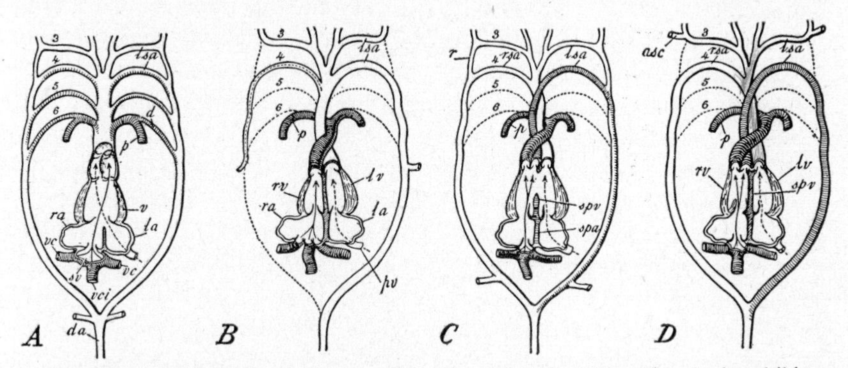

Fig. 266. Diagram of the heart and aortic arches in tetrapods. *A,* Amphibian; *B,* a mammal; *C,* typical modern reptiles; *D,* crocodilian. Ventral views; the heart (sectioned) is represented as if the chambers were arranged in the same plane; the dorsal ends of the arches are arbitrarily placed at either side. Solid arrows represent the main stream of venous blood; arrows with broken line the blood coming from the lung. Vessels with aerated blood are unshaded; those leaving heart with venous blood, hatched. The two vessels at the top of each figure are the internal carotid (laterally) and external carotid (medially). In amphibians without a ventricular septum the two blood streams are somewhat mixed; subdivision of the arterial cone tends to bring about partial separation, but some venous blood is returned to the dorsal aorta. In mammals ventricular separation is complete, the arterial cone subdivided into two vessels, and the arches are reduced to the left systemic and pulmonary. The mammalian condition has apparently arisen directly from the primitive type preserved in the Amphibia, for in modern reptiles the conus arteriosus shows a division into three vessels, rather than two; one, returning venous blood back to the body, leads only to the left fourth arch. In crocodilians the ventricular septum is nearly complete, and the elimination of the left fourth arch would give the avian condition. *asc,* Anterior subclavian; *d,* ductus Botalli; *da,* dorsal aorta; *la,* left atrium; *lsa,* left systemic arch; *lv,* left ventricle; *p,* pulmonary artery; *pv,* pulmonary vein; *r,* portion of lateral aorta remaining open in some reptiles; *ra,* right atrium; *rsa,* right systemic arch; *rv,* right ventricle; *spa,* interatrial septum; *spv,* interventricular septum; *sv,* sinus venosus; *v,* ventricle; *vc,* anterior vena cava; *vci,* posterior vena cava. (From Goodrich.)

third, fourth, and sixth of the series; arch V persists in urodeles, but is absent in frogs and all adult amniotes. In fishes the dorsal ends of all these arches connected with the uninterrupted dorsal aorta. Variations in these connections, however, are already present in amphibians. Even in fishes the blood flowing up through the third arch tended to pass forward alone toward the head region, and cranial blood supply becomes the sole function of this arch in tetrapods. The part of the dorsal aorta between arches III and IV tends to disappear in adult tetrapods, thus completely separating cranial and body segments of the original aorta. However, this connection still persists in the adult in the Apoda, some urodeles and even in reptiles—Sphenodon and many lizards—where it is known as the *carotid duct.* The third arch and the associated anterior part of the dorsal aorta is called the *internal carotid artery;* the ventral aortic stem leading forward from the base of the fourth arch becomes the *common carotid* (cf. Fig. 269).

The fourth arch is always a large, bilaterally developed vessel in lower tetrapods; it is termed the *systemic arch* and is the main channel for blood flowing from heart to body. The fifth arch, on the other hand, tends to "lose out" in tetrapods; it persists in urodeles, but is present in no other land vertebrates beyond the embryonic stage, and even in the embryo is usually small and transient in appearance, and may fail to develop at all. Why the fourth arch is selected as the blood channel to the body rather than the seemingly shorter and more direct route via arch V is an unsolved puzzle.

As we have noted, the lung is supplied with blood via arch VI. The lung is functionless in the amphibian larva (as in the embryos of amniotes). During the larval period the major part of the blood in this arch travels straight on upward into the dorsal aorta as part of the main blood stream to the body. But when, in the adult amphibian (or in the amniote at birth or hatching), the lung begins to function, the dorsal part of this arch should, for the sake of efficiency, be eliminated to prevent admixture of blood streams. This connection, the *ductus arteriosus* (or duct of Botalli), persists in the adult in urodeles and Apoda, in Sphenodon, the alligator, and some turtles; in frogs and all amniotes except the reptiles mentioned, it disappears after metamorphosis or birth.

The original ventral aorta, as a result of the processes just described, tends to be reduced to a channel of supply for arches III, IV, and VI. The last, as a bearer of venous blood to the lung, merits a separate channel from the heart. We find that in living amphibians the ventral aorta as such is done away with; it has been split to its base to separate the *pulmonary trunk* from the remainder of the original aorta.

In urodeles there is a relatively complete and "fishy" arch system, but in anurans, as a result of the changes discussed, we find an arch system of a sort characteristic of a majority of reptiles and, indeed, more advanced than in many of that group. In this pre-amniote stage we have remaining (1) a carotid system, including the anterior end of the ventral aorta, the third arch, and the anterior end of the paired aorta; (2) a paired systemic arch, including the major basal stem of the ventral aorta, the fourth pair of arches, and continuing posteriorly into the dorsal aorta; (3) a pulmonary system, including a basal stem split off from the ventral aorta, part of the sixth arch, and a *pulmonary artery* leading from arch to the lung. Variant relics of more primitive stages which may persist in other amphibians include the carotid duct, connecting arches III and IV dorsally, the arterial duct, completing the dorsal end of arch VI, and the fifth arch retained in urodeles.

Amniote Aortic Arches. In the amniotes further developments in the aortic arch system have to do mainly with variations in the fourth arch. In earlier stages the arch structures were in general strictly bilaterally symmetrical; in the amniotes asymmetry appears.

It is probable that in the ancestral reptiles, as in the frog, there were two ventral trunks from the heart, one to the lungs, the other a common trunk to both systemic arches and both carotids. In living reptiles, however, we find that three vessels, rather than two, open forward from the heart (Fig. 266, *C, D*). These include (1) the pulmonary arch; (2) a separate tube leading only to the left systemic arch; and (3) one for the

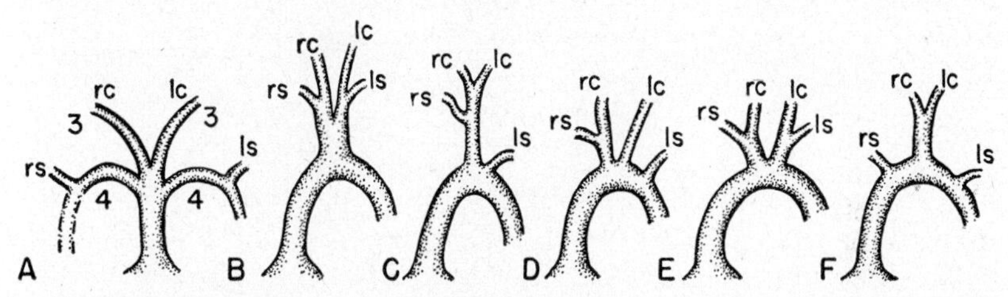

Fig. 267. Diagrams, in ventral view, to show variations in the branching of the main blood vessels from the mammalian aortic arch. *A*, Embryonic condition, with ventral trunk of aorta and third (carotid) and fourth pairs of arches, of which the right fourth arch is later lost beyond the point of departure of the subclavian. By differential growth of the vessels the various arrangements shown in *B* to *F* are brought about (*D* is the human type). *lc*, Left carotid, *ls*, left subclavian, *rc*, right carotid, *rs*, right subclavian arteries. (After Hafferl, in part.)

right systemic arch. These are so situated that in the ventricular cavity of the heart, incompletely divided in most reptiles, the vessel for the left fourth arch tends to receive mainly venous body blood and hence has little function as an oxygen carrier. When, in crocodiles, the two ventricles are completely separated, the left fourth arch definitely receives its blood from the right ventricle. This recirculation of "venous" blood is obviously highly inefficient; and the birds, descended from reptiles related to the crocodiles, have eliminated this useless vessel completely (Fig. 264, *G*). Apart from the pulmonary stem from the right ventricle, there remains only a single trunk leaving the heart—that from the left ventricle. This carries aerated blood to both the carotids and thence to the head, and to the body by a single systemic arch—the right member of the original pair.

The history of the mammalian arches seems to have been simpler (Figs. 264, *H*, 266, *B*). Mammal ancestry diverged from that of other reptiles at an early date, and there is no reason to believe that the system of three heart orifices seen in modern reptiles was ever present in the mammalian line. Presumably, as in anurans, the early reptilian ancestor of mammals had, beside a pulmonary arch, a single trunk leading from the left side of the ventricle to the two carotids and the two systemic arches. But a double systemic arch is unnecessary and inefficient; somewhere along the line of development to mammals, the right fourth arch disappeared from the pic-

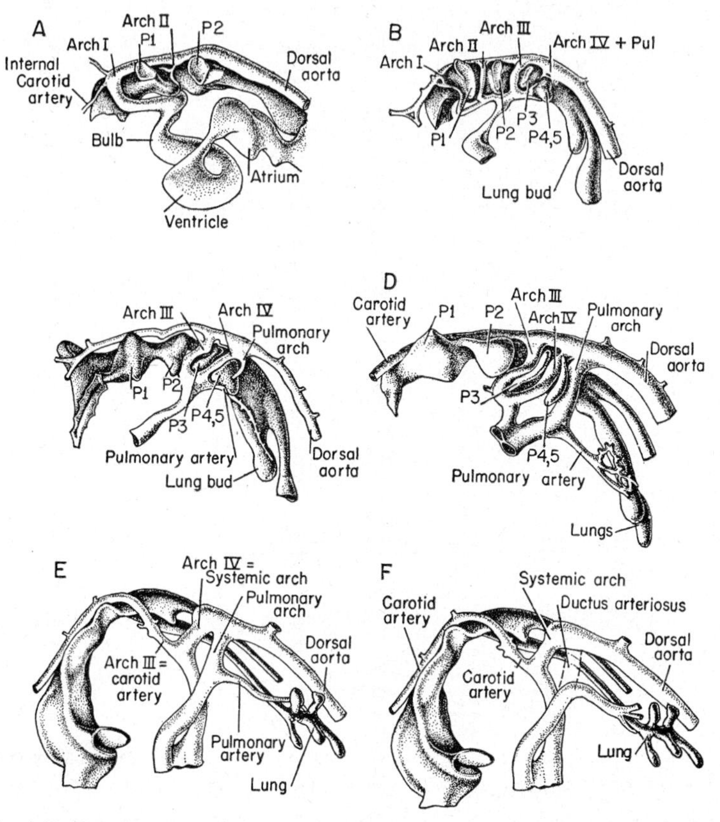

Fig. 268. Development of the aortic arches of a mammal (Homo). The outlines of the gut cavity, gill pouches, and lung buds are shown in addition to blood vessels; in *A*, the cavities within the heart are included. Arch I is developed. *B*, Arch I already reduced, II and III formed. *C*, Arch II reduced, IV (systemic) formed, VI (pulmonary) arch and pulmonary artery forming. *D*, Pulmonary arch well developed. *E*, Carotid arch (III) separated dorsally from aorta, pulmonary arch becoming distinct at root from ventral aorta. *F*, Diagram to show reduction at birth of upper end of arch VI (ductus arteriosus). p^1 to p^5 = pharyngeal pouches. (After Streeter.)

ture (except that its base remained as a connection with the subclavian artery to the arm), and the entire blood supply to the trunk follows the curve of the left member of the embryonic pair. The mammal, like the bird, has simplified the systemic blood supply; but the two groups are in contrast as to which member of the pair of fourth arches is retained.

In mammals there is great variation in the way in which the pairs of carotids and subclavians branch off from the arch of the aorta (Fig. 267).

The carotids may leave the aorta separately, by a common stem, or jointly with the neighboring subclavian; and as a final variant, all four vessels may depart as a single, large trunk.

The embryonic development of the aortic arches of a mammal recapitulates to a considerable degree the phylogenetic story outlined above (Fig. 268). The first blood channels established from heart to body in the embryo are a pair of anteriorly placed arches which, subsequent history shows, are the first pair of the typical fish series. Gill pouches develop in

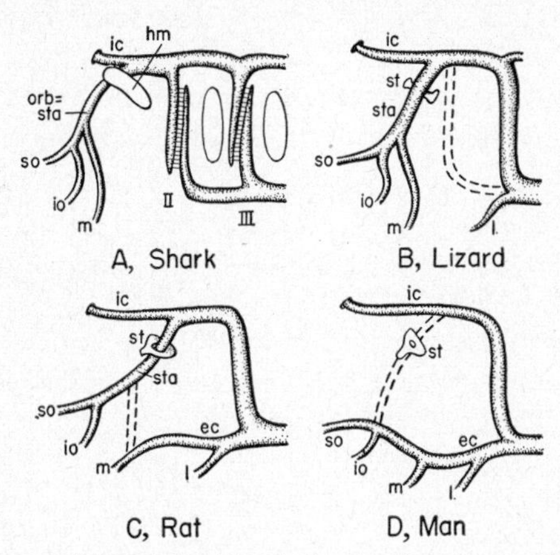

Fig. 269. Diagrams of the left side of the head to show the evolution of the carotid system. In a primitive fish stage (*A*) the direct forward continuation of the aorta is the internal carotid artery, which enters the braincase near the pituitary. This gives off a major branch, the orbital artery, which, passing close to the hyomandibular, supplies most of the more superficial features of the skull and jaws. (An accessory blood supply to the head from the spiracle is omitted from the diagram.) In many tetrapods (*B*) a similar situation persists, the orbital artery being commonly called the stapedial, since it passes close to or through the stapes (= hyomandibular). However, a small branch present near the root of the carotid extends forward, and as the external carotid of mammals may take over part (*C*) or all (*D*) of the functions of the stapedial. *ec,* External carotid; *hm,* hyomandibular; *ic,* internal carotid; *io,* infraorbital artery; *l,* lingual artery; *m,* mandibular artery; *orb,* orbital artery; *so,* supraorbital artery; *st,* stapes; *sta,* stapedial artery; II, III, second and third aortic arches. In *B* to *D* modified aortic root = common carotid.

series behind this point, and successive aortic loops are formed between them as arches II, III, and IV. As the more posterior arches develop they tend, each in turn, to become the most prominent pair; synchronously, the more anterior arches tend to be reduced in importance, so that the first two cease to be functional in a short time. Behind arch IV, arch VI presently develops, but the fifth arch is sometimes absent and at the best is small and transitory. In the embryo, arch VI runs directly upward to the dorsal aorta, and little blood passes through the pulmonary artery branching from it. At birth, however, the upper part of this arch—the ductus arteriosus—becomes almost immediately occluded, and the full flow of blood passes to the lung.

Blood Supply to the Head (Fig. 269). Though the main dorsal flow of arterial blood is in a posterior direction, to the trunk, tail, and limbs, an important element in the circulation, if a minor one in quantity, is a forward flow to the head region.

The anterior part of the dorsal aortic system is a paired vessel which, in the skull region, primitively runs along either side of the braincase, receiving the efferent branchial arteries from more anterior members of the gill series. Posteriorly, this vessel is considered a paired part of the dorsal aorta; anteriorly, it is an obvious homologue of the *internal carotid artery* of higher vertebrates above the fish level. If we follow this artery forward, we find that after giving off a major branch—*orbital artery* or temporal artery—to supply much of the general face and jaw region, its main stem passes upward into the braincase in front of the pituitary. Here it furnishes the main blood supply for the brain, gives off an artery to the eyeball, and connects with the efferent pseudobranchial artery mentioned earlier.

The condition just described is a primitive one, which holds, with little variation, among the various fish groups. In amphibians the system tends to be modified in relation to the disappearance of the internal gill system; all that remains of the aortic arch system anterior to the fourth arch is the proximal part of the carotid system, and the pseudobranchial artery of fishes is absent.

In typical reptiles and amphibians the carotid gives off near its base, while ventral in position, a small *external carotid* (lingual) *artery* which supplies tissues of the tongue and throat (and may exceptionally reach the lower jaw region). The main carotid stem passes upward and forward as the internal carotid, to the middle ear region of the skull. Here the internal carotid artery continues forward toward its opening into the braincase; but it gives off, close by the stapes, a major branch, a large *stapedial artery,* which supplies all the structures of the outer parts of the head and most of the area of the jaws. This last structure is identical with the orbital artery of the fish ancestors.

In primitive mammals and all mammalian embryos the arterial system of the head persists in much the same fashion. In most mammals, however, major changes occur. The originally small external carotid grows forward and taps the branches of the stapedial artery. As a result the external carotid is large in most mammals, the stapedial artery reduced or absent; a new and shorter route has developed for supplying blood to the face and jaws.

Blood Supply to the Body and Limbs (Figs. 261–263, p. 438; 270). In every vertebrate the major blood supply to the trunk, tail, and limbs is furnished by the dorsal aorta and its branches. Its most anterior, cranial, part may be paired, but farther back the aorta becomes a median vessel, lying beneath notochord or vertebrae and above the root of the mesentery;

posteriorly, it continues as a median *caudal artery*. The branches of the aorta tend to be of two types: (1) median ventral branches running downward in the mesentery to the gut, and associated structures; and (2) paired lateral branches to the outer areas of the body and to the limbs.

The median ventral vessels may be numerous in certain lower vertebrates and in embryos; in general, however, there is a somewhat variable tendency for the concentration of these vessels into a few main trunks, usually including a *celiac artery* to the stomach-liver region, and one or more *mesenteric arteries* to the intestines.

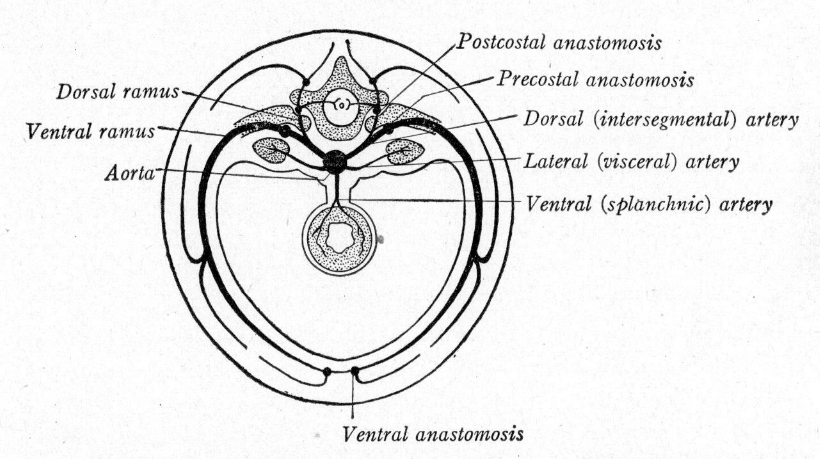

Fig. 270. Diagrammatic cross section of the body of a higher vertebrate to show various types of branches which may be given off by the aorta. Most prominent are median ventral branches descending in the mesentery to the gut, and so forth, and paired intersegmental arteries, the main ventral ramus of which descends the flanks between the myomeres or adjacent to successive ribs. Longitudinal anastomoses may occur between successive segments in various regions. (From Arey.)

Of paired vessels, short lateral branches are present for the gonads and the kidneys. Separate from these is a series of paired segmental blood vessels which run laterally from the aorta to supply the trunk musculature, the skin, and the spinal cord. At several points there may be longitudinal connections between the paired vessels of successive segments. These may form lengthwise arterial channels, supplementary to that afforded by the aorta, in the dorsal region of the body and the skin of the back and flanks. These channels may, further, afford the opportunity for originally segmental arteries to coalesce into a smaller number of large vessels stemming from the dorsal aorta, and in higher vertebrates the segmental arrangement is increasingly obscured.

In the embryonic development of paired fins or legs, there is seen, at an early stage, a network of small arteries continuous with the segmental arteries of the neighboring region of the trunk (Fig. 283, p. 470). During ontogeny one or another of these channels tends to become dominant and to form a main channel from aorta to limb. In the pectoral limb this main trunk is usually given its mammalian name of *subclavian artery,* although it is improbable that it is exactly the same vessel in every case. Giving off branches to the shoulder and chest region, this same main trunk is termed the *axillary artery* as it enters the limb, and the *brachial artery* as it proceeds down the "arm." The point of departure of the subclavian from the primitive aortic trunk seems to have been rather variable; in consequence, its relation to carotids and main aorta varies in tetrapod types (Figs. 262, 263, 267, pp. 439 and 445). In some cases the subclavians arise from the unpaired posterior stem of the dorsal aortic trunk, and this condition is seen, for example, in salamanders. In a majority of reptiles and in frogs they take origin somewhat farther forward from the paired portion of the dorsal aorta of the embryo, and in the adult they thus arise from the paired systemic arch. In mammals the right systemic arch is done away with, but the base of the arch is retained as a proximal part of the right subclavian. In crocodiles the origin of the subclavians is still farther forward, from a point near the top of aortic arch III, and hence in the adult the subclavians and carotids have a common trunk; the same situation persists in their avian relatives (Fig. 266, *D*, p. 443).

In the pelvic appendage, as in the pectoral, there is a tendency for the concentration of the blood supply to the limb into a single vessel, usually termed the *iliac artery,* but variable in position and details of embryological origin. A *femoral artery* typically supplies the anterior or inner side of the thigh. In land vertebrates with well-developed limbs the iliacs are large branches of the aorta; in forms such as frogs and typical mammals in which the tail (and consequently the caudal artery) is reduced, the iliacs appear essentially as a bifurcation of the aorta.

Embryonic Arteries (Figs. 284, 285, p. 471). In the embryo of all vertebrates the primary blood circulation begins in the walls of the gut and its outgrowths, where food materials are found. The blood must be returned to these same areas to complete the circuit; hence the first large branch or branches to develop from the dorsal aorta go to such regions. In forms with mesolecithal eggs the primary veins take origin from the gut walls; hence the major early arteries are those which descend the mesentery to the gut—that is, homologues of the celiac, mesenteric, and the like. In forms with large amounts of yolk the first veins are those from the yolk sac; consequently the *vitelline arteries,* branching from dorsal aorta out over the yolk are large structures in the early embryo. Early paired and numerous, they tend to fuse to a single trunk, comparable to a mesenteric artery, as the body takes shape. They persist till birth or the absorption of the yolk in Chondrichthyes, reptiles, and birds; in mammals, where the yolk sac is empty, they disappear at an earlier stage. In reptiles and birds a vascular circuit early develops to and from the allantoic "lung," and in mammals the same vessels, following the allantoic stalk, supply the placenta. Paired *umbilical arteries* leading to the allantois appear early in the development of all amniotes; in mammals a single fused umbilical artery remains prominent in the umbilical cord until birth, affording the sole embryonic source of blood to the placenta.

VENOUS SYSTEM

The veins—by definition, vessels bringing blood from the tissues to the heart—have a complicated and variable arrangement. If, however, their embryonic history is studied, it is seen that they can be logically sorted out into a small number of systems. The discussion of the venous system here will be based largely on the developmental story. So treated, we may consider the following principal components of the venous system (Fig. 271):

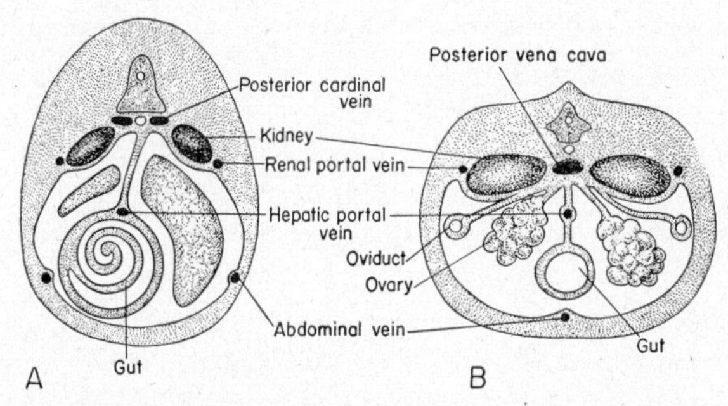

Fig. 271. *A,* Cross section of the abdominal region of a shark, to show the position of the main veins. *B,* The same in a urodele amphibian.

1. A *subintestinal system* of veins flowing forward below the gut, and in the adult forming the *hepatic portal system* to the liver and the *hepatic veins* from liver toward the heart.

2. Veins situated dorsally or dorsolaterally to the celom or gut and carrying blood toward the heart from the dorsal parts of the body and head (and generally from the paired limbs as well); they include the *cardinals* or the *venae cavae* which replace them, and their affluents.

3. A relatively minor group, the *abdominal vein* or veins, draining the ventral part of the body wall.

4. In lung-bearing forms, the *pulmonary veins* from lungs to heart.

The first of these four components forms the drainage of the gut tube and its outgrowths; it is essentially a visceral venous system. The second and third components are, on the contrary, somatic venous systems, draining the outer wall of the body structure.

Hepatic Portal System and Hepatic Veins (Figs. 261–263, pp. 438, 439). The hepatic portal system, found in all vertebrates, is composed of veins which collect blood from the intestine and transport it to the capillary network of the liver. Functionally the system is of great importance, for the intestinal capillaries take in all the food materials absorbed from the gut (with the exception of fats), and the presence of a hepatic portal system guarantees that the liver has "first chance" at such materials, for their storage or transformation, before they are turned out into the general circulation.

Beyond the liver the blood from the intestine is re-collected into a *hepatic vein* or veins. In most fishes this is a large median vessel which empties directly into the sinus venosus of the heart. In the choanate fishes and all tetrapods, however, we find that, as discussed later (p. 456), part of this hepatic vein system has been utilized in the formation of the posterior vena cava, which carries the main blood stream forward from the dorsal part of the trunk to the heart. In consequence, the term "hepatic vein" is restricted in these forms to the vessel or its subdivisions which empty from the liver into the terminal part of the posterior vena cava.

Generally the first blood vessels to appear in the embryo in forms with a mesolecithal type of egg are a pair of veins which form in the floor of the gut and coalesce into a single channel running forward ventrally as a *subintestinal vein* (Fig. 284, p. 471). From the far anterior end of this trunk develop the ventral aorta and heart—structures with which we are not concerned here. Posteriorly, this trunk gives rise to the hepatic and hepatic portal veins.

In large-yolked types, as elasmobranchs, reptiles, and birds, where the early embryo is spread out over the yolk surface and there is no formed midventral line, the blood channels which here correspond to the embryonic subintestinal vein are paired structures which collect blood from the surface of the yolk mass. As the body of the embryo takes shape, the parts of these vessels which lie within it fuse to form a typical subintestinal vessel; those parts remaining external to the body persist as large and important embryonic structures, the *vitelline* or omphalo-mesenteric *veins* (Fig. 285, p. 471). Similar veins develop in mammals in early stages, although the yolk is absent. With the reduction and disappearance of the yolk sac during development, the parts of this system external to the embryo disappear.

In early stages the subintestinal or vitelline system runs forward below the gut directly to the heart. Presently, however, the liver grows out ventrally from the gut; with its growth, liver and vein materials become intermingled; the vein trunk breaks up into small vessels and finally into a liver capillary system, with the resulting formation of a separate portal trunk posteriorly and a hepatic vein anteriorly (Fig. 286, p. 473). In typical fishes no further important development occurs; in higher vertebrates a branch of the hepatic vein reaches dorsally along the mesenteries to tap the posterior cardinal system and form the anterior part of the posterior vena cava (p. 456). The hepatic portal vein remains a large and important vessel, collecting blood not only from the intestine, but from the stomach, pancreas, and spleen as well, for conduction forward to the liver.

Dorsal Veins—Cardinals and Venae Cavae. The principal blood drainage from the "outer tube" of the body is cared for by important longitudinal vessels situated dorsally above the gut and mesenteries. In lower vertebrates these veins are the cardinals; in higher forms modification of these vessels produces the venae cavae.

In the embryo of every vertebrate paired veins appear at an early stage in the tissues above the celomic cavity, one on either side of the midline

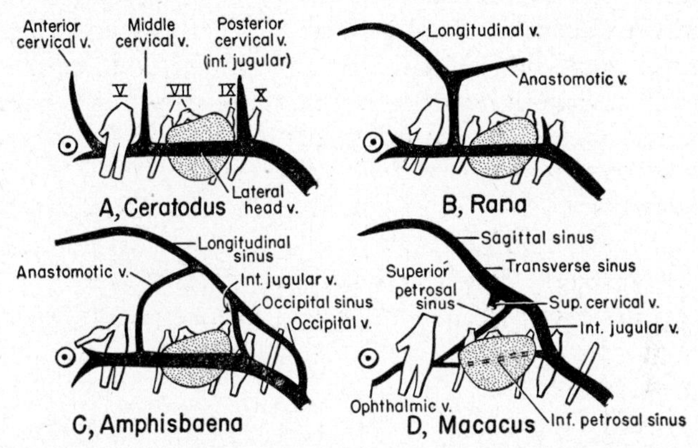

Fig. 272. Diagrams of the left side of the head region to show stages in the evolution of the venous drainage. Roots of some of cranial nerves are indicated; the position of the eye is shown, and the otic capsule is stippled. In lower vertebrates the main drainage is by a lateral head vein, which forms in the orbital region and runs backward to become the anterior cardinal. This primitively receives several successive veins from the interior of the skull. A series of sinuses develops within the braincase; the lateral head vein is abandoned in mammals, and blood from the orbital region enters the sinus system, all finally draining from the skull as the internal jugular vein. *A,* Lungfish; *B,* frog; *C,* lizard; *D,* mammal (macaque monkey). (After van Gelderen.)

(Fig. 285). These are the primitive *cardinal veins*. The *posterior cardinals* run along the trunk on either side of the dorsal aorta to a point in the body wall dorsal to the heart. Paired *anterior cardinals* begin as head veins on either side of the developing braincase and run back dorsally on either side of the neck region to meet their posterior mates. From this point of junction on either side a major vessel descends to enter the sinus venosus of the heart; this is the *common cardinal* (or duct of Cuvier). This characteristic embryonic cardinal system is retained in almost diagrammatic form in the adult of elasmobranchs (Fig. 261, p. 438). We may discuss separately the history of the anterior and posterior parts of this system.

In all vertebrate classes except the mammals the main stem of each anterior cardinal (or anterior vena cava) is the *lateral head vein* (Fig. 272). This characteristically arises in the space behind the orbit, where it may be present as an expanded sinus, and receives vessels from various areas of the anterior part of the head; it connects, in most vertebrates below the mammalian level, with its fellow of the opposite side, by an interorbital vein located behind the hypophysis. Traveling posteriorly along the side of the braincase and ear region, the lateral head veins receive tributaries from the brain and, continuing backward, enter the common cardinals. In lungfishes and tetrapods (where the posterior vena cava has taken the place of the posterior cardinals) the common cardinals are incorporated in the trunk of the anterior cardinals, including in their course, therefore, the openings of the subclavian veins from the pectoral limbs. With this modification the anterior cardinals come to resemble the vessels termed anterior venae cavae in mammals, and are frequently called by this name.

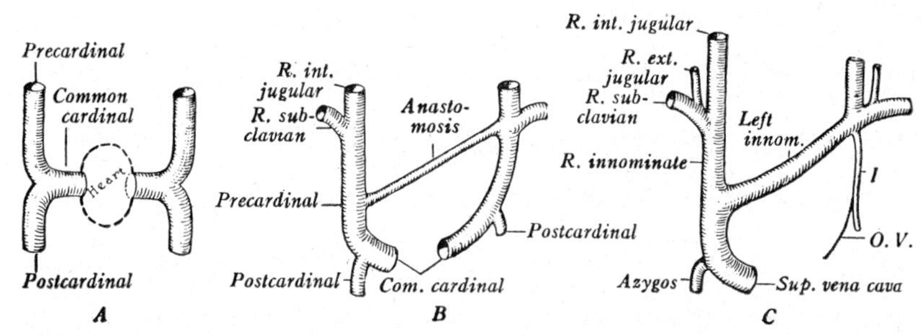

Fig. 273. Ventral views of the veins anterior to the heart region to show the formation, in man and certain other mammals, of a single anterior (or superior) vena cava from the two anterior cardinals. An intercostal vein (*I*) and a small vein from the wall of the left atrium (*OV*, oblique vein) are persisting vestiges of the original left anterior cardinal. (From Arey.)

In mammals there is an important change in the cranial circulation. A system of intercommunicating venous sinuses is set up within the expanded brain cavity. The lateral head veins disappear, and the blood from the deep vessels of the front part of the head enters the braincase, to leave posteriorly on either side, after collection of the venous blood from the brain, as the *internal jugular veins*. As these travel posteriorly they are joined by vessels, the *external jugulars,* which gather blood from the more superficial parts of the head, to form *common jugulars;* joining the subclavian vein, the further course of each of these vessels to the heart is a trunk termed an *anterior vena cava*. Such veins are easily recognizable as the old anterior cardinals and common cardinals, except that intracranial channels supplant the lateral head veins which originally formed the most anterior part of the cardinal stems.

In many mammals (including man) the terminal part of the system is further modified (Fig. 273). In the embryo a cross connection develops between the two trunks, a short distance anterior to the heart, and the left common cardinal disappears. All the blood from the left cardinal then flows over to enter the heart through the right cardinal, so that in the adult only a single anterior (or superior) vena cava is present.

The story of the *posterior cardinals* is more complex (Figs. 274, 275). It begins with a pair of rather simple dorsal trunk vessels draining forward

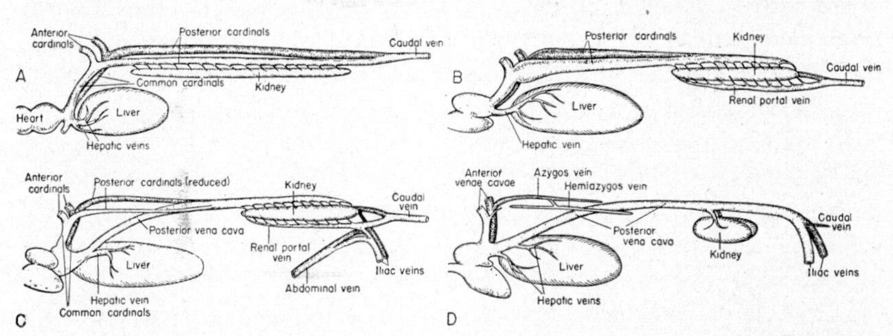

Fig. 274. Diagrams in lateral view to show the evolution of the posterior cardinals and their transformation into a posterior vena cava. *A*, Lamprey (larva); *B*, typical fish condition; interjection of renal portal system. *C*, Lungfish or primitive tetrapod; a shortened route to the heart is established by utilizing part of the hepatic vein system in the initiation of a posterior vena cava. *D*, Mammal; the renal portal eliminated. (Vessels of the right side are shown in deeper shading.)

into the heart via the common cardinals; it ends in mammals with the draining of the same region by a single but complex vessel, the posterior vena cava. In between lies a considerable history.

In cyclostomes the posterior cardinals are simple paired vessels, receiving blood from the tail, kidneys, gonads, and dorsal parts of the body

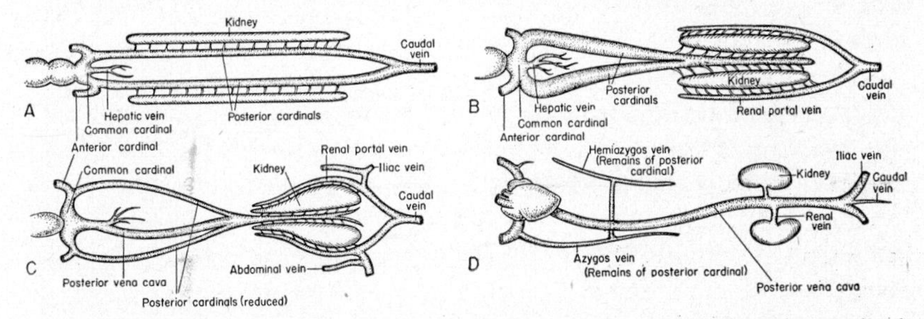

Fig. 275. Diagrams in ventral view to show the evolution of the posterior cardinals and their transformation into a posterior vena cava. Stages as in Figure 274.

musculature and running forward uninterruptedly to the common cardinals. In the Chondrichthyes, however, a new feature appears in the system which was destined to persist upward into the reptilian stage—the development of a *renal portal system*. Seemingly in order to "guarantee" that a fair amount of blood would pass through the opisthonephric kidney, the blood from the posterior part of the trunk and from the tail, which passes forward in the original cardinal channels, does not go directly to the heart, but is instead shunted to the kidneys to pass through a network of capillaries around the kidney tubules. After this passage it re-collects to course forward in the truncated posterior cardinals.

The actinopterygians exhibit a specialized condition in that only part of the blood from the tail enters the renal portal; some blood may run forward directly into the cardinals, and (most unusual for an adult, although not for the embryo) a branch of the caudal vein may run downward into the hepatic portal trunk.

In the Choanichthyes, as shown by the living lungfishes, a progressive change initiates the development of a *posterior vena cava.* A branch of the hepatic veins draining the liver extends upward past that organ along a pulmonary fold (cf. p. 480) to the dorsal side of the body cavity and taps the right posterior cardinal vein some distance back of its entrance into the common cardinal. Once this is done, blood from the right cardinal follows this new channel to the heart; further, since there are cross connections in the kidney region between the two posterior cardinals, that from the left vessel follows the same course. As a result, in higher tetrapods the anterior ends of both posterior cardinals lose connection with their more posterior trunks; they remain only as small and somewhat variable *azygos veins,* draining the blood from a part of the flank into the anterior venae cavae. The new major trunk, from kidney via this short circuit to the heart, may be properly termed the posterior vena cava.

The lungfish situation continues relatively unchanged through the amphibians and reptiles. Among members of the latter class, however, the renal portal system begins to degenerate, and some of the portal blood may pass through the kidney without entering the tubule circulation. In birds and mammals the renal portal system is abandoned. The blood from hind legs (and tail when present) passes directly forward into the posterior vena cava; further, of the paired renal portal channels, only that of the right side is retained. The abandonment of the renal portal system may perhaps be associated with the fact that the higher blood pressure in amniotes renders it unnecessary to "compel" the passage of blood through the kidney; arterial blood, under better pressure, will "voluntarily" take this course.

The posterior (inferior) vena cava of the mammal is a single structure of seeming simplicity, running forward dorsally along the back of the body cavity from the pelvic region and finally downward past liver to heart. As its history shows, however, it is a composite patchwork structure, including part of the old right posterior cardinal and vessels replacing it (such as the renal portal vein) and, anteriorly, a branch of the original hepatic vein. In its development from the primitive posterior cardinal system there have been three main steps: (1) development of a renal portal system; (2) tapping of the cardinals by the hepatic vein and consequent degeneration of much of the anterior ends of the cardinal system; (3) subsequent abandonment, in mammals as in birds, of the renal portal.

The development of the mammalian circulation shows a complicated history which recapitulates this phylogenetic story in an elaborate (and somewhat variable) fashion.

Abdominal Veins. In the Chondrichthyes longitudinal veins—*abdominal veins*—are present low down in the body wall on either side (Figs. 261, 271, *A*, pp. 438, 451). They drain blood from the lower part of the trunk musculature; the small veins from the pelvic fins enter the back end of these veins, and the pectoral fin veins empty into or near their anterior ends. Abdominal veins are absent in actinopterygians. In the lungfishes we find, instead of paired vessels, a median ventral abdominal vein, which is presumably homologous. In amphibians (Figs. 262, 271, *B*) and reptiles the median abdominal persists, despite the reduction of the trunk musculature; but anteriorly, instead of entering the heart directly, it becomes incorporated into the liver sinus system with the hepatic portal. In some of these lower tetrapods the blood from the tail may flow into the abdominal vein; but there is a connection with the kidneys so that hind leg blood as well as blood from the tail may also flow forward through the renal portal system. In birds and mammals the abdominal vein disappears in the adult; hind legs and tail drain into the dorsal vein system.

The abdominal vein system persists in the embryos of all amniotes, even when absent in the adults, for the significant reason that the veins from the allantois empty into the abdominals. These veins in reptiles and birds are important for bringing oxygenated blood from the allantoic "lung" into the body; in placental mammals they are the carriers of the entire food supply from the placenta (Fig. 285, p. 471). These are in mammals termed the *umbilical veins,* since, fused into one vessel for part of their course, they form the one venous afferent in the umbilical cord of the late embryo. As in the abdominal vein of adult amphibians and reptiles, the umbilical veins drain into the liver (Fig. 286, p. 473); paired in the body of the early embryo, only one vein may persist at a later stage, and beyond birth the whole system disappears in birds and mammals.

We may mention here a pair of anterior veins in fishes which perform for the anterior end of the body the function of the abdominal veins in the trunk; they are found, in sharks and other lower vertebrates, as paired ventrolateral structures draining the walls of the "throat" region and running back into the common cardinals (Fig. 261). They are frequently termed external jugulars, but appear not to be homologous with the vessels of that name in mammals.

Limb Veins (Figs. 261–263, p. 438). The pectoral fins and tetrapod pectoral limbs are drained by a large vessel usually termed by its mammalian name, the *subclavian vein.* This may be associated with the anterior end of the abdominal vein; it may enter directly the common cardinal; it may join the anterior cardinal or the jugular vein which replaces it.

The main pelvic limb vessel is usually termed the *iliac vein*. It may follow either of two courses toward the heart: it may enter the abdominal vein system ventrally, or it may join the posterior end of the cardinal system; or—the most common condition in nonmammalian forms—it may connect with both. Through its cardinal connection it may enter, with the caudal vein, the renal portal system; the forked iliac may act as a connection between abdominal and renal portal systems and serve to equalize the pressure in the two. In mammals the two iliacs join to form the posterior vena cava.

Pulmonary Veins. These veins are, of course, absent in living lower fishes, in which lungs are nonexistent. In the actinopterygian Polypterus, where a functional lung exists, the pulmonary veins empty into the hepatic vein; hence their blood mixes with the general blood stream from the body.

In the lungfishes, however, the pulmonary trunk by-passes the sinus venosus and enters the atrial cavity of the heart directly and at the left side. This separate course to the heart of the aerated blood from the lungs persists in all tetrapods. The entrance of the pulmonary veins into this heart region is a causal factor in the subdivision of the atrium and the final subdivision of the entire heart.

LYMPHATICS

Lymphatic Vessels. In all except the lowest vertebrate groups we find, supplementary to the venous system, a second series of vessels returning from the tissues to the heart—the lymphatic system. Although paralleling the veins in many functions (and often paralleling them topographically), the lymphatics differ from them in major respects. A fundamental difference is that the lymphatics are not connected in any way with the arteries, but arise from their own capillaries which are blind at their tips (Fig. 260, p. 437). Although the nature of the tips of the lymphatics was once debated, it is now generally agreed that they are closed throughout. There is thus no arterial pressure behind the fluid contained in the lymphatics; the flow of materials in them is generally sluggish and brought about only by body movements.

The contained liquid, the *lymph,* gains entrance into the lymphatic vessels by diffusion through their walls and hence is essentially similar to the general tissue juice and (except for the absence of blood proteins) to the plasma of the blood vessels into which the lymphatics empty. There are, in most cases, few blood corpuscles in the lymph, since there is no flow of corpuscle-bearing blood into their distal ends; ameboid white blood cells, however, may make their entrance from adjacent tissue spaces. The lymphocytes owe their name to the fact that in mammals the main site of their formation is in the lymph nodes placed in the course of the lymphatic vessels. This, however, is a mammalian peculiarity, and in other vertebrate groups there is no such association of the lymphocytes with the lymphatic system.

Related to the low pressure under which lymph travels is the fact that the walls of lymphatics are thin (they are consequently difficult to find and dissect unless specially injected). Even the largest of lymph vessels have but a thin coat of musculature and connective tissue. Like veins, the lymphatics in birds and mammals are supplied with valves preventing a backward flow of the lymph. In amphibians the lymphatic system is supplied with *lymph hearts*—small, two-chambered, muscular structures which lie at points where lymph vessels empty into veins and actively pump lymph onward into the general circulation. These hearts are usually but few in number, but in the Apoda there may be 100 or so in two paired series. A pair in the pelvic region is prominent in both anurans and urodeles, and persists in reptiles and ratite birds. No lymph hearts are, however, found in other birds or in mammals.

In the Agnatha and Chondrichthyes there are no lymphatics, although the presence of thin-walled sinusoids which drain into veins, but have little or no arterial connections, are suggestive of a stage in the evolution of these structures. Lymphatics are present, however, in bony fishes and all tetrapods. They are highly developed in amphibians, where they are remarkably abundant in the subcutaneous tissues as a protection against potential desiccation; owing to the development of the lymph hearts, amphibians have an active lymph circulation.

Lymphatic vessels in higher vertebrates penetrate most of the body tissues; liver, spleen, and bone marrow are exceptions. They are highly developed in the gut region. The intestinal lymphatics are of great importance in metabolism in that, while carbohydrate and protein food materials pass into the hepatic portal vein, much of the fats enters the circulation by way of the lymphatics; this is perhaps associated with the relatively large size of the molecules involved and difficulties encountered by them in entering against pressure the intestinal capillaries of the blood vessels.

In their arrangement the lymphatic vessels (Fig. 276) vary greatly from group to group. Terminally, the lymphatics enter the veins—generally the dorsal veins of the cardinal and vena cava vessels. Primitively, the points of entrance may have been numerous and even more or less metameric; conditions of this sort prevail in urodeles and apodous amphibians, and even in amniotes small lymphatics enter the veins at a variety of points. In general, however, there tends to be a concentration of lymph discharge in three areas: (1) Anterior. There are always one or more main entrances into the anterior cardinals or jugulars near the heart (where pressure is, of course, lowest and unopposed entrance easiest). (2) Middle. Entrances along the course of the posterior cardinals or posterior vena cava are variable; they are rare in mammals, numerous in urodeles. (3) Posterior. In fishes and birds (but not mammals) there may be a major flow of lymph into the veins in the pelvic region.

The fatty lymph, or *chyle,* leaving the intestine runs up the mesenteries in vessels termed *lacteals* from their milky appearance when distended.

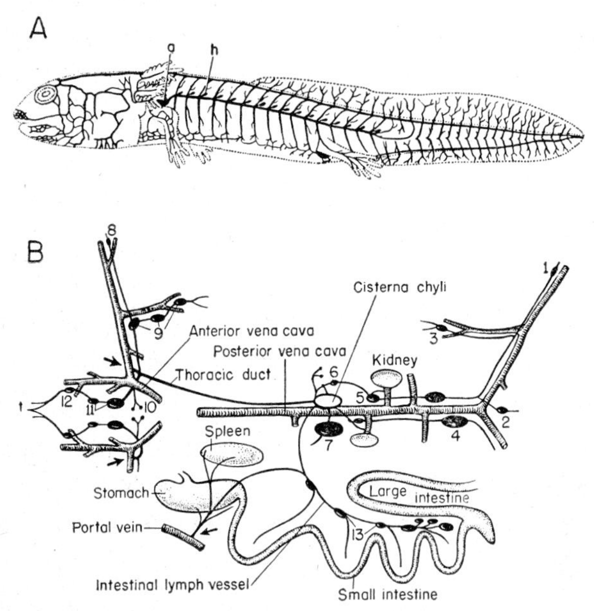

Fig. 276. *A,* Side view of a salamander, showing superficial lymph vessels. Dorsal, lateral, and ventral longitudinal vessels are present; a series of lymph hearts (*h*) is present along the lateral vessel. Lymph from this vessel enters the venous circulation through an axillary sac (*a*); lymph from the ventral vessel enters through an inguinal sac. *B,* Diagram of the deep vessels of the lymphatic system of the rat, anterior end at left. Lymphatics in solid black; the neighboring veins also shown. Nodes are numbered according to the region in which they lie: *1,* knee, *2,* tail; *3,* inguinal; *4,* lumbar; *5,* kidney; *6,* nodes about the cisterna chyli; *7,* intestinal node: *8,* elbow; *9,* axilla; *10,* thoracic; *11,* cervical; *12,* submaxillary; *13,* mesenteric nodes; *t,* plexus of lymphatics around tongue and lips. Arrows indicate the point of entrance of lymph into the veins near the junction of jugular and subclavian and into the portal vein. (*A* after Hoyer and Udziela; *B* after Job.)

There tends to develop a pair of longitudinal lymphatics, *thoracic ducts,* along the back of the body cavity, paralleling the posterior cardinals or vena cava, with anterior and sometimes posterior openings. These ducts serve as collectors for the lacteals. Paired ducts are common in fishes, reptiles, and birds, but in some fishes and reptiles and in mammals generally the pair tend to fuse into a single asymmetrical duct; a large cistern —*cisterna chyli*—may develop along its course in the lumbar region.

The lymphatic vessels frequently run parallel to the veins and have been thought by some to represent a part of the embryonic venous system. They appear, however, to be of independent origin, although lymphatics may grow outward during development from the veins into which they are to empty.

The functional explanation of the development of lymphatics would appear to be suggested by the fact that they are most highly developed in tetrapods, in which the gills have been eliminated from the arterial circuit, and the blood, hence, courses at a higher pressure than in lower fishes. In consequence, entrance of tissue liquids into the capillaries, against pressure, is more difficult than in the original fish condition. The lymphatics offer a relatively low pressure system of drainage into the veins, where the blood pressure is at its lowest.

THE HEART

Some type of muscular pump is necessary for the efficient circulation of the blood. A variety of heart structures are found in invertebrates, and in Amphioxus there is a whole series of paired "heartlets." In true vertebrates, however, the heart is a single structure, situated ventrally and well anteriorly in the trunk, sucking in posteriorly venous blood from all regions of the body and pumping it anteriorly, in lower vertebrates, to the aortic arch system and the gill circulation. Primitively, it consisted of four successive chambers, termed, from back to front, sinus venosus, atrium, ventricle, conus arteriosus; in advanced groups the first and last lose their identity, and atrium and ventricle tend to subdivide.

The heart is situated ventral to the gut in a special anterior region of the celomic cavity; it is in the adult attached to the walls of this cavity only at the points of entrance and departure of the blood vessels and thus is able to change its shape readily during its powerful pumping movements. This pericardial cavity (Fig. 279, p. 463; also pp. 477, 478) is usually completely separated from the general body cavity in the adult, but connections persist in the cyclostomes and Chondrichthyes.

The heart is essentially a series of expansions developed along the course of a main blood trunk; its histological structure, although much modified, is essentially comparable to that seen in other blood vessels. There is a thin internal lining, the *endocardium;* an outer covering is a thin, mesodermal epithelium, the *epicardium,* similar to that lining the rest of the celom. The main bulk of the heart consists of connective tissues and muscle—the *myocardium.* The connective tissue may be compact (particularly about the ventricles) and become essentially a type of skeleton; in some instances there may even be a development of cartilage or bone. There are frequently cross strands of connective tissue across the cavity of the ventricle (and sometimes of the atrium as well), preventing undue expansion under pressure. The muscular tissue has been discussed earlier; it is in some regards similar to striated muscle, but presumably originated as a specialization of smooth muscle tissue. Between the heart chambers and at the points of entrance or exit of blood vessels lie a series of heart *valves,* similar basically to those in veins or lymphatics, but usually powerful and of complex structure. These yield freely to the forward propulsion of blood, but prevent a backflow when a chamber contracts. Frequently there are tendinous strands of tissue which limit the extent to which the valves may be pushed backwards into (or beyond) the orifice to be closed; in some higher vertebrates the valves may even be furnished with small muscles.

The heart is, of course, supplied by blood vessels which ramify in its muscular walls. Though small, these vessels are of crucial importance; we are familiar with the fact that occlusion of the little coronary artery

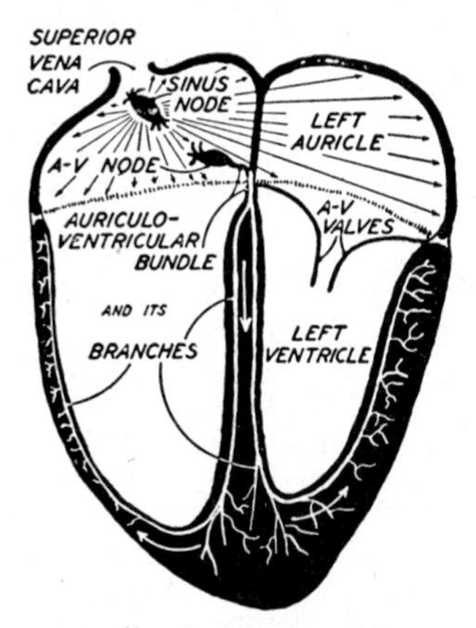

Fig. 277. Diagram of a mammalian heart opened from the ventral surface to show the conducting system. (From Carlson and Johnson, The Machinery of the Body, University of Chicago Press.)

which supplies the heart in man may result in sudden death through heart stoppage.

Heart Beat. Fibers from the autonomic nervous system reach the heart (at the sinus venosus or atrium) and may affect the rhythm of its beat; the heart, however, is essentially "on its own," as is shown by the fact that heart muscle will continue a rhythmic contraction even when cultured apart from the body. In lower vertebrate classes the four successive chambers present contract in sequence in postero-anterior order: the beat starts with a contraction of the sinus venosus, which is thought to be stimulated as a result of its periodic expansion by blood full of carbon dioxide. The wave of contraction of the sinus musculature stimulates that of the atrium to activity and so on down the remaining chambers. In amniotes there tends to develop a unique conducting system for this stimulus, the *sinoventricular system*—essentially a local simulation of a nervous system (Fig. 277). It consists of strands or aggregations of fibers which are specialized muscle fibers—specialized not for contraction, but carriage of an impulse. In a mammal a *sinus node* of these tissues, situated in the right atrium, initiates the beat. The stimulus is in the main transferred to a second node in the septum between the atria, whence a fiber bundle—the *atrioventricular bundle*—descends to disperse among the muscles of the ventricles and set up a contraction of these chambers. The asymmetri-

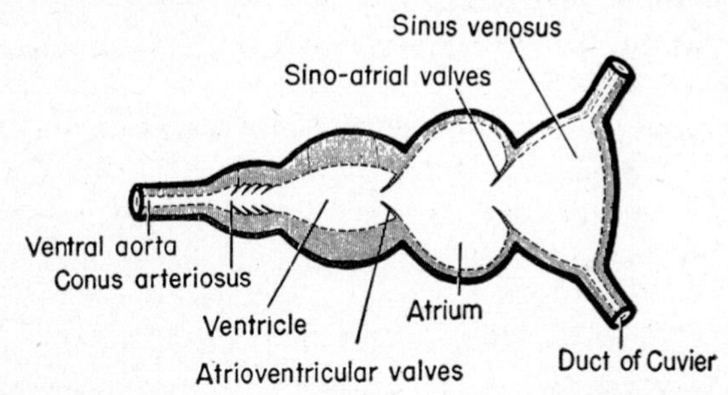

Fig. 278. Diagram of the chambers of the primitive vertebrate heart. (After Ihle.)

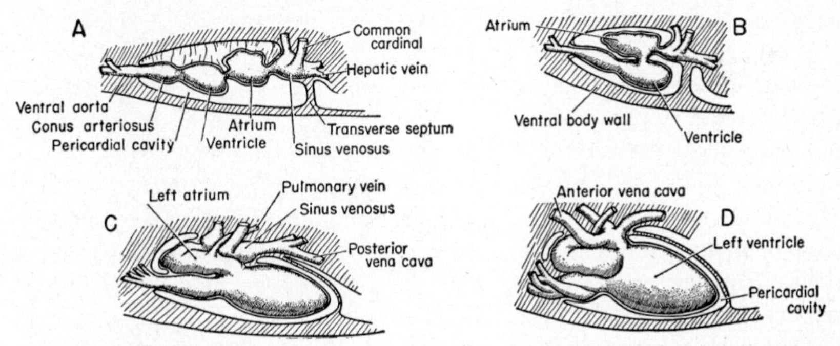

Fig. 279. Diagrammatic views of left side of heart in various vertebrates, to show its position in the pericardial cavity and phylogenetic modification of the heart chambers. *A,* Hypothetical ancestral condition, found essentially repeated in embryos (cf. Figure 278). The four primitive chambers are in line anteroposteriorly, and a dorsal mesentery is still present. *B,* Selachian stage; the mesentery is gone; the atrium has pushed forward above the ventricle, but the sinus venosus is still posteriorly placed. *C,* Amphibian stage; the sinus and accompanying blood vessels have moved anteriorly. *D,* Amniote stage; sinus and conus arteriosus have lost their identity; the heart attaches to the walls of the pericardium only anteriorly. (After Goodrich, 1930.)

cal point of origin of the beat is connected with the fact that the sinus venosus, in which the stimulus began in the primitive heart, is here incorporated in the right atrium.

The Primitive Heart. The heart of typical fishes is a single tube consisting of four consecutive chambers. The heart of a bird or mammal is also four-chambered, but the four chambers do not correspond to the original four; it is, rather, a double pump, with two chambers in each of its two parts. Great changes have occurred in the history of the heart, changes correlated with the shift from gill breathing to lung breathing and the necessary changes associated in the course of the circulation.

In the more primitive vertebrate heart (Figs. 278, 279, *A*) the four chambers are:

1. *Sinus venosus,* a thin-walled sac, with little muscular tissue, essentially a place for the collection of venous blood, lying posteriorly, in the region of the septum between heart cavity and general celom. It receives the hepatic vein or veins, posteriorly, and laterally the paired common cardinal veins.
2. *Atrium* (or auricle), the next anterior chamber, still relatively thin-walled and distensible.
3. *Ventricle,* thick-walled; the main contractile organ of the heart.
4. *Conus arteriosus,** thick-walled, but smaller in diameter and frequently lined by several sets of valves.

These four chambers are arranged in postero-anterior series in the embryo of lower vertebrates. But during development the anterior part of the heart tube tends to fold back ventrally (and somewhat to the right) in an S-shaped curve, thus combining length with compact structure. As a result, the more posterior heart chambers tend, even in a fish heart, to be situated dorsal to the anterior ones, and the atrium may be anterior as well as dorsal in position with regard to the ventricle. The ventricle tends to be so placed that it forms a pocket with a "free" posterior end and with both atrial entrance and arterial exit via the conus placed near one another anteriorly. The folded position makes it difficult to visualize heart construction, and frequently (as in Fig. 266, p. 443) diagrams of heart structure arbitrarily represent the organ as if "pulled out" into its embryonic longitudinal arrangement.

The primitive type of heart here described is typically developed in the cartilaginous fishes, and most other fishes have a heart similarly constructed. There may be, however, minor modifications. In cyclostomes, for example, the sinus venosus tends to be reduced in size; in teleosts the conus is much reduced and practically absent and is more or less replaced by an arterial bulb.

Development of Double Heart Circuit (Fig. 266). In the lungfishes, and more fully in the amphibians, a problem arises connected with the substitution of lungs for gills as breathing organs. The heart now receives blood of two different types: "spent" venous blood from the body, and "fresh," oxygenated blood from the lungs. These two streams should be kept separate, as far as possible, and sent to two different destinations—the venous blood to the lungs, the "fresh" blood to the body. Beyond the heart a splitting of the arterial channels is made for these two destinations.

* Often confused with the *bulbus arteriosus,* developed as a proximal part of the ventral aorta just in front of the heart. The term *bulbus cordis* is frequently used for the embryonic equivalent of the conus in mammals.

But how to keep the two streams distinct in a single-baralleled pump? The perfect solution of this difficulty was not attained until the avian and mammalian stages were reached, but the lungfishes and amphibians are already far along toward attainment of a separation of the two blood streams.

In the lungfishes we find that the blood coming from the lungs by way of the pulmonary veins does not enter the sinus venosus, as does the typical venous blood (and, indeed, the pulmonary veins of Polypterus), but enters the atrium directly and separately. The sinus venosus, thus, is restricted to the typical venous blood circuit; the sinus remains large in lungfishes, but is reduced in size in amphibians and is absorbed into the atrial structure in amniotes.

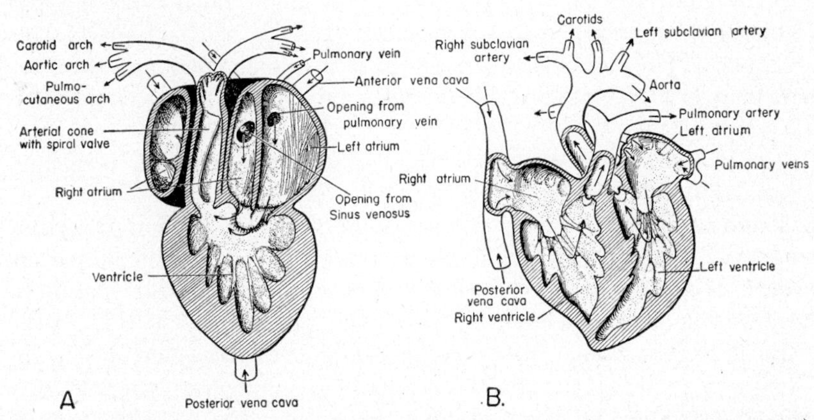

Fig. 280. Diagrammatic section through the heart of *A*, a frog; *B*, a typical mammal. (Partly after Jammes.)

A second change lies in the gradual subdivision of the originally single atrium. The pulmonary venous trunk enters the left side of the atrial chamber. A partition develops which cleaves the atrium in two in such a fashion that the left chamber contains only pulmonary blood; the right (with which the sinus becomes confluent), typical venous blood. In lungfishes there is a partial septum between the two sides of the atrium; in some (but not all) amphibians the division is complete. In both groups, however, there is still a single opening from the atrial chamber into the ventricle. Primitively, this was guarded by a pair of valves. In lungfishes a special plug of tissue from the posterior (originally ventral) wall of the heart helps close this opening (now functionally double); in amphibians four valves are present, although the opening is still single.

But atrial subdivision is in vain if the two blood streams are to meet and mix in the ventricle. In lungfishes and amphibians (Figs. 266, *A*, 280, *A*) the ventricle is single, but complete mixture of the two blood streams is prevented by adaptations of one sort or another. Modern lungfishes have developed a partial septum, essentially a continuation of that in the atrium, along the posterior wall of the ventricle. But this seems to be an independent development in this group, not present in tetrapod ancestors, for amphibians have failed to develop a septum. There is, however, a functional separation of blood streams to some extent in amphibians. The spongy, cavernous structure of the cavity of the ventricle tends to prevent too free a mixing of the blood received from the two auricles, and the arrangement of the openings from these chambers is such that the "venous" blood is poured into the ventricle in advance of that from the lung.

A division of the ventricle takes place in reptiles which is comparable to that of lungfish (Fig. 266, *C*, p. 443). Here, however, it is better developed and nearly completely separates the ventricle into two chambers. In typical reptiles a gap persists near the points of entrance and exit of the blood streams, so that some admixture can still take place. The crocodilians have, technically, a complete ventricular septum, but even here there is a gap, with possible leakage, at the base of the arterial trunks (Fig. 266, *D*).

The two most progressive vertebrate classes, the birds and mammals (Figs. 266, *B*, 280, *B*), have completed the ventricular septum and at long last have completely separated the two blood streams along the length of the major heart chambers. This development has obviously been brought about independently in the two cases, since mammals and birds have evolved independently from primitive reptiles.

But there is no point in establishing separate heart circuits unless the blood, when it leaves the heart, is directed into appropriate arterial channels—aerated blood to head and body, venous blood to the lung. We have earlier (p. 444) noted subdivisions which occur in the ventral arterial trunk and separate these two streams. We must now, finally, consider the history of the most anterior region of the heart, that of the *conus arteriosus,* which is the primitive connecting piece between ventricle and aorta.

Primitively, as seen in a shark, the well-developed conus was a valved, tubular structure with a single terminal opening into the ventral aorta, and the aorta itself was a single stem. The aorta, as noted previously, becomes subdivided in tetrapods so that at least the pulmonary arteries of the sixth aortic arch are separated from the more anterior arches which circulate blood to head and body. There remains as a problem the course of the blood through the conus arteriosus, most anterior of heart chambers. In lungfishes and amphibians we find that there occurs a change in the nature of the conus. Originally it contained simple, typical heart valves; in these primitive lung bearers there develops, instead, a twisted "spiral valve" which runs the length of the conus. This valve partially subdivides the conus for its entire length into two channels (Figs. 266, *A*, 280, *A*), and in the South American lungfish and amphibians the conus is completely subdivided for part of this distance. The structure and functioning of the heart are such that the venous blood is directed mainly into one of these two channels, and the oxygenated blood into the other; distally, the two channels tend to direct their streams mainly into, respectively, the pulmonary arch and the arches leading to trunk and head. Thus in lungfishes and amphibians, despite the incomplete division of the heart, there is a moderate degree of separation of the two blood streams.

In amniotes the conus arteriosus is reduced and not identifiable in the adult. In part it may be incorporated into the ventricles, in part subdivided into the roots of the discrete aortic vessels, three in reptiles, two in birds and mammals (Fig. 266, *B-D*).

The introduction of the lung into the blood circuit in advanced fishes threw out of order the simple and efficient heart plan of primitive vertebrates and furnished a "problem" which vertebrates found difficult to

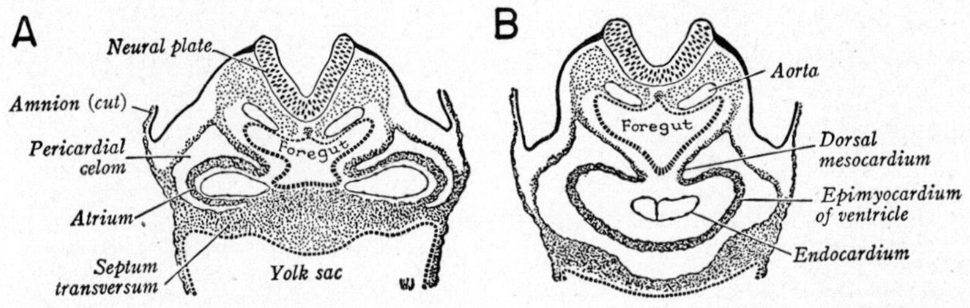

Fig. 281. Cross sections of a mammalian embryo to show an early stage in heart formation, before the fusion of the two subintestinal vessels from which the heart forms. In the atrial region (*A*) the two are still widely separated; farther forward, in the ventricular region (*B*), the two tubes are apposed, inside a single pericardial sac, but not yet fused. (From Arey.)

solve. Lungfishes, amphibians, and reptiles even today have not solved it completely, although their partial solutions are satisfactory enough to allow them to survive. In birds and mammals alone is the complete solution seen and complete separation of circuits attained. The result is surprising in its efficiency. The single pump of the original heart has become a double one; each half of the heart performs effectively its own distinct task.

Heart Development (Fig. 281). We have noted that the first blood vessels to develop in the typical vertebrate embryos are paired vitelline vessels, which fuse to form a subintestinal vein running forward beneath the gut. The heart is a development in the course of this longitudinal tube strategically situated, in the embryo and in the adult of lower vertebrates, near its anterior end, just behind the gill region. The necessity for a circulatory pump develops at an early stage of embryonic history, and in large-yolked types heart formation usually begins even before the two vitelline veins have fused; it may thus appear in transitory fashion as a paired structure. Celomic cavities develop within the lateral plate mesoderm lying on either side of the embryonic heart. The peritoneal linings of the two cavities come to surround the heart tube; they supply it with its epicardial outer coat, furnish mesenchyme, from which the heart muscle later forms, and form dorsal and ventral mesenteries—*mesocardia*. A ventral mesocardium, if developed at all, is short-lived, and the dorsal mesocardium disappears at an early stage, so that a single pericardial cavity is created. Posterior to the heart a transverse septum is presently formed (p. 477) which in most vertebrates shuts off the pericardial cavity from the general celom. With this septum is associated the entrance into the subintestinal vessel of a pair of veins—the common cardinals—descending from the dorsal regions of the body. Their entrance causes an enlargement at the back of the heart region which becomes the sinus venosus. Meanwhile the more anterior part of the heart tube shows a series of enlargements and intervening constrictions which delimit, in linear order from back to front, atrium, ventricle, and conus; in amniotes indications of such divisions may appear even before the pair of heart rudiments is fused. Simultaneously there begins the elongation and twisting of the tube noted in the description of the adult chamber, so that the ventricle comes to lie below and in behind the atrium, and somewhat to the right, rather than in its morphologically primitive anterior position.

With these developments the heart has attained the general structural features found in typical fishes, and needs little but valvular development to reach the adult piscine condition. In higher vertebrates, however, further differentiation is needed. We have noted phylogenetically a series of progressive stages in the subdivision and rearrangement of heart chambers in adjustment to lung breathing and the consequent necessity of separating two blood streams. In the various tetrapod classes ontogenetic events essentially recapitulate the phylogenetic stages. In a mammalian heart, for example, principal developments include subdivision of both atrium and ventricle and of the original single opening between them; absorption of the sinus venosus into the right atrium and the segregation of the pulmonary veins so that they open into the left atrium; subdivision of the conus arteriosus into a pulmonary trunk leading from the right ventricle and a major aorta leading from the left. As noted later, however, the different breathing mechanisms before and after birth in amniotes cause certain embryological developments in the heart which have no phylogenetic significance.

CIRCULATION

Blood Circuits. In earlier sections of this chapter the components of the blood circuit have been described piecemeal. We shall here briefly review the general evolutionary history of the circulation as a whole with especial reference to blood pressures and capillary nets (Fig. 282). As in any passage of liquid through vessels, the friction of the liquid on the vessel walls tends to lessen the pressure given by the "pump," and the capillaries are, of course, the parts of the system in which the loss of pressure is the greatest.

In fishes, in general, every drop of blood which leaves the heart must pass through at least two capillary systems before its return to that organ, for blood first passes through a capillary network in the gills before it reaches the aorta and passes to the capillary system of the ordinary body tissues. But much of the blood passes, in addition, through a portal system—the hepatic portal in all forms, and a renal portal system in fishes above the cyclostome level. In this situation the blood passes through

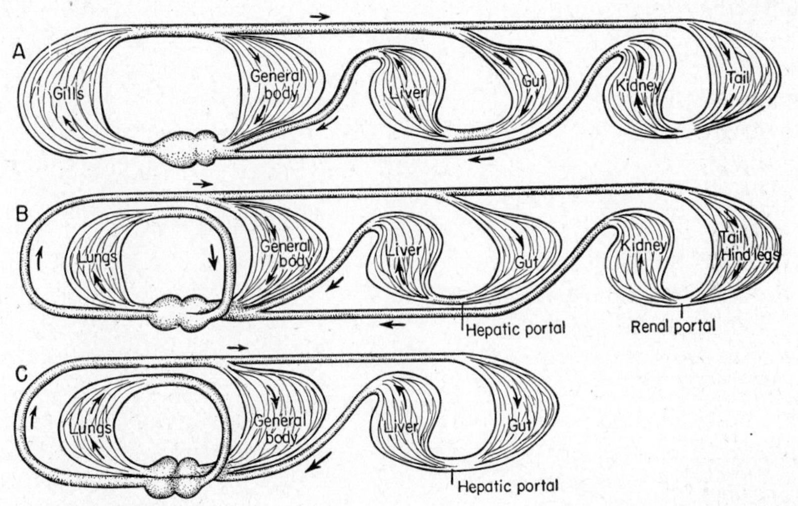

Fig. 282. Diagrams showing the general nature of the blood circuits and capillary networks encountered in *A*, a typical fish; *B*, a terrestrial amphibian with elimination of gill circulation and introduction of pulmonary circuit; *C*, a mammal, with elimination of renal portal system.

three, rather than two, capillary systems, losing much pressure in each. In consequence, blood circulation in fish is relatively sluggish.

With the introduction of the lung circuit into the circulatory system and the abolition in adult amphibians of gill breathing, circulatory efficiency is greatly promoted. The gill capillary system is eliminated; hence all body tissues are reached arterially directly with little loss of pressure. Blood from the lungs and from much of the body tissues makes the complete circuit back to the heart with but a single capillary system en route. Blood to the posterior end of the body or to the intestine, however, still has to pass through a portal system, renal or hepatic, on its return; but even here there are only two capillary drops, rather than three.

With reduction of the renal portal system in reptiles and its abolition in both birds and mammals, the efficiency of the circulation is further increased. In these last classes only the intestinal circulation passes two capillary networks; blood to every other region goes directly from and to the heart, with the only capillary intervention that in the tissues supplied.

The substitution of lungs for gills has, seemingly by a happy accident, caused, in the long run, the development of a more efficient circulatory system, and even the elimination of the renal portal may be, perhaps, a direct result of the increase in arterial pressure. On the other hand, lymphatic development in higher vertebrates may perhaps be a response to the increased difficulties in tissue drainage with increased pressure in the vessels of the blood-vascular system.

Embryonic Circulation (Figs. 284, 285). In earlier sections of this chapter we mentioned the formation of this or that blood vessel in the

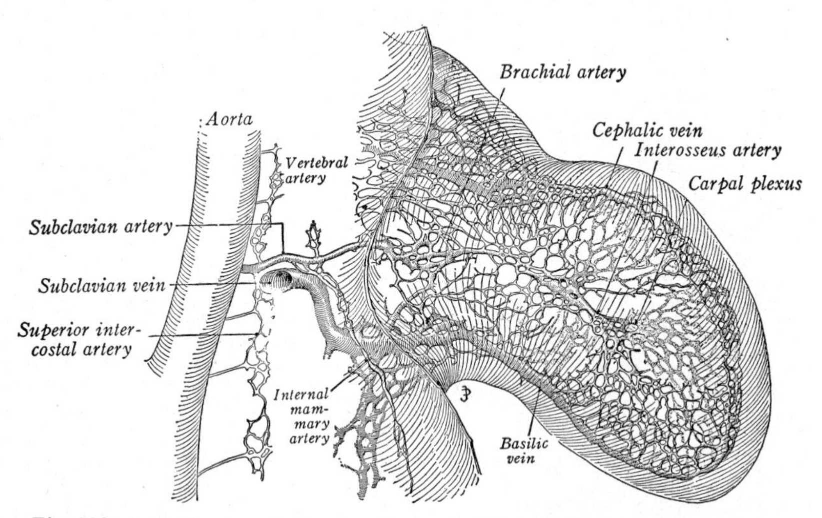

Fig. 283. An early stage in the development of the forelimb of a pig embryo, to show the manner of formation of patterns in limb circulation. There is a network of interweaving small vessels from which the main adult vessels develop. Choice of one definitive channel or another allows for the occurrence of variants as anomalies. (From Woollard.)

embryo. Here we may attempt to gain a general picture of the development of the circulatory system, although this necessarily involves some repetition.

The most generalized pattern of blood vessel development is that in such lower vertebrates as amphibians and lungfishes, in which the egg is not heavily yolked, normal body form is soon attained, and the picture is not complicated by the presence of intricate extra-embryonic membranes. The basic plan of the vertebrate circulation is one in which (in contrast with the annelid scheme) the blood flows forward ventrally beneath the gut, turns dorsally, and then runs backward above the gut tube. The earliest blood vessels to appear in an amphibian or lungfish are portions of the ventral part of this system. Beneath the yolk-laden gut small vessels form and run forward, assembling into a pair of vitelline veins. These fuse, anteriorly at least, to form a subintestinal vessel. The most anterior part of this vessel forms the ventral aorta; the succeeding section early develops into a heart structure. Posterior to this point the ven-

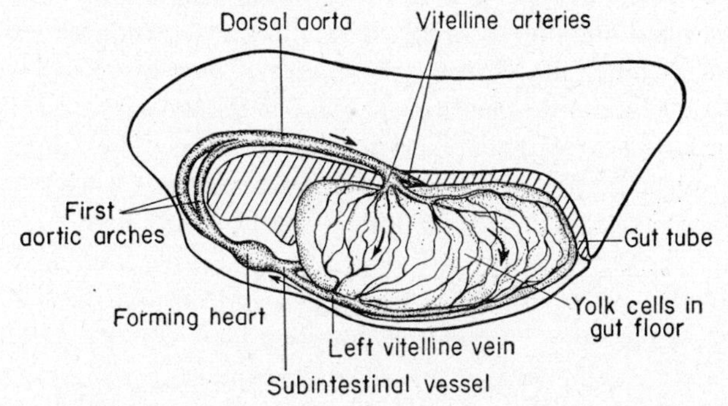

Fig. 284. Diagram of the general circulation in a young frog tadpole. The food supply still lies in the yolky gut cells, and the vitelline circulation is of great importance.

tral vessel is soon invaded by liver tissue and disrupted here into a capillary system. Between this liver "portal" and the heart the ventral trunk becomes the hepatic vein; posterior to it there develops the hepatic portal system, which continues to drain the gut in the adult.

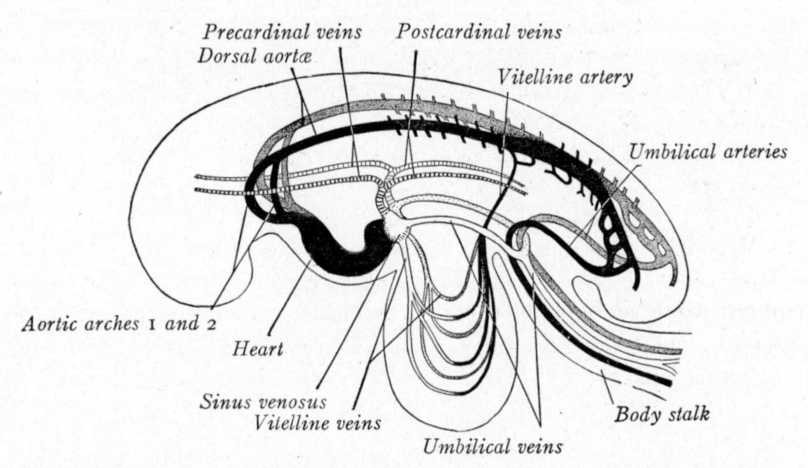

Fig. 285. The circulatory vessels in a mammalian embryo. (From Arey, after Felix.)

But before the ventral vessels have proceeded far in their differentiation there appears a dorsal arterial trunk which, coursing backward, completes the return circuit of the blood to the gut walls. From the front end of the subintestinal vessel, blood vessels curve upward around either side of the foregut to the dorsal part of the body. Here, below notochord and nerve cord, they turn backward above the gut and send off branches by which the blood descends around the digestive tube to enter once more the vitelline system. The pair of loops ascending around the foregut are the first of the aortic arches of the pharynx (Fig. 268, p. 446). As development proceeds, additional aortic loops develop more posteriorly. These may break up along their course into capillary systems as the gills develop; the anterior arches may atrophy or undergo modification even in early stages of development. The dorsal vessel is the dorsal aorta, and its anterior continuation becomes the carotids. At its first appearance the aorta may be a paired structure, and its derivatives remain paired in the most anterior cephalic region in the adult; for most of the body length the two vessels fuse, however, at an early stage to form a single median longitudinal aortic vessel. Its first branches, which descend around the gut walls, are paired, to begin with; but as the gut diminishes in size relative to the body as a whole, these vessels fuse to descend the mesentery as median arteries which become the mesenteric, celiac, and so on, of the adult.

With the development of the subintestinal and aortic vessels and their connections the primary circulatory system is completed. This, it will be noted, is essentially a visceral circulation, associated almost entirely with the gut tube region. With growth of the mesodermal structures and of the nervous system there develops the necessity for circulatory vessels in the outer body tube—a "somatic" circulation. The afferent arterial supply in this new system is the dorsal aorta, which sends out lateral branches into the trunk musculature and upward to the neural tube. From the capillary networks developed at either side of the aorta arise return venous vessels as the paired cardinal veins, anterior and posterior; these gain a discharge into the heart by forming a descending trunk, a common cardinal, down either side of the thoracic region. In higher vertebrates the anterior cardinals are modified to some degree to form the anterior (superior) vena cava—double or single. The posterior cardinals, we have noted, undergo important changes to become the posterior (inferior) vena cava; these changes include, in every embryo, the interpolation of a renal portal system, which is eliminated again in the adult stage of higher amniotes. Other elements in the "somatic" circulation include an arterial supply and venous return from the fins or limbs, and, in lower vertebrates, the development of abdominal veins which drain the ventral part of the body wall.

In other vertebrate types—elasmobranchs, amniotes—we find an extra-embryonic circulation composed of vessels which run from and to the yolk sac and, in amniotes, connect the embryo with allantois or placenta. Their presence complicates the embryonic story (Fig. 285).

In elasmobranchs and amniotes alike there is at first no "floor" to the gut cavity; in consequence, the paired vitelline veins arise on either side from the roof of the yolk sac. There can be no formation of a single ventral trunk until a relatively late stage in embryonic development; the heart and ventral vessels persist for a time as paired elements. The return vessels from the dorsal aorta are likewise persistently paired structures as vitelline arteries. A highly developed system of vitelline veins and returning vitelline arteries is present throughout embryonic life in elasmobranchs, reptiles, and birds. In mammals, despite the absence of yolk in the yolk sac, similar vessels early develop and persist for some time.

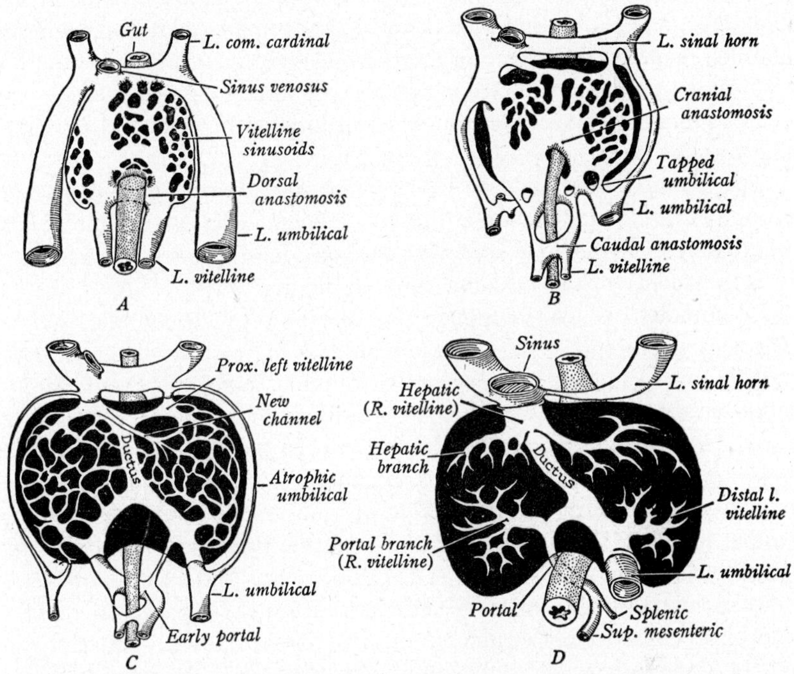

Fig. 286. Diagrams of the liver region of human embryos at successive stages (4.5, 5, 6, and 9 mm. in length), seen from the ventral surface, to show developmental changes in the vitelline and umbilical veins. The gut tube is represented. In *A* the vitelline vessels from the yolk sac are well developed and pass through the liver tissue; in later figures their transformation into the portal system is seen. The umbilical veins (from the placenta) are already well developed in *A*, but run directly to the sinus venosus. In later stages this blood flow is diverted to the liver circulation, much of it flowing through this organ via a hepatic duct. The right umbilical vein is reduced; the left persists until birth, when this vessel and the hepatic duct undergo reduction. (From Arey.)

In amniotes a second series of embryonic vessels of note are those which extend along the stalk of the allantois, which in reptiles and birds serves as an embryonic lung. Paired allantoic or umbilical veins enter the body of the embryo and course forward in the ventral part of the lateral body walls; they appear to represent, within the body, the lateral abdominal veins of lower vertebrates. In the early embryo these veins enter the sinus venosus directly, as do the abdominal veins of a shark. In later development, however, they turn inward anteriorly to enter the liver and become part of its portal system (Fig. 286). Apparently, however, the flow of blood of the combined vitelline and umbilical veins is too great for proper handling by the liver capillaries, and in embryonic amniotes much of the blood passes through the liver by a direct channel—the *ductus venosus.* Paired allantoic or umbilical arteries arise from the dorsal aorta posteriorly and, in strong contrast to the vitelline vessels, descend to the allantoic stalk in the body walls rather than via the gut. In mammals the allantoic stalk becomes the connection between placenta and embryo, and the allantoic vessels are of the utmost importance, bringing food as well as oxygen to the embryo; in this class of vertebrates the yolk sac circulation diminishes with the gradual absorption of the yolk, and the vitelline vessels disappear.

The allantoic or umbilical vessels of amniotes persist until birth, but abruptly cease their function at that time. They are shut off at the points where they leave the body, and their internal courses are soon obliterated, as is the ductus venosus channel through the liver.

In amniotes oxygen is supplied to the embryo by the umbilical veins. The abrupt entrance of the lung circuit into the picture at the time of hatching or birth calls for special modifications in the embryonic circulation. In the embryo the pulmonary arteries and veins develop at a fairly early stage, but the blood for the most part by-passes these vessels. We have noted that the upper part of the pulmonary arch, from which the pulmonary arteries proper take origin, remains open in amniote embryos so that blood which takes this course from the heart need not enter the lungs, but instead may proceed to the dorsal aorta and enter the general body circulation. This vessel is the ductus arteriosus—paired in reptiles and birds, present only on the left side in mammals (Fig. 268, *E, F,* p. 446). At the other end of the circuit we find that in the amniote heart the left atrium has no proper blood supply in the embryo, for the only vessels entering it are the pulmonary veins, which carry almost no blood. This situation is remedied by the fact that in all amniotes the septum between the two atria is imperfect in the embryo: blood from the right atrium may cross over and fill the left atrium sufficiently to enable it to function properly.

At birth there is a sharp change in the lung circuit. The ductus arteriosus closes rapidly in birds and mammals and most reptiles, so that all blood entering the pulmonary arch is forced into the lungs. The return of this blood to the heart furnishes the left atrium with its proper share of the blood stream, and the opening between the two atria is soon closed.

CHAPTER XIV

BODY CAVITIES

IN THE VERTEBRATES, as in all the more highly organized invertebrate types, most of the body organs are not surrounded by solid tissues or mesenchyme, but are situated within the bounds of body cavities, more properly *celomic cavities*. These are fluid-filled spaces in which the viscera are placed in a situation relatively free from ties, and more or less at liberty to move freely during their functional activity and to change the more readily in size or shape during growth processes. In earlier chapters body cavities have been given incidental mention in the description of various organs. We may here give a brief résumé of their arrangement, for the moment considering, so to speak, the hole rather than the doughnut.

Development of the Celom. Such cavities are formed in the mesodermal tissues, and their linings are a mesodermal epithelium, the *peritoneum*. We have noted that in Amphioxus, as in echinoderms, the celomic cavities arise in part, at least, from pouches pinched off from the primitive gut as mesodermal somites (Figs. 47, 48, p. 98). In true vertebrates, however, this presumably primitive method of celom formation is abandoned, and the body cavities are formed by a splitting of mesodermal sheets after that tissue has separated from other germ layers. Although seemingly celomic cavities may appear in the somites or the kidney-forming tissue, these are ephemeral, and the term "celom," in all late embryonic and adult stages, is confined to the cavities in the more lateral and ventral parts of the mesoderm, the lateral plate.

In forms developing from a mesolecithal type of egg the lateral plate of mesoderm at an early stage extends down the flanks of the body on either side so that the two sheets presently meet, or come close to meeting, in the ventral midline (Fig. 49, *B*, p. 99). In large-yolked types the lateral plates are at first directed outward laterally over the yolk, and are continuous with the mesoderm of the extra-embryonic regions (Fig. 43, p. 93); only later, with the assumption of body form, do these sheets meet ventrally. Each of the two lateral sheets is a continuum, to begin with, from the head to the hinder end of the trunk region, and never shows any indication of segmentation. At first each plate is solid, but presently it splits in two (Fig. 287). The cavity, liquid-filled, between the two layers thus formed is the embryonic celomic cavity; its outer and inner walls, apart from giving rise to mesenchyme and other materials, are destined to form, respectively, the parietal and splanchnic layers of the peritoneum. The *somatic peritoneum* (or parietal peritoneum) forms the inner surface of the great external "tube" of the body, termed in the embryo the *somatopleure,* in which the somatic skeleton, somatic musculature, kidneys, and gonads presently develop. The *splanchnic peritoneum* forms the outer covering of the digestive tract and its outgrowths, which constitute the *splanchnopleure.*

In this fashion there is formed a pair of celomic cavities, one on either side, with a large gut cavity between them. The diameter of the gut, behind the pharyngeal region, becomes relatively small as development proceeds, and the two celomic cavities come to lie adjacent to one another both above and below the gut tube (Fig. 2, *B, D,* p. 6). At this stage the right and left celoms are separated by longitudinal thin sheets of tissue,

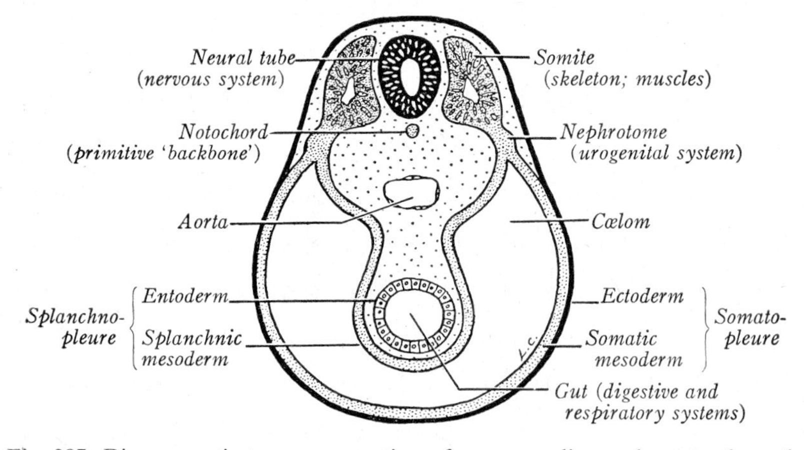

Fig. 287. Diagrammatic transverse section of a mammalian embryo to show the relations of the mesoderm. (From Arey.)

bounded on either side by peritoneum; these are the dorsal and ventral mesenteries. The *dorsal mesentery* is a persistent structure, containing nerves and blood vessels supplying the gut. The *ventral mesentery,* however, is of little functional importance in the adult and usually disappears for most of its length. With this disappearance the two celomic cavities are merged.

In the early embryo the sheet of lateral plate mesoderm containing the celomic cavity is continuous along the length of the gut (Fig. 49, *C*). The anal or cloacal region is the definitive posterior boundary of the trunk, and neither the celom nor the viscera associated with it extend into the tail. Anteriorly, the situation is complicated by the development of the gill system of the pharynx. The developing gill slits disrupt the lateral plate in the side walls of the pharyngeal region, so that in this part of the embryo the celom is confined to the ventral part of the throat, below the pharynx (cf. Fig. 153, p. 255).

Pericardial Cavity. The most anterior and ventral region of the embryonic celom becomes the pericardial cavity, in which the heart develops. In the early embryo of all vertebrates the pericardial cavity opens widely at the back into the general abdominal cavity. Soon, however, there forms a *transverse septum* which partially or completely separates the two (Figs. 288, 289).

This septum owes its origin to the fact that the common cardinal veins from either side of the body push across to enter the back end of the heart tube at the sinus venosus, bringing folds of the celomic wall with them. The septum as thus formed blocks off the lower part of the passage between pericardium and general celom, but leaves a gap above. In most

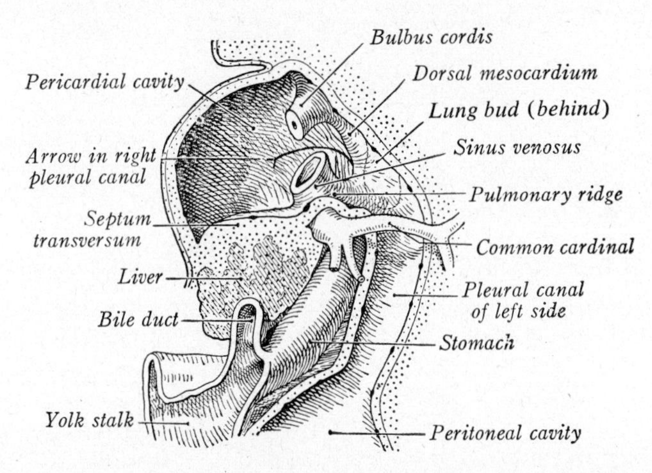

Fig. 288. Lateral view of a dissection of the trunk of a human embryo 3 mm. in length, showing pericardial cavity (heart removed), pleural, and peritoneal cavities. The transverse septum, with the developing liver, has formed behind the heart, but otherwise all body cavities are in open communication. (From Arey, after Kollman.)

vertebrates the transverse septum is completed dorsally in the adult; but in hagfishes, elasmobranchs, and a few lower ray-finned fishes a small dorsal opening between heart cavity and general celom persists into the adult stage. In the early embryo the liver bud, growing forward below the gut, enters the transverse septum and expands within its substance, so that in lower vertebrates the main liver attachment is to the posterior surface of the septum. In many higher forms the liver tends in the adult to pull away again from the transverse septum, but remains attached to it by a relatively narrow *coronary ligament*.

In most groups of land vertebrates the disappearance of a functional gill system and the development of a neck have been associated with a marked change in the relative position of the pericardial cavity. It has been pushed back into the floor of the thorax, where it lies, protected by the coracoids or sternum, diagonally below the anterior end of the main body cavity. Above it in paired recesses of the celom lie the lungs. Separating heart and lung cavities, a *pleuropericardial membrane,* bounded on either surface by celomic epithelium, extends forward from the top of the transverse septum, of which it is considered to be an extension (Fig. 289).

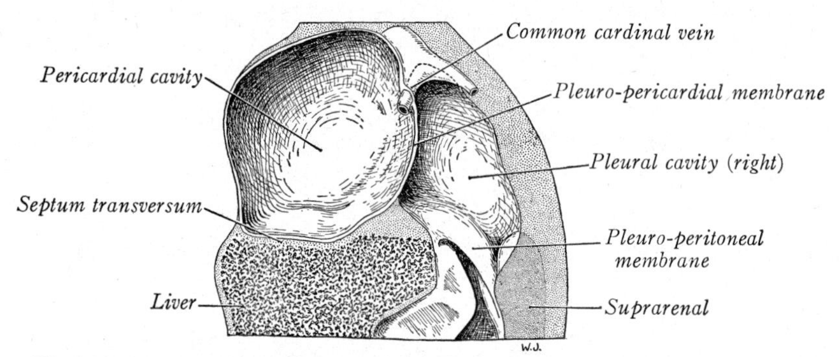

Fig. 289. A lateral view similar to that of Figure 288, but a later stage (16 mm.). In addition to the transverse septum, pleuropericardial and pleuroperitoneal membranes are present. The last, together with the septum, are the primary elements forming the diaphragm. (From Arey, after Frazer.)

General Body Cavity. With the separation of the pericardial cavity, there remains in the majority of vertebrates a single great celomic cavity occupying, with its enclosed organs, most of the trunk region. This is partially subdivided into right and left halves by the mesentery system.

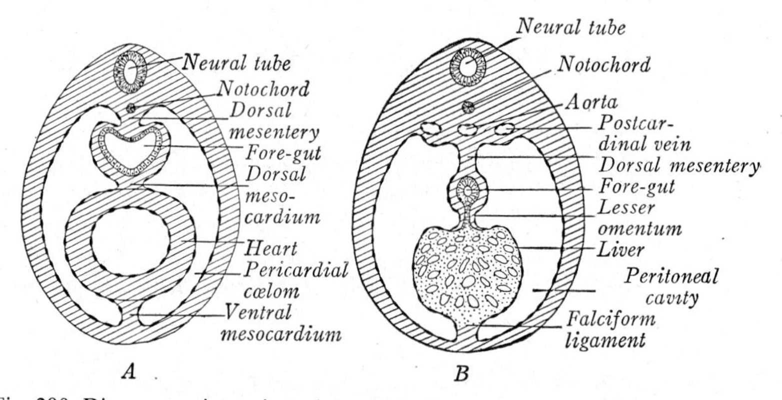

Fig. 290. Diagrammatic sections through the heart and liver regions of an amniote embryo to show the relations of the mesenteries. (From Arey, after Prentiss.)

The ventral mesentery, we have noted, usually disappears for most of its length (It is persistent in a few types of fishes). A short posterior segment may remain as a ligament tying in the bladder with the rectum. Anteriorly, however, it persists in connection with the liver, which arises in the midst of the primitive ventral mesentery. The part of the ventral mesentery connecting stomach and liver forms the gastrohepatic ligament or *lesser omentum;* that part below the liver is the *falciform ligament* (Fig. 290).

The dorsal mesentery is a simple, continuous longitudinal sheet of tissue in the early embryo. But in later stages the development of the gut tube and its outgrowths into a series of distinctive organs results in a parallel development of distinct mesenteric areas. The term "mesentery" is sometimes used in a narrow sense to apply merely to that part of the sheet supporting the small intestine (when such a structure is developed), and special terms such as mesocolon, mesorectum, and so forth, may be used for other parts. Of special importance is the mesogaster, or *greater omentum,* supporting the stomach. The dorsal mesentery system is present in all vertebrates, since it is along the mesenteries that arteries, lymphatics, and nerves descend to the gut. In reptiles and mammals the dorsal mesentery remains unbroken; in birds and amphibians, however, there may be gaps along its length, and in some fish groups it is broken up into discontinuous segments. In teleosts and land vertebrates in which a coiled small intestine is developed, the mesentery is, of course, folded in a complicated fashion, and parts of the mesentery may fuse with one another in a confusing way (Figs. 291, 294).

The stomach, as developed in most groups of jawed vertebrates, swings to the left side of the body, and that part of the dorsal mesentery forming the greater omentum swings with it. In consequence, the right part of the celomic cavity adjoining the stomach encroaches on the left side of the body. This encroachment is exaggerated by the fact that generally the greater omentum (in which the spleen is embedded) does not descend directly from the dorsal body wall to the stomach, but tends to fold still further to the left. Ventrally, this folded area is, of course, closed off by the expanse of the liver, tied to the stomach by the lesser omentum. As a result this area of the right celom becomes essentially a pouch, the *omental bursa,* opening out to the right above the liver. In many lower vertebrates it is not especially developed and may open broadly to the right side (Fig. 291).

Dorsally, various organs not part of the gut system may project into the body cavity, as we have noted in earlier chapters (Figs. 2, *D,* 237, pp. 6, 396). The testes or ovaries formed from the genital ridges of the embryo, usually expand into the body cavity, surrounded by its peritoneum, and may be supported by special mesenteries—*mesorchium* or *mesovarium.* Farther laterally the developing kidneys of the embryo also expand into the body cavity as longitudinal swollen ridges, and in the adult the kidney bulges to a varied degree into the celom from the dorsal side. The oviducts arise in the kidney ridges and may persist there, but are often suspended by a special paired mesenteric structure, the mesosalpinx.

Lung Pockets. The lungs, as developed in certain bony fishes and in land animals, push backward from the pharynx along either side of the esophagus and extend posteriorly into the body cavity. Here they lie above the pericardium on either side of a forward extension of the mesentery, termed the *mediastinum,* in which the esophagus is embedded. In lower tetrapods the lungs are usually supported by separate paired mesenteries, *pulmonary folds,* extending down to them from the dorsal wall of the body cavity and continuing downward to tie into the upper surface

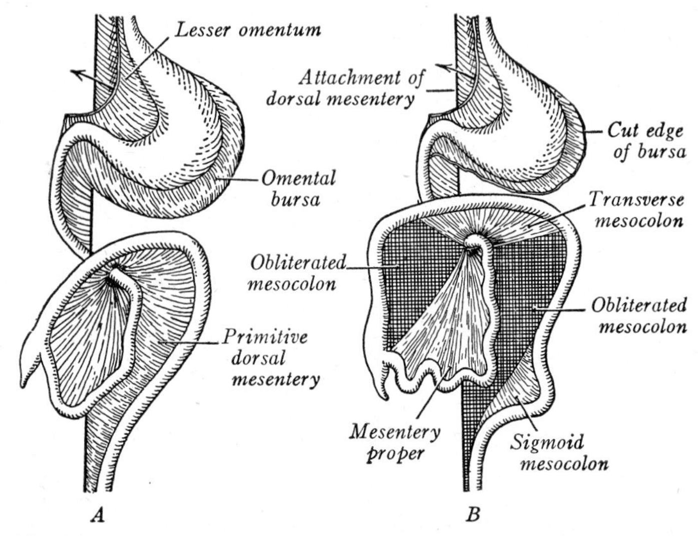

Fig. 291. Diagrammatic ventral views of the gut and mesenteries of a mammal in *A,* an embryo and, *B,* essentially adult conditions. *A* shows the general type of folding which the mesenteries must undergo because of the asymmetrical position of the stomach and twisting of the intestine. As shown in *B,* this folding may result in obliteration or fusion of parts of the mesentery. In *A* the bursa is a structure of minor dimensions; in many mammals the enlarged bursa would extend down, covering over much of the intestine, but has been cut off short in *B.* The opening from the bursa to the celom of the right side (epiploic foramen) is shown by an arrow; the double line "above" the arrow (i.e., ventral to it) is the cut attachment of the lesser omentum to the liver. For lateral views of the same structures, see Figure 294. (From Arey.)

of the liver on either side (Fig. 292). As a result there is, in amphibians and reptiles, a pair of *pulmonary recesses,* little blind celomic pouches pushed forward on either side of the mediastinum, with the lungs and pulmonary folds lying lateral to them. The left fold—and consequently the left pouch—is usually little developed. The right fold, however, is conspicuous in the embryo of every tetrapod and frequently remains so in the adult. Its greater development may, perhaps, have been associated originally with the asymmetrical development of the stomach region; its persistence is definitely associated with the fact that the posterior vena cava utilizes it to make its short circuit down into the liver and thence to the heart (cf. p. 456). Since the right pulmonary fold is large, the right pulmonary recess is likewise well developed. It opens back into the omental bursa and forms part of this recessed cavity.

In amphibians and a majority of reptiles the dorsal and anterior recesses which contain the lungs are still parts of the general body cavity. In some reptiles the lungs may be more or less buried in the body wall dorsally, with little or no projection into the celom. In still others, however, there may be a partial or even complete closure of the lung pockets from the remainder of the celom, making *pleural cavities,* much as in birds and mammals.

The Celom in Birds. The great development of air sacs in birds is associated with a complicated subdivision of the body cavity which appears to promote the efficiency of the breathing apparatus (Fig. 293). The pericardium forms a separate anterior chamber, and posteriorly a transverse septum divides off a large cavity containing intestine, kidneys, and gonads.

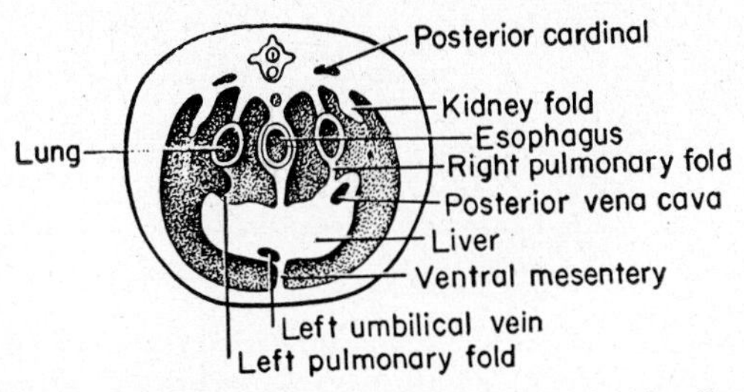

Fig. 292. Diagrammatic cross section through the body of a lizard in the region of the lungs, to show the pulmonary folds, between which and the esophagus lie pulmonary recesses. (After Goodrich.)

Between these two extremities lies the area of the lungs, liver, and stomach; here three paired longitudinal septa, more or less horizontal in position, bring about the development of four pairs of celomic cavities. The uppermost, pulmonary septum shuts off the two lungs into individual pleural cavities; this septum somewhat parallels the development of the diaphragm of the mammals, described below, but is of independent origin. Another, lower septum causes the formation of a lateral pair of cavities in which are air sacs (these are not true celomic cavities). Farther ventrally is the area where the liver expands on either side of the gut; here a horizontal septum running lengthwise along either side of the liver forms separate celomic cavities dorsal and ventral to that organ.

Mammalian Celomic Cavities; Diaphragm. Quite another type of celomic development has occurred in mammals. There are, of course, no air sacs of avian type and none of the fantastic subdivision of the main body cavity which accompanies that development. The major features are (1) a high degree of development of the omental bursa and, especially, (2) the presence of pleural cavities closed off by the diaphragm.

We have noted the presence to a variable degree in other vertebrates of a pouch of the right celomic cavity between stomach and liver and the greater omentum associated with this condition. In mammals this recess is much developed; the greater omentum becomes a huge fold extending down as an apron over much of the ventral surface of the abdominal cavity (Fig. 294). The opening to the bursa (on the right side) is closed except for the small *epiploic foramen* (foramen of Winslow, Fig. 291).

In mammals the lungs, as has been noted in discussing those organs, are enclosed in paired *pleural cavities,* recesses in the thorax which are shut off from the major abdominal (or peritoneal) cavity. The partition between thoracic and abdominal parts of the celom is the *diaphragm;* movements of this partly muscular structure, together with rib movements, are responsible for the typical breathing action of mammals.

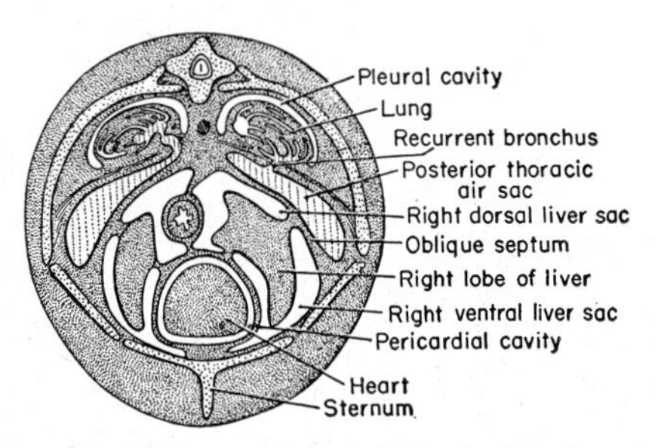

Fig. 293. Diagram of transverse section through the thorax of a bird to show especially the subdivision of the celomic cavity. In addition to the pericardial cavity, the pleural cavities and the sacs dorsal and ventral to the liver are shown; the main intestinal celom is too far posterior to be included in this section. The thoracic air sacs are hatched. (After Goodrich.)

In the adult mammal the diaphragm has the appearance of a relatively simple partition, tendinous in its central part, but with a sheet of muscle all about its periphery; it extends across the body below the base of the rib basket and above the liver and stomach to divide the thoracic region from the abdomen. Despite its apparent simplicity, the diaphragm is complicated in its mode of development and in its structure; we shall here give the barest outline of this story.

We have noted in connection with the formation of the pericardium the transverse septum, developed in front of the liver, which closes off the pericardial cavity from the main celom. This is the principal ventral component of the diaphragm. In typical reptiles, however, there are, above the top of the septum (as noted above) paired forward extensions of the peritoneal cavity which contain the lungs. In mammals these pockets expand downward on either side of the heart. As a result the lung cavities may approach one another ventrally, leaving only a median strand of tissue, a ventral mediastinum, connecting the pericardial sac with the ventral body wall (Fig. 200, p. 336). To close off these enlarged lung pockets from the abdominal celom as pleural cavities, adjuncts to the transverse septum are demanded (Fig. 295). The more medial parts of the opening are closed by mesodermal folds growing dorsally from the top of the transverse septum. Laterally, *pleuroperitoneal membranes* (Fig.

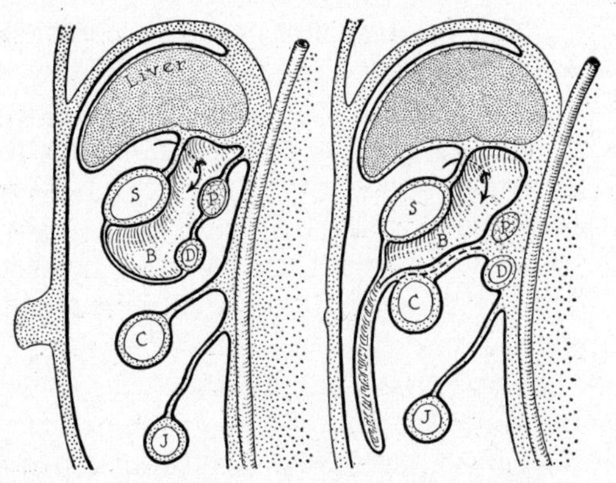

Fig. 294. Diagrams of embryonic and adult conditions of the omentum and bursa in a mammal; longitudinal sections, seen from the left, with head end above. In *A* the bursa is small; the entrance (epiploic foramen) is indicated by an arrow. In *B* the dorsal mesentery, as the greater omentum, has become a long fold. The diagrams further show how the mesenteries of the gut may be fused together (as is here that of the transverse colon to the greater omentum), or obliterated (as is that of the duodenum, in the diagram): *B*, omental bursa; *C*, transverse colon; *D*, duodenum; *J*, jejunum; *P*, pancreas, *S*, stomach. For comparable ventral views, see Figure 291. (From Arey, after Kollman.)

289, p. 478) fold inward from the body wall to close much of the opening, and, as the lungs continue to expand, the body wall itself takes part in forming the lateral areas of the diaphragm beyond the base of the original pleuroperitoneal membranes. A small opening persists for a time far dor-

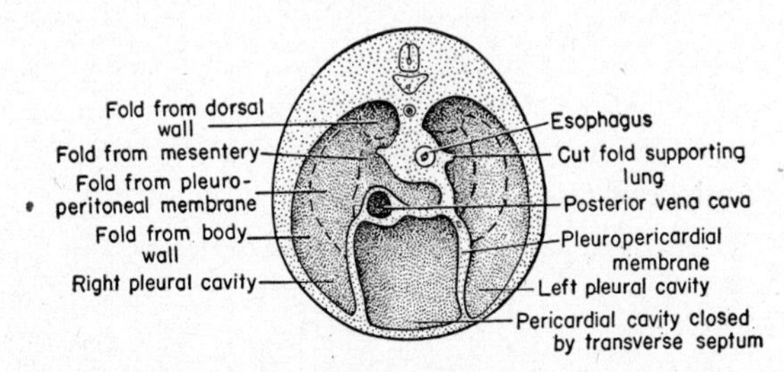

Fig. 295. Anterior view of the diaphragm region of an embryonic mammal, to show the various elements which make up the diaphragm. (Heart and lungs have been removed to show the posterior walls of the pleural and pericardial chambers.) (After Broman, Goodrich.)

sally on either side; here a final closing fold develops from the dorsal mesentery or the mesentery supporting the lung itself.

The diaphragm is supplied with a sheet of muscle, striated and subject to voluntary control. It is obviously a derivative of the axial musculature

(p. 263), although its embryological origin is not clear. It is of interest that the phrenic nerve, which innervates this musculature, is formed from cervical nerve roots. Surely these muscles became associated with the lung cavities while they were still situated more anteriorly in "neckless" early tetrapods, and have followed the lungs posteriorly in their migration to the thorax.

Abdominal Pores. In the ancestral fishes there appear to have been present a pair of small abdominal pores connecting the celom with the exterior in the cloacal region (Fig. 249, p. 415). Such openings are present in cyclostomes and most elasmobranchs, and occur in all the various groups of bony fishes, although most teleosts have lost them. In land vertebrates the pores have been lost in most cases, but may be found in turtles and crocodilians. In cyclostomes these pores serve as exits for the sperm and eggs. In all gnathostomes, however, special ducts have taken over the transportation of the sex cells, and the only function which we can assume the abdominal pores to have is that of getting rid of surplus liquid from the celomic cavity.

CHAPTER XV

SENSE ORGANS

ALL CELLS, one may believe, are capable of receiving sensations and responding to sensory stimuli. If, however, in a vertebrate the reaction to a sensation is one that should be performed by the organism as a whole, its reception is in vain unless there is some channel of communication between the sensory receptor and the organs—muscles or glands—which should make the appropriate response. Such communication may be made by hormonal action, but in general the mechanism used is that of the nervous system. The tips of nerve fibers are themselves capable of stimulation, but in general the reception of sensation in vertebrates is the function of specialized *sensory cells,* generally grouped in organs of lesser or greater degree of complexity. These are attuned to physical or chemical stimuli of specific types and are associated with nerves which relay these stimuli to specific centers in the brain or to the nerve cord.

Anatomists divide such nerves, sensory in nature, into two groups: the *somatic sensory nerves,* carrying impulses of a sort which in ourselves reach the level of consciousness, from the "outer tube" of the body—the skin and body surfaces and the muscles; and *visceral sensory nerves,* whose impulses, seldom reaching our consciousness, arise from the viscera. The physiologist customarily classifies sensory receptors in a fashion which fits readily into the neurological scheme. *Exteroceptors* are those sensory structures of the skin and special senses which receive sensations from the outer world; *proprioceptors* are those situated in the striated voluntary muscles; *interoceptors* are those located in the internal organs. The first two of these correlate with the somatic system of sensory nerves, the third with the visceral sensory nerve components of the anatomist.

SIMPLE SENSE ORGANS

One ubiquitous sensation—that of *pain*—seems not to need any special organ for its reception. Pain, as it appears in our own perception, is not specific, but may be due to a variety of physical and chemical causes. This sensation is produced, as far as known, by direct stimulation of the end fibers of sensory nerves.

Sense Corpuscles. Other "simple" sensations are apparently received by small sensory structures present in almost every part of the body—skin, muscles, viscera. Most are of such tiny size that their study lies in the realm of the histologist with his microscope rather than that of the dissector with his scalpel. Their appearance and structure vary greatly (Fig. 296). In general we see a series of specialized cells with which are associated tangled skeins of terminal nerve fibrils. Most common are varied types of corpuscles in which the sensory cells are arranged in concentric layers, onion-fashion. Such corpuscles are commonly present in the connective tissues, but similar sensory structures may be present in epithelia (special tactile cells in the mammalian epidermis, for example).

With small and varied bodies of this sort, accurate determination of function is difficult. Some knowledge can, of course, be gained from our

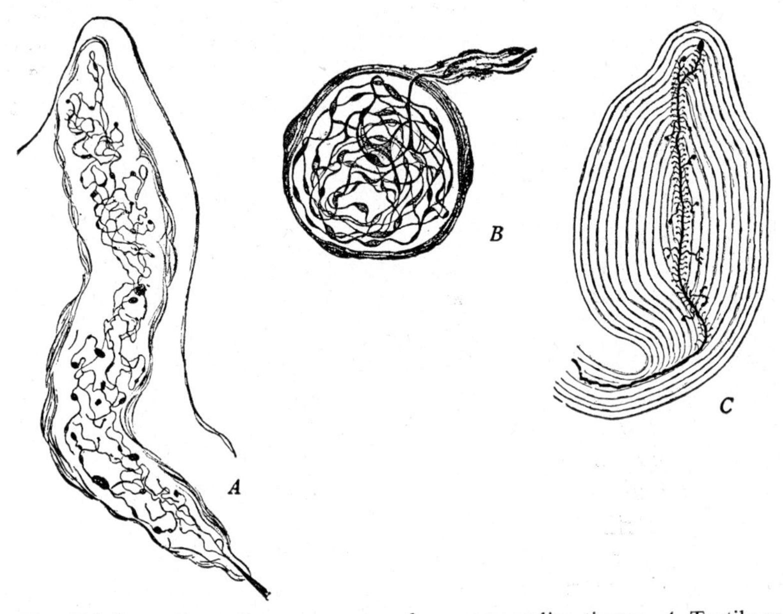

Fig. 296. Some types of sensory organs from mammalian tissues. *A*, Tactile corpuscle (Meissner's corpuscle) from the connective tissue of the skin; *B*, an end bulb of Krause, from the conjunctiva of the human eye; *C*, a pacinian corpuscle. (From Ranson, after Dogiel, Sala, Böhm-Davidoff, Huber.)

own sensory experience, and can be extrapolated to give us some idea of the sensory situation in other mammals and, with lesser chance of correctness, in other vertebrate classes. Among mammals, some four types of sensations, in addition to pain, may be distinguished in the dermis—*warmth, cold, touch, pressure*—all apparently associated with corpuscles of different sorts. A number of types of interoceptive corpuscles are present in and among the viscera, but it is generally difficult to tell the nature of the sensations to which they respond.

Proprioceptive sensations, of the physiologist's terminology, are those obtained from sensory structures located in the striated muscles and tendons as *muscle spindles* (Fig. 297) and *tendon spindles*. In lower vertebrate classes we find that sensory nerve fibrils may twist about individual muscle fibers or spread among a group of tendon fibers. In mammals muscle spindles consist of several muscle fibers surrounded by a maze of sensory nerve endings and enclosed in an elongate connective tissue sheath. These muscle and tendon structures are the seat of "muscle sense." They furnish to the central nervous system a report on the state of contraction of the muscles—data necessary before any further action of the muscle can be properly "ordered" by way of motor nerves. In addition, these spindles are the source of the animal's information concerning the position in space of the various parts of the body—knowledge, as we are ourselves well aware, which can be furnished without the aid of other sensory structures, but strictly confined (in the absence of contact with other objects) to parts of the body containing striated muscles or their tendons.

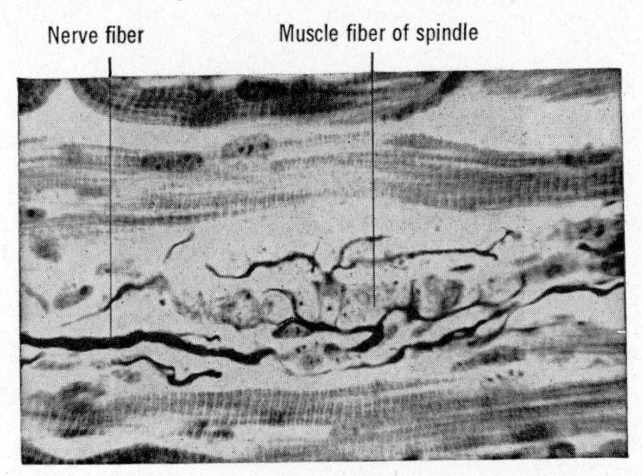

Fig. 297. A muscle spindle. (Gardner, Anat. Rec., vol. 83, 1942, courtesy Wistar Institute of Anatomy and Biology.)

Taste. Most of the simpler senses are responses to physical stimuli; taste (and smell) is a response to chemical stimulation. The organs of taste are simple. They are *taste buds,* which usually consist of barrel-like collections of elongated cells sunk within the ectodermal epithelium (Fig. 298). The bud's external cells, forming the barrel staves, are sustentacular in function; the taste cells proper fill the interior of the tiny structure. Each taste cell has a sensitive, hairlike projecting process. Taste buds are for the most part confined to the mouth, and in mammals are in great measure concentrated on the tongue. They may be more widely distributed, however. Even in mammals they may be found in the pharynx and its outgrowths (as in the larynx). In fishes they may develop on the outer surface of the body. In a sturgeon, for example, they are abundant on the under side of the projecting rostrum, so that as the fish glides over the bottom it can obtain a foretaste of potential food before the mouth reaches it. In some catfishes there is a tremendous spread of taste buds over the surface of the body, giving a phenomenal possibility of pleasant (or unpleasant) gustatory sensations. It must be noted that much of what we casually think of as taste is actually a smelling of mouth contents (foods do not "taste" as well when a head cold clogs the nose). All taste buds look alike, but in such animals (including ourselves) as have been experimentally tested, four distinct types, at least, are present, giving sensations of sour, salty, bitter, and sweet. These appear to be responses to specific types of ions present in solution in the liquids of the mouth (or body surface).

In sections which follow are described the more prominent of the complicated sensory structures of vertebrates, including nose, eye, ear, and lateral line organs. The first three of these are familiar as part of our

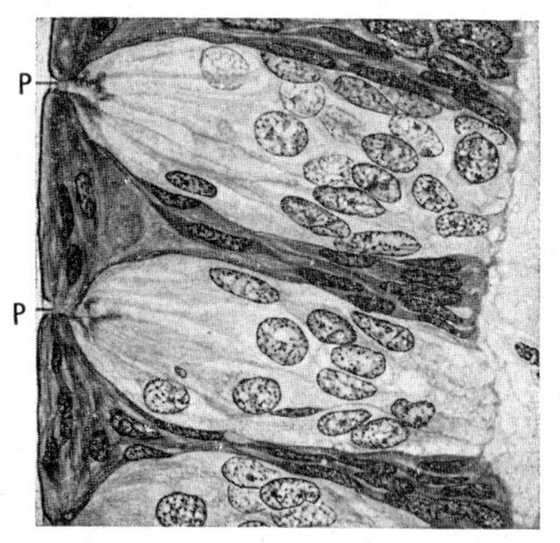

Fig. 298. Two taste buds from a rhesus monkey: *P*, pores. (From Maximow and Bloom.)

own sensory apparatus; the last is a sense which we do not ourselves possess, but is so distinctively developed in fishes that it cannot be overlooked, nor its general import missed.

Presumably there is a variety of still other sensory structures in other vertebrates, particularly the lower classes of them, which give responses of types unfamiliar to us and hence difficult for us to understand. One such sense, about which we do have data, lies in the *pit organ* of the so-called pit vipers among snakes, of which the rattlesnake is a familiar (although not popular) example. Between eye and nose on either side of the face is a pit filled with a peculiar, highly vascular tissue with numerous nerve endings. Experiments show that this is a heat-sensitive organ; a rattlesnake can perceive the movement of a moderately warm body past its head at a distance of several feet—a sensory power extremely useful to an animal which makes much of its living by capturing warm-blooded rodents and is not too well endowed with "normal" sense organs.

THE NOSE

In the Choanichthyes and tetrapods the nose, with an internal opening, has become associated with breathing, but its primary function, of course, is that of olfactory reception. In certain vertebrate groups smell is relatively unimportant; it is not in general highly developed in teleost fishes, and is rather feebly developed in birds, in marine mammals, and in higher primates, including man; most are groups in which the eyes have become the dominant sense organs. These cases, however, are exceptional; among vertebrates in general smell is in many ways the most important of all senses, and from cyclostomes and sharks up through the more primitive bony fishes, amphibians, reptiles, and even among most mammals, smell

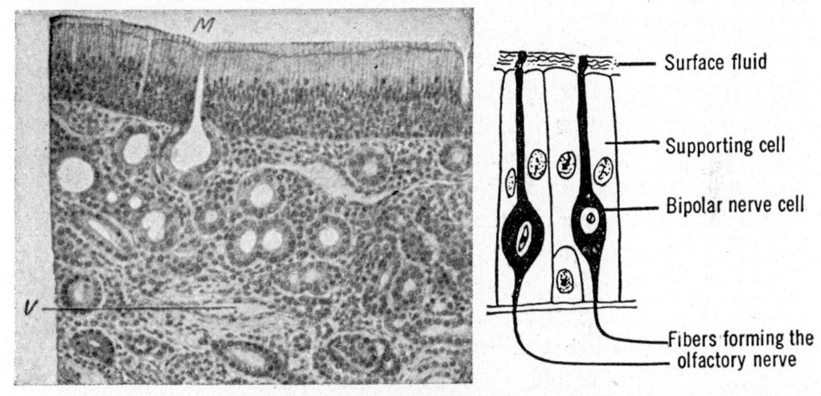

Fig. 299. *Left,* a section through the olfactory mucous membrane of a mammal, showing at the top the sensory epithelium, with mucous glands in the underlying tissue: *M,* opening of a gland, *v,* a vein. (From Maximow and Bloom, after Schaffer.) *Right,* diagram to show arrangement of supporting and sensory cells and "nerve" fibers. (From Gardner.)

is a main source of information concerning the outside world. That we ourselves have this sense feebly developed should not blind us to its general importance; the fact that, as we shall later see, the most highly developed brain centers arise in a brain area primarily connected with smell is testimony to the part olfaction plays among the vertebrates.

In a majority of fishes the nasal structures consist of a pair of pockets placed well anteriorly on the head and without any internal opening to the mouth. Typically, they are lateral in position; in sharks they are placed on the under surface of the snout. The external opening of each sac is paired, as in typical teleosts, where water flows inward through an anterior and out through a posterior opening, or partially subdivided into two openings by a fold of skin, as in sharks. Within the nasal sac is a highly folded epithelium of simple columnar type. This epithelium contains, in all vertebrates, *olfactory cells*, interspersed with supporting elements (Fig. 299). On its exposed surfaces each olfactory cell bears a radiating brush of short, hairlike sensitive processes. In one remarkable feature these cells differ from any other receptor organs in vertebrates. All others depend upon nerve fibers to relay inward the sensations received. The olfactory cells, on the contrary, do their own work; a long fiber extends from each cell into the olfactory bulb of the brain. This type of organization, with the sensory cell itself serving for nerve conduction as well, is found in various instances among invertebrates and even in Amphioxus, and is suggestive of the antiquity of the sense of smell in vertebrate history.

The surface brushes of the olfactory cells are sensitive to minute amounts of chemical materials, derived from distant objects, which come to be present in solution in the liquid covering the nasal epithelium. Despite much work on the part of physiologists, we have little data as to the precise nature of olfactory sensations of different sorts; our lack of success in analysis may be due in considerable measure to our lack of subjective appreciation of the delicate nuances of smell perceived by vertebrates more fortunate in their olfactory powers.

The jawless vertebrates present a puzzling structural situation. In contrast to that of all other living vertebrates, the cyclostome olfactory organ is a single median pouch, situated either at the tip of the snout (hagfishes) or far back on the upper surface of the head (lampreys) (Figs. 13, 189, pp. 36, 320). A further peculiarity, already noted, is the fact that the hypophyseal sac is combined in a common opening with the nostril. In the hagfish the nasal pouch breaks through internally into the pharynx as a water passage to the gills.

Is the cyclostome dorsal single-pouched structure primitive or specialized? The answer is none too clear. In the ancient ostracoderms two out of three main groups show the lamprey situation; but in a third group the nostrils were presumably ventral and paired, so that both situations were present at an early stage in vertebrate history. Some embryological evidence can be adduced for a belief that the cyclostome condition is not primitive. In the embryo lamprey, we have seen (Fig. 169, p. 294) that the nasal pocket is at first ventral in position and only later migrates dorsally; and in the larval lamprey the nasal sac is bilobed, although not distinctly paired.

In earlier chapters we commented upon the evolution, in the Choanichthyes and their tetrapod descendants, of a type of nostril with an opening into the roof of the mouth as well as to the exterior, and the utilization of this passage as an adjunct to air breathing. As a result the olfactory epithelium tends to be restricted to a part only of the lining of this passage. In typical fishes the nasal sac is filled with water; in tetrapods, air replaces water, but the air-borne chemical particles are absorbed into a film of liquid which covers the olfactory mucous membranes. Specific glands are generally developed which function in keeping the membranes moist.

In amphibians the main air channel is usually simply constructed. Each small *external naris* opens into a somewhat elongated sac; postero-ventrally, a large opening, the *choana,* or internal naris, opens directly into the front part of the roof of the mouth. In fishes practically the entire surface of the olfactory pouch is covered with sensory epithelium, which is thrown into numerous small folds. These folds are absent in amphibians; the surface of the main nasal sac is nearly smooth, and part of the surface is covered by a nonsensory epithelium.

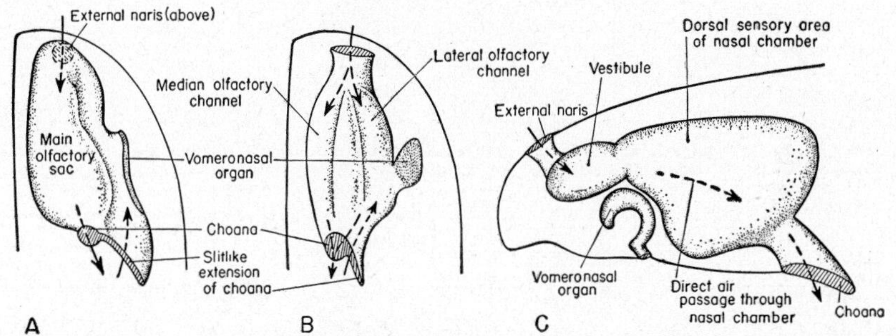

Fig. 300. *A,* Ventral view of anterior part of the left side of the palatal region of the salamander, Triton, with the nasal channels shown as solid objects, remainder as transparent structures (cf. Fig. 170, *A*, p. 295). *B,* Similar view of the toad Pipa. *C,* Longitudinal section of the nasal region of a lizard, cut somewhat to the right of the midline, to show the cavities of the nasal apparatus. In the embryo lizard the vomeronasal organ was a lateral pocket of the main nasal channels, as in an amphibian; in the adult (as in many mammals) this organ has separated to open independently into the roof of the mouth by a nasopalatine duct. Arrows show the main air flow inward in all figures and the outward flow toward the vomeronasal organ in the amphibians. (*A* after Matthes; *B* after Bancroft; *C* after Leydig.)

In typical reptiles the nasal region begins to assume a more complicated structure (Fig. 300). The air passage is longer, the choana more posteriorly placed, and there is usually a distinct, small anterior sac, or *vestibule,* through which inspired air passes before it reaches the main olfactory chamber. Of this main chamber, the ventral part, the direct air passage to the choana, is usually lined by a nonsensory epithelium. The expanded dorsal part contains the sensory area; here we find typically developed a swollen, curved expansion, a *concha* or *turbinal* from the lateral (never the medial) wall of the chamber; this increases the surface over which sensory epithelium may form. In the chelonians there is some development of a secondary palate and, in relation to this, a development of a short air duct leading from the nasal chamber back to the choana.

In the crocodilians, with a long snout and an elongate secondary palate, the nasal chamber, containing three conchae, is itself considerably elongated, and there is a long tube leading back to the pharynx.

In birds, smell appears to play but a small role in sensory reception. The nasal structures, built on the general reptilian plan, are of modest size, and the olfactory epithelium is generally restricted to a small area of the main nasal chamber. The birds, however, retain the three nasal conchae seen in their crocodilian relatives, and these may develop into complicated scrolls. The external nares, generally situated at the base of the beak, are of varied form and, as in fowls and doves, are often protected by a horny structure something like a small eyelid.

The nasal structures reach the peak of their development, as far as size is concerned, among the mammals (Fig. 171, p. 295). In lower vertebrates they are generally confined to a short stretch of territory at the anterior end of the head; in typical mammals (higher primates are exceptional) they push backward to the orbital region, and occupy roughly half of the skull length. Their size is augmented through the fact that there has developed in mammals a secondary palatal structure. As a result, the original vault of the mouth is included in the nasal area; this mouth region anteriorly forms the ventral part of the main nasal cavity and posteriorly forms the *nasopharyngeal duct,* carrying air backward to the pharynx.

These new additions to the nasal area function primarily as aids to respiration. There is, however, a considerable development of olfaction. We have seen that in reptiles there tended to be a concentration of sensory epithelium in the upper part of the nasal chamber. This is true in mammals as well. The lower part of the system, forming a straight passage from external naris to nasopharyngeal duct and choana, is an air passage; the sensory areas are concentrated in the upper and posterior parts of the chamber.

The nasal chamber is, however, by no means a simple sac. We have noted the development in reptiles and birds of folded ingrowths from the lateral wall of the nasal chambers, the conchae or turbinals. In mammals these are developed to a high degree as coiled scrolls of bone, covered with epithelium, which subdivide the air passage into a series of semitubular channels. Ventrally, these structures act as an "air conditioner," filtering and warming incoming air; dorsally and posteriorly, they afford, as in reptiles, additional sensory surface.

In placental mammals there occur, generally, extensions of the nasal air cavities into the adjacent bones as *sinus pneumatici* (plural), frequently present in the adjacent maxillary, frontal, and sphenoid regions, and reaching to the occiput in the high-domed head of the elephant. These sinuses effect a lightening of the skull; other attributes are perhaps all too familiar to some readers.

In many tetrapods there is a specialized part of the olfactory system termed the *vomeronasal organ* (organ of Jacobson, Fig. 300), which in most cases appears to have as its main function the picking up of olfactory sensations from the food in the mouth cavity. It is seen in its simplest form in urodele amphibians. There a partially separated grooved channel runs along one side of the main nasal sac and opens out into the roof of the mouth in a slit continuous with the choana. This groove bears an area of olfactory epithelium quite separate from that of the main part of the nose and served by a distinct branch of the olfactory nerve. In the other two amphibian orders this pocket is nearly completely separated, except at the choana, from the rest of the nasal system; the vomeronasal organ comes to lie in a blind pouch.

In Sphenodon this organ forms a club-shaped blind pouch which opens into the choana. In lizards and snakes the separation from the normal nasal structures is complete; the two vomeronasal organs are separate pouches which open independently into the roof of the mouth anterior to the choanae. The organ is not distinct in chelonians. Adult crocodilians have lost it, and it is absent in birds as well.

Among mammals the vomeronasal organ is absent in man and other higher primates, in many bats and in various aquatic forms. In almost all other groups of mammals, however, it persists as a functional organ. It is a small, cigar-shaped structure, buried in the floor of the nasal region toward the midline. In many rodents it opens into the main nasal cavity. In all other cases, however, the vomeronasal organs connect independently with the mouth, much as in lizards and snakes, by paired *nasopalatine ducts* which pierce through the secondary palate and appear to retain their original mouth-smelling functions.

A peculiar development is seen in many lizards and snakes. The tongue, cleft into two prongs and darting in and out of the mouth, serves as an accessory olfactory organ. When the tongue is withdrawn into the mouth, the tips are inserted into the vomeronasal pockets; chemical particles which adhered to them in the air are transferred to the moisture on the sensory epithelium of these pouches.

THE EYE

The vertebrate body is subjected constantly to radiations which may vary from the extremely short but rapid waves of cosmic rays and those from atomic disintegration to the long, slow undulations utilized for radio transmission; in figures, from waves with as high a frequency as some quintillion vibrations per second and a length of a fraction of a billionth of a centimeter, to others with only a few hundreds or thousands of oscillations per second and a theoretical wavelength measurable in miles. Many of these radiations may affect protoplasm, but specific sensitivity to them appears to be limited to a narrow band part way between the two extremes;

for knowledge of other wavelengths we must resort to mechanisms which transform their effects into terms comprehensible to our own limited senses. It is not unreasonable to find that the animal band of sensitivity corresponds in great measure to the range of radiations reaching the earth from the sun, since that body is the source of the vast bulk of the radiations that normally reach us. Of this band, the slower waves are received

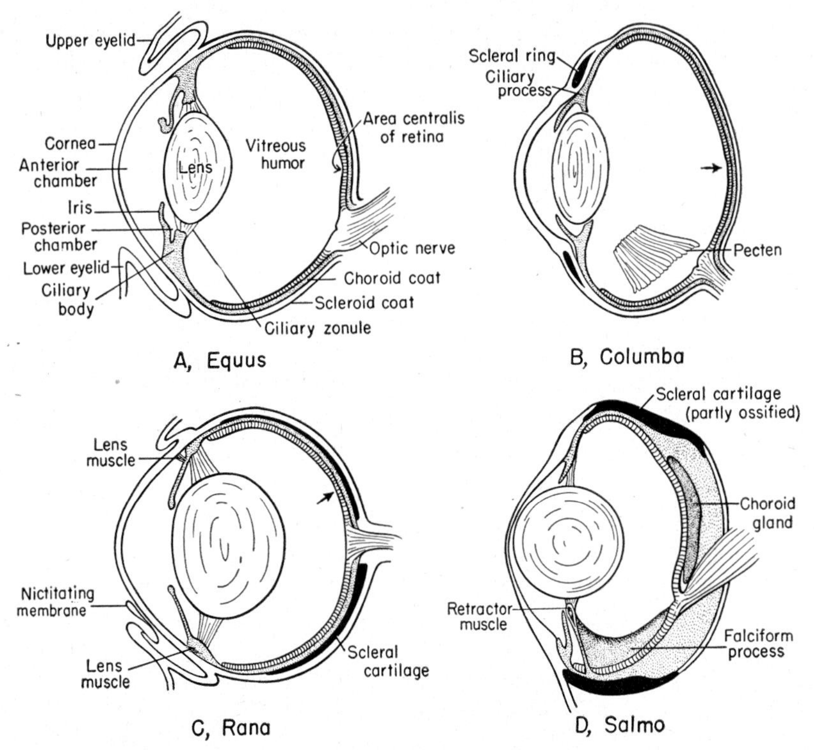

Fig. 301. Diagrammatic vertical sections through the eye of *A*, a horse; *B*, a dove; *C*, a common frog; *D*, a teleost (salmon). Connective tissue of sclera and cornea unshaded; scleral ring or cartilage black; choroid, ciliary body, and iris stippled; retina hatched. Arrows point to the fovea. In *B* is shown the pecten, lying to one side of the midline. In *D* the section is slightly to one side of the choroid fissure through which the falciform process enters the eyeball. (After Rochon-Duvigneaud, Walls.)

as heat, the faster, shorter waves perceived as light—the process of *photoreception.*

Specific sensitivity to light is widespread through the animal and plant kingdoms. In many animals such sensitivity is concentrated in special cells, or cell clusters, frequently with associated pigment, which form "eye spots." Often there is evolutionary progress to an organized photoreceptive structure, with a lens for concentrating light upon sensitive cells contained in a closed chamber. Many such organs are merely receptors for light "in bulk;" with better organization and the arrangement of the sensory cells in a definite pattern so that they receive light from different external areas, true vision results. Well-developed eyes with many common features, but surely independently evolved, are found in forms as far apart as molluscs, arthropods of various sorts, and vertebrates.

The general composition of the vertebrate eye (Figs. 301, 302) may first be noted before consideration of details. Leaving aside for the moment various accessories, we find that the essential structure is the roughly spherical *eyeball,* situated in a recess, the *orbit,* on either side of the braincase, and connected with the brain by an *optic nerve* emerging from its internal surface. The eyeball has a radial symmetry, with a main axis running from inner to outer aspects. Internally, there is a set of chambers filled with watery or gelatinous liquids. In the interior, well toward the front, lies a *lens.* The walls of the hollow sphere are formed primarily of

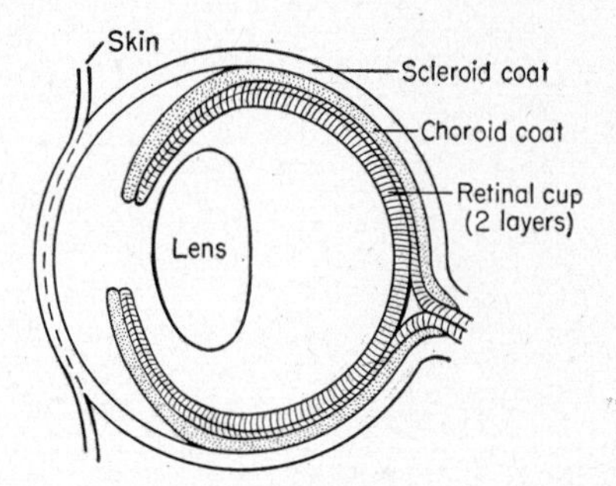

Fig. 302. Diagrammatic section of an eye to show the arrangement of the successive embryonic layers.

three layers, in order, from the outside inward, the *scleroid* and *choroid* coats, and the *retina.* The first two are of mesenchyme origin and are essentially supporting and nutritive in function; the retina (actually a double layer) includes the sensory part of the eye system where visual stimuli are received and transferred to the brain via the optic nerve. At the outer end of the eyeball the scleroid coat is modified to form, with the overlying skin, the transparent *cornea.* Anteriorly, choroid and retinal layers are fused and modified. Opposite the margins of the lens the conjoined layers usually expand to form a *ciliary body,* from which the lens may be suspended. Beyond this point they curve inward parallel to the lens to form the *iris,* but leave a centrally situated opening, the *pupil.*

The general nature of the operation of the eye is commonly (and reasonably) compared to that of a simple box camera. The chamber of the eyeball corresponds to the dark interior of the camera box. In the eye the lens (and in land vertebrates the cornea as well) operates, like that of the camera, in focusing the light properly upon the sheet of sensory materials in the back of the chamber; the retina functions in reception of the image as does the camera film. The iris of the eye is comparable to the similarly named diaphragm of the camera, regulating the size of the eye opening.

Development (Fig. 303). Embryologically, the most important functional parts of the eye arise from the ectoderm (including neurectoderm), but the mesoderm also enters prominently into the picture.

The first indication of an eye comes at about the time of completion of the brain tube. Outward from the forebrain area on either side grow

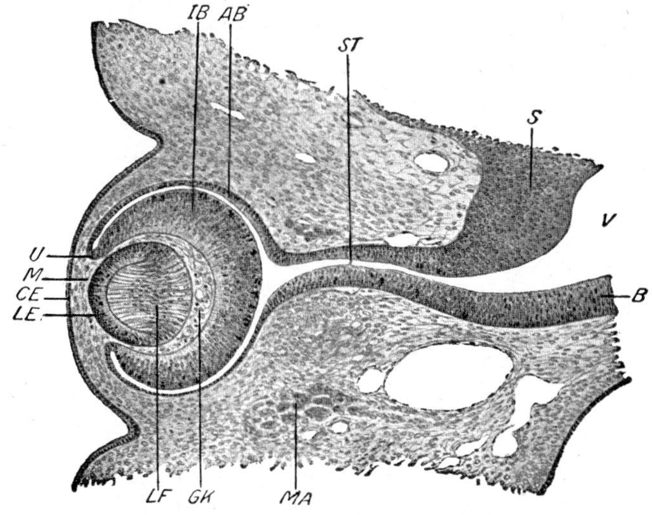

Fig. 303. Section through the developing eye of a mouse embryo. The retina is still connected with the brain by a stalk, *st*, but has become folded into a double-layered cup with a persistent cleft between the two layers; the outer layer, *AB*, remains thin; the inner layer, *IB*, is beginning to differentiate. The lens has sunk from the surface into the cavity of the retinal cup, and at this stage shows an outer epithelial layer, *LE*, and an inner part, *LF*, in which fibers are developing. At *GK* there is a beginning of the vitreous body, but there is as yet no indication of the liquid-filled chambers external to the lens, and there is no definite evidence of the development of the choroid and scleroid coats from the mesenchyme. *B* and *S*, ventral and lateral walls of the brain cavity, *V*; *MA*, developing eye muscle; *U*, border of the optic cup; *M*, mesenchyme; *CE*, epithelium of cornea. (From Maximow and Bloom, after Schaffer.)

spherical vesicles which remain connected by a stalk with the brain. As each *optic vesicle* develops, its outer hemisphere folds into the inner to form a double-layered *optic cup;* ventrally, however, there is a slit in the cup, the *choroid fissure,* through which blood vessels enter during development (normally this fissure is later obliterated). The optic cup becomes the retina, which is thus primarily a two-layered structure; from the retina come part of the ciliary body and iris.

The lens is the second main eye structure to put in its appearance. As the optic vesicle approaches the surface, the overlying ectoderm of the skin thickens, and a spherical mass or pocket of ectoderm detaches itself to sink into the orifice of the cup and forms the lens body. It has been proved in a number of amphibians that the stimulus for lens formation is provided by the approach of the optic vesicle to the epidermis. If the optic vesicle in these forms is removed experimentally, no lens forms; if the vesicle is transplanted to (say) the flank, formation of a lens from the ectoderm of that region occurs. (In certain other forms, however, induction by the optic vesicle is not necessary for lens formation.)

The mesoderm contributes the further structures of the eyeball. From the mesenchyme arises an inner sheathing layer, primarily vascular, to enclose the retinal cup as the choroid, and to form much of the ciliary body and iris. Subsequently an external coat of connective tissue forms a complete sphere about the organ as the sclerotic coat (or sclera) and cornea. These sheaths correspond in great measure to wrappings (men-

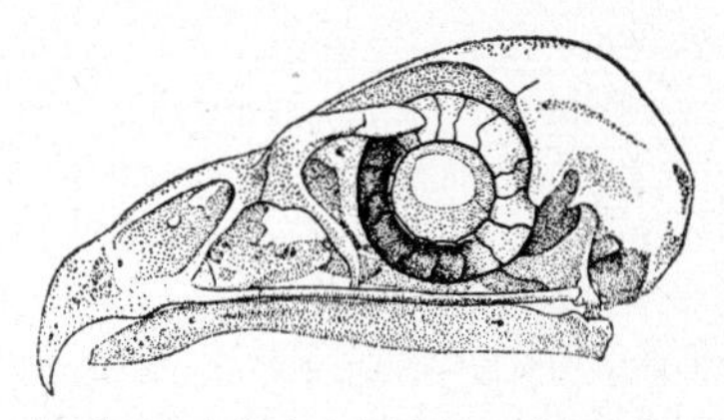

Fig. 304. A skull of a bird (Aquila), showing the scleral ring in place. (From Edinger.)

inges) which surround the brain—a condition which we might reasonably expect, since the developmental picture shows that the retina is in reality a peculiarly developed part of the brain itself.

Sclera and Cornea. The most external eye sheath is that of the scleroid coat, or sclera, a stiff external structure which preserves the shape of the eyeball and resists pressure, internal or external, which might modify this shape. In cyclostomes and mammals, at the two extremes of the scale of living vertebrates, it consists entirely of dense connective tissue. In most groups, however, this is reinforced by the development of skeletal tissues. Often there is a cartilaginous cup enclosing much of the eyeball posteriorly. Further protection is afforded by the frequent presence of a scleral ring (Fig. 304), a series of bony plates lying in the sclera in front of the equator. Fossil evidence shows that such a ring was present in ancestral vertebrates and continued onward along all the main lines of vertebrate evolution. It has, however, disappeared in many groups, and is found today only in actinopterygian fishes, reptiles, and birds. Primitively, it appears, the ring consisted in fishes of four plates, but is reduced in living ray-finned forms to two elements(or to one or none). In the crossopterygian ancestors of land vertebrates and the older fossil amphibians the number of plates was greatly increased, often to a score or more, and a high count is still present in reptiles and birds.

The external part of the sphere of sclerotic tissue is the translucent *cornea,* through which light enters the eyeball. Primitively, the cornea appears to have lain beneath the skin as a distinct structure, and this is still the case in lampreys. In all gnathostomes, however, the scleral coat and skin fuse inseparably in the adult; the skin component of the cornea and the sensitive skin area beneath the lid folds constitute the *conjunctiva.*

The refractive index of the cornea—its ability, that is, to deflect the course of light waves—is practically the same as that of water. Hence in primitive water-dwelling vertebrates the cornea has no power to act as a lens structure. In air, however, the curved cornea does much of the work of focusing and relieves the lens of much of this task. Its importance is shown by the fact that in ourselves many defects such as astigmatism calling for optical correction are due to departures of the cornea from its proper form as a segment of a sphere.

Choroid. The inner of the two mesodermal layers of the eyeball is that of the choroid coat. This contrasts strongly with the sclera, for it is a soft material containing numerous blood vessels which are particularly important for the sustenance of the underlying retina. The choroid is pigmented and absorbs most of the light reaching it after penetrating the retina. But in addition, the choroid of many vertebrates, from elasmobranchs to a variety of mammals, includes a light-reflecting device, most familiar to us as seen in the ghostly eyes of a night-prowling cat illuminated by automobile headlights. This phenomenon is due to the *tapetum lucidum,* generally developed in nocturnal terrestrial animals and in fishes which live below the surface. With plenty of light coming to the eye, internal reflection is not needed and is in fact harmful, since it may confuse the details of the visual image. Where light is scarce, however, conservation of light rays, turned back to the retina by this mirror, more than makes up for the disadvantages.

Tapeta are variously constructed. In one type, common in hoofed mammals, the inner part of the choroid develops as a sheet of glistening connective tissue fibers which act as a mirror of sorts. In a second type, exemplified in carnivorous mammals and many teleosts, the cells of this part of the choroid form an epithelium filled with fiber-like crystals of guanine (a material already mentioned in connection with skin pigmentation). Although the tapetum when present is normally formed in the choroid, some teleosts develop a guanine mirror in the pigment layer of the retina.

Iris. This structure, universally present, is formed by a combination of the modified anterior segments of both choroid and retinal layers of the eyeball. Both layers have in this region lost their most characteristic functions—vascular supply and photoreception, respectively. In attenuated form, they join to form an anterior covering for the lens and retinal cavity, and outline the pupil—the restricted opening through which light penetrates into the inner recesses of the eyeball. The iris is universally pigmented. Its inner (retinal) layer, for example, gives in man a blue effect, while brown pigments which may mask the blue are added by the outer (choroid) component. In a number of fish groups the iris is fixed in dimensions, except where forward or backward movement of the eyeball may affect its distention and the consequent size of the pupil. In sharks, however, and in tetrapods generally, muscle cells are present—striated fibers in reptiles and birds, smooth in amphibians and mammals. Arranged in circular and radial patterns, these fibers may contract or expand the pupillary opening. This gives the iris the function of a camera diaphragm; autonomic nerve stimulation causes expansion of the pupil in dim light for maximum illumination, contraction of the pupil in bright light for protection of the retina and for more precise definition. Nocturnal forms tend to have a slit-shaped pupil which closes more readily to exclude bright light.

The nature of the muscle fibers in the iris is of great theoretical interest. They are formed, when present, from the retinal component of the iris. This is a derivative of the neural ectoderm of the embryo. But muscle, smooth or striated, is supposedly formed by the mesoderm alone, and an ectodermal origin of muscle fibers is unorthodox, improper, and shocking to a right-minded embryologist. However, these cells have all the attributes of muscle fibers, structurally and functionally. The embryo eye has never read a textbook on embryology, and in its innocence violates the rules laid down for it.

The Lens and Accommodation. In land vertebrates light rays are bent as they enter the cornea, which consequently does much of the work of focusing. In fishes this is not the case, and the entire task is thus performed by the lens itself. In consequence, we find that the fish lens has a spherical shape, which gives it the highest possible power, and is situated far forward in the orbit to afford the maximum distance for the convergence of rays on to the retina. In land vertebrates, in which the cornea relieves it of part of its optical task, the lens is less rounded (the primate lens is exceptionally flat) and is situated farther back in the eye cavity. The lens is formed of elongated fiberlike-cells, arranged in a complicated pattern of concentric layers; it nevertheless has excellent optical properties. It is firm in shape and, in lower vertebrates, resistant to distortion.

In cyclostomes the lens is not attached to the walls of the eyeball; pressure from the vitreous humor behind it and the cornea in front fix it in place. In all other groups, however, the lens is attached peripherally by a belt or zonule of some sort. This may be (as in elasmobranchs) a circular membrane or zonule, shaped like a washer; in most cases, however, the suspensory structure consists of a circular belt of zonule fibers. The area of attachment of these fibers is in a zone of conjoined choroid and retinal layers, the *ciliary body,* lying opposite the equator of the lens. The inner surface of this body is an epithelium formed by retinal tissue; its substance is derived from the choroid.

As every user of a camera is aware, it is impossible to obtain exact "definition" of objects at varied distances without adjustment of the lens focus. Such adjustment in the eye is termed *accommodation.* The eyes of most vertebrates are capable of accommodation; but, curiously, it is attained in a different fashion in almost every major group. The methods used may be broadly classified as follows:

A. Lens moved to achieve accommodation:
   1. Fixed position for near vision; moved backward to accommodate for distant objects (lampreys and teleosts)
   2. Fixed position for far vision; moved forward to accommodate for near objects (elasmobranchs, amphibians)

B. Shape of lens modified; fixed form for distant objects, expanded shape for near objects (amniotes)

In lampreys and teleosts the lens is normally at its most forward position and is thus adjusted for nearby objects. For distant vision the lens is, in lampreys, pushed back by a flattening of the cornea, which it directly underlies, and this flattening is accomplished by an external somatic muscle peculiar to this animal. In teleosts there is an unusual situation in that the embryonic fissure in the floor of the eye remains open. Through it projects an elongated vascular structure, the *falciform process* (Fig. 301, *D*, p. 494), serving a nutritive function for the interior of the eyeball. From the front edge of this process there takes origin a small *retractor lentis muscle* (of mesodermal origin) which attaches to the lower edge of the lens and pulls it backward.

In elasmobranchs the lens is fixed for distant vision; for close sight it is swung forward (most strongly suspended by a dorsal ligament) by a forward pull of a small muscle-like protractor structure attached to the ventral rim of the lens. This muscle is, like those of the iris, of ectodermal (retinal) origin, and thus presents another worry for the embryologist.

The amphibian lens movement is the same as that of sharks—a forward pull of the lens for close-up focus. All amphibians have a small ventral muscle which pulls forward on the lens, and in anurans there is a second, dorsal muscle as well. These muscles form in the substance of the ciliary body and hence are (respectably) of mesodermal derivation.

In all amniotes the second principal mode of accommodation is present. The lens is somewhat flexible and capable of changing its shape from a flattened condition adjusted to far vision to more rounded contours for near sight. These changes in shape are accomplished by smooth muscles of mesodermal origin situated in the ciliary body. But despite this basic agreement among amniotes, we find that (perversely) reptiles and birds, on the one hand, and mammals on the other, do the trick in two different ways. In typical reptiles and in birds the greatly developed ciliary body has a ring of padlike processes which extend inward to gain a contact with the periphery of the lens. When the muscle fibers of the ciliary body contract, these processes push in on the lens and force it to bulge into the more rounded shape suitable for near vision.

In reptiles the zonule fibers are unimportant. In mammals, however, the ciliary body does not touch the lens; this is suspended by the fibers, which attach at the back edge of the ciliary region (Fig. 305). The pull

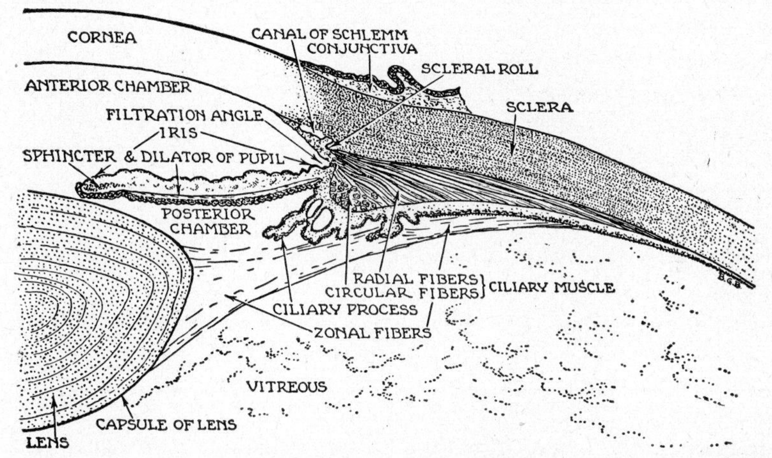

Fig. 305. Details of the anterior segment of the human eye. (From Weymouth, after Maximow and Bloom.)

of the fibers holds the lens, under strain, in a flattened condition. Contraction of the ciliary muscles pulls the region to which the fibers attach inward, closer to the lens; the fibers relax, and the elastic lens, released from tension, takes a more "natural" rounded shape. This type of mechanism is an efficient method in many regards, but is effective only if the lens retains its elasticity. In man, as older persons know, it stiffens with age, accommodation diminishes, and without artificial aids a book can be read only at arm's length, if at all.

Why this great variety of methods to attain a single result? One may more than suspect that it is due to the absence of any device of this sort in the ancestral vertebrates (some fish still have none today). Over the long course of hundreds of millions of years the various main vertebrate groups have all solved the problem of accommodation, but have solved it independently of one another, each in its own fashion.

Cavities of the Eyeball. Much of the eyeball is, as far as function goes, merely a blank space which needs only be filled by some substance which will not block or distort light rays passing through. This filling is in the form of liquids, here known under the old-fashioned name of humors. The principal cavity of the eyeball, between lens and retina, is filled by the *vitreous humor,* a thick, jelly-like material. In front of the lens is the *aqueous humor,* a thinner watery liquid, as it needs to be in lower vertebrates where the lens moves to and fro in this area. The cavity filled by aqueous humor between the cornea and iris is termed the *anterior chamber* of the eye; the *posterior chamber* is not, as one might think, that filled by the vitreous humor, but the area—never of any great dimensions—which the aqueous humor may occupy between iris and lens (Fig. 305).

Intrusive structures are sometimes found in the eye cavity occupied by the vitreous humor. We have already noted the intrusion of the falciform process into this region in teleosts. In reptiles a *papillary cone* projects into this cavity from the region of attachment of the optic nerve. This cone is a highly vascular structure which presumably aids in the supply of nutriment to the retina. In birds it has developed into the *pecten* (Fig. 301, *B,* p. 494), a characteristic structure of this class. The pecten is undoubtedly, like its reptilian homologue, a source of nutritive supply, but its usual shape, with pronounced parallel ridges, leads to the belief that it is, in addition, a visual aid. It is suggested that the shadows of these ridges, falling on the retina, act as a grille ruling, and that small or distant objects are more readily discerned as their images pass from one component to another on this grille. The presence of a pecten may be responsible, in part at least, for the high visual powers attributed to many birds.

Retina. All other parts of the eye are subordinate in importance to the retina; their duty is to see that the light rays are brought in proper arrangement and focus to this sensory structure for reception and transfer inward to the nervous system. We have noted that embryologically the retina is a double-layered structure; in the adult, however, the two layers are fused. The outer one is thin and unimportant; it contributes nothing but a set of pigment cells to reinforce the pigment present external to it in the choroid. The complex sensory and nervous mechanisms of the retina all develop from its inner layer.

The nature of the retina varies greatly from form to form and from one part to another of a single retina. Frequently, however, a sectioned retina has the general appearance seen in Fig. 306. Externally, there is the thin pigmented layer; just inside this, a zone showing perpendicular striations; inside this, again, three distinct zones containing circular objects recognizable as cell nuclei. Special methods of preparation reveal the nature of this stratification. The striated zone contains the elongated sensory portions of the light-receiving cells, the rods and cones; the outer nuclear zone contains the cell bodies and nuclei of these structures. The nuclear zone next in order is that of the bipolar cells, which transmit the impulses inward from the rods and cones, and of other retinal nerve cells which appear to establish cross connections between retinal elements.

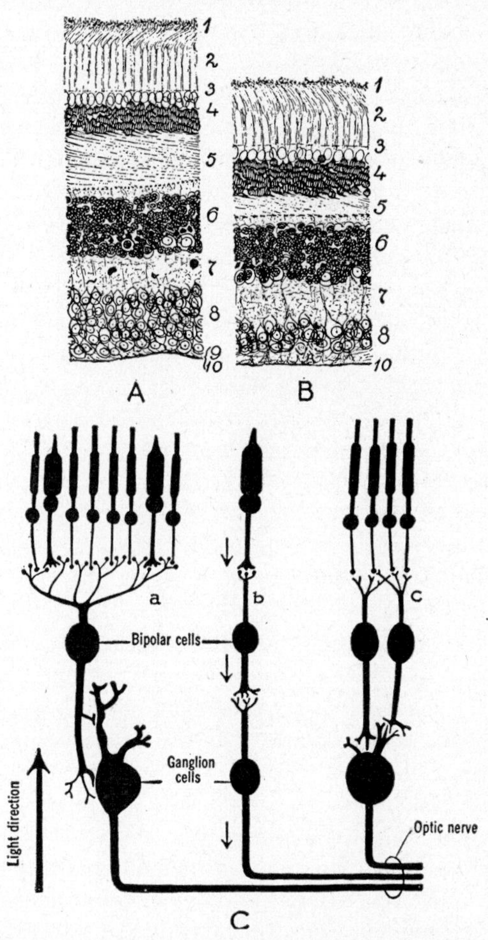

Fig. 306. *A, B,* Sections through the central area of the retina of a rhesus monkey. The orientation is such that the internal surface through which light enters is at the bottom. Successive layers are: *1,* pigment layer; *2,* outer part of rod and cone cells; *3,* basement layer through which rods and cones project; *4,* inner parts of rod and cone cells, with nuclei; *5,* layer of fibers connecting rods and cones with bipolar cells; *6,* bipolar cells; *7,* layer of fibers connecting bipolars with ganglion cells; *8,* ganglion cells; *9* and *10,* nerve fibers and inner membrane bordering the vitreous humor. *C,* Diagram of the "wiring" arrangement of the retina. Cone cells frequently have individual paths to the brain as in *b;* rods usually discharge in groups as in *c;* in some cases cones may be included in a common discharge route with other cones and rods, as in *a.* (*A* and *B* after Maximow and Bloom; *C* from Gardner.)

The innermost nuclei are those of ganglion cells which pick up the stimuli from the bipolar elements and send fibers along the optic nerve to the brain.

The *rods* and *cones* are the actual photoreceptors of the system. They receive their names from their usual shape in mammals. Each cell includes a sensory tip directed outward toward the choroid, a thickened section and (beneath a membrane lying at this level of the retina) a basal piece containing the nucleus. In most mammalian eyes the two types are readily distinguishable: the rod cell is slenderly built throughout; the cone cell has generally a short broad tip and broad "body." In other groups, however, rods and cones vary greatly in shape and are sometimes difficult to tell apart.

One is immediately impressed by the fact that rod and cone cells in the vertebrate retina are *pointing the wrong way!* In a "logically" constructed retina, the tips of the photoreceptors should point toward the light source, and the parts of the retina which carry the impulses inward toward the brain should be farther removed. Some invertebrate eyes are so built; not those of vertebrates. Various theories have been made to account for this anomalous situation—theories based mainly on the fact that the retina is a brain outgrowth. One may assume, for example, that visual cells first appeared in the floor (or side wall) of the brain cavity and were able to function with upwardly (and inwardly) pointing tips, owing to a non-opaque build of the head in ancestral chordates. When these areas became outfolded optic vesicles, the retinal elements simply retained their original positions.

Considerable data are currently at hand regarding the chemistry of photoreception in the rods, at least. Their tips contain a substance known as *visual purple* (chemically termed "rhodopsin"), which consists of a protein plus a substance similar to vitamin A and derived from it. Light reception causes a breaking down of the visual purple into its two components; in the dark they are recombined. It is probable that a similar chemical situation holds for the cones.

Rods and cones differ markedly in their functions, as one can determine from his own eyes, where the cones are concentrated in the center of the field of vision, the rods situated, in the main, peripherally. (1) Good illumination is necessary before the cones come into play; rods are effective in faint light, and it is calculated that rod stimulation can be produced by a single quantum, the smallest theoretical unit of energy. At night one can often catch a glimpse of a faint star in the margin of the field of vision, but fail to see it if looking directly at it. (2) Cones give good visual details; rods give a more blurred picture. For accurate detail we focus the cone-bearing center of the eye on the object; things seen peripherally with the rods are blurred and indistinct. (3) Cones give color vision; rods give black-and-white effects only. In one's visual field, peripheral objects are gray and colorless.

The difference in acuity of vision between cones and rods is explained by the nerve fiber connections of these two types, noted below. We do not know what makes the difference between the high and low "thresholds" of cone and rod cells.

Cell morphology offers no aid in the interpretation of color vision. The visual radiations are measured in terms of wavelengths based on angstrom units of 1/10,000,000 of a millimeter. The longest, and slowest, vibrations perceptible to the eye are those on the order of 700 such units, producing the sensation of the color red (longer waves—infra-red—are discerned as heat). Shorter and shorter waves give color sensations of orange, yellow, green, blue, indigo, and, at about 400 units, violet; beyond this lies the invisible ultraviolet region. The dominant modern theory holds that the cones are of three main types which pick up sensations in the red, green, and violet regions of the spectrum, and that intermediate color sensations are due to a combination within the brain of the relative proportion of stimulation of two or all of these cone types. Thus, stimuli of red and green-perceiving cones in various proportions produce orange or yellow sensations; green and violet produce shades of blue; violet and red produce purple—a color not in the spectrum at all. It is possible that there are three types of photoreceptive materials in the cone cells, or that in addition to a photoreceptive chemical they contain liquids acting as color filters of three sorts; but there is as yet no proof of this, and all cones in a given animal appear to be anatomically of a single type.

Color vision is widespread in vertebrates, but far from universal. Obviously there is little if any color vision in forms in which cones are rare or absent. But it is none too certain that color vision is present, or if present is similar to our own, even where cones are found in the retina. Observation and experiment on living animals are necessary to determine whether color sensitivity is present. Many teleosts have color vision, as do many reptiles and most birds. There is, however, no sure proof of color vision in amphibians. In mammals as a whole relatively few of those tested show much response to color; the ancestral mammals, it is believed, were mainly nocturnal forms, with rods predominating in the eyes. The high color sensitivity of higher primates, including man, is an exceptional situation in mammals; to a dog or cat the world is probably gray, or has at the most faint pastel tints.

The distribution of rods and cones in different animals and in different retinal regions is highly variable. As would be expected, rods dominate in nocturnal animals, or fishes dwelling in deeper waters; cones are more abundant in forms active by day. Since, however, members of a given order or class may differ widely in their habits, the distribution of rod and cone types of retinas does not sort out well taxonomically. In an "average" vertebrate probably not over 5 per cent or so of the total photoreceptive cells are cones. Retinas containing rods alone are found in many sharks and some deep sea teleosts, and a relatively few nocturnal forms in the amniote classes. Eyes which, on the other hand, have cones only are also known in a few amniotes of each class. A majority of reptiles and birds have eyes relatively rich in cones and hence mainly suited for daytime life.

Rods and cones may both be present in any retinal region. Often, however, in forms in which both types of cell are common, there develops a distribution such as that in our own eyes. In this pattern the cones are but sparingly present over most of the retinal area and concentrated, to the practical exclusion of rods, in a central region in which vision is most acute, and perception of detail best developed. This is the *area centralis* (sometimes termed macula lutea because, exceptionally, the area has a yellow tinge in man). Frequently there is present here a *fovea* as well, a depression in the surface of the area centralis from which blood vessels and the inner layers of cells have been cleared away.

In many birds there is, curiously, a development of two central areas in each eye, one centrally situated, the other well toward the posterior or outer region. The bird's eye normally is pointed well to the side, and the primary central area is aimed at this lateral visual field. In bird flight, vision straight ahead is of great importance; the secondary center gives each eye a perception of detail in the anterior part of each visual field.

Inward from the rods and cones is a layer which consists for the most part of *bipolar cells;* and, well to the inner surface of the retina, a third cell layer is that of *ganglion cells.* The bipolar cells have short processes serving to connect with rods and cones on the one hand and with ganglion cells on the other; the latter cells are neurons from which long fibers relay on into the brain the impulses initiated by light reception. Both bipolar and ganglion layers include in addition peculiar types of cells which appear to make cross connections between the various direct pathways. It appears that often one cone cell alone connects with a given bipolar cell, and only one such bipolar cell connects with an associated ganglion cell; thus each cone may have an individual pathway to optic nerve and brain. On the other hand, a considerable number of rods converge into a single bipolar cell. In consequence, the brain obtains no information as to which of the group of rods has been stimulated, a condition which accounts for the lack of precision in rod vision as compared with that of cone areas.

Optic Nerve. Although the brain and nerves are the subject of the next chapter, we may perhaps permit ourselves to violate proper order and discuss the central connections of the eye in relation to vision. The ganglion cells of the retina produce long nerve fibers which travel inwardly along the course of the original optic stalk to the forebrain region. The fibers originate on the inner surface of the retina, converge to a point opposite the attachment of the stalk, and there plunge through the substance of the retina. At this point there can, of course, be no rods or cones, and there is, hence, a *blind spot* in the field of vision. In ourselves, at least, this blank area is "fudged" by the brain, which fills it in with the same materials seen in the surrounding region.

While the connection of retina with brain proper is customarily called a nerve, it differs from any normal nerve in two regards. Here, somewhat as in the olfactory nerve, we are dealing with fibers running inward toward the brain rather than outward from cells in or near that organ. Further, since the retina is, from an embryological point of view, properly a part of the brain itself, the optic "nerve" is rather to be thought of not as a true external nerve, but as a fiber tract connecting two brain regions.

Reaching the floor of the forebrain, the optic nerves enter via the X-shaped *optic chiasma* (Fig. 307). We find that frequently fiber bundles cross from one side to the other within the substance of the brain, without any immediately obvious reason, and this is the case also with the optic nerves. In the crossroads of the chiasma, in primitive vertebrates, all the fibers of the right optic nerve cross to the left side of the brain, and vice versa; this crossing of fibers is termed a *decussation.* The two sets of fibers in lower vertebrates run to the paired optic lobes of the midbrain; in mammals they are relayed to special areas in the gray matter of the cerebral hemispheres. Details are little known in lower vertebrates, but it is presumed that in them, as is known to be the case in mammals, the points in the brain receiving sensations from the different parts of the retina are arranged topographically in such a fashion as to give a brain pattern—and a resulting mental "picture"—reproducing the arrangement of objects as "seen" by the retina.

In a majority of vertebrates the eyes are directed nearly straight laterally, and the two fields of vision are partially or totally different; the brain builds up two separate pictures of two quite separate views. In a number of higher animal types, however—such as birds of prey and many mammals—the eyes are turned forward, the two fields of vision overlap to a greater or lesser degree, and the two sets of impressions transmitted to the brain are more or less the same. Most notable is the case in primates, where, from Tarsius to man, the two fields of vision are practically identical.

In such cases the formation of two duplicate mental pictures seems an unnecessary procedure. Nevertheless, this is done, as far as can be discovered, in nonmammalian forms with overlapping visual fields. In mammals, however, we find a new development. *Stereoscopic vision* appears. The field of vision is unified; sensations from all common objects received by the two eyes are superposed. As a result such a form as man gains, by the slight differences in point of view of the two eyes, effects of depth and shape of objects otherwise impossible of attainment.

An important anatomical phenomenon concerned with this development is *incomplete decussation* at the optic chiasma. In mammals—and in mammals alone—we find that, for parts of the field of the two eyes which overlap, fibers from the areas which in both retinas view the same objects go to the same side of the brain. In consequence, certain groups of fibers do not cross (i.e., decussate), but turn a right angle at the chiasma, to accompany their mates from the opposite eye. In man, for example, where the overlap is nearly complete, all the fibers from the left halves of both retinas enter the left side of the brain, and all fibers from the right halves of the retinas enter the right side of the brain (Fig. 307). As a result, the visual area of each brain hemisphere builds up a half-picture of the total visual field as a "double exposure;" by further complicated interconnections between the hemispheres, the two halves of the picture are welded together to emerge into consciousness as a single stereoscopic view.

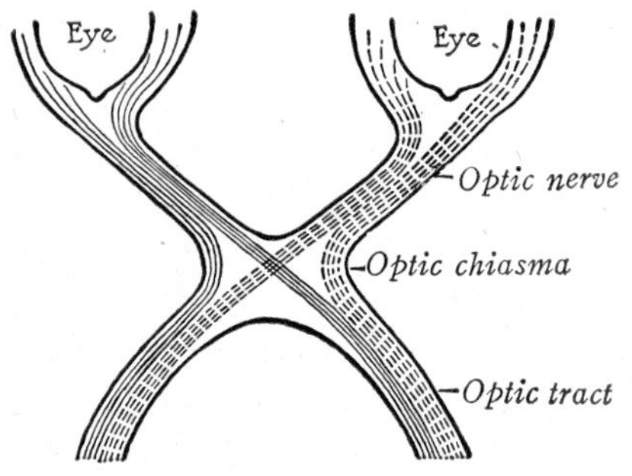

Fig. 307. Diagram of the optic chiasma in a mammal with good stereoscopic vision. All fibers from the corresponding half of each eye pass to the same side of the brain. (From Arey.)

Accessory Structures. External to the eyeball are various visual aids. We have described earlier the six striated eye muscles which move the eye. This muscle group may form additional muscles upon occasion; notably in many tetrapods a retractor muscle which pulls the eyeball inward. We may note that in some vertebrates there develops in the orbit a derivative of the jaw musculature which, when contracted, bulges with the effect of popping the eye outward.

In fishes there may be present folds of skin around the margins of the orbit, but there is little further development of *eyelids* in most fishes. In some teleosts, however, the cornea may be partly covered by fixed skin folds, and in some sharks there are upper and lower lids, the latter movable.

All tetrapods have the lids movable in some fashion. A dry cornea becomes opaque; the lids, by closing at intervals, moisten and clean the corneal surface. Upper and lower lids are always present; they are usually thick and opaque. In most groups the lower lid is more prominent than the upper; in mammals the upper lid is the major structure. In reptiles and birds generally, and in many mammals, there is a third eyelid, the *nictitating membrane,* a transparent fold of skin lying deep to the other eyelids and drawn over the cornea from anterior (or medial) to posterior margins.

No muscles were originally present in the region of the eyelids, and various methods have been evolved to effect their movements. In many tetrapods, as noted above, the eyeball can be withdrawn and again protracted outward; such motions may produce a passive closing and opening of the lids. The operation of the nictitating membrane is effected by the development of a tendon which passes inward from it around the eyeball and becomes associated in variable fashion with the eyeball muscles; a similar tendon may arise in connection with the lower lid. Slips from the eyeball muscles proper in some cases grow forward to attach to the lids. In mammals use is made of a new development in that class. We have noted that facial muscles have here grown forward over the head; a ring of these muscle fibers operates as a sphincter to close the eye (Fig. 166, p. 290).

In snakes there lies external to the cornea a second superficial eye-covering of transparent tissue. This pair of protective "spectacles" is formed from a fusion of the eyelids, giving the snake its unwinking stare.

In land vertebrates are developed *lacrimal glands,* whose function is to furnish liquid for moistening the cornea. As seen in some urodeles, there is primitively a row of small glands running the length of the eyeball within the lower lid. In most groups, however, there is a concentration of glandular development at either the front (medial) or the back (lateral) margin of the orbit. Characteristic of the former position is the *harderian gland,* which tends to secrete a rather oily liquid, in contrast to the true lacrimal gland. The latter is typically developed at the back or outer corner of the eye. In amphibians, reptiles, and birds the harderian glands are usually prominent, the lacrimal glands often poorly developed or absent; in mammals, on the other hand, the lacrimal gland proper is characteristic, and harderian glands relatively rare.

A minor if useful adjunct to the tetrapod eye is the *tear duct* (lacrimal duct), not developed in fishes, which carries surplus fluid from the anterior or medial corner of the eye to the nasal cavity. The lacrimal fluid has thus an accessory function in moistening the nasal mucosa and, in lizards and snakes, of reaching the mouth via the nose and forming an addition to the salivary fluid. (The duct is lacking in turtles.)

Eye Variations. Our previous comments on optic structures have given some faint idea of the great amount of adaptive variation seen in vertebrate eyes as a whole. Throughout the vertebrate series the basic eye pattern persists with remarkable consistency; nevertheless, entire volumes could be (and have been) devoted to an account of the wealth of changes rung on the basic eye structures in different groups. A few additional notes on variations may be given here.

Eyeballs are generally nearly spherical, but there are striking departures from this. "Tube eyes"—deep and narrow—are seen in deep-sea teleosts on the one hand, and in owls on the other. Both these forms operate in dim light; in this eye type all available illumination is concentrated on a narrow patch of sensory retina to bring the light intensity up to the minimum necessary for visibility. Typical day birds, in contrast with owls, have an expanded retinal surface, making perception of small details possible.

Eyes are relatively small in large animals, relatively large in little forms. This one might reasonably expect, for if an eye of a certain size furnishes satisfactory vision for a given animal, there is no selective advantage in a larger eye even if general body size has increased.

Eyes have degenerated and lost much of their structural niceties in a variety of forms from hagfishes to mammals. Such degeneration is, of course, usually correlated with a mode of life in which light is dim or absent. Among marine teleosts, those living at rather considerable depths universally have large and specialized eyes which make the most of the faint illumination present; but in the great deeps some fishes have, so to speak, abandoned the struggle, and the eyes have degenerated. Cave life, too, has led to eye degeneration in various fishes and amphibians, some closely related to forms with normal eye structure. In land vertebrates burrowing forms usually exhibit a similar tendency.

Median Eyes. Our attention thus far has been exclusively directed to the paired laterally placed eyes common to all vertebrates. In their early history, however, the ancestral vertebrates had a third eye, medially situated on the forehead and directed upward. In the oldest ostracoderms (Fig. 16, p. 38) a socket which obviously contained such an eye is universally present and often conspicuous, although always much smaller than the paired eyes. This eye opening was generally present in placoderms; it is found in the Devonian in primitive representatives of all three major groups of bony fishes and is especially common in crossopterygians (Fig. 131, p. 224). In land animals this third eye persisted; it was present in almost all the older amphibians and early reptiles, including most of the mammal-like types (Figs. 130, *A*, 138, *A*, *B*, pp. 218, 232). But by Triassic times median eyes had apparently gone out of fashion. Even in fishes they disappeared at an early stage in most groups, and in modern amphibians, most reptiles, and all birds and mammals they are likewise absent. Today we find median eyes present only in lampreys, on the one hand, and Sphenodon and some lizards on the other (Fig. 353); buried beneath the skin, they can presumably do little more than detect the presence or absence of light, although a miniature cornea, lens, and retina may be developed. In frogs there may be a vestigial external structure; in all other groups the remains of such organs are no more than glandular materials in the braincase.

A functional interpretation of this story can be made in terms of the habits of the vertebrate ancestors. Presumably they were bottom-dwelling mud strainers. In this mode of life, knowledge furnished by a median eye of what threatened from above was vital. With the development of a more active mode of life, vertical vision lost its importance.

Such dorsal eye structures are brain outgrowths, springing from the roof of the diencephalon. There is, however, a curious quirk in the story: the median eye, it appears, is not the same structure throughout, but may develop from either of two outgrowths, the *parapineal organ* (or parietal organ) and *pineal organ*. Both are present in many forms and, although adjacent, are readily distinguishable through differences in the position of their stalk attachments. In lampreys both form eyelike structures, but the pineal is the functional element. In Sphenodon and lizards, however, the parapineal is the functional eye. The situation suggests that possibly the remote ancestor of the vertebrates may have had paired dorsal eyes as well as paired lateral ones.

Despite loss of eye function, the pineal organ persists in higher vertebrates as a glandular structure. It has been suspected of being an endocrine gland, but there is little better evidence that it functions in this fashion than there is for the suggestion, made by Descartes some centuries ago, that it is the seat of the soul.

LATERAL LINE ORGANS

A highly developed sensory system of a type quite unknown in land vertebrates is that of the lateral line organs (Figs. 308, 309) found in

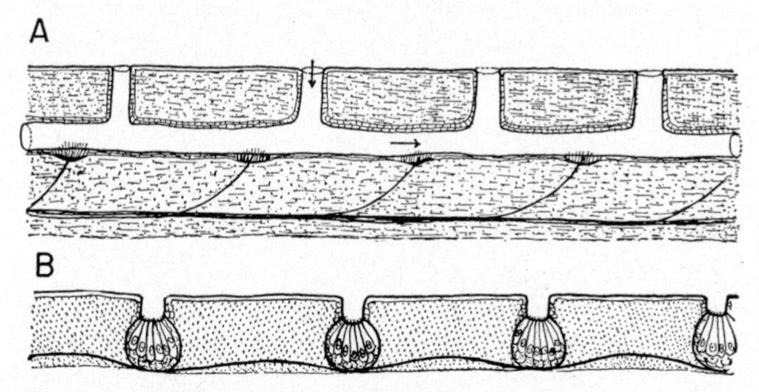

Fig. 308. *A,* Longitudinal section through a lateral line canal of a fish, showing sensory areas and accompanying nerve. Arrows show water current. *B,* Groups of neuromasts occurring in isolated fashion in the skin. (From Dean.)

fishes and in aquatic larval amphibians. The receptors are sensory cells termed neuromasts; these may be found distributed in isolated fashion in the skin, but are generally distributed along a series of canals or grooves on the head and body.

A main element of the system is the *lateral line* in the narrower sense of the term. In typical fishes this is a long canal running the length of either flank. The canal continues forward onto the head, where similar canals form a complex pattern. Generally, a major canal runs forward over the temporal region, then curves downward and forward beneath the orbit to the snout as an infraorbital canal. A second cranial canal is supraorbital in position in its forward course. The canals of the two sides may have a transverse connection across the occiput; a canal generally crosses the cheek and then runs downward and forward along the lower jaw. Other accessory canals may be present on the head and gill region, and in some cases there are accessory body lines. Isolated neuromasts, *pit organs,* may also be found on the head, often in linear arrange-

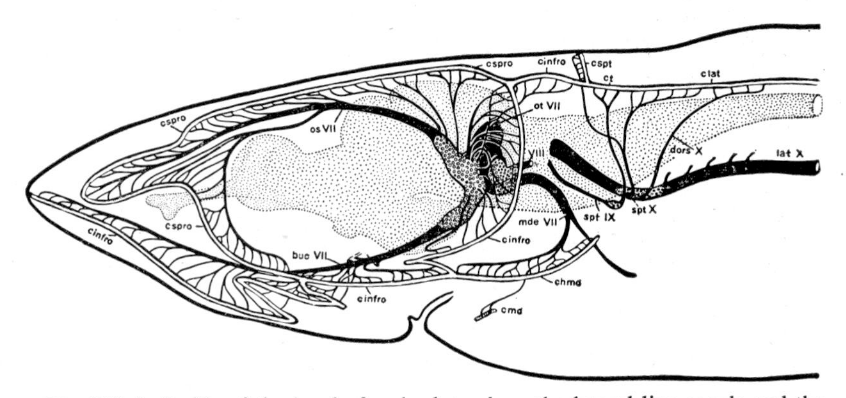

Fig. 309. Left side of the head of a shark to show the lateral line canals and the nerves supplying them. Pit organs are not shown: *buc VII,* buccal ramus of nerve VII; *chmd,* hyomandibular canal; *cinfro,* infraorbital canal; *clat,* lateral line canal proper; *cmd,* mandibular canal; *cspro,* supraorbital canal; *cspt,* supratemporal (or occipital) canal; *ct,* temporal canal; *dors X,* dorsal ramus of nerve X; *lat X,* lateral line ramus of nerve X; *mde VII,* external mandibular ramus of nerve VII; *os VII,* superficial ophthalmic ramus of nerve VII; *ot VII,* otic ramus of nerve VII; *spt IX,* supratemporal ramus of nerve IX; *spt X,* supratemporal ramus of nerve X. (From Norris and Hughes.)

ments. In elasmobranchs there are, in addition, deep-lying tubular, slime-filled *ampullae* of similar nature.

In most fishes the lateral line organs are contained in closed canals, opening at intervals to the surface. In many bony fishes this canal system may be sunk within the substance of the head plates and within or below the body scales; in other bony fishes and in elasmobranchs generally the canals are superficially situated. In a few cartilaginous fishes open grooves take the place of canals, and the same appears to have been true among the ancient fossil amphibians. In modern amphibians the neuromasts lie in the skin in more or less isolated clusters, although preserving in part a linear arrangement.

The sense organs of the lateral line system are the *neuromasts,* consisting of bundles of cells, often with much the appearance of taste buds. The neuromast cells are elongated, and each bears a projecting hairlike structure. Lacking comparable structures ourselves, the nature of the sensations registered has been difficult for us to determine. It appears that they respond, by bending of their sensory processes, to waves or currents in the water. A fish seldom has visible "landmarks" against which to measure its progress through the water; sensory structures of this sort will at least tell it of water movements and the pressure of currents against its body and thus be of great aid in locomotion.

The neuromasts and their nerve cells are derived embryologically from thickenings of the ectoderm on either side of the head, termed *placodes;* these form external to the neural tube after this has closed, but are nevertheless generally considered to be outlying parts of the nervous system. The lateral line placodes become associated with three of the cranial nerves. Much of the system of head canals is formed from an anterior placode associated with cranial nerve VII. A short segment in the temporal region is formed from a small placode related to nerve IX. A posterior placode forms the posterior part of the head system and grows backward the length of the trunk; it is associated with cranial nerve X (the vagus), which sends a branch along the flank to accompany it.

Although the lateral line system is present in larval amphibians and some adults which remain water dwellers, it was lost, it seems, in the earliest reptiles. Many reptiles and mammals have returned to a water-dwelling life, but, useful as it would be to them, this sensory system, once lost, never reappears.

THE EAR

First thoughts as to the primary anatomical or functional aspects of the vertebrate ear are liable to be misleading when based on familiar human features. One tends to think, when the word is mentioned, of the ornamental pinna of the mammalian external ear, or, perhaps, of the middle ear cavity behind the drum, with its contained ear ossicles. These items, however, are entirely lacking in fishes; the basic ear structures of all vertebrates are those of the internal ear, the sensory organs buried deep within the ear capsule. We naturally think of hearing as the proper ear function; but in the ancestral vertebrates audition was apparently unimportant and perhaps absent; equilibrium was the basic sensory attribute of the "auditory" organ.

The Ear as an Organ of Equilibrium. Before considering the hearing function, which becomes increasingly important as we ascend the scale of vertebrates, we may discuss the ear as an organ of equilibrium, a basic and relatively unchanged function from fish to man. Equilibrium is a type of sensation produced by the internal ear alone; all accessory ear structures are related to the hearing sense and need not concern us for the moment.

In a variety of fishes, amphibians, and reptiles the paired internal ears are built upon a relatively uniform pattern in which most of the structures present are related to equilibration (Fig. 310). The *membranous labyrinth* consists of a series of sacs and canals contained within the otic region on either side of the braincase. These form a closed system of cavities, lined by an epithelium and containing a liquid, not dissimilar to that of the interstitial fluid, termed the *endolymph.*

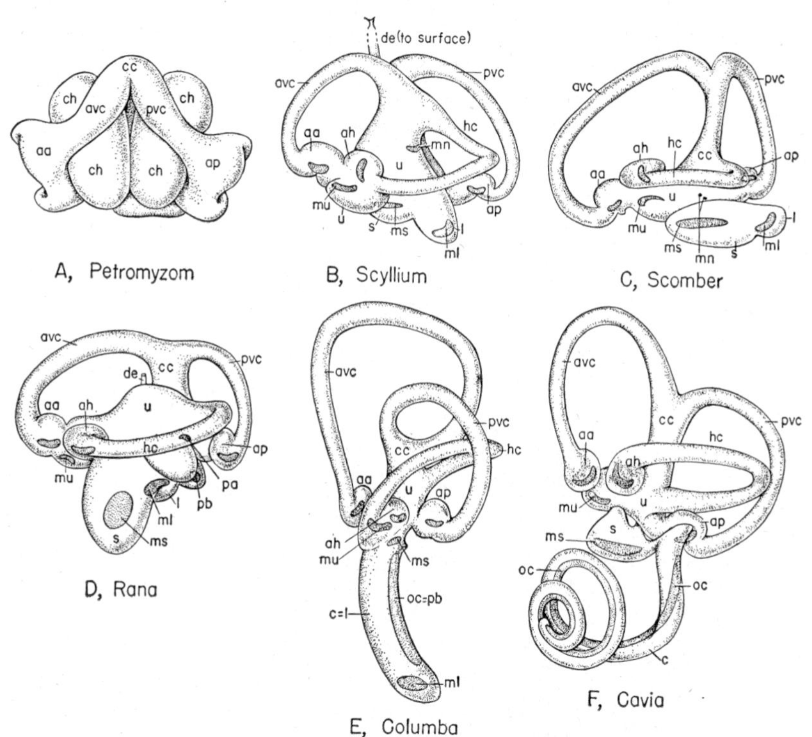

Fig. 310. Membranous labyrinth of *A*, lamprey; *B*, shark; *C*, teleost; *D*, frog; *E*, bird; *F*, mammal; all external views of the left ear. Sensory areas are shown (except in *A*) as if the ear were transparent. *aa,* Ampulla of anterior canal; *ah,* ampulla of horizontal canal; *ap,* ampulla of posterior canal; *avc,* anterior vertical canal; *c,* cochlear duct; *cc,* crus commune with which both vertical canals connect; *ch,* chambers in the lamprey ear lined with a ciliated epithelium; *de,* endolymphatic duct; *hc,* horizontal canal; *l,* lagena; *ml,* macula of lagena; *mn,* macula neglecta; *ms,* macula of sacculus; *mu,* macula of utriculus; *oc,* organ of Corti; *pa,* papilla amphibiorum; *pb,* papilla basilaris; *pvc,* posterior vertical canal; *s,* sacculus; *u,* utriculus. (After Retzius.)

Two distinct, major saclike structures are generally present in jawed vertebrates, the *utriculus* and the *sacculus,* usually connected by a narrow channel. From a point on or near this connecting duct there generally arises a slender tube, the *endolymphatic duct,* which usually extends upward and inward to terminate within the braincase as an *endolymphatic sac.* In general, the sacculus lies below the utriculus. In both there is found a large oval "spot" consisting of a sensory epithelium associated with branches of the auditory nerve; these areas are the *utricular macula* and *saccular macula.* The utricular macula lies on the floor of that sac, in a horizontal plane; the saccular element is typically in a vertical plane on the inner wall of the sacculus.

The sensory cells of these maculae (and, indeed, of the entire internal ear) are comparable to the neuromasts of the lateral line system. During development there forms over the combined tips of the sensory hairs of these cells a gelatinous membrane. In the utricular and saccular maculae this material becomes a thickened structure in which there is deposited a mass of crystals of calcium carbonate termed an *otolith.* Its nature varies. It may remain a rather amorphous mass of crystals. In ray-finned fishes, however, the otoliths develop into compact structures. The saccular otolith is generally the better developed; in teleosts it is so large as practically to fill the sacculus and reflect in its outlines the shape of that cavity. This shape varies from genus to genus and from species to species; a single tiny otolith will often furnish positive identification of the fish which bore it.

The function of these maculae is readily deduced. They are organs of equilibrium, static "balancing" organs which give information as to the tilt of the head (and in fishes, of course, of the whole body). Somewhat similar organs are found in a number of invertebrate types. As the head is bent in one direction or another, the pressure of the otoliths on the sensory hairs which support them is increased in one area or another of the macula; these areal differences are presumably registered in the brain, with which all sensory ear structures are connected by way of the auditory nerve.

The two major maculae are sufficient to register the static position of the head in space at any given moment, but they do not give information as to movement of the head or body. Such data are furnished by another series of structures, the *semicircular canals.*

These tiny tubes are dorsal elements of the endolymphatic system, springing out from the utriculus and connected at either end with this sac. In every jawed vertebrate, without exception, three such canals are present; each of the three lies in a plane at right angles to the other two, so that one is present for each of the three planes of space. Two canals lie in vertical planes, an *anterior vertical* (or superior) *canal,* angled forward and outward, and a *posterior vertical canal,* running backward and outward from the upper surface of the utriculus; a *horizontal canal* extends laterally from that body. Often the two vertical canals arise at their proximal ends by a common stalk, a *crus commune,* from the upper surface of the utriculus. Each semicircular canal has at one end a spherical expansion, an *ampulla.* The vertical canals bear these ampullae at their distal ends, anteriorly and posteriorly; the horizontal canal (for no known reason in particular) has its ampulla anteriorly placed. Within each ampulla is a sensory area, usually elevated, termed a *crista.* Here we find again the familiar hair cells, or neuromasts; their tips are embedded in a common membrane or mass of gelatinous substance.

The arrangement of the canals and their sensory structures makes it seem clear that their function is to register movement of the animal in the several planes of space; they furnish a dynamic register of equilibrium, as contrasted with the registration of static equilibrium in the maculae. One would at first assume that the sensory effects would be caused by a flow of endolymph through the canals as the head moves, affecting the sensory hairs of the cristae; movements in various directions would cause varied combinations of flow in the three canals. It has, however, been pointed out that these canals are so tiny that no free flow of liquid can occur through them, and some have argued that, apart from the ampullae, the canals themselves can have no function. But the fact that no jawed vertebrate, however specialized or degenerate in other regards, has ever lost a canal, argues strongly for a belief that they are not superfluous ornaments, but have a real and important function. Movement of the animal must at least cause some sort of jar or vibration to be set up in the canals and impinge on the ampullae.

The organization of the parts of the ear dealing with equilibrium is in general constant through the jawed vertebrates, high or low. There are, however, certain variables in the system. As in many other features, the cyclostomes exhibit an unusual and puzzling condition. The lamprey has but two semicircular canals, equivalent, apparently, to the vertical canals of other vertebrates, and, in addition, a peculiar system of sacs beneath them; the hagfish has but a single canal, presumably representing a combination of these two. As in the nose, we cannot be sure whether this is a primitive or a degenerate condition. The lamprey type of ear (as well as nostril arrangement) is known to have been present in some of the most ancient fossil vertebrates. In cartilaginous fishes, sacculus and utriculus are merely two lobes of a common cavity from which canals and endolymphatic duct arise. The posterior vertical canal may be associated, at its ampullar end, with the saccular region in elasmobranchs. In these forms the crus commune is commonly absent, and in skates the canals are connected only by narrow ducts with the remainder of the canal system.

The endolymphatic duct usually terminates in a sac of modest proportions lying within the braincase. In Chondrichthyes—presumed to be primitive in this regard—the endolymphatic ducts may extend upward to open on the top of the head. The endolymphatic sac is absent in some teleosts. In some bony fishes, the reptiles, and particularly in the amphibians, the endolymphatic sacs may be of great size; those of the two sides may connect either above or below the brain in some lower tetrapods, and in the typical frogs may extend the length of the cavity of the spinal cord as well.

Origin of the Vertebrate Ear. Embryologically, the internal ears first appear, like lateral line organs, as a pair of placodes on either side of the head (Fig. 311). These sink inward to form a pair of sacs, which for some time may retain a tubular connection with the exterior (as do the endolymphatic ducts in the adult shark). Typically, each sac then subdivides into utricular and saccular portions; from the former the canals are gradu-

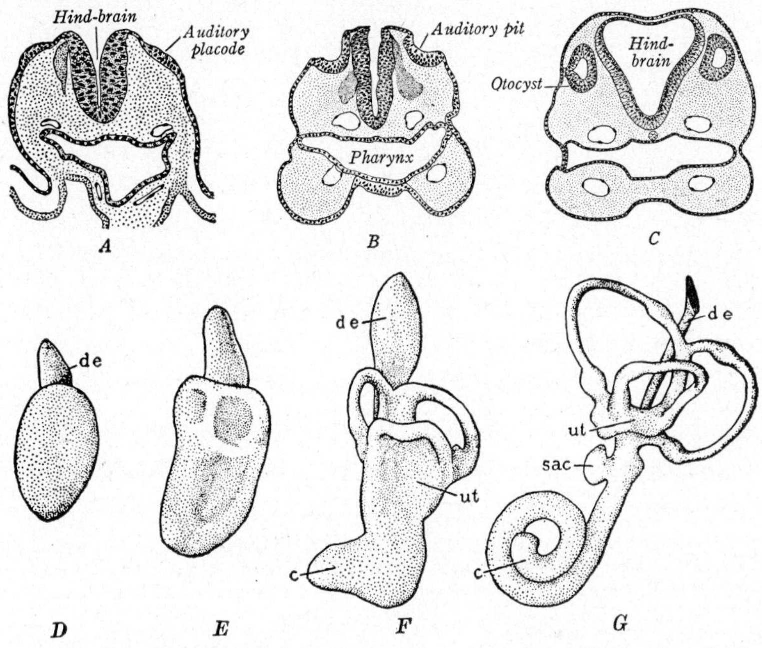

Fig. 311. Diagrams to show the development of the internal ear. *A* to *C*, Cross sections of the head of early mammalian embryos; an ectodermal placode sinks inward on either side to form an otic vesicle. *D* to *G*, Successive stages in the development of the various parts of the membranous labyrinth from the otic vesicle: *c*, cochlear duct; *de*, endolymphatic duct; *sac*, sacculus; *ut*, utriculus. (*A* to *C* from Arey; *D* to *F* after His and Bremer.)

ally separated off, and from the latter arise accessory structures discussed later.

This embryological story, together with the nature of the sensory endings of the ear, suggests that the internal ear is a specialized, deeply sunk part of the lateral line system. Like the lateral line neuromasts, the hair cells of the internal ear respond to pressure or to movements in the liquid-filled spaces in which they lie. The nerves from the two sets of organs are closely associated (as will be noted in the next chapter) to form an acoustico-lateralis system. And functionally also the two sets of sensory structures are closely knit; together they offer a primitive fish the major part of the data by which its locomotion is regulated.

Hearing in Fishes. Whether or not fish hear is a topic that was long disputed. A considerable body of evidence, however, indicates that hearing is certainly present in many forms, at least. There is no positive evidence as to the location in the endolymph of the sensory receptors for "hearing" vibrations. However, fishes have accessory sensory areas for which no equilibrium function is apparent and which thus may come under suspicion here (Fig. 310, *B, C*). Fishes and a variety of land vertebrates have a small and obscure sensory spot, the *macula neglecta,* generally located in the posterior part of the utriculus. It may have some auditory function, but there is no proof of this, and the macula neglecta certainly has nothing to do with the development of hearing in tetrapods. More promising as a possible seat of hearing is the *lagena,* usually present as a pocket-like depression in the floor of the sacculus near its posterior end. This contains a *lagenar macula,* frequently distinct in fishes from that of the sacculus, and in bony fishes commonly carrying a well-formed otolith. As will be seen, this macula does not become the developed hearing organ of higher vertebrates; the lagenar pocket, however, is definitely concerned with later advances in hearing, and it is not impossible that the lagenar macula may have some primitive auditory function.

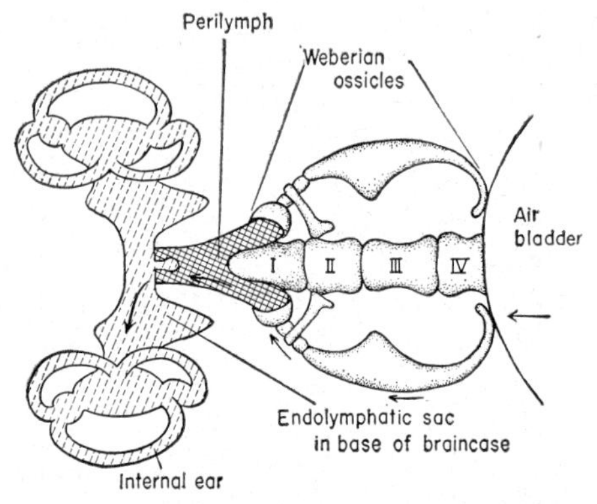

Fig. 312. Diagrammatic horizontal section of the posterior part of the head and anterior part of the body of a teleost with weberian ossicles. Vibrations in an anterior subdivision of the air bladder set up corresponding vibrations in a series of small ossicles which in turn set up waves in a perilymphatic sac. This, again, sets up vibrations in an endolymphatic sac at the base of the braincase. Arrows show the course of transmission of the vibrations. Roman numerals indicate the vertebrae from which the weberian ossicles are derived. (After Chvanilov.)

Fishes in general lack the various accessory devices by which in higher vertebrates sound waves reach the internal ear. But, as shown by studies on men with hearing deficiencies, the conduction of vibrations through the head skeleton to the internal ear may produce some degree of hearing. In most fishes water vibrations, to be heard, must set up head vibrations, and these in turn produce endolymphatic vibrations which can be picked up by the hair cells of the internal ear.

Some bony fishes, however, have accessory structures which parallel in a way the "hearing aids" found in land vertebrates, although evolved quite independently and along other lines. In these fishes, it appears, the

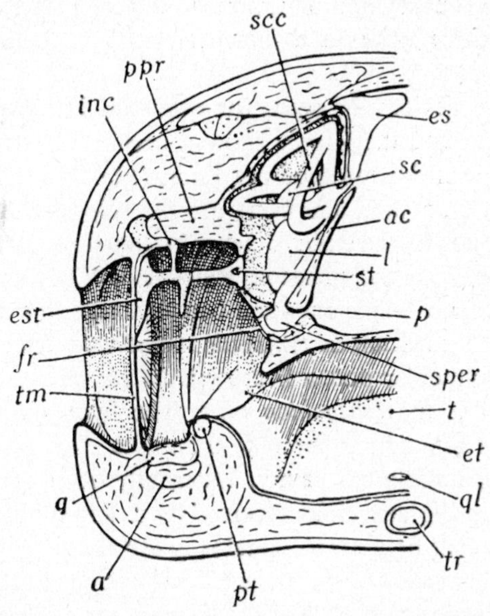

Fig. 313. Posterior view of the left side of a lizard head to show the auditory apparatus. A shallow external depression leads to the ear drum (*tm*). Internal to the drum, the stapes is seen, divided into two parts, the "extracolumella" (*est*), and columella or stapes proper (*st*); processes from the former articulate above with the skull (*inc*) and below with the quadrate anterior to the middle ear cavity. This cavity opens by a broad eustachian tube (*et*) to the throat. The internal ear is shown in diagrammatic fashion. Other abbreviations: *a*, articular bone of lower jaw (= malleus); *ac*, inner wall of auditory capsule; *es*, endolymphatic sac; *fr*, fenestra rotunda; *gl*, glottis; *l*, lagena; *p*. perilymphatic duct connecting inner ear with brain cavity; *ppr*, paroccipital process of otic region; *pt*, pterygoid; *q*, quadrate (= incus); *sc*, sacculus; *scc*, semicircular canals; *sper*, position of perilymphatic sac; *t*, tongue; *tr*, trachea. (From Goodrich, after Versluys.)

swim bladder is utilized as a device for the reception of vibrations. In herring-like teleosts this air bladder sends forward a tubular extension which comes to lie alongside part of the membrane system of either ear and can thus induce endolymph vibrations. In a group of teleosts, termed the Ostariophysii, which includes the catfishes, carp and relatives, in which hearing is unquestionably developed, another method is used. Processes of the most anterior vertebrae develop on either side as four small detached bones termed the *weberian ossicles* (Fig. 312). These articulate in series to form a chain extending from the air bladder forward to the ear region. The ossicles operate somewhat in the fashion of the little ear bones of a mammal, transmitting air bladder vibrations to the liquids of the internal ear system.

The Middle Ear in Reptiles and Birds. Hearing is an important sense in the tetrapods. A different problem in sound reception is found in land vertebrates as contrasted with the fish situation. The sounds to be heard are relatively faint air waves which can have ordinarily little direct effect in setting up endolymph vibrations in the internal ear. Devices for amplification of these waves and their transmission to the internal ear are a necessity.

Basically, these devices, as seen, for example, in many reptiles, are simple. Sphenodon and the lizards show a middle ear structure not far removed from a primitive tetrapod condition (Fig. 313). The spiracular gill cleft and the hyomandibular bone are the sole elements concerned. The spiracular pouch of the embryo never breaks through to the surface; the corresponding surface depression, if developed, becomes an *external auditory meatus.* The thin membrane between external depression and pouch becomes the eardrum or *tympanic membrane,* which picks up air vibrations. The spiracular pouch is the *middle ear cavity;* its connection with the pharynx is the *eustachian tube.* The hyomandibular bone of fishes changes its function to become a rodlike *stapes* (or columella) which crosses the middle ear cavity. Externally, it attaches to the eardrum and picks up the vibrations received by it. Internally, an expanded footplate of the stapes fits into an opening in the ear capsule, the *fenestra ovalis;* through the agency of the stapes, air vibrations are carried inward to set up vibrations in the liquids of the internal ear and eventually reach its sensory structures.

The fish spiracular opening lies high up on the side of the head (Fig. 314, *A*), the position in which the eardrum lay in the older fossil amphibians (Fig. 314, *B;* the eardrum position in the skull is shown by the presence of an otic notch, Fig. 314, *E*). In amniotes, however, the drum has moved farther down the side of the head, lying close to the jaw joint in most reptiles and birds (Fig. 314, *F*). Primitively, the drum was nearly flush with the surface of the head. It is still in this position in anurans and turtles; in some reptiles and in birds, however, it attains some protection by being placed at the inner end of a short external meatus.

The eardrum is in general a thin membrane capable of rapid vibration in response to air waves. It contains a sheet of fibrous tissue and is bounded outside and inside by epithelia derived from the skin and from the lining of the first gill pouch, respectively. Its presence is due to the "failure" of the spiracular pouch to break through to the surface. The pouch itself expands lateral to the braincase to form the air-filled middle ear cavity—a cavity which, in crocodiles, birds, and mammals, may extend its air spaces into the neighboring regions of the skull (as in the mammalian mastoid). Primitively, as in anurans and many lizards, the middle ear cavity opens broadly into the pharynx; in most amniotes, however, the connection is narrowed to form the eustachian tube. In many amphibians (as noted later) the drum and middle ear cavity have degenerated, as is true of snakes. A specialization of the turtles and snakes is the development of a secondary outer wall of the ear capsule around the shaft of the

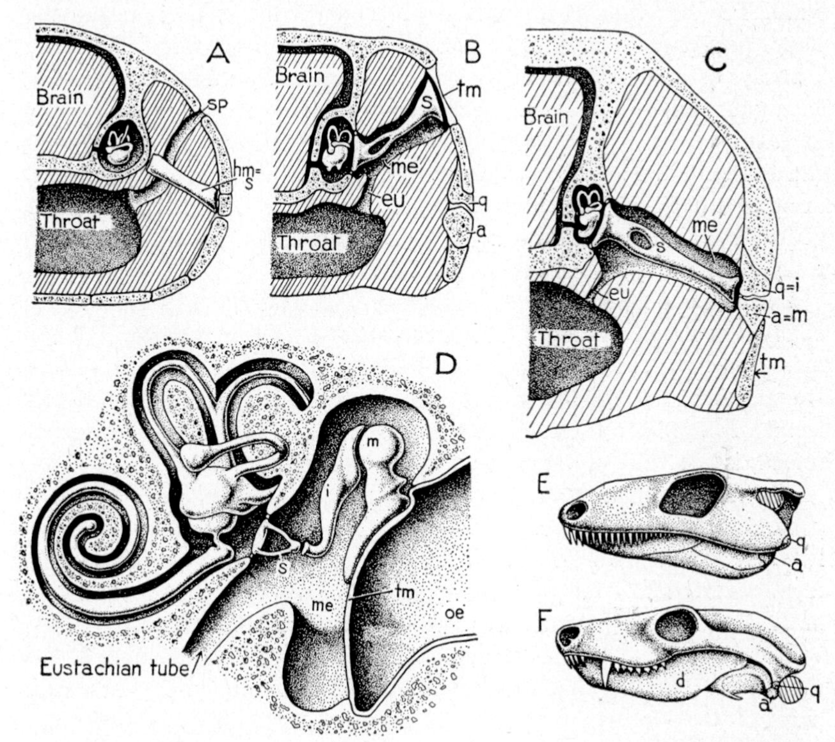

Fig. 314. Diagrams to show the evolution of the middle ear and auditory ossicles. Diagrammatic sections through the otic region of the head of *A*, a fish; *B*, a primitive amphibian; *C*, a primitive reptile; *D*, a mammal (showing the ear region only); *E*, side view of the skull of a primitive land vertebrate; *F*, of a mammal-like reptile to show the shift of the eardrum from the otic notch of the skull to the region of the jaw articulation. *a*, Articular; *d*, dentary; *eu*, eustachian tube; *hm*, hyomandibular; *i*, incus; *m*, malleus; *me*, middle ear cavity; *oe*, outer ear cavity; *q*, quadrate; *s*, stapes; *sp*, spiracle; *tm*, tympanic membrane (in *C*, posterior to the plane of section; cf. *F*). (From Romer, Man and the Vertebrates, University of Chicago Press.)

stapes; into the space beneath this is an extension of the fluid-filled cavities of the inner ear.

The fish hyomandibular lies just behind the spiracular pouch; in typical tetrapods it comes to lie within the bounds of the expanded middle ear cavity. In the fish ancestors of the tetrapods the hyomandibular, once part of the gill apparatus, functioned as a jaw prop, bracing the quadrate on the otic region of the skull. In land vertebrates jaw suspension is adequately taken care of by the skull itself, and the hyomandibular, as the stapes, functions in transmitting vibrations from eardrum to internal ear. The stapes, essentially a rodlike structure, has its main outer attachment to the eardrum; medially, there has developed, underneath an original attachment to the braincase, an opening into the internal ear capsule, the fenestra ovalis. In reptiles and birds the stapes often has a complicated build. The structure as a whole is often called the *columella,* and the stapes proper considered to be the proximal part of the shaft. The latter is always ossified, sometimes pierced by an arterial foramen, and ends in an expanded footplate in the oval window. The outer part, often termed the *extracolumella,* may remain cartilaginous and may have a complicated structure. The direct outer termination lies, of course, on the eardrum, but there may be extra processes connecting, as in fishes, with the quadrate region of the jaw articulation, with the skull dorsally, and, ventrally, with the hyoid arch (of which the stapes was originally a part).

External and Middle Ear in Mammals (Fig. 314, *D*). In mammals the external ear region is well developed. There is a deep, tubular external meatus. In some reptiles the meatus may be surrounded by cartilage; this is expanded in mammals, and in addition there is almost always a projecting ear *pinna* of elastic cartilage; this structure, often elaborately folded, is generally of value as a collector of sound waves.

The mammalian middle ear is, as noted in an earlier chapter (p. 240), generally enclosed in a bony bulla. Instead of a single auditory bone, the stapes, it contains an articulated chain of three auditory ossicles leading from eardrum to oval window—*malleus, incus,* and *stapes* (hammer, anvil, and stirrup)—named in relation to their fancied shapes.

The origin of this series of auditory ossicles was long debated. It was thought at one time that they might be due to a subdivision into three parts of the reptilian columella or stapes. Emybryology, comparative anatomy, and paleontology combined have, however, revealed the true story. The inner element, the stapes, although much shortened, is equivalent to the whole columellar apparatus of a lower tetrapod. The other two elements are homologous with the articular and quadrate bones which in lower vertebrates form the jaw joint. Mammals have evolved a new joint system for the jaw, and the older structures of this region have become superfluous. They have been put to new use. The reptilian eardrum lies close to the old jaw joint. The articular has kept an attachment to the membrane and become the malleus. The quadrate, connecting in reptiles with articular on the one hand, and stapes on the other, has become the incus, and the homologue of the jaw articulation of a reptile lies in mammals at the joint between two ear ossicles. These bones afford a good example of the changes of function which homologous structures can undergo. Breathing aids have become feeding aids and finally hearing aids.

The Internal Ear in Reptiles (Fig. 315). In tetrapods, as already indicated, the parts of the internal ear devoted to equilibrium remain little changed from the fish condition. The auditory apparatus, however, gradually develops into structures which attain such size and importance that the older regions of the sacs and canals are often termed (rather slightingly) the *vestibule* of the internal ear.

In fishes, we have noted, little or nothing can be suspected of an auditory function except the macula of the lagena. This macula persists in all tetrapods, except mammals above the monotreme level, but never

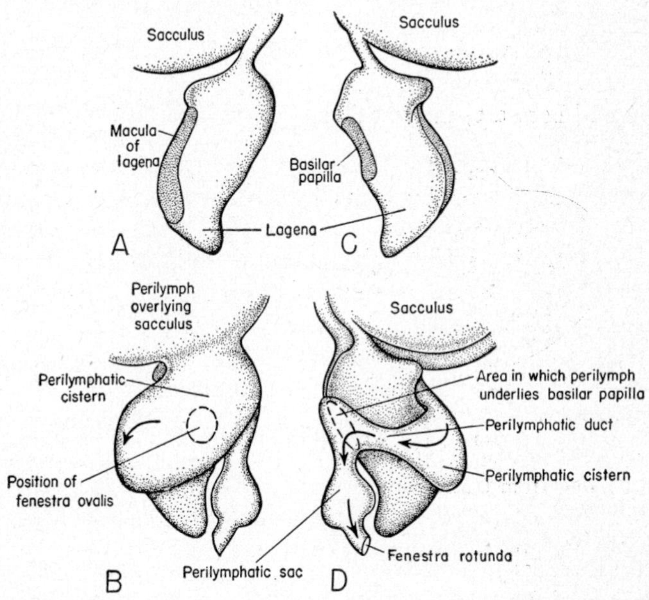

Fig. 315. The ear of a late embryo of a lizard (Lacerta). *A*, Left ear, lateral view of the membranous labyrinth in the floor of the sacculus and the lagena. *B*, The same with the perilymphatic system shown in addition. *C*, *D*, Medial views comparable to *A* and *B*, respectively. Arrows indicate course of conduction of vibrations from stapes to basilar papilla and on to the "round window" at the distal end, beyond the perilymphatic duct.

develops to any degree; if it has an auditory function in land vertebrates, this function cannot be significant. The lagena itself, however, is important as a basal structure in the development of the cochlea.

The sensory area vital in the development of hearing in tetrapods is that termed the *basilar papilla* (Fig. 310, *D*, *E*, p. 514). This area is unknown in fishes, but is characteristically developed in many amphibians and in typical reptiles. In these forms it is an area of hair cells, covered by a common membrane, situated on the posterior wall of the sacculus or at the base of the lagena. It is here that vibrations, brought in from without by the stapes, are received.

In general the sacs and canals of the inner ear are not adherent to the skeletal walls of the auditory capsule that surround them. They are separated from them by spaces filled with liquid and crossed by connective tissue strands. This liquid is the *perilymph* (Fig. 316). It surrounds the membranous labyrinth and thus is quite distinct from the endolymph within. The perilymph has no important function in fishes, nor has it in the vestibular area of tetrapods. From the perilymph, however, there develops in land vertebrates a conduction system which leads from the fenestra ovalis to the basilar papilla and forms the last link in the chain of structures by which potentially audible vibrations reach the auditory sensory areas.

Inside the oval window in lower tetrapods there develops a large *perilymphatic cistern,* against which the stapes plays (Fig. 315). Vibrations received here are, in typical reptiles, carried around the lagena to

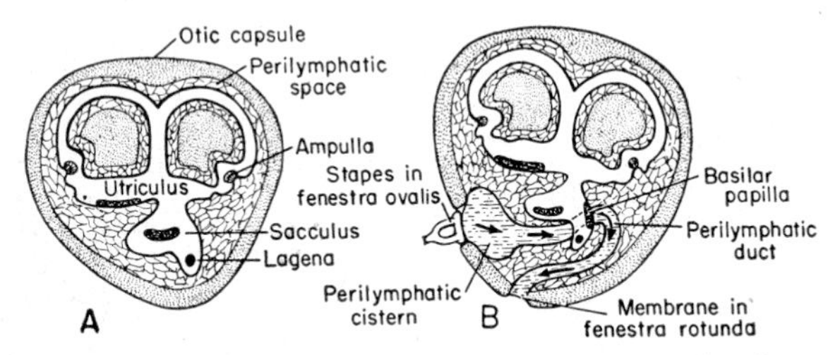

Fig. 316. *A,* Schematic section through the ear capsule of a fish, to show the perilymphatic space surrounding the membranous labyrinth containing endolymph. *B,* Similar scheme of a tetrapod, in which a part of the perilymph area is specialized to conduct sound from the fenestra ovalis to and past the auditory sensory area. (After de Burlet.)

its posterior border in a perilymph-filled canal. This duct passes just outside the area of the basilar papilla, and is separated from the base of its sensory cells only by a thin *basilar membrane.* Vibrations in this membrane agitate the hair cells and at long last, in this roundabout fashion, the sensory organ is reached. This situation—the auditory sensory structure agitated by vibrations of the membrane at its base—is a fundamental feature in the construction of the hearing apparatus in all amniotes. We shall find it repeated, and described in much the same words, in birds and mammals.

For vibrations set up by the stapes in the perilymphatic system and carried along by the perilymphatic duct, some release mechanism, allowing a corresponding vibration, must be set up at the far end of the perilymph system. In many amphibians this consists of a *perilymphatic sac,* which projects from the auditory capsule into the braincase. In some amphibians and in reptiles (except the turtles) a further development takes place. There is a large, rather tubular foramen (primarily for the vagus nerve) in the braincase wall which runs out from a point near this sac. We find that the perilymphatic duct system may, instead of terminating in a sac, run outward through this foramen—or a separate one formed close by, the *fenestra rotunda* (Fig. 313, p. 519)—and terminate in a membrane which vibrates in phase with the impulses received through the fenestra ovalis at the other end of the perilymph system. Primitively, this membrane is buried in the tissues at the margin of the skull; in more advanced forms the round window opens back into the middle ear chamber.

Development of the Cochlea. Both birds and mammals have greatly extended their hearing ability by the evolution of a *cochlea*—a highly developed structure for auditory reception. The crocodilians, related to the birds, demonstrate the manner in which the cochlea is developed.

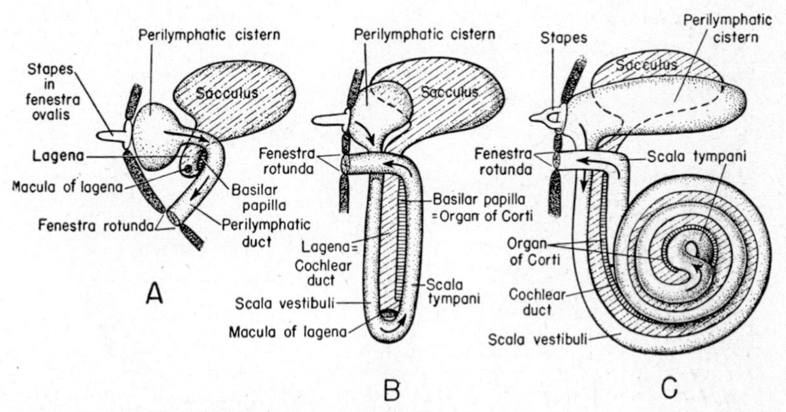

Fig. 317. Diagrammatic sections through the saccular region to show the evolution of the cochlea. *A,* Primitive reptile with a small basilar papilla adjacent to the perilymphatic duct. *B,* The crocodile or bird type; the lagena has elongated to form a cochlear duct, the basilar sense organ with it, and a loop of the perilymphatic duct follows the cochlear duct in its elongation. *C,* The mammalian type; the cochlea is further elongated and coiled in a fashion economical of space.

Three structural features are involved: the lagena, the perilymphatic duct, and the basilar papilla.

In the crocodiles and birds (Fig. 317) the finger-like lagena has become expanded into a long but straight tube, filled of course with endolymph; this is the *cochlear duct* (or scala media). At the tip persists the original macula of the lagena. The important sensory structure, however, is the basilar papilla, which is here greatly expanded into an elongate area running the length of the cochlear duct. Thus developed, the papilla can be recognized as the equivalent of the mammalian *organ of Corti.* It has a complicated series of sensory and supporting cells over which folds a membranous flap. With the lengthening of this sensory organ the perilymphatic duct, which remains closely applied to its base, expands in a double loop. The part of this tube leading inward from the oval window is termed the *scala vestibuli* (so called because the fenestra ovalis is considered to lie in the vestibular part of the inner ear). The distal limb of the tube, leading to the round window with its membrane, or "tympanum," is the *scala tympani.* It is in this distal segment of the loop that the perilymphatic duct underlies the sensory organ.

These modifications of the primitive reptilian system result in the development, in birds and their crocodilian relatives, of a true, although simply constructed, cochlea. This consists of three associated tubular structures—one endolymphatic, and two filled with perilymph—and, in

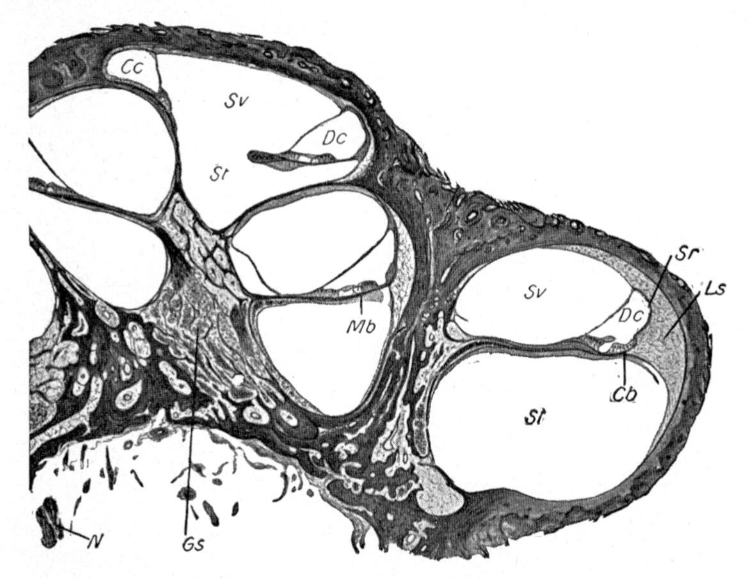

Fig. 318. Diagrammatic section through a mammalian cochlea. Three complete sections through the duct system are shown, and at the top the terminal point at which vestibular and tympanic scalae connect (cf. Fig. 319). *Cb,* Basal crista; *Cc,* a small distal pocket of the cochlear duct (cecum cupulare); *Dc,* cochlear duct; *Gs,* spiral ganglion of cochlear nerve; *Ls,* spiral ligament attaching basilar membrane to outer wall of bony labyrinth; *Mb,* basilar membrane; *N,* cochlear nerve; *Sr,* vascular epithelium; *St,* scala tympani; *Sv,* scala vestibuli. (From Maximow and Bloom, after Schaffer.)

the central duct, the organ of Corti, separated from the distal perilymphatic tube—the scala tympani—only by the thin basilar membrane. Crudely one may visualize the formation of this type of cochlea by imagining that he has grasped the region of the lagena and the perilymphatic duct crossing its base and pulled this area ventrally. This pull has stretched the lagena out into a long blind tube as the cochlear duct; the perilymphatic duct, attached at both ends, has pulled out as a double-looped structure, the two scalae.

The mammalian cochlea (Figs. 318, 319), we believe, has developed independently of that of birds, but in closely parallel fashion. In monotremes it is still a simple, uncoiled structure, readily comparable to that of birds. The typical mammalian cochlea is, however, much more elongated than that of birds or crocodiles. If it were to remain straight, it could not be comfortably accommodated within the ear capsule, and in relation to this we find that (as the name implies) the mammalian cochlea of three tubes has been coiled into a tidy spiral structure.

We cannot here discuss in detail the complicated microscopic structure of the organ of Corti, which includes a complex system of sensory and supporting cell types and a covering membranous flap. A structural feature of importance is the basilar membrane which underlies it, and separates it from the scala tympani. As in lower tetrapods, vibrations of this

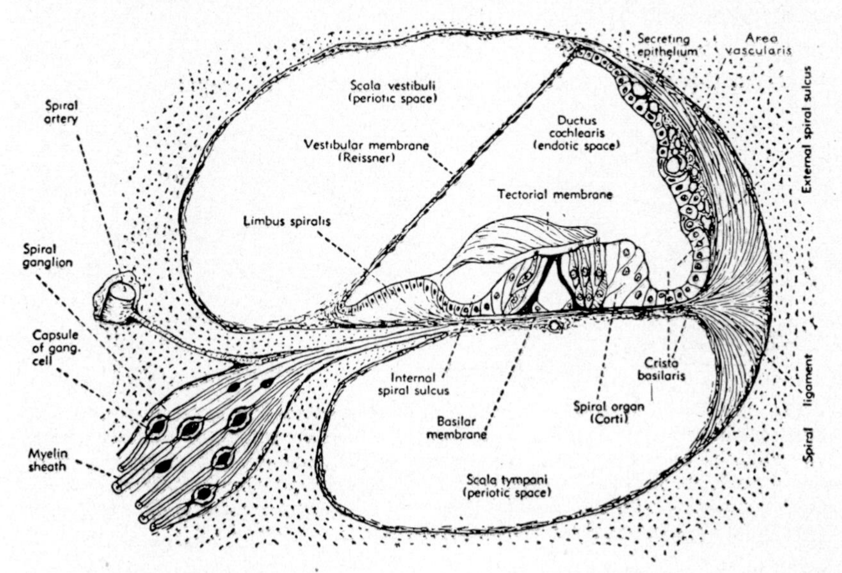

Fig. 319. A much enlarged section through a mammalian cochlea to show details of the organ of Corti (cf. Fig. 318). (Ruch from Fulton-Howell.)

membrane aroused by air waves brought in by the perilymph system appear to be basically responsible for stimulation of the organ of Corti. The functional "reason" for elongation of the basilar papilla into the formed organ of Corti appears to be the discrimination of sounds of different pitch—i.e., of different wavelengths. Differences in the transverse span of the basilar membrane, which grades in width from one end of the cochlea to the other, appear to result in making the membrane sensitive to different wavelengths at different parts of its extent and thus render tone distinctions possible.

The Middle Ear in Amphibians. We have above omitted any reference to ear construction in living amphibians, for the reason that conditions in these forms are not, in general, primitive. They are specialized and seemingly degenerate in most cases, and are, further, extremely varied. Three main points may be noted initially: (1) the drum and middle ear cavity are often missing; (2) the stapes is frequently reduced or absent; (3) a second ossicle, the operculum, is frequently present in the oval window.

A majority of frogs and toads have a well-developed middle ear region, with a large, superficially placed eardrum and a good stapes* carrying vibrations inward from drum to oval window. But in urodeles and the Apoda, and even in a number of anurans, there is no drum or middle ear cavity, and typical sound transmission is impossible. The stapes may persist in the absence of drum and cavity, and is well developed in the Apoda and some salamanders. Its outer end articulates with the quadrate region of the skull, and it may in some fashion pick up for transmission vibrations in this region of the head. In a great number of both urodeles and anurans, there is present a second ear ossicle, the *operculum,* peculiar

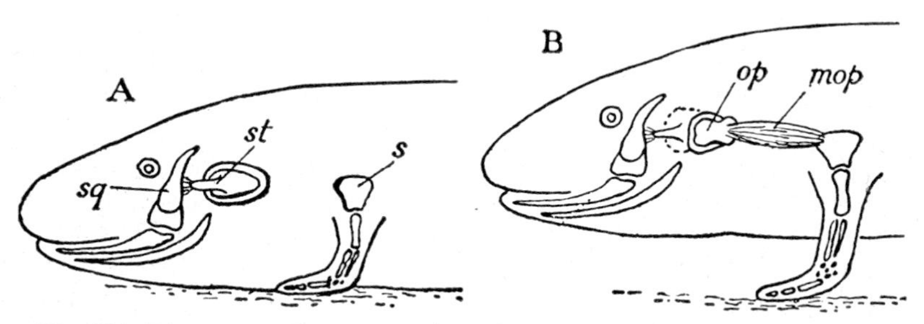

Fig. 320. Diagram to show mechanism of communication between exterior and inner ear in urodeles. *A,* An aquatic form in which the stapes, or columella, picks up vibrations by a ligamentous attachment to the squamosal. *B,* A type in which the stapes is reduced, and the operculum picks up ground vibrations through a muscular connection with the shoulder girdle. *mop,* Opercular muscle; *op,* operculum; *s,* scapula; *sq,* squamosal; *st,* stapes. (After Kingsbury and Reed.)

to amphibians. This is a flat plate fitting into the oval window. It may be present in the fenestra in company with a well-developed stapes, as in typical frogs and toads; in many forms, however, the stapes is reduced, and the operculum replaces it functionally as a hearing aid. The operculum has no homologue in the skeleton of other groups of vertebrates; it appears to be a separate part of the wall of the otic capsule itself. In most urodeles, and in those anurans in which the eardrum is lost, the operculum is connected with the shoulder blade by a small *opercular muscle* (a specialized axial muscle) (Fig. 320). It is suggested that when this muscle is tensed, ground vibrations are carried from the limb along the muscle to the operculum and thence to the inner ear.

The Inner Ear in Amphibians. In inner ear as in middle ear, the amphibians were passed over earlier because they tend to exhibit specializations which obscure the general story. There is typically developed in modern amphibians, as in reptiles, a perilymphatic cistern, perilymphatic

* The stapes, or columella, is sometimes termed the plectrum in anurans.

ducts, and a perilymphatic sac or release mechanism at the base of the skull. But though many amphibians have a basilar papilla, this may be absent. Instead, there is developed a special amphibian sensory structure, the *papilla amphibiorum,* which appears to play the main role in sound reception in this class of vertebrates (Fig. 321). This is situated near the upper margin of the sacculus. It is structurally similar to the simpler types of basilar papillae. However, the perilymphatic duct does not lie beneath the base of this structure; instead, it approaches it, so to

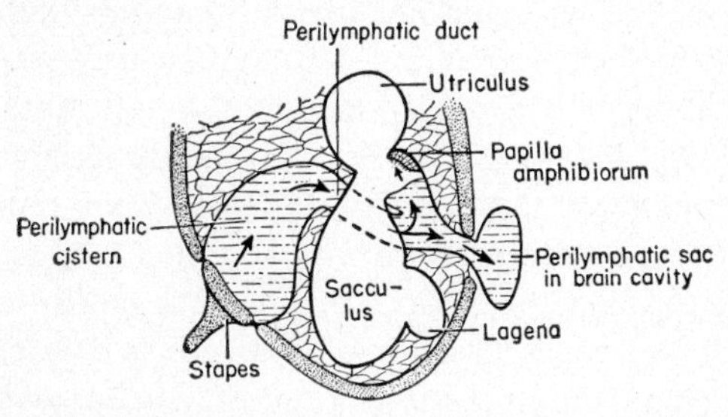

Fig. 321. Schematic section of the internal ear in a salamander. The basilar papilla is absent here, but is found in addition to the papilla amphibiorum in anurans. (After de Burlet.)

speak, laterally, being separated from the endolymph in a neighboring region by a thin membrane through which vibrations are transmitted to the latter fluid. (The same condition holds for the relations of the perilymph to the basilar papilla in amphibians.)

Evidently, auditory structures were in an experimental stage in the ancestral amphibians; the ancestors of the modern amphibians chose less progressive paths in ear evolution, as in many other features, than did those of their ancient relatives who gave rise to the amniotes.

CHAPTER XVI

THE NERVOUS SYSTEM

IN A PROTOZOAN the single cell in itself receives sensations and responds to them. In higher, metazoan organisms there tends to be, to an increasing degree, a differentiation between cells specialized for the reception of sensations—*receptors*—and those which make the appropriate response—the *effectors*. In simple multicellular organisms the relations between these two types of cells may remain relatively simple, and cells receiving sensations may, by their physical and chemical activities, arouse their neighbors to respond. Even in the vertebrates the circulation of hormones in the blood stream is a retention of such essentially primitive methods of stimulation. But such a method of arousing responses tends to be slow, is often nonspecific, and is relatively ineffective for swift and precise reaction. With the development of a nervous system there are formed special tissues designed to transmit stimuli from sensory structures directly and rapidly to specific end organs—the muscles or glands which furnish the response.

In primitive metazoans, such as coelenterates, the nervous system may be a diffuse network of cells and fibers spread through the tissues. But in most animals of any degree of complexity, the system is a well-integrated, well-organized structure. Instead of scattered fibers we find collections of fibers formed into nerve trunks, and centers where transfer of nerve impulses between fiber systems takes place, much as a telephone system brings its numerous lines into major cables and central exchanges. In most groups a dominant center, a brain in some fashion or other, makes its appearance, usually in a situation strategically located with regard to major sense organs. In bilaterally symmetrical animals there is usually a brain structure situated anteriorly and one or more main longitudinal nerve trunks. In the vertebrates there is a well-organized tubular brain and a single dorsal hollow nerve cord—the spinal cord—running backward from this along the body. These form the *central nervous system*. Running outward from brain and spinal cord are numerous paired nerves, along which ganglia—clusters of nerve cells—are found at specific regions; these nerves and ganglia constitute the *peripheral nervous system*.

STRUCTURAL ELEMENTS

The Neuron. Nervous tissues, in the brain, spinal cord, and ganglia, contain numerous cell bodies. Prominent in the nervous system, however, are slender but elongate nerve fibers, making up the nerves and much of the substance of the central nervous system as well. It was once thought that the cell bodies and the fibers, conveyers of the nervous impulses, were separate structures, independent of one another. Most nerve fibers are exceedingly elongate, and connections between them and cell bodies were not often observed. But as it gradually became clear, through the decades, that all the vital materials of the body were cellular, it was realized that the fibers are universally processes of cells rather than independent structures. The basic units of the nervous system are *neurons;* each consists of a cell body and its processes, long or short.

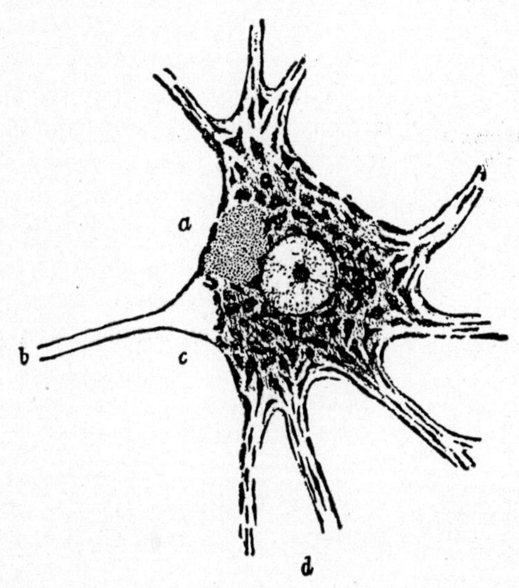

Fig. 322. The cell body of a motor neuron from the spinal cord of an ox; *a,* a mass of pigment. At *b* is the axon, with a clear cytoplasmic area (the "axon hillock"), *c,* at its base; the remaining processes, *d,* are dendrites. Much of the cytoplasm is filled with readily stainable masses of material, the Nissl bodies. The large nucleus contains a prominent cluster of chromatin, the nucleolus. (From Herrick, after von Lenhossék.)

Most of the cell bodies of the neurons (Fig. 323, p. 532) are situated within the walls of the central nervous system, a relatively small number in the peripheral ganglia. The shape of the neuron body is variable; in many cases, however, it is stellate, owing to the fact that a number of processes extend out from the cell body. With appropriate stains, microscopic preparations show in the protoplasm of the nerve cell various characteristic structures not present in other tissues. Some are illustrated in Figure 322. In the adult there is never seen any evidence of mitosis in the neurons, indicating, as a notable peculiarity of the nervous system, that growth has been completed, for the most part at least, by about the time of birth. In consequence, destruction of nerve cells through injury or disease is a permanent loss; cell processes may regenerate if the cell body is intact, but a nerve cell, once lost, cannot be replaced.

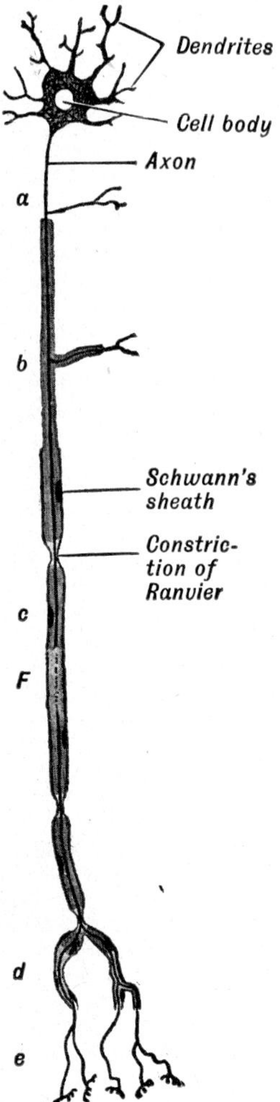

Fig. 323. A peripheral motor neuron: *a,* portion of axon without sheaths; *b,* axis cylinder covered by myelin sheath; *c,* neurolemma (Schwann's sheath) added externally; *d,* myelin sheath ends, axis covered only by neurolemma; *e,* naked fiber endings; *F,* broken line indicating the greater length of the fiber omitted. (From Maximow and Bloom.)

Nerve Fibers (Fig. 323). Extending outward from the cell body of the neuron are slender processes—typically, numerous dendrites and a single axon—along which the nerve impulses pass; the cell body is perhaps to be regarded mainly as a center for the maintenance and nutriment of these vitally important fibers. Short processes, usually slender, numerous, and branching, and most frequently carrying impulses inward toward the cell body, are termed *dendrites,* from their treelike appearance. The *axon* is a relatively stout and elongate process, sometimes (in large animals) as much as several yards in length, which generally carries impulses away from the region of the cell body. The axons perform the bulk of the work of nerve impulse transmission; as nerve fibers they form, collected in bundles, the peripheral nerves and the tracts in the central nervous system.

Functionally, the most important part of a nerve fiber (Fig. 323) is its central structure, the *axis cylinder,* a thin strand of protoplasm continuous with that of the cell body. Its appearance is homogeneous in unstained materials; appropriate stains, however, show the presence of threadlike longitudinal *neurofibrils,* each of which is shown by the electron microscope to consist of a bundle of exceedingly minute tubules. Every fiber lying outside the brain or spinal cord is covered by a thin sheathing membrane, the *neurolemma.* This is not part of the neuron itself, but is formed by separate cells which lie in series along the course of the fiber. A majority of long nerve fibers both within and without the central nervous system have a second type of covering, a thick layer of fatty material interposed between axis cylinder and neurolemma—the *myelin sheath.* This sheath gives a white, glistening appearance to nerves composed of myelinated fibers in contrast to a gray tone characteristic of nerves or fiber bundles lacking it. The myelin is, it seems, produced by the individual neurolemmal sheath cells bounding the fiber; and the boundary between successive sheath cells is marked by a constriction, a *node of Ranvier*. Here the myelin sheath is interrupted, and the neurolemma dips inward to the axis cylinder.

If nerve fibers are cut, the part distal to the cut degenerates, and the proximal region may show evidence of damage. But frequently the fiber regenerates, growing out again from the stump still connected with the cell body. It appears that it is aided in seeking out again its former path by the persistence of the sheath cells which surrounded the former axis cylinder. Fiber degeneration is a feature of great value in neurological work. Particularly in the central nervous system it is difficult to trace groups of long fibers through from origins to termini. However, a group of problematical fibers may be cut and allowed to degenerate. If the material is stained in appropriate fashion and sectioned, microscopic study enables one to distinguish the degenerating fibers from their normal neighbors and to trace them to their destination.

The Nervous Impulse. Investigation of the nature and properties of the impulses transmitted by nerve fibers has been the object of a vast amount of physiological work. There is, as yet, no agreement as to the general nature of the nerve impulse. By analogy one tends to compare nerve transmission with an electrical impulse, and it can be demonstrated that as an impulse travels along a fiber there is a change of the electric potential at the fiber surface: a "ripple of leakiness" in the surface membrane. Chemical changes occur in the axis cylinder, and some argue that the electrical phenomena are merely incidental to rapid metabolic processes analogous to those in muscle contraction. The passage of a nerve impulse may on this hypothesis be crudely compared to the burning of a powder fuse, with the notable difference, however, that the "burnt" material is quickly restored, so that a second impulse may follow after a short interval.

A nerve impulse, though rapid, cannot be compared for speed with an electric impulse. On the whole, mammalian fibers give speedier transmission than do those of lower vertebrates—about twice the speed of frog fibers, for example. But even in mammals the highest recorded speeds are at the rate of only 100 yards per second, and some mammal fibers have a speed of only about a foot a second. The speed of transmission along some fibers is not inconsiderable; nevertheless, there is a distinct time lag between stimulation and response, and there are potential difficulties in coordination in such a large animal as an elephant, for example, because of the time necessarily consumed by the stimulus in even the simplest reflex in travelling many yards of fiber length.

A nerve impulse is anonymous and nonspecific. The nature of a sensation "felt" in the brain depends upon the centers which received it, not on any difference in the type of impulse received. Could the "wiring" be changed, nerve impulses from the nose, for example, if received in an auditory center, would give a hearing effect, and vice versa; and we are all familiar with the fact that a blow on the eye causes one to "see stars."

Nerve impulses normally travel in only one direction along the fibers. But the fiber is equally capable of transmitting impulses in either direction, as can be shown experimentally. The unidirectional transmission actually found in the working nervous system is due to the pattern of fiber connections; neurons are "polarized."

A nerve impulse is an "all or none" phenomenon. There are no "strong" or "weak" impulses; either the fiber reacts fully, or it does not react at all. True, there may be variations in the strength of the nervous impulses sent along a nerve, but these variations are due to two factors of other sorts. One is that there may be differences in the number of fibers in a

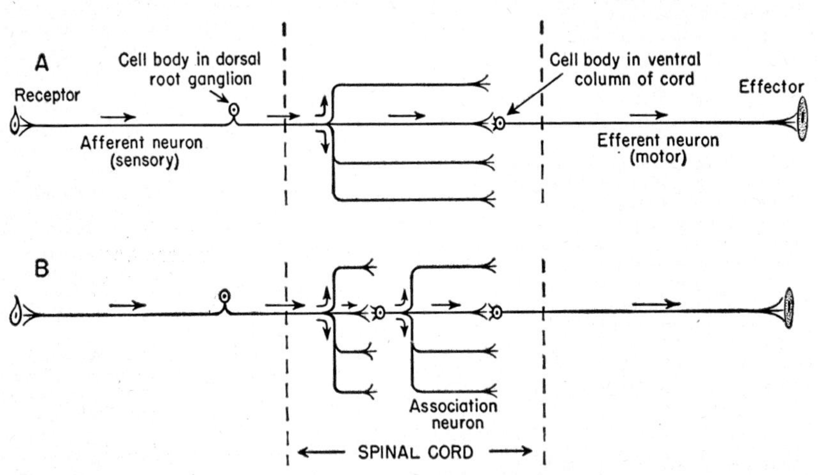

Fig. 324. Diagrams to show simple reflexes. Area between broken lines is part of arc lying within the cord (cf. Fig. 325). *A*, Two-neuron reflex; *B*, association neuron interpolated, increasing the number of possible paths.

nerve which are stimulated at one time or another. A second is that impulses do not ordinarily come singly, but follow one another in a rapid series, so that their effect on a muscle, for example, may be cumulative.

The Synapse. Never, among vertebrates, does a single neuron span the entire distance between the sensory organ (receptor) and the motor or glandular structure stimulated (effector). The action takes place through a chain of neurons, always two and usually more. The point of transfer between successive neurons is termed a *synapse*. When studied under the microscope, the end of the axon of the first of the two neurons is seen to break up into fine fibrils which intertwine with the dendrites of the second neuron or else twine about the cell body of that element. Many once believed that there is an actual protoplasmic connection between the neurons at the synapse; currently, however, it is generally thought that no such direct continuity exists, although the processes may be closely appressed. The timing of nerve transmission shows that a distinct interval is taken up by the bridging of synapses, although the time involved is only a tiny fraction of a second. According to one theory the time lag is due to an electrical situation, comparable to the building up of enough current for a "spark" to bridge the gap. Another belief, however, is that the phenomenon is a chemical one: that the tips of the distal fibrils of a neuron give off a tiny amount of a chemical which crosses the synaptic gap and stimulates the next fiber. Such a liquid is termed a *neurohumor*.

If synaptic stimulation between neurons is chemical, it may be reasonably concluded that the stimulation of nerve fibers by receptor organs is, in general, of a chemical nature and that (at the other end of the chain) the action of the last neuron on the effector is due to the giving off of a neurohumor. This last seems to be definitely proved in certain cases; we

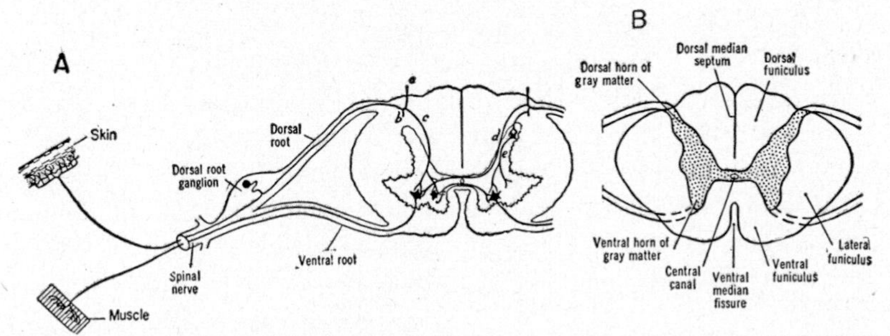

Fig. 325. *A*, Diagram of mammalian spinal cord and nerve to show path of reflex arc. A sensory fiber entering by the dorsal root may send branches (*a*, *b*) up and down the cord. At various levels the sensory fiber may connect with motor neurons of the same side (*c*) or opposite side of the cord (*d*) or with association neurons (*e*, *f*). *B*, Diagrammatic section of a mammalian spinal cord, to show distribution of white and gray matter and funiculi. (After Gardner.)

have noted the effect of such neurohumors on chromatophores in the skin. At least two chemical compounds appear to be given off by nerve endings, one a substance similar to adrenalin; the second, acetylcholine.

The Reflex Arc. Before considering anatomical structure in the system, we may note the general nature of a simple type of nerve action of the sort known as a *reflex* (Figs. 324, 325, *A*). Reflex actions are those of such type as the "automatic" withdrawal of a bare foot that has trod on a tack, or of a finger that has touched a hot stove. In general a sensory stimulus is picked up from receptor cells and carried in toward the central nervous system by a long *afferent* nerve fiber. The cell body of the *sensory neuron* to which this fiber belongs lies, along with many others, in a ganglion close to cord or brain; a fiber, however, continues on from the cell body into the central nervous system. Here, normally, it does not connect simply with one successive neuron, but branches so that a stimulus may be caused in whole series of neurons, with which the sensory neuron synapses. Conversely, each neuron with which it connects may receive impulses from numerous afferent fibers, so that a considerable amount of interplay between receptors and effectors may take place.

In the simplest of reflexes the neurons stimulated may be *efferent*—usually *motor neurons*. Their cell bodies lie within the central nervous system, and long axons run out from them to the effector end organs (usually muscle fibers). But even a simple reflex is usually one stage more complicated than this, and is a three-neuron rather than a two-neuron chain. In this the afferent fibers entering the central nervous system do not synapse directly with motor cells, but with neurons contained entirely within the central nervous system. These *association neurons,* as they are frequently termed, send out, like the afferents, branched processes which synapse with numerous motor cells. This further multiplies the number of possible responses which a sensory impulse may cause and, conversely, also increases the number of sensory impulses which might produce any given motor effect. The interposition of a single association neuron in the system is as far as any normal spinal reflex usually goes in complexity, but gives to a considerable degree a clue to the way in which more complicated brain mechanisms may have been evolved. By the intercalation of still further association neurons one can comprehend the building up of higher association centers, into which a wide variety of sensory impressions may drain, and from which may come a wide variety of responses.

SPINAL NERVES

As contrasted with the central nervous system, the peripheral nervous system, which we shall first consider, is simply constructed. It consists essentially of the nerves which penetrate to almost every region of the body: groups of fibers along which the sensory "raw material" of nervous action is brought into the spinal cord and brain by afferent pathways, and out through which resulting efferent impulses pass, as the result of central actions, to affect muscular or glandular structures. Included, too, in the peripheral system are the ganglia found along the course of the nerves and containing the cell bodies of peripheral neurons. We shall treat of spinal nerves, relatively simple and uniform in structure, before considering the more specialized nerves of the cranial region.

The typical spinal nerves of most vertebrates (Figs. 325, *A*, 326) are paired structures present in every body segment. Each nerve arises from the spinal cord by two roots, dorsal and ventral (posterior and anterior in the specialized human terminology). The *ventral root* runs straight outward from the ventral margin of the side wall of the cord; the *dorsal root,* which bears a prominent ganglion, enters the cord on its side wall. In most vertebrates the ventral and dorsal roots are united to form the main trunk of the nerve leaving the canal of the vertebral column. Emerging from the vertebral column, the nerve divides into a variety of branches, or rami. Neglecting for the time a ramus running toward the viscera, we may note the frequent presence of a major *dorsal ramus* supplying fibers to the dorsal part of the axial musculature and the skin of the back, and a *ventral ramus* to the more lateral and ventral parts of the skin and musculature of the body wall.

The main spinal nerve trunk and its principal branches carry both afferent and efferent fibers. The two roots, however, show a sharp division of functions. The ventral roots carry only outward bound impulses through efferent fibers, whose cell bodies are contained within the cord (Fig. 327). In all typical spinal nerves of higher vertebrate groups the dorsal root, in contrast, mainly carries afferent fibers with incoming impulses. The cell bodies of these neurons lie in the dorsal root ganglion; they are unipolar in structure, sending out one stem which, within the

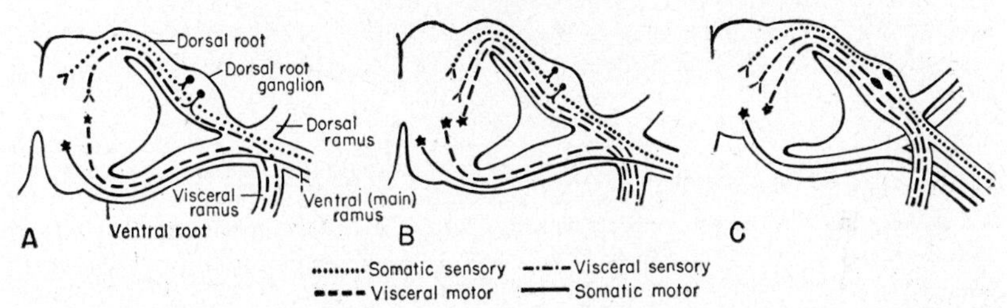

Fig. 326. Diagram to show the distribution of nerve components in dorsal and ventral roots of spinal nerves. *A,* Typical mammalian situation: sensory fibers in dorsal root, motor fibers exclusively in ventral. *B,* More primitive type, particularly common in lower vertebrates: visceral motor fibers present in dorsal root. *C,* The lamprey condition, presumably most primitive for vertebrates: separate dorsal and ventral nerves, with visceral motor elements in dorsal root (cf. cranial nerves). (The dorsal ramus, left blank in the diagrams, may include the same components as the ventral ramus; visceral fibers are present to some degree in the dorsal and ventral rami in higher vertebrates.)

limits of the ganglion, branches in a T-fashion; one process brings impulses inward from sensory structures, a second carries them centrally into the cord.

Spinal Nerve Plexuses.* In many typical segments of the body each spinal nerve is a discrete structure, without connections with its neighbors fore or aft; it innervates the trunk muscles which have arisen from the myotome pertaining to this segment, and on the sensory side supplies a corresponding strip of skin, although often with some overlap with the nerves of the adjacent segments. In certain regions, however, there is an interweaving of branches of a number of successive spinal nerves to form a plexus (Fig. 328). In forms with well-developed paired limbs there is associated with each a well-developed nerve plexus—the *brachial plexus* and *lumbosacral plexus* for front and hind legs, respectively. As a result

* In Latin, singular and plural forms of the word plexus are identical in spelling; we use an anglicized plural here for the sake of clarity.

we find that the muscles of any limb region may be supplied by fibers from several spinal nerves. This condition is perhaps to be associated

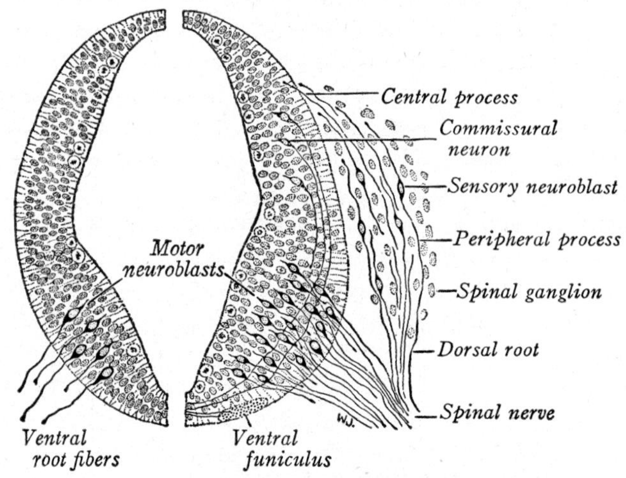

Fig. 327. Sections of the spinal cord of early mammalian embryos. *Left,* axons are growing outward from motor nerve cells. *Right,* association or commissural neurons are developing within the cord; sensory neurons are developing externally from neural crest cells (cf. Fig. 46, p. 97). These sensory neuroblasts are at this stage bipolar, i.e., with separate proximal and distal processes; later the two processes fuse proximally to give a unipolar condition to the mature ganglion cell. (From Arey.)

functionally with the development of complex limb reflexes, and structurally with the fact that the limb musculature in tetrapods has lost all trace of its original segmental nature.

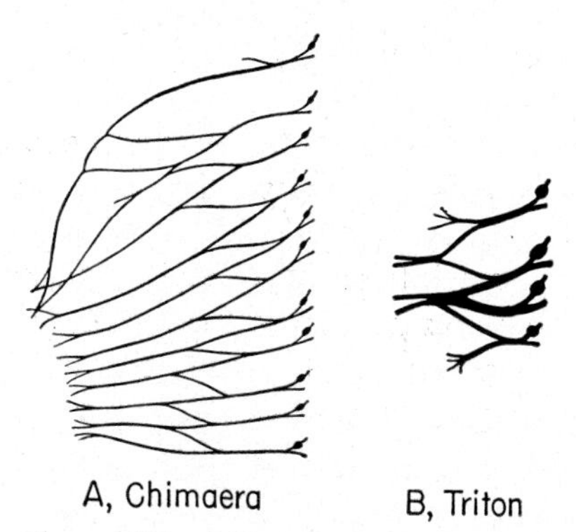

Fig. 328. *A,* Nerve plexus of the left pelvic fin of a chimaera, showing the interchange of fiber bundles between the series of spinal nerves concerned (nerve roots at right; spinal cord not shown). *B,* Similar view of the nerve plexus of the pelvic limb of a newt (distal part of the limb not shown).

Constancy of Innervation; Nerve Growth. Even in complicated limb plexuses we find that there is a high degree of constancy in the innervation of a given muscle in various animals by twigs from seemingly comparable nerve stems—a feature of practical value in the study of muscle evolution. This and other facts tended to create the doctrine that innervation is an absolutely constant feature, that a given muscle is always, in every generation and over long evolutionary lines, innervated by the same nervous elements and by similar pathways. Actually, however, there are notable (although relatively rare) cases where such a doctrine cannot be maintained. Any discussion of this problem leads to a consideration of nerve growth.

The peripheral nerves are derived embryologically from two sources. The fibers of efferent neurons emerge from cell bodies within the cord and grow out from the cord toward the muscles which they innervate. In the vertebrate ancestors the afferent fibers may have had cell bodies in the cord; this condition occurs in Amphioxus, and in cyclostomes, too, we find some cell bodies of afferent neurons in the dorsal part of the cord rather than in the dorsal root ganglia. In all higher vertebrates, however, these cells do not develop in the cord, but in the neural crest, which we have noted in an earlier chapter to have been given off at the point of closure of neural tube from skin ectoderm. Cells from the neural crest migrate downward to form the spinal ganglia and grow fibers both inward to the cord and outward to the periphery.

As can be seen experimentally, peripheral nerve fibers grow out as long processes which finally reach their end organs—sensory structures, muscle fibers, or gland cells. How do they do this? Some have assumed that there is a sort of specific, mystic affinity between a special nerve fiber and a special organ to which it attaches, so that the nerve fiber "finds its mate." That there is a degree of specific association is certain, for efferent fibers do not attach to sensory structures, nor afferent fibers to muscle cells. Experimental work demonstrates that the presence of materials to be innervated does attract nerve fibers in some broad way; but such work also furnishes good evidence against precise specific affinities. For example, a salamander limb bud, moved in the embryo to an unusual position, may become connected with nerves; but these may be nerves arising from body segments quite different from those which normally supply it. Though the manner of growth of nerve fibers is not completely understood, it seems certain that in great measure fibers tend to push out from the nerve cord along paths of least resistance in the surrounding materials. Since in successive generations the topography of a region will tend to be the same, given nerve fibers will tend to follow similar courses and innervate similar structures. Essential constancy of innervation may thus be explained without the necessity of postulating precise specificity.

Nerve Components and Spinal Nerve Composition (Fig. 326). Highly useful in the study of both peripheral and central nervous systems is the doctrine of nerve components. This points out that both afferent and efferent types of nerve connections can be divided into two types—*somatic* and *visceral*. We have, in discussing sensory structures, noted this type of subdivision. *Somatic afferent fibers* are those carrying impulses in from the skin and the sense organs of muscles and tendons—the exteroceptive and proprioceptive groups of the physiologist. *Visceral afferents,* on the other hand, carry sensations from sensory endings (interoceptive) in the gut and other internal structures, including blood vessels. On the motor side, *somatic efferents* are the motor fibers to the striated voluntary muscles of the body and limbs derived (directly or, at any rate, historically) from myotomes—the somatic musculature (cf. Chap. IX). *Visceral efferents,* on the other hand, carry stimuli primarily to the muscles of the gut (the visceral musculature) and to other smooth muscles and glands in various regions. There are thus four components present in most spinal nerves: the two afferents in the dorsal root, two efferents in the ventral. All four components are present in the trunk of a typical spinal nerve as it emerges from the vertebral column. Beyond this point, however, the larger and more characteristic trunks are composed mainly of the two types of somatic fibers; visceral fibers turn ventrally in a visceral ramus, and their further distribution will be discussed in the next section.

The root arrangement of the spinal nerve described above is usually assumed to be that present in vertebrates as a whole. As a matter of fact, however, as we descend the scale, we find increasing evidences of a different structure which gives us a clue to some of the peculiar features of the cranial nerves studied later.

In mammals both afferent components are present dorsally in every known case, and both efferent components are in general confined to the ventral root. But in some few cases, at least, among mammals, birds, and reptiles, efferent visceral fibers are carried by the dorsal root as well as the ventral, and in amphibians and jawed fishes, where investigated, visceral fibers in the dorsal root appear to be common (Fig. 326, *B*). In the higher vertebrates the two roots emerge from the column at about the same level of the cord, one above the other; but in the lower fishes there tends to be an alteration of dorsal and ventral roots, and in sharks the union of the two roots is less intimate than in higher forms. The lowest evolutionary stage appears to be that seen in the lampreys (Fig. 326, *C*). Here dorsal and ventral roots do not connect, and they are quite separate nerves. Further, as far as the evidence goes, the visceral efferent fibers are confined to the dorsal roots; these roots thus carry three components, and the ventral root is purely somatic motor in nature. Still further, dorsal and ventral roots alternate in position in lampreys.

Probably this lamprey condition is primitive; in Amphioxus, as in lampreys, dorsal and ventral roots are separate, and the union of the two may not have occurred until some stage in early fish history. That the visceral efferent fibers were primitively carried by the dorsal root and only later tended to shift to emerge by the ventral root is suggested by the series of stages cited; further, we see a retention of this seemingly primitive three-to-one split of components in various cranial nerves in all vertebrates.

In the alternating position of the separate dorsal and ventral spinal nerves of the lamprey, the ventral roots emerge opposite the middle of each body segment, and the dorsal roots are intersegmental in position (Fig. 329). This is a logical, primitive situation as regards somatic components, at least. The sole function of the primitive ventral root is the innervation of somite musculature, and the most effective position for the emergence of a group of fibers is opposite the middle of a somite. On the other hand, the main function of the sensory component is innervation of the skin; the body surface is best reached by a path between successive somites, with an intersegmental position a logical point of origin

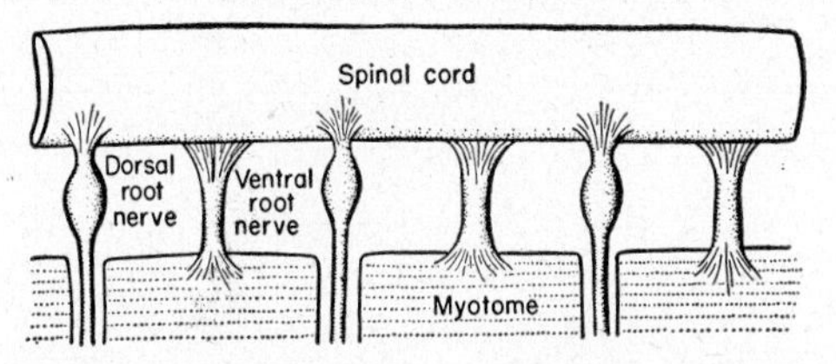

Fig. 329. Diagram of the spinal cord and nerves of the left side of a lamprey seen in dorsal view (anterior end to the left), to show the alternating arrangement of separate dorsal and ventral spinal nerves, related to intermyotomic spaces and myotomes.

of the nerve from the cord. Spinal nerves are segmentally arranged; but it would seem that this segmental arrangement is not inherent in the nervous system, but has, historically, been imposed upon it by the mesodermal somites. We shall see, in the case of the brain, a partial imposition of a second type of "segmental" nerve arrangement due to the presence of the gill structures.

VISCERAL NERVOUS SYSTEM

Apart from special nerve components associated with cranial nerves (discussed later), the visceral nerve connections, both afferent and efferent, are of a distinctive sort. As we are well aware from personal experience, visceral sensations and motor responses to them are not generally associated with the higher centers of the brain. We "know" little of what our viscera feel and have little control over them; they live their own lives essentially independent of the rest of the body.

The afferent pathways of the visceral system call for little remark. Visceral sensory endings are present in most internal organs, blood vessels, and so forth. Myelinated fibers from these ascend to the spinal cord and brain through special visceral nerves, discussed below, or through the vagus nerve of the cranial system, which extends over most of the length of the gut cavity.

The Autonomic System (Figs. 330, 331). More interesting, because more complicated in structure and function, are the outgoing fibers which innervate the smooth muscles and glands of the body, both deep and superficial. This system of efferent fibers was formerly called the sympathetic system; the term "sympathetic," however, is now commonly restricted to one subdivision of the visceral efferent,* and the efferents as

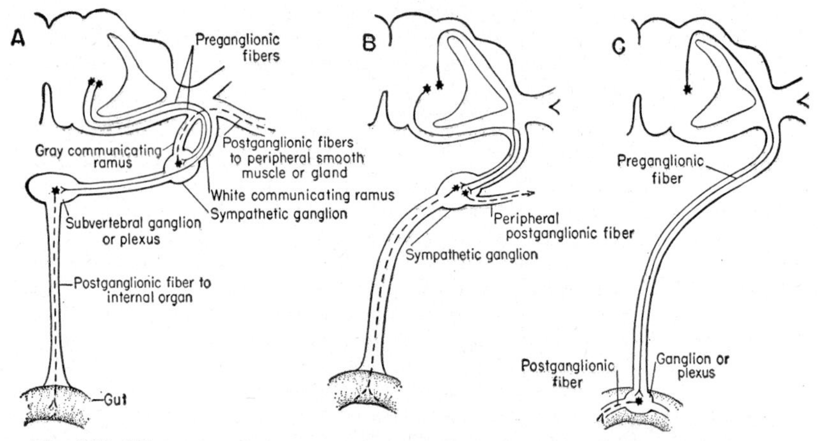

Fig. 330. Diagrammatic cross sections to show the path of autonomic fibers. *A*, Sympathetic (thoracolumbar) distribution in a mammal, with autonomic ganglia both in a lateral chain and in a subvertebral position. Preganglionic fibers may be relayed in either position and run either to superficial structures via the major nerve trunks or to the viscera, in either case with a long postganglionic neuron. *B*, Sympathetic distribution as found in many lower vertebrates; there is little development of a sympathetic chain, and no distinction of ganglia into two groups; fibers to peripheral structures course independently or with blood vessels, rather than with the major (somatic) nerve trunks. *C*, Course of parasympathetic fibers; the preganglionic fiber makes the entire run from cord or brain to a point in or near the organ concerned, where there is a relay to a short postganglionic neuron.

a whole are usually referred to as the *autonomic system,* with reference to the generally self-governing nature of the reflexes affecting smooth muscles and glands. A classification is as follows:

Visceral nervous system { Afferent; Efferent = Autonomic { Sympathetic; Parasympathetic

The fiber pattern followed by autonomic impulses from the spinal cord differs in one notable regard from that described earlier as typical for

* But there is still wide variation in usage of the term sympathetic. By various authors it may be defined (*a*) narrowly, as here; (*b*) as equivalent to autonomic; (*c*) as equivalent to the entire visceral ("vegetative") system, both afferent and efferent.

motor impulses to striated muscles (including the striated muscles of the visceral system in the head and branchial region). Somatic motor impulses are carried from cord to effector by the long axon of a single neuron. The typical visceral efferent is, in contrast, a two-neuron chain. The first

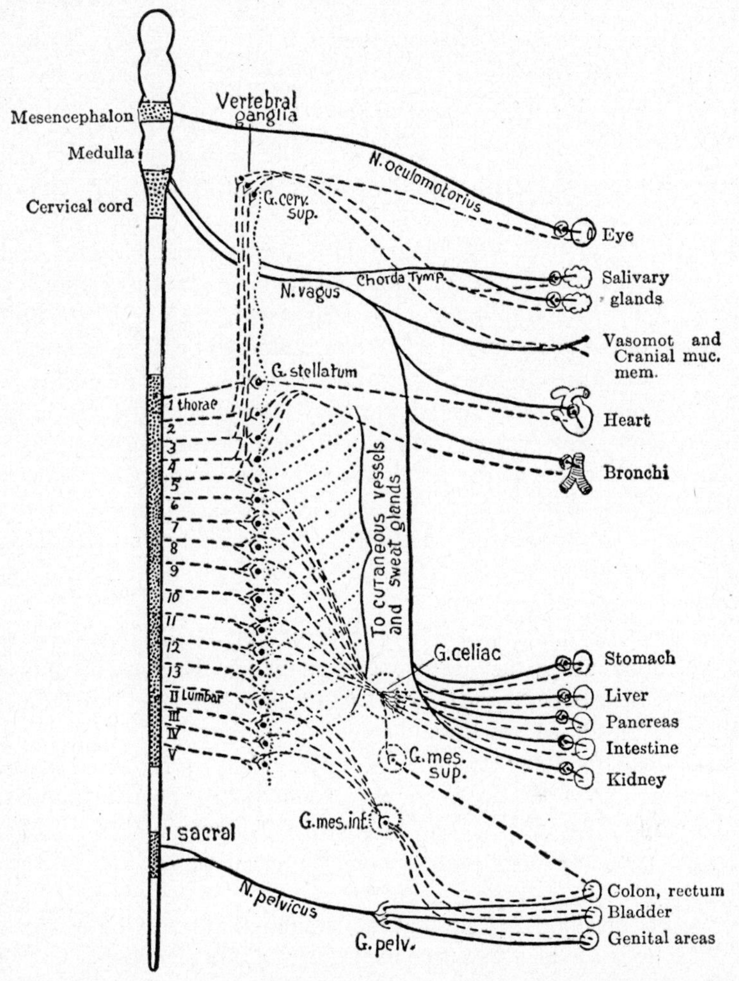

Fig. 331. Diagram of the distribution of the nerves of the autonomic system. Parasympathetics in full line; sympathetics in broken line; course of fibers to superficial structures from ganglia of sympathetic chain in dotted line. Relay of sympathetics takes place in sympathetic chain ganglia (including superior cervical) or subvertebral ganglia (celiac, superior or inferior mesenteric). (Modified from Meyer and Gottlieb, from Kuntz.)

neuron is one whose cell body lies in the cord, termed a *preganglionic neuron*. Its axon is typically myelinated. This axon, however, does not extend the full length of the pathway to the effector organ, muscular or glandular. At some point along its course it enters a ganglion of the autonomic system. Here the impulse is relayed to a second, *postganglionic neuron,* whose axon (usually lacking a myelin sheath) completes the passage to the end organ. Although this second neuron is often far removed ventrally from the region of the cord, it appears to be derived (like dorsal root ganglion cells) from neural crest cells or the cord itself; these descend in the embryo along the developing nerves.

A combination of physiological and anatomical investigations indicates that, in higher vertebrates at least, the autonomic system is sorted out into two subdivisions, termed (1) the *sympathetic* (in a narrow sense) or *thoracolumbar system* and (2) the *parasympathetic* or *craniosacral system* (Fig. 331). The two differ both functionally and topographically.

Stimulation of true sympathetic nerves tends to increase the activity of the animal, speed up heart and circulation, slow down digestive processes, and, in general, to make it fit for fight or frolic. On the other hand, the action of the parasympathetic tends to slow down activity in general, but promotes digestion and a "vegetative" phase of existence. This functional difference is associated with known differences in the way in which these systems affect the end organs. In both cases "neurohumors" are produced by the tips of the nerve fibers. In the sympathetics proper the material produced is similar to, if not identical with, *adrenalin,* a stimulating hormone produced by the medulla of the adrenal gland (see p. 546). In the case of the parasympathetic, the substance given off is the chemical known as *acetylcholine.*

Anatomically, the two autonomic types can be distinguished as to (*a*) the point along the nerve course at which the relay to the postganglionic fiber occurs, and (*b*) the point along the length of the neural tube at which the nerves leave the central nervous system. In true sympathetic fibers the relay typically takes place close to the spinal cord, and the preganglionic fibers are in consequence short, the postganglionic long. In the parasympathetic system, in contrast, the preganglionic axons are long, the postganglionics short, and the relay occurs in or close to the organ concerned.

Topographically, it is seen that in a mammal, at least, the sympathetic system proper (as its alternative name suggests) is confined at its origins to the thoracic and lumbar regions of the cord. The parasympathetic fibers emerge from the central nervous system both anteriorly and posteriorly. Anteriorly, they are associated with certain cranial neves, most particularly the important vagus nerve; posteriorly, they are present in the spinal nerves of a few segments in the sacral region. From their origins both sets of nerves spread out widely, so that glands and smooth muscle* receive, in every part of the body, a double innervation from the two antagonistic systems.

* Including heart muscle as a specialization from the smooth type.

The true sympathetic fibers in mammals leave the vertebral canal in the main trunk of the nerve of each segment. Just outside the vertebra, however, they leave this trunk and turn ventrally in a short *visceral ramus* (white ramus, ramus communicans) ending in a small *sympathetic ganglion*. In most vertebrate groups the ganglia on either side of the vertebral column are connected by strands of fibers to form a *sympathetic chain* along which preganglionic fibers may course some distance before reaching the level at which they terminate. In the sympathetic ganglia occurs the relay to postganglionic neurons in the case of impulses destined for the smooth muscles of peripheral blood vessels and for smooth muscles and glands in the skin. In mammals such an outward-bound axon rejoins the main spinal nerve trunk from the ganglion by way of an accessory communicating ramus, gray because of the absence of a myelin sheath in its fibers. In many lower vertebrates, however, the peripheral sympathetic nerves pursue an independent course, running outward along the blood vessels.

But the sympathetic fibers relayed in the ganglia of the chain are only a part of the sympathetic system. Most of those bound for the deep internal organs pass uninterruptedly through these ganglia and turn ventrally and inward. Beneath the column, sympathetic fibers from the two sides meet in an interlacing network of fibers and ganglion cells, the *subvertebral ganglia* or subvertebral plexuses (such as the celiac plexus, lying beneath the anterior lumbar vertebrae of mammals). Here these fibers are relayed to postganglionic cells whose axons extend downward, mainly via the mesenteric system, to reach the various gut organs.

The greater part of the parasympathetic distribution is by way of the vagus nerve, a major branch of which follows the gut tube for much of its length; apart from minor elements associated with other cranial nerves, the remaining parasympathetic nerves come from a few segments in the sacral region. In this system there are no proximal ganglia; long, myelinated preganglionic fibers extend the entire distance to, or nearly to, the various viscera; the relay takes place in plexuses generally embedded in the organ concerned.

Our knowledge of the autonomic system has been gained chiefly from a study of mammals. Autonomic structure in lower vertebrates is relatively poorly known, for the reason that both anatomical and physiological studies are needed to elucidate conditions in any animal, and both types of work are difficult in this system. The ganglionic chain can be followed down in phylogeny as far as the bony fishes. In lower vertebrates the chain often extends from head to tail, connecting with most or all of the cranial and spinal nerves along its entire course. In the Chondrichthyes ganglia are present in the chain region, but the chain itself is not developed. Subvertebral ganglia and plexuses are in general little developed in lower vertebrates. It appears probable that in lower groups the two autonomic components are less well sorted out than in mammals. In cyclostomes there appears to be little development of autonomic nerves in the trunk, and most of the connections between viscera and central nervous system occur, it seems, through the long vagus nerve from the brain.

Although we have emphasized the relative independence of the autonomic system, it will be noted that, as here described, every reflex, nevertheless, passes through the cord or brain stem and is thus subject to potential influence from other parts of the nervous system. It has been found, however, that in certain instances local stimulation of a part of the gut tube may produce a nervous response even if nerves to or from the central nervous system are cut. Such reflexes appear to be purely local, with both ends of the arc contained in the nerve network of the gut walls. This leads one to speculation. The main components of the nervous system are somatic and are concerned with "external affairs"—sensations received from without, and motor responses to them. Perhaps in the ancestral vertebrates internal affairs in the gut were regulated independently. The nervous system of the viscera may have been, to begin with, an independent local nerve network, unconnected with cord, brain, and somatic nerves, and the present central connections of the visceral system may be secondary. There is, however, no proof of such a hypothesis; this lies in the realm of pure speculation.

Paraganglia and the Adrenal Medulla. In various instances there are described clusters of cells here and there in the vertebrate body, particularly along the region near the dorsal aorta, which are termed *chromaffin cells* because of the readiness with which they stain with chromium salts. Embryologically, they arise from cells migrating downward along the path of the visceral nerve rami, and are thus identical in origin with the postganglionic neurons of the sympathetic system. In certain lower vertebrates, small masses of such cells are present between the kidneys along the dorsal wall of the body cavity. They are appropriately termed *paraganglia,* since they are embryologically identical with the sympathetic ganglia, which they adjoin. In tetrapods occasional cell clusters of this type persist, but most of the chromaffin material is concentrated into a compact mass of tissue which forms part of the *adrenal gland* capping the kidney. In mammals the chromaffin cells are concentrated in the center of the adrenal body forming its medulla; in lower tetrapods they are more diffuse, and interspersed with the cortical component (Fig. 332).

The product of these cells is *adrenalin* (or epinephrine), familiar for its stimulating effect on body activity—an effect almost exactly that caused by the sympathetic nervous system. The cells of the adrenal medulla do not look like nerve cells, for they lack any fiber processes. But actually they are homologous with postganglionic sympathetic neurons;

hence it is not surprising that they produce a neurohumor identical with that of their homologues. The difference is that, whereas the true postganglionic sympathetic cell produces only a tiny amount of adrenalin which affects only structures immediately adjacent, the mass of adrenal cells produces a large quantity of this material, which has a "shotgun" effect on all parts of the body when carried about by the circulatory system.

If the medullary cells of the adrenal are equivalent to the postganglionic

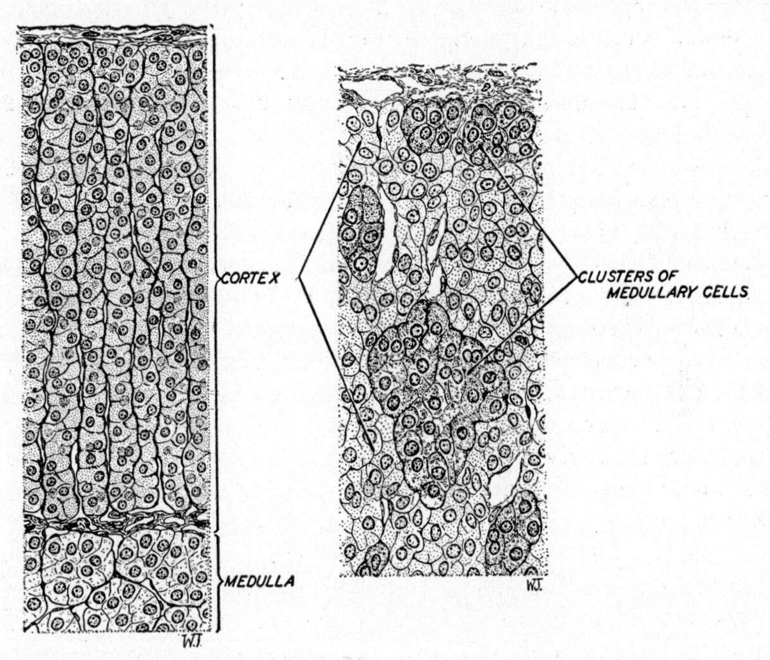

Fig. 332. A section through part of the adrenal gland (outer surface above) in a mammal (rat), with division into cortical and medullary layers, and a reptile (Heloderma), in which the two tissues are intermingled. (From Turner.)

cells of the sympathetic fiber system, one would theoretically expect that the autonomic innervation of the adrenal medulla would be by way of a single neuron rather than two. Theory and findings agree; this is indeed the case.

CRANIAL NERVES

In the head region of vertebrates there is present a special series of varied nerves which are, particularly at first sight, difficult to compare with those of the body (Figs. 333, 334). They deserve close consideration and analysis. The cranial nerves were first studied in man and were early given names and numbers based on their mammalian functions and positions. Although, as we shall see, the human arrangement does not hold throughout, we shall introduce the study of cranial nerves by listing them:*

I. *Olfactory:* sensory, from the olfactory epithelium.
II. *Optic:* sensory, from the eye.
III. *Oculomotor:* innervates four of the six eye muscles.
IV. *Trochlear:* to the superior oblique muscle of the eye (sometimes termed the trochlear muscle).
V. *Trigeminal:* a large nerve with three main branches, mainly bringing in somatic sensations from the head, with motor fibers to the jaw muscles.
VI. *Abducens:* to the posterior rectus muscle, which abducts the eye.
VII. *Facial:* partly sensory, but mainly important as supplying the muscles of the face.
VIII. *Acoustic:* sensory, from the internal ear.
IX. *Glossopharyngeal:* a small nerve, mainly sensory, and innervating (as the name implies) much of the tongue and pharynx.
X. *Vagus:* a large nerve, both sensory and motor, which (as the name suggests) does not restrict itself to the head region, but runs backward to innervate much of the viscera—heart, stomach, and so forth.
XI. *Accessory:* a nerve accessory to the vagus.
XII. *Hypoglossal:* a motor nerve to the tongue muscles.

One can memorize such a list of cranial nerves and their functions, but no student of the nervous system can stop at this point. We have here a series of nerves which are amazing in variety and seemingly haphazard in distribution, and one cannot but attempt to "make sense" out of them. Is there any logic in their distribution? Can they be grouped in any sort of natural categories?

Stimulus to attempt this is added through the fact that comparative study shows that the fixed, orthodox mammalian scheme just presented is not found in all vertebrates. Even the number of cranial nerves varies. The hypoglossal, for example, is absent as such in fishes, and the accessory is in lower vertebrates an integral part of the vagus; on the other hand, most vertebrates have anteriorly a small "terminal" nerve absent in man, and one main branch of the mammalian trigeminal appears originally as a quite distinct element.

A clue to a possible classification of cranial nerves lies in a consideration of nerve components (cf. Table III). We have noted that, in the postcranial region, nerve components of four types are present. In the head region three further categories may be distinguished. In addi-

* The initial letters of the names of the cranial nerves are the initial letters of the words of the following choice bit of poetry: "On old Olympus' towering top a Finn and German viewed a hop."

tion to the ordinary smooth visceral muscles found in the trunk, the head and throat have special striated visceral muscles connected with the gill bar system, and a *special visceral motor* category has been established

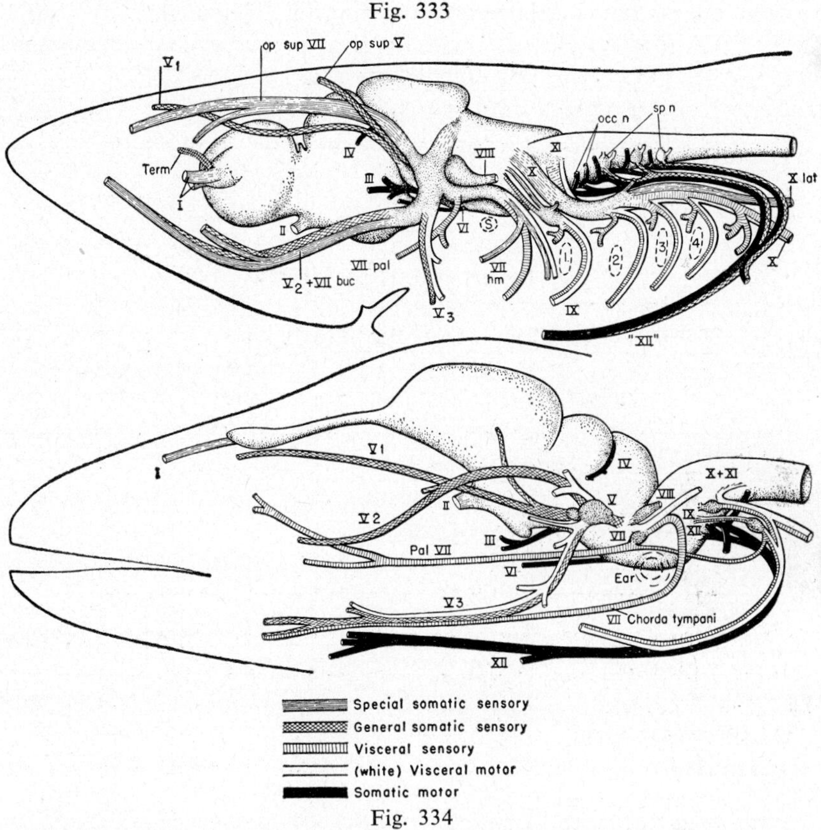

Fig. 333. Diagram of the distribution and components of the cranial nerves of a shark (Squalus). Roman numerals refer to cranial nerves. *buc,* Buccal; *hm,* hyomandibular; *lat,* lateral line trunk; *occ n,* occipital nerves; *op sup,* superficial ophthalmic; *pal,* palatine; *S,* position of spiracle; *sp n,* anterior spinal nerves; *Term,* terminal nerve; *1* to *4,* position of gill slits; "*XII*", trunk of conjoined occipital and anterior spinal nerves corresponding to amniote hypoglossal XII. (After Norris and Hughes.)

Fig. 334. Diagram of the distribution and components of the cranial nerves of a lizard (Anolis). V_1, ophthalmic (profundus) ramus of trigeminal; V_2, maxillary division; V_3, mandibular division of trigeminal; *Pal VII,* palatine ramus of facial. (Data from Willard, Watkinson.)

for nerve components supplying them. A *special visceral sensory* component is distinguished for fibers from the taste organs. Further, there are found in the head special somatic sensory structures—nose, eye, and acustico-lateralis organs; their nerves merit a *special sensory* category. We can group cranial nerves into three types which (as indicated by the double vertical rulings in the table) show essentially clean-cut distinctions as to the components present. These three types are (*a*) *special sensory nerves* of somatic type—I, II, VIII, and lateral line nerves; (*b*) *dorsal root* or *branchial nerves* containing sensory and special visceral motor components associated with the branchial region; and (*c*) *ventral*

TABLE III

Table of Nerve Components of Cranial Nerves

Nerve Types	Special Sensory	Branchial (Dorsal)					Ventral
Components	Special Somatic Sensory	General Somatic Sensory	General Visceral Sensory	Special Visceral Sensory	Special Visceral Motor	Visceral Motor (Autonomic)	Somatic Motor
0. Terminalis		X					
I. Olfactory	X						
II. Optic	X						
III. Oculomotor						(X)	X
IV. Trochlear							X
V_1. Profundus		X				(X)	
$V_{2,3}$. Trigeminal proper		X			X		
VI. Abducens							X
VII. Facial	L	(X)	X	X	X	X	
VIII. Acoustic	A						
IX. Glossopharyngeal	L	(X)	X	X	X	X	
X and XI. Vagus (and accessory)	L	X	X	X	X	X	
XII. Hypoglossal							X

Proprioceptive fibers (muscle sense) are not included. *L*, Lateralis sensory components of lower vertebrates (in X, the vagus, alone in amphibians); *A*, acoustic component of tetrapods. Components in parentheses: variable or negligible. The three areas between vertical double-ruled lines indicate the components proper to each of the three nerve types. Except for the usual presence of autonomic fibers accompanying the oculomotor nerve, the distinctions are clear-cut.

root nerves, containing almost exclusively somatic motor fibers. The first category is peculiar to the cranial region, the other two are comparable to the dorsal and ventral roots of spinal nerves of lower vertebrates, and especially to the separate dorsal and ventral spinal nerves of lampreys and Amphioxus.

Special Sensory Nerves. In all vertebrates the three main sense organs—nose, eye, and ear—are innervated by special nerves; in primitive aquatic vertebrates we find in addition nerve trunks, intimately associated with that for the ear, which innervate the lateral line organs.

Olfactory (I). As mentioned previously (p. 490), the olfactory is not a typical nerve. A normal sensory nerve has the cell bodies of its neurons located in a ganglion close to the spinal cord or brain. In this case, however, the nerve fibers are not formed by nervous system elements at all; they are, instead, formed by the cells of the sensory epithelium, and run inward to the olfactory bulb at the front of the brain.*

In mammals the olfactory is not a compact nerve; rather, it consists of a number of short fiber bundles which pass back from the olfactory organ through the cribriform plate of the ethmoid bone. In many other vertebrates, nose and brain are farther apart, and a distinctive pair of olfactory nerves is formed. In typical amphibians and reptiles and in many mammals, with a distinct vomeronasal organ (p. 493), a discrete branch of the olfactory nerve develops for the innervation of this structure.

Optic (II). The optic nerve, entering the brain at the chiasma, was discussed in connection with the eye (p. 506). It is, properly, not a nerve at all, but a brain tract, since the retina is formed as an outgrowth of the brain tube. As in the olfactory nerve, we find here that the fiber-forming neurons are situated peripherally—in the ganglion cell layer of the retina.

Acoustic (VIII). More normal in construction is the acoustic nerve, connected with the internal ear. Here the nerve fibers are formed by true ganglion cells; these are situated peripherally, in the otic capsule, close to the sensory structures. The acoustic nerve is not, of course, visible in an external dissection of the braincase region, since its fibers run directly from the capsule into the endocranial cavity, where it enters the medulla oblongata at a far dorsal position. Two main branches are present, each with a separate ganglion. The *vestibular nerve* serves the anterior parts of the system of canals and sacs; the *cochlear nerve* is so named because in mammals it arises from the auditory organ of the cochlea as well as the posterior part of the organs of equilibrium.

Lateral Line Nerves (cf. Fig. 309, p. 512). The lateral line organs of primitive water dwellers are, we have noted, intimately related to the auditory sense; hence it is reasonable to find that the nerves innervating these structures are closely associated with the acoustic nerve. In fishes a pair of stout lateral line nerves flank the acoustic nerve anteriorly and posteriorly on either side of the medulla. These nerves enter the braincase with certain of the branchial nerves and, therefore, frequently are reckoned as part of these nerves. This association, however, seems to be merely one of "convenience," and the essential independence of the lateral line nerves should be emphasized.

* Prior discussion of cranial nerves is necessary for a proper consideration of brain structure, but, conversely, some mention of prominent brain landmarks is necessary in the description of cranial nerves.

The *anterior lateral line nerve* enters the skull in company with the facial nerve and supplies the lateral line organs of much of the head region as well as those of the lower jaw; its main branches—superficial ophthalmic, buccal, hyomandibular—are often quite independent of the branches of the facial nerve. Some fibers of the *posterior lateral line nerve* may fuse with the glossopharyngeal (IX), but most or all of the posterior nerve accompanies the vagus. It innervates the neuromast organs of the occipital region, but primarily forms a stout trunk accompanying the lateral line the length of the body.

Branchial Nerves. We have noted that in the spinal nerves of most vertebrates, dorsal and ventral roots unite to form a single trunk, but cited evidence suggesting that primitively (as in Amphioxus and the lamprey today) the two were quite distinct. The dorsal roots contain all sensory components and in addition appear to have carried originally part, and perhaps all, of the visceral motor fibers as well. If Table III is in-

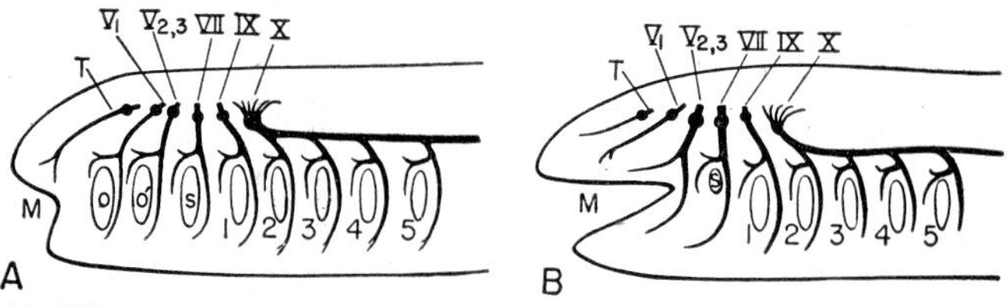

Fig. 335. Diagrams showing the distribution of branchial (dorsal root) cranial nerves. *A*, Hypothetical primitive condition, with typical nerves to each of two anterior gill slits lost in jawed vertebrates, and a terminal nerve to anterior end of head. *B*, Condition in jaw-bearing fishes. *M*, Mouth; *O, O′*, anterior gill slits lost in gnathostomes; *s*, spiracular slit; *T*, terminal nerve; *1* to *5*, typical gill slits of gnathostomes.

spected, it will be seen that a large series of cranial nerves appear to belong to this category of dorsal root nerves, all containing sensory fibers and in most (but not all) cases visceral motor components as well. Such are the terminal nerve, profundus nerve, trigeminus proper, facial, glossopharyngeal and vagus (Fig. 335).

These dorsal cranial nerves all, however, differ in various regards from dorsal root nerves of the trunk. Most of the differences are obviously due to their close association with the gill system, prominently developed in this region of the body in lower vertebrates. The presence of the gill slits tends to bring about a serial pattern of distribution of these nerves, which is enhanced by the serial pattern arrangement of the striated branchial muscles which they innervate.

Most typical of these dorsal root or branchial nerves is perhaps the glossopharyngeal in fishes (Fig. 336), which is associated with the first typical gill slit. Like all members of this series, it arises from the upper part of the medulla oblongata and bears a prominent ganglion on its sensory root. A major trunk, containing both sensory and motor fibers, runs down behind this slit as a *post-trematic ramus*. A smaller, *pretrematic ramus* runs down in front of the gill slit, and a *pharyngeal ramus* turns inward and forward beneath the lining of the pharynx. These two branches contain visceral sensory fibers only. A small *dorsal ramus*, often absent, may bring in somatic sensations from the skin.

Posterior to the first, each of the gill slits has a similar pattern of nerve branches; all, however, unite to connect with the brain via a single large nerve, the vagus. Anteriorly, the development of the jaw apparatus has greatly modified in gnathostomes the branchial nerve pattern which may have been present in the ancestral jawless vertebrates. The facial nerve, associated with the spiracular slit, is not too atypical, but has reduced or lost its somatic sensory component. Jaw development, it is thought, may have eliminated two more anterior gill slits. The trigeminal nerve complex appears to have been originally associated with this lost gill region, and the little terminal nerve may be a vestigial anterior member of this series.

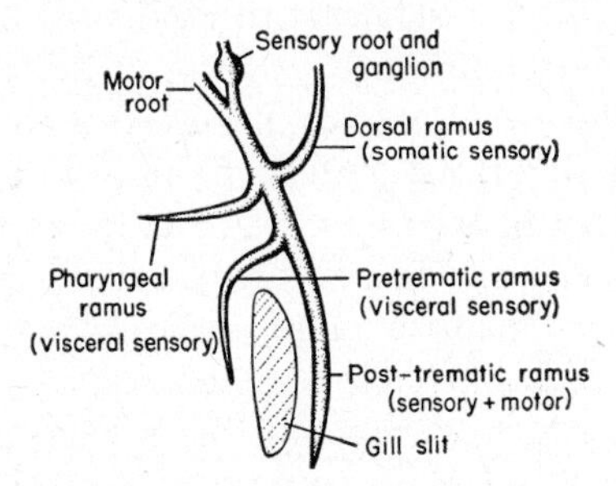

Fig. 336. Diagram showing the composition of a typical branchial nerve.

These dorsal root nerves may be described in anteroposterior sequence.

Terminal Nerve. In many vertebrates, representing every class except cyclostomes and birds, there is found a tiny nerve which parallels the olfactory in its course inward from the nasal epithelium; its fibers run back to the diencephalon. Little is known of its nature; it is apparently sensory in function, but has nothing to do with olfactory reception. Quite possibly it is the last remnant of a most anterior member of the branchial series which innervated the primitive small mouth area.

Profundus Nerve (V_1). The *ophthalmicus profundus* nerve* is a stout trunk which runs forward through the deeper part of the orbit and functions as a somatic afferent, receiving sensations from the skin of the "snout" region. In mammals this nerve is intimately associated with the trigeminus and counted as one of its three main trunks. In lower vertebrates, however, it has frequently a separate ganglion and may emerge independent of the trigeminus from the brain. In consequence, it is gen-

* In mammals termed simply the ophthalmic; in most vertebrates, however, it is termed the deep ophthalmic to distinguish it from a more superficial nerve in the same general region.

erally held that it was originally a separate nerve of the branchial series, associated with an anterior gill slit which has been lost with the development of the jaws and expansion of the mouth region.

Trigeminal Nerve (V_2, V_3). The trigeminal nerve proper is believed to have been originally associated, like its profundus component, with an anterior gill slit lost with the expansion of the mouth gape. This nerve has not, however, been reduced; it has become the main nerve of the mouth region in all jawed vertebrates. Two main branches, somewhat comparable to pretrematic and post-trematic rami of a typical branchial nerve, are always present in the trigeminus proper: a *maxillary ramus* (V_2) to the upper jaw region, and a *mandibular ramus* (V_3) to the lower jaw. Both branches include somatic sensory elements which supply much of the surface of the head and mouth cavity. Fibers innervating the jaw muscles are associated with V_3. (The trigeminal has no visceral sensory component.)

Facial Nerve (VII). This is the nerve proper to the spiracular gill slit. Its main, post-trematic trunk in fishes is the *hyomandibular nerve,* which descends behind the spiracle alongside the hyomandibular cartilage or bone. This supplies the branchial muscles proper to the hyoid arch and contains visceral sensory fibers as well. An anterior branch is purely visceral sensory in nature, and includes a *palatine ramus* which corresponds to the pharyngeal rami of more posterior nerves. As we have noted, the anterior lateral line nerve is associated with the facial. The visceral sensory fibers innervate much of the mouth, including part of the taste bud series.

In tetrapods the nerve preserves its various branches in modified form. The main ventral trunk, after looping through the middle ear region, turns forward as the *chorda tympani* to give a sensory innervation to taste buds and lower jaw. In mammals, we have noted, the musculature of the hyoid arch has become expanded over the surface of the head as the facial muscles, and branches of the facial nerve spread over skull and face to innervate them.

Glossopharyngeal Nerve (IX). As noted, this small nerve is associated in fishes with the first typical gill slit; it is never of any great size or importance. A somatic sensory element is absent except in certain fishes and amphibians. A small amount of striated visceral musculature of the throat and larynx and part of the salivary glands are innervated by this nerve, and it carries visceral sensations from the back part of the mouth and tongue and part of the taste buds.

Vagus Nerve (X, XI). The vagus is the largest and most versatile of all cranial nerves. The *accessory nerve* of mammals is essentially a part of the vagus which continues its area of origin back from the medulla into the cervical region of the spinal cord, and in lower groups appears to be an integral part of the vagus. In fishes the posterior lateral line nerve emerges from the braincase with the vagus. Usually there is a minor skin sensory component from the neck region; for the most part, however, the vagus is a visceral nerve. The vagus nerve supplies, in fishes, a typical series of branches to each of the gills behind the first, innervating the pharynx and the striated gill muscles. This branchial part of the vagus survives, in modified form, in the tetrapod pharynx and its musculature. In every vertebrate the main trunk of the vagus runs on backward into the body as an *abdominal ramus* which extends in all forms to the stomach and heart, to the lungs, when present, and in some cases may run the entire length of the gut. This nerve carries visceral sensations to the brain and (as noted previously) is the main route of the fibers of the parasympathetic system.

Somatic Motor Nerves. The nerves in this category—III, IV, VI, and XII of the human series—are highly comparable in most regards to ventral roots of spinal nerves. They are almost entirely composed of somatic motor fibers which emerge ventrally from the brain stem. Their arrangement is closely correlated with the distribution of the myotomes in the head region. As we have seen earlier (p. 264; Fig. 153, p. 255), the myotome succession seen in the trunk is interrupted anteriorly in the otic region; farther forward, three further somites form the muscles of the eyeball. Nerves III, IV, and VI innervate the eye muscles formed by these somites (Fig. 156, p. 264); nerve XII, when present, supplies musculature arising from somites formed in the occipital region.

Since both the branchial and the somatic motor nerves of the brain show a serial arrangement, it is but natural that students of the cranial nerves have attempted to arrange them, theoretically, in pairs corresponding to dorsal and ventral roots in trunk segments. Such attempts, however, have been none too successful; and on consideration it is clear that no such arrangement need ever have been present. The serial arrangement of the somatic motor nerves is a true segmental one, corresponding to the segmental arrangement of the somites which they innervate. The linear arrangement of the dorsal, branchial nerves, on the other hand, is based on the arrangement of the gill slits, which develop in the embryo quite independently of the somites.

Oculomotor (III). The greater part of this nerve, which arises ventrally from the midbrain, consists of somatic motor fibers innervating the four eye muscles developed from the first cranial somite—the superior, anterior, and inferior rectus, and the inferior oblique. (An exception is the innervation in cyclostomes of the inferior rectus muscle by the abducens nerve). Autonomic fibers, partly derived from nerve V_1 and hardly to be considered an integral part of the oculomotor, accompany the nerve to the eyeball and enter it to innervate the smooth muscle fiber of the ciliary body and iris; they are associated with accommodation and pupil reflexes.

Trochlearis (IV). The trochlearis is a small nerve with a most unusual course. It arises ventrally within the midbrain on either side; but instead of emerging directly outward, its fibers turn upward within the substance of the brain and cross to emerge dorsally on the opposite side. It innervates solely the superior oblique muscle of the eyeball, derived from the second head somite.

Abducens (VI). This small nerve emerges ventrally from the anterior end of the oblongata and supplies the posterior rectus muscle, derived from the third head somite. A muscle retracting the eye is developed in many forms from the same somite and is also innervated by the abducens.

Hypoglossal (XII). In fishes (particularly among the sharks) the posterior end of the skull—and hence of the cranial nerve series—is not a fixed point. The number of potential vertebrae which are incorporated in the occipital region of the skull is variable; hence we find a variable number of would-be spinal nerves actually emerging from the occiput. Such *occipital nerves* tend to lose their dorsal roots and are mainly motor nerves supplying muscles formed from the somites of the occipital region. In amniotes the condition is stabilized, and back of the vagus-accessory region there is a final cranial nerve, the hypoglossal, usually formed from three roots which presumably represent three body segments fused into the occiput. There is no hypoglossal nerve in modern amphibians. But fossil amphibians and crossopterygians show a foramen for it in the skull; it would seem that a definitive hypoglossal was formed in the ancestors of the tetrapods and that its absence in frogs and salamanders today is one of the many degenerate characters of these forms.

The myotomes in the region close to the back of the skull migrate in the embryo backward and downward around the gill region to form the hypobranchial muscles in fishes and the tongue musculature of higher vertebrates (Fig. 153, p. 255). Trunks of the occipital nerves of fishes follow this migration route to innervate the hypobranchial muscles, and the hypoglossal nerve of amniotes follows a similar path, back, down, and then forward around the pharyngeal region to innervate the tongue musculature.

CENTRAL NERVOUS SYSTEM—ACCESSORY ELEMENTS

Besides the functionally important neurons, other types of cells are developed as part of the nervous system. We have already noted the neurolemmal cells sheathing nerve fibers; within the central nervous system are various other cell types termed, in a broad sense, the *neuroglia*. In most body organs supporting functions are performed by connective tissue. This is not present in the central nervous system, and the neuroglial cells take its place as protective and supporting structures.

The neuroglial elements arise embryologically as a part of the true nervous tissues, from which they become differentiated during development. They are found throughout the brain and spinal cord as relatively small but numerous elements scattered among the neurons. Typically, there is a rather small cell body from which radiate numerous fine processes, giving the neuroglia a starlike appearance. One type, the *ependymal cells,* retain the epithelial character true of all the tissue of the embryonic nerve tube, and persist in a layer of epithelium (the ependyma), frequently ciliated, lining the cavities of the brain and spinal cord. Outside the central nervous system neuroglial structures in a broad sense include not only the cells of the neurolemma, but also supporting cells present in nerve ganglia.

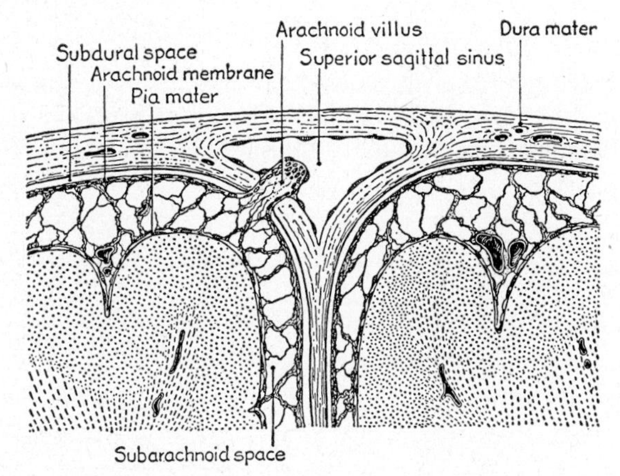

Fig. 337. Section of a portion of the brain of a mammal to show the meninges. The section is taken through the partition (falx cerebri) between the two cerebral hemispheres. The superior sagittal sinus is a prominent venous channel following a longitudinal course backward between the hemispheres; sections of smaller blood vessels are seen in meninges and brain. The subarachnoid space is occupied by cerebrospinal fluid; the arachnoidal villi offer a minor means of transfusion of material between this fluid and the blood. (After Weed.)

The brain and spinal cord, enclosed within the braincase and the neural arches of the vertebrae, are further protected and sustained by one or more wrappings, or *meninges,* of mesenchymal origin (Fig. 337). In most fishes there is but a single meninx of compact connective tissue, external to which a loose mucous or fatty connective tissue lies between cord or brain and neural arches or braincase. In all tetrapods two meninges are present. The outer, the *dura mater,* is a stout sheath, connected by numerous slender ligaments with an inner membrane applied to the brain or cord. In mammals there is a separation of the softer, inner wrapping into two delicate structures, an outer *arachnoid* and an inner, vascular *pia mater,* the two separated by a fluid-filled *subarachnoid space,* which is crossed by a cobweb of delicate tissue threads.

The vertebrate nervous system is basically a hollow tube, and this tubular arrangement persists in the adult; within the brain are the ventricles and in the spinal cord a central canal. Flowing through these cavities, and also found between arachnoid and pia mater in the brain wrappings, is a clear liquid, the *cerebrospinal fluid*. This is similar in composition to the interstitial fluid or the perilymphatic liquid of the ear. Food or other materials reach the cerebrospinal fluid from the blood through the medium of special vascular structures, the choroid plexuses of the brain (p. 564). These are more impervious to large molecules or other structures of any magnitude than are ordinary capillary walls, and thus serve as a barrier against the introduction of foreign materials into the brain tissues.

SPINAL CORD

The spinal cord (Figs. 325, 327, pp. 535, 538; 338, *A*), extending the length of the body, is a little-modified adult representative of the nerve tube formed in the early embryo. It still contains, as did that of the embryo,

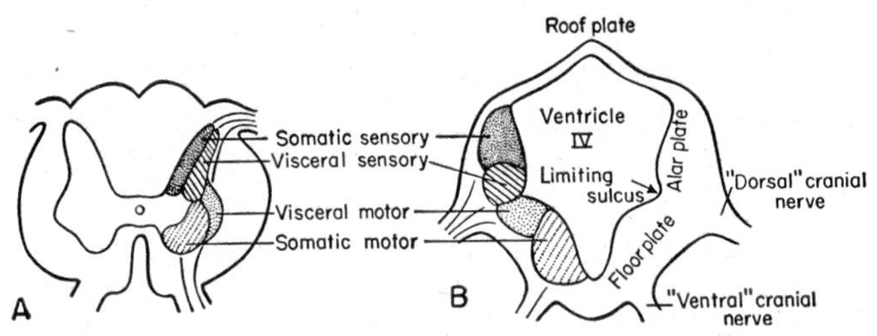

Fig. 338. Diagrams showing the distribution of the sensory and motor columns in the spinal cord, *A*, and the medulla oblongata, *B*. The medulla is that of an embryo; the "white matter" fiber tracts external to the gray are much more massive in the adult. The plate of tissue lying below the limiting sulcus is termed the floor plate; from this, motor centers arise. The sensory region above is the "wing" or alar plate. (Partly after Herrick.)

a centrally situated, fluid-filled canal; this, however, has become relatively tiny in diameter, owing to the great growth of the walls of nervous tissue surrounding it.

The spinal cord is subcircular or oval in section in lower vertebrates; in higher groups, however, its walls have tended to expand in bilateral fashion, and pronounced grooves may be present in the midline dorsally and ventrally. Two layers of material can be readily distinguished, a central area of *gray matter* surrounding the canal of the cord, and a peripheral region of *white matter*. The former consists mainly of cell bodies, the latter of countless myelinated fibers coursing up and down the cord.

Primitively, the gray substance was, it seems, arranged in a fairly even fashion about the central cavity; in most vertebrates, however, there is a bilaterally symmetrical arrangement, as seen in section, into an H-shape, or that of a butterfly's wings. Such a pattern gives the appearance of a pair of "horns" on either side, dorsally and ventrally. Actually, of course, each "horn" is merely a section of a longitudinal structure, and we should speak rather of a dorsal column and a ventral column.

The *ventral column* is the seat of the cell bodies of the efferent or motor neurons of the spinal nerves, their axons passing out through successive ventral nerve roots. The number of neurons in any given part of the cord will, naturally, vary with the volume of musculature at that level of the body, and in land vertebrates the ventral column (indeed, the entire cord) may be much expanded in the regions supplying the limbs. Most of the motor supply is to somatic muscles; but visceral efferents are present over much of the length of the cord. The cell bodies of these latter neurons are situated above and lateral to those of the somatic motor type, and sometimes are distinguishable as a *ventrolateral column*.

The *dorsal column* of the gray matter is associated with the dorsal, sensory root of the spinal nerves; it is the seat of the cell bodies of association neurons through which impulses brought in from sense organs are relayed and distributed. These neurons send out processes which may ascend and descend the cord to connect with motor neurons of the same side, may cross to connect with motor neurons of the opposite side of the cord, or, still further, may ascend the cord to the brain. The arrangement of various clusters of these sensory association cells in the dorsal column is complex and variable, but in some cases (particularly in certain embryos) it appears that we can distinguish a larger series associated with somatic sensory reception, situated dorsally and medially, and a smaller, visceral sensory group situated more ventrally and laterally. There thus appear to be in the gray matter four areas on either side related to the four nerve components, the four being in sequence, from dorsal to ventral: somatic sensory, visceral sensory, visceral motor, and somatic motor. It is of interest that the same arrangement is found in the gray matter of the brain stem (Fig. 338, *B*).

The *white matter* is composed, as has been said, of innumerable myelinated fibers. These include ascending and descending fibers of sensory nerve cells which enter the cord through the dorsal nerve roots, and of similar fibers taking origin from association cells. Present as well are fibers which carry sensory stimuli forward to the brain, and fibers returning from brain centers to act on motor neurons. Fibers of these latter categories are especially abundant in higher vertebrate groups, in which the trunk loses the semi-autonomous nature which it has in fishes and comes more directly under the influence of the brain. Topographically, the white matter is more or less subdivided by the dorsal and ventral "horns" into major areas: dorsal, lateral, and ventral *funiculi*. Anatomical and physiological investigation enables one to distinguish in these areas *fiber tracts* with the varied types of connections noted above. These tracts, however, vary greatly in nature and position from form to form and cannot be described here in detail.

THE BRAIN

In all chordates, as well as in the more highly organized invertebrates, we find a concentration of nervous tissues at the anterior end of the body in the form of a brain or brainlike structure of some sort. Such a concentration is to be expected. In an actively moving, bilaterally symmetrical animal this region is that which first makes contact with environmental

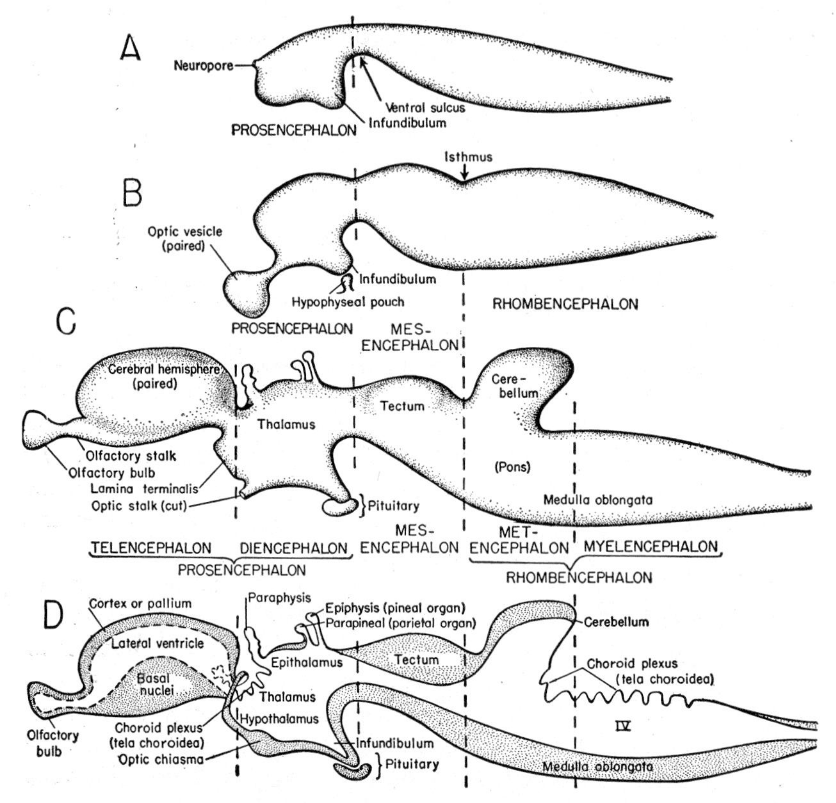

Fig. 339. Diagrams to show the development of the principal brain divisions and structures. *A*, Only prosencephalon (primitive forebrain) distinct from remainder of neural tube. *B*, Three main divisions established. *C*, More mature stage in lateral view. *D*, The same in median section. (Partly after Bütschli.)

situations to which response must be made and, in consequence, is that in which main sensory structures come to be situated.

Primitively, we may believe, the vertebrate brain was merely a modestly developed anterior region of the neural tube in which, in addition to facilities for local reflexes to the head and throat region, special sensory stimuli were assembled and "referred for action" to the semi-autonomous body region via the spinal cord. Within the vertebrates, however, there has occurred a strong trend for the concentration in the brain of command over bodily functions (except for the simplest reflexes), with the development of a series of complex, intercommunicating brain centers. We have noted, in discussing the elementary composition of the nervous system, the way in which the intercalation of an association neuron into the simple reflex arc greatly broadens the field of possible responses to a sensory stimulus and, conversely, greatly increases the variety of stimuli which may excite a specific motor response. The brain pattern is essentially an elaboration of this principle—the interposition of further series of neurons between primary areas of sensory reception and final motor paths. These intermediate neuron groups are clustered in functional centers. In such centers afferent impulses may be correlated and integrated for appropriate responses, or motor mechanisms coordinated; on still higher levels there may develop association centers of whose activity memory, learning, and consciousness may be the products.

Brain Development (Fig. 339). The general topography of the brain and its parts is best understood through a consideration of its development. The brain develops rapidly in the embryo—much more rapidly than almost any other organ—and there is early established a generalized structural pattern upon which the numerous variations seen in the adult brains of different groups are superposed. In early stages the future brain is merely an expanded area of the neural tube. With continued increase in size, its anterior end tends to fold downward. The point of flexure is marked by a *ventral sulcus*. This distinguishes a median, terminal sac-like structure, the primitive forebrain or *prosencephalon,* from the remainder of the tube. Somewhat later a second flexure, in a reverse direction, is found more posteriorly at the *isthmus*. This separates the midbrain, or *mesencephalon,* from the primitive hindbrain, the *rhombencephalon,* which becomes the *medulla oblongata* of the adult. Posteriorly, the rhombencephalon tapers gradually, without any abrupt change, into the spinal cord.

Prosencephalon, mesencephalon, and rhombencephalon are the three primary subdivisions of the brain. The three vesicles which constitute them in early stages are still recognizable in the adult, where they are termed collectively the *brain stem*—that part of the brain which is presumably phylogenetically the oldest and in which are persistently located centers for many simple but basically important reactions within the nervous system.*

At the three-vesicle stage midbrain and hindbrain regions are simple in construction, but in the primitive forebrain region special structures appear early. Most notably, the optic vesicles (whose history was dis-

* The cerebellum and the cerebral hemispheres are "new additions" not developed at this stage and not included in the brain stem. Also excluded by neurologists in defining the brain stem are certain structures, such as the pons (p. 572), lodged in the brain stem, but intimately connected with the functioning of cerebellum or hemispheres.

cussed in the last chapter) push out at either side; the optic stalks, in which the optic nerves later develop, remain attached anteriorly to the base of the primitive forebrain at the *optic chiasma*. More posteriorly, there is a down-growing median projection from the diencephalon, the *infundibulum*. Concomitantly a pocket of epithelium, the *hypophyseal pouch* (Rathke's pouch) grows upward from the roof of the embryonic mouth. In later stages modified infundibular tissues and those derived from the pouch combine to form the pituitary gland. Dorsally there grows presently from the roof of the primitive forebrain a series of median processes, the paraphysis and a median "eye stalk" (sometimes two) which are described elsewhere.

Two further great developments occur to transform the tripartite brain stem into a brain of five regions. The midbrain shows little important change except for paired dorsal swellings which form the *tectum,* but both hindbrain and forebrain regions become subdivided. In the hindbrain a dorsal outgrowth develops from the roof of the front of the rhombencephalon; this becomes the *cerebellum*. The region of the medulla oblongata beneath it is little changed in most vertebrates, but in mammals is expanded into the structure there termed the *pons*. Pons and overlying cerebellum are distinguished as the *metencephalon* from the more posterior part of the oblongata, the *myelencephalon*.

Still more important is the development anteriorly of paired outgrowths from the primitive forebrain. These are hollow pockets of brain tissue (at first somewhat analogous to the optic cups) which grow forward toward the nasal region; from them develop the *cerebral hemispheres* and, still farther anteriorly, the *olfactory bulbs*. These structures constitute the *telencephalon,* the anterior terminal segment of the brain; the unpaired part of the forebrain is the *diencephalon*. It must be noted, however, that, by definition, the most anterior tip of the median forebrain pocket, between the foramina leading to the paired vesicles of the hemispheres, is reckoned as part of the telencephalon.

The principal brain structures of the adult may be tabulated according to the divisions established in the embryo:

Prosencephalon	Telencephalon	Cerebral hemispheres, including olfactory lobes, basal nuclei (corpus striatum) and cerebral cortex (pallium); olfactory bulbs
	Diencephalon	Epithalamus; thalamus; hypothalamus; and appendages
Mesencephalon		Tectum, including optic lobes (corpora quadrigemina in mammals); tegmentum; crura cerebri (cerebral peduncles) in mammals
Rhombencephalon	Metencephalon	Pons of mammals (part of medulla oblongata); cerebellum
	Myelencephalon	Part of medulla oblongata

Ventricles (Figs. 340, 359, p. 586). The original cavity of the embryonic neural tube persists in the adult brain in the form of a series of cavities and passages filled with cerebrospinal fluid. A cavity, or *lateral*

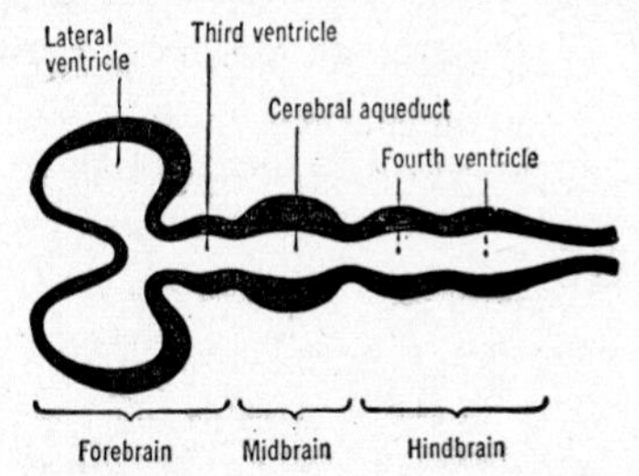

Fig. 340. Diagram showing position of brain ventricles. (From Gardner.)

ventricle, is present in each of the cerebral hemispheres; these connect posteriorly through an *interventricular foramen* (foramen of Monroe)

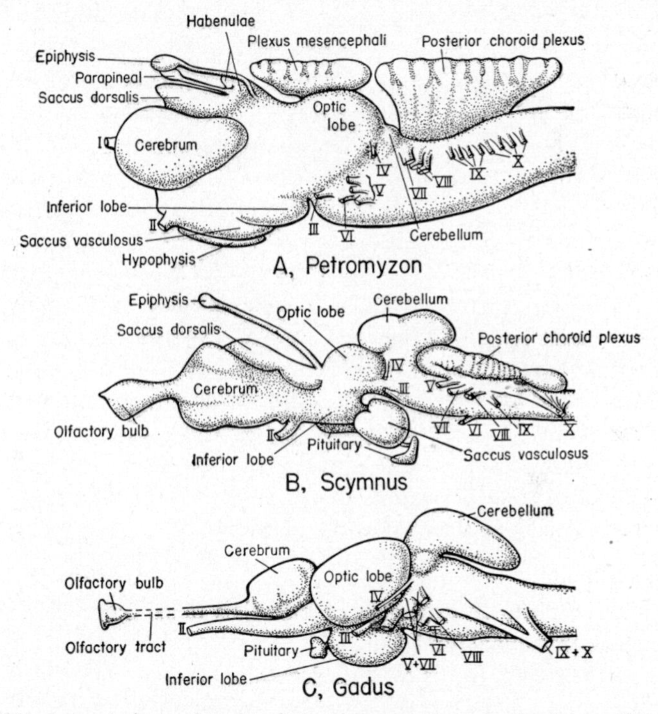

Fig. 341. Lateral views of brain of *A,* a lamprey; *B,* a shark; *C,* a codfish. In the lamprey an exceptional condition is the development of a vascular choroid area, the plexus mesencephali, on the roof of the midbrain. (After Bütschli, Ahlborn.)

with a median *third ventricle,* situated in the diencephalon. Within the midbrain there is in lower vertebrates a ventricle-like cavity, but in amniotes this becomes a narrow channel termed the *cerebral aqueduct.*

Within the medulla oblongata is a *fourth ventricle;* this tapers posteriorly into the canal of the spinal cord. Over most of the extent of the brain the ventricles are surrounded by thick walls of nervous tissue. The walls are commonly thin, however, in two roof regions, one at the junction of the hemispheres with the diencephalon, the other forming the roof of the fourth ventricle. In each of these areas there develops a *choroid plexus* (or tela choroidea)—a highly folded area of richly vascular tissue.

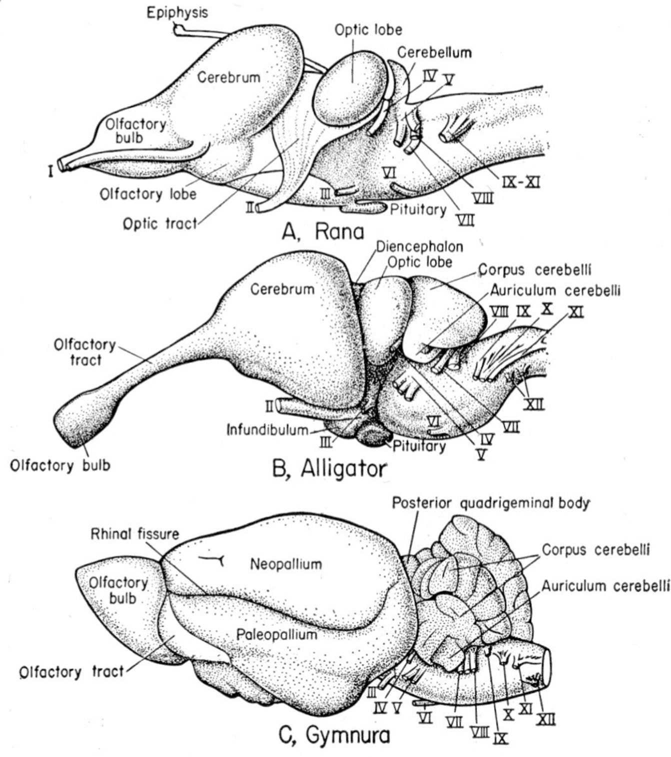

Fig. 342. Lateral views of brain of *A*, a frog; *B*, an alligator; *C*, an insectivore (representing a primitive mammalian type). (After Bütschli, Clark, Crosby, Gaupp, Wettstein.)

Through these plexuses exchange of materials takes place between the blood and cerebrospinal fluid.

Brain Architecture. In the present elementary account of the vertebrate brain our attention will be mainly centered on external features and gross structures (Figs. 341–349). But while such superficial aspects of brain anatomy are significant, an adequate understanding of the working of the brain can no more be gained from them than a knowledge of the working of a telephone system can be had from an acquaintance with the external appearance and room plan of the telephone exchange. What is of importance in a telephone system is the wiring arrangements and switchboards; in a brain it is the centers in which various types of activities occur and the tracts of fibers connecting these centers.

It may well be that the brain "wiring" was primitively much like that of the spinal cord—a general crisscross of fibers interconnecting all areas. Such types of connection still persist in the brain of all vertebrates as the *neuropil.* Generally, however, there is a strong tendency for the clustering together of nerve cells of specific functions in centers, and the assembling of fibers with like connections into definite bundles.

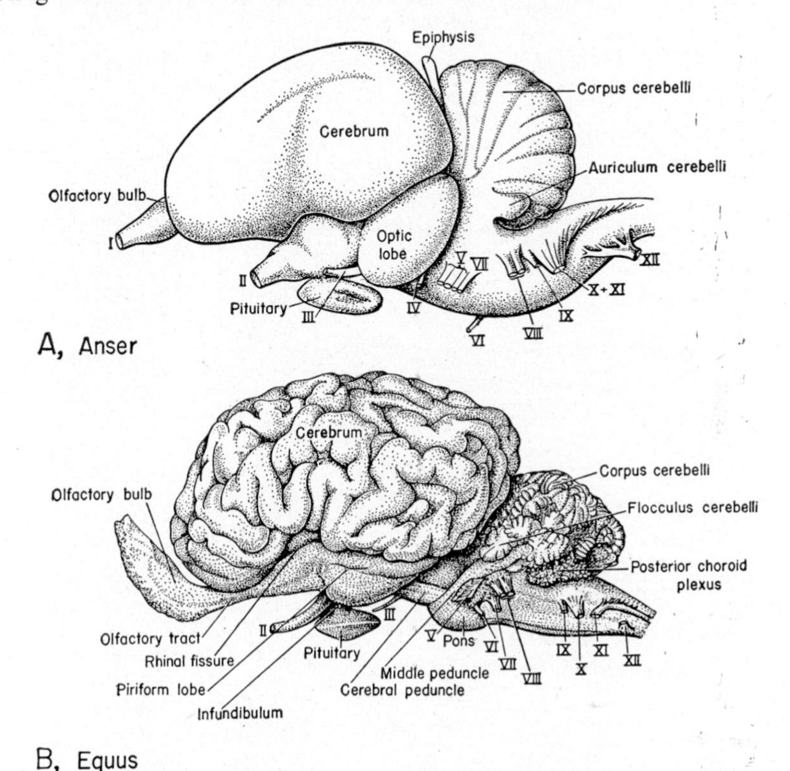

Fig. 343. Lateral views of brain of *A,* a goose; *B,* a horse. (After Bütschli, Kuenzi, Sisson.)

Cell centers are generally termed *nuclei,* although certain special centers have special names. Nuclei, or centers, in a broad sense, range in size from tiny clusters of cells embedded in the gray matter of the brain stem and discernible only on microscopic study, to such massive structures as the cerebellum. Fiber bundles connecting nuclei with one another are in general termed *tracts;* the fibers constituting such a tract are, of course, axons of neurons whose cell bodies lie in the nucleus of origin. A tract is generally given a compound name, the two parts of which designate its origin and termination; thus the corticospinal tract carries impulses from the cortex of the cerebral hemispheres to the motor cells of the spinal cord.

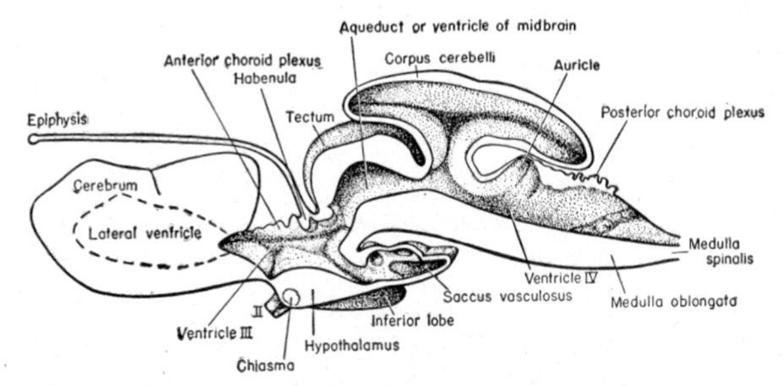

Fig. 344. Median section of the brain of a shark (Scyllium). Unshaded areas are those sectioned. (After Haller, Burckhardt.)

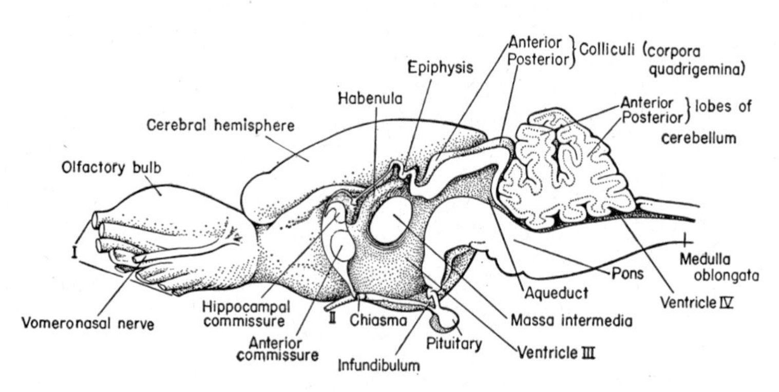

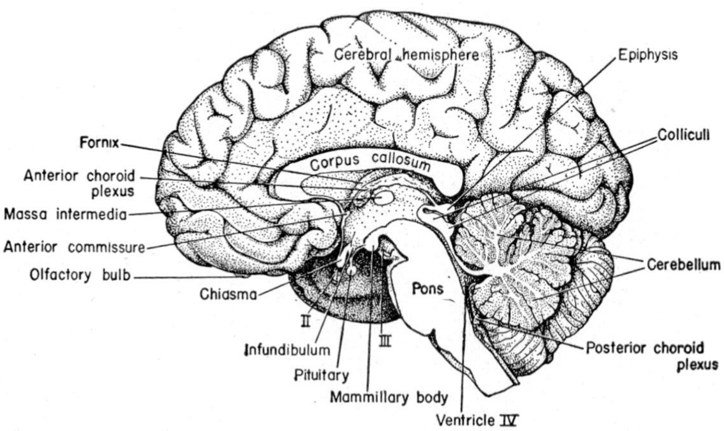

Fig. 345. Median section of the brains of *A,* an opossum; *B,* man. Unshaded areas are those sectioned. The internally bulging side walls of the diencephalon may meet and fuse in the midline, forming a "massa intermedia," which, however, has no functional importance. (*A* after Loo.)

Despite the general tendency of brain cells and fibers to organize into clean-cut centers and tracts, a primitive condition persists throughout in the *reticular system*. This is a band of interlacing cellular and fibrous material associated with the motor columns in the brain stem and anterior part of the spinal cord, and particularly well developed in the anterior part of the brain stem. This system appears to be important in motor coordination. Still further, the reticular network appears to function in carrying stimuli downward from the anterior brain regions to the motor centers of the oblongata and cord. This function is especially important

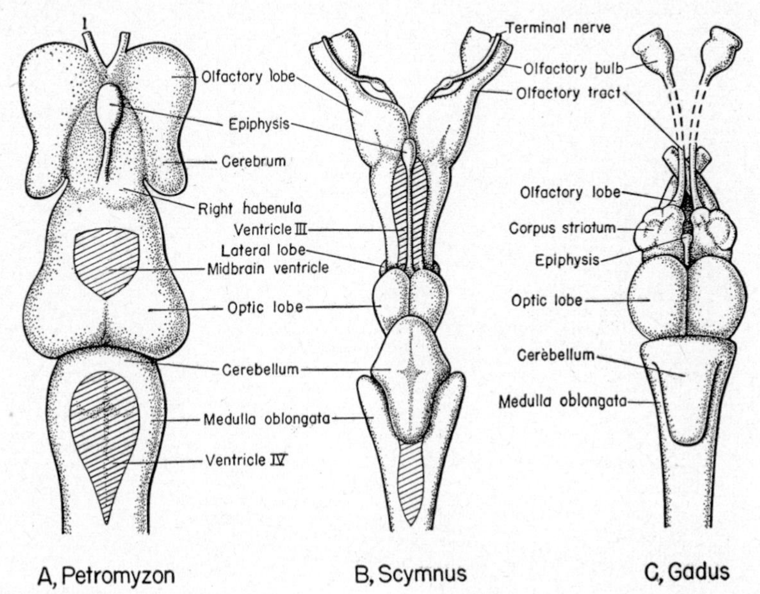

Fig. 346. Dorsal views of the brain of *A*, a lamprey; *B*, a shark; *C*, a teleost (codfish). Hatched areas are those in which a choroid plexus has been removed, exposing the underlying ventricle. (After Bütschli, Ahlborn.)

in lower vertebrates, in which motor tracts giving the brain control over trunk activities are poorly developed (cf. Figs. 361 and 362, page 590).

The brain is constructed on a bilaterally symmetrical pattern; in consequence, cross connections must be established in order that an animal may not have a literally dual personality. We have noted that even in the spinal cord there are numbers of association neurons whose fibers cross to the opposite side, and in the brain such connections are numerous. In the anterior regions of the brain there are a number of *commissures,* fiber tracts which connect corresponding regions of the two sides. In addition, there are frequent cases where fiber tracts in their course along the brain cross over from one side to the other (i.e., decussate), sometimes without apparent reason. We have described such a decussation in the case of the optic nerves (which are really brain tracts). Another example is recalled by the well-known fact that movements of one side of the body are controlled by the gray matter of the brain hemisphere of the opposite side—a situation caused by a decussation of the corticospinal tracts of the two sides as they pass backward along the brain stem.

Medulla Oblongata. Approach to the study of brain architecture is best made by first considering those parts of it which are simplest in construction and most closely resemble the spinal cord. The brain stem, including the three primary vesicles of the embryo, is simpler than its specialized outgrowths, the cerebral hemispheres and cerebellum. Even in the stem, however, the anterior part appears to have been the seat of complex ner-

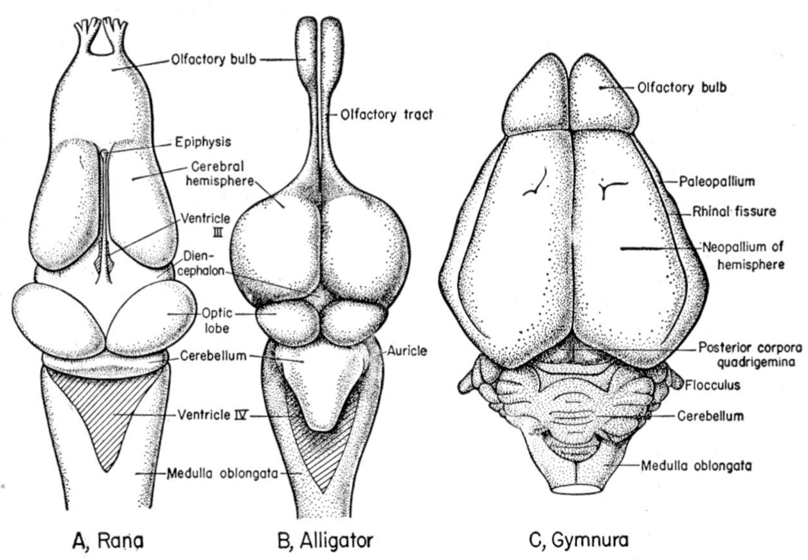

Fig. 347. Dorsal views of the brain of *A*, a frog; *B*, an alligator; *C*, a tree shrew. Hatched areas are those in which a choroid plexus has been removed, exposing the underlying ventricle. (After Gaupp, Crosby, Wettstein, Clark.)

vous centers from the beginning of vertebrate history. It is in the hindbrain region, the medulla oblongata, that we find the closest structural approach to the cord. Further, from the oblongata and the adjacent parts of the midbrain arise all the cranial nerves except the atypical ones from nose and eye.

The medulla oblongata is, particularly in lower vertebrates, closely comparable to an anterior section of the spinal cord, enlarged, however, through the expansion of the central canal to form the fourth ventricle with the columns of gray matter of either side widely separated dorsally as a consequence. Dorsally for most of its length the roof of the oblongata is thin, membranous, and infolded to form the posterior choroid plexus; anteriorly, the oblongata is covered by the cerebellum and fiber tracts connected with it.

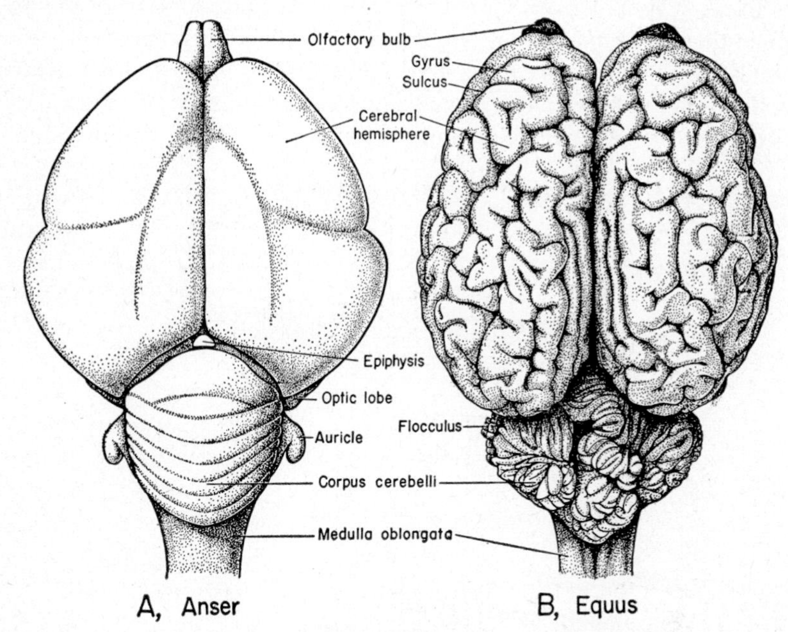

Fig. 348. Dorsal views of the brain of *A*, a goose; *B*, a horse. (After Bütschli, Kuenzi, Sisson.)

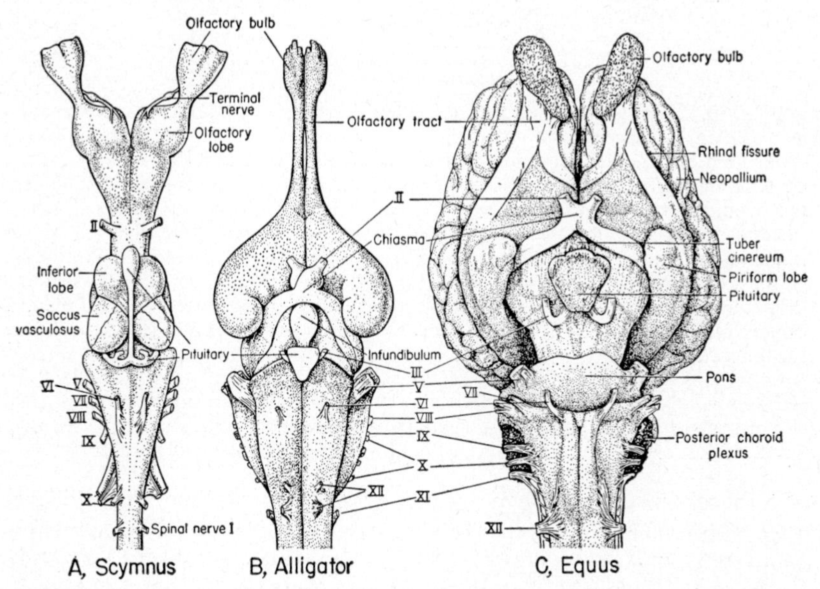

Fig. 349. Ventral views of the brain of *A*, a shark; *B*, an alligator; *C*, a horse. (After Bütschli, Wettstein, Sisson.)

In the gray matter of the oblongata and the posterior part of the midbrain there is found a series of columns or nuclei of gray matter basically similar to those present in the cord (Figs. 338, *B*, p. 558; 350). In that region we have noted the presence of a dorsal column, containing association centers for the reception and distribution of sensory impulses, and a

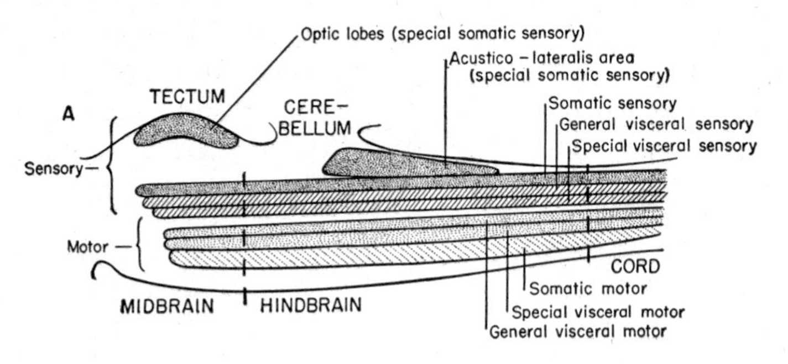

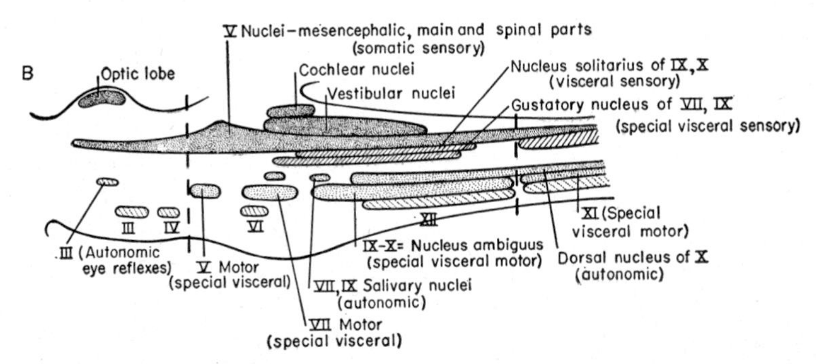

Fig. 350. Diagrams of midbrain and hindbrain regions in lateral view to show the arrangement of sensory and motor nuclei. *A*, Hypothetical primitive stage, in which brain stem centers were continuous with one another and with the columns of the cord. Even at such a stage, however, it would be assumed that special somatic centers would have developed for eye and ear. The brain region includes a special visceral motor column for the branchial muscle. *B*, Comparable diagram of the mammalian situation. The somatic sensory column is still essentially continuous (almost entirely associated with nerve V), but the other columns are broken into discrete nuclei. The visceral sensory column includes both a general visceral nucleus (many for efferent fibers from the viscera via the vagus) and a special nucleus for the important sense of taste. Of efferent visceral nuclei, there are small anterior ones for autonomic eye reflexes and the salivary glands, and a large nucleus for parasympathetic fibers to the viscera via the vagus. There are important branchial motor nuclei for *V*, *VI*, and *IX*, *X* (ambiguus). The somatic motor column includes small nuclei for eye muscles anteriorly and a hypoglossal nucleus posteriorly.

ventral column containing motor neurons; we have further noted evidence that both columns may be subdivided into somatic and visceral components. A similar situation is found in the oblongata and the posterior part of the midbrain, although the widely expanded ventricle makes for a somewhat different appearance. A horizontal groove, the *sulcus limitans*, runs along the inner surface of the brain stem on either side and separates a dorsal sensory region from a ventral motor region. Further, here, as in the cord, each of these two can be subdivided into somatic and visceral components.

In the embryo each subdivision appears to be formed as an essentially continuous fore and aft column. Most ventrally is a somatic motor column. Dorsal to this, but still below the limiting sulcus, are visceral motor columns for both branchial muscle and autonomic efferents. Above the sulcus are, in order, visceral sensory and somatic sensory columns, the primary areas of reception for sensations from gut and skin, respectively. The ventralmost column is that from which the ventral cranial nerves—III, IV, VI, XII—arise; the others are the areas of central connection of the dorsal root or branchial nerves—including V, VII, IX, X.

In the adults of lower vertebrate groups much of the embryonic continuity of the column is preserved; in higher types, however, there is a strong trend for a breaking down of the columns into discrete nuclei for the various nerves concerned. Thus in mammals the somatic motor column is fragmented into (1) several small anterior nuclei in the midbrain and in the anterior end of the oblongata for the eye muscle nerves, and (2) a more posterior nucleus for the hypoglossal. The special visceral motor column for branchial musculature is broken up into separate special visceral motor nuclei for nerves V, VII, and IX and X (*nucleus ambiguus*). Small autonomic nuclei are present in the midbrain for eye reflexes, farther back in the oblongata for salivary glands, and still farther back for the autonomic component of the vagus. In vertebrates generally the nuclei associated with the normal visceral sensory fibers remain concentrated in a *nucleus solitarius* of the oblongata, but a parallel *gustatory nucleus* is present for the special visceral sense of taste. The somatic sensory column remains as a single elongate nucleus which, associated with the trigeminal nerve, extends much of the length of the brain stem and even back into the cord.

The series of motor and sensory columns or nuclei give us all the elements required for reflex circuits in the brain between sensory reception and the responding effector organs of the head and gill region. But brain mechanisms are not built solely on such a simple plan. In addition to ordinary sensory receptors in the skin, muscles, and gut tube, such as are found throughout the body, we have in the head region special sensory organs. Centers must be present for the reception of sensations from these organs; and higher centers must be built up for the association and correlation of these sensations before final "directions" can be issued to the motor columns of brain stem and cord. Much of this apparatus is situated elsewhere in the brain. But even in the relatively simple brain stem region here considered we find the primary area of reception of one of the main sensory systems—the acustico-lateralis system.

The lateral line organs and the ear, a special development from them, are somatic sensory structures; as such, sensations from them were primitively received in the somatic sensory column of the oblongata. So special are they, however, that a specific center, an *acustico-lateralis* area, often of considerable size, develops above the normal somatic sensory column in the anterior part of the oblongata, below the cerebellum. In land vertebrates the lateralis system disappears, but acoustic nuclei persist; in mammals there are distinct centers for the vestibular and cochlear parts of the ear.

The gray matter of the oblongata is, like that of the cord, sheathed externally by white matter composed of longitudinally arranged fibers. This sheathing is increasingly thick in higher vertebrate groups, for with greater control by brain centers over bodily activities in these forms, the volume of fiber connections between brain and cord via the oblongata is vastly augmented. Further expansion of the outer layers of the oblong-

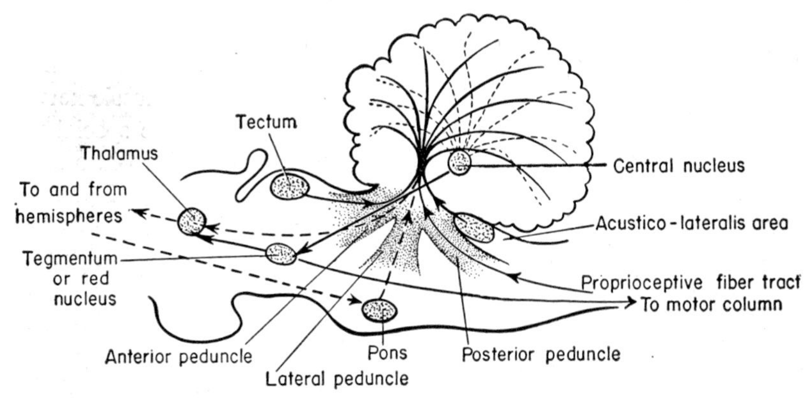

Fig. 351. Diagram to show the main connections of the cerebellum. The connections with the cerebral cortex, peculiar to mammals, are shown in broken line.

ata is due to fiber tracts connecting with the cerebellum; in mammals a mass of cerebellar fibers which bridge over the base of the oblongata anteriorly causes so prominent a swelling that this region is termed the *pons*.

Cerebellum (Figs. 351, 352). Rising above the brain stem at the anterior end of the medulla oblongata is the cerebellum, a brain center, often of large size, which is of extreme importance in the coordination and regulation of motor activities and the maintenance of posture. The cerebellum acts in a passive, essentially reflex fashion in the maintenance of equilibrium and body orientation. Its function in regulating muscular activity may be compared to that of "staff work" in the movement of an army. To carry out the general orders of an army commander it is necessary that there be in hand information as to the position, current movements, condition, and equipment of the bodies of troops concerned. A "directive" (from the higher brain centers of the hemispheres or anterior part of the brain stem) for a muscular action—say, the movement of a limb—cannot be carried out efficiently unless there are available data as to the current position and movement of the limb, the state of relaxation or contraction of the muscles involved, the general position of the body, and its relation to the outside world. Such data are assembled in the cerebellum and synthesized there, and resulting "orders" are issued by efferent pathways which render the movement effective.

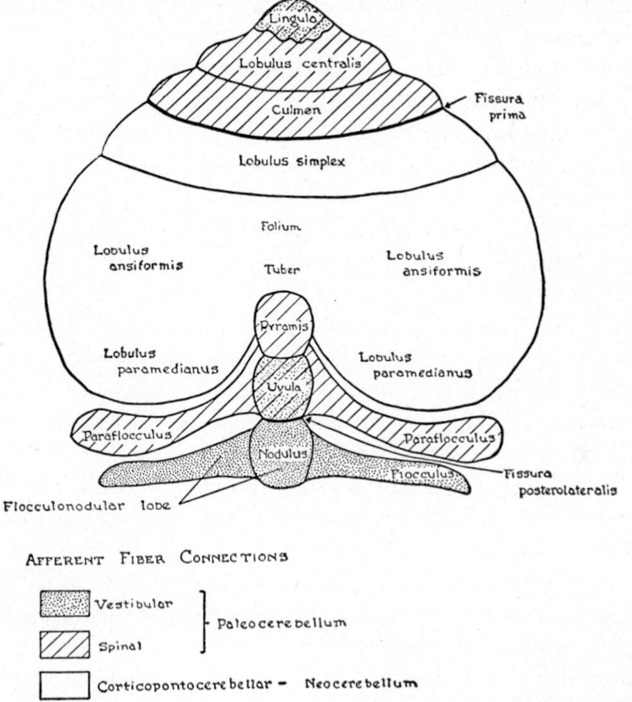

Fig. 352. Diagram of a surface view of a mammalian cerebellum (showing details not discussed in the text). The stippled and hatched portions, associated with equilibrium and with muscle sensations, are the phylogenetically oldest parts of the cerebellum; the white area is a mammalian addition associated with the cortex of the cerebral hemispheres. (From Fulton, after Larsell.)

The data upon which the cerebellum acts are derived from two main sources. One is the acoustic area of the medulla—acustico-lateralis area of lower vertebrates—in which are registered equilibrium sensations from the ear and lateral line sensations. The base of the cerebellum lies immediately above this area, and it is generally accepted that the cerebellum originated as a specialized part of this sensory center. The second main source of cerebellar data is the system of sense organs found throughout the muscles of the body. Into the cerebellum are directed fibers of this proprioceptive system carrying data regarding the position of the body parts and the state of muscle tension.

In addition to these two primary sources, the sensory picture assembled in the cerebellum is rounded out by additional fiber relays from skin sensory areas, from the optic centers, and, in lower vertebrates, even from the nose. Information, further, is supplied regarding muscular movements which are directed by higher brain centers, but upon which the cerebellum has influence. In lower vertebrates these data are furnished by fibers from the midbrain, where such impulses in great measure originate; in mammals, where the cerebral cortex dominates motor functions, strong fiber tracts connect cortex with cerebellum via the pons.

After integration of data by the cerebellum, outgoing fiber tracts carry impulses forward and downward on either side of the brain to the side walls of the midbrain (the tegmental region), whence they continue (in part after a relay there) to the appropriate motor nuclei of head or body or, in mammals, to the thalamus and thence to the cerebral cortex.

In forms in which the cerebellum is well developed the fiber tracts leading to and from it are prominent features in the architecture of the brain stem, particularly of the oblongata. In mammals, for example, these fibers form three pairs of pillar-like structures, the *cerebellar peduncles* (or brachia). Anterior peduncles, mainly composed of efferent fiber tracts, connect cerebellum and midbrain. A middle, or lateral, pair of peduncles rise straight upward along the sides of the metencephalon from the pons; these carry fibers running from cerebral cortex to cerebellum, which cross and are relayed ventrally in the pons. Posterior peduncles rise up through the oblongata and carry proprioceptive fiber bundles from the cord.

The cerebellum varies greatly in size and shape from group to group. Its degree of development is correlated with the intricacy of bodily movements; it is large and elaborately constructed in many fishes and in birds and mammals, little developed in cyclostomes and amphibians. In these last groups it is little more than a pair of small centers lying just above the acustico-lateralis nuclei and seemingly developed as a specialized part of these sensory areas. This most ancient part of the cerebellum persists in all groups as the *auricles* (or flocculi), especially concerned with equilibrium and closely connected with the inner ear.

Even in lampreys and amphibians, however, there is some development of a median cerebellar region between the two auricles above the front end of the fourth ventricle. To this come fibers from the organs of muscle sense and from sensory centers in the brain for integration and correlation with stimuli from the organs of equilibrium in the ear. This median structure, the *corpus cerebelli,* constitutes the main mass of the cerebellum in typical fishes, reptiles, and birds. In mammals the rise of the cerebral cortex to command of motor functions is correlated with the development of powerful tracts from motor cortex to cerebellum and the appearance of large paired and convoluted structures—"cerebellar hemispheres"—which make up much of the bulk of the mammalian organ.

In contrast to almost every other area of the central nervous system except the cerebral hemispheres, the cerebellum is a region in which the gray cellular material is superficially placed as a cortex, the white matter internal. In a well-developed cerebellum the gray matter is spread out as a surface sheet which is often highly convoluted and thus gains greater area. Beneath this is the white matter, consisting of a fan-shaped radiation of incoming fibers spreading out in all directions to the surface, and of fibers returning from the gray matter. The latter are relayed through a central nucleus on their way to the midbrain. The cerebellar cortex contains cells of several distinctive types between which sensory data are interchanged by a complex fiber network. There is evidence of some degree of localiza-

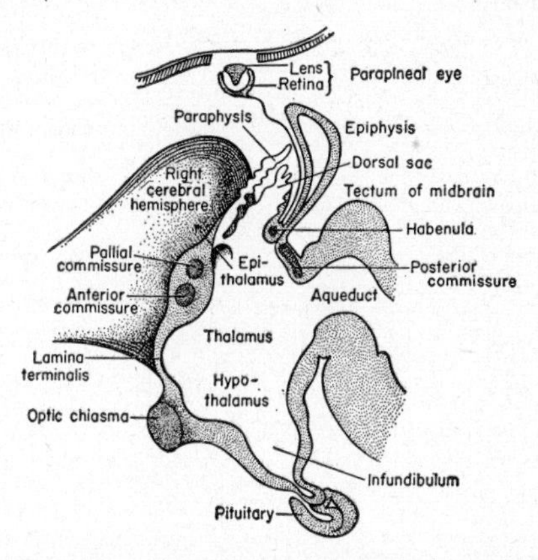

Fig. 353. A median section through the diencephalon of a lizard to show, particularly, the various outgrowths. The arrow indicates the position of the interventricular foramen leading from ventricle III into the right lateral ventricle. (After von Kupfer, Nowikoff.)

tion of function in the cerebellum, but in many respects the entire corpus cerebelli appears to act as an integrated unit.

Midbrain and Diencephalon. In contrast with the posterior part of the brain stem, the mesencephalon and diencephalon show specialized functional features in vertebrates of all classes. Sensory and motor nuclei connected with cranial nerves extend some distance into the midbrain from the oblongata. Farther forward, however, such structures are absent. The anterior brain stem centers have no direct connection with afferent or efferent stimuli apart from those of the optic nerve. This region serves two main functions. In higher vertebrates, more particularly, it is a principal way station between "lower" brain areas and the cerebral hemispheres. In all groups it is important as a locus of centers of nervous correlation and coordination. In mammals and birds these latter functions are overshadowed by those of the hemispheres, but in many lower groups the most highly developed association mechanisms lie in the midbrain and thalamus. In a mammal, destruction of the hemispheres results in functional disability; a frog, on the other hand, can go about its business in essentially normal fashion without cerebral hemispheres as long as the brain stem is intact.

The topography of the midbrain and diencephalon and their annexes may be described before dealing with the functions and connections of the various nuclei of their gray matter (Fig. 353).

The midbrain is that part of the brain tube traversed by the cerebral aqueduct; the diencephalon is a region lying about the third ventricle. In mammals the diencephalon is buried below and between the expanded hemispheres. Diencephalon and midbrain lie on the lines of communication between anterior and posterior brain areas. Fiber tracts of white

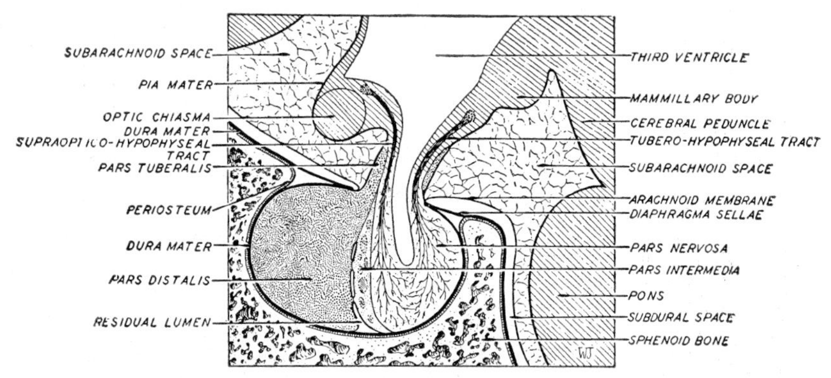

Fig. 354. Section through a human pituitary and the adjacent structures at the base of the brain. (From Turner.)

matter form the greater part of the walls and floor of the midbrain; its sides are formed by great motor fiber bundles (the pyramidal tracts descending from the hemispheres) which form the *cerebral peduncles* (crura cerebri).

In the midbrain the gray matter above the aqueduct is greatly thickened to form the mesencephalic *tectum;* thinner areas of gray matter in the side walls are termed the *tegmentum.*

Of the areas bounding the third ventricle, the anterior wall, the *lamina terminalis* in which the brain stem ends, is considered to form part of the telencephalon; the remainder forms the diencephalon. Accessory structures, described below, are present in the roof and floor of the diencephalon; the main substance of this brain subdivision consists of the lateral walls of the ventricle, which contain in their gray matter a host of nuclei. These wall areas are termed collectively the *thalamus,* because of the fancy that the diencephalon forms a "couch" upon which rest the great cerebral hemispheres of higher vertebrates. Dorsal and ventral parts—roof and floor—of the diencephalon are called the *epithalamus* and *hypothalamus.* Thickened parts of the hypothalamus may cause swellings of the brain surface as *inferior lobes* in fishes (Fig. 341, p. 563) or *lateral lobes* in amphibians and reptiles. The thalamus proper is further divided, on the basis of the centers contained within it, into a *dorsal thalamus* and a *ventral thalamus*.

The roof of the third ventricle is for the most part a thin, non-nervous structure. In it develop a varied series of outgrowths. Most anteriorly, in that part of the roof which pertains to the telencephalic region, there grows upward in embryos of most groups a thin-walled sac, the *paraphysis*,

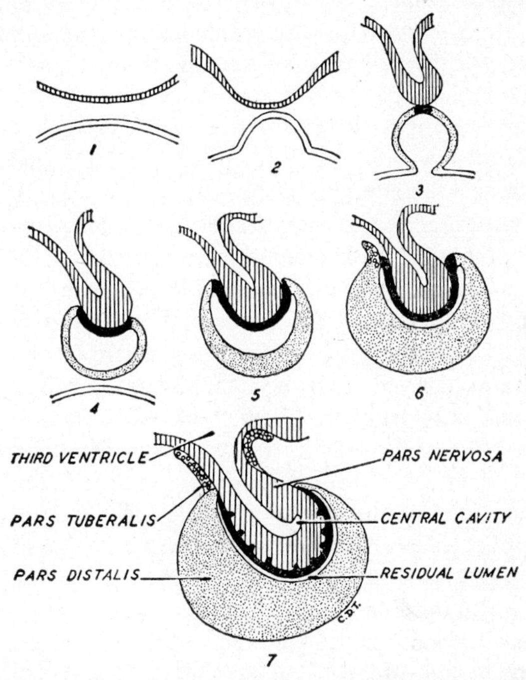

Fig. 355. Diagrams showing stages in the development of the mammalian pituitary from brain tissue (hatched) and from the hypophyseal (Rathke's) pouch. As will be seen, only the pars nervosa (neural lobe or posterior lobe) is formed from neural material; the three other parts (including the pars intermedia, in solid black) are derived from the pouch epithelium. (From Turner.)

of unknown function; this structure disappears in the adult in most cases. Adjacent to it, in the anterior part of the diencephalic roof, is the region of the anterior choroid plexus. In higher vertebrate classes this is typically invaginated into the cavity of the ventricles; in lower groups, however, it is more generally an extroverted structure, a *saccus dorsalis*. Still more posteriorly in the diencephalic roof there may develop one or both of a pair of median, stalked eyelike structures, the parietal and pineal organs, discussed in the preceding chapter (p. 510).

In the floor of the diencephalon the *optic chiasma* is a prominent feature anteriorly. Posteriorly, we find in most fishes the *saccus vasculosus*, a thin vesicle which may reach considerable size. Its walls contain sensory structures, but its function is unknown. Possibly it may register fluid pressure, internal and external; the sac is generally best developed, it may be noted, in actively swimming oceanic fishes (Fig. 341, p. 563).

Below the diencephalon lies the *pituitary gland* or *hypophysis cerebri* (Figs. 354, 355), most important of all the endocrine structures of the body; its hormones have been noted previously (p. 424). Embryologically, this has a dual origin (Figs. 168, 339, pp. 292, 560). Downward from the embryonic diencephalon extends a hollow, finger-like process, the infundibulum. Upward from the embryonic mouth there grows an ectodermal pocket, the hypophyseal pouch (Rathke's pouch). This latter usually closes over in adult life, but remains open in a few fishes (notably the cyclostomes, where, as we have seen, it combines with the nasal sac). From both these embryonic structures there proliferate masses of tissue which unite to form the adult pituitary.

A prominent part of the pituitary in all vertebrates is the *pars distalis,* locus of origin of most of the known hormones and formed from the hypophyseal pouch. In most types a *pars tuberalis* is present, an upgrowth of tissue surrounding the stalk of the infundibulum. The tissue arising from the infundibulum constitutes a *pars nervosa.* There frequently develops a *pars intermedia,* derived from the hypophyseal pouch, but tending to fuse with the nervous part of the gland.*

The eyes appear to have been responsible in great measure for the development, in lower vertebrates, of important association centers in the anterior region of the brain stem (just as the acustico-lateralis system is related to cerebellar development in the hindbrain and the olfactory sense, as we shall see, to cerebral development). The optic nerves enter ventro-anteriorly at the chiasma. In all vertebrates except mammals, however, their fibers do not tarry in the diencephalon, but proceed, almost without exception, upward and backward to the roof of the midbrain, where the primary visual center is located. The midbrain roof, the tectum, became, early in vertebrate history, the seat of an important association center; it is the dominant brain region, it would appear, in fishes and amphibians, but became of lessened importance in amniotes with increased development of the cerebral hemispheres (cf. Figs. 361–363, pp. 590, 591).

This center developed in connection with visual reception. A greater part of it develops as a pair of *optic lobes* (corpora bigemina), varying in size with the importance of the eyes, and becoming particularly large

* The pituitary is frequently considered to consist of two lobes, but the terminology (based on the structure in man) is variable and sometimes inconsistent. The anterior lobe includes the pars distalis, with or without the pars tuberalis; the posterior lobe is basically the pars nervosa, with the pars intermedia sometimes included. If the gland is to be considered as consisting of two lobes, a more reasonable terminology is a division into a glandular lobe, including the three parts arising from the mouth pouch, and a nervous lobe, composed solely of the infundibular derivative.

in birds and many teleosts. The histological pattern of the optic lobes of nonmammalian vertebrates is complex, with successive layers of cell and fiber areas, giving them a structure broadly comparable to the gray matter of cerebellum or cerebrum. It appears probable that a visual pattern is here laid out in a fashion similar to that developed in our own case in the cerebral gray matter.

The presence of this visual center in lower vertebrates was, it would seem, responsible for the attraction to the tectal region of stimuli from other sensory areas. Fiber paths lead hither from the acustico-lateralis area, from the somatic motor column, and from the olfactory region via the diencephalon, and connections with the cerebellum are developed. Sensory stimuli from all somatic sources are here associated and synthesized, and motor responses originated. Primitively, these motor stimuli were relayed, it would seem, by way of the reticular formation lying more ventrally in the tegmental region of the midbrain, and a variety of nuclei are developed in this region in one vertebrate group or another. In amphibians and reptiles, however, direct motor paths are developed, with the formation of a definite tract from the tectum to the motor columns of the brain stem and cord.

In fishes and amphibians the tectum appears to be the true "heart" of the nervous system—the center which wields the greatest influence on body activity. In reptiles and birds the tectum is still an area of great importance, but is rivaled, and in birds overshadowed, by the development of the basal nuclei of the hemispheres.

In mammals the midbrain has undergone a great reduction in relative importance; most of its functions have been transferred to the gray matter of the cerebral hemispheres, and most of the sensory stimuli which are integrated here in lower vertebrates are, instead, projected to the hemispheres in mammals. Most auditory and other somatic sensations which reach the midbrain are, in mammals, relayed onward by way of the thalamus to the hemispheres. It is especially notable that few of the optic fibers in mammals follow the original course to the midbrain; most are interrupted in the thalamus, and visual stimuli are shunted forward to the hemispheres. The tectum retains little function except that of serving in a limited way as a center for visual and auditory reflexes. In mammals it takes the form of four small swellings of the midbrain roof, the *corpora quadrigemina*. Of these the anterior pair deal with visual reflexes and represent the main part of the tectum of lower vertebrates. In amphibians, with the development of hearing functions in the ear, secondary auditory centers develop on either side of the midbrain adjacent to the optic tectum. These develop into the posterior pair of elements of the corpora quadrigemina, which attain considerable size in some groups of mammals; they function as a relay station for auditory stimuli on their way to the thalamus and the cerebral hemispheres.

The region on either wall of the midbrain termed the *tegmentum* is essentially a forward continuation of the motor areas of the oblongata. As such it functions as a region in which varied stimuli from diencephalon, tectum, and cerebellum are coordinated and transmitted to the motor nuclei of the brain stem. In lower vertebrate groups it consists in general of a rather diffuse series of nerve cells and fibers of the reticular system (p. 567). In some instances, however, well-defined nuclei may be formed; we may note, for example, the presence in this region in mammals of the *red nucleus* through which are forwarded efferent impulses from the cerebellum.

Of diencephalic regions, the epithalamus is of little importance as a brain center. We may note the constant presence here in all vertebrates of the *habenular body,* a group of small nuclei through which olfactory stimuli pass on their way back from hemispheres to brain stem.

The hypothalamus also contains olfactory centers, notably the *tuber cinereum* and, in mammals, the *mammillary bodies* adjacent to the pituitary region. Its main importance, however, is that of a visceral brain center; other major brain centers are almost exclusively somatic in their activities. Most of the visceral nervous functions are carried out by reflexes in the cord or medulla, but the hypothalamus is in all vertebrate classes a region with control over certain, at least, of the body's visceral activities. A number of nuclei are present in this region. To them pass stimuli from olfactory and taste organs as well as sensations from various visceral structures of the body. These nuclei have efferent connections with the pituitary and, posteriorly, with motor centers for visceral movement, and the like. The range of their regulatory functions is incompletely known. It is, however, of interest that (for example) temperature regulation and sleep in mammals are controlled by the hypothalamus.

The thalamus proper is in lower vertebrate classes an area of modest importance. Its ventral part may be considered an anterior outpost of the motor column of the brain stem; it is in every class a motor coordinating center, and is further a relay center on the motor path from the basal nuclei of the hemispheres back to the brain stem.

In lower vertebrates the dorsal region of the thalamus appears to be merely an anterior extension of the correlation areas connected with the tectal region of the midbrain. Its importance increases proportionately with the increased development of association centers in the cerebral hemispheres. Even in vertebrates with a lowly brain organization, a certain amount of sensory data may be passed forward from the brain stem by relay in the thalamus, to the cerebral hemispheres for synthesis there with olfactory stimuli. With a high degree of development of association centers in the hemispheres of amniotes, there is a great development of nuclei in the dorsal thalamus for relay purposes. In the mammalian stage (with the practical abandonment of the tectal association center) the thalamus reaches the height of its development. In mammals all somatic sensations are assembled in the gray matter of the cerebral cortex; and all this sensory material (excepting, of course, olfaction) is relayed to it via the dorsal thalamus. Conspicuous elements in its structure are the *lateral geniculate body,* whence optic stimuli are projected upward to the hemispheres; and the *medial geniculate body,* for the relay of auditory stimuli (Fig. 363, p. 591).

Cerebral Hemispheres. The evolution of the cerebral hemispheres is the most spectacular story in comparative anatomy. These paired outgrowths of the forebrain began their history, it would seem, simply as loci of olfactory reception. Early in tetrapod history they became large and important centers of sensory correlation; by the time the mammalian stage is reached, the greatly expanded surfaces of the hemispheres have become the dominant association center, seat of the highest mental faculties. The development of centers of such importance in this anterior, originally olfactory, segment of the brain emphasizes the importance of the sense of smell in vertebrates. The acustico-lateralis system and vision, are, as we have seen, senses upon which important correlation mechanisms were erected early in vertebrate history, but in the long run smell has proved dominant. It is of little account in higher primates, such as ourselves. But in following down the ancestral line from mammals to their early vertebrate ancestors, it appears that this sense has been throughout a main channel through which information concerning the outside world has been received. It is thus but natural that its brain centers should form a base upon which higher correlative and associative mechanisms have been built.

Presumably the brain tube of the earliest vertebrates was, like that of Amphioxus, a single, unpaired structure all the way forward to its anterior end; the cavity of the telencephalon was a median, unpaired terminal ventricle, as it is today in the early embryo of every type. In all vertebrates the most anterior part of the wall of the third ventricle is considered as belonging to the telencephalon, but in land vertebrates most of that brain segment consists of distinct paired structures, the lateral ventricles and the tissues surrounding them. Most fishes—cyclostomes, Chondrichthyes, actinopterygians—show a transitional condition, for there is, for much of the length of the end brain, a single ventricular cavity which bifurcates only distally.

In even the lowest of living vertebrates, the cyclostomes, each half of the end brain is subdivided into two parts, the *olfactory bulb* and the hemisphere. The bulb is a terminal swelling in which the fibers of the olfactory nerve end; its size varies with the acuteness of the sense of smell. Here is located the primary olfactory nucleus. From its cells, fibers pass back (sometimes with a relay en route) to be distributed to the various regions of the hemisphere. Depending on the configuration of the head and braincase, bulb and hemisphere may be in close contact (as they are, for example, in cyclostomes) or may be well separated, with a distinct *olfactory tract* between them.

Primitively, as seen in cyclostomes, the hemisphere is merely an *olfactory lobe*—an area in which olfactory sensations are assembled and relayed to the diencephalon for correlation with other sensory impulses (Figs. 356, *A*, 357, *A*). At this level of development few if any fibers ascend from the stem to the hemispheres for correlation there. In general, fiber paths from olfactory areas follow two routes: ventrally to end in

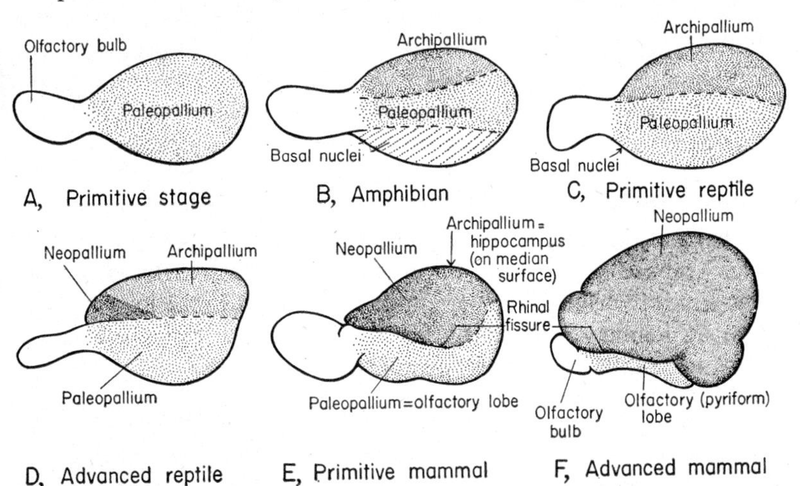

Fig. 356. Diagrams to show progressive differentiation of the cerebral hemisphere (cf. Fig. 357). Lateral views of left hemisphere and olfactory bulb. In *A* the hemisphere is merely an olfactory lobe. *B*, Dorsal and ventral areas, archipallium (= hippocampus) and basal nuclei (corpus striatum) are differentiated. *C*, The basal nuclei have moved to the inner part of the hemisphere. *D*, The neopallium appears as a small area (in many reptiles). *E*, The archipallium is forced to the median surface, but the neopallium is still of modest dimensions, and the olfactory areas are still prominent below the rhinal fissure (as in primitive mammals). *F*, The primitive olfactory area is restricted to the ventral aspect, and the neopallial areas are greatly enlarged (as in advanced mammals).

visceral centers in the hypothalamus (tuber cinereum, mammillary bodies of mammals), or dorsally to the habenulae in the epithalamus and thence to the motor areas of the brain stem.

A higher stage, although still primitive, is that seen in the amphibians (Figs. 356, *B*, 357, *B*). Here most of the tissues of the hemispheres can be divided regionally into three areas, of interest because of their history in more progressive types. All three areas receive olfactory stimuli and exchange fibers with one another; all three discharge to the brain stem. Ventrally lies the region of the *basal nuclei,* essentially equivalent to the corpus striatum of mammals, and destined in higher stages to move into the central parts of the hemisphere. The basal nuclei form a correlation center at an early evolutionary stage and become increasingly more important in this regard in more advanced groups. Sensory impulses are "projected" upward into the basal nuclei from the thalamus for correla-

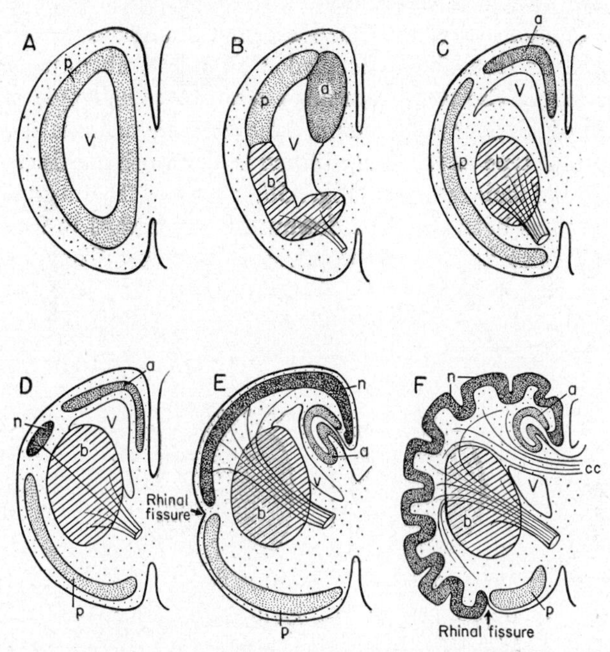

Fig. 357. Diagrammatic cross sections of left cerebral hemisphere to show stages in the evolution of the corpus striatum and cerebral cortex. *A*, Primitive stage, essentially an olfactory lobe; gray matter internal and little differentiated. *B*, Stage seen in modern amphibians. Gray matter still deep to surface, but differentiated into paleopallium (= olfactory lobe), archipallium (= hippocampus), and basal nuclei (= corpus striatum), the last becoming an association center, with connections from and to the thalamus (indicated by lines representing cut fiber bundles). *C*, More progressive stage, in which basal nuclei have moved to interior, and pallial areas are moving toward surface. *D*, Advanced reptilian stage; beginnings of neopallium. *E*, Primitive mammalian stage; neopallium expanded, with strong connections with brain stem; archipallium rolled medially as hippocampus; paleopallial area still prominent. *F*, Progressive mammal; neopallium greatly expanded and convoluted; paleopallium confined to restricted ventral area as pyriform lobe. The corpus callosum developed as a great commissure connecting the two neopallial areas. *a*, Archipallium; *b*, basal nuclei; *cc*, corpus callosum; *n*, neopallium; *p*, paleopallium; *V*, ventricle.

tion with olfactory sensations; descending fibers carry impulses from the basal nuclei to centers in the thalamus and midbrain tegmental region.

The gray matter of all parts of the hemisphere except the basal nuclei tends progressively to move outward toward the surface, and thus becomes the *cerebral cortex,* or *pallium* ("blanket"). In the amphibian stage the gray matter is still largely internal, but these terms may nevertheless be used in the light of later history. A band of tissue along the lateral surface of the hemisphere is termed the *paleopallium.* This area remains almost purely olfactory in character, and the paleopallial region is that of the olfactory lobes in higher stages. Dorsally and medially lies the *archipallium,* which is antecedent to the hippocampus of mammals. This area is to a minor extent a correlation center in all land vertebrates, with ascending fibers from the diencephalon as well as fibers from olfactory bulbs and lobes. The tract from this region to the habenula is the main component of the fiber bundle termed the *fornix* in mammals.

An aberrant type of forebrain is seen in the dominant fishes of today, the teleosts and their relatives (Fig. 358, *A*). Here the gray matter of the hemispheres has been crowded downward and inward to form massive structures bulging up into the ventricles from below. They include both the basal nuclei or striatum and tissues above it representing the pallial areas; the roof of the hemispheres is but a thin, non-nervous membrane.

The reptile hemispheres are advanced over the amphibian type in

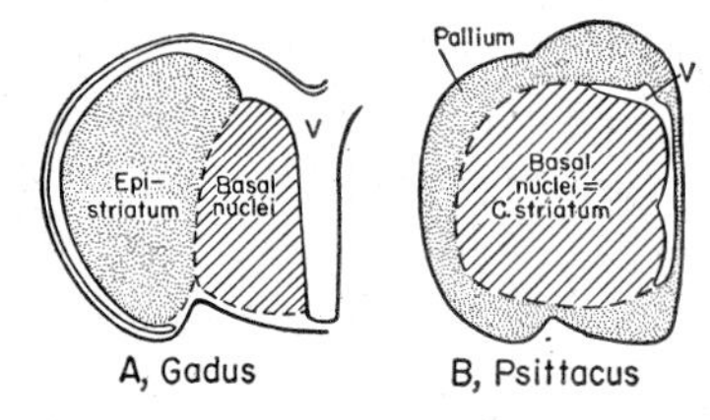

Fig. 358. Aberrant types of forebrains. Section of left hemispheres of *A*, a teleost (codfish); *B*, a bird (parrot). In teleosts the roof of the ventricles is only a membrane; the gray matter has been pushed downward and inward to join the basal nuclei (or striatum) as an epistriatum. In birds the roof of the ventricles is little developed; the basal nuclei, on the other hand, are enormously developed. *v*, Ventricle. Basal nuclei hatched.

both complexity of organization and relative size compared with other brain parts. Roof and side walls of the hemispheres show in the main an essentially primitive arrangement of the pallial areas, but some of the gray matter has spread outward toward the surface, and some reptiles even show the beginning of the neopallial development which is of crucial importance, as will be seen, in the evolution of the mammalian brain. The basal nuclei have moved inward to occupy a considerable area in the floor. Strong projection fiber bundles run to the basal nuclei from the thalamus and back from them to the brain stem; the basal nuclei are obviously correlation centers of importance.

In birds (Fig. 358, *B*) the hemispheres are further enlarged. Their evolutionary development, however, has taken place in a fashion radically different from that seen in mammals. The various pallial areas are present in birds, but are little developed, and the outer walls of the hemispheres are relatively smooth and thin; this is associated with the fact that the primitive pallium was almost exclusively associated with olfaction—a sense of little importance in birds. On the other hand, the basal nuclei, corresponding to the mammalian corpus striatum, are of enormous size. The midbrain tectum is still important as a center in which visual stimuli are correlated with other sensations, but the basal nuclei are, to judge by the pattern of fiber tracts, the dominant association center regulating body activities.

Our knowledge of the mode of functioning of the basal nuclei is imperfect, but some idea of this can be gained by contrasting the actions of birds with those of mammals. In birds we see a complex series of action patterns which may be called forth to meet a great variety of situations. They are all, however, essentially stereotyped; the actions are innate, instinctive. The bird, its brain dominated by its basal nuclei, is essentially a highly complex mechanism with little of the learning capacity which in mammals is associated with the development of the expanded cerebral cortex.

The first faint traces of mammalian cortical development are to be seen in certain reptiles (Figs. 356, *D,* 357, *D*). In the hemispheres of these forms we find, between paleopallium and archipallium, a small area of superficial gray matter of a new type, that of the *neopallium.* Even at its inception it is an association center, receiving, like the basal nuclei, fibers which relay to it sensory stimuli from the brain stem.

The evolutionary history of the mammal brain is essentially a story of neopallial expansion and elaboration. The cerebral hemispheres have attained a bulk exceeding that of all other parts of the brain, particularly through the growth of the neopallium, and dominate functionally as well. This dominance is apparent in mammals of all types, but is particularly marked in a variety of progressive forms, most especially in man. In even the more primitive mammals (Figs. 356, *E,* 357, *E*) the neopallium has expanded over the roof and side walls of the hemispheres. It has crowded the archipallium on to the median surface above; the paleopallium is restricted to the ventrolateral part of the hemispheres, below the rhinal fissure—a furrow which marks superficially the boundary between olfactory and nonolfactory areas of the cortex. With still further neopallial growth (Figs. 356, *F,* 357, *F*) the archipallium is folded into a restricted area on the median part of the hemispheres, where it remains obscurely as the hippocampus,* and the olfactory lobes which include the paleopallium come to constitute but a small ventral hemisphere region. The corpus striatum persists, but its mammalian functions are not clear.

As it develops in mammalian evolution, the neopallium assumes newer and higher types of neural activity in correlation and association and also takes over much of the functions previously exercised by brain stem centers and much of those of the basal nuclei. The midbrain tectum loses its

* So called because of its fancied resemblance (in section) to a sea horse, with its coiled tail.

former importance and is reduced to a reflex center. Auditory and other somatic sensations are relayed to the thalamus, most of the optic fibers are intercepted there, and all are projected from thalamus to hemispheres by great fiber tracts. Thalamic connections of this sort for the basal nuclei had evolved in lower vertebrate groups, and in birds, we have seen, powerful projection tracts evolved in connection with this dominant brain region. In mammals, however, the greater part of these fibers plunge on through the basal nuclei—here termed the *corpus striatum**—to radiate out to the neopallial surface. With all the sensory data thus made available, the appropriate motor "decision" is made by the cortex. As men-

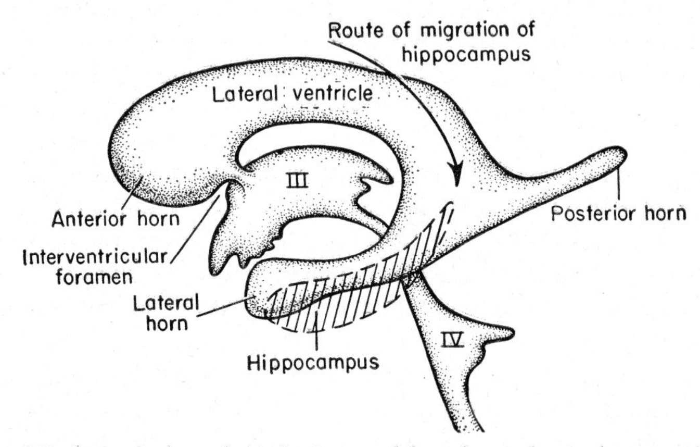

Fig. 359. Lateral view of the brain ventricles of an advanced mammalian type (Homo) in lateral view. The ventricles are represented as solid objects, the brain tissue being removed. With expansion of the cerebral hemisphere, the lateral ventricle has expanded backward to a posterior horn in the occipital lobe, and downward and forward laterally to a lateral horn in the temporal lobe. With this backward and downward expansion, various shifts in position of brain parts occurred. The hippocampus, which developed dorsally on the median surface of the hemisphere (cf. Fig. 357, *F*) has been rotated, in advanced mammals, backward and downward into a ventral position near the midline.

tioned earlier (p. 574), one set of stimuli is sent from cortex to cerebellum by way of the pons for appropriate regulatory effects. The main motor discharge, however, is by way of the *pyramidal tract,* a fiber bundle which extends directly, without intervening relay, from the neopallial cortex to the motor nuclei and columns the length of brain stem and cord. (It is to be noted that this cortical control does not include the visceral system; this has retained its highest centers in the diencephalon.)

With expansion, the cerebral hemispheres tend to cover and envelop the other brain structures. In a primitive mammalian brain (Fig. 345, *A,* p. 566) the cerebrum leaves much of the midbrain exposed; but in a majority of living mammals (as in the horse, Fig. 343, *B*, p. 565) the

* The name is due to the striated appearance caused by the passage of the fiber bundles through the gray matter of the basal nuclei.

midbrain and part of the cerebellum are overlapped, and in man (Fig. 345, *B*) cerebral expansion is so great that all other brain parts are concealed in any but a ventral view. The paired ventricles and the "old-fashioned" areas associated more purely with olfaction—olfactory lobe, hippocampus, and related tracts and nuclei—have been shifted and distorted, with the growth of the mammalian hemispheres, into patterns which are difficult to compare with those of lower and more simply built brains (Fig. 359).

Since the neopallium is essentially a thin sheet of material, underlain by the white fibrous mass of the cerebrum, simple increase in hemisphere bulk fails to keep cortical expansion in step with increase in fiber volume; surface folding is necessary. In small or primitive mammals the neopallial surface is often smooth; in large or more progressive types the surface is generally highly convoluted—thrown into folds which greatly increase the surface area. The folds are termed *gyri,* the furrows between, *sulci.* These are prominent landmarks on the brain surface, and it was once believed that in some cases they were structural boundary markers for specific cortical areas. Further study, however, shows that (apart from the rhinal fissure and a central sulcus in primates) there is no fixed relationship between the convolution pattern and the structural subdivisions of the cortex.

In man, most particularly, the hemispheres are often described as composed of a series of lobes—a *frontal lobe* anteriorly, a *parietal lobe* at the summit of the hemisphere, an *occipital lobe* at the posterior end, and a lateral *temporal lobe.* These terms are, however, purely topographic, and have no precise meaning as regards the architecture or functioning of cortical areas.

The gray matter of the neopallium has a complex histological structure with, in placentals, as many as six superposed cellular layers and masses of intervening fibrils, in contrast to paleopallial and archipallial regions, in which but two cell layers are usually present. The white matter internal to the gray includes, besides a fan of cortical connections to lower brain regions, a great interweaving meshwork of fibers which connect every part of the cortex with every other. An *anterior commissure* (Fig. 353, p. 575) connecting olfactory portions of the two hemispheres is present in all vertebrates, and other commissural fibers are present in the fornix; to connect the neopallial structures a massive new commissure, the *corpus callosum* (Fig. 345, *B,* p. 566), develops in placental mammals.

The complex "wiring" system connecting all parts of the cortex with one another would suggest that the gray matter is essentially a unit, equipotent in all its parts for any cerebral activity. This is to a considerable extent true; experiments show that in laboratory animals a good part of the neopallium may be destroyed without permanently interfering with normal activity, and the results of injury and disease conditions show that the same holds for certain areas of the human brain. On the other hand, it is clear that certain cortical areas are normally associated with specific functions (Fig. 360). We have previously discussed the "old-fashioned" cortical regions devoted mainly to olfactory sensations—the paleopallium and archipallium, represented in mammals in the pyriform lobe and hippocampus, respectively. Regional differentiation is present in the neopallium as well. The front part of the hemispheres includes a motor area. The posterior part is associated with sensory perception. Special regions are associated with eye and ear, in parietal and temporal lobes, while the areas for sensations received from the skin and proprioceptive organs are situated farther forward, close to the motor area. In man and other higher primates a *central sulcus* which crosses the top of the hemisphere from medial to lateral surfaces proves to be the precise boundary between motor

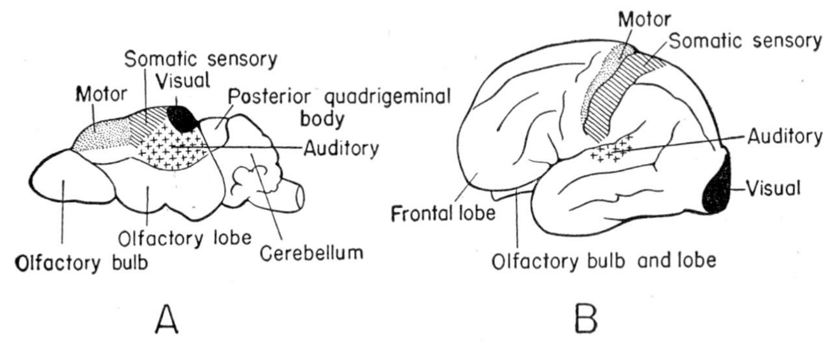

Fig. 360. Lateral view of the brain of *A*, a shrew; *B*, man; to show cortical areas.

and sensory divisions. (In other mammals the boundary is less sharp.) Along the front margin of this sulcus are specific motor areas for each subdivision of the body and limbs, arranged in linear order; along the back margin is an exactly parallel arrangement of loci for sensory reception from the various members. In many mammals nearly the entire surface of the neopallium is occupied by areas associated with specific sensory or motor functions. In man, particularly, however, we find that these specific functional areas occupy only a relatively small part of the neopallium. Between them have developed large "blank areas" of the gray matter. Obviously they are far from blank, and are reasonably considered to be association areas of the highest and most generalized type.

Size of cerebral hemispheres might be thought of as giving a clue to the mental abilities of a mammal. This is true, but subject to qualification. If the amount of cortical surface present is related in any way to intelligence, it is obvious that of two brains of the same size, one with convoluted hemispheres is better than one with a smooth surface. The bulk of the animal concerned affects brain size, presumably because of the need of greater terminal areas for the increased sensory and motor connections, but the brain increase is not absolutely proportionate to body bulk, and large animals tend to have relatively small brains without, it seems, any loss of mental powers. That absolute brain size is not a criterion of intellect is indicated by the fact that the brain of a whale may have five times the volume of that of a man. Nor is the proportion of brain size to body size a perfect criterion, for small South American monkeys may have a brain one-fifteenth or one-twentieth of body weight, while the brain weight of an average man is but one-fortieth that of the body.

Brain Patterns—Summary. In earlier paragraphs of this chapter we described the principal structural and functional elements of the brain and certain of their interconnections. The more important features may be summarized here.

In the posterior part of the brain stem there is present a series of sensory and motor columns closely comparable to those of the spinal cord. As in the cord, direct reflexes may take place between them. The history of brain evolution, however, has been mainly one of the development of higher centers of coordination and association, interposed between sensory and motor areas. In such centers sensory data are assembled and synthesized and resultant motor stimuli sent out to the motor columns of brain stem and cord. These centers have been in the main built up about areas primarily concerned with the special senses of the head region—acustico-lateralis, visual, and olfactory.

1. The primary area of reception for equilibrium and lateral line stimuli lies in the sensory columns of the medulla oblongata; above this developed the cerebellum, important in all vertebrates. This organ initiates no bodily movement, apart from adjustments in posture, but insures that motor directives initiated in other centers be carried out in proper fashion. It may receive fibers from all the somatic senses, but is principally informed by the acustico-lateralis centers adjacent and by fibers from the proprioceptive system of the muscles. Its outgoing, regulatory stimuli are for the most part sent to the midbrain and thence distributed to the motor areas. In mammals, where the cortex assumes direct control over most motor responses, powerful circuits are established connecting cerebellum and cortex in both directions (Fig. 351, p. 572).

2. In lower vertebrates the main centers dominating nervous activity are situated in the anterior regions of the brain stem. (*a*) A great center of coordination in which motor activity is initiated is established in the tectum of the midbrain, in which the primary visual center is situated (Fig. 361). To this region are relayed also stimuli from the nose, ear, and other somatic senses; from it are sent out stimuli to the motor columns. As we ascend the vertebrate scale the tectal area becomes rivalled and then exceeded by the association centers of the cerebral hemispheres; in mammals most of the sensory data once assembled in the midbrain are relayed to the neopallium via the thalamus, and the mesencephalon is

reduced to a reflex center. (*b*) The tectal centers are somatic in nature; corresponding centers for visceral sensations and visceral motor responses

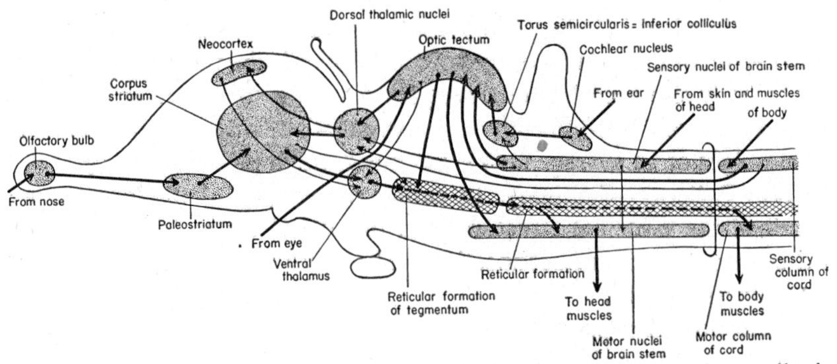

Fig. 361. Diagram of the main centers and "wiring" arrangement of a reptile, in which the tectal region of the midbrain plays a dominant role; the corpus striatum (basal ganglia) is of some importance as a correlation center, but the neocortex (neopallium) is unimportant. The reticular formation of the brain stem (cross hatched) is important in carrying motor impulses to nuclei of the stem and cord. In this oversimplified diagram only a limited number of paths between somatic receptors and effectors are included; visceral centers and paths are omitted, as are cerebellar connections (shown in Figure 351, p. 572).

were early established farther anteriorly and ventrally, in the hypothalamus. This situation remains little changed throughout the vertebrate series.

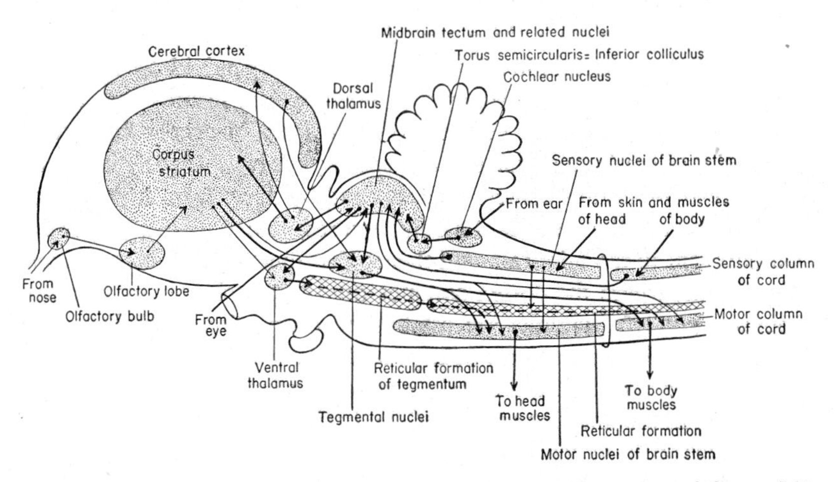

Fig. 362. A "wiring diagram" of a bird brain, comparable to that of Figure 361. The midbrain tectum is still of importance, but the corpus striatum is the dominant center in many regards.

3. As the vertebrate scale is ascended, the cerebral hemispheres, originally only a center for olfactory sensation, have become more and more important as association centers. (*a*) First of cerebral areas to gain importance is that of the basal nuclei, the corpus striatum. Fiber tracts from

the thalamus relay somatic sensations to this body, and return fibers carry motor stimuli back to the midbrain and thence to the motor columns. In reptiles the corpus striatum is a prominent structure which rivals the older tectal center in importance, and in birds the corpus striatum is a large, complex, and dominant center (Fig. 362). (*b*) In mammals, however, a different development has occurred. From the gray matter of the

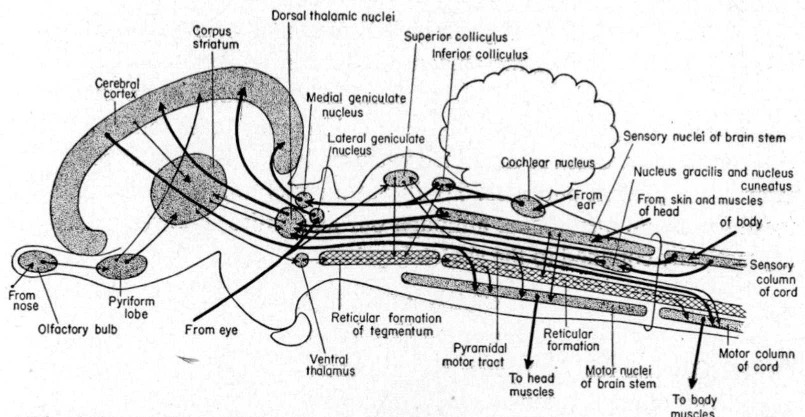

Fig. 363. A "wiring diagram" of a mammalian brain comparable to Figures 361 and 362. The midbrain tectum is reduced to a minor reflex center, and the corpus striatum is relatively unimportant; most sensory impulses are projected "upward" to the cerebral cortex, whence a direct motor path (pyramidal tract) extends to the motor centers of brain stem and cord.

cortex develops a new, greatly expanded correlation and association area, the neopallium. This assumes the greater part of the higher functions once concentrated in the tectum or corpus striatum, gains a complete array of somatic sensory data through projection fibers from the thalamus, and develops direct motor paths to the motor columns of the brain stem and spinal cord (Fig. 363).

APPENDIX I

A SYNOPTIC CLASSIFICATION OF VERTEBRATES

THE CLASSIFICATION given here is presented primarily for the purpose of allowing the student to place in their proper position the forms discussed in the text. In consequence, no attempt is made to list the genera of vertebrates, and in many cases families, suborders, and even orders are neglected when such subdivisions of groups lack interest for present purposes. To fill out the picture, the more important fossil forms are mentioned, although our anatomical knowledge of them is practically confined to the skeletal system.

CLASS AGNATHA

(Jawless Vertebrates)

Orders Osteostraci, Anaspida, Heterostraci. These constitute the ostracoderms of the Silurian and Devonian periods. Representatives of these three groups are shown in Figure 15 (p. 38); Cephalaspis-like forms of the Osteostraci are also shown in Figures 16 and 190 (pp. 38, 322).

Order Cyclostomata. The living cyclostomes (Figs. 12, 13, pp. 34, 36).

Suborder Myxinoidea. Hagfishes.

Suborder Petromyzontia. Lampreys.

CLASS PLACODERMI

(Primitive jawed vertebrates; extinct and confined to the Paleozoic)

Order Acanthodii. Spiny "sharks" (Fig. 17, *A*, p. 39).

Order Arthrodira. The arthrodires; jointed-necked fishes (Fig. 17, *B*).

Order Antiarchi. Related to arthrodires, but with peculiar bony "arms" (Fig. 17, *C*).

Order Macropetalichthyida.

Order Stegoselachii. The last two orders are modified arthrodire relatives which are suspected of being transitional, with reduction of bony armor, to the sharklike fishes.

CLASS CHONDRICHTHYES

(Cartilaginous Fishes)

Subclass Elasmobranchii

(Sharks and Related Forms)

Order Cladoselachii. Primitive Paleozoic sharks (Fig. 18, *A*, p. 40).

Order Selachii. Typical sharks, Paleozoic to Recent, with claspers, narrow-based fins, and so forth (Fig. 18, *B*).

Order Batoidea. The skates and rays (Fig. 19, p. 41).

Subclass Holocephali

Differing from sharks in having an operculum, upper jaws fused to skull, and so on.

Order Bradyodonti. Paleozoic forms, poorly known, mainly represented by tooth plates.

Order Chimaerae. The chimaeras (Fig. 18, *C*).

CLASS OSTEICHTHYES

(The Higher Bony Fishes) (see Fig. 22)

Subclass Actinopterygii

(Ray-Finned Fishes)

Superorder Chondrostei. Primitive ray-finned fishes, with heterocercal tails, and so forth, represented by fossil palaeoniscoids (Fig. 25, *A*, p. 47), mainly Paleozoic, and three modern types: Polypterus (Fig. 25, *B*), the sturgeons (Fig. 26, *B*, p. 47), and the paddlefish (Fig. 26, *A*).

Superorder Holostei. Dominant ray-finned forms of the Mesozoic, with abbreviate heterocercal tails, and so on; living forms include the gar pike and Amia (Fig. 27, p. 49).

Superorder Teleostei. The dominant fishes of Cenozoic and Recent times, with a homocercal tail. Include many thousands of forms, classed in a number of orders (Fig. 28, p. 50).

Subclass Choanichthyes

(With Internal Nostrils)

Order Crossopterygii. Forms broadly ancestral to land vertebrates. Mainly Paleozoic fossils; one aberrent living form (Fig. 23, p. 45).

Order Dipnoi. The lungfishes, including three living genera. Many similarities to ancestral types, but aberrant in teeth, skulls, and so forth (Fig. 24, p. 46).

CLASS AMPHIBIA

(Tetrapods, but without development of amniote type of egg)

Subclass Apsidospondyli

(Amphibians in which vertebral centra were primitively formed from cartilaginous blocks in the embryo [cf. Fig. 85, p. 155])

Superorder Labyrinthodontia. The stem Amphibia, extinct, but dominant in late Paleozoic and Triassic times. Includes many dozens of fossil genera arranged in several orders.

Superorder Salientia, Order Anura. The frogs and toads; living forms have highly specialized limbs and shortened trunks, and so on. Vertebrae formed mainly from neural arches; original elements of centra reduced or absent.

Subclass Lepospondyli

(Vertebral centra formed by direct spool-shaped ossification around notochord [Fig. 85])

Orders Aistopoda, Nectridia, Microsauria. These are late Paleozoic fossils; the last consists of small forms from which the two living orders may have descended.

Order Urodela. The salamanders and newts, with a normal body form, but many degenerate characters.

Order Apoda (or Gymnophiona). Wormlike, burrowing types.

CLASS REPTILIA

(Amniotes, but without advanced avian or mammalian characters [cf. Fig. 31, p. 56])

Subclass Anapsida

(Without a Temporal Opening)

Order Cotylosauria. Archaic "stem reptiles" of the late Paleozoic and Triassic.

Order Chelonia (Testudinata). The turtles.

Subclass Ichthyopterygia

Order Ichthyosauria. The ichthyosaurs, with a body highly specialized for marine life.

Subclass Synaptosauria

(Extinct groups, characterized by a single temporal opening high up on the side of the cheek)

Order Protorosauria. Includes various obscure Permian and Mesozoic reptiles.

Order Sauropterygia. The plesiosaurs and their relatives; marine Mesozoic reptiles swimming by means of limbs transformed into powerful paddles.

Subclass Lepidosauria

(Diapsid type of temporal region; without archosaur specializations)

Order Eosuchia. Permian and Triassic ancestral diapsids.

Order Rhynchocephalia. The living Sphenodon of New Zealand and fossil relatives.

Order Squamata. Lizards and snakes. Related to the last, but temporal arches reduced.

Subclass Archosauria

(Ruling reptiles. Diapsid temporal region, with specializations tending toward bipedal life [*cf. Fig. 32, p. 57*]*)*

Order Thecodontia. Triassic ancestors of dinosaurs, birds, and others.

Order Crocodilia. Crocodiles and alligators; degenerate amphibious survivors of archosaur group.

Order Pterosauria. Extinct flying reptiles with membrane wing.

Order Saurischia. "Reptile-like" dinosaurs with triradiate pelvis. Carnivores and large amphibious forms.

Order Ornithischia. "Birdlike" dinosaurs with tetraradiate pelvis; herbivorous. Bipeds (including duckbills), armored and horned quadrupeds.

Subclass Synapsida

(Lateral temporal opening; forms leading to mammals. Extinct)

Order Pelycosauria. Primitive Permian mammal-like reptiles, close to stem reptiles.

Order Therapsida. Advanced mammal-like forms of late Permian and Triassic.

CLASS AVES

(Winged archosaur descendants, with feathers, temperature control, and so forth)

Subclass Archaeornithes

(Primitive Jurassic fossil birds with many reptilian characters)

Subclass Neornithes

(All Other, "Modernized," Birds)

Superorder Odontognathae. Toothed birds of the Cretaceous.

Superorder Palaeognathae. Including mainly the ostrich-like birds, or ratites, with relatively primitive structures.

Superorder Neognathae. All remaining birds, arranged in some twenty-three orders, but all essentially similar in most anatomical features.

CLASS MAMMALIA

(Animals with hair, nursing habit, brain of advanced type, and so on)

Subclass Prototheria

(Egg-Laying Mammals)

Order Monotremata. The duckbill and spiny anteater of the Australian region.

Subclass Theria

(Mammals Bearing the Young Alive)

Infraclass and Order Pantotheria. Typical small Jurassic fossil mammals, probably ancestral to remaining groups. (There are two further Jurassic orders of obscure relationships.)

Infraclass Allotheria, Order Multituberculata. An extinct group of primitive but aberrant mammals, perhaps comparable in habits to later rodents. Jurassic to Eocene.

Infraclass Metatheria, Order Marsupialia. The pouched mammals, as opossum and many Australian types. Young born alive, but at immature stage.

Infraclass Eutheria. The higher mammals, with an efficient placenta.

Order Insectivora. Ancestral placentals and modern descendants, such as shrews, moles, hedgehog.

Order Chiroptera. Bats.

Order Primates. Essentially an arboreal offshoot of the primitive placental stock.

Suborder Lemuroidea. Tree shrews and lemurs.

Suborder Tarsioidea. Tarsius and extinct relatives, transitional between lemurs and monkeys.

Suborder Anthropoidea. Monkeys, apes, and man.

INFRAORDER PLATYRRHINI. South American monkeys, with nostrils opening sideways.

Family Hapalidae. Marmosets.

Family Cebidae. Typical South American monkeys.

INFRAORDER CATARRHINI. Old World monkeys, apes, and man; nostrils open downward.

Family Cercopithecidae. Old World monkeys.

Family Simiidae. Manlike apes.

Family Hominidae. Men.

Order Carnivora. The carnivores.

Suborder Creodonta. Extinct archaic carnivores.

Suborder Fissipedia. Modern land carnivores.

INFRAORDER EUCREODI. Extinct ancestors of modern types.

INFRAORDER AELUROIDEA. The "cats" and relatives.

Family Viverridae. Civets, mongoose, and the like; primitive Old World aeluroids.

Family Hyaenidae. Hyaenas.

Family Felidae. Cats, lions, tigers, and others.

INFRAORDER ARCTOIDEA. The dogs and relatives.

Family Mustelidae. Primitive. Weasels, skunks, badgers, otters, and so on.

Family Canidae. Dogs, wolves, foxes.

Family Procyonidae. Raccoons, pandas, kinkajous.

Family Ursidae. Bears.

Suborder Pinnipedia. Marine carnivores: seals, sea lion, walrus.

Orders Amblypoda, Dinocerata, Embrithopoda, Astrapotheria, Litopterna, Notoungulata. Extinct orders of ungulates, mainly of archaic character, the last three characteristic of South America.

Order Hyracoidea. The conies of Africa and Syria; rabbit-like in habits, but actually ungulates. This order and the next two are related, and probably of African origin.

Order Proboscidea. The elephants and fossil relatives, mammoths and mastodons.

Order Sirenia. The sea cows—manatee and dugong; an aquatic offshoot of an ungulate stock.

Order Condylarthra. Primitive extinct ungulates, possibly ancestral to next order.

Order Perissodactyla. Odd-toed ungulates.

Suborder Hippomorpha.

Family Equidae. Horses, asses, zebras.

Family Titanotheriidae. Titanotheres—large, ungainly, horned, fossil forms.

Family Chalicotheriidae. Extinct forms related to the last two—but with claws, not hoofs.

Suborder Tapiromorpha.

Family Tapiridae. Tapirs.

Family Rhinocerotidae. Rhinoceroses.

Order Artiodactyla. Even-toed ungulates.

Suborder Palaeodonta. The earliest and most primitive fossil artiodactyls.

Suborder Suina. Relatively primitive types with simple stomachs, including the following living families as well as several extinct ones:

Family Suidae. Pigs of the Old World.

Family Dicotylidae. Peccaries of the New World.

Family Hippopotamidae. The hippopotamus.

Suborder Ruminantia. Cud chewers, with complex stomach, and selenodont teeth.

INFRAORDER TYLOPODA. Primitive cud chewers, including, in addition to early extinct families:

Family Camelidae. Camels, llamas.

Family Oreodontidae. The oreodons, short-legged ruminants abundant in North American fossil deposits.

INFRAORDER PECORA. Advanced ruminants, mostly with horns or antlers, including, besides extinct forms:

Family Tragulidae. Chevrotains, tiny, hornless, deerlike animals of the tropical Old World.

Family Cervidae. The deer tribe.

Family Giraffidae. The giraffe and okapi of Africa.

Family Antilocapridae. The American pronghorn.

Family Bovidae. The cattle family, mainly Old World forms, including bison, sheep, goats, and numerous types of antelopes.

Order Edentata. So-called "toothless" mammals, developed in South America.

Suborder Pilosa. Hairy edentates.

INFRAORDER GRAVIGRADA. Extinct ground sloths.

INFRAORDER TARDIGRADA. Family Bradypodidae. Tree sloths.

INFRAORDER VERMILINGUA. Family Myrmecophagidae. South American anteaters.

Suborder Loricata. Armored edentates.

Family Dasypodidae. Armadillos.

Family Glyptodontidae. The extinct giant glyptodonts.

Order Pholidota. The Old World pangolin, an anteater, but not closely related to the last order.

Order Tubulidentata. The aardvark of Africa; an anteater, but not related to the preceding orders.

Order Cetacea. Whales.

Suborder Archaeoceti. Extinct ancestral whales.

Suborder Odontoceti. The toothed whales, porpoises, dolphins.

Suborder Mysticeti. Whalebone whales.

Order Rodentia. Gnawing animals (except rabbit group).

Suborder Sciuromorpha. Squirrels, gophers, woodchuck, beaver, and others.

Suborder Hystricomorpha. Guinea pig and many other South American rodents, porcupines.

Suborder Myomorpha. Rats and mice.

Order Lagomorpha. Hares and rabbits; gnawing forms, but not closely related to last order.

SCIENTIFIC TERMINOLOGY

In anatomical terminology common Latin (or Greek) words are used as such for any part of the body for which the ancients had a name. For numerous other structures, scientific names have been invented (1) by using, in a new sense, some classical word which seemed to be descriptive of the part concerned, or (2) commonly, by combining Greek or Latin roots to form a new compound term. The student frequently attempts to memorize such terms without understanding their meaning and with consequent mental indigestion. We give here the roots from which many of these descriptive terms and compounds are derived, as an aid to comprehension. As will be seen, some names formed by the anatomists are rather fanciful or far-fetched; some are none too appropriate. This list is not intended, of course, as a glossary or dictionary of scientific words. We have not, for instance, included common names of bones and muscles. Most of the terms used in this book are defined or discussed in the text. For a wider vocabulary, use of a standard biological or medical dictionary* is recommended, but the larger editions of Webster and the like are satisfactory in most regards. Abbreviations: *F.*, French; *G.*, Greek; *L.*, Latin; *NL.*, "New" Latin; *Sp.*, Spanish.

A-, ab. L. prefix implying separation.
Abdomen. L., from *abdere* (?), to hide.
Abducens. L., *ab,* away, + *ducens,* leading.
Abductor. L., *ab,* away, + *ducere,* to lead.
Accessory. L., *accessorius,* supplementary.
Acetabulum. L., *acetabulum,* vinegar cup.
Acelous, G., *a,* not, + *koilos,* hollow.
Acoustic. G., *akoustikos,* pertaining to hearing.
Acrania. G., *a,* without, + *krania,* heads.
Acrodont. G., *akron,* height or extremity, + *odous,* tooth.
Acromion. G., *akron,* height or extremity, + *omos,* shoulder.
Ad. L. prefix, to, toward, at, or near.
Adductor. L., *ad,* to, + *ducere,* to lead.
Adrenal. L., *ad,* near, + *renes,* kidneys.

* Such as Dorland, W. A. N.; American Illustrated Medical Dictionary. 21st edition. Philadelphia, W. B. Saunders Company, 1948.

Alisphenoid. L., *ala,* wing, + G., *sphen,* wedge, + G., *eidos,* form.
Allantois. G., *allas,* sausage, + *eidos,* form, appearance.
Alveolus. L., *alveolus,* little cavity.
Ambiguus. L., *ambiguus,* uncertain, changeable.
Amnion. G., *amnion,* fetal membrane.
Amphi-. G. prefix, on both sides; hence, around or double.
Amphibia. G., *amphi,* double, + *bios,* life.
Amphicelous. G., *amphi,* both, + *koilos,* hollow.
Amphioxus. G., *amphi,* both, + *oxys,* sharp.
Amphiplatyan. G., *amphi,* both, + *platys,* flat.
Amphistylic. G., *amphi,* both, + *stylos,* pillar.
Ampulla. L., *ampulla,* a flask or vessel swelling in the middle.
A-, an-. G., prefix (alpha privative), without or not.
Ana-. G. prefix, on, upward, throughout, frequently, or reinforcing a meaning.
Analogy. G., *ana,* according to, + *logos,* due ratio; hence, proportionate.
Anamniota. G., *an,* without, + *amnion,* fetal membrane.
Anapsid. G., *an,* not, + *apsis,* arch.
Anastomosis. G., *anastomoein,* to bring to a mouth, cause to communicate.
Ankylosis. G., *ankylosis,* stiffening of the joint.
Annulus. L., *anulus* (*annulus*), a ring.
Anura. G., *an,* without, + *oura,* tail.
Anus. L., *anus,* fundament.
Apo. G. prefix, from.
Apoda. G., *a,* without, + *poda,* feet.
Aponeurosis. G., *apo,* from, + *neuron,* tendon.
Apophysis. G., *apo,* from, + *physis,* growth.
Appendicular. L., *appendere,* to hang upon.
Arachnoid. G., *arachnes,* spider, + *eidos,* shape or likeness.
Arch-, archi-. G. prefix, first or chief; hence, primitive or ancestral.
Archenteron. G., *arch,* first, + *enteron,* intestine, gut.
Archipallium. G., *archi,* first, + *pallium,* cloak.
Archipterygium. G., *archesthai,* to begin, + *pterygion,* a little wing.
Arcualia (pl.). L., *arcualis,* bow-shaped.
Arrector. L., *arrigere,* to raise.
Arytenoid. G., *arytaina,* jug, + *eidos,* shape or likeness.
Astragalus. G., *astragalos,* ankle bone, used as a die (commonly pl.).
Atrium. L., *atrium,* court, entrance hall.
Auditory. L., *audire,* to hear.
Auricle. L. dim., *auricula,* external ear.
Auto-. G. prefix, *autos,* self.
Autonomic. G., *autos,* self, + *nomos,* law.
Autostylic. G., *autos,* self, + *stylos,* pillar.
Axial. L., *axis,* axle of a wheel, the line about which any body turns.

Azygos. G., *a,* not, + *zygon,* yoke; hence, unpaired.
Basal. L., *basis,* footing or base.
Basi-. L. prefix, pertaining to the base.
Basibranchial. L., *basis,* base, + *branchiae,* gills.
Basihyoid. L., *basis,* base, + G., *hyoeides,* Y-shaped.
Bi-. L. prefix, two, twice, or double.
Biceps. L., *bis,* twice, + *caput,* head.
Bilateral. L., *bi,* two, + *latus,* side.
Blasto-. G. prefix, bud, germ, or sprout.
Blastocele. G., *blastos,* germ, + *koilos,* hollow.
Blastoderm. G., *blastos,* germ, + *derma,* skin.
Blastodisc. G., *blastos,* germ, + *diskos,* a round plate, quoit.
Blastomere. G., *blastos,* germ, + *meros,* part.
Blastopore. G., *blastos,* germ, + *poros,* passage, opening.
Blastula. L. dim. of G. *blastos,* germ.
Brachial. L., *brachialis,* belonging to the arm.
Brachium (pl. **-ia**). L., *brachium,* arm, especially, the forearm.
Branchial. L., *branchiae,* or G., *branchia,* gills.
Branchiostegal. G., *branchia,* gills, + *stegein,* to cover.
Bronchus. G., *bronchia,* end of windpipe.
Buccal. L., *bucca,* cheek.
Bulbus. L., *bulbus,* bulb, swollen root.
Bunodont. G., *bounos,* mound, + *odous,* tooth.
Cecum. L., *caecus, -a, -um,* blind.
Calcaneum. L., *calcaneum,* heel.
Callosum. L., *callosus, -a, -um,* thick-skinned.
Calyx (pl. **calices**). L., *calyx,* husk, cup-shaped protective covering.
Caninus. L., *caninus,* pertaining to a dog.
Capillary. L., *capillaris,* pertaining to the hair.
Capitulum. L. dim. (*caput*), small head.
Caput (pl. **capita**). L., *caput,* head.
Carapace. NL., *carapax,* bony or chitinous covering.
Cardiac. G., *kardiakos* (*kardia*), pertaining to the heart.
Cardinal. L., *cardinalis,* pertaining to a door hinge.
Carnassial. F., *carnassier,* carnivorous (L., *caro, carnis,* flesh).
Carnivorous. L., *caro,* flesh, + *vorare,* to devour.
Carnosus. L., *carnosus,* fleshy.
Carpus. G., *karpos,* wrist.
Cartilago (pl. **-agines**). L., *cartilago,* gristle, cartilage.
Caudal. L., *cauda,* tail.
Cava. L., *cavus, -a, -um,* hollow.
Cavernosus. L., *caverna,* a hollow or cave.
Cephalic. G., *kephale,* head.
Cephalo-. G. combining form, *kephale,* head.

Ceratobranchial. G., *keras,* horn, + *branchia,* gills.
Ceratotrichia. G., *keras,* horn, + *thrix,* hair.
Cerebellum. L. dim., *cerebrum,* brain.
Cerebrum. L., *cerebrum,* brain.
Cervical. L., *cervix,* neck.
Chiasma. G., *chiasma,* figure of X.
Choana (pl. **-ae**). G., *choane,* funnel.
Choledochus. G., *chole,* bile, + *dochos* (*dechomai*), container.
Chondrichthyes. G., *chondros,* cartilage, + *ichthys,* fish.
Chondro-. G., combining form, cartilaginous.
Chondroblast. G., *chondros,* cartilage, gristle, + *blastos,* shoot or germ.
Chondroclast. G., *chondros,* cartilage, + *klaein,* to break.
Chorda. G., *chorde,* string of gut, cord.
Choroid. G., *chorion,* skin, + *eidos,* likeness.
Chorion. G., *chorion,* skin.
Chromaffin. G., *chroma,* color, + L., *affinis,* showing affinity for.
Chromatophore. G., *chroma,* color, + *pherein,* to bear.
Chromosome. G., *chroma,* color, + *soma,* body.
Chyme. G. *chymos,* juice.
Ciliary. L., *cilium* (pl. *cilia*), eyelash.
Circum-. L. prefix, around.
Cloaca. L., *cloaca,* sewer, drain.
Cnemial. G., *kneme,* lower leg.
Cochlea. L., *cochlea* (G., *kochlias*), spiral, snail shell.
Celiac. G., *koilia,* belly.
Celom (**e**). G., *koiloma,* a hollow.
Colon. L., *colon,* great gut.
Columella. L. dim. (*columna*), little pillar.
Concha. L., *concha,* bivalve, oyster shell.
Constrictor. L., *constringere,* to draw together.
Commissure. L., *commissura* (*cum* + *mittere*), connection.
Condyle. G., *kondylos,* knuckle.
Conjunctiva. L., *conjunctivus,* connecting.
Coprodeum. G., *kopros,* dung, + *hodaios,* pertaining to a way.
Corium. G., *chorion,* skin, leather.
Cornea. L., *corneus,* horny.
Coronary. L., *coronarius,* pertaining to a wreath or crown; hence, encircling.
Cortex. L., *cortex,* bark, rind.
Cortical. L., *cortex, -icis,* bark, rind.
Cosmin. G., *kosmos,* orderly arrangement, + *eidos,* form.
Costa. L., *costa,* rib.
Costal. L., *costa,* rib.
Cranial. G., *kranion,* skull.

Cribriform. L., *cribrum,* sieve, + *forma,* form.
Cricoid. G., *krikos,* ring, + *eidos,* form.
Crista. L., *crista,* crest.
Crus (pl. **crura**). L., *crus,* leg.
Ctenoid. G., *kteis* (gen. *ktenos*), a comb, + *eidos,* form.
Cuneiform. L., *cuneus,* wedge, + *forma,* shape.
Cutis. L., *cutis,* skin.
Cycloid. G., *kyklos,* circle, + *eidos,* form.
Cystic. G., *kystis,* the bladder, a bag or pouch.
Cytoplasm. G., *kytos,* cell, + *plasma,* plasma, anything molded.
De-. L. prefix signifying down, away from, deprived of.
Deciduous. L., *deciduus* (*de* + *cado*), falling off.
Decussatio. L., *decussatio,* crosswise intersection.
Decussation. L., *decussatio,* intersection of two lines, as in Roman X.
Deferens. L., *de,* away, + *ferens,* carrying.
Dens. L., *dens, dentis,* tooth.
Depressor. L., *de,* down, + *premere,* to press.
Dermal. G., *derma,* skin.
Dermatome. G., *derma,* skin, + *temnein,* to cut.
Di-. G., *dis,* twice; hence, twofold or double.
Di-, dia-. G. prefix, through, between, apart, across.
Diaphragm. G., *diaphragma,* partition, midriff.
Diaphysis. G., *dia,* between, + *physis,* growth.
Diapophysis. G., *dia,* apart, + *apophysis,* outgrowth.
Diapsid. G., *di,* double, + *apsis,* arch.
Diarthrosis. G., *dia,* through, + *arthroun,* to fasten by a joint.
Diastema (pl. **-ata**). G., *diastema,* interval.
Digit. L., *digitus,* finger.
Diphycercal. G., *diphyes,* twofold, + *kerkos,* tail.
Diplospondylous. G., *diploos,* double, + *spondyle,* vertebra.
Dipnoi. G., *di,* double, + *pnein,* to breathe.
Distal. L., *distare,* to stand apart.
Dorsal. L., *dorsum,* back.
Duct. L., *ducere,* to lead or draw.
Duodenum. L., *duodeni,* twelve (meaning twelve fingerbreadths).
E-, ex-. L. prefix, out, out of, from.
Ectepicondyle. G., *ek,* out, + *epi,* upon, + *kondylos,* knuckle, knob.
Ectoderm. G., *ektos,* outside, + *derma,* skin.
Effector. L., *efficere,* to bring to pass.
Efferent. L., *ex,* out, + *ferre,* to bear.
Ejaculatory. L., *e,* out, + *jacere,* to throw.
Ek-, ekto-. G. prefix, out of, from, outside.
Embolomerous. G., *en,* in, + *ballein,* to throw, + *meros,* portion or part.
En-, endo-. G. prefix, in, within.

Endocardium. G., *endon,* within, + *kardia,* heart.
Endochondral. G., *endon,* within, + *chondros,* cartilage.
Endocrine. G., *endon,* within, + *krinein,* to separate.
Endoderm. G., *endon,* within, + *derma,* skin.
Endolymph. G., *endon,* within, + L., *lympha,* water.
Endometrium. G., *endon,* within, + *metra,* womb.
Endoneurium. G., *endon,* within, + *neuron,* nerve.
Endoskeleton. G., *endon,* within, + *skeletos,* dried up.
Endostyle. G., *endon,* within, + *stylos,* pillar.
Endothelium. G., *endon,* within, + *thele,* nipple.
Entepicondyle. G., *en.,* in, + *epi,* upon, + *kondylos,* knuckle.
Enzyme. G., *en,* in, + *zyme,* leaven.
Ependyma. G., *ependyma,* upper garment.
Epiaxial. G., *epi,* on, + *axis,* center line.
Epibranchial. G., *epi,* on, + *branchia,* gills.
Epicardium. G., *epi,* on, + *kardia,* heart.
Epicondyle. G., *epi,* on, + *kondylos,* knuckle.
Epidermis. G., *epi,* on, + *derma,* skin.
Epididymis. G., *epi,* on, + *didymoi,* testicles.
Epiglottis. G., *epi,* on, + *glotta,* tongue.
Epimere. G., *epi,* on, + *meros,* part.
Epineurium. G., *epi,* on, + *neuron,* nerve.
Epiphysis. G., *epi,* on, + *physis,* growth.
Epiploic. G., *epiploon,* caul, omentum.
Epithalamus. G., *epi,* on, + *thalamos,* chamber.
Epithelium. G., *epi,* on, + *thele,* nipple.
Erythrocyte. G., *erythros,* red, + *kytos,* cell.
Esophagus. G., *oisein* (*phero*), to carry, + *phagein,* to eat.
Ethmoid. G., *ethmos,* sieve, + *eidos,* form.
Excretion. L., (*excretus*) *ex,* out, + *cernere,* to sift.
Exocrine. G., *ex,* out, + *krinein,* to separate.
Extensor. L., *ex,* out, + *tendere,* to stretch.
Extrinsic. L., *extrinsecus,* on the outside.
Facialis. L., *facies,* face.
Falciform. L., *falx,* sickle, + *forma,* shape.
Falx. L., *falx, falcis,* sickle.
Fascia (pl. **-iae**). L., *fascia,* band.
Fiber. L., *fibra,* string, thread.
Fibril. NL., *fibrilla,* a little thread.
Filoplume. L., *filum,* thread, + *pluma,* soft feather, down.
Filum. L., *filum,* thread.
Fimbria. L., *fimbriae,* threads, fringe.
Firmisternal. L., *firmus,* steadfast, strong, + G., *sternon,* chest.
Fissure. L., *fissura* (*findo*), a cleft.

Flagellum (pl. **-a**). L., *flagellum,* a little whip.
Flexor. L., *flexus,* bent.
Flocculus. NL. dim. (*floccus*), a tuft of wool.
Follicle. L., *folliculus,* a small bag.
Fornix. L., *fornix,* arch or vault.
Fovea. L., *fovea,* small pit.
Frontal. L., *frons, frontis,* forehead, brow.
Fundus. L., *fundus,* bottom.
Funiculus. L., *funiculus,* a slender rope.
Gametes. G., *gametes,* spouse.
Ganglion. G., *ganglion,* a swelling under the skin.
Gastralia (pl.). G., *gaster,* belly.
Gastrula. NL. dim. (from G., *gaster,* stomach).
Geniculate. L., *geniculatus,* with bent knee.
Genital. L., *genitalis* (*gigno*), pertaining to birth.
Germinal. L., *germen,* bud, germ.
Germinative. L., *germen,* bud, germ.
Glans. L., *glans,* acorn.
Glenoid. G., *glene,* socket, + *eidos,* form.
Glomerulus. L. dim. of *glomus,* a ball.
Glomus (pl. **glomera**). L., *glomus,* ball.
Glossopharyngeus. G., *glossa,* tongue, + *pharynx,* throat.
Glottis. G., *glottis,* mouth of the windpipe.
Gluteus. G., *gloutos,* rump.
Gnathos. G., *gnathos,* jaw.
Gnathostomata (pl.). G., *gnathos,* jaw, + *stoma,* mouth.
Gonad. G., *gone,* seed.
Granulocytes. L., *granulum,* small grain, + G., *kytos,* cell.
Granulosus. L., *granulosus,* full of grains.
Granulum. L., *granulum,* small grain.
Guanin (**e**). Sp., *guano,* dung of sea fowl.
Guanophore. Sp. from Peruvian, *huanu,* dung, + G., *pherein,* to bear.
Gubernaculum. L., *gubernaculum,* helm.
Gular. L., *gula,* throat.
Gyrus (pl. **gyri**). G., *gyros,* a turn.
Habenula. L., *habena,* strap.
Haemal, hemal. G., *haima,* blood.
Hamatum. L., *hamatus,* hook-shaped.
Hemi. G. prefix, signifying half.
Hemibranch. G., *hemi,* half, + *branchia,* gills.
Hemichordata. G., *hemi,* half, + *chorde,* string.
Hemipenis. G., *hemi,* half, + L., *penis,* penis.
Hemisphere. G., *hemi,* half, + *sphaira,* ball.
Hemocytoblast. G., *haima,* blood, + *kytos,* cell, + *blastos,* germ.

Hemoglobin. G., *haima,* blood, + L., *globus,* sphere.
Hemopoietic. G., *haima,* blood, + *poietikos,* creative.
Hepatic. L., *hepar,* liver.
Hetero-. G. combining form signifying other, different.
Heterocercal. G., *heteros,* other, + *kerkos,* tail.
Heterocelous. G., *heteros,* other, + *koilos,* hollow.
Heterodont. G., *heteros,* other, + *odous,* tooth.
Heterotopic. G., *heteros,* other, + *topos,* place.
Hippocampus. G., *hippos,* horse, + *kampos,* sea monster.
Histology. G., *histos,* web, + *logos,* discourse, account.
Holo-. G. combining form signifying whole.
Holoblastic. G., *holos,* whole, + *blastos,* germ.
Holobranch. G., *holos,* whole, + *branchia,* gills.
Holocephali. G., *holos,* whole, + *kephale,* head.
Holonephros. G., *holos,* whole, + *nephros,* kidney.
Holostei. G., *holos,* whole, + *osteon,* bone.
Homo-. G. combining form signifying one and the same.
Homocercal. G., *homos,* same, + *kerkos,* tail.
Homolecithal. G., *homos,* same, + *lekithos,* yolk.
Homology. G., *homos,* same, + *logos,* ratio.
Homothermous. G., *homos,* same, + *thermos,* hot.
Hormone. G., *hormaein,* to excite.
Humor. L., *humor,* moisture, fluid.
Hyaline. G., *hyalos,* glass.
Hyoid. G., *hyoeides,* U-shaped.
Hyomandibular. G., *upsilon* (Y-shaped letter) + L., *mandibula,* jaw.
Hyostylic. G., *upsilon* (Y-shaped letter) + *stylos,* pillar.
Hypaxial. G., *hypo,* under, + L., *axis,* center line, axis.
Hypo-. G. prefix signifying under, below.
Hypobranchial. G., *hypo,* below, + *branchia,* gills.
Hypoglossal. G., *hypo,* under, + *glossa,* tongue.
Hypomere. G., *hypo,* under, + *meros,* part.
Hypophysis. G., *hypo,* under, + *physis,* growth.
Hypothalamus. G., *hypo,* under, + *thalamos,* chamber, couch.
Hypsodont. G., *hypsos,* height, + *odous,* tooth.
Hypural. G., *hypo,* under, + *oura,* tail.
Ileum. G., *eilein,* to wind or turn.
In-. L. prefix signifying not; also signifying in, into, within, toward, on.
Incisor. L., *incisus* (*incidere*), cut.
Incus. L., *incus,* anvil.
Inductor. L., *inducere,* to lead on, excite.
Infra-. L. prefix signifying below, lower than.
Inframeningeal. L., *infra,* below, + G., *meninx,* membrane.
Infraparietal. L., *infra,* below, + *paries,* wall.

Infraspinous. L., *infra,* below, + *spina,* spine.
Infundibulum. L., *infundibulum,* a funnel.
Inguinal. L., *inguina,* groin.
Integument. L., *in,* over, + *tegere,* to cover.
Inter-. L. prefix signifying between, among.
Intercalated. L., *inter,* between, + *calare,* to call.
Intercostal. L., *inter,* between, + *costa,* rib.
Intermaxillary. L., *inter,* between, + *maxilla,* jaw.
Interrenal. L., *inter,* between, + *renes,* kidneys.
Interstitial. L., *inter,* between, + *sistere,* to set.
Intervertebral. L., *inter,* between, + *vertebra,* joint.
Intestine. L., *intus,* within.
Intrinsic. L., *intrinsecus,* inward.
Invagination. L., *in,* in, + *vagina,* sheath.
Invertebrate. L., *in,* not, + *vertebratus,* jointed.
Iridocyte. G., *iris, -idos,* rainbow, + *kytos,* cell.
Iris. G., *iris,* rainbow.
Ischiofemoral. G., *ischion,* hip, + L., *femur,* thigh.
Ischium (pl. **-ia**). G., *ischion,* hip.
Iso-. G. prefix signifying equal.
Isolecithal. G., *isos,* equal, + *lekithos,* yolk.
Isomer. G., *isos,* equal, + *meros,* part.
Jugal. L., *jugum,* yoke.
Jugular. L., *jugularis* (*jugulum*), pertaining to the neck.
Jejunum. L., *jejunus,* empty (of food).
Labial. L., *labialis* (*labia*), pertaining to the lips.
Lacerate. L., *lacerare,* to tear.
Lacrimal. L., *lacrima,* tear.
Lagena. L., *lagena,* flask.
Lamina (pl. **-ae**). L., *lamina,* thin plate.
Larva (pl. **-ae**). L., *larva,* ghost, mask.
Larynx. G., *larynx,* upper part of windpipe.
Lateral. L., *lateralis* (*latus*), pertaining to a side.
Lepidotrichia. G., *lepis,* scale, + *thrix,* hair.
Leukocyte. G., *leukos,* white, + *kytos,* cell.
Levator. L., *levare,* to raise.
Ligamentum. L., *ligamentum,* a bandage.
Lipid. G., *lipos,* fat, + *eidos,* resemblance.
Lipo-. G. combining form signifying fat.
Lipophore. G., *lipos,* fat, + *phoros* (*phero*), bearing.
Lobus. G., *lobos,* lobe.
Lophodont. G., *lophos,* ridge, + *odous,* tooth.
Lucidum. L., *lucidus* (*lux*), full of light, clear.
Lumbar. L., *lumbare,* apron for the loins.

Luteum. L., *luteus,* yellow.
Lymphocyte. L., *lympha,* water, + G., *kytos,* cell.
Macrophage. G., *makros,* large, + *phagein,* to eat.
Macula. L., *macula,* spot, stain.
Malleus. L., *malleus,* hammer.
Mammillary. L., *mamillaris* (*mamma, -ae*), of or in the breast.
Marginal. L., *marginalis* (*margo*), bordering.
Marsupium. L., *marsupium,* pouch.
Mastoid. G., *mastos,* breast, + *eidos,* resemblance.
Matrix. L., *matrix* (*mater*), womb, groundwork, or mold.
Meatus (pl. **-us**). L., *meatus,* passage.
Medial. L., *medialis* (*medius*), pertaining to the middle.
Mediastinum. L., *mediastinus,* servant, drudge.
Medulla. L., *medulla,* marrow, pith.
Melanin. G., *melas, melanos,* black.
Membrane. L., *membrana,* skin.
Meninx (pl. **meninges**). G., *meninx,* membrane.
Mes-, meso-. G. prefix signifying middle.
Mesencephalon. G., *mesos,* middle, + *en,* in, + *kephale,* head.
Mesenchyme. G., *mesos,* middle, + *en,* in, + *chymos,* juice.
Mesentery. G., *mesos,* midway between, + *enteron,* gut.
Mesocardium. G., *mesos,* middle, + *kardia,* heart.
Mesoderm. G., *mesos,* middle, + *derma,* skin.
Mesolecithal. G., *mesos,* middle, + *lekithos,* yolk.
Mesonephros. G., *mesos,* middle, + *nephros,* kidney.
Mesopterygium. G., *mesos,* middle, + *pterygion,* little wing.
Mesorchium. G., *mesos,* middle, + *orchis,* testis.
Mesovarium. G., *mesos,* middle, + L., *ovarium,* ovary.
Meta-. G., prefix meaning after, next; denoting change of time or situation.
Metabolic. G., *metabole,* change.
Metacarpus. G., *meta,* after, + L., *carpus,* wrist.
Metamere. G., *meta,* after, + *meros,* part.
Metamorphosis. G., *meta,* signifying change, + *morphe,* form.
Metanephros. G., *meta,* after, + *nephros,* kidney.
Metapleura. G., *meta,* after, + *pleura,* side.
Metapodial. G., *meta,* after, + *pous, podos,* foot.
Metatarsus. G., *meta,* after, + L., *tarsus,* ankle.
Metencephalon. G., *meta,* after, + *enkephalos,* brain.
Molar. L., *molaris* (*mola*), pertaining to a millstone.
Monocyte. G., *monos,* single, + *kytos,* cell.
Mucus. L., *mucus,* snivel, slippery secretion.
Multangulum. L., *multus,* many, + *angulus,* angle.
Myelencephalon. G., *myelos,* marrow, + *enkephalos,* brain.
Myelin. G., *myelos,* marrow.

Myo-. G. combining form signifying muscle.
Myocardium. G., *mys, myos,* muscle, + *kardia,* heart.
Myocomma. G., *mys, myos,* muscle, + *komma,* implying separation.
Myodome. G., *mys, myos,* muscle, + L., *domus,* house.
Myomere. G., *mys, myos,* muscle, + *meros,* part.
Myotome. G., *mys, myos,* muscle, + *tome,* a cutting.
Naris (pl. **-es**). L., *naris,* nostril.
Neopallium. G., *neos,* youthful, new, + L., *pallium,* cloak.
Nephridia. G., *nephridios,* belonging to the kidneys.
Nephrotome. G., *nephros,* kidney, + *tome,* a cutting.
Neural. G., *neuron,* nerve.
Neurenteric. G., *neuron,* nerve, + *enteron,* gut.
Neurilemma. G., *neuron,* nerve, + *lemma,* husk, sheath.
Neuro-. G. combining form signifying nerve.
Neuroglia. G., *neuron,* nerve, + *gloia,* glue.
Neurohumor. G., *neuron,* nerve, + L., *humor,* fluid.
Neuromast. G., *neuron,* nerve, + *mastos,* round hill.
Neuron. G., *neuron,* sinew, tendon; equivalent of L. *nervus;* whence, nerve.
Neuropil. G., *neuron,* nerve, + *pilos,* felt.
Nictitating. L., *nictare,* to wink.
Nidamental. L., *nidamentum,* materials for a nest.
Node. L., *nodus,* knot.
Notochord. G., *noton,* back, + *chorde,* cord.
Nuchal. L., *nucha,* nape of the neck.
Obliquus. L., *obliquus,* slanting.
Oculomotor. L., *oculus,* eye, + *motor,* mover.
Odontoblast. G., *odous,* tooth, + *blastos,* shoot, germ.
Olecranon. G., *olekranon,* point of the elbow.
Olfactory. L., *olere,* to smell, + *facere,* to make.
Omasum. L., *omasum,* paunch.
Omentum. L., *omentum,* adipose membrane enclosing the bowels.
Omphalo-. G. combining form (*omphalos*) signifying the navel.
Ontogeny. G., *onta,* things that exist, + *gennan,* to beget.
Operculum. L., *operculum,* lid.
Ophthalmic. G., *ophthalmos,* eye.
Opistho-. G., prefix signifying backward, behind.
Opisthocelous. G., *opisthe,* behind, + *koilos,* hollow.
Opisthonephros. G., *opisthe,* behind, + *nephros,* kidney.
Optic. G., *opsis,* sight.
Osseous. L., *os* (pl. *ossa*), bone.
Ossicle. L., *ossiculum,* small bone.
Osteoblast. G. *osteon,* bone, + *blastos,* germ.
Osteocyte. G., *osteon,* bone, + *kytos,* cell.

Otic. G., *otikos,* belonging to the ear.
Otolith. G., *ous, otos,* ear, + *lithos,* stone.
Ovum (pl. **ova**). L., *ovum,* egg.
Oxyphil. G., *oxys,* acid, + *philos,* friend.
Paleontology. G., *palaios,* ancient, + *onta,* existing things, + *logos,* science.
Paleopallium. G., *palaios,* ancient, + L., *pallium,* covering.
Pallium. L., *pallium,* cloak.
Palma. L., *palma,* the (open) hand.
Palpebra. L., *palpebra,* eyelid.
Pancreas. G., *pan,* all, + *kreas,* flesh.
Papilla. L., *papilla,* pimple.
Para-. G. prefix signifying alongside of, near.
Parabronchii. G., *para,* beside, + *bronchos,* windpipe.
Paracentrum. G., *para,* near, + *kentron* (L. *centrum*), center.
Parachordal. G., *para,* beside, + *chorde,* cord.
Paraganglion. G., *para,* along, beside, + *ganglion,* knot.
Paraphysis. G., *para,* beside, + *physis,* growth.
Parapsid. G., *para,* beside, + *apsis,* arch.
Parathyroid. G., *para,* near, + *thyreos,* oblong shield, + *eidos,* form.
Parencephalon. G., *para,* beside, + *enkephalon,* brain.
Parietal. L., *paries,* wall.
Parotid. G., *para,* near, + *ous, otos,* ear.
Pecten. L., *pecten,* comb.
Pectoral. L., *pectoralis* (*pectus*), belonging to the breast.
Pedunculus. L., *pediculus,* a little foot.
Pelvic. L., *pelvis,* basin.
Peri-. G. prefix meaning around.
Pericardial. G., *peri,* around, + *kardia,* heart.
Perichondrium. G., *peri,* around, + *chondros,* cartilage.
Perichordal. G., *peri,* around, + *chorde,* cord.
Perilymph. G., *peri,* around, + L., *lympha,* fluid.
Perimysium. G., *peri,* around, + *mys,* muscle.
Periosteum. G., *peri,* around, + *osteon,* bone.
Peristalsis. G., *peristaltikos,* clasping and compressing.
Peritoneum. G., *peritonaion,* membrane containing the lower viscera.
Phallic. G., *phallikos,* pertaining to the penis.
Pharynx. G., *pharynx,* throat.
Photophore. G., *phos, photos,* light, + *pherein,* to bear.
Phylogeny. G., *phylon,* race, + *gennan,* to beget.
Pineal. L., *pinea,* pine cone.
Pinna (pl. **-ae**). L., *penna, pinna,* feather; hence, wing.
Pisiform. L., *pisum,* pea, + *forma,* shape.
Pituitary. L., *pituita,* slime, phlegm.

Placenta. L., *placenta,* a flat cake.
Placode. G., *plax,* plate, + *eidos,* likeness.
Planta. L., *planta,* sole of the foot.
Plastron. F., *plastron,* breastplate.
Platybasic. G., *platys,* broad, flat, + L., *basis,* base.
Plectrum. G., *plektron,* hammer.
Pleuro-. G. combining form signifying the side.
Pleurocentrum. G., *pleura,* side, + *kentron* (L., *centrum*), center.
Pleurodont. G., *pleura,* ribs, side, + *odous,* tooth.
Plexus (pl. **-us**). L., *plexus,* plaiting, braid.
Pneumatic. G., *pneumatikos,* pertaining to breath.
Poikilothermous. G., *poikilos,* changeful, + *thermos,* heat.
Pons. L., *pons, pontis,* bridge.
Portal. L., *porta* (pl. *-ae*), gate.
Porus. G., *poros,* passage.
Prae-, pre-. L. prefix signifying before, in front.
Premolar. L., *pre,* in front, + *molaris,* molar.
Prepuce. L., *praeputium,* foreskin.
Primordial. L., *primordium,* beginning.
Pro-. G. or L. prefix signifying before, in front of, or prior.
Procelous. G., *pro,* in front, + *koilos,* hollow.
Proctodeum. G., *proktos,* anus, + *hodaios,* pertaining to a way.
Profundus. L., *profundus,* deep.
Pronator. L., *pronare,* to bend forward.
Pronephros. G., *pro,* before, + *nephros,* kidney.
Proprioceptor. L., *proprius,* special, + *capere,* to take.
Prosencephalon. G., *pros,* before, + *enkephalos,* brain.
Prostate. L., *pro,* in front, + *stare,* to stand.
Protonephros. G., *protos,* first, + *nephros,* kidney.
Protoplasm. G., *protos,* first, + *plasma,* form.
Proximal. L., *proximus,* next.
Pseudobranch. G., *pseudes,* false, + *branchia,* gills.
Pterygoid. G., *pteryx,* wing, + *eidos,* likeness.
Pterylae. G., *pteron,* feather, + *hyle,* a wood.
Pubis (pl. **-es**). L., *pubis,* mature.
Pulmonary. L., *pulmo,* lung.
Pygal. G., *pyge,* rump.
Pygostyle. G., *pyge,* rump + *stylos,* pillar.
Pylorus. G., *pylouros,* gate-keeper.
Pyriform. L., *pirum,* pear, + *forma,* shape.
Quadriceps. L., *quattuor,* four, + *caput,* head.
Quadrigeminus. L., *quadrigeminus,* fourfold, four.
Radial. L., *radius,* rod, spoke.
Receptor. L., *recipere,* to take back, receive.

Rectus. L., *rectus,* straight.
Remiges (pl.). L., *remex,* rower.
Renal. L., *renes,* kidneys.
Rete. L., *rete,* network.
Reticulum. L., *reticulum,* a little net.
Retina. L., *rete,* net.
Retractor. L., *retrahere,* to draw back.
Retrices. L., *retro,* back, + *cedere,* to go.
Rhachitomous. G., *rhachis,* spine + *temnein,* to cut.
Rhinal. G., *rhis,* nose.
Rhombencephalon. G., *rhombos,* kind of parallelogram, + *enkephalos,* brain.
Rostrum (pl. **-a**). L., *rostrum,* beak.
Rotator. L., *rotare,* to whirl about.
Ruminare. L., *ruminari,* to chew the cud.
Sacculus. L., *sacculus,* a little bag.
Sacrum. L., *sacer,* sacred.
Sagittal. L., *sagitta,* arrow.
Salpinx. G., *salpinx,* trumpet.
Sarcolemma. G., *sarx,* flesh, + *lemma,* husk, skin.
Scala. L., *scala,* staircase.
Sclera. G., *skleros,* hard.
Sclerotic. G., *skleros,* hard.
Sclerotome. G., *skleros,* hard, + *temnein,* to cut.
Scrotum. L., *scrotum* (*scrotum*), skin.
Sebaceous. L., *sebum,* tallow, grease.
Selenodont. G., *selene,* moon (hence, crescent) + *odous,* tooth.
Seminiferous. L., *semen,* seed, + *ferre,* to bear.
Septum. L., *saeptum,* fence.
Sinus (pl. **-us**). L., *sinus,* curve, cavity, bosom.
Somatic G., *soma* (pl. *somata*), body.
Somatopleure. G., *soma,* body, + *pleura,* side.
Somite. G., *soma,* body, + suffix *-ite,* indicating origin.
Spermatozoon (pl. **-a**). G., *sperma,* seed, + *zoon,* animal.
Sphenoid. G., *sphen,* wedge, + *eidos,* likeness.
Sphincter. G., *sphingein,* to bind tight.
Spina. L., *spina,* thorn.
Spiracle. L., *spiraculum,* air hole.
Splanchnic. G., *splanchna,* viscera.
Splanchnopleure. G., *splanchna,* viscera, + *pleura,* side.
Stapes. L., *stapes,* stirrup.
Stereospondylous. G., *stereos,* solid, + *sphondylos,* vertebra.
Stomodeum. G., *stoma,* mouth, + *hodaios,* pertaining to a way.
Stratus (pl. **strata**). L., *stratus,* layer.

Striatum. L., *striatus,* grooved, streaked.
Styloid. G., *stylos,* pillar, + *eidos,* likeness.
Sub-. L. prefix signifying under, beneath, near.
Subcostal. L., *sub,* under, + *costa,* rib.
Sublingual. L., *sub,* under, + *lingus,* tongue.
Subunguis. L., *sub,* under, + *unguis,* nail.
Subvertebral. L., *sub,* under, + *vertebra,* joint.
Sulcus. L., *sulcus,* furrow.
Supinator. L., *supinare,* to bend backward.
Supracostal. L., *supra,* above, + *costa,* rib.
Supraspinatus. L., *supra,* above, + *spina,* thorn.
Sym-, syn-. G. prefix signifying with or together.
Sympathetic. G., *syn,* together, + *pathein,* to suffer.
Symphysis. G., *syn,* together, + *physis,* growth.
Synapse. G., *syn,* together, + *haptein,* to fasten.
Synarthrosis. G., *syn,* together, + *arthron,* joint.
Synsacrum. G., *syn,* together, + L., *sacer* (*os sacrum*), sacrum.
Syrinx. G., *syrinx,* pipe.
Tabular. L., *tabula,* board, table.
Talonid. L., *talus,* heel, + G., *eidos,* form.
Tapetum. L., *tapete,* carpet.
Tarsus. G., *tarsos,* sole of the foot.
Tectum. L., *tectum* (*tego*), roof.
Tegmentum. L., *tegumentum,* a covering.
Tela. L., *tela,* web.
Telencephalon. G., *telos,* end, + *enkephalos,* brain.
Telolecithal. G., *telos,* end, + *lekithos,* yolk.
Temporal. L., *temporalis,* belonging to time.
Temporal. L., *tempora,* the temples.
Tendon. L., *tendere,* to stretch.
Tentorium. L., *tentorium,* tent.
Terminalis. L., *terminare,* to limit.
Testis. L., *testis,* testicle.
Tetrapod. G., *tetra,* four, + *pous,* foot.
Thalamus. G., *thalamos,* chamber or couch.
Thecodont. G., *theke,* case, sheath, + *odous,* tooth.
Thorax (thoracis). G., *thorax,* breastplate, breast.
Thrombocytes. G., *thrombos,* clot, + *kytos,* cells.
Thymus. G., *thymos,* sweetbread.
Thyroid. G., *thyreos,* shield, + *eidos,* form.
Trabecula (pl. **-ae**). *L., trabecula,* a little beam.
Trachea. G., *tracheia,* windpipe.
Triceps. L., *tres,* three, + *caput,* head.
Trigeminus. L., *trigeminus,* born three together.

Triquetrum. L., *triquetrus,* three-cornered, triangular.
Trochanter. G., *trochos,* wheel, pulley.
Trochlea. G., *trochilia,* pulley.
Trophoblast. G., *trophe,* nourishment, + *blastos,* shoot, germ.
Tropibasic. G., *trope,* a turning, + L., *basis,* base.
Tuberculum. L., *tuberculum,* a small hump.
Tunica. L., *tunica,* undergarment.
Turbinal. L., *turbo,* a top, anything that spins or shows turning.
Tympanic. L., *tympanum,* drum.
Umbilical. L., *umbilicus,* navel.
Unciform. L., *uncus,* hook, + *forma,* shape.
Uncinate. L., *uncinatus,* furnished with a hook.
Urea. G., *ouron,* urine.
Urodeum. G., *ouron,* urine, + *hodaios,* pertaining to a way.
Urodela. G., *oura,* tail, + *delos,* evident.
Uropygial. G., *orros,* end of os sacrum, + *pyge,* rump.
Urogenital. G., *ouron,* urine, + L., *genitalis,* genital.
Urostyle. G., *oura,* tail, + *stylos,* pillar.
Uterus. L., *uterus,* womb.
Utriculus. L., *utriculus,* small skin or leather bottle.
Vagus. L., *vagus,* wandering.
Valvula. L., *valvula,* a little fold or valve.
Vas. L., *vas,* vessel.
Vascular. L., *vasculum,* small vessel.
Ventral. L., *venter,* belly.
Ventricle. L., *ventriculus,* little cavity, loculus.
Vermiform. L., *vermis,* worm, + *forma,* shape.
Vesicle. L., *vesicula,* a small bladder.
Vestibulum. L., *vestibulum,* entrance court.
Vibrissa (pl. **-ae**). L., *vibrissa,* hair in the nostril.
Villus (pl. **villi**). L., *villus,* shaggy hair.
Visceral. L., *viscera,* entrails, bowels.
Vitelline. L., *vitellus,* yolk of egg.
Vitreus. L., *vitreus,* of glass; hence, transparent.
Viviparous. L., *vivus,* living, + *parere,* to beget.
Vomer. L., *vomer,* ploughshare.
Xiphiplastron. G., *xiphos,* sword, + F., *plastron,* breastplate.
Zygapophysis. G., *zygon,* yoke, + *apophysis,* process of a bone.
Zygomatic. G., *zygoma,* cheekbone.

LATIN WORD ENDINGS

Although scientific terms are often used in English form, some knowledge of the use of these words in Latin form is desirable. Latin is a highly inflected language, with a variable series of terminations for nouns and adjectives expressing not only singular and plural numbers, but also genders (of an artificial nature) and a variety of cases; still further, there are several different systems of forming such terminations ("declensions"). Fortunately, however, nearly all use of scientific terms involves only two cases—nominative and genitive. Less than a score of endings affixed to the word root will cover most instances.

Adjectives (which must agree in gender, number, and case with their nouns) are "declined" according to one of the two following schemes, for each of which a common adjective is used as an example (the ending, attached to the root, is in boldface).

First and Second Declension (Combined)

	Masculine	*Neuter*	*Feminine*
Nominative singular	magn**us**	magn**um**	magn**a**
Nominative plural	magn**i**	magn**a**	magn**ae**
Genitive singular	magn**i**		magn**ae**
Genitive plural	magn**orum**		magn**arum**

Third Declension

	Masculine and Feminine	*Neuter*
Nominative singular	grand**is**	grand**e**
Nominative plural	grand**es**	grandi**a**
Genitive singular	grand**is**	
Genitive plural	grand**ium**	

Most nouns follow one of these same schemes. Thus, *fibula* is a feminine noun of the first declension and is declined *fibula, fibulae, fibulae, fibularum; humerus* is a masculine noun of the second declension, declined *humerus, humeri, humeri, humerorum; sternum, sterna, sterni, sternorum,* a neuter noun of the second declension; *cutis, cutes, cutis, cutium* (skin), a feminine noun of the third declension.

There are, however, two complications: (1) in the third declension most nouns have a short form for the nominative singular, a longer root for the other case endings. Thus *femur* (third declension neuter) becomes *femora,* and so on, in other cases; other typical examples are *meninx, meninges; foramen, foramina; caput, capita.* (2) A few nouns used anatomically belong to a further declension—a fourth declension. Of masculine words of this gender—*plexus* and *meatus* are examples—the plural spelling is the same as the singular; hence the English form is preferable for common use. A common neuter noun of this declension is *cornu* (horn), declined *cornu, cornua, cornus, cornuum.*

APPENDIX III

REFERENCES

BELOW ARE listed a few of the more useful general works, or works on special topics or animal types. No attempt is made to give references to the numerous original papers from which the data in such works are assembled. To look further into the literature of any special topic, these two publications are most useful:

1. Zoological Record, 1864-date. London.
   Each annual volume lists all papers published during the year concerning each class of vertebrates, and follows this with classified lists of those papers which deal with various topics in anatomy, embryology, and so forth.
2. Biological Abstracts, 1926-date. Philadelphia.
   A voluminous journal which attempts to abstract and index all papers published on any field of biology.

GENERAL

3. Bolk, L., and others: Handbuch der vergleichenden Anatomie der Wirbeltiere. 6 vols. Berlin and Vienna, 1931-1938.
   A comprehensive work on vertebrate anatomy by many specialists; includes extensive bibliographies.
4. Ihle, J.E.W., and others: Vergleichende Anatomie der Wirbeltiere. Berlin, 1927.
   A substantial volume on comparative anatomy by Dutch authors, translated into German. A new edition, in Dutch, was published in 1947.
5. Goodrich, E. S.: Studies on the Structure and Development of Vertebrates. London, Macmillan Company, 1930.
   A stimulating discussion of many anatomical problems by a first-rate authority.
6. Bronn, H.G., and others: Klassen und Ordnungen des Thier-Reichs. Abteilung VI [Vertebrates]. Leipzig and Heidelberg, 1874-1938.
   A voluminous work by various authors, published in parts, some old, some as yet incomplete, which gives great attention to the anatomy of the various vertebrate groups as well as to classification and distribution.
7. Kükenthal, W., and Krumbach, T., editors: Handbuch der Zoologie. Berlin and Leipzig, 1923-date.
   A work similar to the last in scope; incomplete.
8. Parker, T. J., and Haswell, W. A. A Textbook of Zoology. 6th ed., New York, Macmillan Company, 1940, vol. 2.
9. Turner, C. D.: General Endocrinology. Philadelphia, W. B. Saunders Company, 1948.
   Includes much anatomical and histological data.

HISTOLOGY

10. Maximow, A., and Bloom, W.: A Textbook of Histology. 5th ed. Philadelphia, W. B. Saunders Company, 1948.
    One of several good histologies based on human material.

11. Krause, R.: Mikroskopische Anatomie der Wirbeltiere. Berlin and Leipzig, 1923.
A comparative histology of the vertebrates, with sections on each class.

EMBRYOLOGY

12. McEwen, R. S.: Vertebrate Embryology. Revised ed. New York, 1949.
13. Brachet, A., Dalcq, A., and Gérard, P.: Traité d'Embryologie des Vertébrés. 2d ed. Paris, Masson & Cie, 1935.
14. Huettner, A. F.: Fundamentals of Comparative Embryology of the Vertebrates. 2d ed. New York, Macmillan Company, 1949.
15. Hertwig, O.: Handbuch der vergleichenden und experimentallen Entwickelungslehre der Wirbeltiere. Jena, 1901–1906, 3 vols.
A comprehensive, well-illustrated work; somewhat out of date in certain aspects.
16. Kerr, J. G.: Text-Book of Embryology. Vol. II. Vertebrata with the Exception of Mammals. London, Macmillan Company, 1919.
17. Arey, L. B.: 1946. Developmental Anatomy. 5th ed. Philadelphia, W. B. Saunders Company, 1946.
Primarily mammalian and human.

OSTEOLOGY

18. Reynolds, S. H.: The Vertebrate Skeleton. 2d ed. Cambridge, 1913.
19. Romer, A. S.: Vertebrate Paleontology. 2d ed. Chicago, University of Chicago Press, 1945.
20. DeBeer, G. R.: The Development of the Vertebrate Skull. Oxford University Press, 1937.
21. Gregory, W. K.: Fish Skulls: A Study of the Evolution of Natural Mechanisms. Tr. Am. Philos. Soc., *23:* 75-481, 1933.
22. Williston, S. W.: The Osteology of the Reptiles, W. K. Gregory, ed. Cambridge, 1925.
23. Jayne, H.: Mammalian Anatomy. Part I. The Skeleton of the Cat. Philadelphia, 1898.
24. Flower, W. H.: An Introduction to the Osteology of the Mammalia. 3d ed. London, 1885.
An old but useful little book.

SENSE ORGANS

25. Walls, G. L.: The Vertebrate Eye and Its Adaptive Radiation. Cranbrook Institute of Science, Bull. No. 19, 1942.
26. Rochon-Duvigneaud, A.: Les Yeux et la Vision des Vertébrés. Paris, 1943.
27. Retzius, G.: Das Gehörorgan des Wirbelthiere. Morphologischhistologiche Studien. Stockholm, 2 vols., 1881-1884.

NEUROLOGY

28. Kappers, C. U. A., Huber, G. C., and Crosby, E. C.: The Comparative Anatomy of the Nervous System of Vertebrates, Including Man. New York, Macmillan Company, 2 vols., 1936.
A mine of information on comparative neurology, but difficult to work for one not a neurologist.
29. Herrick, C. J.: An Introduction to Neurology. 5th ed. Philadelphia, 1934.
30. Papez, J. W.: Comparative Neurology. New York, Thomas W. Crowell Company, 1929.
31. Krieg, W. J. S.: Functional Neuroanatomy. Philadelphia, Blakiston Company, 1942.
On the human brain, but with comparative discussions and excellent figures.
32. Tilney, F.: The Brain from Ape to Man. New York, Paul B. Hoeber, Inc., 2 vols., 1928.
Includes lower primates.

LOWER CHORDATES

33. Delage, Y., and Hérouard, E.: Traité de Zoologie Concrète. Vol. 8. Les Procordés. Paris, 1898.
Detailed and well-illustrated description of structure and development of lower chordates.

FISHES

34. Goodrich, E. S.: A Treatise on Zoology, edited by E. Ray Lankester. Part IX. Vertebrata Craniata, Fasc. I. "Cyclostomes and Fishes". London, 1909.
A mine of data on fish anatomy; badly indexed, however.

35. Daniel, J. F.: The Elasmobranch Fishes. 3rd ed. Berkeley, University of California Press, 1934.
Shark anatomy.

36. Norman, J. R.: A History of Fishes. 2d ed. London, Ernest Benn, Ltd., 1936.
Life history, habits, and so on, as well as structure.

AMPHIBIA

37. Noble, G. K.: The Biology of the Amphibia. New York, McGraw-Hill Book Company, Inc., 1931.
38. Francis, E. T. B.: The Anatomy of the Salamander. Oxford University Press, 1934.
39. Ecker, A., Wiedersheim, R., and Gaupp, E.: Anatomie des Frosches. 2d ed. Braunschweig, 3 vols., 1888-1904.
A thorough account of frog anatomy, which has passed through the hands of three successive authors.

40. Holmes, S. J.: The Biology of the Frog. 4th ed. New York, Macmillan Company, 1927.

BIRDS

41. Pycraft, W. P.: A History of Birds. London, 1910.
Includes anatomy. A good account in No. 7 by Stresemann; a brief description of the fowl in No. 47.

MAMMALS

42. Weber, M., Burlet, H. M. de, and Abel, O.: Die Säugetiere. 2d ed. Jena, 2 vols., 1927-1928.
A standard work on mammalian anatomy and classification.

43. Howell, A. B.: Gross Anatomy: A Brief Systematic Presentation of the Macroscopic Structure of the Human Body. New York, D. Appleton-Century Company, Inc., 1939.
More detailed accounts of human anatomy are available in the larger standard texts, such as Cunningham, Gray, and Morris.

44. Flower, W. H., and Lydekker, R.: An Introduction to the Study of Mammals, Living and Extinct. London, 1891.
Old, but still useful.

45. Greene, E. C.: Anatomy of the Rat. Tr. Am. Philos. Soc. (N.S.), *27*:1, 1935.
46. Bradley, O. C., and Grahame, T.: Topographical Anatomy of the Dog. 4th ed. New York, Macmillan Company, 1943.
47. Sisson, S., and Grossman, J. D.: The Anatomy of the Domestic Animals. 3rd ed. Philadelphia, W. B. Saunders Company, 1938.
A comprehensive account of horse anatomy; ox, sheep, pig, and dog are covered more briefly.

48. Reighard, J. E., and Jennings, H. S.: Anatomy of the Cat. 3rd ed. New York, Henry Holt & Company, Inc., 1935.

49. Howell, A. B.: Anatomy of the Wood Rat. Baltimore, Williams & Wilkins Company, 1926.

50. Clark, W. E. LeGros: Early Forerunners of Man. Baltimore, Williams & Wilkins Company, 1934.
A discussion of the anatomy of lower primates.

51. Hartman, C. G., and Straus, W. L., Jr., ed.: The Anatomy of the Rhesus Monkey. Baltimore, Williams & Wilkins Company, 1933.

52. Davison, A., and Stromsten, F. A.: Mammalian Anatomy, with Special Reference to the Cat. 7th ed. Philadelphia, 1937.

Index